Personal Copy
Bruce D. Martin
2/92

New Concepts in Global Tectonics

New Concepts in Global Tectonics

Edited by

Sankar Chatterjee and Nicholas Hotton III

Texas Tech University Press

Copyright 1992 Texas Tech University Press

All rights reserved. No portion of this book may be reproduced in any form or by any means, including electronic storage and retrieval systems, except by explicit, prior written permission of the publisher, except for the brief passages excerpted for review and critical purposes.

This book was set in 10 on 12 Baskerville and printed on acid-free paper that meets the guidelines for permanence and durability of the Committee on Production Guidelines for Book Longevity of the Council on Library Resources.

Jacket photogaph by NASA: Gemini XI earth-sky view. This photograph of the Indian subcontinent was taken by Gemini XI on 14 September 1966 at an altitude of 650 kilometers. The Himalayan Mountains, 3,700 kilometers away, are just visible on the horizon. The Indian Ocean is at the bottom of the photograph, at left center is the Arabian Sea, and at upper right is the Bay of Bengal. India's paleoposition in the Pangea reconstruction is the single most contested issue in plate tectonics.

Manufactured in the United States of America

Library of Congress Cataloging-in-Publication Data
New concepts in global tectonics / edited by Sankar Chatterjee and
  Nicholas Hotton.
      p. cm.
    Includes bibliographical references.
    ISBN 0-89672-269-4
    1. Plate tectonics. 2. Paleobiogeography. I. Chatterjee,
Sankar. II. Hotton, Nicholas.
QE511.4.N48 1992
551.1'36—dc20   91-32304
                                                         CIP

92 93 94 95 96 97 98 99 00 / 9 8 7 6 5 4 3 2 1

Texas Tech University Press
Lubbock, Texas 79409-1037 USA

Dedication to Curt Teichert

This volume is a tribute to Curt Teichert, who has contributed significantly to the knowledge and understanding of paleontology, stratigraphy, paleobiogeography, polar geology, and Gondwana studies on a truly global scale as both researcher and teacher. Curt's remarkably full and diverse career spans 60 years, and the vast range of his contributions is reflected in 250 publications, which include a series of long monographs. His editing of the monumental *Treatise of Invertebrate Paleontology* is only one of his outstanding accomplishments.

Curt was born 8 May 1905 in Koenigsberg, Prussia. He received his Ph.D. in geology from the University of Koenigsberg in 1928, and a D.Sc. from the University of Western Australia in 1944. He emigrated to Australia in 1937, where he taught at the University of Western Australia and worked as assistant chief goverment geologist at Victoria Mines department. In 1954 he joined the U.S. Geological Survey to head various laboratories. He left the USGS in 1964 to become Regents Distinguished Professor of geology at the University of Kansas. After his nominal retirement in 1977, Curt moved to the University of Rochester as adjunct professor of geological sciences, and his publications continue to attest to his high level of activity.

Curt has led several expeditions to different parts of the world from Greenland to Gondwana. He has earned many national and international awards, among them the David Syme Prize, University of Melbourne (1949); the Raymond Cecil Moore Medal, Society of Economic Geologists, Paleontologists, and Mineralogists (1982); and the Paleontological Society Medal (1984). For more than half a century he has been an inspiring teacher and stimulating associate for those lucky enough to know him. This volume is dedicated to Curt Teichert with appreciation, respect, and love.

Contents

Introduction   ix

Plate Tectonics: An Overview

Plate tectonics and continental drift in geologic education
    Paul D. Lowman, Jr.   **3**

Geometry of Plate Tectonics

A kinematic perspective on finite relative plate motion, provided by the first-order cycloid model
    Vincent S. Cronin   **13**

An experimental study of ridge offsets in relation to trends of ridge and transform faults, using wax models
    S. K. Ghosh   **23**

A kinematic model for the evolution of the Indian plate since the Late Jurassic
    Sankar Chatterjee   **33**

Continental Tectonics and Orogeny

Intracratonic tectonism: key to the mechanism of diastrophism
    A. C. Grant   **65**

The tectonic framework of Australia
    Vadim Anfiloff   **75**

Conventional plate tectonics and orogenic models
    S. E. Cebull and D. H. Shurbet   **111**

Geological evidence for a simple horizontal compression of the crust in the Zagros Crush Zone
    Mansour S. Kashfi   **119**

Constraints to major right-lateral movements, San Andreas fault system, central and northern California, USA
    B. D. Martin   **131**

Tectonics of the Ocean Basins

Origin of midocean ridges
    Arthur A. Meyerhoff, William B. Agocs, Irfan Taner, Anthony E. L. Morris, and Bruce D. Martin   **151**

Paleoland, crustal structure, and composition under the northwestern Pacific Ocean
    D. R. Choi, B. I. Vasil'yev, and M. I. Bhat   **179**

Past distribution of oceans and continents
    J. M. Dickins, D. R. Choi, and A. N. Yeates   **193**

Paleomagnetism

Rotating plates: new concept of global tectonics
    K. M. Storetvedt   **203**

Reykjanes Ridge: quantitative determinations from magnetic anomalies
    William B. Agocs, Arthur A. Meyerhoff, and Karoly Kis   **221**

Paleobiogeography

Paleofloras, faunas, and continental drift: some problem areas
    Charles J. Smiley   **241**

Paleozoogeographic relationships of Australian Mesozoic tetrapods
    R. E. Molnar   **259**

Global distribution of terrestrial and aquatic tetrapods, and its relevance to the position of continental masses
    Nicholas Hotton III   **267**

Alternatives to Plate Tectonics

Has the Earth increased in size?
    H. G. Owen   **289**

Earth expansion theory versus statical Earth assumption
    G. O. W. Kremp   **297**

Surge tectonics: a new hypothesis of Earth dynamics
    Arthur A. Meyerhoff, Irfan Taner, Anthony E. L. Morris, Bruce D. Martin, W. B. Agocs, and Howard A. Meyerhoff   **309**

Endogenic regimes and the evolution of the tectonosphere
    V. V. Beloussov   **411**

Global change: shear-dominated geotectonics modulated by rhythmic Earth pulsations
    Forese-Carlo Wezel   **421**

The contracting-expanding Earth and the binary system of its megacyclicity
    L. S. Smirnoff   **441**

Introduction

The judgment of contemporaries is always subject to severe limitations. The long-range judgment of posterity is unknown.

S. Chandrasekhar

Modern plate-tectonic theory is a profoundly integrative concept that has reinvigorated the earth sciences as evolution reinvigorated biology. The general concepts of plate tectonics are simple. The outer shell of the Earth, the lithosphere, is segmented into a mosaic of rigid, shifting plates. These plates are created at oceanic ridges, consumed at oceanic trenches, and slide past one another along transform fault boundaries. This dynamic model links a great range of geological phenomena, from the age and composition of sea floors to the rise of mountains, the past distribution of flora and fauna, and the occurrence of various deposits of economic value. Most earthquakes, volcanoes, and deformations are concentrated along the network of plate boundaries. Earth's magnetic field, because of its recurrent reversals, acts as a chronometer, accurately timing these events.

After some 25 years of acceptance, plate tectonics has acquired a conventionality comparable to that of Darwinian evolution. Yet many areas remain unexplained, and new evidence continues to appear, requiring modification of current theory. Although there is a consensus that some sort of thermal convection may drive the plates, the mechanism behind plate motion is still controversial and poorly understood. It is ironic that Wegener's hypothesis of continental drift was rejected largely because it lacked a plausible driving mechanism, yet plate tectonics is widely accepted today despite a similar lack.

Paleomagnetic pole determination for the paleoposition of continents is one of the weaker links in the chain of evidence for plate tectonics. Rock magnetism is subject to modification by weathering, later magnetism, and metamorphism. It provides clues to paleolatitude, but no information regarding longitude at the time of magnetization. In consequence, we are unable to say whether two continents, correctly oriented and placed at the proper latitude, were contiguous or were separated by a significant seaway. The issue is further complicated because the magnetic pole, not the rotational pole, of Earth is the point of reference. The two poles do not coincide at present, and there is considerable evidence for slow migration of the magnetic pole over long distances, independent of plate migration. With increasing age, the frame of reference furnished by paleomagnetic data becomes progressively less satisfactory. Young marine magnetic anomalies provide strong support for the sea-floor spreading model and have resolved the paradox of old continents and young oceans. The sequence of normal and reverse polarity can provide relative dates, but few sequences are available prior to the Upper Jurassic. As a result, our confidence in the utility of plate-tectonic models diminishes with rocks older than 200 million years.

There are other observations that do not conform closely to current plate-tectonic theory. Plate tectonics assumes that large lithospheric plates behave as rigid units that deform only at their boundaries. The intraplate deformation in the central Indian Ocean basin is a good example of deviation from a basic tenet of plate tectonics, and the common association of extensional tectonics with convergent plate boundaries is another.

Perhaps the time has come to acknowledge that there are inconsistencies in plate tectonics. Whether such inconsistencies are fundamental defects, or merely loose ends that need tying up in a basically sound hypothesis remains to be seen as the theory evolves. With such ideas in mind, a mixed group of earth scientists gathered to discuss new concepts in global tectonics at the Smithsonian Institution following the International Geological Congress in 1989. It was an unusual convention because of the diversity of opinions, which ranged from devotion to the received wisdom in the field of plate tectonics to unreconstructed continental fixity. It provided a basis for interaction that is not easily available otherwise. This volume is the result of that discussion.

The 23 chapters that comprise this volume are divided into seven sections. Within each section some chapters are strongly supportive of plate tectonics, some are strongly critical, and others are neutral. An entire section is devoted to alternative models to plate tectonics. We chose the title "New Concepts in Global Tectonics" to avoid confusion with conventional plate tectonics and to embrace different aspects of earth dynamics. We hope that the reader will find these papers stimulating and thought-provoking as well as heretical.

The lead paper, by P.D. Lowman, Jr., provides a balanced overview of the role of plate tectonics in geologic education. Lowman argues that the current approach to the teaching of geology is one-sided, so heavily based on plate tectonics that it may stifle further research into alternative aspects of tectonics. He recommends that other models of earth dynamics should be taught along with plate tectonics and comparative planetology, so that students can appreciate the complexities of tectonic problems of the Earth's crust. This eclectic approach is intended to encourage students to understand fundamental questions and uncertainties of global dynamics, which will stimulate penetrating inquiry into the way the Earth works.

The chapters in the next section concern the geometry of plate tectonics. Much of the beauty of plate tectonics lies in the exactness and simplicity of the geometry of movement. Euler's poles play a central role in the geometry of plate tectonics and continental reconstruction. Given Euler's theorem, plate motion is described as the rotation of one plate relative to another about a pole on the surface of the Earth, requiring the location of the pole and the angular velocity of rotation about this pole. However, it is becoming apparent that plate motion may not be as simple and elegant as was originally conceived, and may require more complex geometry such as the cycloid relative motion model presented by V.S. Cronin in this volume. Cronin's model is more robust, shows more predictive capabilities, and should change our understanding of variations of relative motion across plate margins through time. S.K. Ghosh documents experimentally the geometric and genetic relationships of ridge offsets in relation to trends of ridges. With the use of wax models, he explains why the movement of transform faults is opposite to the sense of ridge offsets, as predicted by J.T. Wilson. S. Chatterjee proposes a new kinematic model for the motion of the Indian plate since the Late Jurassic, and suggests that India maintained an overland connection with other landmasses during its transfer from Antarctica to Asia. At the K-T boundary, a large meteorite impact might have triggered the rifting between India and the Seychelles, initiated the main eruptive phase of Deccan volcanism, and created a large oval crater about 900 km long and 800 km wide. The crater was split by the newly formed Carlsberg Ridge, and the two halves then drifted apart. Today, one part of the crater is attached to the Seychelles and the other to the western part of India.

Modern development of plate tectonics is mainly a result of the study of oceanic areas. Successful explanations of the tectonics of oceanic crust have been blurred by failures to explain the tectonics of continental interiors. For example, buoyant continental crust can "float" on the underlying mantle to form mountain ranges, as in Asia. Intraplate magmatic activity, reflected in small plutons, plugs, laccoliths, dikes, diatremes, kimberlites, and some extrusive rocks, seems to have no relation to plate-boundary activity. Intraplate vertical movement, manifest in the development of basins and arches within continents, likewise appears unrelated to activity at plate margins. Vertical movement of continental crust is fundamental to the formation of continental topography, but its nature and cause remains inexplicable by plate tectonics.

Various mechanisms for continental tectonics and orogeny are discussed in the next section. A.C. Grant shows convincingly the genesis of the Nares Strait between Greenland and North America by a vertical tectonic mechanism instead of conventional plate tectonics. In a comprehensive review, V. Anfiloff elaborates a new model of vertical crustal displacement and thermal anomaly for the tectonic framework of Australia. This model invokes three systematic and one random process during intracratonic tectonism. Systematic processes include rifting of crustal blocks by compression, upwelling of magma into rift valleys, and channelling of compression along these igneous bodies. The random process is the local chelogenic heat engine driven by large bubbles of mantle-derived magma. The compressive force for intracratonic tectonism might have come from a contracting Earth (see Meyerhoff et al. on surge tectonics). S.E. Cebull and D.H. Shurbet point out succinctly the increasing difficulty of plate tectonics in explaining noncollisional (subductive) orogeny such as that of the Andes and much of the Pacific rim, because of the lack of a trigger mechanism. M.S. Kashfi questions the plate-tectonic interpretation of the Zagros Crush Zone in the Middle East as a suture of a former continent-to-continent collision between Arabian and Eurasian plates. He marshalls a great deal of field data to show that the evidence of subduction along this putative suture zone is nonexistent, and that the basement under the Zagros is continental. The geologic and biostratigraphic continuity across the Zagros Mountains indicates geographic proximity of Arabia and Iran before the obliteration of Tethys. B.D. Martin speculates that there are three critical regions along the San Andreas Fault (SAF) in California where the major movement has been vertical rather than horizontal (right-lateral strike slip) since the Cenozoic. The recent Loma Prieta earthquake of 17 October 1989, agrees with Martin's hypothesis of vertical movement along some sectors of the SAF system.

In the plate-tectonic model, the ocean floor is considered relatively young, mainly basaltic, and is formed by sea-floor spreading along the midocean ridge. As a corollary, it is held that no crust flooring the modern oceans is older than about 200 million years. The three chapters in the section on the tectonics of the ocean basins are provocative and question the basic tenets of sea-floor spreading. A.A. Meyerhof, W.B. Agocs, I. Taner, A.E.L. Morris, and B.D. Martin find no support of spreading along the midocean ridges. They provide an exciting account of linear faults, fractures, and fissures demonstrated by side-scanning imagery, that extend for thousands of kilometers parallel with midocean ridge systems and are not predicted by plate tectonics.

These linear structures resemble those of glaciers and lava tunnels, and suggest that principal movement is parallel to the ridges instead of perpendicular to them. Moreover, Meyerhoff et al. observe that the age of the ocean floor does not increase systematically from the ridge crest, as required by plate tectonics. This raises a serious question abut the utility of linear magnetic anomalies for plate reconstruction. D.R. Choi, B.I. Vasil'yev, and M.I. Bhat conclude from combined evidence of seismic stratigraphy, dredging, and provenance study that the basement rock beneath the NW Pacific Ocean is essentially Precambrian continental shield overlain by younger geosynclinal sediments and volcanics. They identify various paleolands in the NW Pacific that bordered Japan during the Paleozoic and Early Mesozoic. These conclusions argue against the young age and basaltic composition of the entire ocean floor, and raise questions about the validity of the subductive and accretionary model of the Pacific plate. A similar conclusion is reached by J.M. Dickins, D.R. Choi, and A.N. Yeates. These authors document the claim that considerable areas of the present oceans were land or relatively shallow seas before the mid-Cretaceous; the great depths of modern oceans were established relatively recently, during the Neogene.

The next section critically assesses paleomagnetic evidence commonly used in plate reconstruction; the data presented are not congruent with the sea-floor spreading concept. K.M. Storetvedt synthesizes the Mesozoic-Cenozoic paleomagnetic data and proposes a novel hypothesis of "rotating plates" that is fundamentally different from conventional plate motions. He endorses the oceanization concept of Beloussov and restricts the use of plates to continental masses. Such plates, according to Storetvedt, show opposing rotational movements across the South Atlantic and Indian oceans and the Alpine-Himalayan mountain belt during K-T boundary time, and may be strongly linked to the Earth's rotation. Plate rotation generated a transpressive tectonic regime in the oceanic domain, which culminated along the midocean ridge system to form a new class of Alpine foldbelts. With the termination of opposing plate motion in the Late Eocene, the midocean ridge system entered a tensional phase, producing magmatic activity and a central median valley. W.B. Agocs, A.A. Meyerhoff, and K. Kis analyze quantitatively 15 magnetic profiles across the Reykjanes Ridge, Iceland. They conclude that linear magnetic anomalies on midocean ridges are not necessarily a "taped record" of normal and reversed polarity during sea-floor spreading, as is generally assumed in plate tectonics, but can be explained more effectively by differences in magnetic-susceptibility contrasts.

On the paleontological side, India fails to show the endemism expected as a consequence of its putative voyage across the Southern Ocean, whereas Australia shows endemism inconsistent with its putative attachment to Gondwana. The chapters in the next section show the application of fossil evidence to problems involving paleobiogeography and continental drift. C.J. Smiley traces distribution patterns of Late Paleozoic, Mesozoic, and Cenozoic terrestrial plants and tetrapods, and of contemporaneous marine faunas, and concludes that they do not support the concept of moving plates on a large scale. Instead, these data are consistent with continental and polar stability, and suggest that North America, India, and the Australia-New Guinea block moved very little from their Paleozoic to their present positions. R.E. Molnar reviews the highly endemic Early Cretaceous terrestrial fauna of Australia, holding that it arose in isolation from South America and Africa during this interval. He emphasizes that some Australian Mesozoic taxa are relicts, which indicates that the factors that led to their extinction overseas did not affect Australia. N. Hotton examines the global cosmopolitanism of a wide variety of taxa of Permian and Triassic terrestrial tetrapods, and finds better support for a Pangean hypothesis than for continental separation like that of the present. However, he also reports large numbers of endemic families, some of them obviously relicts of earlier distribution, indicating local isolation that was probably a consequence of separation by distance and climate.

A central element of plate tectonics is the assumption that the radius of the Earth has remained constant throughout geologic history, so that the creation of lithosphere at the ridge must be compensated by destruction at the trench. Paleomagnetic data tend to suggest that the Earth's radius has not changed significantly since the Paleozoic, but these results are not precise enough to discount contraction or expansion of a few percent during the last 400 million years. The six chapters in the final section provide alternative models to plate tectonics, hypothesizing the Earth as expanding, contracting, static, or oscillating. H.G. Owen argues cogently for a slowly expanding Earth to explain continental breakup and sea-floor spreading data. He uses precise cartographic methods and satellite laser ranging data to suggest that Earth, before the breakup of Pangea (180-200 Ma), had attained about 80% of its modern diameter, and has gained the remainder since that time. G.O.W. Kremp favors the concept of a rapidly expanding Earth as advocated by S.W. Carey and K. Vogel. In this model, Earth had attained only 60% of its modern diameter by 200 Ma. The slowing down of Earth's rotation may be linked to Earth expansion. An opposing view, that of a contracting Earth, is proposed by A.A. Meyerhoff, I. Taner, A.E.L. Morris, B.D. Martin, W.D. Agocs, and H.A. Meyerhoff. These authors present a new and elegant hypothesis of earth dynamics, termed "surge tectonics," which they document with a wealth of geological and geophysical evidence. Meyerhoff et al. have mapped bands of high heat flow and microseismicity beneath every conceivable large tectonic feature of Jurassic or younger age—midocean ridges, aseismic submarine ridges, oceanic rises, linear island and seamount chains, eugeosynclines and volcanic arcs, foldbelts, wrench fault systems, long linear to curvilinear valleys, actively rising mountain chains and plateaus, and intracratonic basins. They identify these bands as a global network of interconnected surge channels containing fluid magma. Lithosphere compression, generated by a cooling and contracting Earth, is the prime mover of this geodynamic. As this compression increases through time in a geotectonic cycle, it causes the magma to move in a surge channel and eventually to rupture it, so that contents of the channel surge bilaterally upward and outward to initiate tectogenesis. The authors claim that every important structure of the lithosphere, every major tectonic process, is genetically linked to surge tectonics. On the basis of a static Earth model, V.V. Beloussov postulates a mechanism for the evolution of the tectonosphere (lithosphere) by endogenic processes such as tectonism, magmatism, and metamorphism. Both constructive and destructive processes operate in the formation of the tectonosphere. An older stage (pre-Mesozoic) is constructive, involved in the formation of continental crust across the Earth's surface, and is closely related to calc-alkaline magmatism. A younger stage (post-Mesozoic) involves destruction of continental crust and its replacement with oceanic crust (oceanization) by tholeiitic magmatism. F.-C. Wezel presents an integrated approach to explain geodynamic processes in terms of a pulsating rhythm, alternating between contraction and expansion of the lithosphere. Contraction produces an array of simultaneous global events such as normal geomagnetic polarity, marine transgression, diastrophism, oceanization, flysch deposition, flooding, climatic amelioration, and diversity of species. Expansion, on the other hand, creates reversal of the geomagnetic field, marine regression, acid plutonism, the morphotectonic phase of mountain building, and a decline in global temperatures. Shifts of contraction to expansion and back cause climatic and magnetic instabilities that may trigger biotic crisis. All geological phenomena are manifestations of a continuous oscillation between destruction and regeneration of continental crust, and may be linked to Earth's rate of rotation. Wezel believes that the lithosphere is crisscrossed by megashears along which active tectonism takes place. Finally, L.S. Smirnoff expands the concept of a contracting-expanding Earth in relation to its parameters and cosmological context. Several parameters of Earth, such as its radius, gravitational acceleration, and angular velocity, possibly change through time, in concert with the geologic cyclicity of tectonism, magmatism, and sedimentation.

Collective works are difficult to publish, the more so when the topic is as controversial as global tectonics. To assure that contributions met a high standard of quality, each manuscript was peer-reviewed by at least two referees, and closely edited for uniformity of style. For their help in meeting this requirement, we gratefully acknowledge the efforts of the following: D.L. Baars, J.F. Bonaparte, A.J. Boucot, S.E. Cebull, A.R. Crawford, V. S. Cronin, J.M. Dickins, E. Erlich, W. Glenn, A.C. Grant, G. Gray, M. Hill, M. Kamen-Kaye, A. Kaplan, G.O.W. Kremp, P.D. Lowman, Jr., N.J. McMillan, A.A. Meyerhoff, G. Mitra, H.G. Owen, J. Pierce, T. Rich, D.A. Russell, H. Schorn, N.C. Steenland, K.M. Storetvedt, D. Tucker, S. Uyeda, P.J. Wyllie, F.-C. Wezel, and T. Yeates. For special help and editing we thank Beth Davidow-Henry and Sanghmitra Kundu. We would also like to thank the editorial staff of Texas Tech University Press, especially Carole Young and Wendell W. Broom, for their support, encouragement, and patience. The cover photograph of India was provided by NASA. Financial support for this project was provided in part by Texas Tech University and the Smithsonian Institution. Special thanks are due J.M. Dickins, Curt Teichert, and A.J. Boucot for the initial planning of the conference.

Sankar Chatterjee and Nicholas Hotton III
30 September 1991

Plate Tectonics: An Overview

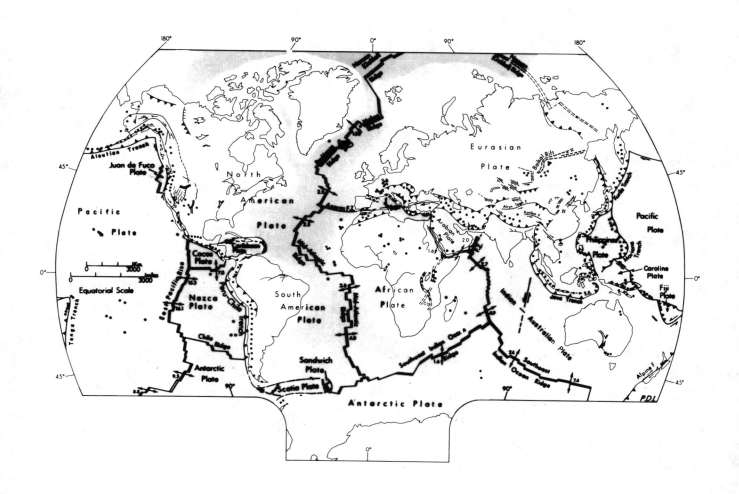

Plate tectonics and continental drift in geologic education

Paul D. Lowman, Jr., Goddard Space Flight Center, Geology and Geomagnetism Branch (Code 622), Greenbelt, Maryland 20771 USA

ABSTRACT

The undesirable effects of plate-tectonic theory on geologic education at all levels, kindergarten through graduate school is discussed. The theory is a useful pedagogical tool in being coherent, simple, visual, and easily taught. However, these strengths are outweighed by weaknesses: deceptive simplicity, limited scope, and possibly fundamental incorrectness. These problems are already slowing geologic progress by inducing narrow, rigid viewpoints, and discouraging research by convincing students that the main problems are solved. Suggested remedies include: teaching plate tectonics as a theory requiring continual testing; inclusion of plausible alternative theories (contracting Earth, expanding Earth, surge tectonics, oceanization, and plate tectonics without continental drift); devoting adequate time to Archean, extraterrestrial, and intraplate geology, and to anomalies not explained by plate tectonics (e.g., New Madrid earthquake).

INTRODUCTION

There is currently a remarkable and dismaying difference between the intellectual climate in astrophysics and geology. Astrophysics is alive with controversies and new theories stimulated by recent discoveries: quasars, superluminal quasars, the missing mass, homogeneity of the cosmic 3° background radiation, and other first-order problems. Geology was similarly alive in the 1960s when plate tectonics suddenly blossomed. However, it has recently, with few exceptions, become a bland mixture of descriptive research and interpretive papers in which the interpretation is a facile cookbook application of plate-tectonics concepts: sutures, terrane accretion, opening of the Atlantic, and passive margins used as confidently as trigonometric functions. This situation appears to stem largely from the degree to which plate tectonics has saturated geologic education.

The theory of plate tectonics and its presumed corollary of continental drift within the last 15 years have come to dominate geologic education on all levels from grade school through graduate school, as a "master plan into which everything we know about the earth seems to fit" in the words of a leading text (Hamblin, 1978). The purpose of this paper is to point out some of the undesirable effects of this dominance and to suggest remedies.

PEDAGOGICAL STRENGTHS OF PLATE-TECTONIC THEORY

It is helpful to outline the reasons for the dominance of plate-tectonic theory in education by summarizing its very real strengths as a teaching device. Plate tectonics has come to mean almost anything relevant to regional structure and crustal dynamics. In its original sense it consists of three elements: the ridge (or spreading center), the trench (or subduction zone), and the transform fault (Morgan, 1968). Plates are the relatively rigid and inactive segments of lithosphere bounded by some combination of these three elements (Fig. 1). Proponents of the theory generally believed that most tectonism and volcanism occur at plate boundaries, although these are often wide and ill-defined (e.g., the Basin and Range Province). Continental drift is almost universally considered a consequence of plate movement, apparently circumventing the geophysical problems of motive force and crustal resistance for which Wegener (1929) had no answer.

Plate-tectonic theory is an enormously powerful and helpful conceptual framework for teaching geology, whatever else may be said about it. First, it is fundamentally simple and easy to understand, even by students in the lower primary grades. An indirect benefit of this simplicity is that the theory is easy for teachers to assimilate, an important factor because many grade and high school earth science teachers have no background in geology, being pressed into service for lack of anyone else. A second pedagogical strength of plate tectonics is that the concept is easily visualized, with devices such as continent jigsaw puzzles, and easily dramatized with photographs or videotapes of volcanic eruptions, earthquakes, and the like. Orbital photography has been used by the author (Lowman, 1985a) to illustrate the theory, taking advantage of its global coverage and synoptic perspective.

Perhaps the most important educational strength of plate-tectonic theory, especially for university level teaching, is its fundamental coherence and broad base of apparently supporting evidence. Orogenesis was taught in graduate level courses even in the late 1950s as a process whose cause was an unsolved problem. In contrast, plate-tectonic mechanisms appear to provide both general and specific origins for orogenic belts and many types of geosyncline (a term nearly banished by plate-tectonic theory). Furthermore, there are several independent lines of evidence that can be called upon to support the theory. Recent long-range geodetic measurements in the Pacific Basin in particular appear to demonstrate sea-floor spreading and, by implication, subduction under the Andes. These fundamental

Chatterjee, S., and N. Hotton III, eds. *New Concepts in Global Tectonics.*
Texas Tech University Press, Lubbock, 1992, xii + 450 pp.

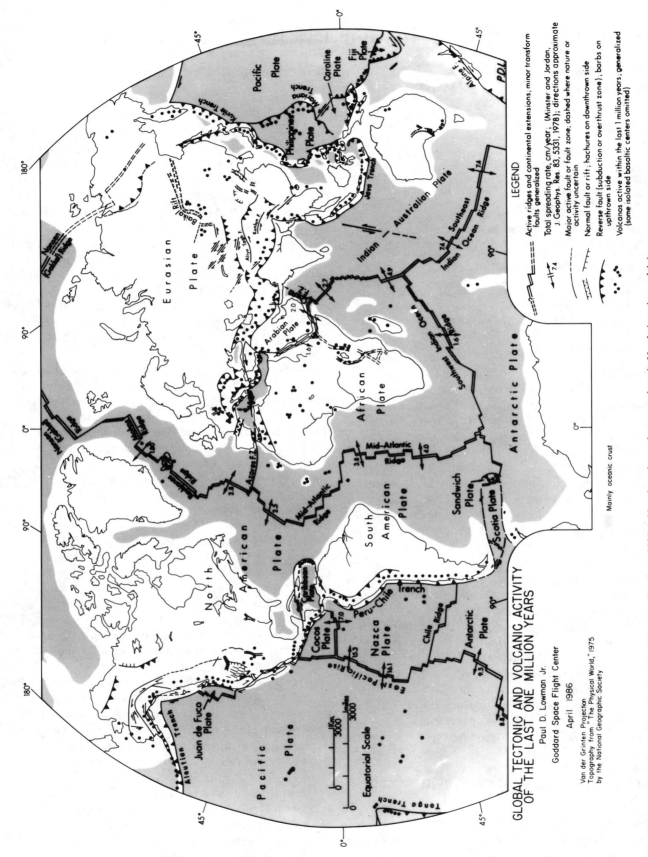

Figure 1. Global tectonic and volcanic activity map (Lowman, 1985a). Note broad areas between plates in North America and Asia.

strengths are the main reason for acceptance of plate tectonics at the upper levels of geologic education, and by the profession as a whole. Let us now consider its weaknesses.

PEDAGOGICAL WEAKNESSES OF PLATE-TECTONIC THEORY

This paper is a critique of plate tectonics as a teaching approach, not as a scientific theory. The latter topic has been covered by the author elsewhere (Lowman, 1985b, 1985c), and by Meyerhoff, Beloussov, and others cited. However, viewed from an education viewpoint, plate-tectonic theory appears to have several weaknesses.

Deceptive simplicity

The very strength of plate-tectonic theory, as just discussed, is also a major weakness. The plate maps and attendant explanations found in all modern texts are over-simplified, in that they concentrate on features that actually may be well-explained by the theory, such as the midocean ridges, transform faults, and young circum-Pacific mountain belts. But they omit, because plate tectonics does not explain them, many major crustal features and characteristics, such as the following.

The structural symmetry of North America, first described in a prescient paper by Keith (1928), seems increasingly real as our geologic knowledge increases. The tectonic map of North America by King (1969) showed that most major features of the North American Cordillera can be matched by similar features in the Appalachian foldbelt if we allow for age and superimposed features such as the Basin and Range Province. On each side of the continent there are thin-skinned miogeosynclinal overthrusts, chains of batholiths, and metamorphic belts interpreted by King as eugeosynclinal. Orthodox plate-tectonic theory explains the Appalachian foldbelt as resulting from repeated opening and closing of the Atlantic, culminating with a climactic collision of North America with Eurasia. This collision was followed immediately by the classic Wegenerian drift episode. Regardless of its inherent plausibility (which the writer regards as low), we should ask how such a series of events could have occurred in the eastern Pacific Basin, assuming a constant radius Earth. There is no Eurasia with which western North America could have collided or rifted from to produce essentially analogous Phanerozoic orogenic belts. Plate-tectonic theory thus fails to explain Keith's "symmetry."

It should be said at once that the newer theory of terrane accretion (Ben-Avraham et al., 1981; Williams and Hatcher, 1982), involving collisions of smaller crustal units, does not encounter this objection and appears inherently more plausible. But proposing a culminating megacollision, so to speak, to produce an orogenic belt essentially similar to the Cordillera appears unnecessary.

The list of features or phenomena not explained by plate-tectonic theory could be greatly enlarged. The widespread Proterozoic rhyolitic volcanism of the North American midcontinent region has no obvious relation to subduction. The 1811 New Madrid earthquake, in the middle of a continent supposedly moving westward as part of a single plate, is inexplicable by plate tectonics. (A recent review by Johnston and Kanter [1990] discussed the enigmatic nature of the entire class of "stable" crust earthquakes.) Leading plate-tectonics advocates agree that our understanding of intracontinental basins (for example, Michigan, Hudson Bay, Williston) is unsatisfactory (Dickinson, 1974; Bally, 1989). Even Phanerozoic foldbelts, confidently explained as the result of continental or arc-continent collisions in plate-tectonics treatments, do not fit as well as commonly assumed. The foldbelts of central Australia, for example, appear to have formed on normal continental crust and, moreover, to have evolved over 600 million years (Lambeck, 1986), completely contrary to plate-tectonic interpretations. Papers given by Australian geologists at the 1989 International Geological Congress were notable for treatments in which plate-tectonic concepts were hardly mentioned in relation to rifts, foldbelts, and volcanism.

To summarize the problem of deceptive simplicity, plate-tectonic theory is easy to teach and easy to learn in part because it simply ignores several major categories of structure and crustal characteristics, concentrating on what is actually a rather narrow (though important) range of features. It would be well to remember A. N. Whitehead's advice: Seek simplicity, and distrust it.

Induced overconfidence

The apparent simplicity of plate-tectonic theory is producing whole generations of students who believe that they really understand the way the Earth works (Wyllie, 1976), the title, ironically, of one of the few textbooks to present a balanced treatment of tectonic theory. As Anita Harris (quoted by McPhee, 1983) put it: "I get all heated up when some sweet young thing with three geology courses tells me about global tectonics, never having gone on a field trip to look at a rock." Similar overconfidence is evident also in popular science writing. Most popular accounts of geologic events such as earthquakes casually refer to, for example, the supposed collision of Africa with Europe, unaware that two leading proponents of plate-tectonic theory (Burke and Wilson, 1972) concluded, from the absence of hot-spot trails, that Africa has not moved appreciably for the last 25 million years or so. Opening of the Atlantic is referred to in even the scientific literature as confidently as if the event had happened during the 1969 Atlantic City G.S.A. meeting. It is interesting to note that this overconfidence is felt primarily by young geologists and others with limited knowledge of geology. Leading plate-tectonics advocates, such as Dewey, MacKenzie, Heirtzler, Sengor, Molnar, and Atwater frequently call attention to the complexities and unexplained problems of tectonics. Unfortunately, their awareness of these difficulties rarely finds expression in textbooks or popular science presentations.

Effects of this growing overconfidence can be foreseen, one being a failure of coming generations of geologists to recognize problems. It is widely realized, even among nonscientists, that ability to solve problems is far more common than ability to recognize them in the first place. A specific example of the bad effect of overconfidence is related to the problem of how continental crust was formed (Lowman, 1989a, 1989b). The writer's interest in this subject was aroused as an undergraduate in 1953, when many leading scientists, such as Wilson, Kay, Vening Meinesz, Umbgrove, and Hills were publishing papers and books on the origin of continents. No alert student could fail to realize that here was a fruitful field for research. But today's geology students are exposed to masterful expositions of the lateral accretion theory in its plate tectonic incarnation that give no hint of the theory's critical weaknesses. These treatments, furthermore, cover all levels of geologic education, and few students have the background or the confidence to question the authoritative views of, for example, Windley (1984), Bally (1989), or Hoffman (1989). The origin of continental crust may have been solved by plate-tectonic theory. But if it has not, students raised on this theory will probably not do so, because they have been taught essentially that there is no longer a problem. They are confident that plate tectonics has answered not only this but most other great questions of physical geology.

Restricted scope

Despite Hamblin's (1978) description of plate tectonics as a master plan, it is becoming increasingly clear that there are many things that do not fit this plan. To put it somewhat differently, plate tectonics is most useful for, if not actually restricted to, ocean basins, Andean-type margins, and transform faults in explaining geologic phenomena currently observable.

The problem of restricted scope already has been touched upon by implication in previous sections, but deserves further discussion. There are entire fields in geology to which plate tectonics has little if any application. The origin of intracratonic basins has been recognized for decades as essentially untreatable by a plate tectonic approach (Dickinson, 1972; Bally et al., 1989). The formation of Precambrian dyke swarms appears to bear no relation to recognizable plate boundaries. Furthermore, flood basalts cover much of the Moon, Mars, and Venus, representing basaltic magmatism that was clearly independent of plate tectonics (Lowman, 1989a). The evolution of Archean granite-greenstone terranes, in contrast to Proterozoic foldbelts that can be explained reasonably by some form of plate tectonics (Kroner, 1985; Hoffman, 1988), has appeared until very recently unapproachable by plate tectonic concepts. It was only at the 1989 Montreal meeting of the Geological Association of Canada that Archean geology was interpreted extensively in terms of plate tectonics, and one must wonder if this was not because most geologists now active in the field know no other approach.

Many classic problems in geomorphology, such as the origin of peneplains (still controversial, as summarized by Bloom, 1978), appear untouched by plate-tectonic theory. The processes of continental hydrothermal ore deposition, with a few notable exceptions, have not been clarified by plate tectonics. It is significant that a leading text on economic mineral deposits published long after the new global tectonics revolution of the mid-1960s (Jensen and Bateman, 1979) has only three index entries under "plate tectonics." The formation of continental crust is widely believed to have been essentially a plate tectonic process by which island arcs and other terranes are accreted. However, the thin-skinned structure, chemistry, relative ages, and structural relation of greenstone belts to adjacent granitoids argue strongly against such accretion as a means of forming Archean crust (Fig. 2). Furthermore, other planets (Fig. 3) on which plate tectonics could never have been effective have analogous crusts (Lowman, 1989b), implying that planetary differentiation is quite possible without plate tectonics.

Potential fundamental falsity

The most serious charge against plate tectonics as an educational device is also the simplest: the theory may be wrong. This suggestion would be dismissed by most western geologists, students, and recent graduates. But ruling theories have been overthrown before, as the words of A. N. Whitehead (Whyte, 1956) remind us: "By the middle of the 1890s there were a few tremors, a slight shiver, but no one sensed what was coming. By 1900 Newtonian physics were demolished. Done for! It had a profound effect on me. I have been fooled once, and I'll be damned if I'll be fooled again."

It is at least conceivable that a similar fate may await plate-tectonic theory in its present form. It must be

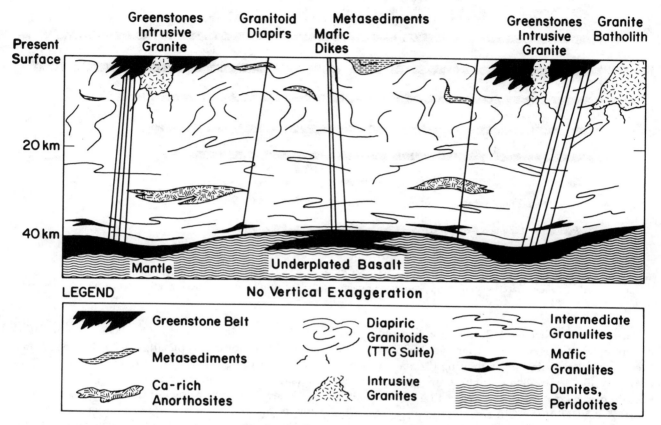

Figure 2. Structural and lithologic cross section of typical Archean crust (e.g., Superior Province, Candian Shield). Note shallow root zones of greenstone belts, implying that lateral accretion of such cannot account for formation of entire thickness of crust (Lowman,1989b).

remembered that even neglecting the discoveries that await us in space exploration, there is a largely unseen planet under our feet. Seismic tomography (Anderson, 1987) and reflection profiling, for example, are revealing totally unexpected features of the mantle and crust, such as the apparent thin-skinned structure of the eastern U.S. piedmont. Dense networks of ultra-precise geodetic measurements are being established with the Global Positioning System and similar space-related techniques, and we should have a true picture of crustal deformation within ten years or so.

These new techniques may of course confirm plate tectonic theory. Data from satellite laser ranging and radio interferometry strongly support sea-floor spreading in the Pacific Basin. But if the theory is overturned, an entire generation of geologists will have been seriously misled. Physics recovered from the 1900 revolution, and presumably geology could recover from a similar event. But it is unfair to students not to prepare them for this possibility.

WHAT IS TO BE DONE?

Several remedies for the foregoing difficulties can be suggested.

Teach plate tectonic theory—This fairly obvious approach was followed with respect to continental drift by leading geology teachers of the 1940s: teach plate tectonics as a stimulating, useful, and interesting theory, but only a theory, requiring continual examination and testing. Students at upper levels should be reminded that although Einstein's general relativity theory was verified immediately by Eddington's 1919 eclipse observations, physicists 70 years later still test it at every opportunity. Plate tectonics should similarly be tested repeatedly, and students should be taught to expect this.

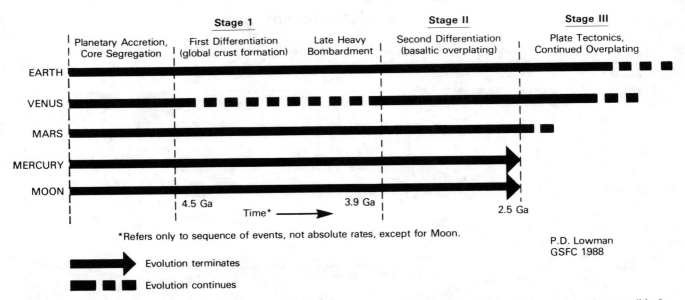

Figure 3. Crustal evolution in Moon, Mars, Mercury, Venus, and Earth (Lowman, 1989b). First differentiation considered responsible for formation of earliest continental crust, well before plate tectonic stage. Analogous crusts on other bodies imply that plate-tectonics is not necessary for differentiation.

Teach other theories—Although no one would guess it from current western texts, there are serious alternatives to plate tectonics for explaining the way the Earth works (Wyllie, 1976). These include the classic contraction theory; its opposite, the expanding Earth (Carey, 1981); oceanization (Beloussov, 1989); surge tectonics (Meyerhoff et al., this volume); Wegener's original theory; or modified versions of plate tectonics (Lowman, 1985a). Each of these has been published in reviewed journals, and although they obviously can not all be correct, each has elements of truth. Wyllie (1976) is almost alone among present textbook authors in giving something like adequate space to non-plate tectonic theories; it is recommended that others emulate him.

Discuss anomalies—Perhaps the most damaging tactic that can be used in scientific controversies is to ignore contradictory evidence, anomalies, and dissenting views. For example, the deep structure of continents, a critical problem for orthodox plate-tectonic theory, has been known to geophysicsts since MacDonald's (1964) paper, and Jordan (1979) repeatedly presented new evidence supporting the concept. Yet virtually all American geology texts continue to show diagrams illustrating continent-bearing plates moving on a lithosphere of uniform thickness. Similarly, several lines of evidence indicate that slab pull must be the driving force for plate movement (Lowman, 1985c), but the question of what drives plates with nonsubductable (continental) leading edges is universally ignored. To tell students about such anomalies is not simply picking out toads and smoads, as the saying goes, but warning them of potentially fatal flaws in the conventional wisdom.

Cover all aspects of tectonics—By this recommendation is meant simply devoting adequate attention in geology courses to aspects not easily explained by plate tectonic theory, even if the theory is proven correct. An obvious example is comparative planetology. The Moon somehow was thoroughly differentiated early in its history, clearly with no help from plate tectonics. Mars also may have been so differentiated. The geology of Venus, with foldbelts remarkably similar to those of the Tethyan foldbelt, can be explained by plate tectonics only if we apply Kroner's (1981) ensialic version, a theory so different as to represent a wholly new approach. Similarly, Archean geology, occupying nearly half the geologic life span of Earth, should be given adequate attention not in spite of its intractability with plate tectonics but because of this intractability. Recent texts such as those by Cloud (1987) and Nisbet (1987) give adequate space to the Archean; teachers should take advantage of these.

SUMMARY

The theory of plate tectonics has given earth sciences for the first time a master plan of compelling simplicity, beauty, and coherence, one which has had an enormously stimulating effect on both formal education and popular presentations of geology. But as taught in most geology courses today, it can have the opposite effect: it can stifle research on the great problems of tectonics, by convincing students that these problems have already been solved. This effect was best summarized in Gilbert Highet's (1950) concise classic, *The Art of Teaching:*

> One of the worst depressants in school learning is the feeling that everything is already known and filed away, that knowledge is all dead wood it does them good to feel that, by hard work, they will go further; and one way to inculcate this feeling is to show them problems which the best brains have so far failed to solve. They may never solve them. They may not even try but they will always benefit from having learned that human knowledge is expanding, and that its expansion is a stimulus to our will power, our brains, and our co-operation

Geology teachers at all levels, as well as science popularizers, should try to avoid this pitfall, to broaden their treatments, and to remember that the theory of plate tectonics was not delivered on Mount Sinai carved in stone.

REFERENCES CITED

Anderson, D.L., 1987, Global mapping of the upper mantle by surface wave tomography, *in* Fuchs, K., and Froidevaux, C., eds., Composition, structure, and dynamics of the lithosphere-asthenosphere system, Geodynamics Series, v. 16, American Geophysical Union, p. 89-97.

Bally, A.W., 1989, Phanerozoic basins of North America, *in* Bally, A.W., and Palmer, A.R., eds., The geology of North America: an overview, Geological Society of America, p. 397-446.

Ben-Avraham, Z., Nur, A., Jones, D., and Cox, A., 1981, Continental accretion: from oceanic plateaus to allochthonous terranes: Science, v. 213, p. 47-54.

Beloussov, V.V., 1989, Endogenic regimes: Interactions between the upper mantle and crust (abstract): Proceedings of the 28th International Geological Congress, p. 1-118.

Bloom, A.L., 1978, Geomorphology: Englewood Cliffs, N.J., Prentice-Hall, 510 p.

Burke, K.C., and Wilson, J.T., 1972, Is the African plate stationary? Nature, v. 239, p. 387-389.

Carey, S.W., 1981, The expanding Earth: A symposium: Hobart, Tasmania, University of Tasmania, 423 p.

Cloud, P.C., Jr., 1987, Oasis in space: San Francisco, Freeman, 357 p.

Dickinson, W.R., 1974, Plate tectonics and sedimentation, *in* Dickinson, W.R., ed., Tectonics and sedimentation, Society of Economic Paleontologists and Mineralogists Special Publication 22, p. 1-27.

Hamblin, W.K., 1978, Earth's dynamic systems: Minneapolis, Burgess, 459 p.

Highet, G., 1950, The art of teaching: New York, Random House, 259 p.

Hoffman, P.F., 1988, United plates of America, the birth of a craton: Early Proterozoic assembly and growth of Laurentia: Annual Reviews of Earth and Planetary Science, v. 16, p. 543-603.

——, 1989, Precambrian geology and and tectonic history of North America, *in* Bally, A. W., and Palmer, A. R., eds., The geology of North America: an overview, Geological Society of America, p. 447-512.

Jensen, M.L., and Bateman, A.M., 1979, Economic mineral deposits (3rd ed.): New York, John Wiley, 593 p.

Johnston, A., and Kanter, L.R., 1990, Earthquakes in stable continental crust: Scientific American, v. 262, p. 68-75.

Jordan, T., 1979, The deep structure of continents: Scientific American, v. 240, p. 92-107.

Keith, A., 1928, Structural symmetry in North America: Geological Society of America Bulletin, v. 39, p. 321-386.

King, P. B., 1969, Tectonic Map of North America, 1:5,000,000: Washington, U.S. Geological Survey.

Kroner, A., 1981, Precambrian plate tectonics, *in* Kroner, A., Precambrian plate tectonics: Amsterdam, Elsevier, p. 57-90.

Lambeck, K., 1986, Crustal structure and evolution of the central Australian basins, *in* Dawson, J.B., Carswell, D.A., Hall, J., and Wedepohl, K.H., eds., The nature of the lower continental crust: Geological Society Special Publication 24: Oxford, Blackwell Scientific Publications, p. 133-145.

Lowman, P.D., Jr., 1985a, Geology from space: a brief history of orbital remote sensing, *in* Drake, E.T., and Jordan, W.M., eds., Geologists and ideas: a history of North American geology: Centennial Special Volume 1, Geological Society of America, 525 p.

——, 1985b, Plate tectonics with fixed continents: a testable hypothesis: Journal of Petroleum Geology, v. 8, p. 373-388, continued in v. 9, p. 71-87, 1986.

——, 1985c, Mechanical obstacles to the movement of continent-bearing plates: Geophysical Research Letters, v. 12, p. 223-225.

——, 1989a, Origin of the continental crust: an example of differentiation in silicate planets (abstract): Proceedings of the 28th International Geological Congress, p. 2-329.

——, 1989b, Comparative planetology and the origin of continental crust: Precambrian Research, v.44, p. 171-195.

MacDonald, G.J.F., 1964, The deep structure of continents: Science, v. 142, p. 921-929.

McPhee, J., 1983, In suspect terrain: New York, Farrar, Straus, and Giroux, 210 p.

Morgan, W.J., 1968, Rises, trenches, great faults, and crustal blocks: Journal of Geophysical Research, v. 73, p. 1959-1982.

Nisbet, E.G., 1987, The young Earth: An introduction to Archean geology: Boston, Allen and Unwin, 402 p.

Wegener, A., 1929, The origin of continents and oceans, English translation by J. Biram, 1966: New York, Dover Publications, 246 p.

Whyte, W.H., Jr., 1956, The organization man: Garden City, N.Y., Doubleday, 471 p.

Williams, H., and Hatcher, R.D., Jr., 1982, Suspect terranes and accretionary history of the Appalachian orogen: Geology, v. 10, p. 530-536.

Windley, B.F., 1984, The evolving continents (2d ed.): New York, John Wiley, 399 p.

Wyllie, P.J., 1976, The way the Earth works: New York, John Wiley, 296 p.

ACKNOWLEDGMENTS

This paper owes much to discussions with Art Meyerhoff, Bruce Martin, Jim Heirtzler, Barbara Christy, Ruth Freitag, John Dewey, Kevin Burke, and Hank Williams, though few if any agree with my views on plate tectonics. A review by S. E. Cebull clarified several unclear or poorly expressed sections. I wish to express very belated thanks to Larry Warner, University of Colorado, for giving me a solid grounding in tectonic theory as it was in 1957 and an open mind on the subject, accomplishments that are becoming increasingly rare today.

Geometry of Plate Tectonics

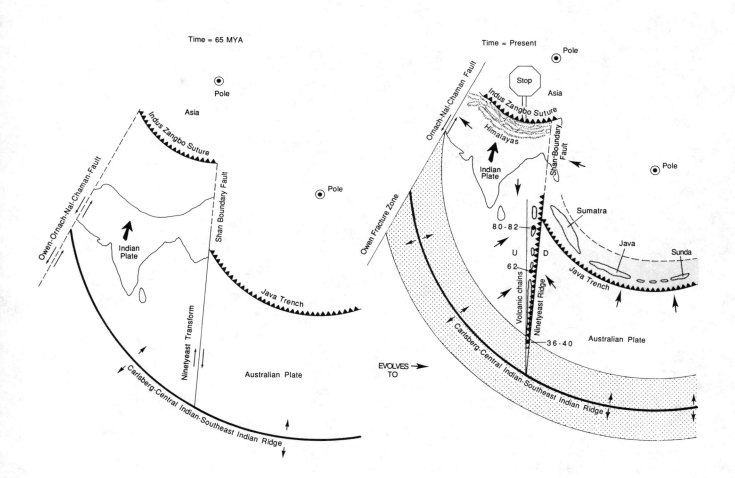

A kinematic perspective on finite relative plate motion, provided by the first-order cycloid model

Vincent S. Cronin, Department of Geosciences, University of Wisconsin–Milwaukee, P.O. Box 413, Milwaukee, Wisconsin 53201 USA

ABSTRACT

Assume Earth to be a sphere, in the interest of obtaining a tractable model of the finite relative motion of plates (used herein as relative motion). Relative motion cannot be along a line, because a plate cannot diverge from Earth's surface. The simplest curve that may be a path of relative motion on a sphere is a figure of rotation around one axis: a circle. In a given three-plate system, it is conceivable that plates A and B could move along a circular path relative to plate C; however, plates A and B would not move along circular paths relative to each other unless the three poles of relative motion are coaxial throughout the displacement. Solutions currently available for the instantaneous relative motion of plates indicate that the poles of relative motion are not coaxial in any three-plate system, so circular relative motion trajectories are atypical. Models that assume every plate moves around a set of finite relative motion axes that are fixed to the plate are not generally valid. Hence, transform faults should not be assumed to be surfaces along which there is neither convergence nor divergence, and fracture zones should not be assumed to have a small-circle shape.

The next simplest curve that can be a path of relative motion is a figure of rotation around two axes: a two-axis spherical cycloid. Imagine a point on plate A rotates around the pole to plate A and is observed from plate B while the observer rotates around the pole to plate B. The trajectory of the point on plate A would be a two-axis spherical cycloid as observed from plate B during relative motion. Instantaneous, plate-specific motion characteristics can be determined using instantaneous, relative plate motion data. The first-order cycloid model (CYC1) assumes (1) Earth is a unit sphere; (2) plates are rigid spherical shells, except adjacent to plate edges; (3) there is relative motion between all plates under consideration; (4) the axial vector that describes individual plate motion is constant for all plates under consideration; and (5) the angular relationships between all axial vectors that describe individual plate motion are constant. Even though the CYC1 model assumptions are quite restrictive, CYC1 can be employed usefully to gain an understanding of many boundary processes dependent upon relative motion. The variation in velocity and direction of relative motion can be predicted using CYC1, which also has been used to reproduce the observed sigmoid long-wavelength shapes of fracture zones between the African and South American plates.

It is not possible to uniquely define relative motion curves that are more complex, involving more than two axes of rotation, without either (1) a complete, detailed knowledge of the nature and temporal variation in plate dynamics or (2) a complete, detailed, empirical knowledge of the kinematic history of relative motion. The former is not currently available, and the latter would negate the need for relative motion modeling. The CYC1 model permits useful modeling of relative motion, even in the absence of complete dynamic or empirical kinematic data. When additional data are available, the cycloid model can be refined to improve accuracy. The cycloid model is particularly useful in applications that can utilize present-day instantaneous relative-motion data along with data from oceanic fracture zones and marine magnetic anomalies.

PREFACE

Several of the participants in the workshop on New Concepts in Global Tectonics found it difficult to accept plate tectonics, in whole or in part. I find plate tectonics to be a useful paradigm for many branches of research in the geosciences, and I am persuaded by the preponderance of evidence that indicates that Earth's surface is a mosaic of lithospheric plates that are in motion with respect to one another, substantially as described by Wegener (1929). Plate tectonics provides an excellent framework with which to understand the interrelationships among the thermal, magmatic, structural, seismic, sedimentary, and other geological processes associated with plate motion.

What, then, is plate tectonics? This question is asked in response to statements such as "Plate tectonics cannot explain this" or "This is the cornerstone theory of plate tectonics." The first statement describes plate tectonics as an oracle, whereas the second implies that there are theoretical points which, if refuted, would cause the edifice of plate-tectonic theory to collapse. Thousands of technical papers have been written on various aspects of plate tectonics in the years since Harry Hess published his exercise in geopoetry (Hess, 1962). We might define plate-tectonics as the summation of hypotheses and theories that are founded in physical laws and that explain the observed motion of lithospheric plates through the mechanisms of subduction and sea-floor spreading. The breadth of this literature and the lack of a good textbook summary of modern plate-tectonic theory can be obstacles hindering education in tectonics, although several fine historical treatments of the early development of plate tectonics have been written (e.g., Tarling and Tarling, 1971; Hallam, 1973; Uyeda, 1978; Glen, 1982; Weiner, 1986; Allegre, 1988) along with an excellent early textbook (Le Pichon et al., 1973).

Those who do not accept plate tectonics often focus on the viability of "cornerstone theories" to promote alternative global models. The actual cornerstone of plate tectonics is the collection of observations that unambiguously shows that, among other things, (1) Earth's outer surface is composed of an interlocking array of lithospheric plates, (2) most earthquake and magmatic activity occurs along the margins of plates, (3) all plates are in motion relative to one another, (4) plates diverge

---

Chatterjee, S., and N. Hotton III, eds. *New Concepts in Global Tectonics.* Texas Tech University Press, Lubbock, 1992, xii + 450 pp.

along midocean ridges, where new crust forms to fill the gaps, in a process known as sea-floor spreading, (5) mantle lithosphere is resorbed as it sinks into the deeper mantle in subduction zones, (6) the bathymetry of ocean basins is related to the variation in age and thermal structure of the plates, (7) the apparent offset of midocean-ridge segments is opposite to the actual sense of slip along the intervening transform fault, (8) oceanic fracture zones emanate from both ends of a given transform fault, (9) collision of continental lithosphere is a mechanism for the generation of mountain ranges, and (10) the oldest continental crust is Archean in age, whereas the oldest oceanic crust that is still in an ocean basin is Jurassic.

Any model devised to account for the characteristics of the entire lithosphere through time must be physically possible and robust enough to explain the observational data. The ever-evolving plate-tectonic model appears to be the only model currently available that can explain the observational data and make useful predictions. If another model is proposed that can improve our understanding of Earth processes, then all geoscientists should be open and interested; however, I am not aware of any model currently available that is able to explain observational data regarding the lithosphere as successfully as the plate tectonic model. Without compromising scientific skepticism, plate tectonics is accepted as a paradigm in modern geosciences.

INTRODUCTION

Finite relative motion is the motion of one object as observed from another object over an extended period of time. In the context of plate tectonics, finite relative plate motion is the motion of one plate as observed from another plate through a finite time interval. If the history of motion and deformation along a plate boundary is of interest, then the finite relative motion of plates should be investigated to understand the boundary conditions that existed along the plate margin throughout its evolution. For example, it is not sufficient to know the relative motion of the Pacific and North American plates today to extrapolate the history of the San Andreas Fault Zone. Rather, understanding will come from knowledge of the history of finite relative plate motion for the entire time period from the initiation of the San Andreas Fault Zone through today.

There is a "ruling theory" regarding the kinematics of finite relative plate motion, to use the terminology of Chamberlin (1897). This ruling theory is termed the *small-circle model*, because it suggests that the motion of one plate as observed from another plate is circular, around a pole of relative motion that is fixed to both plates (e.g., Bullard et al., 1965; McKenzie and Parker, 1967; Morgan, 1968; Le Pichon, 1968; Heirtzler et al., 1968; Cox, 1973; Cox and Hart, 1986). The research described herein was initiated because the small-circle model has mathematical and conceptual shortcomings when it is extended from instantaneous to finite relative motion (McKenzie and Morgan, 1969; Cox, 1973; McKenzie and Parker, 1974; Dewey, 1975a, 1975b; Cox and Hart, 1986). In addition, this model is not particularly helpful in predicting some of the observed phenomena related to finite relative plate motion, although it provides many significant insights regarding the tectonics of the ocean floor during the early years of plate tectonics.

The method of multiple working hypotheses (Chamberlin, 1897) is employed to evaluate several potential kinematic models for finite relative plate motion. It would be preferable to use dynamic models constrained by abundant empirical data to generate accurate kinematic models of plate motion; however, adequate global-scale dynamic models and comprehensive empirical data sets are not yet available in plate tectonics. Hence, first-order kinematic modeling is still of value in generating hypotheses regarding the evolution of plate boundaries.

A note on terminology and general assumptions—The term *relative motion* is used to signify the motion of one lithospheric plate as observed from another plate during an extended or finite period of time. In contrast, *instantaneous* relative motion refers to the relative motion of plates during a very small time interval, as might be measured by very long baseline interferometry or other short-term direct measurement. Earth is assumed to be a sphere in first-order kinematic models, and a plate is considered to be a rigid piece of a spherical shell except where the plate dips into the sublithospheric mantle in a subduction zone. The *axis of rotation* for a given plate or plate pair is assumed to pass through the center of Earth. A *pole* is one of the two points at which an axis of rotation pierces the surface of Earth.

MULTIPLE WORKING HYPOTHESES

What shape might a path of relative motion be? The simplest hypothesis is that relative motion is along a straight line. Another hypothesis is that relative motion is along a circle, which is a figure of rotation around one axis. A third hypothesis is that relative motion is along a figure of rotation around two axes, or a cycloid. Other hypotheses might be that relative motion is ellipsoidal (that is, motion around two foci), or along *n*-axial cycloids, or entirely irregular to the extent that first-order kinematic models are meaningless. In the interest of simplicity, consideration is given only to the first three hypotheses. Which of these hypotheses are admissible, and which is best able to account for the observed phenomena of plate motion and boundary interactions?

Linear motion hypothesis

If Earth were flat, the simplest shape a path of relative motion could be is a line: a linear translation. A vector of instantaneous relative motion would be parallel with the path of finite relative motion. Earth is not flat, however. If the relative motion path of a particle currently on Earth's surface is a line, then the particle would leave Earth's surface after an infinitesimal displacement, because a line can intersect a sphere in, at most, two points (one, if the line does not pierce the sphere's surface). Plates are constrained to move along the spherical surface of Earth, except at subduction zones where lithosphere bends and sinks into the upper mantle, so plates cannot move along linear paths of relative motion. Hence, the linear motion hypothesis is not admissible, and is not valid for relative plate motion.

Circular motion hypothesis

The first-order small-circle relative-motion model assumes that the relative motion paths of plates have a circular shape. For the relative motion path of each particle on an observed plate to be circular, the plate must move around a single axis of rotation as viewed from another plate. The pole corresponding to this axis of finite relative motion is termed herein the relative pole. The relative pole must remain fixed with respect to both the observed and the observer's plate throughout the relative motion if the path of relative motion is to be a circle. If the position of the *relative pole* varies, as observed from either plate, the path of relative motion will be a curve that is more complex than a circle.

Implications—The consequences of circular relative motion are easily visualized (Fig. 1). If the trace of a particular transform fault is circular and concentric with small circles around the relative pole, then there will be neither convergence or divergence along that fault during finite relative plate motion (McKenzie and Parker, 1967). A given transform fault will remain a fixed distance from the relative pole during finite displacements. Fracture zones that emanate from either end of a given transform fault will have a small-circle shape, and will be symmetric across the ridge axis. The direction of relative motion at a given position along a plate boundary will be constant, so the direction of divergence along a midocean ridge and the direction of convergence along a subduction zone or a mountain range will be constant. Hence, the stability of triple junctions can be determined using simple vector calcula-

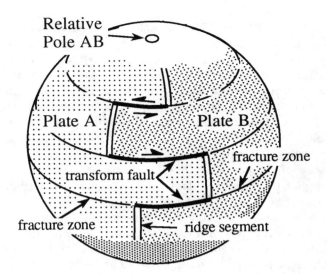

Figure 1. Typical depiction of the small-circle relative-motion model. Motion of Plate B relative to Plate A is around Relative Pole AB, which is fixed to both Plates A and B. The trace of each transform fault and the two corresponding fracture zones is a small circle around the relative pole.

tions in plane geometry, using the plane tangent to Earth at the triple junction (McKenzie and Morgan, 1969).

Limitations—The limitations of the first-order small-circle model are apparent when the relative motion of more than two plates is considered. Imagine a three-plate system, in which plate A, plate B, and plate C are in motion with respect to one another (Fig. 2). In any three-plate system, there will be three axes of relative motion: between plates A and B (axis AB), between plates B and C (axis BC), and between plates A and C (axis AC). The small-circle model requires that axis AB is fixed with respect to both plates A and B, axis BC is fixed to plates B and C, and axis AC is fixed to plates A and C, just as a bicycle tire must be fixed to its axle if all parts of the tire are to move in a circular path around the axle.

The problem with the small-circle model arises because plate A must be fixed to both axes AB and AC. Hence, axes AB and AC must be fixed to each other, if neither is in motion relative to plate A. Continuing the

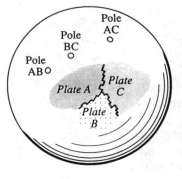

Figure 2. The three-point problem: three plates (A, B, and C) cannot simultaneously be in motion relative to each other and maintain a fixed distance to their corresponding relative poles (AB, BC, and AC) if the poles are not in the same location. Relative poles generally cannot be fixed to the corresponding plates during finite motion.

logic, axes BC and AB must be fixed to plate B as well as to each other, and axes BC and AC must be fixed to plate C as well as to each other. Hence, axes AB, BC, and AC must all be fixed to each other, and by extension, plates A, B, and C must be fixed to the three axes and to each other. If plates A, B, and C are fixed to each other, there is no relative motion and the initial condition that the three plates are in motion relative to one another is violated. It can be shown that the only case in which the small-circle model is valid for a given three-plate system is when the three axes of relative motion are coaxial (that is, axis AB = axis BC = axis AC).

The observation that the small-circle model assumptions are not generally valid for finite relative motion in systems of three or more plates has been called the three-plate problem (Cox, 1973). The best estimates of the present-day positions of relative poles for major plates on Earth show that the relative poles are not coincident (Minster and Jordan, 1978; DeMets et al., 1985, 1990). If the axes of relative motion in any given three-plate system are not coaxial, then the axes and the corresponding relative poles must be in motion with respect to the plates in the system. If the relative pole is in motion with respect to a given plate, the path of relative motion cannot be circular. Hence, the first-order small-circle model is not a generally valid model for finite relative plate motion.

Higher-order small-circle model—Higher-order models based on the small-circle relative-motion model typically utilize a finite-difference approach (e.g., Atwater, 1970; McKenzie and Sclater, 1971; Dewey et al., 1973; Sclater and Fisher, 1974; Dewey, 1975a; Engebretson et al., 1985). The relative-motion path of a particle on a given plate has a shape that can be divided into a series of circular segments that combine to form a complex curve. Each circular segment of the relative-motion path represents motion during a discrete time interval, and has a corresponding relative pole. This pole is alternately termed a stage pole, a finite-difference pole or a differential pole. As the complex shape of the path of relative motion is divided into an increasingly greater number of circular segments, the accuracy of the finite-difference approximation improves. Conversely, as the time interval used to calculate a stage pole increases, the divergence of the stage pole from the pole of relative motion for any time within that interval generally increases.

The accuracy of the finite-difference method is dependent upon the quantity, quality, and chronologic distribution of empirical data that is available to constrain the determination of relative motion paths. These data are potentially available through the analysis of the magnetic, structural, and petrological fabric of the oceanic crust. During time intervals for which these data are not available, as in future time or during the time before the genesis of the oldest ocean basin, the finite-difference method cannot be used to model relative motion.

Reconsideration of the small-circle model—In a given three-plate system, it is conceivable that plates A and B could move along a circular path relative to plate C; however, plates A and B would not move along circular paths relative to each other unless the three poles of relative motion are coaxial throughout the displacement. Solutions that are currently available for the instantaneous relative motion of plates indicate that poles of relative motion are not coaxial in any three-plate system, so circular relative-motion trajectories are atypical. Paths of relative plate motion are generally complex curves.

Models that assume that every plate moves around a set of finite relative-motion axes that are fixed to the plate are not generally valid. If the small-circle model generally is not valid for finite relative plate motion, then the assumed consequences of circular relative motion must be reexamined. Transform faults should not be assumed to be surfaces along which there is neither convergence nor divergence. Transform faults are generally in motion with respect to the corresponding relative pole. Fracture zones should not be assumed to have a small-circle shape, and are generally not symmetrical across the ridge axis. Neither the direction nor the velocity of relative motion of a point on one plate generally will be constant as observed at a given point on another plate. The simple vector calculations regarding the kinematic stability of triple junctions generally are not valid over a finite time interval, because of the variation in relative motion at any point along a given plate boundary through time.

Cycloid motion hypothesis

The next simplest shape of a relative-motion trajectory, after the small circle, is a figure of rotation around two axes: a cycloid. The shape of the path of a point on one rotating object as observed from another rotating object is a cycloid, if the two axes are a fixed distance from one another. The trajectories of the planets as observed from Earth would be cycloids if planetary orbits were circular rather than elliptical. In the small-circle model, the single axis of relative motion was presumed to be provided by an axis of instantaneous relative motion that can be determined by direct observation. What might constitute the dual axes utilized by a cycloid-based relative-motion model?

The perception of motion requires displacement between the observer and the observed object. Description of motion requires the specification of a frame of reference from which the motion is observed. Each lithospheric plate is observed to be in motion with respect to each other plate. Without specifying the frame of reference at this point in the discussion, let us assert that each plate is in motion, and that the motion of each plate across a spherical Earth can be described as a rotation around a plate-specific axis. The observed relative motion of plates is a result of two, simultaneous, plate-specific motions

$$_A\omega_B = \omega_B - \omega_A$$

where ω_A is the vector that describes the motion of plate A, and ω_B describes the motion of plate B, and $_A\omega_B$ is the vector that describes the motion of plate B as observed from plate A (Cronin, 1988b).

The observed relative motion of plates is the result of a unique set of plate-specific motions. The set of plate-specific vectors cannot be obtained from analysis of relative-motion vectors alone; however, a variety of geological phenomena can be used to constrain plate-specific motion models. For example, Minster and Jordan (1978) used hot-spot trails to constrain plate-specific motion in the AM1-2 model. The two axes of relative motion utilized by the cycloid model are the two axes of plate-specific motion, vectorially described by ω_A and ω_B for the plate A-plate B system.

First-order model—The assumptions of the first-order cycloid model (CYC1) are essentially the same as the assumptions of the small-circle model, except the poles of relative motion are not assumed to be fixed to the corresponding plates. The assumptions are: (1) Earth is a unit sphere; (2) plates are rigid spherical shells, except adjacent to plate edges; (3) there is relative motion between all plates under consideration; (4) the axial vector that describes individual plate motion is constant for all plates under consideration; and (5) the angular relationships between all axial vectors that describe individual plate motion are constant.

Cycloid relative motion can be easily conceptualized. Imagine each plate is a patio umbrella (Fig. 3). Umbrella A has the letter A painted on it, and umbrella B has a B painted on it (Fig. 3a). The umbrellas' poles are sunken in concrete, so the poles are always the same distance apart. Next, umbrella A is rotated 60° clockwise, and umbrella B is rotated 50° counterclockwise (Fig. 3b). Figure 3b is the view of the umbrellas from a vantage point on a balcony above the patio—a frame of reference that is fixed relative to the axes of the umbrellas. The same rotations observed from umbrella A would yield the configuration in Figure 3c.

What would the finite rotation look like if it was viewed by an observer fixed to umbrella A? No part of umbrella A would appear to have moved as the umbrellas rotate, because the observer is fixed to umbrella A, but both B and the axis of umbrella B would appear to have moved from their initial position. The finite motion of B as observed from A can be broken down into two components of motion that occur at the same time. First, the pole to umbrella B appears to rotate 60° in a

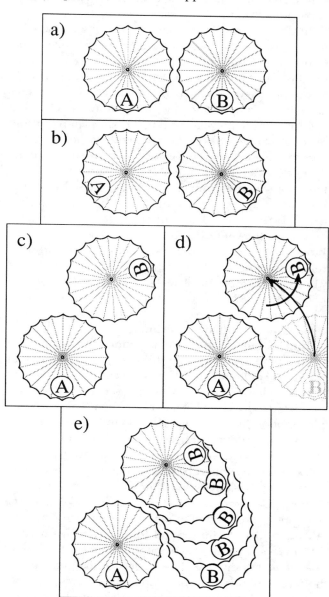

Figure 3. Conceptualization of first-order cycloid motion. **a**—Umbrellas A and B in their initial state. **b**—Umbrella A rotates 60° clockwise, and umbrella B rotates 50° counterclockwise, observed in a frame of reference that is fixed to the poles of the umbrellas. **c**—Configuration of the two umbrellas after rotation, as observed from umbrella A. **d**—Umbrella B pole traces a circular path counterclockwise around the pole of umbrella A, as observed from point A. Concurrently, umbrella B rotates around its pole. **e**—Point B traces a 2-axis cycloid as observed from umbrella A during finite relative motion.

counterclockwise direction around the pole to umbrella A as observed from umbrella A (Fig. 3d). Second, the painted B rotates 50° in a counterclockwise direction around the pole of umbrella B. Combining these two simultaneous, circular motions, the finite path of the painted B relative to an observer on umbrella A is a figure of rotation around two axes: a cycloid.

In CYC1, an observer is fixed on the observer's plate, so the pole to the observer's plate is also fixed with respect to the observer. CYC1 assumes that the pole to the observer's plate is a constant distance from the pole to the observed plate. As the observer's plate rotates around its plate-specific pole, the pole to the observed plate appears to move in a circular path around the pole of the observer's plate. Concurrently, a point on the observed plate is rotating around its plate-specific pole. The path of a point on the observed plate is therefore a figure of rotation around two axes, as viewed from the observer's plate (Cronin, 1986, 1987a, b, c, d, e, f, g, 1988 a, b, c, 1989 a, b, c). The mathematics used in CYC1 are explained elsewhere (Cronin, 1987f, 1988b, 1991 a, b). It does not matter whether the discussion concerns the relative motion of planets, umbrellas, bugs on bicycle wheels, or lithospheric plates—if a point on one rotating body is observed from a point on another rotating body, where the axes are fixed to one another, the path of relative motion will be a cycloid.

Implications—What are the implications of the cycloid model, and how do they compare with those of the small-circle model? The small-circle model would predict relative motion around a small circle. CYC1 predicts a more complex curve, and hence a more complex motion history along plate boundaries. In Figure 4a, the hypothetical trajectories of two points on the Pacific plate relative to the North American plate are shown, as determined by CYC1. One of the two reference points has the same latitude and longitude as Los Angeles, California, and its trajectory is marked by dots. The points are plotted at 5 Ma time intervals, and the AM1-2 model of Minster and Jordan (1978) is used in CYC1 to specify plate-specific motion vectors. Dots along the same path that are closer together, like those at the bottom of the cusp in the center of Figure 4a, indicate higher relative velocities. The trajectory of the other point, at Point Mendocino, is marked with a solid curve. Note that CYC1 does not yield concentric trajectories, even though the points being observed are rigidly coupled to one another on the same plate. The two curves have the same wavelength and frequency, but differ in amplitude and phase. Hence, finite simple shear without convergence or divergence is not possible on any plate boundary along which there is cycloid relative motion. For example, the San Andreas fault may

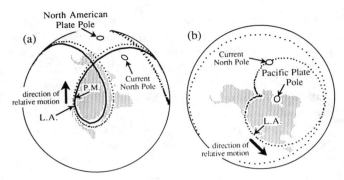

Figure 4. Cycloid relative motion paths along the Pacific-North American plate boundary, calculated using CYC1 (Cronin, 1988b) and AM1-2 model of Minster and Jordan (1978). Dots represent different relative positions, plotted at 5 Ma intervals. Present position of North America is indicated in halftone. L.A. = Los Angeles; P.M. = Point Mendocino. **a**—Dotted curve is the predicted trajectory of Los Angeles; solid curve is trajectory of Point Mendocino. Both are assumed rigidly coupled to the Pacific plate, and are observed from the North American plate. **b**—Dotted curve is the predicted trajectory of Los Angeles, assumed rigidly coupled to the North American plate, and observed from the Pacific plate.

be a transform fault, but CYC1 indicates that there must be systematic variations in convergence, divergence, and velocity across its trace through time (Cronin, 1986, 1987b, 1987d, e).

The cycloid in Figure 4b traces the path of one of the same points used in Figure 4a, with the same initial longitude and latitude as Los Angeles, only this time we assumed that Los Angeles is fixed to the North American plate. This curve is the hypothetical path of a point on the North American Plate as observed from the Pacific Plate, determined using CYC1. By changing the frame of reference, we have generated very different, but related, cycloid curves. Notice that the dots are plotted at 5 Ma intervals as before, but the dots are much more closely spaced near the center of the illustration, in the tight curve or cusp, than away from the center. The cycloid model predicts that the velocity and direction of relative plate motion changes systematically during finite displacements.

The following is a list of some of the more significant implications of CYC1 (Cronin, 1987g, 1988a, b), along with comparisons to the first-order small-circle model.

1. CYC1 predicts that the direction of relative motion varies systematically through time, whereas the small-circle model assumes the direction of relative motion to be constant around a fixed relative pole.

2. CYC1 predicts that the velocity of relative motion varies systematically through time, whereas the small-circle model does not predict a variation in plate velocity through time. The velocity of relative motion of a given point on an observed plate is often assumed to be constant during finite displacements.

3. CYC1 predicts a systematic variation in the distance from a plate to any of its relative poles through time. The relative poles generally have meaning only at a specified instant in time, but not over finite time intervals. One of the basic tenets of the small-circle model is that the relative pole is fixed to both plates in a two-plate system.

4. CYC1 predicts a systematic variation in the distance from a given transform fault to the corresponding relative pole. In contrast, the small-circle model assumes that the distance from a transform fault to the corresponding relative pole is constant.

5. If transform faults are in motion with respect to the relative pole, so too are the ridge segments on either side of a transform fault (Cronin, 1987c). Therefore, there must be variation in the rate of spreading along a given ridge segment through time. The small-circle model does not predict any motion between ridge segments and the corresponding relative pole.

6. Transform faults are predicted generally to be in motion with respect to the plates that they bound, as well as to the corresponding plate-specific poles of rotation. The small-circle model suggests that the motion of a transform fault is around the corresponding relative pole, as observed from one of the plates that the fault bounds.

7. CYC1 predicts that each point on a rigid plate traces a different path of finite motion relative to another plate. The paths traced by points on the same plate have the same wavelength and frequency, but the differ in amplitude or phase. The small-circle model predicts paths that are concentric circles around the relative pole.

8. CYC1 predicts a systematic change in the direction and velocity of relative motion at a given point along a plate boundary through time. In fact, it predicts an oscillation or wobble of one plate relative to another (Cronin, 1987a). The small-circle model does not predict this type of oscillation, because all points on one plate are assumed to move in concentric circles relative to another plate. The geometry of boundary interactions is assumed to be constant in the small-circle model.

9. CYC1 predicts a systematic increase in convergence or divergence along a given transform fault during finite displacement because of the cycloid wobble effect (Cronin, 1987b, 1987d, e). The small-circle model considers transform faults to be lines of pure slip along which crust is conserved (McKenzie and Parker, 1967). This characterization is valid only when the fault is circular and the paths of finite relative motion are concentric with the fault. Paths of finite motion are not generally circular, so transform faults are not generally lines of pure slip along which crust is conserved, regardless of whether the path of finite motion is cycloidal or another non-circular shape.

10. If there is a systematic increase in convergence or divergence along transform faults, then there may be a maximum stable length for a transform fault at any given location along the margin of a two-plate system (Cronin, 1989b). Beyond that length, the fault would be unstable and tend to segment, for mechanical reasons related to factors such as the rheological properties of the faulted media, the strain rate and the variation in direction of relative motion. The small-circle model places no constraints on the length of transform faults.

11. CYC1 makes no explicit assumptions regarding the shape of a transform fault. Fault shape is a function of fault dynamics. The small-circle model assumes that transform faults should be small circles concentric with the relative pole.

12. CYC1 considers a fracture zone on a given plate to follow the flow line of the corresponding transform fault as viewed from that plate (Cronin, 1987c, 1988b, c). The transform fault is in motion with respect to both plates and the system relative pole. Hence, the shape of oceanic fracture zones is rather complex—a figure of rotation around several axes. This complex shape is not symmetrical across the ridge axis. CYC1 does not consider fracture zones to follow the flow line of a point on one plate as observed from another plate. In preliminary tests (Cronin, 1988b, c), CYC1 was successful in modeling the asymmetry and sigmoid long-wavelength shape of observed fracture zones in the South Atlantic, which had been mapped using SEASAT radar altimetry (Shaw, 1987). It may be possible to use fracture zone shape to help constrain plate-specific motion characteristics.

In contrast, the small-circle model predicts that fracture zones are circular, concentric around the relative pole and hence, symmetric across a ridge axis. The small-circle model would agree that fracture zones are flow lines of transform faults, but would add that they are also flow lines of plates.

13. If there are systematic variations in the direction and velocity of relative motion at any given point along a plate boundary through time, as predicted by CYC1, then the geometric stability of a boundary cannot be described with static vector circuits. It can be shown using CYC1 that, of the 125 possible configurations of triple junctions, geometric stability throughout a finite time interval is possible only under very unusual circumstances. CYC1 provides a way to assess the rate of change of the geometry of a triple junction (Cronin, 1988b, 1989a). The small-circle model suggests that many triple junction configurations are stable, and that

stability can be assessed using simple vector circuits as described by McKenzie and Morgan (1969).

Limitations—The small-circle model is limited in its utility because it is not valid for modeling finite relative motion in n-plate systems in which the relative poles are not coaxial. Higher-order modeling is generally necessary, requiring an abundance of empirical data to constrain the finite-difference methods. Where suitable empirical data are not available, the small-circle model cannot be extended with confidence. CYC1 does not share this limitation, because it is generally valid for finite relative motion within the constraints of the model assumptions. Hence, the cycloid hypothesis for the finite relative motion of plates is admissible, although this judgement is based upon kinematic arguments and not upon dynamic modeling.

The principal limitation for CYC1 is the uncertainty in quantifying plate-specific motion characteristics through time. Instantaneous relative motion is directly observable, so the present-day positions of all relative poles can easily be determined along with the corresponding velocities (e.g., Minster and Jordan, 1978; DeMets et al., 1985, 1990). In contrast, the present-day positions of all plate-specific poles and the corresponding angular velocities cannot be determined directly or simply calculated by inversion of the relative-motion data. Plate-specific motion has been investigated relative to the trail a hot spot might make on the overriding lithosphere if the hot spot is fixed in the sublithospheric mantle (e.g., Wilson, 1965; Dietz and Holden, 1970; Morgan, 1971, 1972a, 1972b; Minster et al., 1974; Minster and Jordan, 1978; Gripp and Gordon, 1990). Alternately, plate-specific motion may be quantifiable using the idea that no net torque is exerted on the lithosphere (Solomon and Sleep, 1974; Solomon et al., 1975, 1977; Jurdy, 1978; Davis and Solomon, 1981; Jurdy and Gordon, 1984; Gordon and Jurdy, 1986). Another method for determining plate-specific motion is under development, utilizing relative-motion data along with data from oceanic fracture zones. The shape of the fracture zone along with the record of the variation in transform fault length that is contained in offset magnetic anomalies along fracture zones may provide an adequate recording of both relative- and plate-specific motion in the past.

CONCLUSIONS

Three hypotheses for relative plate motion are examined. Linear relative motion on a sphere is impossible. The small-circle hypothesis is not generally valid for finite relative plate motion. Predictions of boundary interactions based upon the invalid small-circle model should be re-evaluated. Of the three hypotheses evaluated, only the cycloid model is kinematically admissible for finite relative motion within the limitations imposed by the CYC1 model assumptions. The implications of cycloid relative motion with respect to plate boundary interactions are significant, and should fundamentally alter our understanding of variations in relative motion across plate margins through time. Full utilization of CYC1 and higher-order modeling based upon the cycloid model awaits the further development of methods to define plate-specific motion characteristics through time.

REFERENCES CITED

Allegre, C., 1988, The behavior of the Earth, continental and seafloor mobility: Cambridge, Harvard University Press, 272 p.

Atwater, T., 1970, Implications of plate tectonics for the Cenozoic tectonic evolution of western North America: Geological Society of America Bulletin, v. 81, p. 3513-3536.

Bullard, E.C., Everett, J.E., and Smith, A.G., 1965, The fit of the continents around the Atlantic, in Blackett, P.M.S., Bullard, E.C., and Runcorn, S.K., eds., A symposium on continental drift: Philosophical Transactions of the Royal Society of London, Ser. A, v. 258, p. 41-75.

Chamberlin, T.C., 1897, The method of multiple working hypotheses: Journal of Geology, v. 5, p. 837-848.

Cox, A., (Compiler), 1973, Plate tectonics and geomagnetic reversals: San Francisco, W.H. Freeman, 702 p.

——, and Hart, R.B., 1986, Plate tectonics, how it works: Blackwell Scientific Publications, Palo Alto, California, 392 p.

Cronin, V.S., 1986a, Cycloid tectonics: Relative motion on a sphere: Geological Society of America, Abstracts with Programs, v. 18, p. 575.

——, 1987a, The contribution of cycloid wobble to plate boundary strain: Geological Society of America, Abstracts with Programs, v. 19, p. 149.

——, 1987b, Kinematics of transform plate-boundary faults re-evaluated [abs.], in Hilde, T.W.C., and Carlson, R.L., eds., Proceedings of the 1987 Geodynamics Symposium: Geodynamics Research Institute, Texas A&M University, p. 84-86.

——, 1987c, Cycloid tectonics: Predicted changes in the position and morphology of midocean ridge segments, transform faults, and oceanic fracture zones [abs.]: EOS (American Geophysical Union Transactions), v. 68, no. 16, p. 408.

——, 1987d, Cycloid tectonics: The kinematics of transform faulting re-evaluated [abs.]: American Association of Petroleum Geologists Bulletin, v. 71, p. 544.

——, 1987e, Cycloid tectonics: Are transform plate-boundary faults "lines of pure slip" along which "crust is conserved"?: Geological Society of America, Abstracts with Programs, v. 19, No. 7, p. 631.

——, 1987f, Cycloid tectonics: A first-order matrix solution for finite relative plate motion [abs.]: EOS (American Geophysical Union Transactions), v. 68, p. 1473.

——, 1987g, Cycloid kinematics of relative plate motion: Geology, v. 15, p. 1006-1009.

——, 1988a, Reply to comment on "Cycloid kinematics of relative plate motion: Geology, v. 16, p. 473-474.

——, 1988b, Cycloid tectonics: A kinematic model of finite relative plate motion, [Ph.D. thesis]: College Station, Texas A&M University, 118 p.

——, 1988c, Cycloid tectonics: Fracture zones as flow lines of transform faults: EOS (American Geophysical Union Transactions), v. 69, p. 1415.

———, 1989a, Cycloid tectonics: Geometric stability of triple junctions during finite relative plate motion: Geological Society of America, Abstracts with Programs, v. 21, no. 4, p. 8.

———, 1989b, Cycloid tectonics: Is there a maximum stable length for a transform fault? Geological Society of America, Abstracts with Programs, v. 21, no. 4, p. 8.

———, 1989c, A kinematic model for the evolution of a transtensional basin from a transform fault whose initial displacement is purely strike-slip [abs.]: American Association of Petroleum Geologists Bulletin, v. 73, p. 1029.

———, 1991a, The cycloid relative-motion model and the kinematics of transform faulting: Tectonophysics, v. 187, p. 215-249.

———, 1991b, Corrections to "The cycloid relative-motion model and the kinematics of transofrm faulting": Tectonophysics, v. 192, no 3/4, in press.

Davis, D.M., and Solomon, S.C., 1981, Variations in the velocities of the major plates since the Late Cretaceous: Tectonophysics, v. 74, p. 189-208.

DeMets, C., Gordon, R.G., Stein, S., Argus, D.F., Engeln, J., Lundgren, P., Quible, D.G., Stein, C., Weinstein, S.A., Wiens, D.A., and Woods, D.F., 1985, NUVEL-1: A new global plate motion model and dataset: EOS (American Geophysical Union Transactions), v. 66, p. 368-369.

DeMets, C., Gordon, R.G., Argus, D.F., and Stein, S., 1990, Current plate motions: Geophysical Journal International, vol. 101, p. 425.

Dewey, J.F., 1975a, Finite plate implications: some implications for the evolution of rock masses at plate margins: American Journal of Science, v. 275-A, p. 260-284.

———, 1975b, Plate tectonics: Reviews of Geophysics and Space Physics, v. 13, p. 326-332.

———, Pitman, W.C., III, Ryan, W.B.F., and Bonnin, J., 1973, Plate tectonics and the evolution of the Alpine System: Geological Society of America Bulletin, v. 84, p. 3137-3180.

Dietz, R.S., and Holden, J.C., 1970, Reconstruction of Pangaea: Breakup and dispersion of continents, Permian to present: Journal of Geophysical Research, v. 75, p. 4939-4956.

Engebretson, D.C., Cox, A., and Gordon, R.G., 1985, Relative motions between oceanic and continental plates in the Pacific basin. Geological Society of America, Special Paper 206, 59 p.

Glen, W., 1982, The road to Jaramillo: Critical years of the revolution in Earth science. Stanford, California, Stanford University Press.

Gordon, R.G., and Jurdy, D.M., 1986, Cenozoic global plate motions: Journal of Geophysical Research, v. 91, p. 12,389-12,406.

Gripp, A.E., and Gordon, R.G., 1990, Current plate velocities relative to the hotspots incorporating the NUVEL-1 global plate motion model: Geophysical Research Letters, v. 17, p. 1107-1112.

Hallam, A., 1973, A revolution in the Earth sciences. London, Oxford University Press, 127 p.

Heirtzler, J.R., Dickson, G.O., Herron, E.M., Pitman, W.C., III, and Le Pichon, X., 1968, Marine magnetic anomalies, geomagnetic field reversals, and motions of the ocean floor and continents: Journal of Geophysical Research, v. 73, p. 2119-2136.

Hess, H.H., 1962, History of ocean basins, in Engel, A.E.J., James, H.L., and Leonard, B.F., eds., Petrological studies: A volume in honor of A.F. Buddington: Geological Society of America, p. 599-620.

Jurdy, D.M., 1978, An alternate model for early Tertiary absolute plate motion: Geology, v. 6, p. 469-472.

Jurdy, D.M., and Gordon, R.G., 1984, Global plate motions relative to the hot spots 64 to 56 Ma: Journal of Geophysical Research, v. 89, p. 9927-9936.

Le Pichon, X., 1968, Sea-floor spreading and continental drift: Journal of Geophysical Research, v. 73, p. 3661-3705.

———, Francheteau, J., and Bonnin, J., 1973, Plate tectonics: New York, Elsevier, 300 p.

McKenzie, D.P., and Morgan, W.J., 1969, The evolution of triple junctions: Nature, v. 224, p. 125-133.

———, and Parker, D.L., 1967, The north Pacific: An example of tectonics on a sphere: Nature, v. 216, p. 1276-1280.

———, and Parker, D.L., 1974, Plate tectonics in ω space: Earth and Planetary Science Letters, v. 22, p. 285-293.

———, and Sclater, J.G., 1971, The evolution of the Indian Ocean since the late Cretaceous: Geophysical Journal of the Royal Astronomical Society, v. 24, p. 437-528.

Minster, J.B., Jordan, T.H., Molnar, P., and Haines, E., 1974, Numerical modelling of instantaneous plate tectonics: Geophysical Journal, Royal Astronomical Society, v. 36, p. 541-576.

———, and Jordan, T.H., 1978, Present-day plate motions: Journal of Geophysical Research, v. 83, p. 5331-5354.

Morgan, W.J., 1968, Rises, trenches, great faults, and crustal blocks: Journal of Geophysical Research, v. 73, p. 1959-1982.

———, 1971, Convection plumes in the lower mantle: Nature, v. 230, p. 42-43.

———, 1972a, Plate motions and deep mantle convection, in: Shagam, R., ed., Studies in Earth and space sciences: Geological Society of America, Memoir 132, p. 7-22.

———, 1972b, Deep mantle convection plumes and plate motions: American Association of Petroleum Geologists Bulletin, v. 56, p. 203-213.

Sclater, J.G., and Fisher, R.L., 1974, Evolution of the east-central Indian Ocean, with emphasis on the tectonic setting of the Ninetyeast Ridge: Geological Society of America Bulletin, v. 85, p. 683-702.

Shaw, P.R., 1987, Investigations of relative plate motions in the South Atlantic using Seasat altimeter data: Journal of Geophysical Research, v. 92, p. 9363-9375.

Solomon, S.C., and Sleep, N.H., 1974, Some simple physical models for absolute plate motions: Journal of Geophysical Research, v. 79, p. 2557-2567.

Solomon, S.C., Sleep, N.H., and Richardson, R.M., 1975. On the forces driving plate tectonics: Inferences from absolute plate velocities and intraplate stress: Geophysical Journal, Royal Astronomical Society, v. 42, p. 769-801.

Solomon, S.C., Sleep, N.H., and Jurdy, D.M., 1977, Mechanical models for absolute plate motions in the early Tertiary: Journal of Geophysical Research, v. 82, p. 203-212.

Tarling, D.H., and Tarling, M.P., 1971, Continental drift: London, G. Bell and Sons, 112 p.

Uyeda, S., 1978, The new view of the Earth: San Francisco, California, W.H. Freeman and Co., 217 p.

Wegener, A., 1929, The origin of continents and oceans (4th revised edition, translated by John Biram; English edition, 1966): New York, Dover Publications, 226 p.

Weiner, J., 1986, Planet Earth: New York, Bantam Books, 370 p.

Wilson, J.T., 1965, Submarine fracture zones, aseismic ridges and the International Council of Scientific Unions line: Proposed western margin of the east Pacific ridge: Nature, v. 207, p. 907-910.

ACKNOWLEDGMENTS

This work evolved from conversations with A.O. Woodford, M.L. Hill, and C.L. Drake. Later discussions with many others, including R.L. Carlson, T.W.C. Hilde, J.T. Pitts, J.H. Spang, N.L. Carter, S. Uyeda, and C.R. Scotese, further stimulated these studies. I would also like to thank the anonymous reviewer who contributed to the final version of this work. I gratefully acknowledge the financial assistance provided by the following: University of Wisconsin–Milwaukee, Texas A&M University, Amoco Production Company, Exxon Company USA, and Sigma Xi.

An experimental study of ridge offsets in relation to trends of ridge and transform faults, using wax models

S. K. Ghosh, Department of Geological Sciences, Jadavpur University, Calcutta 700 032, India

ABSTRACT

Oceanic ridge-transform patterns show certain geometrical peculiarities that any theory of transform faulting must explain. The magnitude and sense of ridge offset depends upon the long-range trend of the ridge with respect to the direction of spreading. Structures similar to these have been simulated in two sets of experiments. In one of these, a thin film of wax, solidifying on the surface of molten wax, is subjected to tension. The hot-wax film under tension develops a fine lineation that controls the initiation of a set of transform faults. In the other set of experiments a thin layer of molten wax is brushed on a slab of pitch, and the pitch slab is subjected to tension. A set of parallel incisions is made on the wax layer along the direction of tension. The resulting steplike pattern of ridge and transform faults on the wax layer is remarkably similar to the patterns on the ocean floor.

In light of these experiments it is suggested that the newly formed lithosphere in spreading zones develops sets of mechanically weak planes that ultimately control the orientation of the transform faults. The nature of ridge offset is greatly controlled by the gross shape of the zone of spreading. The experiments suggest that there may be some transcurrent movement associated with the transform faults. The net effect of spreading and transcurrent movement is that the instantaneous movement on transforms is opposite to the sense of ridge offsets.

RIDGE-TRANSFORM GEOMETRY

One fundamental problem of plate tectonics is to explain the pattern of ridge offsets associated with transform faults. Oceanic ridges and transform faults form a steplike pattern with a side-stepping or offset of ridge segments. Wilson (1965) indicated that ridge offsets were not produced by transcurrent faulting. Rather, the steplike pattern was inherited from the shape of the initial break in the continent. He further implied that the orientation of the transform faults could have been inherited from old planes of weakness in the continents. The sense of movement of the transform faults is opposite to that of the ridge offsets. This sense is the most characteristic feature of transform faults.

The ridge-transform pattern does not show the haphazard shape of a ragged surface of splitting. The pattern is dominated by the following geometrical characteristics.

1. Figure 1 shows two hypothetical patterns. In Figure 1a, the long-range trend of the ridge is linear, the ridge segments and the ridge trend are more or less at a right angle to the transforms and there is a large range in the magnitude of ridge offsets. In Figure 1b, the overall shape of the ridge is curved, the ridge segments are not everywhere at a right angle to the transforms and the magnitude of the ridge offset is consistently small, even where the long-range trend of the ridge is at a low angle to the transforms. The oceanic ridge-transform pattern does not conform with any of these two hypothetical patterns. The oceanic pattern shows a clear relationship between the relative magnitude of ridge offsets and the long-range trend of the ridge. Where the gross orientation of the ridge is at a high angle to the transforms, the offsets are generally small. Where the overall orientation of the ridge is at a low angle to the transforms, the offsets are generally large, whatever the angle between the transforms and the individual ridge segments is.

2. A sawtooth pattern is obtained by joining the ridge segments and the connecting transform faults. Where the ridge and the transforms are not mutually perpendicular, the sawtooth pattern almost always shows obtuse

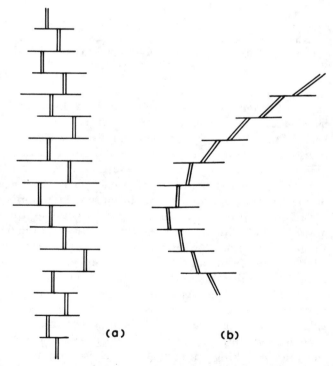

Figure 1. Two hypothetical ridge-transform patterns. In (a) the long-range trend of the ridge is linear and at a right angle to the transforms. The offsets of ridge segments show a large variation of magnitude. In (b) the offsets are constant irrespective of the angle between the transforms and the long-range trend of the ridge. None of these patterns are characteristic of the ocean floor.

Chatterjee, S., and N. Hotton III, eds. *New Concepts in Global Tectonics.*
Texas Tech University Press, Lubbock, 1992, xii + 450 pp.

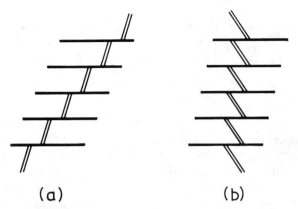

Figure 2. Sawtooth pattern formed by ridge segments and transform faults. The ridge-transform pattern generally makes obtuse angles as in (**a**). The pattern of acute angles as shown in (**b**) is rare.

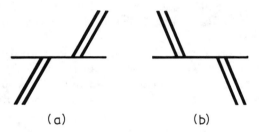

Figure 3. Sense of offset of oblique ridge segments. While looking across a transform fault, if the ridge segment is tilted towards the right as in (**a**), ridge offset is nearly always dextral. If the ridge segment is tilted towards the left as in (**b**), the ridge offset will be sinistral.

EXPERIMENTS TO INITIATE TRANSFORM FAULTS

Why do transform faults develop at all? Experiments of tension fracturing or of normal faulting in isotropic material do not show such patterns of splitting. However, in the series of experiments outlined below, such faults formed readily when the material was mechanically anisotropic. In these experiments, the fracture that separated the fragments of the stiff plates is described as a ridge, and the line midway between the separated walls is designated as a ridge crest. These terms are merely for the convenience of description; the terms do not have a topographic connotation.

In one series of experiments, a thin coating of plaster of paris was applied with a brush to the surface of a rectangular slab of pitch (Fig. 4a). The brush marks on the plaster sheet were fine grooves and ridges, which gave rise to a mechanical anisotropy in the plaster sheet when the material dried and became brittle (see Ghosh, 1988). When the underlying pitch slab was extended parallel to the brush marks, a boxlike pattern of fracture was produced in the plaster sheet (Fig. 4b) with the short

angles (Lachenbruch and Thompson, 1972; Fujita and Sleep, 1978), as shown in Figure 2a; acute angles, as in Figure 2b, are rare. Consequently, if we look across a transform fault, and the ridge segment on the far side is inclined towards the right-hand side (Fig. 3a), we can correctly predict that the ridge offset will be dextral. Similarly, if the ridge segment on the far side tilts towards the left, the offset will be sinistral (Fig. 3b). Along the approximately linear trend of the southwest Indian Ridge, for example, the individual ridge segments show a leftward tilt on the far side of the transform faults and a consistent sinistral side-stepping throughout the entire length of the ridge. As a consequence, the sense of curvature of ridge arcuation is related to the sense of ridge offsets. For example, when the successive ridge segments veer towards the left, as in the northern part of the Carlsberg Ridge, the ridge offsets are dominantly sinistral.

3. Where the gross orientation of the ridge makes an arcuate pattern, the array of transform faults weakly converges towards the concave side of the arcuation (see Collette, 1974, from the equatorial and north Atlantic).

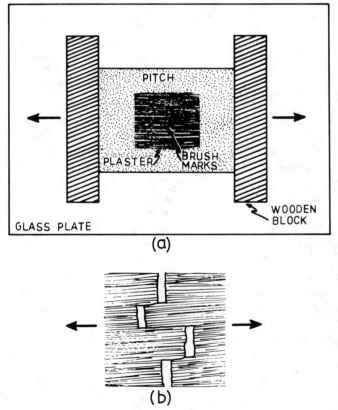

Figure 4. **a**—Experimental set-up for development of transform faults in plaster-of-paris sheet. The sheet, on the surface of a slab of pitch resting on a glass plate smeared with a soapy solution, is painted with a brush. The slab of pitch is extended by pulling apart the wooden blocks. **b**—Typical fracture pattern in plaster sheet. The transform faults have developed along weak planes parallel to the brush marks.

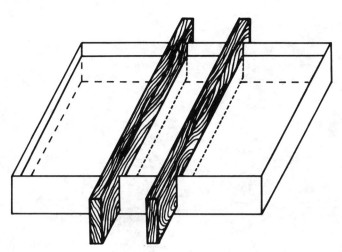

Figure 5. Experimental set-up for solidifying wax model showing metal tray with molten wax and wooden paddles. The paddles with grooves, resting on the long edges of the tray, are moved apart on a direction parallel to the long edge of the tray to produce fractures on the thin film congealing on the wax surface.

side-stepping tension fractures across the brush marks being connected by fractures dominantly parallel to the brush marks. Separation across the tension fractures produced a structure similar to that of offset ridges and the connecting transform faults when the two were more or less perpendicular. Here the development of the transform faults clearly was controlled by the presence of lines of weakness parallel to the brush marks of the plaster plate.

Similar orthogonal ridge-transform patterns were obtained by Oldenburg and Brune (1972) by pulling a thin crust of solidifying wax over the surface of molten wax. According to Oldenburg and Brune, the transform faults in these experiments were produced because of the lack of shear strength of the solidifying wax in the direction of pulling. These experimental results were reinterpreted by Vroman (1976) who suggested that the jagged edge of a tension fracture, while moving through the liquid wax, hindered its congealing so that the lines that connected the kinks became zones of impaired strength. Vroman concluded that the transform faults were created in a material with Newtonian properties.

I repeated experiments similar to that of Oldenburg and Brune (1972). The experimental set-up is illustrated in Figure 5. Two wooden paddles were placed in the central part of a tray of molten paraffin wax that was dyed pink. The experiments were performed in an air-conditioned room with an air temperature of 18° C. The best results were obtained when the temperature of the wax was a little above its melting point. The paddles were pulled at a rate of less than 2 mm per second away from each other. The frozen film on the wax surface immediately developed a fine linear structure parallel to the direction of extension. It was later found that when the hot congealed wax was subjected to an extension, it developed a fine fibrous structure parallel to the direction of extension. This is clearly visible when a thick slab of frozen, warm wax breaks under tension; the fine fibrous structure becomes visible on the jagged fracture surface. It is this linear structure that controlled the development of transform faults in the experiments. When the congealing wax film in the experiment was subjected to tension, it developed short segments of tension fractures not aligned. The terminations of these fractures were immediately joined up by straight fractures parallel to the direction of tension. When the wax plates were pulled away, the movement along these latter fractures was similar to that of transform faults.

The color contrast between the frozen and molten wax was so small that the pattern of fractures did not come out well in the photographs of the initial experiments. In addition, holding the camera at a low angle to bring out the contrast distorted the plane view of the fracture pattern. However, the structures can be seen well and can be photographed if the frozen surface of the wax is dusted with talc powder from time to time. Although the absorption of the powder by the hot wax gives the model surface a splotchy appearance, the ridge segments and the transform faults can be seen clearly. Moreover, because of the variable absorption of the powder by the hot wax, the traces of successive positions of older fracture walls were preserved in some of the models—somewhat like the frozen pattern of linear magnetic anomalies.

The development of the transformlike structures in these experiments depended, as was noted by Oldenburg and Brune (1972), on a delicate balance between the rate of pulling of the paddles and the rate at which the frozen film developed on the surface of the molten wax. The transform faults grew only when the film was very thin or was just forming. When the transform faults formed in the solidifying wax, the nature of side-stepping of the tension fractures depended largely on the geometry of the newly formed wax film with respect to the direction of the pull. At the beginning of the experiment, a thickness of a few millimeters of wax was allowed to congeal on the surface. This crust was then cut along a straight or a curved line and the paddles were slowly pulled apart. As the thick layer of frozen wax was pulled apart, the surface of the molten wax was exposed. The pulling was continued slowly enough to keep pace with the growth of a thin film on the exposed surface of the molten wax. Depending on the shape of the initial incision, the shape of the newly formed thin film was either curved or more or less straight. When the orientation of this newly formed film was roughly perpendicular to the direction of pulling,

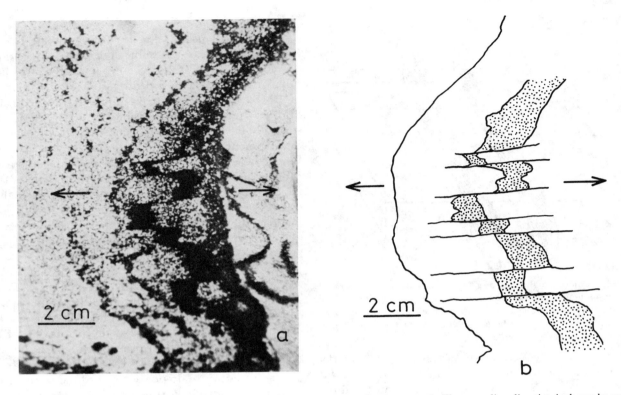

Figure 6. **a**—Fracture pattern on solidifying wax model. The initial zone of spreading was curved. The spreading direction is shown by arrows. **b**—Sketch of central fractures of Figure 6a. The stippled portions show areas of molten wax exposed by separation walls of the fracture. The initial curved fracture was smoothly curved and its walls have moved apart to become a part of the frozen crust. The current position of one of these walls is still visible (the curved line on the left) as a dark line on the wax crust. The current spreading zone (stippled) is discontinuous because of development of transform faults parallel to the direction of tension.

the fracture segments had no consistent sense of side-stepping (Fig. 6a and b). Along an oblique zone of the thin film, however, the side-stepping was dominantly dextral or sinistral depending upon whether the zone of the thin film as a whole was tilted towards the right or towards the left. In all of these experiments, the ridge segments were perpendicular to the transform faults even when the gross orientation of the ridge was oblique.

Figure 7 shows a model in which about a 2-mm-thick crust was allowed to form on the surface of the molten wax. An incision by a razor blade was then made in this crust at an angle of about 70° to the direction of pulling (shown by the arrows). When the paddles were slowly pulled apart a narrow zone of molten wax was exposed. The zone was oblique (tilted towards the right) to the direction of pulling. With continued pulling the thin film congealing on this narrow zone developed fine grooves and ridges parallel to the direction of tension. Continued pulling gave rise to a series of dextral side-stepping ridge segments (fractures) connected by fine lines of fracture. As in a transform fault the

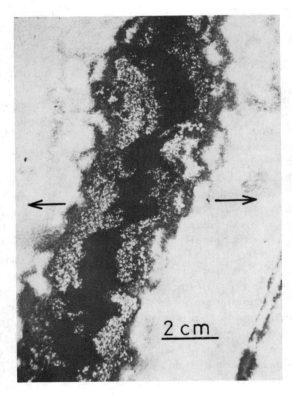

Figure 7. Fracture pattern on wax film solidifying over molten wax. Arrows show direction of extension. The dark area in center represents surface of molten wax exposed by fracturing of wax crust. The gross orientation of this spreading zone is tilted towards the right. Along the transform faults the ridge offsets are dextral.

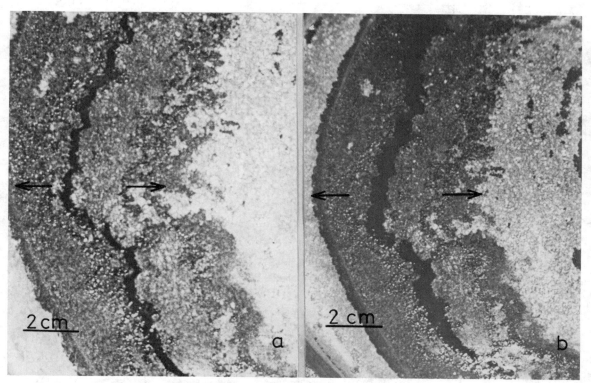

Figure 8. Curved spreading zone in solidifying model. As viewed across the transform faults the spreading zone is tilted towards the right in the upper part of the model and the offset in this part of the zone is dextral. Arrows indicate the direction in which the wooden paddles have been moved. (a) and (b) are different stages of development of the same model.

movement of the thin-wax film between the ridge crests was opposite to the sense of offset of ridge crests. Figures 8a and 8b show two stages of another experiment in which the initial incision through the wax crust was arcuate. Note that the ridge crests show a dextral sidestepping in the upper part of the figure where the gross trend of the ridge is tilted towards the right.

The experiments described above indicate that a pattern of side-stepping ridge segments connected by transform faults can develop in a layer that is mechanically anisotropic, with planes of weakness subparallel to the spreading direction. The sense of side-stepping of the experimental ridge segments is consistent with patterns found in ocean floors. However, the experiments with molten wax failed to simulate oblique ridge segments. Moreover, in the ocean floor, the largest offsets are found where the gross orientation of the ridge is distinctly oblique to the transform faults. Such was not necessarily the case in the experiments with the freezing wax. The transform faults in these experiments did not show a fanning when the general orientation of the ridge was arcuate.

The behavior of the freezing wax under tension is not exactly similar to that of the oceanic crust in the spreading zone. Nevertheless, it is likely that there is some mechanism that causes the formation of mechanically weak planes parallel to the direction of spreading in the ocean floor. A likely process is the development of thermal contraction joints as suggested by Turcotte (1974) and Collette (1974). If the initiation of transform faults is indeed controlled by the presence of mechanically weak planes along which slip is possible, then there is no reason why this mechanical property will cease to exist beyond the side-stepping portions of the ridge crests (i.e., beyond the "active" segments of the transform faults). What kind of movement do we expect if the planes that permit sliding extend beyond ridge off-sets? The experiments of freezing wax do not clarify this problem. The ridge-transform pattern in these experiments changes too quickly, and the movement along a single plane of discontinuity cannot be followed for a long time.

The next series of experiments was of a different type. The experimental set-up is shown in Figure 9. In these experiments, molten wax was brushed along a curved zone on the surface of a rectangular slab of pitch. The pitch slab was placed on a glass plate smeared with a soapy solution. Two opposite vertical faces of the pitch slab were attached to wooden blocks, and the other two faces were held in position by glass plates smeared on the inside with a soapy solution. The wax layer on the surface of the pitch contained a number of incisions parallel to the direction of extension (Fig. 9). When the pitch slab was extended by pulling apart the wooden blocks, a series of side-stepping tension fractures developed in the wax plate. Each fracture terminated against the lines of incisions in the wax plate. The sense of side-stepping was

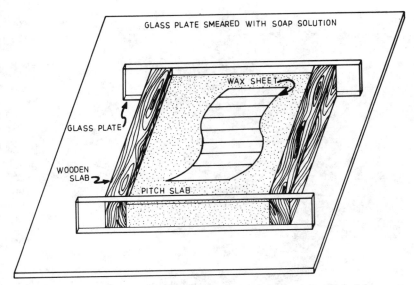

Figure 9. Experimental set-up for fracturing of wax plates overlying pitch slab.

controlled essentially by the overall shape of the wax plate (Fig. 10) and was consistent with the pattern of ridge offsets in the ocean floor.

In some of these experiments displacement along an incision took place because of the continuous separation of the walls of the tension fracture and, because the tension fractures had an initial offset, displacement along the incisions was similar to that of a transform fault. In addition, in some of the experiments there was a small displacement along the incisions in the same sense as that of the offsets. For these models there was a small transcurrent movement opposed to the sense of transform faulting. For any small interval of time, the displacement due to transcurrent movement was smaller than that resulting from separation across a tension fracture. The total movement along the planes of weakness (the incisions) remained opposite to the sense of offsets of the tension fractures.

On some of the models, in addition to the set of incisions parallel to the direction of extension, the wax plate was also cut by a single smoothly curved incision congruent with the plate boundary. When the pitch slab underlying the wax plate was pulled apart, separation across the curved line of incision was associated with slip along the parallel set of incisions (Fig. 11a and b; Fig. 12a and b). The middle line of the zone of separation is analogous to the ridge crest. The sense of transcurrent movement or the sense of offset of the line of separation was either dextral or sinistral depending on whether the line of separation was inclined towards the right or the left as viewed across the planes of weakness.

In other words, the sense of offset was such that it tended to reduce the angle between the extension direction and the gross orientation of the line of separation. The largest offsets occurred where this angle was smallest. There was little or no offset where the gross orientation of the line of separation was at a right angle to the direction of extension. Because the offsets in these experiments were caused entirely by the transcurrent movements, the magnitude of offset increased in the course of time. It should be noted, however, that the transcurrent movement along the planes of weakness was accompanied by a much faster movement resulting from the separation across segments of the curved incision (i.e., across the ridge segments). The sense of movement resulting from the separation or spreading was opposed to that of the transcurrent movement.

Figure 10. Curved wax plate on pitch slab. The direction of extension of the pitch slab is along the shorter edge of the figure. The wax plate was dissected by incisions parallel to the spreading direction. Tension fractures terminate against the incisions and show systematic offsets, the sense of which is controlled by the shape of the wax plate. Bar at lower right represents 2 cm.

Figure 11. **a**—The curved wax plate initially had a central incision congruent with the plate boundary. In addition, it had parallel incisions along the direction of extension of the pitch slab. The model shows an initial stage of deformation. **b**—Same model at an advanced stage of deformation. Note that the maximum offsets are where the plate as a whole makes a low angle with the spreading direction. The zone here is tilted towards the left and the offsets are sinistral. Note also the fanning of the transform faults.

RELATIVE CONTRIBUTION OF TRANSCURRENT MOVEMENT

The last series of experiments showed that when the gross trend of the ridge was at an angle to the direction of bulk extension, separation across the ridge was associated with some transcurrent movement. If the magnitude of separation or spreading across the ridge crest is s' and the magnitude of the offset caused by the transcurrent movement is s (Fig. 13), the following analysis will show that in the usual case s' will be larger than s.

Let us assume that (1) a viscous material underlying the stiff plates is being deformed by progressive uniaxial extension, and (2) the overlying plates are dissected by mechanically weak planes parallel to the direction of extension. On one side of the ridge there are two parallel strips of the plate each of width T and length ι_0 and separated by a plane of weakness. The centers of the strips lie on a line that makes an initial angle θ_0 with the extension direction (Fig. 13). After a time t the centers of the strips will make an angle θ with the extension direction. The change in the angle from θ_0 to θ will depend upon the magnitude of the extension undergone by the viscous material underneath. Under uniaxial extension an initial point (x_0, y_0) is shifted to the position (x, y) with

$$x = x_0 \exp(\dot{\xi} x\, t)$$
$$y = y_0 \qquad (1)$$

where $\dot{\xi} x$ is the rate of natural or logrithmic strain in the direction of extension. From equation (1) we find that

$$\frac{x}{y} = \frac{x_0}{y_0} \exp(\dot{\xi} xt)$$

or

$$\cot \theta = \cot \theta_0 \exp(\dot{\xi} x\, t). \qquad (2)$$

Differentiating with respect to time t, we find

$$-\csc^2 \theta \cdot \dot{\theta} = \cot \theta_0 \exp(\dot{\xi} x\, t) \times \dot{\xi} x\,.$$

Replacing $\cot \theta_0$ in terms of $\cot \theta$ with the help of equation (2) we find, after simplification,

$$\dot{\theta} = -\dot{\xi} x \sin \theta \cos \theta \qquad (3)$$

Because of separation across the ridges, the line segment $2\iota_0$ is changed to 2ι, where $2\iota - 2\iota_0 = s'$. Hence,

$$s' = 2\iota_0 \{\exp(\dot{\xi} x\, t) - 1\} \qquad (4)$$

Differentiating with respect to t we have

$$\dot{s}' = 2\iota_0 \dot{\xi} x \exp(\dot{\xi} x\, t) \qquad (5)$$

Figure 12. a—Wax plate with central curved incision after a small amount of deformation. The spreading of the underlying pitch slab is parallel to the short edge of the figure. **b**—Same model at an advanced stage of deformation. Note that the transform faults are converging towards the concave side of the curved plate.

or, since

$$\iota/\iota_0 = \exp(\dot{\xi} \times t),$$
$$\dot{s}' = 2\iota \, \dot{\xi} \, \xi \quad (6)$$

Because of the change in the angle from θ_0 to θ, there will be a slip s along the line of weakness separating the two stiff strips. In the triangle ABC of Figure 13, the angle BAC is θ and the angle ABC is $(\theta_0 - \theta)$, so that

$$\frac{AC}{\sin(\theta_0 - \theta)} = \frac{BC}{\sin \theta}$$

or, because $AC = s/2$ and $BC = T/(2 \sin \theta_0)$,

$$s = T(\cot \theta - \cot \theta_0). \quad (7)$$

Differentiating with respect to t, we have, after simplification with the help of equation (3),

$$\dot{s} = T \cot \theta \times \dot{\xi}_x \quad (8)$$

From equations (6) and (8) we find

$$\frac{\dot{s}'}{\dot{s}} = \frac{2\iota}{T} \tan \theta \quad (9)$$

Because ι is much larger than T and θ is generally large, $\dot{s}'$ is, in ordinary circumstances, significantly larger that $\dot{s}$. Because between ridge crests the transcurrent movement is opposite to the spreading, the sense of the total movement along the planes of weakness will be the same as that caused by spreading, s', at any one instant. In other words, despite the transcurrent movement between the strips of rigid plates, the sense of movement between the two offset ridge segments will be the same as the one that is found in transform faults.

DISCUSSION

The experiments described above are broadly of two types. In the freezing-wax model the tensile strength was in the overlying plate, whereas the substratum behaved in a passive manner and did not produce a significant drag on the undersurface of the stiff plate. The fracture and separation of the plate fragments were caused by this drag. In both sets of experiments, structures similar to transform faults were produced because of the presence of a mechanical anisotropy that favored slip on a set of planes parallel to the direction of bulk extension. Both sets strongly suggest that the presence of a set of mechanically weak planes is an essential factor for the development of transform faults in the ocean floor.

The thickness of the experimental plates was so small that the effect of gravity on them was negligible and hence the fractures across which the plates were separated were tension fractures. For thicker plates the tension would have caused rifting by normal faults.

The ridge-transform patterns of the experiments are strikingly similar to those of the ocean floor. In both situations a consistent sense of ridge offsets is clearly controlled by the gross orientation of the ridge with respect to the direction of spreading. This gross orientation, in the experiments as well as in the ocean floor

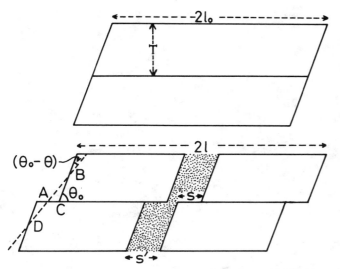

Figure 13. Relation among extension, spreading, and offset. DB joins the midpoints of the shorter edges of the parallelograms. See text for details.

(Mid-Atlantic Ridge, for example), depends on the shape of the initial fracture.

In light of these experiments, it is reasonable to suggest that on both sides of the initial fracture a narrow zone of lithosphere developed whose properties set it apart as a distinct tectonic unit. The sense of ridge offsets is controlled by the orientation of this zone with respect to the direction of bulk extension across the ridge. For example, dextral ridge offsets are expected if this zone trends in the northeasterly direction where the extension across the ridge is roughly east to west. Such geometrical control is evidently present for the Mid-Atlantic Ridge along with its transform faults. Here, the relevant geometric factor is the gross shape of the Atlantic rather than that of the entire African or South American plate.

In the experiments, the offsets were produced in two ways. In the first set of experiments, en echelon tension fractures developed, each of which terminated against transverse planes of weakness. The joining up of the tension fractures by the transverse planes gave rise to a single sawtooth fracture. The sense of side-stepping was entirely controlled by the geometry of a mechanically distinct zone of spreading. In some of the experiments in which the planes of easy sliding cut across the entire width of the plate on both sides of the ridge, spreading or separation was accompanied by a sliding motion along the planes of weakness. This motion caused a progressive increase of the ridge offsets. Because the rate of transcurrent movement was relatively small, the overall sense of movement between the ridge crests was, in conformity with transform faulting, opposite to that of the ridge offsets.

In the second mode of offset development, the ridge is represented by a single continuous fracture cut across by a set of parallel planes of weakness. Under bulk extension parallel to the strike of these planes, there is a separation across the fracture along with the development of offsets of the fracture walls themselves. If the initial ridge was curved, it is represented by a side-stepping system of ridge segments with their local trends changing in accordance with the gross shape of the ridge. Although the ridge offsets in these experiments are produced entirely by transcurrent movement, the instantaneous movement between two offset ridge crests is in the opposite sense, in agreement with what is expected in transform faulting. Evidently, the results of these experimental situations, or modes of offset development, have a relevance for ocean-floor tectonics only if there is evidence of transcurrent movement along the extensions of oceanic transform faults.

A comparison of the experimental structures (Fig. 11 and 12) with those in the ocean floor implies that, if transcurrent movement along the planes of weakness takes place in oceanic lithosphere, a weak fanning of the transform faults and their extensions should occur. In the experiments with pitch slabs, there was a slight shortening of the model perpendicular to the direction of extension. In all likelihood this bulk shortening was also a relevant factor in causing a fanning of transform faults.

The major conclusions are summarized below. These conclusions are necessarily tentative because of inherent problems of extrapolating the laboratory results to the natural situation.

1. The ridge-transform pattern has a characteristic geometry. Such a consistent pattern, with an obtuse angle of the sawtooth shape, does not develop during splitting of an isotropic plate.

2. The experiments strongly suggest that the sea-floor spreading process results in a transversely anisotropic oceanic lithosphere and that the presence of a set of transverse, mechanically weak planes is an essential factor for the initiation of transform faults.

3. The sense of ridge offset is controlled by the shape of a mechanically distinct zone of spreading.

4. The experiments suggest that there may be some transcurrent movement associated with the transform faults. The net effect of spreading and transcurrent movement is such that the instantaneous movement on transforms is opposite to the sense of ridge offset.

REFERENCES CITED

Collette, B.J., 1974, Thermal contraction joints in a spreading sea floor as origin of fracture zones: Nature, v. 251, p. 299-300.

Fujita, K., and Sleep, N.H., 1978, Membrane stresses near mid-ocean ridge-transform intersections: Tectonophysics, v. 50, p. 207-221.

Ghosh, S.K., 1988, Theory of chocolate tablet boudinage: Journal of Structural Geology, v. 10, p. 541-553.

Lachenbruch, H.H., and Thompson, G.A., 1972, Oceanic ridges and transform faults: their intersection angles and resistance to plate motion: Earth and Planetary Science Letter, v. 15, p. 116-122.

Oldenburg, D.W., and Brune, J.N., 1972, Ridge transform fault spreading pattern in freezing wax: Science, v. 178, p. 301-304.

Turcotte, D.L., 1974, Are transform faults thermal contraction cracks? Journal of Geophysical Research, v. 79, p. 2573-2577.

Vroman, A.J., 1976, A tentative explanation of the origin of "transform faults" (with a critical remark on the term): Tectonophysics, v. 30, p. T11-T16.

Wilson, J.T., 1965, A new class of faults and their bearing on continental drift: Nature, v. 207, p. 343-347.

A kinematic model for the evolution of the Indian plate since the Late Jurassic

Sankar Chatterjee, Department of Geosciences, Texas Tech University, Lubbock, Texas 79409-1053 USA

ABSTRACT

The paleopositions of India and its travel paths during the Mesozoic and Cenozoic remain a riddle in conventional plate tectonics. Most reconstructions show peninsular India separating itself from Gondwana, and as an island continent drifting northward for more than 100 my toward its eventual collision with the mainland of Asia. However, the lack of endemism among Indian Cretaceous terrestrial biota is clearly inconsistent with the island continent hypothesis. On the contrary, these fossils show their closest affinity with those of Laurasia and Africa, indicating that India maintained overland connections with these landmasses during this period.

A new model for the tectonic evolution of the Indian plate from the Early Jurassic to the present is proposed that is paleontologically well constrained and corroborated by magnetic anomalies, basement ages, and fracture zone lineations. The configurations are reconstructed manually by moving models of continents across a 16" diameter globe. Smith and Hallam's fit of the Early Jurassic Gondwana is taken as the starting point. The separation chronology of Madagascar from the Africa and Indo-Seychelles blocks is constrained by the Late Jurassic-Late Cretaceous magnetic anomalies in the Somali and Mascarene basins. During this period the conjoined Africa and Indo-Seychelles blocks moved northward, while Madagascar moved relatively southward. This initial spreading phase reflects the relative counterclockwise rotational motion that rotated the Indo-Seychelles block relative to Africa. The pivot of rotation appears to have been situated at Socotra near the tip of Somalia. Late Cretaceous reconstruction places the western coast of India against Arabia. During this period a narrow Tethys Sea, comparable to the present Mediterranean, lay between India and Asia. Paleontologic evidence supports the idea that during the Late Cretaceous terrestrial animals were moving freely between India and Asia across the Kashmir corridor.

Although the sequence of events of the Indian plate motion is well constrained since the K-T boundary time, what triggered this motion is poorly understood. The synchrony of the initiation of spreading from the Carlsberg Ridge, the development of the Owen Fracture Zone and the Amirante Arc, the eruption of the Deccan flood basalts, and their close spatial association indicate that these features may have been caused by a single physical event. It has been proposed recently that a large meteorite collided with the Earth during this crucial time, but the actual site of the putative impact remains elusive. Various tectonic and volcanic features, when restored at the Seychelles-India boundary at 65 Ma, reveals a large oval impact scar, the Shiva Crater. The western rim of the crater survives in the Amirante Arc, south of the Seychelles. The eastern rim is concealed by Deccan lava near the Bombay coast but has a surface expression along the Panvel Flexure. The Shiva Crater is about 800 km long and 700 km wide, and shows the morphology of a complex impact scar. It is estimated that a 40 km diameter asteroid could have created the Shiva Crater and the Carlsberg Ridge by excavating and shattering the lithosphere respectively. The Carlsberg Ridge generated new ocean floor between India and the Seychelles. Lava erupted from this ridge may have flooded the crater to form a lava lake similar to lunar mare. The crater was split by the newly-formed Carlsberg Ridge and the two halves then drifted away from each other. Today, one part of the crater is attached to the Seychelles, the other part to the western coast of India. The westerly slope of the pre-Deccan topography possibly represents the collapsed rim structure associated with downfaulting. The radial and asymmetric fractures on the northeast side of the crater exterior became the main feeder channels for the extrusion of the Deccan flood basalts.

The main vector of the Indian plate motion since the K-T boundary time is essentially north-northeast as determined from the strike of the Owen Fracture Zone rather than northward as has been generally assumed. Continued sea-floor spreading along the Carlsberg Ridge resulted in plate reorganizations in the Indian Ocean, with the development of the Chagos-Laccadive-Mascarene Plateau in the west and the Ninetyeast Ridge in the east. The tectonic history of the Ninetyeast Ridge is far more complex than the popular hot-spot model. An alternative model for the oblique convergence between Indian and Australian plates is proposed here to explain the origin of this linear volcanic chain. The Ninetyeast Ridge is interpreted as an ancient transform fault that evolved into a trench boundary because of a convergent strike-slip motion. As a result, a linear volcanic chain is produced with age gradient mimicking hot-spot tracks. The spreading of the Carlsberg Ridge led to the strike-slip displacement of India from Arabia along the Owen-Ornach-Nal-Chaman fault system, followed by India's convergence with Asia, the closure of the Tethys, the origin of the Himalayas, and the plate reorganizations in the Indian Ocean.

The north-northeast penetration of India into Asia is seen as a major factor in the asymmetric plan pattern of the Himalayas. In the west the convergent motion of India was accommodated in the Kirthar-Sulaiman Range along the Ornach-Nal-Chaman fault system. The collision between India and Asia is considered to have resulted in spectacular tectonic deformation and lateral extrusion of eastern Asia. However, the Baikal Rift and the Shansi Graben may have played an important part in producing the east-west trending strike-slip faults in this region. The detailed chronology of the collision and plate reorganizations in the Indian Ocean supports the hypothesis that ridge-push is the dominant driving mechanism of the Indian plate.

INTRODUCTION

The paleoposition of India (Indo-Pakistani subcontinent) in Gondwana, its past plate motion and its dispersal route have been the subject of speculation and controversy for several decades. In most plate-tectonic reconstructions (Dietz and Holden, 1970; McKenzie and Sclater, 1973; Johnson et al., 1976; Norton and Sclater, 1979; Barron and Harrison, 1980; Smith et al., 1981; Scotese et al., 1988; Patriat and Segoufin, 1988; Royer and Sandwell, 1989), India is shown as an island continent located far out in the Tethys Sea during its flight from Madagascar to Asia (Fig. 1). According to these models, it took more than 100 million years from the time that India separated from Gondwana (Late Jurassic, approximately 160 Ma) until India made subaerial contact with Asia (Late Eocene, approximately 40 Ma). Such

---

Chatterjee, S., and N. Hotton III, eds. *New Concepts in Global Tectonics.*
Texas Tech University Press, Lubbock, 1992, xii + 450 pp.

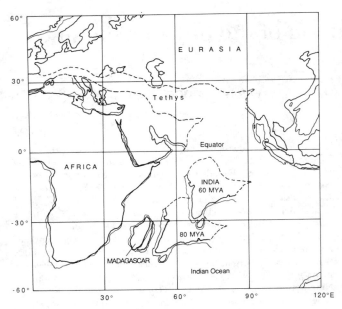

Figure 1. Reconstructed position of India during the Late Cretaceous (Santonian, 80 Ma) and the Paleocene (60 Ma) according to Smith et al., 1981. Note the total isolation of Peninsular India during this period.

an extended period of isolation should have produced a highly peculiar endemic biota, but Cretaceous fossils from India do not show the evolutionary effects of isolation. Contrariwise, Late Cretaceous Indian biota clearly show Eurasian and African affinities (Chatterjee and Hotton, 1986; Sahni and Bajpai, 1988; Jaegger et al., 1989; Briggs, 1989). It seems reasonably evident that India was not an island continent during its northward journey, but maintained an overland communication with Asia and Africa that allowed terrestrial vertebrates to cross back and forth.

The evolution of the Indian plate is unusually complex in several respects. First, the intraplate deformation in the central Indian Ocean is a well-known example of a violation of the basic tenets of plate tectonics (Weissel et al., 1980). Second, a large asteroid impact may have contributed significantly to the Indian plate motion since the K-T boundary time. It has been speculated that the K-T impact occurred at the Indo-Seychelles plate to initiate the rifting between the Seychelles and India.(Hartnady, 1986; Alt et al., 1988; Chatterjee, 1990a, 1990b). The impact was so powerful that it created an enormous crater, about 900 km long, and produced a midocean ridge, rift basins, and volcanism. The Indo-Seychelles rift margin documents for the first time the possible correlation of a giant impact to sea-floor spreading.

The determination of the past motion of India relative to Africa is particularly important when considering the breakup of Gondwana. The details of the physical and biological relationships between India and its adjacent continents during its flight present a challenge to conventional plate tectonics. Fracture zones, transform fault lineations, magnetic anomaly bands, linear volcanic chains, and basement ages of the Indian Ocean floor have been used in the past to chart the successive positions of India. The tectonic style of the Himalayan mountain belts also preserves a record of the direction of Indian plate motion during its convergence to Asia.

One of the important path tracers of the Indian plate motion is the lineation of the Owen Fracture Zone, trending roughly in the north-northeast–south-southwest direction. If in its northward journey India stayed closer to the Somalia-Arabian coast than is depicted in most plate reconstructions, and if the Owen Fracture Zone marked the passage of its western margin, then a connection between India and Africa-Arabia appears to be plausible (Briggs, 1987). The Owen Fracture Zone provides a tight constraint on the paleospreading direction of the Indian plate.

A reconstruction of the Indian plate motion is attempted here on a constant-radius Earth by refitting the outlines of the interacting plates. The new reconstruction is paleontologically well constrained and offers a mechanism for the kinematic evolution of the Indian plate, the reorganization in the Indian Ocean, and the eruption of the Deccan flood basalts at the K-T boundary. The reinterpretation of the sea-floor spreading data and tectonic features suggest that India was never an island continent during its journey. These data provide an improved version of India's paleogeography, which is quite different from the usual plate reconstructions.

PALEONTOLOGIC CONSTRAINTS

The Cretaceous (145–65 Ma) is among the longer periods of the Earth's history, and is divisible into two phases, early and late. Because of extensive marine incursions, Early Cretaceous terrestrial life is poorly known, represented in some beds in Europe, North America, and China. In India, Early Cretaceous record of vertebrate fossils is blank. Various continental formations such as Chikiala, Gangapur, Nimar, and Jabalpur have potential for future discovery of dinosaurs and other terrestrial vertebrates, but so far, no systematic search has been made. In contrast, Late Cretaceous vertebrates, especially dinosaurs, are well known throughout the world. This discussion will be restricted to the Late Cretaceous Indian biota to understand its continental connections during this period.

In India, the Deccan Traps constitute an important marker bed for correlating the K-T boundary events. These thick lava flows, though confined to a narrow interval of time, were not extruded in one singular event. Intercalated with flows are fluviatile and lacustrine sediments (such

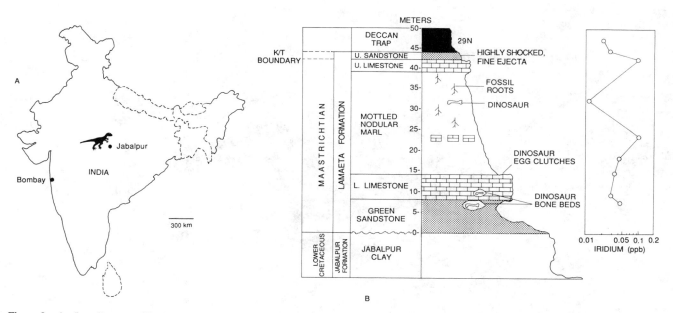

Figure 2. **A**—Locality map of India showing Jabalpur site. **B**—Geological section of the Late Cretaceous Lameta Formation at Bara Simla Hill, Jabalpur, showing stratigraphic positions of dinosaur fossils and eggs, and the ejecta layer at the K-T boundary; corresponding Iridium profile on the right column.

as Lamenta and Takli formations) containing abundant plant and animal remains including a varied dinosaur fauna of Maastrichtian age. These fossils provide important clues to the Cretaceous paleobiogeography of India.

The Lameta dinosaurs have been known for over a century, most of them coming from a single quarry on the western slope of Bara Simla Hill, Jabalpur, in central India (Fig. 2). Although they are represented by fragmentary material, they show a great taxonomic diversity. Huene and Matley (1933) identified very large sauropods (titanosaurids), a variety of theropods (both coelurosaurs and carnosaurs), and an ankylosaur, *Lametasaurus*. Chatterjee (1978) described a partial skull of a carnosaur *Indosuchus* from the Lameta Formation, which may belong to the new theropod family Abelisauridae (Bonaparte and Novas, 1985). Berman and Jain (1982) reported a titanosaurid braincase from the Lameta Formation of Dongargaon, about 320 km south of Jabalpur. Late Cretaceous dinosaur egg clutches have been reported from the Lameta Formation in the Kheda district, Gujrat (Srivastava et al., 1986). Biostratigraphic evidence, drawn from dinosaurs, pelobatid frogs, charophytes, nonmarine ostracodes, and the selachian *Igdabatis*, clearly indicates a Late Maastrichtian age for the Lameta Formation (Courtillot et al., 1986; Sahni and Bajpai, 1988; Jaegger et al., 1989).

Recently we have collected well-preserved dinosaur material from the Bara Simla quarry, representing partially articulated skeletons of several individuals including cranial elements. Preliminary study indicates that titanosaurids, ankylosaurids, and abelisaurids are present in the new collection (Fig. 3). Both titanosaurids and abelisaurids are found to be identical to those of France, recently reported from Late Cretaceous deposits (Buffetaut, 1989; Le Loeuff et al., 1989). The Lameta ankylosaurids represent typical Laurasiatic forms with distinctive tail club. Similarly, the Lameta sauropod eggs are very similar in microstructure and gross morphology to those reported from the Maastrichtian of France (Vianey-Liaud et al., 1987). The dinosaurs from India provide evidence of faunal interchange between India and Eurasia across the Tethys Sea during the Late Cretaceous

Similar biotic correspondence comes from the recent study of microfossils. Sahni (1984) concluded that the terrestrial fossil assemblages from the infra- and intertrappean sediments, such as pelobatid frogs, boid snakes, anguid lizards, pelomedusid turtles, nonmarine ostracodes, and charophytes show no evidence for faunal endemism, which contradicts the popular model of an oceanically isolated Indian subcontinent. On the contrary, they show closer affinities with contemporaneous Laurasian and African fossils. There are striking similarities between the freshwater fish fauna (*Lepidotes, Lepisosteus, Igdabatis*), boid snake (*Madtosia*), and pelomedusid turtles of the Late Cretaceous of India and Niger indicating India-Africa communication at that time (Courtillot et al., 1986; Sahni et al., 1987). The anguid lizards from the infratraps are extremely similar to those of the Lance Formation of North America

Figure 3. Life restorations of the Late Cretaceous Lameta dinosaurs. In the foreground is the ankylosaurid, *Lametasaurus*; on the left is a herd of titanosaurids, *Titanosaurus*, on the right is the abelisaurid theropod, *Indosuchus*.

(Sahni et al., 1987). These data are now reinforced by the discovery of paleoryctid mammals from the Takli Formation of Asifabad (Prasad and Sahni, 1988). Paleoryctids are considered typical Laurasiatic mammals and have been reported from the Late Cretaceous of Mongolia and North America. This faunal similarity implies that India was connected with Asia and Africa during the Late Cretaceous. Asia, in turn, was connected to North America at that time across Beringia, which allowed almost continuous highway for dispersal of terrestrial and freshwater fauna for a long time (Lillegraven et al., 1979).

The floral record also suggests that India was not an island continent during the Cretaceous. Smiley (1974) noted that 50 percent of the genera of Indian Cretaceous flora are recorded from several areas across Laurasia. A typical North American Cretaceous palynomorph, *Aquilapollenites*, has been recovered from the Lameta Formation (Sahni and Bajpai, 1988). Similarly, Kar and Singh (1986), in their investigation of the palynology of Cretaceous sediments of Meghalaya, noted a strong similarity to a Middle Cretaceous assemblage from the Peace River in northwestern Canada. Sahni (1989) also mentioned the discovery in the Takli Formation of some representative of a Mongolian charophyte genus.

When a landmass becomes completely isolated as an island, endemic species are very common, especially among terrestrial vertebrates. The larger an island, and the longer the period of isolation, the greater the probability that a resident population will persist long enough to develop into a distinct endemic species. The strongly endemic fauna of Australia and the Seychelles is clear evidence that these landmasses were isolated throughout the Tertiary period. India, if it really was an isolated island continent for more than 100 millions of years during its journey (Barron, 1987), should have evolved a distinctive endemic fauna during its isolation, and we surely should find some evidence of it, if in fact it ever existed. Instead, we find that almost all Indian taxa were possessed in common with Africa and Laurasia, indicating close biotic links. The lack of endemism among Indian vertebrates during the Cretaceous period is clearly incompatible with the island continent hypothesis (Colbert, 1973; Chatterjee, 1984; Chatterjee and Hotton, 1986; Sahni, 1984; Briggs, 1989). Another paleobiogeographic paradox is posed by the occurrence of typical Laurasiatic biota in India. It is suspected that terrestrial corridors between Asia and India were established sometime during the Cretaceous to allow floral and faunal exchange. Any reconstruction of India during this time should be congruent with these important paleobiogeographic constraints.

TECTONIC SETTING OF THE INDIAN OCEAN

The Indian Ocean is the most complex and least understood of the Earth's major oceans. Its complex tectonic history is slowly unfolding from the physiography of the ocean floor, marine geophysical measurements, and deep sea drilling results (Sclater and Fisher, 1974; Schlich, 1982; Royer et al., 1989). The sea floor of the Indian Ocean is dominated morphologically by a

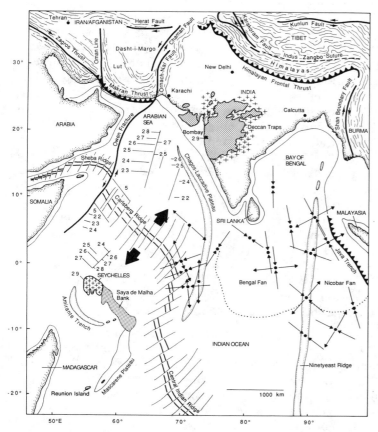

Figure 4. Major tectonic features of the Indian Ocean and Asia. Compiled from McKenzie and Sclater (1973), Fisher et al. (1974), Molnar and Tapponnier (1977), Powell (1979), Weissel et al. (1980), De Jong (1982), Schlich (1982), Wezel (1988), and Royer et al. (1989).

system of active midocean ridges that mark the boundaries of several plates (Fig. 4). The ridges form the Rodriguez Triple Junction at 25° S, 70° E, and resemble an inverted Y. One arm of the inverted Y, the Southwest Indian Ridge (SWIR) separates Africa and Antarctica and connects with the Mid-Atlantic Ridge. The other arm, the Southeast Indian Ridge (SEIR) separates the Indo-Australian plate from Antarctica and joins with the East Pacific Rise. North of the triple junction, the Central Indian Ridge (CNIR) runs almost due north as a series of en echelon spreading centers and fracture zones before turning northwest as the Carlsberg Ridge (CR). The northern end of the Carlsberg Ridge and the Sheba Ridge is offset about 300 km by the Owen Fracture Zone. The Sheba Ridge enters the Gulf of Aden and connects with the rifts of Africa (Laughton and Tarmontini, 1969). The Owen Fracture Zone, a plate boundary between India and Arabia (Gordon and DeMets, 1989), extends for hundreds of kilometers in a north-northeast–south-southwest direction and connects northward with the Ornach-Nal-Chaman (ONC) strike-slip fault along the Baluchistan Arc (Abdel-Gawad, 1971; Molnar and Tapponnier, 1977).

An extraordinary feature of the Indian Ocean is the large number of long, linear, aseismic ridges and plateaus, as well as several microcontinents, scattered throughout the basin. The most striking of these submarine features is the Ninetyeast Ridge running along the 90° E meridian, marking the limit between the western and eastern Indian Ocean. It is more than 4,500 km long, roughly 50–100 km wide, and its relief above the surrounding ocean floor averages 2 km (Sclater and Fisher, 1974; Peirce, 1978). It is the longest linear feature in the Indian Ocean, and its tectonic history is more complex than it first appears. The Ninetyeast Ridge has commonly been referred to as an aseismic ridge, although it has been pointed out by Stein and Okal (1978) that it is seismically very active and forms a plate boundary between India and Australia. Seismic studies in the Bay of Bengal (Curray and Moore, 1971; Curray et al., 1982; Weissel et al., 1980) revealed highly deformed oceanic crust and overlying sediments, stretching from the Chagos-Laccadive Ridge to the Ninetyeast Ridge and from 5° N to 10° S. This intraplate deformation implies that the India-Australian plate is under severe compressive stress and that the Ninetyeast Ridge may be a precursor to a new covergent boundary between separate Indian and Australian plates (Le Pichon and Heirtzler, 1968; Minster and Jordan, 1978).

In the west-central Indian Ocean Basin lies the Chagos-Maldive-Laccadive Ridge that extends nearly 3,000 km due north along the 73° E meridian and intersects with the Deccan Traps. This ridge is bordered easterly by the Chagos Trench. South of this ridge lies the Mascarene Plateau, a 2,000-km-long arcuate ridge stretching from the Seychelles Island to the volcanic chains of Mauritius and Reunion. The Mauritius Trench borders the Mascarene Plateau eastward.

The Indian Ocean shows three distinct phases of ocean-floor spreading history. The first phase of spreading occurred along the SWIR extending from the Late Jurassic to the Late Cretaceous. It separated Africa-Madagascar-India from Antarctica-Australia. During its northward journey, Madagascar lagged behind Africa and India to produce the bulk of the Somali Basin and Mascarene Basin. The second phase of spreading took place at the K-T boundary time along the Carlsberg Ridge (C-CNIR), when the microplate Seychelles was rifted from India. India began to converge with Asia following the contraction of the intervening Tethyan Sea, resulting in the formation of the Himalayan mountain chain. This converging motion still continues today. Finally, from the Eocene to the present, the spreading of the SWIR began with the separation of Australia from Antarctica. These three phases of spreading have greatly modified and reorganized the Indian Ocean floor. For the sake of clarity, I did not represent in Figure 4 all the magnetic anomalies, especially the poorly defined M-series.

The margins of India are inactive, the spreading of the C-CNIR and the movement of the Indian plate being absorbed in the folding and intracratonic subductions of the Himalayas. The only active trench in the Indian Ocean is the Java Trench absorbing ocean crust generated farther south along the SWIR.

Reconstruction methods

The reconstruction of the Indian plate is done manually by moving models of continents across a 16" diameter globe (National Geographic, physical) in which the physiography of the ocean floor is accurately portrayed. This method does not involve any external reference frame and therefore provides fairly accurate means of reconstructing past plate configurations. The five continental fragments (India, Madagascar, Africa, Seychelles, and Arabia) with continental slope outlines at 1,000 m isobaths were molded in raised relief with epoxy sculpting putty (sculpall) to form segments of rigid spherical shells on the globe (Fig. 5). Individual continental fragments can be fixed at any position over the globe with a little wax placed on the inner surface of the fragment, thus acting as a fastener. The spreading pattern of the globe on the present-day Earth was used to reconstruct the relative plate motions since the Late Jurassic. Weijermars (1989) pointed out that manual constructions of the movement of tectonic plates on globe are simple and fairly accurate. The interactive computer graphics developed by the Paleoceanographic Mapping Project (POMP) at the University of Texas were very helpful to cross-check the sequence of Indian plate motion.

At any given time, an accretionary plate boundary is made up of a combination of ridge segments and transform faults. As sea-floor spreading progresses, the spreading history will be preserved in the ocean floor as a series of magnetic anomalies that are symmetrically disposed on either side of the spreading axis. The tectonic fabric chart of the Indian Ocean with magnetic anomaly data (Royer et al., 1989) was useful for continental reconstructions that were produced by the successive closing of the anomaly bands relative to the ridge axis.

Minster and Jordan (1978) noted that Indian plate motion data could be improved if India and Australia are assumed to lie on separate plates divided along a hypothetical plate boundary following the Ninetyeast Ridge. Gordon et al. (1990) proposed a diffuse plate boundary between rigid Indian and Australian plates. To reconstruct India with respect to Australia requires an accurate interpretation of the origin of the Ninetyeast Ridge. To eliminate these ambiguities, data along only the C-CNIR were used. Note that the anomalies adjacent to Ninetyeast Ridge and the Mesozoic anomalies adjacent to Australia (Markl, 1974) have not been used to define the relative motion of India with respect to Australia. In this discussion the geological time scale from Early Jurassic times to the present is adopted from Harland et al. (1982).

MESOZOIC TECTONIC RECONSTRUCTIONS

Early Jurassic (Pliensbachian, 200 my)

The initial fit of Gondwana is poorly constrained by data from the sea floor. Consequently, initial reconstructions are dependent on geologic data such as the correlation of concurrent depositional facies, basement lineations, and the distribution of shield areas. The geometric fit of Gondwana is controversial, the main uncertainties being the locations of Madagascar and India. The shapes of the continental edges of the east coast of Africa, Madagascar, India, Australia, and Antarctica are not quite so well suited to fitting as the circum-Atlantic continents.

Smith and Hallam's fit (1970) is taken as the starting point of Gondwana reconstruction at the Early Jurassic prior to its breakup. In this fit, the northwest part of India was attached to Somalia-Arabia. The location of Madagascar in a northerly position adjacent to Tanzania, Kenya, and southern Somalia is recently confirmed by the discovery of M-series magnetic anomalies in the Somali Basin (Rabinowitz et al., 1983). However, the Seychelles microcontinent was part of India at that time (Davies, 1968; Norton and Sclater, 1969; Chatterjee and Hotton, 1986; White and McKenzie, 1989a), and the reconstruction is modified accordingly to accommodate the Seychelles into the small gap north of Madagascar. The Laccadive-Chagos Ridge and the Mascarene Plateau are absent because they are of Tertiary to Recent age (Figs. 6A and 6B).

Late Cretaceous (Maastrichtian, 70 my)

The geometry of plate movements and spreading axes in the period between the initial splitting and the Late Cretaceous remains a major problem in the history of the breakup of Gondwana because of the long duration (approximately 34 my) of the Cretaceous Quiet Zone (118–84 Ma) and the generally poor quality of M-series anomalies. For this reason, no satisfactory model has been proposed to explain the sequence of dispersal of Gondwana plates during this period.

In the western Indian Ocean, three plates—Africa, Madagascar, and Indo-Seychelles—were interacting with each other since the Middle Jurassic. Identification of a sequence of east-west trending magnetic anomalies (M9–M25) in the western Somali Basin (Rabinowitz et al., 1983) indicates a much earlier separation (Middle Jurassic, approximately 160 Ma) of Madagascar from Africa than previously considered, placing the separation at about the same time as the initial breakup of Gondwana. The anomalies are symmetric about ancient ridge segments. The magnetic data suggest that Madagascar moved south relative to Africa along the Davie Ridge Transform Fault and maintained its present position with Africa at the time of the formation of anomaly M9 (Early Cretaceous, approximately 130 Ma). The motion of Madagascar in relation to Africa was a combination of

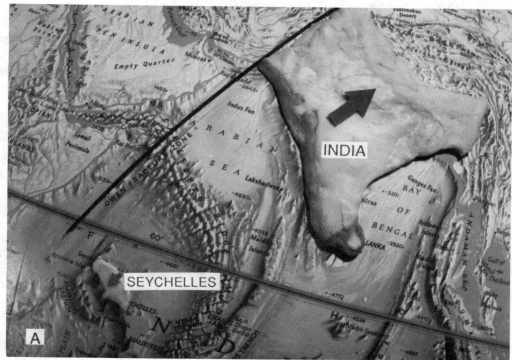
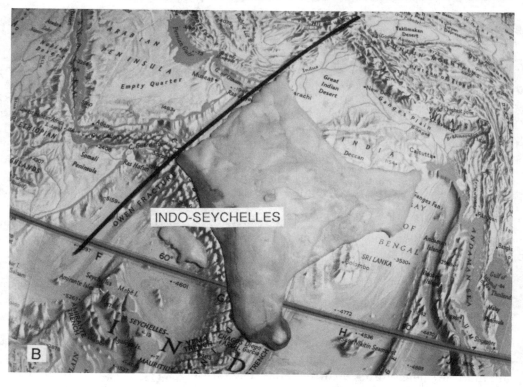

Figure 5. Reconstruction of the India-Seychelles continental block on a National Geographic physical globe. **A**—Present postion of India and the Seychelles. **B**—India-Seychelles fit at the K-T Boundary time relative to the Carlsberg Ridge.

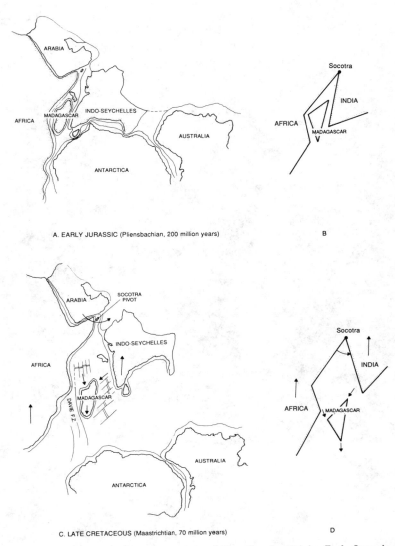

Figure 6. Mesozoic reconstruction of East Gondwana. **A**—Early Jurassic reconstruction before the breakup of Gondwana. **B**—Sketch map showing the interrelationships of Africa, Madagascar, and India in the same time. **C**—Late Cretaceous configuration of East Gondwana; note the divergent strike-slip separation of Madagascar and its southward displacement relative to the African and Indo-Seychelles blocks; Indo-Seychelles block rotated counterclockwise in relation to Africa around the Socotra pivot. **D**—Sketch map showing the southward displacement of Madagascar relative to Africa and India during the Late Cretaceous.

strike-slip and extension (transtension). The distribution of marine Jurassic rocks provides further support for the early breakup of Madagascar. The faunal and lithological successions of East Africa, Madagascar, and Kutch in north-west India are strikingly similar and show faunal provincialism in this region (Smith and Hallam, 1970).

To this is added the effect of paleontologic constraint, which requires continental connection between Africa and Indo-Seychelles during the Cretaceous. The Africa and Indo-Seychelles blocks maintained their original ties along Socotra-Karachi while breaking away from the rest of the Gondwana, moved northward in concert, and simultaneously opened the extreme South Atlantic and southeastern Indian Ocean. The early history of sea-floor spreading between Africa and Antarctica can be reconstructed using the identifications of Mesozoic magnetic anomalies (M22–M0) in the Mozambique Channel and their counterparts to the south, in the vicinity of Dronning Maud Land (Simpson et al., 1979). During this time, the motions of India relative to Antarctica were not well constrained. The occurrence of Late Cretaceous anomalies (31–34) in the Mascarene Basin along a fossil ridge defines the relative transtension motion between Madagascar and India (Schlich, 1982). The combination of these two transtension motions is shown by the southward displacement of Madagascar in relation to the Africa and Indo-Seychelles blocks, producing the bulk of the Somali Basin and Mascarene Basin. At the same time, these divergent motions led to an anticlockwise rotational movement of Indo-Seychelles block in relation to Africa; the pivot of rotation appears to have been around the Socotra region at the tip of Somalia (Fig. 6 C and D).

Cretaceous-Tertiary boundary (65 my)

The reconstruction map for the Maastrichtian and the K-T boundary are almost identical. At the K-T boundary time, India began to separate from the Seychelles with the development of the Carlsberg Ridge. Because the separation of the Seychelles from India is well constrained by magnetic anomaly data, we can reconstruct the paleoposition of India backward from the present to the K-T boundary time.

Today, the Seychelles is separated from the western coast of India by 2,800 km because of spreading of the Carlsberg Ridge. The Carlsberg Ridge shows symmetrical magnetic anomalies of 5, 23, 24, 25, 26, 27, 28, and 29 on either side of the crest between India and Seychelles (Fig. 4). Presumably the successive pairs of magnetic anomalies were generated at the ridge axis and moved apart on rigid plates recording the history of the opening of the Arabian Sea. Prerifting reconstruction requires a single rotation to close the Arabian Sea. Seychelles may have behaved as an independent microplate during this rifting event (Vine and Livermore, 1985). The lineament of the Owen Fracture

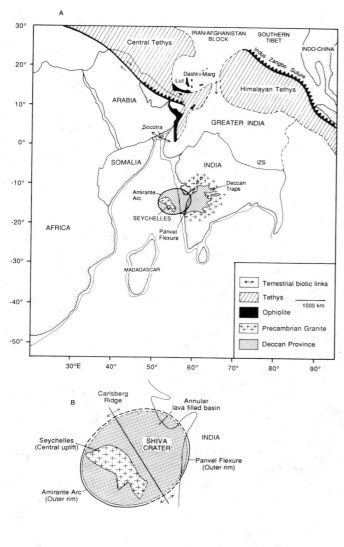

Figure 7. **A**—India-Seychelles fit at the K-T boundary time showing the location of the Shiva Crater; possible dispersal routes of land vertebrates between India-Africa and India-Asia are shown by arrow; the Tethys Sea bordering the Indian subcontinent at the K-T boundary time is reconstructed. **B**—Morphology of the Shiva Crater exhibits a distinct central uplift in the form of the Seychelles granitic core; an annular trough filled with lava to form a mare; and a slumped outer rim. The crater is somewhat oval shaped because of oblique impact along southwest-northeast trajectory. The impact simultaneously created the Carlsberg Ridge and the Shiva Crater.

Zone, which defines the strike-slip motion between India and Somalia-Arabia, is used as a guide to determine the direction of the Indian plate motion since the K-T boundary time. Reversing the motion, India fits close to Somalia-Arabia with an intervening Somali Basin as in the Maastrichtian reconstruction; Karachi and Socotra maintained physical and biotic links.

The reconstruction places the western coast of India against the Seychelles-Saya de Malha Bank and shows matching geological provinces (Fig. 7A). The largely submerged continental block that bears the Seychelles Islands contains enormous flood basalt deposits in the submarine plateau of the Saya de Malha deposits, which were erupted as part of the Deccan volcanism about 65 Ma (Meyerhoff and Kamen-Kaye, 1981, Backman et al., 1988). The link between magmatism on the Seychelles and in the Deccan Plateau is emphasized by the matching geochemistry (White and McKenzie, 1989b). Similarly, the Late Proterozoic Mahe Granite of the Seychelles (Baker and Miller, 1963) is isochronous (700 ± 50 my) with the Siwana-Jalor Granite (Auden, 1974) of western India. The conjoined Seychelles-India block forms a huge, continuous Deccan Province (Fig. 7).

INDIA-SEYCHELLES RIFTING, K-T IMPACT EVENT, AND DECCAN VOLCANISM

The rifting of India-Seychelles

How does a continent split open, allowing an ocean to form in the gradually widening rift? What physical and chemical processes are involved? These are some of the controversial issues in plate-tectonic theory debated for many years. Structurally, a rift is essentially a downfaulted block or graben. All rifts exhibit anomalous crustal and upper mantle profiles, usually interpreted as the result of asthenospheric diapirism or lithospheric thinning. The interplay between lithospheric thinning and magmatism is central to several physical models of continental rifting.

Two alternative views of the rifting process are known, active and passive. Active rifting begins in the asthenosphere with the development of a thermal anomaly, a hot spot, which might be responsible for thinning the lithosphere from below, causing uplift, and initiating the continental rift (Morgan, 1972, 1981; Burke and Wilson, 1976; Bonatti, 1987; Hooper, 1990). Passive rifting, on the other hand, is produced entirely as a result of lithosphere extension, the source of this horizontal stress being attributed to the forces arising from the interaction of lithosphere plates along their mutual boundaries (Forsyth and Uyeda, 1975; White and Mckenzie, 1989a, 1989b). In this view, the asthenosphere is essentially passive; it simply rises to fill the gap left by the thinning lithosphere. Intraplate extensional tectonic effects in the form of graben, rifts, and sedimentary basins are widespread in the major continental plates. It appears that such conditions can readily be met, particularly at times of unusual plate configurations when continental intraplate extensional stress is at a maximum resulting from the trench suction (Forsyth and Uyeda, 1975). A horizontal deviatoric tension of 100–200 MPa is required to rift the continental lithosphere (Kusznir et al., 1986).

A new model of lithospheric thinning by meteoritic impact is proposed here to explain the origin of the Carlsberg Ridge. In this model, the lithosphere could be shattered by a projectile of considerable size (≥ 10 km) to initiate midoceanic ridge (MOR). Asteroids strike the Earth at an average speed of 25 km/sec and transfer considerable kinetic energy to the target rocks (Shoemaker, 1983). The pressures exerted on the meteorite and the target rocks can exceed 100 GPA; temperatures can reach several thousand degrss Celsius (Grieve, 1990). Such a hypervelocity impacting body penetrates the target rocks to two to three times its radius (Grieve, 1987). An asteroid of 40 km diameter would produce cratering and associated tectonic rebound-collapse effects sufficient to shatter the 80-km-thick lithosphere that could form plate boundaries and continental rifts. The proposed giant K-T impact at the India-Seychelles boundary may have created the Carlsberg Ridge and continental rifting by initiating a crack in the stressed lithosphere (Alt et al., 1988; Chatterjee 1990a, 1990b). This concept opens up a new field of research to investigate whether plate tectonics may be influenced by impacts of large bodies.

A possible K-T impact site

It is clear that Earth has been bombarded throughout its history by extraterrestrial objects of various sources, sizes, and compositions (Clube and Napier, 1982). Incontrovertible evidence of large cosmic collisions is the typical occurrence of circular craters that are associated with considerable local structural disturbance and shock metamorphism (French and Short, 1968). Because of the dynamic nature of the terrestrial lithosphere, where such forces as erosion, volcanism, deposition, orogeny, and plate tectonics constantly restructure the surface, impact structures are often quite quickly obscured, unlike the more static surfaces of the Moon, Mercury, and Mars, for example. To date, over 120 impact craters have been recognized on the Earth's surface. They range in size from approximately 100 m to 150 km and in age from Proterozoic to Recent (Grieve, 1987, 1990). Other craters may be submerged under oceans and remain accessible or undetected. Scientific interest in the role of impact in geological and biological evolution has been enhanced by several developments in recent years. Among the most prominent of these are the hypothesis of Alvarez et al. (1980) concerning terminal Cretaceous extinction, and lunar and planetary exploration by manned and unmanned spacecraft. As interest in bombardment mounts, previously unknown or cryptic sites frequently are recognized.

Hypervelocity impacts can have a large range of effects that depend on the strength and density of the projectile and the nature of the target material. The most obvious result of the larger collisions is seen in the spectrum of crater sizes and morphologies. Craters associated with meteorites and shock metamorphic effects are testimony of an impact event, but when large craters are deeply eroded or buried, the evidence of impact is obscured or blurred. Such evidence may be identified indirectly from shock metamorphic effects on the target rock and ejecta components, as well as distinctive geochemical signatures attributable to a particular type of meteoritic projectile. These signatures may be preserved locally near the impact site, or globally at a particular stratigraphic level containing ejecta fallout. Together they may provide clues to the nature of the target material and the impacting body. Impact craters of this obscure nature are the most controversial and require additional information for verification.

Such is the case of the putative K-T impact. Ever since Alvarez et al. (1980) proposed that a 10 km asteroid collided with the Earth at a velocity of 25 km/second, producing a biotic crisis at the K-T boundary, investigators have searched painstakingly for evidence of a crater marking the point of collision. Such a crater could have been 150 km or more in diameter (Grieve, 1982). One of the strongest criticisms levied against a terminal Cretaceous impact event is the failure to locate a possible impact site commensurate in age and size (Hallam, 1987; Officer et al., 1987). There are number of candidates for the K-T impact site, none of which is very compelling. The 35 km Manson structure in north-central Iowa has been mentioned as a possible K-T impact site (French, 1984; Anderson and Hartung, 1988). However, the Manson structure is apparently too small to account for globally integrated estimates of ejecta at the K-T boundary, or for the dramatic effects upon the biosphere. Twin impact structures in the Kara Sea in the northern USSR, the Kara (diameter, 60 km) and Ust-Kara (diameter, 25 km), have been proposed as possible impact sites (Koeberl et al., 1988), but recent geochronologic age suggests that these structures are older than the K-T boundary event (Koeberl et al., 1990). Hildebrand and Boynton (1990) placed the K-T impact location in the Colombian Basin between Colombia and Haiti on the basis of seismic data and deep-sea drilling program (DSDP) core samples, but the putative structure is not only under water, but buried by 1,000 m of sediments and is subject to other interpretations, such as tectonic origin or a change in the thickness of the oceanic crust. This idea of oceanic impact is seemingly incompatible with evidence of shocked quartz found in K-T boundary rocks in western North America (Bohor et al., 1987). Bohor and Seitz (1990) speculated that the impact site was near Cuba, about 1,350 km from the site proposed by Hildebrand and Boynton, but the Cuban site is ruled out as point of collision (Dietz and McHone, 1990).

Not only the specific site of K-T impact, but also its general location, whether continental or oceanic, remains controversial. Trace element, basalt spherules, and isotopic studies of the highly altered boundary-layer components tend to support the oceanic impact hypothesis (Gilmore et al., 1984; Hildebrand and Boynton, 1990). On the other hand, the presence of shocked quartz at several K-T boundary sites would indicate a terrestrial site (Bohor et al., 1987). The apparent contradiction can be reconciled if the impact was at a continental margin involving both oceanic and continental crust.

Hartnady (1986) suggested that the Amirante Basin, south of the Seychelles may be a possible K-T impact site. The basin has a subcircular shape of about 300 km in diameter, bounded on the northeast by the Seychelles Bank and is partially ringed on the southwest by the semicircular structure of the Amirante Arc. Hartnady favored an oceanic impact site and cited the presence of extensive chaotic slump structures at a K-T boundary section on the East African continental margin as a consequence of tsunami effects. However, the morphology of the Amirante Basin is enigmatic. It is semicircular in outline, preserving half of a supposed crater rim. What happened to the other half of the crater?

Alt et al. (1988) remedied the deficiency of the Amirante Basin model as the point of collision. They suggested that the western rim of the crater survives in the Amirante Arc, as Hartnady (1986) proposed, but the eastern rim lies along the western coast of India, hidden by the overlying Deccan Traps. They speculated that the impact was forceful enough to create not only an enormous crater approximately 600 km in diameter, but also to cause pressure-release melting in the asthenosphere. Basalt then filled the crater basin to form an immense lava lake, the terrestrial equivalent of a lunar mare.

Chatterjee (1990a, 1990b) elaborated this K-T impact scenerio at the India-Seychelles rift margin, and identified the eastern rim of the crater along the Panvel Flexure, near the Bombay coast. The Panvel Flexure is an arcuate segment of the crater, about 120 km long, on the Deccan Traps, and is difficult to explain in terms of conventional tectonics. It is marked by a line of hot springs, dikes, deep crustal faults, and seismicity (Kaila et al., 1981; Powar, 1981). The Panvel Flexure may represent the eastern rim of the crater in the form of a collapsed rim structure. It exercises a tectonic control on the attitude of the Deccan lavas. To the east of the flexure, the basaltic flows are horizontal; to the west of the flexure, the basalatic flows dip west to west-southwest at 50°–60° toward the coast. The abrupt change of dip along the flexure axis may indicate the slope of the crater wall, which is now concealed by Deccan lavas. By completing the circle combining the Amirante Arc and the Panvel Flexure, the extent of the crater can be extrapolated (Fig. 7B). I suggest this impact structure be named Shiva Crater, after Lord Shiva, the Hindu god of destruction.

If real, the restored Shiva Crater is the largest, most spectacular and most complex of terrestrial craters known, and provides unique opportunities for comparisons with lunar craters. It shows the structure and morphology of a giant, complex crater, oval in outline (Fig. 7B) with (1) a collapsed outer ring, 900 km long and 800 km wide; (2) an annular trough, filled with a lava lake like a lunar mare; (3) a distinct central uplift, represented by the Seychelles; and (4) a highly dispersed ejecta blanket with shock quartz (discussed later). This crater exceeds the dimensions of other impact features observed on Earth, but corresponds in diameter with the Orientale Basin of the moon. The smooth, flat, marelike surface is particularly well developed at the Amitante Basin.

It is estimated that a 40 km diameter asteroid, about the size of the Amor object Ganymed could have created the Shiva Crater and the Carlsberg Ridge. Lava generated from the ridge flooded the crater to form a lava lake similar to a lunar mare. The crater was split by the newly forming Carlsberg Ridge, and the two parts were rifted apart and drifted away from each other. Today one part of the crater is attached to the Seychelles, the other to the western coast of India (Fig. 8).

Although hypervelocity impact normally creates circular craters, impact at an angle often generates elongate craters such as the Messier and Schiller craters of the Moon (Wilhelms, 1987). Craters formed by oblique impact mimic their shape (Gault and Wedkind, 1978). The impact that produced the Shiva Crater was probably oblique along a southwest-northeast trajectory as evident from the general oval shape. It might have created asymetric and radial fracture patterns on the northeast side of the crater exterior through which Deccan lava erupted (Chatterjee, 1990a, 1990b). A similar fracture halo extending radially from the crater rim is known from the Manicougan Crater of Canada (Grieve et al., 1988).

Evidence of shock metamorphism

Although the morphology of the Shiva structure is suggestive of impact origin, the critical piece of evidence in support of this model comes from the occurrence of

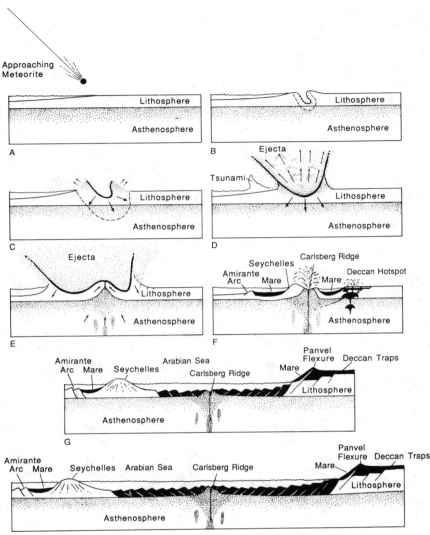

Figure 8. Schematic sequence of events that formed the Shiva Crater and the subsequent rifting between Seychelles and India according to meteorite impact concept. **A**—A giant meteorite, about 40 km across, approaches from the southwest at the India-Seychelles continental margin about 65 million years ago. The target rock was partly oceanic (basaltic) and partly continental (Late Proterozoic granite with a veneer of quartz or sandstone). **B**—The meteorite collides at the continental margin at a velocity of 25 km/s at an oblique angle. Immediately after the impact, a crater begins to form. The shock wave spreads out followed by fracturing. **C**—These waves compress, melt, vaporize, and excavate rocks, and create a cavity. **D**—A large transient cavity is formed; a steam "fireball" ejects condensed meteorite vapor to a high altitude; the crater on the sea floor begins to collapse and large tsunamis are generated at the oceanic edge. **E**—Uplift of the floor produces a central peak, somewhat asymmetrically placed with fractures in the lithosphere to form midoceanic ridge (MOR) and rift basins. **F**—Lava generated from the MOR fills the crater to form the mare basin; lava erupted from the rift basins forms the Deccan Traps; the rim of the crater collapses to form the final crater. **G**—With the spreading of the ridge, the Mare Shiva splits apart, and India and the Seychelles begin to drift away from each other with the development of the Arabian Sea. **H**—The Shiva Crater at the present time; sea-floor spreading has separated the two parts of the Shiva Crater 2800 km apart.

shock metamorphic minerals at the K-T boundary layer of the Jabalpur section. If significant portions of boundary layer display signs of shock metamorphism, then correlation with the impact event can be assumed. Many of the effects of extremely high transient pressures and temperatures, observed in minerals and rocks found at meteorite craters, have never been observed in normal geologic environments (Chao, 1967). Shock metamorphic effects in the silicic target rocks reflect the progressive breakdown of crystallographic order due to the passage of the shock wave, and include so-called planar features, solid-state phase change, thermal decomposition, melting and finally vaporization (French and Short, 1968; Stoffler, 1971; Grieve, 1990). Shock metamorphism differs from traditional endogenic metamorphism in the scales of pressures, temperatures, and time. The range of pressures achieved by shock is immense. Moreover, the rise and fall of pressure are extremely rapid, with the result that shock metamorphism is a nonequilibrium process, and it produces a wide range of physical, mineralogical, and chemical effects.

In Bara Simla Hill of Jabalpur section, the K-T boundary interval is represented by a thick (2.7 m) sandstone layer at the top of the Lameta Formation (Fig 1). The sandstone horizon is entirely barren; sudden disappearance of megaflora and *Aquilapollenites* at this horizon is the primary means for stratigraphic placement of the K-T boundary. The boundary layer is marked by shocked mineral anomalies; it is overlain by a thin flow of the Deccan Trap showing normal polarity (Courtillot et al., 1988). This flow is inferred as chron 29N as it overlies the boundary layer.

Shocked quartz has been reported from the K-T boundary layer of the Lameta Formation of Jabalpur (Basu et al., 1988). However, the shock expression in these grains has been questioned by others (Sharpton et al., 1989). Recently, Necip Guven and H. Todd Schaef (pers. comm.) confirmed the evidence of shock metamorphism of the clastic minerals (10 microns) from the K-T boundary layer of Jabalpur by analytical electron microscopy using SEM and EDS modes. Clay-free mineral grains were prepared to determine their shock metamorphic effects. The residues consist primarily of silica, but also trace amount of metallic particles. The mineral grains were digested with HF (20%) for 5 minutes, and were coated with carbon and gold. The surface textures of the grains were examined by SEM, and their chemistries were simultaneously determined by Kevex 8000 microanalyzer. The X-ray spectrum of the shocked minerals shows pure silica composition with only Si- and O-lines. The quartz grains showed multiple sets of shock-induced planar features indentical to those reported from other K-T sections (Bohor et al., 1987). It should be emphasized that the magnitude of the shocked mineral spike at the Jabalpur impact layer is enormous as compared to other K-T boundary sections in the world.

In a typical large impact, ejecta blanket contains a substantial fraction of the total mass excavated during the formation of an impact crater. The principal mass of ejecta from the crater may be found beyond the rim for a great distance, depending upon the impact force and the ballistic trajectory. The unconsolidated, 2.7-m-thick sandy layer of Jabalpur, which is about 800 km radial distance from the crater rim, may represent the distal facies composed of early high-speed, far-traveled, highly shocked, and finely crushed ejecta. This postulated ejecta layer would represent the thickest K-T boundary section anywhere in the world, and was probably emplaced ballistically. The exceptionally thick boundary layer at Jabalpur also may indicate that impact would have taken place somewhere on or near the Indian subcontinent. The clay fraction of this boundary layer at Jabalpur is smectitic with a subdued iridium anomaly (0.029 ppb), but contains metallic particles of titanium, iron, chromium, and cobalt with shock-induced microdeformations. The protected Deccan flow at the top of the boundary layer has preserved the ejecta components without any sedimentary mixing or large detrital influx.

Origin of the Deccan volcanism

Rifting of the Indian-Seychelles blocks and associated fracturing were associated with extrusion of vast amounts of basaltic magma onto the continental margins. These volcanic rocks include the Deccan flood basalts on land along the west coast of India as well as submarine deposits that form the Amirante Basin and the Saya de Malha Bank around the Seychelles.

The Deccan Traps cover 800,000 km^2 of west-central India, and extend seaward along more than 500 km of coastline of the western Indian Ocean, reaching as far as the continental shelf and beyond (Devey and Lightfoot, 1986). They are dominantly tholeiitic in composition and probably are continuous with oceanic flood basalts; their original extent may well have been more than 2×10^6 km^2 (Krishnan, 1982). Deccan volcanism is considered to be one of the largest continental flood basalt deposits in the Phanerozoic (Courtillot et al., 1986).

Many flood-basalt plateaus are closely associated with the initiation and early development of rifted continental margins; the Deccan Trap is no exception. There always has always been controversy whether the plume or rifting was the initiating factor for the Deccan volcanism. The Deccan flood basalts seem to be related in some way to the detachment of the Seychelles platform from the western part of India. The breakup of the Seychelles block from India appears to be contemporaneous with the major epidsode of Deccan eruptions. Naini and Talwani (1982) recognized that the oldest sea floor between the Seychelles and India formed during 29R at the K-T boundary time. Recent radiometric and paleomagnetic data suggest that the bulk of Deccan lavas was erupted rather rapidly (approximately 1 my) between 64 and 68 Ma, during anomaly 29R, and was coincident with the K-T boundary event (Courtillot et al., 1988; Duncan and Pyle, 1988). Paleontologic data corresponds well with the geochronologic and paleomagnetic age of the Deccan Traps. Sediments immediately below the Deccan flows contain two distinctive index fossils: *Abatomphalus mayaroensis*, a plankton, and *Aquilapollenites*, a palynomorph; both thrived at the topmost Maastrichtian in other parts of the world, but disappeared at the K-T boundary (Courtillot, 1990; Sahni and Bajpai, 1988). The fossil record suggests that the Deccan volcanism began during the very last stage of the Cretaceous. Similarly the flood basalt deposits of the Saya de Malha Bank and the Amirante Basin as well as the basaltic dikes at the Seychelles appear to be contemporaneous with the Deccan Traps (Fisher et al., 1968; MacIntyre et al., 1985; Backman et al., 1988).

Despite decades of extensive geological work, the origin of the Deccan volcanism remains elusive. Morgan (1981) proposed that the Deccan flood basalts were the first manifestation of the stationary hotspot that subsequently produced the volcanic ridge underlying the Laccadive, Maldive, and Chagos islands; the Mascarene Plateau; and the young volcanic islands of Mauritius and Reunion. Although a hot-spot model is very attractive in explaining the Deccan flood basalt volcanism and linear volcanic chains of the western Indian Ocean, there are distinctions in both trace element and isotope geochemistry between present-day Reunion eruptives and those of the Deccan province; the likely source of the Deccan volcanism is similar to a midocean ridge's source rather than the source of the Reunion hot-spot (Mahoney, 1988).

It is still unknown what triggers the disturbances in the mantle to form hot-spots. A positive energy source may come from impact. It has been speculated whether the Reunion hot-spot was itself generated by an impact to cause pressure relief melting (Alvarez et al., 1982; Alt et al., 1988; Basu et al., 1988; Rampino and Stothers, 1988; Chatterjee, 1990a, 1990b). The synchrony of initiation of the spreading from the Carlsberg Ridge, the emplacement of the flood basalts at the Deccan Plateau, Saya de Malha Bank, and Amirante Basin, as well as their close spatial association around a crater basin indicate that the array of simultaneous tectonic and volcanic features might have been triggered by a single physical cause—a giant impact. Should correlation between Deccan volcanism with the K-T boundary be accepted (Jaeger et al., 1989), the impact-triggered mechanism seems to be an attractive hypothesis for the origin of the Deccan volcanism. Critical to this correlation of impact volcanism hypothesis is the absolute age and duration of Deccan volcanic rocks. If the basal flows of the Deccan volcanism are found to be older than the K-T event, the preexistence of a separate volcanic source such as the Reunion hot-spot would be a strong possibility. Paleomagnetic data tend to suggest that Deccan volcanism probably began during an interval of normal magnetic polarity (30N), reached the major peak of eruptive activity in the next reversal interval (29R), then waned in a final, normal interval (29N) (Courtillot, 1990). If this observation is corroborated by other geochronologic evidence, Deccan volcanism might have started a few hundred thousand years earlier than the K-T event. In that case, the impact might have been close enough to the Reunion hot-spot to activate the major phase of the volcanic outbursts during the K-T boundary. Impact might have been the catalyst rather than the proximate cause of the Deccan eruption. Much of the Western Ghat lava pile appears to have deposited during a strong epoch of reversed magnetic polarity (29R), a fact that contributes the strongest evidence for rapid, voluminous extrusion, possibly triggered the impact.

Biswas (1982) has documented highly fractured western rift basins of India, radially arranged (Fig. 9). In addition, several workers postulated the existence of basement rifts, the Kirudvadi and the Koyna, in the inland region (Kaila et al., 1981; Powar, 1981). These rift basins tend to converge near Bombay. Fractures probably formed by the cratering event served as channelways for extrusion of Deccan lavas on the crater exterior. Many workers contend that the main eruptive source locations for Deccan volcanisms lie in Western Ghat in the Igatpuri and Kalsubai areas, and the distal portions of the province (e.g., central part of India) and were formed by flows traveling downrange to distances as great as crater length (Mahoney, 1988; Subbarao et al., 1988). The vast basalt flooding within the Shiva Crater may represent a possible terrestrial analog for a lunar mare, whereas the distribution of Deccan Traps outside the crater exterior may be analogous to the overflowing lava lake (Alt et al., 1988). The fingerlike lobes of the Deccan exposures at the northeast side of the Shiva Crater exterior are similar to the flow patterns of impact melt at several lunar craters (Howard and Wishire, 1985). Both compound flows and ash beds along the west coast and the considerable thickening of the volcanic pile generally indicate relative proximity to sites of eruption. All these observations help identify the western part of the Deccan Plateau as the possible site of impact. The pronounced gravity high along the western coast around the eastern rim of the crater is interesting (Fig. 9). Glennie (1951) suggested that this gravity anomaly was caused by a thick intrusion

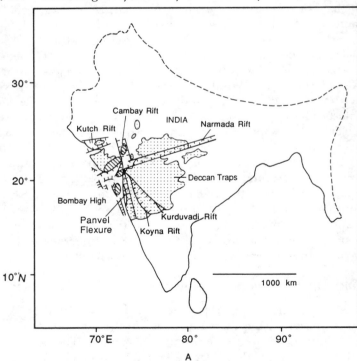

Figure 9. Rift basins of India outside the crater rim that acted as main feeder channels for extrusion of Deccan lava; areas of positive gravity anomalies along the crater rim are shown in hatched pattern (compiled from Kaila et al., 1981; Powar, 1981; Biswas, 1982; Mahoney, 1988).

of mafic material, whereas Qureshy (1981) interpreted it as a gigantic, buried plutonic body. It is tempting to speculate that the buried plutonic body may constitute the meteoritic fragments.

LATE CRETACEOUS PALEOGEOGRAPHY OF INDIA

The reconstruction of the Indian plate at the K-T boundary time involves its interactions with other major continental blocks, namely the Seychelles, Africa, Arabia, Iran-Afghanistan, and southern Tibet. Both geologic and geophysical data are utilized to determine the relative position of India with respect to the other continents (Figs. 6C, 6D, 7). The Red Sea was not opened yet; Arabia was closely apposed to Nubia and Somalia. Africa had separated from Antarctica and Madagascar and had moved northward almost to the present latitudinal position with the development of the Somali Basin. Iran and Afghanistan continental blocks were emergent around the Tethys. Australia was still joined with Antarctica. A large part of the central Indian Ocean was already in existence by the dispersal of Gondwana continents (Smith et al., 1981). Because the future Himalayan mountain chain was part of the Indian plate, Greater India consisted of a greatly expanded northern portion including the spread-out version of the crumpled Himalayas.

India-Africa connection

The tectonics of the Persian Gulf area provide important constraints to the relative movements of India and the surrounding blocks. A crucial area is southern Pakistan near Karachi where the Indian, Arabian, and Iran-Afghanistan plates are joined at a triple junction. The boundary between the Arabian and Indian plates is well defined and consists of the Owen Fracture Zone (Gordon and DeMets, 1989). The boundary between the Indian and Iran-Afghanistan plate is represented by the Ornach-Nal and Chaman (ONC) strike-slip faults. The boundary between the Arabian and Iran-Afghanistan plates is diffuse, complicated, and can be delineated from east to west by the Makran Subduction Zone and the Main Zagros Thrust and strike-slip fault (De Jong, 1982).

The net displacement of the Indian plate from the Carlsberg Ridge is about 1,500 km and is practically in the direction of the strike of the Owen Fracture-Ornach-Nal-Chaman fault systems, which produced a variety of associated folds and faults in the Baluchistan Arc. This whole fault system is an example of a sinistral transform fault of ridge-convex arc type (Wilson, 1965). The relative north-northeast vector motion of the Indian plate since the K-T boundary time, is corroborated by the lineations of the fracture patterns along the Carlsberg Ridge. If the Indian plate is restored against the Seychelles bank at 65 Ma using these strike-slip fault and fracture tracers, the northwest boundary of the Indian plate, as defined by the proto-ONC fault system, becomes adjoined to the continental edge of Arabia. Karachi becomes adjacent to the Socotra microcontinent at the north-east tip of Somalia. At this time, Arabia was juxtaposed to Somalia, and the greater part of the southwestern Arabia was emergent with predominant continental sedimentation (Saint-Marc, 1978). This allowed a dispersal route for terrestrial vertebrates between India and Africa during the Cretaceous (Fig. 7). Sahni (1984) hypothesized a possible filter corridor between India and Africa at that time involving the Mascarene Plateau and the Chagos-Laccadive Ridge, but this model is unacceptable because these volcanic chains came into existence much later, from the Tertiary to Recent time.

The recent discovery of Permo-Carboniferous glacial centers in Saudi Arabia (McClure, 1980) and Oman (Braakman et al., 1982) extends the limit of Gondwana glaciation farther west from India approximately at the same latitude as the Salt Range of Pakistan and the Blaini Boulder Bed in the foothills of the Himalayas, and supports the latitudinal proximity of India and Arabia during the Permian glaciation. Similarly the Cretaceous Nubian Sandstone of Arabia is correlative with the Ahmednagar Sandstone and Nimar Sandstone of western India. The initial strike-slip separation between India and Arabia at the K-T boundary is probably marked by ophiolite outcroppings along the East Arabian coast and on marginal islands (e.g., Masirah Island) along the proto-Owen Fracture Zone (Glennie et al., 1973).

India-Asia connection

The Afghanistan-Iranian belt lying between the Arabian, Turan, and Indian plates was a mosaic of several continental fragments bordered by the Tethys Sea. Regional stratigraphic and tectonic evidence suggests that most of Iran and Afghanistan was an extension of the Arabian platform during the Paleozoic, characterized by platform-type and continental-margin-type sedimentation (Stocklin, 1974). Although data are sparse and imprecise about the Mesozoic paleogeography of this critical area, the central part of Afghanistan includes a continental fragment, the Dasht-i-Margo block, that lay alongside the Lut block of east-central Iran (Crawford, 1972; Takin, 1972; Stocklin, 1974). These two blocks formed a large rectangular microcontinent about the size of Borneo, bordered in the south by the Tethys (Fig. 7). This microcontinent was bounded to the west by the Oman Line, to the south by the Makran Subduction

Zone, to the east by the Ornach-Nal-Chaman Fault, and to the north by the Herat Fault (Fig. 4). Because no Tethyan sediments are known immediately north of this continental block, this microcontinent may have been a peninsula, jutting south from the Turan block and maintaining continental connection with the Eurasian plate (Crawford, 1972). Recent discovery of Jurassic dinosaur footprints from the Lut block, north of Kerman reinforces this hypothesis (Laparrent and Davoudzadeh, 1972). The Iran-Afghanistan microcontinent appears to have formed median masses that influenced the orientation of the surrounding fold belts of the Zagros and Albroz Mountains of the Alpine-Himalayan orogen in western Asia by the obliteration of the Tethys (Stocklin, 1974).

In Late Cretaceous paleogeography, Kashmir would lie proximate to the Iran-Afghanistan microcontinent (Figure 7). A deep gulf from the Tethys Sea occupied the Salt Range, western Sind, and Baluchistan, and had overspread Kutch and Narmada Valley in a scattered way. Most of the Kashmir region was emergent with intense volcanisms at Astor, Burzil, and Dras (Wadia, 1966) and had formed a corridor between the India and Iran-Afghanistan block for the dispersal of land vertebrates (Sahni et al., 1987).

Reconstruction of the Tethys Sea

One of the important consequences of the reconstruction of the Indian plate is the possibility of estimating the size and shape of the Late Cretaceous Tethys before the birth of the Himalayas. Gansser (1964, 1974) pointed out that most of the Himalayan ranges—the Sub-Himalaya, Lower Himalaya, and Higher Himalaya, are made up of Indian shield material. It is only in the 45-km-wide Tibetan Himalaya the Tethyan facies is found that is underlain by the Precambrian crystalline rocks (Chakravorty et al., 1971). Here thick Tethyan sediments were deposited on the continental basement from Proterozoic to Eocene time and form a shallow marine platform facies. In the northern part of the region, however, Mesozoic sedimentary rocks contain abundant flysch with basic dikes and sills, suggestive of continental rifting (Stocklin, 1980). Thus the Tethys in the Himalayan sector was part of the Indian subcontinent; it appears to have been formed primarily on continental crust and may never have been associated with a deep, wide ocean (Dickins, 1987).

The only evidence for oceanic crust is the narrow belt of ophiolite found north of the Tethys Himalaya. Here the ophiolite is associated with Late Cretaceous-Eocene pelagic sediments of relatively small volume and areal extent along the Indus-Zangbo Suture (IZS) (Gansser, 1974). The ophiolite along the IZS is not a continuous feature, being interrupted at several places by structural cross features that are mostly north-south directed, perpendicular to the axis of the Himalayan Mountains (Gansser, 1980). The interpretation of ophiolite at the IZS as a relic of a former vast ocean is questionable (Sonnenfeld, 1978, 1981). The narrow and disconnected nature of outcrops of ophiolite along the IZS and its short geologic span do not establish the existence of a vast oceanic Tethys. The reality of a vast oceanic Tethys appears to be a matter of some contention. Regional stratigraphic and tectonic syntheses tend to favor the existence of a narrow and shallow Tethys during most of the Mesozoic (Meyerhoff and Meyerhoff, 1972; Teichert, 1973; Crawford, 1974, 1979; Carey, 1976; Owen, 1976; Sonnenfeld, 1981; Chatterjee and Hotton, 1986; Dickins, 1987; Smith, 1988).

To understand the paleogeography of the Tethys, how wide it was originally, how its width oscillated with time, or how its position versus climatic belts changed, it is best to look at the most critical area of the Tethyan domain—the eastern Tethys surrounding the Indian plate.

The following factors are considered in the paleogeographic reconstruction of the eastern Tethys Sea at the Late Cretaceous time (Fig. 7A).

1. The Eurasian plate, especially the southern Tibetan (Lhasa) block remained approximately stationary in latitude that extended from 100 to 48 Ma (Allegre et al., 1984).

2. Southern Tibet was continental during the Cretaceous, as evident from the remains of dinosaurs (Zhao, in press) and plant fossils (Allegre et al., 1984), and forming the northern shore of the Tethys.

3. The location of the present IZS Zone is assumed to have coincided with the northern limit of the Cretaceous Himalayan Tethys. The subduction at the IZS was limited as testified by large-scale domal upwarp of the crystalline basement of the leading edge (Valdiya, 1988) and gravity data (Verma and Prasad, 1988). The northern margin of the Greater India formed the southern shore of the Tethys but has been obliterated due to underthrustings of the Himalayas and has to be restored.

4. We do not know where the northern margin of the original Greater India lies with respect to India today, because that margin has been much deformed and shortened in the creation of the Himalayas. The close proximity of the IZS ophiolite to the Kailas (or Kangdese) granites of the Trans-Himalaya and the predominant northward vergence of thrusts in the ophiolites imply substantial postcollisional shortening of the region (Tapponnier et al., 1981). It is generally assumed that there has been several hundred kilometers (500–2,500 km) of crustal shortening between India and

Asia. Part the shortening was absorbed within the Indian plate by intracratonic subductions along the Indus-Zangbo Suture (IZS), Main Central Thrust (MCT), Main Boundary Thrust (MBT), and Main Frontal Thrust (MFT), from south to north. An average of 1,000 km of crustal shortening is assumed here in the northern margin of the Indian plate. Diffuse deformation within Asia has been the other component of accommodating India's northern convergence (Gansser, 1966; Molnar and Tapponnier, 1977; Besse et al., 1984).

The reconstruction shows that the Tethys at the K-T boundary time has been reduced to a long channel, about 1,000 km wide at its maximum. The size and shape of the Himalayan Tethys at this time conforms reasonably well with the present Mediterannean (Figure 7A).

The Himalayas end in the west (near Kashmir) in remarkable syntaxial bend (Wadia, 1966). An understanding of the syntaxes is perhaps the most important key to understanding the configuration of the Tethys and the subsequent tectonics of the Himalayas. Carey (1958) explained this inflection on the Himalayan belt as a consequence of the anticlockwise rotation of India relative to Arabia. He believed that the Cretaceous Tethys Sea was a long, narrow geosyncline, almost straight in trend, and the syntaxes were the result of oroclinal bending due to rotation. We elaborated Carey's orocline model in our previous paper (Chatterjee and Hotton, 1986). During rotational movement, Kashmir might have acted as an indenter to separate the Himalayan Tethys from the Baluchistan Tethys.

In the Cretaceous Tethys configuration (Fig. 7A), a narrow gulf of the Tethys Sea inundated the Baluchistan sector. The *Cardita beaumonti* beds of Sind and the Pab sandstone of Baluchistan represent the Late Cretaceous Tethyan facies in this area. These sediments are largely associated with volcanic tuffs and ophiolite sequences in the Bela and Zhob Valley, contemporaneous with the Deccan Traps (Allemann, 1979). Cretaceous Tethyan domain is also known in western Burma and the Andaman Islands; the eastern parts of Burma were part of the Shan Craton and were continental at that time, marked with extensive volcanisms. The Shan Boundary Fault (SBF) separates the terrestrial from the marine facies of Burma at this time (Bannert and Helmcke, 1981).

The Maastrichtian Tethys Sea was not continuous from west to east as reflected by the diversity of facies and fauna. It was separated by the transversal barriers of Greater India into two separate basins: the Baluchistan Tethys in the northwest, the Himalayan Tethys in the northeast. The separation of the eastern Mediterranean Tethys from the Himalayan Tethys is indicated by the differences of Late Cretaceous marine faunas with the development of faunal provincialism (Kauffman, 1973).

In Kashmir, volcanic activity and partially emergent conditions persisted in many areas from Late Paleozoic time well into the Late Cretaceous. Similarly, much of the area between the eastern Himalaya and northern Burma was emergent during the Cretaceous. Southern Tibet (Lhasa block), lying on the northern shore of the Tethys, was mostly continental at that time and has yielded a spectacular dinosaur faunal succession formed during the Early Jurassic to the Late Cretaceous (Zhao, in press). The Kashmir region of Greater India forms a narrow corridor across the Tethys that allowed faunal migrations between India and Eurasia during the Cretaceous.

Tectonic history of the Ninetyeast Ridge

It is generally believed that the convergence of India into Asia led to the reorganization of the Indian Ocean with the development of the Chagos-Laccadive Ridge to the west, and the Ninetyeast Ridge to the east, but the tectonic evolution of the these ridges remains highly speculative. Sclater and Fisher (1974) suggested that these two parallel ridges are transform faults that facilitated the northward motion of India. Together, these meridional highs resemble train tracks upon which India glided freely toward Asia without interacting with other crustal plates.

Others believe that the Maldive-Laccadive-Mascarene Plateau and the Ninetyeast Ridge represent hot-spot trails similar to the Hawaiian Island chain (Morgan, 1981; Duncan, 1981; Backman et al., 1988; Richards et al., 1989). These two ridges may record the northward motion of India over stationary hot spots at Reunion and Kerguelen respectively (Fig. 10). However, Molnar and Stock (1987) showed that the calculated relative positions of fixed hot spots beneath Reunion and Kerguelen lie several hundred to more than a 1,000 km east from the traces left by them. Thus they questioned the validity of the use of hot-spot tracks to constrain histories of plate motions.

Both transform fault and hot-spot models fail to explain the intraplate deformation in the Indian Ocean bounded by these two ridges. Within the affected regions, the oceanic crust and the overlying sediments are intensely deformed into folds and faults of large dimension (Weissel et al., 1980). If India moved freely northward along these parallel lineaments on its eastern and western boundaries without any interactions, the compressive forces are not typical of the interior of lithosphere plates. The intraplate deformation may be a precursor of a new convergent plate boundary. Neprochnov et al. (1980) identified northeast-striking wrench faults in this deformed region, between the Chagos-Laccadive and Ninetyeast Ridge, indicating that the main vector of the Indian plate motion may be parallel to these faults.

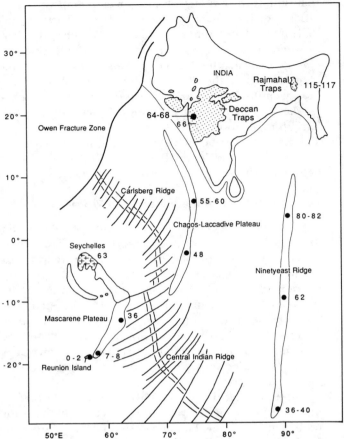

Figure 10. Hot-spot models. The Deccan flood basalts lie at the northern end of the volcanic trace of the Reunion hot spot. The ages (shown in Ma) of the volcanism decrease to the south from the Deccan lavas. A parallel companion volcanic trace follows the Ninetyeast Ridge, and ends with the youngest activity near Kerguelen Island (data from Backman et al., 1988; Duncan and Pyle, 1988).

Fitch (1972) proposed a model for oblique convergence in which transcurrent fault is associated with plate consumption in the same region. This concept has been expanded here to explain the origin of the Ninetyeast Ridge. It invokes a mechanism related to the geometry and kinematics of the Indian plate boundary. This model suggests that the Ninetyeast Ridge might have been formed by the complex interactions of nascent plate boundaries which have produced linear volcanic chains analogous to the hot-spot trails.

Origin of the Ninetyeast Ridge

The Ninetyeast Ridge is the longest and most spectacular tectonic feature in the eastern Indian Ocean. Sclater and Fisher (1974) showed that the Ninetyeast Ridge is largely of extrusive origin and is attached to the Indian plate. They also found that the magnetic anomalies to the west of the ridge age to the north, whereas to the east they age to the south. The ridge becomes progressively younger from north (80 my) to south (36–40 my). Several authors have proposed that the Rajmahal Traps of eastern India (115–117 Ma), the Ninetyeast Ridge and the Kerguelen hot spot located around 50° S together form a hot-spot track system (Peirce, 1978; Duncan, 1981; Backman et al., 1988). Contrary to this, geochemical and isotopic evidence suggests that the Kerguelen hot spot did not feed the Rajmahal Traps (Mahoney et al., 1983).

The origin of the Ninetyeast Ridge is enigmatic. It has been variously interpreted as a horst (Francis and Raitt, 1967); as the result of the convergence of two plates, the Indian plate overriding the Australian plate (Le Pichon and Heirtzler, 1968); as the trace of a hot spot beneath the northward moving Indian plate (Morgan, 1972; Peirce, 1978); as the trace of two hot spots (Luyendyk and Rennick, 1977); as an ancient transform fault (McKenzie and Sclater, 1971); and as a volcanic mass erupted at the junction of a spreading center and a transform fault (Sclater and Fisher, 1974).

The present consensus is that the Ninetyeast Ridge is related to magmatism overlying the Kerguelen hot spot, fixed relative to spin axis (Peirce, 1978; Morgan, 1981; Duncan, 1981). However, the hot-spot model runs into serious difficulties in developing a tectonic model when attempts are made to relate the Kerguelen hot spot to contemporary plate boundaries (Peirce et al., 1989). For example, much of the ridge was produced between 90 and 50 Ma, the period when the Indian plate moved 6,000–10,000 km (Patriat and Achache, 1984). The length of the ridge is about 4,500 km. This indicates that the spreading center must have migrated northward rapidly and would eventually pass over the Kerguelen hot spot which would leave traces on the Antarctic plate. No such hot-spot tracks are known from the Antarctic plate. To explain these inconsistencies, complicated models involving several sources of volcanism have been proposed (Sclater and Fisher, 1974; Luyendyk and Rennick, 1977; Johnson et al., 1976). Moreover, a linear age progression of the Ninetyeast Ridge is not compatible with ages of the ocean crust lying east of the ridge (Royer and Sandwell, 1989). The genesis of the Ninetyeast Transform Fault, which lies immediately east of Ninetyeast Ridge, is difficult to explain with the hot-spot model. Molnar and Stock (1987) pointed out that calculated positions of the Kerguelen hot spot do not coincide with the Ninetyeast Ridge, but they define north-northeast trends about 1,000–1,600 km east of this ridge. This inconsistency, according to them, suggests that hot spots do not define a fixed reference frame.

Stein and Okal (1978) suggested that the traditional interpretation of the Ninetyeast Ridge as being aseismic is not true. The ridge is seismically very active, and is presently a complex zone of deformation within the Indian plate. They recognized left-lateral strike-slip motion in the northern portion of the ridge, which may indicate either a plate boundary or unusual midplate tectonics.

Weissel et al. (1980) showed that the central Indian Ocean is under a large north-south compressive stress regime. Wezel (1988) interpreted the vast area of the Indian Ocean floor, from the south of Sri Lanka to Sumatra as a young Jura-type foldbelt, which he called Indoysian foldbelt. He believed that the oceanic flood basalts in this region might represent the covering of a foundered continental basement. The seismicity, strike-slip motion, and convergent regime along the Ninetyeast Ridge suggest that the origin of the ridge is far more complex than the popular hot-spot model.

The geometry of the Ninetyeast Ridge in relation to its surounding structures may hold the key to unlock its tectonic history. It forms a junction between two plate boundaries—the Java Trench in the north, and the Carlsberg-Central Indian-Southeast Indian (C-CNI-SEI) Ridge in the south. The ridge has an asymmetric cross section with a steep east-facing scarp, an observation that gave rise to the hypothesis of convergence (Le Pichon and Heirtzler, 1968). The model I have presented here predicts the Ninetyeast Ridge, in both time and space, as a compressive feature.

An interesting feature in the Indian Ocean is the geometry of the C-CNI-SEI Ridge and the Java Trench and their interactions. They form concentric arcs of two small circles on the globe, the inner (smaller) arc is the trench, the outer (greater) arc is the spreading ridge; the common center (pole) of these two arcs is around Manila, Phillipines. The Ninetyeast Ridge, a transform fault before evolution, lies almost tangential to the Java Trench. The geometry of the Indian Ocean, according to this model, is an example of tectonics on a sphere.

To visualize interactions of three plate boundaries, I have used a tennis ball as a model of the outer spherical shell of the lithosphere, and a solid inner ball of similar size as a model of the Earth's interior (see Cox, 1973). The tennis ball is cut into two equal halves; only one hemisphere is required for the plate geometry. With a compass, two concentric circles are drawn on the hemisphere about a common center (pole), and the circles are cut along these lines. The hemisphere is now divided into three segments: small, intermediate, and large. The intermediate segment with a central hole is now placed on the inner rubber ball. Let us assume that the great circle of this segment represents the ridge and the small circle as the trench. Three transform plates are created by cutting along the meridians 120° apart and perpendicular to the ridge boundary. Now we have made three separate plates, and three kinds of plate boundaries (Fig. 10).

As the ridge spreads, the three plates move towards the trench along the direction of the transform fault. The leading and trailing edges of each small plate represent trench and ridge boundaries respectively. Because the trench segment is smaller in radius than the ridge segment, the plates will converge towards each other along the transform faults. As a result, each transform evolves into a trench boundary because of convergent strike-slip motion. Transpressive regions will exhibit compressive features, such as subduction and volcanicity. Because the subduction is oblique along this new convergent boundary, a linear volcanic chain is produced with age gradient, oldest near the small circle, and youngest at the large circle, analogous to the hot-spot trails. In consequence of these plate movements, triple junctions such as trench-trench-trench (TTT) are formed at the small circle, and ridge-ridge-trench (RRT) at the great circle (Fig. 11).

The complex series of events described here is the inescapable consequence of the motion of rigid plates whose relative velocity remains constant. The geological history of the central Indian Ocean floor since the Late Cretaceous is in general compatible with the evolution of plates and triple junctions outlined here. Instead of three plates, if we start with two adjacent plates, the Indian and Australian, with the Ninetyeast Ridge as the initial transform between them, the evolution of the surrounding structures become apparent (Fig. 12).

The Ninetyeast Ridge, according to this model, is the result of oblique convergence at a major transform fault between the Indian and Australian plates leading to a linear volcanic chain, similar to a hot-spot trail. Paleolatitude studies (Peirce, 1978) indicate that the ridge formed at a point relatively fixed in latitude. This is considered the most important constraint for a hot-spot origin. However, the fixed paleolatitude of the source does not seem to be a constraint, because spreading along the C-CNI-SEI Ridge concomitant with the backward shift of the volcanic source maintains a position close to that of the paleomagnetic studies. The Indian plate is overriding the Australian plate and is carrying these volcanic chains during its north-northeast convergence into Asia.

The west side of the ridge is encountering resistance due to spreading of the C-CNI-SEI Ridge, and is under compression, as is evident from the principal stress direc-

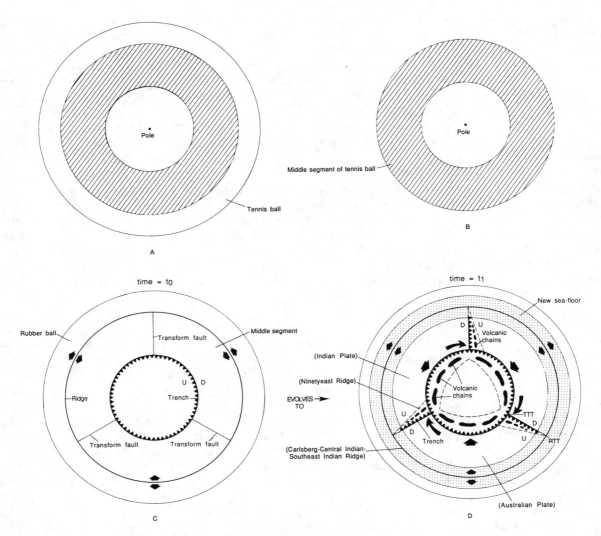

Figure 11. The tennis ball experiment. **A**—Two concentric circles are cut on the hemisphere of the tennis ball about a common center (pole). **B**—The middle segment is separated. **C**—The middle segment is placed on a solid rubber ball and is cut radially into three equal parts to create transform faults; the small circle represents the trench and the great circle the ridge. **D**—As the ridge spreads, each transform evolves into a trench boundary, accompanied by oblique convergence. As a result a series of volcanic chains are produced similar to hot-spot trails. A similar mechanism has been proposed here for the origin of the Ninetyeast Ridge by the interaction of the Indian and Australian plates along the initial Ninetyeast transform fault during the spreading the Carlsberg-Central Indian-Southeast Indian Ridge.

tions (Fig. 4). The east side of the ridge, the Australian plate, is moving northward, and is subducting at the Java Trench. The complex motion between the Indian and Australian plate is accommodated along the Ninetyeast Ridge. Such a situation would give rise to a combination of compression and strike-slip lateral motion (i.e., transpression or convergent strike-slip motion) in the region of the Ninetyeast Ridge, as has been predicted by earlier workers (Le Pichon and Heirtzler, 1968; McKenzie and Sclater, 1971).

INDIA-EURASIA COLLISION CHRONOLOGY

The timing of initial collision between India and Asia has been the subject of much debate. The differences in exact timing seem to stem from two factors: which part of the suture is considered and whether reference is made to the initial collision of the Indian plate carrying part of the Himalayan Tethys, or to the final welding of India to Eurasia with the obliteration of the Tethys. From a paleobiogeographic point of view, the timing of initial contact is important to understand the dispersal corridors.

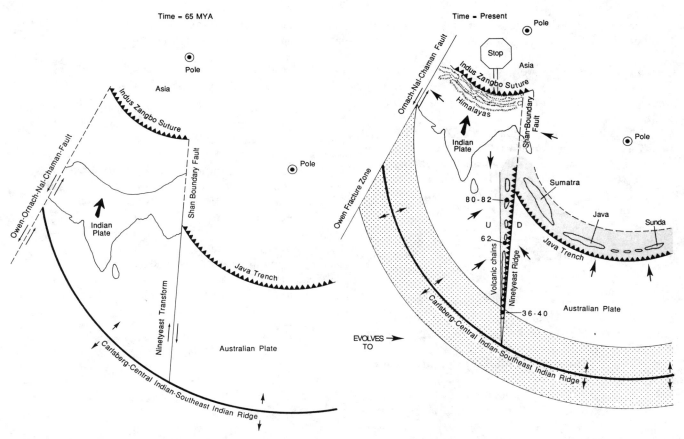

Figure 12. Origin of the Ninetyeast Ridge. **A**—Initial condition at the K-T boundary before spreading of the Carlsberg Ridge; note the concentric arrangement of the spreading ridge and the trench about a common pole, the trench occurs along the inner circle, the ridge along the outer circle. The Indian and the Australian plates were separated at that time by the Ninetyeast transform. **B**—As the Carlsberg-Central Indian-Southeast Indian ridge spreads, the Ninetyeast transform evolves into a trench accompanied by oblique convergence to produce volcanic chains similar to hot-spot trails. The convergence of India into Asia produced the Himalayan orogen. Since the Miocene, there is a growing resistance to shortening across the Himalayas; these resistive forces are now transmitted to the Bay of Bengal Fan area resulting spectacular foldings and faultings of the oceanic lithosphere.

Data from magnetic anomalies in the Indian Ocean and paleomagnetic measurements on the subcontinent have been used to reconstruct India's collision chronology. Analyses of abrupt changes in India's northward movement and in India-Asia relative movement have been interpreted variously in the past as indicating the timing of collision (1) around the Late Paleocene (55 Ma, Powell and Conaghan, 1973; Powell, 1979) or (2) around the Eocene (50 Ma, Besse et al., 1984; Patriat and Achache, 1984; 40 Ma, Molnar and Tapponnier, 1977). However, magnetic lineations can only provide a relatively coarse model of the India-Eurasia convergence. Magmatic activity, tectonism, and stratigraphy along the Indus-Zangbo Suture (IZS) provide better constraints to the sequential movements and suggest a much earlier collision event than generally recognized.

The IZS is the collision zone separating the Precambrian Indian plate from the younger Mesozoic-Cenozoic Trans-Himalaya to the north. The suture zone contains Tethyan ophiolites, blueschists, and granulites. The collision of India was preceded by a short period of Andean-type orogeny. This is evident by the occurrence of a chain of granodioritic batholiths, the Trans-Himalayan granites (Kailasa or Kangdese granites), extending virtually from one end of the range to the other. These Trans-Himalayan granites were formed in response to the subduction of the oceanic component of the Indian plate containing the Tethys beneath southern Tibet (Allegre et al., 1984). Continued convergence eventually brought India-Asia into juxtapostion. Our interest is to determine the timing of the initial contact and eventual suturing. The following geologic evidence from the IZS is important for understanding the collision chronology.

1. Stratigraphic evidence from the IZS zone suggests that Tethyan sediments and ophiolites were subjected to thrusting during Early Cretaceous times indicating the

presence of a subduction zone dipping to the north (Mitchell, 1984).

2. The radiometric age of the Trans-Himalayan Kailas granite (approximately 90–110 Ma) indicates that a subduction zone has been active during the Early Cretaceous between India and Asia (Maluski et al., 1982).

3. The subduction of India at the IZS was minimal and short-lived as evident from the large-scale domal upwarp of the crystalline basement of the leading edge (Valdiya, 1988). Recent gravity field data (Verma and Prasad, 1988) does not support underthrusting of India in the manner envisaged by Powell and Conaghan (1973).

4. The radiometric age suggests that the emplacement of IZS ophiolite took place in the Late Cretaceous (82 ± 6 Ma) during the initial collision of India with Asia (Brookfield and Reynolds, 1981).

5. The Late Cretaceous-Early Paleocene age of the major phase of deformation in southern Tibet (Tapponnier et al., 1986) also favors an early age for the India-Asia collision (Jaeger et al., 1989).

6. The early collision of India corresponds well with the Schlich anomaly of the Mascarene Basin at the time of separation of India from Madagascar.

It becomes clear that both geologic and paleontologic evidence supports the timing of the initial collision of India with Asia in the Cretaceous with the formation of the IZS. This is compatible with the paleontologic constraint. In Late Cretaceous paleogeography (Fig. 7A), most of the Kashmir region was emergent with intense volcanisms at Astor, Burzil, and Dras (Wadia, 1966) and formed a filter bridge between India and Asia for the dispersal of land vertebrates (Sahni et al., 1987).

The convergence of India into Asia continued for a long time after the initial contact (Fig. 13). The complete suturing between India and Asia probably took place in the Paleocene or Early Eocene, when the two continental margins met flush with the extinction of the Tethys. This was followed by the commencement of terrestrial deposition in the Subathu Basin that developed on the depressed northern margin of Greater India. During the Middle Eocene (45 Ma), a broad corridor between India and Asia was established that enabled hordes of Asian mammals to sweep into India. The mammalian fauna from the Charat Series of Pakistan includes taeniodonts, creodonts, titanotheres, anthracotheres, artiodactyls, perissodactyls—all identical to Mongolian, Burmese, and North American taxa (Chatterjee and Hotton, 1986). In the Sabathu Formation of Jammu and Kashmir, Sahni and Khare (1973) recorded a tapiroid, bunodont artiodactyls, a paramyid rodent, and a hyracodont rhinoceratid, all closely related to Asian and North American forms.

NORTH-NORTHEAST CONVERGENCE OF INDIA AND THE TECTONICS OF THE HIMALAYAN ARC

The convergence of India into Asia produced the spectacular Himalayan Arc, along with a series of mountain belts in the east and west (Gansser, 1966, 1974). The tectonic style of the Himalayan Arc and the adjacent regions may provide important clues to the direction of movement of the Indian plate. Peninsular India is almost symmetrical, but the Himalayan mountain chain is highly asymmetric in trend, length, and tectonic style (Crawford, 1974). This asymmetry suggests that the convergent vector of the Indian plate is not due north as is commonly believed, but has a north-northeast direction.

The Himalayan Mountains lie directly northeast of the Indian Shield and form an almost perfect arc for

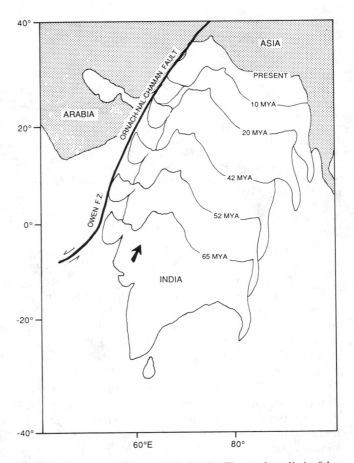

Figure 13. Cenozoic northward drift of India. The northern limit of the Indian continent is taken as the present position of the Indus-Zangbo Suture. The Owen Fracture Zone and the Ornach-Nal-Chaman Fault system marked the passage of the western margin of India and provide a tight constraint on its north-northeast convergence.

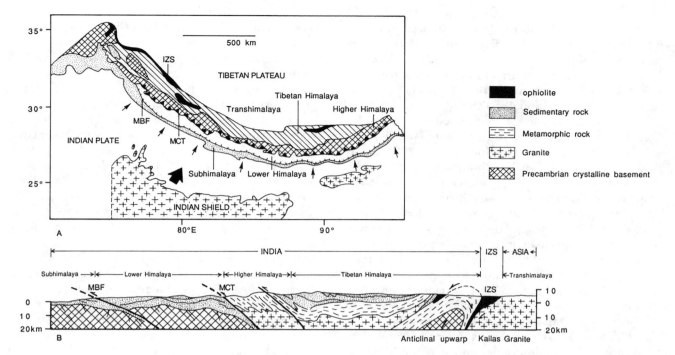

Figure 14. **A**—Morphotectonic zones of the Himalayan Arc, showing its asymmetric plan pattern in relation to Peninsular India. Overthrustings along the Himalayas occur in a direction everywhere locally perpendicular to the arc (shown by small arrow). Large arrow represents the mean direction of convergence of the Indian plate to Asia. **B**— Structural profile across the Himalayan orogen, showing upwarping of the leading edge of the Indian plate along the Indus-Zangbo Suture owing to buoyance resistance, implying that the subduction was limited in extent (modified from Gansser, 1964, 1974; Valdiya, 1988).

2,500 km between longitudes 74° and 96° E, convex toward the south, with an average width of 250 km. They comprise a series of lithologic and tectonic units that run parallel to the mountain belt. On its northern side, it is separated from the Trans-Himalayan zone by the valleys of the IZS. The Himalayan fold thrust belt is bounded on both sides by major strike-slip belts—the Ornach-Chaman Fault in the west, and Shan Boundary Fault in Burma in the east (Abdel-Gawad, 1971). These two strike-slip faults define the vector of convergence of the Indian plate in a north-northeast direction into Asia with the development of the the Himalayan mountain chains (Figs. 13 and 14). There is a gross difference in tectonic style between the Himalayan and Baluchistan Arc on one hand, and the Himalayan and Burmese Arc on the other (Fig. 4). These two terminal arcs were the site of major shearing deformation, associated with foldings, thrustings, and transcurrent faults. The Himalayan Mountains, however, were constructed essentially in a convergent tectonic regime. The major difference in the style of Tertiary deformation east and west of the Himalayas appears to lie in the relative importance of strike-slip faults versus thrusts. Although large overthrusts are preponderant in the Himalayas, major strike-slip faults seem to have played a leading role east and west of the Himalayas with more subordinate thrusts. The asymmetry, structural discontinuity, and different tectonic style in these three sectors of the Himalayan arc system were controlled in part by the north-northeast convergence vector of the Indian plate. The Himalayan orogen is asymmetric and simple, with north-northeast facing orogenic polarity coinciding with the vector of the Indian plate motion (Fig. 14).

The Himalayas were uplifted, folded, and complexly thrusted during a series of orogenetic movements that started in the Late Cretaceous as India moved northward from Madagascar toward Asia. With the spreading of the Carlsberg Ridge about 65 Ma, the convergence of India continued resulting in northeast-southwest crustal shortening. The leading edge of India was inactive, the crustal shortening being absorbed by folding and multiple underthrusting. The IZS was the first tectonic feature that was developed along the axis of the eastern Tethys in response to this convergent motion. Several models have been proposed for the evolution of the Himalayas that are based on the India-Asia collision (Powell and Conaghan, 1973; Gansser, 1974; Graham et al., 1975; Le Fort, 1975; Molnar and Tapponnier, 1977; Patriat and Achache, 1984; Chatterjee and Hotton, 1986). These all recognize that the Himalayan orogeny is composed of distinct cycles; the IZS as the main collisional boundary between India and Asia was formed in the Cretaceous (approximately 80 Ma), followed by the progressive southward migration of thrusting through the Cenozoic.

However, subduction at the IZS was limited in extent. The leading edge of the Indian plate exhibits buoyant resistance to slip beneath the Asian plate as reflected by large-scale anticlinal upwarp of the crystalline basement immediately to the south of the IZS (Valdiya, 1988). After limited subduction, the IZS became immobilized and the India-Asia convergence slowed down considerably. The movement was resumed again in the Miocene. A new zone of weakness developed about 100–200 km farther south of the IZS; the subduction was thoroughly intracontinental. The compression was accommodated along the MCT, pushing up the Himalayas in front of it. The third phase of convergent movement of India started in the Pliocene, and the compression was accommodated along a weaker, more southerly thrust belt, the MBF. Later, in the Pleistocene, the morphotectonic emergence of the Himalayas began with mostly vertical uplift probably due to isostatic emergence, and is still continuing today. The Himalayas are thus slices of the old Indian shield that have stacked over each other by transfer along a series of thrusts migrating southward. Convergent plate motion still occurs between India and Asia, and is evident in particular by the underthrusting of the peninsular shield below the Indo-Gangetic Plain (Figure 14). The increased resistance to shortening across the Himalayas, combined with continued seafloor spreading at the Carlsberg-Central Indian Ridge, implies that the Indian plate is under severe compressive stress. The result is a broad region of intraplate deformation, stretching from the Chagos-Laccadive Ridge to the Ninetyeast Ridge (Weissel et al., 1980).

The geometry of collision zones and vector of compression movement favors the initiation and development of thrust belts within the Indian plate. All the thrust zones in the Himalayas—IZS, MCT, MBT, MFT—are subparallel and form a north-northeast-south-southwest arcuate trend. This indicates that the converging vector of the Indian plate was perpendicular to the direction of these thrust axes, with a mean direction toward N35° E (Fig. 14). The north-northeast motion of the Indian plate may explain the pronounced asymmetrical plan pattern of the Himalayas.

The convergent motion of India to Asia is manifested not only in the Himalayas, but also in the adjacent ranges and platforms. It appears to have compressed and distorted the Earth's crust from the Himalayas to Siberia and from Afganistan to the coast of China (Molnar and Tapponnier, 1977). The most pronounced influences are conjugate sets of gigantic strike-slip faults, trending roughly north-south and east-west respectively. The greatest shearing deformation appears to have occurred along two north-south trending belts at the extremities of the Central Himalayan Arc, coinciding with the two major Himalayan syntaxes (Abdel-Gawad, 1971). In the west, the convergent motion of India was accommodated in the Kirthar-Sulaiman Range along the Ornach-Nal-Chaman fault system.

Molnar and Tapponnier (1977) found that the active tectonic areas of Central Asia, as indicated by current seismicity, formed a number of discrete zones affecting a region up to 4,000 km wide northeast of the Himalayan Front, whereas India is relatively unaffected. They recognized several east-west trending, left-lateral strike-slip faults in East Asia, such as Altyn Tagh, Kunlun, and Kang Ting faults, northeast of the Himalayan chain. The origin of these strike-slip faults is attributed to the eastward propagating extrusions of East Asia resulting from northward penetration of the rigid Indian plate into the more plastic Asian continent. They believed that part of the crustal shortening between India and Asia was achieved by strike-slip movements. The ability of Asia to shorten by lateral displacement is influenced by the boundary conditions of the Asian plate. The vast continental region of Eurasia essentially blocks large-scale movement towards the west. In the east, the presence of continuous subduction zones along the Pacific and Indonesian margins allowed lateral extrusion. The model of extrusive tectonics is attractive and is supported by recent indentation experiments with plasticine (Tapponnier et al., 1982, 1986). The concept of indentation tectonics explains the spatial distribution and various styles of faulting and deformation in Asia. However, difficulties arise when dealing with the unproved assumption that the Asian continent behaved as a "plastic" during convergence of India and was squeezed out laterally like "toothpaste" in a direction perpendicular to the direction of shortening. This concept would violate the basic tenet of the plate tectonic theory—the rigidity of plates. To explain this inconsistency, Molnar (1988) believed that plate tectonics is inadequate to explain many features of continental tectonics. The concept of rigid plate motion, according to him, fails to apply to continental interiors, where buoyant continental crust can detach from the underlying mantle to form orogenic belts and broad zones of diffuse tectonic activity.

A limitation of the plasticine experiments by Tapponnier et al. (1986) is the failure to simulate multiple thrust belts in the Himalayas. In their experiments, the Himalayan-Tibetan belt acted like a "dead zone;" strike-slip faults were produced farther northeast beyond this dead zone. In reality, the Himalayas were the most deformed belt because of the collision of India. Future indentation experiments on plasticine should include the north-northeast vector of the Indian plate motion instead of northward. The asymmetry of the indenter, whose north-northeast edge impinges more obliquely than the north-northwest edge, may explain why deformation occurs over a large area to the north-northeast of the Himalayas. It is likely that the eastern component of the north-northeast vector of the Indian plate has resulted not only in the asymmetric plan pattern of the Himalayas, but also the east-west trending strike-slip faults in East Asia (Chatterjee and Hotton, 1986).

The extrusion model has been criticized by other workers. It does not explain why the extrusion is so young (approximately 5 Ma) in a collision acknowledged to have started at least 55 Ma or earlier (Klootwijk et al., 1985). England and Houseman (1986) pointed out that the indentation model does not take into account the variable thickness of the plastic layer, in which thickening occurs in response to thrusting and thinning, which occurs in response to extension. They have created mathematical versions of India and Asia to model the collision on computer. In their simulations, they see crustal thickening assuming the fundamental role in absorbing the crash of India. They predict that deformation is restricted to the region about 1,000 km north of the Himalayas, and argue against the extrusive model of Tapponnier et al. (1986), in which Indochina has been laterally extended by indentation.

The possible role of the Baikal Rift system at the northern margin of the active tectonic zone and the Shansi Graben at the eastern margin in controlling the tectonics of Asia has never been addressed before. These extensional features, trending roughly northeast-southwest to north-south in direction, can contribute to the formation of several east-west trending strike-slip faults in Asia. The comparatively young age (approximately 5 Ma) of these fault systems is compatible with these nascent rift systems. Molnar and Tapponnier (1977) suggested that these extensional features were the result of the northward convergence of India to Asia. Could the movement of India cause deformation of a region more than 3,000 km away across the Himalayan mountain chain?

From the initial contact in the Cretaceous until the Early Miocene, the northward movement of India was always accommodated within the Indian plate with a progressive southward shift of the active zone of convergence (Gansser, 1964, 1966; Le Fort, 1975; Chatterjee and Hotton, 1986). Since the Miocene, there is a growing resistance to shortening across the Himalayas. With continued sea-floor spreading, these resistive forces are now being transmitted to the interior of the Indian plate as a large horizontal compression, especially in the Bengal Fan area, resulting in foldings and faultings of the oceanic lithosphere (Weissel et al., 1980; Neprochnov et al., 1988). The intraplate deformation may be a precursor of a new convergent plate boundary. Recognition of an east-west trending diffuse plate boundary in this region between India and Australia (Gordon and DeMets, 1990) may indicate the initiation of the latest zone of convergence to accommodate the continued northward movement of India. Interestingly, from north to south, the age of the intraplate subductions decreases. If this interpretation is found to be correct, the principal tectonic features of Asia that are thought to be associated with the continuing northward push of India may be controlled by other factors such as the rift systems of Baikal and Shansi. The surface manifestation of the Neogene convergent movement of India may be confined entirely within the Indian plate.

The forces responsible for continental collision are poorly understood. Among the forces driving India into Asia, the push from the Carlsberg Ridge crest seems to be the most significant. Slab-pull (Forsyth and Uyeda, 1975) cannot be the leading force in the case of continued convergence of India into Asia (Allegre et al., 1984). Subduction along the IZS stopped at least 45 Ma, and since then there is no oceanic lithsopsphere left between India and Asia. Continuing slab-pull force beneath Java and Sumatra may cause northward movement of the Australian plate. The K-T impact may have triggered the initial ridge-push, which is still acting to drive the Indian plate into Asia.

CONCLUSIONS

The analysis of biotic links and dispersal routes provides crucial insights into the travel paths of India during its northward journey. The traditional geophysical reconstruction of India as an island continent for about 100 my is not consistent with its biological relationships. The terrestrial faunal resemblances between India-Asia and India-Africa, together with the lack of endemism within the Late Cretaceous biota of India strongly suggest that India had close terrestrial links with Asia and Africa during this period. A revised reconstruction of India is presented that is paleontologically well constrained and corroborated by marine magnetic anomalies and fracture pattern lineations.

Starting from the Smith and Hallam (1970) fit of the Early Jurassic, successive positions of India were traced. It is proposed that during the Late Jurassic to the Late Cretaceous, the Indo-Seychelles block and Africa remained attached at Socotra-Karachi and moved northward in concert, whereas Madagascar lagged behind and displaced relatively southward. During this period, the Indo-Seychelles block rotated counterclockwise in relation to Africa around a pivot near Socotra. In the Late Cretaceous paleogeography, the Indo-Seychelles block was proximate to Somalia and Arabia. The intervening Tethys Sea between India and Asia was narrow, about the size of the present Mediterranean. Kashmir region in Greater India acted as a dispersal corridor for faunal exchange between India and Eurasia.

At the K-T boundary time, a large impact shattered the lithosphere to initiate continental rifting between India and the Seychelles with the development of the Shiva Crater, Carlsberg Ridge, and Owen Fracture Zone, and resulted in the massive volcanic outbursts of the Deccan Traps. The impact may have decoupled the Indian plate from other Gondwana plates and caused major changes in the velocity and direction of the Indian plate motion. After the initial rifting, India moved away from the Seychelles in a north-northeast direction with the opening of the Arabian Sea and began to converge into Asia.

A model for oblique convergence is expanded here to explain the origin of the Ninetyeast Ridge during the north-northeast movement of India. The volcanic chains at the Ninetyeast Ridge are attributed to the oblique convergence between the Indian and Australian plates along the Ninetyeast Ridge transform fault. These linear volcanic chains superficially resemble hot-spot trails.

The Owen Fracture Zone facilitates direct tracing of the north-northeast convergence of India. This north-northeast motion can be interpreted as the main cause of the pronounced asymmetrical pattern plan of the Himalayas. After the initial contact at the Late Cretaceous, the convergent movement of India was accommodated along the Indus-Zangbo Suture. The complete suturing between India and Asia probably took place in the Paleocene or Early Eocene. Continued convergence produced a series of underthrustings within the Indian plate and led to the formation of the spectacular Himalayan Mountains.

The collision of India is believed to have played an important role in the tectonics of Asia. The most pronounced influences are gigantic east-west trending strike-slip faults that are forcing eastern Asia to slide to the east and out of India's path. This extrusion model, though very attractive, has several shortcomings. India's ongoing push against Asia seems to be confined to the Indian plate and is being absorbed in a new converging zone in the eastern Indian Ocean. Recent spreading centers such as the Baikal Rift and the Shansi Graben may have direct consequences to the formation of the strike-slip faults in Asia.

REFERENCES CITED

Abdel-Gawad, M., 1971, Wrench movements in the Baluchistan Arc and relation to the Himalayan-Indian Ocean tectonics: Geological Society of America Bulletin, v. 82, p. 1235-1256.

Allegre, C.J., et al., 1984, Structure and evolution of the Himalaya-Tibet orogenic belt: Nature, v. 307, p. 17-22.

Alleman, F. 1979, Time of emplacement of the Zhob Valley ophiolites and Bela ophiolites, Baluchistan (preliminary report), in Farah, A. and De Jong, K.A., eds., Geodynamics of Pakistan: Quetta, Geological Survey of Pakistan, p. 215-242.

Alt, D., Sears, J.M., and Hyndman, D.W., 1988, Terrestrial maria: the origins of large basalt plateaus, hotspot tracks and spreading ridges: Journal of Geology, v. 96, p. 647-662.

Alvarez, L.W., Alvarez, W., Asaro, F., and Michel, H.V. 1980, Extraterrestrial cause for the Cretaceous-Tertiary extinction: Science, v. 208, p. 1095-1108.

Alvarez, W., Alvarez, L.W., Asaro, F., and Michel, H.V., 1982, Current status of the impact theory for the terminal Cretaceous extinction: Geological Society of America Special Publication, n. 190, p. 305-315.

Anderson, R.R., and Hartung, J.B., 1988, The Manson impact structure in North-Central Iowa: the largest known impact site in the United States and a K-T boundary astrobleme: EOS (Transactions, American Geophysical Union), v. 69, p. 1291.

Auden, J.B., 1974, Afghanistan-West Pakistan, in Spencer, A.M., ed., Data for orogenic studies: Geological Society of London Special Publication 4, p. 235-253.

Backman, J., et al., 1988, Mascarene Plateau: Proceedings of the Ocean Drilling Program, v. 115, p. 5-15.

Baker, B.H., and Miller, J.A., 1963, Geology and chronology of the Seychelles Islands and structure of the floor of the Arabian Sea: Nature, v. 199, p. 346-348.

Bannert, D., and Helmcke, D., 1981. The evolution of the Asian plate in Burma: Geologische Rundschau, v. 70, p. 446-458.

Barron, E.J., 1987, Global Cretaceous paleogeography—International geologic correlation program project 191: Palaeogeography, Palaeoclimatology, Palaeoecology, v. 59, p. 207-214.

Barron, E.J. and Harrison, C.G.A., 1980, An analysis of past plate motions: the South Atlantic and Indian Oceans, in Davies, P. and Runcorn, S.K., eds., Mechanisms of continental drift and plate tectonics: London, Academic Press, pp. 89-109.

Basu, A.R., Chatterjee, S., and Rudra, D., 1988, Shock-metamorphism in quartz grains at the base of the Deccan Traps: evidence for impact-triggered flood basalt volcanism at the Cretaceous-Tertiary boundary: EOS, Transaction American Geophysical Union, v. 69 (44), p. 1487.

Berman, D.S., and Jain, S.L., 1982, The braincase of a small sauropod dinosaur (Reptile: Saurischia) from the Upper Cretaceous Lameta Group, Central India, with review of Lameta Group localities: Annals of Carnegie Museum, 51, p. 405-422.

Besse, J., Courtillot, V., Pozzi, J.P., Westphal, M., and Zhou, Y.X., 1984, Palaeomagnetic estimates of crustal shortening in the Himalayan thrusts and Zangbo Suture: Nature, v. 311, p. 621-626.

Biswas, S.K., 1982, Western rift basins of India and hydrocarbon prospects: Oil and Gas Journal, v. 80, p. 224-232.

Bohor, B.F., and Seitz, R., 1990, Cuban K-T catastrophe: Nature, v. 344, p. 593.

Bohor, B.F., Modreski, P.J., and Foord, E.E., 1987, Shocked quartz in the Cretaceous-Tertiary boundary clays: evidence for a global distribution: Science, v. 224, p. 705-709.

Bonaparte, J.F., and Novas, F.E., 1985, *Abelisaurus comahuensis*, n.g., n. sp., Carnosaurio del Cretacio Tarido de la Patagonia: Ameghiniana, v. 21, p. 259-265.

Bonatti, E., 1987, The rifting of continents: Scientific American, v. 256(3), p. 97-103.

Braakman, J.H., Levell, B.K., Martin, J.H., Potter, T.L., and van Villet, A., 1982, Late Paleozoic Gondwana glaciation in Oman: Nature, v. 299, p. 48-50.

Briggs, J.C., 1987, Biogeography and plate tectonics: Amsterdam, Elsevier, 204 p.

———, 1989, The historic biogeography of India: isolation or contact?: Systematic Zoology, v. 38, p. 322-332.

Brookfield, M.E., and Reynolds, P.H., 1981, Late Cretaceous emplacement of the Indus Suture Zone ophiolitic melanges and an Eocene-Oligocene magmatic arc on the northern edge of the Indian plate: Earth and Planetary Science Letters, v. 55, p. 157-162.

Buffetaut, E., 1989, Archosaurian reptiles with Gondwana affinities in the Upper Cretaceous of Europe: Terra Nova, v. 1, p. 69-74.

Burke, K.C., and Wilson, J.T., 1976, Hot spots on the Earth's surface: Scientific American, v. 235, n. 2, p. 6-57.

Carey, S.W., 1958, A tectonic approach to continental drift, in Carey, S.W., ed., Continental drift: a symposium: Hobart, University of Tasmania Press, p. 177-355.

———, 1976, The expanding Earth: Elsevier, Amsterdam, 488 p.

Chakravorty, S.C., Nadgir, B.B., Sinha-Roy, S., and Chatterjee, D., 1971, Precambrian rocks of the extrapeninsular India: Indian Geological Survey Records, v. 101, n. 2, p. 100-125.

Chao, E.C.T., 1963, Shock effects in certain rock-forming minerals: Science, v. 156, p. 192-202.

Chatterjee, S., 1978, *Indosuchus* and *Indosaurus*, Cretaceous carnosaurs of India: Journal of Paleontology, v. 52, p. 570-580.

———, 1984, The drift of India: a conflict in plate tectonics: Memoires de la Societe Geologique de France, n. 147, p. 43-48.

———, 1990a, A possible K-T impact site at the India-Seychelles boundary: Abstract, Lunar and Planetary Science Conference XXI, p. 182-183.

———, 1990b, Impact volcanism and dinosaur extinction at the K-T boundary: Abstract, Journal of Vertebrate Paleontology, v. 10(3), p. 17A.

Chatterjee, S., and Hotton, N., III, 1986, The paleoposition of India: Journal of Southeast Asian Earth Sciences, v. 1, p. 145-189.

Clube, S.V.M., and Napier, W.M., 1982, The role of episodic bombardment in geophysics: Earth and Planetary science Letters, v. 57, p. 251-262.

Colbert, E.H., 1973, Continental drift and the distribution of fossil reptiles, in Tarling, D.H., and Runcorn, S.K., eds., Implications of continental drift to Earth sciences: London, Academic Press, v. 1, p. 393-410.

Courtillot, V.E., 1990, A volcanic eruption: Scientific American, v. 263(4), p. 85-92.

Courtillot, V., Besse, J., Vandamme, D., Montigny, R., Jaeger, J.J., and Capetta, H., 1986, Deccan flood basalts at the Cretaceous/Tertiary boundary?: Earth and Planetary Science Letters, v. 80, p. 361-374.

Courtillot, V., Feraud, G., Maluski, H., Vandamme, D., Moreau, M.G., and Besse, J., 1988, Deccan flood basalts and the Cretaceous/Tertiary boundary: Nature, v. 333, p. 843-846.

Cox, A., 1973, ed., Plate tectonics and geomagnetic reversals: San Francisco, Freeman, 702 p.

Crawford, A.R., 1972, Iran, continental drift and plate tectonics: 24th International Geological Congress, section 3, p. 106-111.

———, 1974, The Salt Range, the Kashmir and the Pamic Arc: Earth and Planetary Science Letters, v. 22, p. 371-379.

———, 1979, The myth of a vast oceanic Tethys, the India-Asia problem and Earth expansion: Journal of Petroleum Geology, v. 2, n.1. p. 3-9.

Curray, J.R. and Moore, D.G., 1971, Growth of the Bengal Deep-sea Fan and denudation in the Himalayas: Geological Society of America Bulletin, v. 82, p. 563-572.

Curray, J.R., Emmel, F.J., Moore, D.G., and Raitt, R.W., 1982, Structure, tectonics, and geological history of the Northeastern Indian Ocean, in Nairn, A.E.M., and Stehli, F.G., eds., The ocean basins and margins: New York, Plenum Press, v. 6, p. 399-450.

Davies, D., 1968, When did Seychelles separate from India?: Nature, v. 220, p. 1225-1226.

De Jong, K.A., 1982, Tectonics of the Persian Gulf, Gulf of Oman and southern Pakistan region, in Nairn, A.E.M., and Stehli, F.G., eds., The ocean basins and margins: New York, Plenum Press, v. 6, p. 315-351.

Devey, C.W. and Lightfoot, P.C., 1986, Volcanological and tectonic control of stratigraphy and structure in the Western Deccan Traps: Bulletin Volcanology, v. 48, p. 195-207.

Dickins, J.M., 1987, Tethys—a geosyncline formed on continental crust?, in McKenzie, K.G., ed., Shallow Tethys 2: Rotterdam, A. A. Balkema, p. 149-158,.

Dietz, R.S., and Holden, S.C., 1970, The breakup of Pangaea: Scientific American, v. 223, p. 30-41.

Dietz, R.S. and McHone, J. 1990, Isle of Pines (Cuba) apparently not K-T boundary impact site: Geological Society of America Abstracts with Programs, v. 22 (7), A79.

Duncan, R.A., 1981, Hotspots in the southern oceans—an absolute frame of reference for motion of the Gondwana continents: Tectonophysics, v. 74, p. 29-42.

Duncan, R.A., and Pyle, D.G., 1988, Rapid eruption of the Deccan flood basalts at the Cretaceous/Tertiary boundary: Nature, v. 333, p. 841-843.

England, P., and Houseman, G., 1986, Finite strain calculations of continental deformation 2. comparison with the India-Asia collision zone: Journal of Geophysical Research, v. 91, p. 3664-3676.

Fisher, R.L., Engel, C.G., and Hilde, T.W.C., 1968, Basalts dredged from the Amirante Ridge, western Indian Ocean: Deep Sea Research, v. 15, 521-534.

Fitch, T.J., 1972, Plate convergence, transcurrent faults, and internal deformation adjacent of Southeast Asia and the western Pacific: Journal of Geophysical Research, v. 77, p. 4432-4460.

Forsyth, D.W., and Uyeda, S., 1975, On the relative importance of driving forces of plate motion: Geophysical Journal of Royal Astronomical Society, v. 43, p. 163-200.

Francis, T.J.G., and Raitt, R.W., 1967, Seismic refraction measurements in the Southern Indian Ocean: Journal of Geophysical Research, v. 72, p. 3015-3042.

French, B.M., 1984, Impact event at the Cretaceous-Tertiary boundary: a possible site: Science, v. 226, p. 353.

French, B.M., and Short, N.M., 1968, eds., Shock metamorphism of natural materials: Baltimore, Mono Book Corporation, 644 p.

Gansser, A., 1964, The Geology of the Himalayas: New York, Interscience, 289 p.

———, 1966, The Indian Ocean and the Himalayas, a geological interpretation: Ecologae Geologische Helvetae, v. 52, p. 831-848.

———, 1974, Data for orogenic studies: Himalayas, in Spencer, A.M., ed., Mesozoic-Cenozoic orogenic belts: Geological Society Special Publication 4, p. 267-278.

———, 1980, The significance of the Himalayan suture zone: Tectonophysics, v. 62, p. 37-52.

Gault, D.E. and Wedekind, J.A., 1978, Experimental studies of oblique impact: Lunar and Planetary Science Conference, v. 3, p. 3843-3875.

Gilmore, J.S., Knight, J.D., Orth, C.J., Pilmore, C.L., and Tschudy, R.H., 1984, Trace element patterns at a non-marine Cretaceous-Tertiary boundary: Nature, v. 307, p. 224-228.

Glennie, E.A., 1951, Density or geological corrections to gravity anomalies for the Deccan Traps areas in India: Royal Astronomical Society Geophysics Supplement, v. 6, p. 179-193.

Glennie, K.W., Boeuf, M.G.A., Clark, M.W.H., Moody-Stuart, M., Pilaar, W.F.H., and Reinhart, B.M., 1973, Late Cretaceous nappes in the Oman Mountains and their geologic evolution: American Association of Petroleum Geologists Bulletin, v. 57, p. 5-27.

Gordon, R.G., and DeMets, C., 1989, Present-day motion along the Owen Fracture Zone and Dalrymple Trough in the Arabian Sea: Journal Geophysical Resarch, v. 94, p. 5560-5570.

Gordon, R.G., DeMets, C., and Argus, D.F., 1990, Kinematic constraints on distributed lithospheric deformation in the equatorial Indian Ocean from present motion between the Australian and Indian plates: Tectonics, v. 9, p. 409-422.

Graham, S.A., Dickinson, W.R., and Ingersol. R.V., 1975, Himalayan-Bengal model for flysch dispersal in the Appalachian-Ouchita system: Geological Society of America Bulletin, v. 86, p. 97-116.

Grieve, R.A.F., 1982, The record of impact on Earth: implications for a major Cretaceous/Tertiary impact event: Geological Society of America Special Paper 190, p. 25-37.

———, 1987, Terrestrial impact structures: Annual Review of Earth and Planetary Science, v. 15, p. 245-270.

———, 1990, Impact cratering on the Earth: Scientific American, v. 261(4), p. 66-73.

Grieve, R.A.F., Wood, C.A., Garvin, J.B., McLaughlin G., and McHone. J.F., Jr., 1988, Astronaut's guide to terrestrial impact craters: Houston, Lunar and Planetary Institute, 89 p.

Hallam, A., 1987, End-Cretaceous mass extinction event: argument for terrestrial causation: Science, v. 238, p. 1237-1247.

Harland, W.B., Cox, A.V., Llewellyn, P.G., Pickton, C.A.G., Smith, A.G., and Walters, R., 1982, A geologic time scale: Cambridge, Cambridge University Press, 128 p.

Hartnady, C.H.J., 1986, Amirante Basin, western Indian Ocean: possible impact site of the Cretaceous-Tertiary extinction bolide?: Geology, v. 14, p. 423-426.

Hildebrand, A.R., and Boynton, W.V., 1990, Proximal Cretaceous-Tertiary boundary impact deposits in the Carribean: Science, v. 248, p. 843-847.

Hooper, P.R., 1990, The timing of crustal extension and the eruption of continental flood basalts: Nature, v. 345, p. 246-249.

Howard, K.A., and Wilshire, H.G., 1975, Flows of impact melt at lunar craters: Journal of Research, U.S. Geological Survey, v. 3, p. 237-251.

Huene, F. von, and Matley, C.A., 1933, The Cretaceous Saurischia and Ornithischia of the central Provinces of India: Palaeontologia Indica, v. 21, p. 1-74.

Jaeger, J.-J., Courtillot, V., and Tapponnier, P., 1989, Paleontological view of the ages of the Deccan Traps, the Cretaceous/Tertiary boundary, and the India-Asia collision: Geology, v. 17, p. 316-319.

Johnson, B.D., Powell, M.C., and Veevers, J.J., 1976, Spreading history of the eastern Indian Ocean and Greater India's northward flight from Antarctica and Australia: Geological Society of America Belletin, v. 87, p. 1560-1566.

Kaila, K.L., Murty, P.R.K., Rao, V.K., and Kharetchko, G.E., 1981, Crustal structure from deep seismic soundings along the Koyna II (Kelsi-Loni) profile in the Deccan Trap area, India: Tectonophysics, v. 73, p. 365-384.

Kar, R.K., and Singh, R.S., 1982, Palynology of the Cretaceous sediments of Meghalaya, India: Palaeontographica, v. 202B, 83-153.

Kauffman, E.G., 1973, Cretaceous Bivalvia, in Hallam, A., ed., Atlas of palaeobiogeography: New York, Elsevier, p. 353-383.

Klootwijk, C.T., Conaghan, P.J., and Powell, C.M., 1985, The Himalayan Arc: large-scale continental subduction, oroclinal bending and back-arc apreading: Earth and Planetary Science Letters, v. 75, p. 167-183.

Koeberl, C., Murali, A.V., Sharpton, V.L., and Burke, K., 1988, Geology and geochemistry of the Kara and Ust-Kara Impact structures (USSR): EOS (American Geophysical Union Transactions), v. 69, p. 1293.

Koeberl,, T.M., Sharpton, V.L., Murali, A.V., and Burke, K., 1990, Kara and Ust-Kara impact structures (USSR) and their relevance to the K-T boundary event: Geology, v. 18, p. 50-53.

Krishnan, M.S., 1982, Geology of India and Burma: Delhi, CBS Publishers, 536 p.

Kusznir, N.J., and Park, R.G., 1986, The extensional strength of the continental lithosphere: its dependence on geothermal gradient, and crustal composition and thickness: Geological Society Special Publication, n. 28, p. 35-52.

Laparrent, A.F.de, and Davoudzadeh, M., 1972, Jurassic dinosaur footprints of the Kerman area, Central Iran: Geological Survey of Iran Report 26, p. 5-22.

Laughton, A.S., and Tarmontini, C., 1969, Recent studies of the crustal structure of the Gulf of Aden: Tectonophysics, v. 8, p. 359-375.

Le Fort, P., 1975, Himalayas: the collided ranges; present knowledge of the continental arc: American Journal of Science, v. 275A, p. 1-44.

Le Loeuff, J., Buffetaut, E., Mechin, P., and Mechin-Salessy, A., 1989, A titanosaurid dinosaur braincase (Saurischia, Sauropoda) from the Upper Cretaceous of Var (Provence, France): Comptes Rendus de l'Academie des Sciences de Paris, v. 309, p. 851-857.

Le Pichon, X., and Heirtzler, J.R., 1968, Magnetic anomalies in the Indian Ocean and sea-floor spreading: Journal of Geophysical Research, v. 73, p. 2101-2117.

Lillegraven, J.A., Kraus, M.J., and Bown, T.M., 1979, Paleogeography of the world of the Mesozoic, in Lillegraven, J.A., Kielan-Jaworowska, Z., and Clemens, W.A., eds., The first two-thirds of mammalian history: Berkeley, University of California Press, p. 277-308.

Luyendyk, B.P., and Rennick, W., 1977, Tectonic history of aseismic ridges in the eastern Indian Ocean: Geological Society of America Bulletin, v. 88, p. 1347-1356.

MacIntyre, R.M., Dickin, A.P., Fallick, A.E., and Halliday, A.N., 1985, An isotopic and geochronological study of the younger igneous rocks of the Seychelles: EOS, American Geophysical Union Transactions, v. 66, p. 1137.

Mahoney, J.J., 1988, Deccan Traps, in Macdougall, J.D., ed., Continental flood basalts: Dordrecht, Kluwer Academic Publishers, pp. 151-194.

Mahoney, J.J., Macdougall, J.D., Lugmair, G.W., and Gopalan, K., 1983, Kerguelen hotspot source for Rajmahal Traps and Ninetyeast Ridge?: Nature, v. 303, p. 385-389.

Maluski, H., Proust, F., and Xiao, X.C., 1982, First results of $^{39}Ar/^{40}Ar$ dating of the transhimalayan calcalkaline magmatism of southern Tibet: Nature, v. 298, p. 152-154.

Markl, R.G., 1974, Evidence for the breakup of Eastern Gondwanaland: Nature, v. 251, p. 196-200.

McClure, H.A., 1980, Permian-Carboniferous glaciation in the Arabian Peninsula: Geological Scoiety of America Bulletin, v. 91, p. 707-712.

McKenzie, D.P., and Morgan, W.J., 1969, Evolution of triple junctions: Nature, v. 224, p. 125-133.

McKenzie, D.P., and Sclater, J.G., 1971, The evolution of the Indian Ocean since the Late Cretaceous: Geophysical Journal of the Royal Astronomical Society, v. 25, p. 437-528.

———, 1973, The evolution of the Indian Ocean: Scientific American, v. 228(5), p. 63-72.

Meyerhoff, A.A., and Kamen-Kaye, M., 1981, Petroleum prospects of Saya de Malha and Nazareth Banks, Indian Ocean: American Association of Petroleum Geologists Bulletin, 65, p. 1344-1347.

Meyerhoff, A.A., and Meyerhoff, H.A., 1972, The new global tectonics: major inconsistencies: American Association of Petroleum Geologists Bulletin, v. 56, p. 269-336.

Minster, J.B., and Jordan, T.H., 1978, Present-day plate motions: Journal of Geopshysical Research, v. 83, p. 5331-5354.

Mitchell, A.H.G., 1984, Post-Permian events in the Zangbo 'suture' zone, Tibet: Journal of Geological Society of London, v. 141, p. 129-136.

Miura, Y., 1990, Mineralogical data of shocked quartz grains from K-T boundary: Abstract, Lunar and Planetary Sciences, v. 21, p. 793-794.

Molnar, P., 1988, Continental tectonics in the aftermath of plate tectonics: Nature, v. 335, p. 131-137.

Molnar, P., and Stock, J., 1987, Relative motions of hotspots in the Pacific, Atlantic and Indian Oceans since late Cretaceous time: Nature, v. 327, p. 587-591.

Molnar, P., and Tapponnier, P., 1977, The collision between India and Eurasia: Scientific American, v. 236(4), p. 30-41.

Morgan, W.J., 1972, Deep mantle convection plumes and plate motions: American Association of Petroleum Geologists Bulletin, v. 56, p. 203-213.

———, 1981, Hotspot tracks and the opening of the Atlantic and Indian Oceans, in Emiliani, C. ed., The sea: New York, Wiley, v.7, p. 443-487.

Moore, J.J., 1976, Missile impact craters (White Sand Missile Range, New Mexico) and applications to lunar research: U.S. Geological Survey Professional Paper, v. 812 B, p. B1-B47.

Naini, B.R., and Talwani, M., 1982, Structural framework and the evolutionary history of the continental margin of western India, in Watkins, J.S., and Drake, C.L., eds., Studies in continental margin geology: American Association of Petroleum Geologists Memoir 34, p. 167-193.

Neprochnov, Y.P., Levchenko, O.V., Merklin, L.R., and Sedov, V.V., 1988, The structure and tectonics of the intraplate deformation area of the Indian Ocean: Tectonophysics, v. 156, p. 89-106.

Norton, I.O., and Sclater, J.G., 1979, A model for the evolution of the Indian Ocean and the breakup of Gondwanaland: Journal of Geophysical Research, v. 84(B12), p. 6803-6830.

Officer, C.B., Hallam, A., Drake, C.L., and Devine, J.D., 1987, Late Cretaceous and paroxysmal Cretaceous/Tertiary extinctions: Nature, v. 326, p. 143-149.

Owen, H.G., 1976, Continental displacement and expansion of Earth during the Mesozoic and Cenozoic: Philosophical Transactions of the Royal Scoiety of London, v. A 281, p. 223-291.

Patriat, P., and Achache, J., 1984, India-Eurasia collision chronology has implications for crustal shortening and driving mechanism of plates: Nature, v. 311, p. 615-621.

Patriat, P. and Segoufin, J., 1988, Reconstruction of the Central Indian Ocean: Tectonophysics, v. 155, p. 211-234.

Peirce, J.W., 1978, The northward motion of India since the Late Cretaceous: Geophysical Journal of the Royal Astronomical Society, v. 52, p. 277-311.

Peirce, J.W., et al., 1989, Proceedings of the Ocean Drilling Program, v. 121, p. 517-537.

Powar, K.B., 1981, Lineament fabric and dyke pattern in the westernpart of the Deccan volcanic province, in Subbarao, K.V., and Sukeshwala, R.N., eds., Deccan volcanism and related basalt provinces in other parts of the world: Geological Society of India Memoir 3, p. 45-57.

Powell, C.McA., 1979, A speculative history of Pakistan and surroundings: some constraints from the Indian Ocean, in Farah, A., and De Jong, K.A., eds., Geodynamics of Pakistan: Quetta, Geological Survey of Pakistan, p. 5-24.

Powell, C.McA., and Conaghan, P.J., 1973, Plate tectonics and the Himalayas: Earth and Planetary Science Letters, v. 20, p. 1-12.

Prasad, G.V.R., and Sahni, A., 1988, First Cretaceous mammal from India: Nature, v. 332, p. 638-640.

Qureshy, M.N., 1981, Gravity anomalies, isostacy and crust mantle relations in the Deccan Trap and contiguous regions, India: Memoir Geological Society of India, v. 3, p. 184-197.

Rabinowitz, P.D., Coffin, M.L., and Falvey, D., 1983, The separation of Madagascar and Africa: Science, v. 220, p. 67-69.

Rampino, M.R. and Stothers, B.B., 1988, Flood basalt volcanism during the past 250 million years: Science, v. 241, p. 663-668.

Richards, M.A., Duncan, R.A., and Courtillot, V., 1989, Flood basalts and hot-spot tracks: plume heads and tails: Science, v. 246, p. 103-107.

Royer, J.Y., and Sandwell, D.T., 1989, Evolution of the eastern Indian Ocean since the Late Cretaceous: constraints from geostat altimetry: Journal of Geophysical Research, v. 94, p. 13755-13782.

Royer, J.Y., Sclater, J.G., and Sandwell, D.T., 1989, A preliminary tectonic fabric chart of the Indian Ocean Proceedings Indian Academy of Science, v. 98, p. 7-24.

Sahni, A., 1984, Cretaceous-Paleocene terrestrial faunas of India: lack of endemism during drifting of the Indian plate: Science, v. 226, p. 441-443.

———, 1989, Eurasiatic elements in Indian Cretaceous nonmarine biotas: European Union of Geoscience EUG V, Terra Abstracts, v. 1, p. 253-254.

Sahni, A., and Bajpai, S., 1988, Cretaceous-Tertiary boundary events: the fossil vertebrate, palaeomagnetic and radiometric evidence from Peninsular India: Journal of Geological Society of India, v. 32, p. 382-396.

Sahni, A., and Khare, S.K., 1973, Additional Eocene mammals from the Subathu Formation of Jammu and Kashmir: Journal Palaeontological Society of India, v. 17, p. 31-49.

Sahni, A., Rana, R.S., and Prasad, G.V.R., 1987, New evidence for paleobiogeographic intercontinental Gondwana relationships based on Late Cretaceous-Early Paleocene coastal faunas from Peninsular India, in McKenzie, G.D., ed., Gondwana six: stratigraphy, sedimentology and paleontology: American Geophysical Union Monograph 41, p. 207-228.

Saint-Marc, P., 1978, Arabian Peninsula, in Moullade, M., and Nairn, A.E.M., eds., The Phanerozoic geology of the world II: Amsterdam, Elsevier, p. 435-462.

Schlich, R., 1982, The Indian Ocean: aseismic ridges, spreading centers, and ocean basins, in Nairn, A.E.M., and Stehli, F.G., eds., The ocean basins and margins: New York, Plenum Press, v. 6. p. 51-147.

Sclater, J.G., and Fisher, R.L., 1974, Evolution of the east-central Indian Ocean, with emphasis on the tectonic setting of the Ninetyeast Ridge: Geological Society of America Bulletin, v. 85, p. 683-702.

Scotese, C.R., Graham, L.M., and Larson, R.L., 1988, Plate tectonic reconstructions of the Cretaceous and Cenozoic ocean basins: Tectonophysics, v. 155, p. 27-48.

Sharpton, V.L., Huffman, A.R., Murali, A.V., and Kronenberg, A.K., 1989, Shocked quartz at the base of the Deccan Traps: fact or fiction?: Geological Society of America Abstracts with Programs, v. 21 (6), p. A93.

Shoemaker, E.M., 1983, Asteroid and comet bombardment of the Earth: Annual Review of Earth Sciences, v. 11, p. 461-494.

Simpson, E.S.W., Sclater, J.G., Parsons, B., Norton, I.O., and Meinke, L., 1979, Mesozoic magnetic lineations in the Mozambique Basin: Earth and Planetary Science Letters, v. 43, p. 260-264.

Smiley, C.J., 1974, Analysis of crustal relative stability from some Late Paleozoic and Mesozoic floral records: American Association of Petroleum Geologists Memoir, v. 23, p. 331-360.

Smith, A.B., 1988, Late Paleozoic biogeography of East Asia and paleontological constraints on plate tectonic reconstruction: Philosophical Transactions of the Royal Society of London, v. A 326, p. 189-227.

Smith, A, G., and Hallam, A., 1970, The fit of the southern continents: Nature, v. 225, p. 139-146.

Smith, A.G., Hurley, A.M., and Briden, J.C., 1981, Phanerozoic paleocontinental world maps: Cambridge, Cambridge University press.

Sonnenfeld, P. 1978, Eurasian ophiolites and the Phanerozoic Tethys Sea: Geotectonische Forsungen, v. 56, p. 1-88.

——, 1981, The Phanaerozoic Tethys Sea, in Sonnenfeld, P., ed., Tethys, the ancestral Mediterranean: Stroudsburg, Hutchinson Ross, p. 18-53.

Srivastava, S., Mohabey, D.M., Sahni, A., and Pant, S.C., 1986, Upper Cretaceous egg clutches from Kheda District, Gujrat, India: Palaeontographica, v. A 193, p. 219-233.

Stein, S., and Okal, E.A., 1978, Seismicity and tectonics of the Ninetyeast Ridge area: evidence for internal deformation of the Indian plate: Journal of Geophysical Research, v. 83, p. 2233-2244.

Stocklin, J., 1974, Possible ancient continental margins of Iran, in Burke, C.A., and Drake, C.L., eds., The geology of continental margin: New York, Springer-Verlag, p. 873-887.

——, 1980, Geology of Nepal and its regional frame: Journal Geological Society of London, v. 137, p. 1-34.

Stoffler, D. 1971, Coesite and stishovite in shocked crystalline rocks: Journal of Geophysical Research, v. 76, p. 5474-5488.

Subbarao, K.V., Bodas, M.S., Hooper, P.R., and Walsh, J.N., 1988, Petrogenesis of Jawahar and Igatpuri Formations western Deccan Basalt Province, in Subbarao, K.V., ed., Deccan flood basalts: Geological Society of India Memoir 10, P. 253-280.

Takin, M., 1972, Iranian geology and continental drift in the Middle East: Nature, v. 235, p. 147-150.

Tapponnier, P., et al., 1981, The Tibetan side of the India-Eurasia collision: Nature, v. 294, p. 405-410.

Tapponnier, P., Peltzer, G., Le Dain, A.Y., and Armijo, R., 1982, Propagating extrusion tectonics in Asia: new evidence from simple experiments with plasticine: Geology, v. 10, p. 611-616.

Tapponnier, P., Peltzer, G., and Armijo, R., 1986, On the mechanics of the collision between India and Asia, in Coward, M.P., and Ries, A.C., eds., Collision tectonics: Geological Society Special Publication 19, p. 115-157.

Teichert, C., 1973, Paleozoic Tethyan Ocean: Nature, v. 244, p. 91.

Valdiya, K.S., 1988, Tectonics and evolution of the central sector of the Himalayas: Philosophical Transactions of the Royal Society of London, v. A 326, p. 151-175.

Verma, R.K., and Prasad, K.A.V.L., 1988, Analysis of the gravity field in the Nepalese Himalayaas—Tibetan region in the light of plate tectonics: Tectonophysics, v. 147, p. 59-70

Vianey-Liaud, M., Jain, S.L., and Sahni, A., 1987, Dinosaur egg shells (Saurischia) from the Late Cretaceous intertrappean and Lameta Formation (Deccan, India): Journal of Vertebrate Paleontology, v. 7, p. 408-424.

Vine, F.J., and Livermore, R.M., 1985, A model for the evolution of the Indian Ocean: EOS (Transactions, American Geophysical Union), v. 66, p. 1078.

Wadia, D.N., 1966, Geology of India: London, Macmillan, 536 p.

Weijermars, R., 1989, Global tectonics since the breakup of Pangea 180 million years ago: evolution maps and lithospheric budget: Earth-Science Reviews, v. 26, p. 113-162.

Weissel, J.K., Anderson, R.N., and Geller, C.A., 1980, Deformation of the Indo-Australian plate: Nature, v. 287, p. 284-291.

Wezel, F.C., 1988, A young Jura-type fold belt with the Central Indian Ocean?: Bolletino di Oceanologia Teorica ed applicata, v. 6(2), p. 75-90.

White, R.S., and McKenzie, D.P., 1989a, Volcanism at rifts: Scientific American,v. 261(1), p. 62-71.

——, 1989b, Magmatism at rift zones: the generalization of volcanic continental margins and flood basalts: Journal of Geophysical Research, v. 94(B6), p. 7685-7729.

Wilhelms, D.E., 1987, The geologic history of the moon: U.S. Geological Survey Professional Paper, m. 1348, p. 1-293.

Wilson, J.T., 1965, A new class of faults and their bearing on continental drift: Nature, v. 207, p. 343-347.

Zhao, X., (in press), The Mesozoic vertebrate fossils of Tibet: Vertebrata Palasiatica.

ACKNOWLEDGMENTS

Supported by the National Geographic Society, Texas Tech University and Indian Statistical Institute. I thank A. R. Basu , N. Guven, K. V. Subbarao, and D. K. Rudra for helpful discussions, M. W. Nickell for drafting the figures, and N. Olson for photography. X. Zhao has kindly provided his unpublished manuscript on Tibetan dinosaurs. I thank J. W. Peirce, J. K. Weissel, G. Gray, I. Norton, and Lisa M. Gahagan for numerous suggestions for improvement. Lisa M. Gahagan, the technical coordinator of the Paleoceanographic Mapping Project (POMP), provided valuable database and plate rotation model for producing reconstructions. I am indebted to Necip Guven and H. Todd Schaef for SEM and X-Ray identification of shocked minerals and Carl J. Orth for iridium analysis. I thank D. K. Rudra for field assistance and continued encouragement.

Continental Tectonics and Orogeny

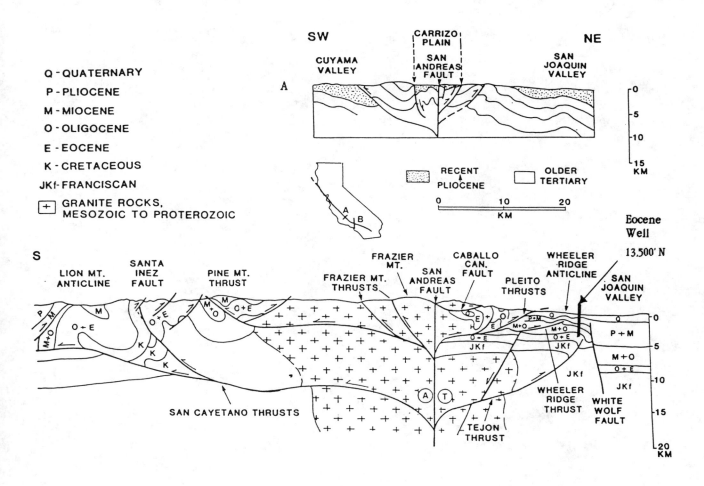

Intracratonic tectonism: key to the mechanism of diastrophism

A. C. Grant, Geological Survey of Canada, Atlantic Geoscience Centre, Bedford Institute of Oceanography, Dartmouth, Nova Scotia, Canada B2Y 4A2

"Now faith is the substance of things hoped for, the evidence of things not seen." Hebrews 11:1.

ABSTRACT

Geological and geophysical observations over the past decade prompt review of the Nares Strait problem, and highlight the need to examine alternatives to plate-tectonic mechanisms for the formation of Baffin Bay, Labrador Sea, and perhaps other oceanic areas as well. Clues to an alternative mechanism lie in the vertical crustal displacements of intracratonic basins (e.g., Hudson Bay Basin). Such displacements are a relatively unambiguous expression of the primary driving mechanism of diastrophism, namely, movement of magmatic fluid that acts to raise or lower the overlying crust. This mechanism is powered by egress of heat from the interior of the Earth, and is modulated by the loading and unloading effects of deposition and erosion. Crustal depression by depositional loading may culminate in reactive uplift and expulsion of altered sediments and crustal material (inversion). The uplift and spreading of this altered mass can form mountains and cause lateral, "thin-skinned" tectonism. This mechanism achieves energy-efficient escape of heat from the Earth through a surficial crust that interacts physically and chemically with both the mantle below and the atmosphere above. Because this mechanism is modulated by erosion and sedimentation, it ultimately may be sensitive to climatic conditions.

INTRODUCTION

The juxtaposition of the Greenland and North American lithospheric plates across Nares Strait (Fig. 1) provides a unique opportunity to test the plate-tectonic theory, because evidence for geologic continuity between the two plates can be compared with displacement required to form Baffin Bay by sea-floor spreading. Geologists who have worked in that area contend there is continuity of Lower Paleozoic geology from Ellesmere Island to Greenland (Fig. 2). In contrast, geophysicists generally regard Baffin Bay as a small ocean basin, formed by processes of Cenozoic sea-floor spreading that moved Greenland away from Baffin Island (Fig. 3). A special session of the 1980 annual meeting of the Geological Association of Canada debated the question "Did Greenland drift along Nares Strait?" Of the 30 papers published from the 1980 meeting (Dawes and Kerr, 1982), 19 express an opinion: 8 papers favor displacement, 11 favor geologic continuity across Nares Strait.

Over the past decade not much has changed—geologists continue to argue for continuity across Nares Strait (e.g., Dawes, 1986; Higgins and Soper, 1989), whereas plate reconstructions based on geophysical in-

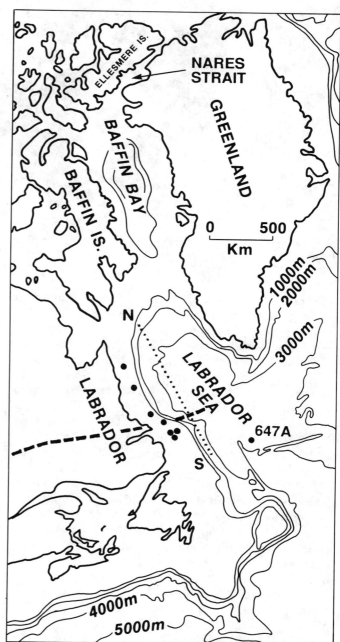

Figure 1. Generalized bathymetric map of Baffin Bay-Labrador Sea region. Dots mark locations of drill sites mentioned in text. Heavy dashed line traces Grenville Front onshore and Cartwright Fracture Zone offshore. Dotted line (N-S) indicates orientation of plane of section in Figure 5.

Chatterjee, S., and N. Hotton III, eds. *New Concepts in Global Tectonics.* Texas Tech University Press, Lubbock, 1992, xii + 450 pp.

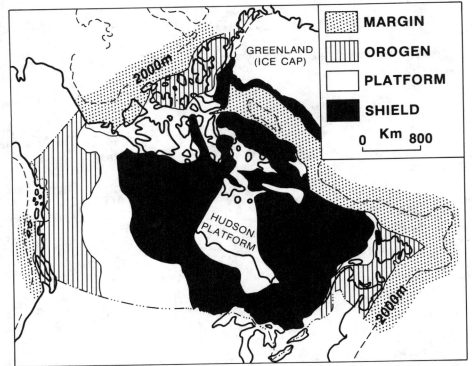

Figure 2. Generalized geological map of Canada and Greenland. Note indicated geologic continuity from Ellesmere Island to Greenland.

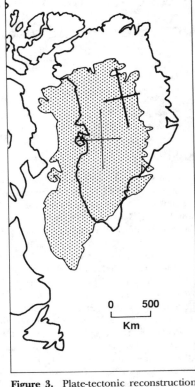

Figure 3. Plate-tectonic reconstruction of Greenland relative to Canada at 92 Ma (stippled), after Roest and Srivastava (1989). Crosses indicate relative translation and rotation.

terpretations require displacements of several hundred km (e.g., Roest and Srivastava, 1989; Fig. 3).

In my view, the geological evidence for continuity across Nares Strait is sound, and geophysical data (e.g., Hood et al., 1985; Jackson and Koppen, 1985) can be reasonably interpreted in support of this view. Stratigraphic evidence from Ellesmere Island and Greenland, as summarized by Higgins and Soper (1989), precludes more than about 25 km of Cenozoic displacement along Nares Strait. In contrast, the plate-tectonic model requires much greater displacement (Fig. 3). I suggest the model be examined.

Labrador Sea is affected by any plate movements prescribed for Baffin Bay (Fig. 3). Recently published ocean drilling program (ODP) results (Srivastava and Arthur, 1989) bear upon the Nares Strait question, because these new observations document problems with plate-tectonic models for the Labrador Sea. In this paper I examine implications of accepting geologic continuity across Nares Strait, and forming Baffin Bay by some mechanism other than lateral movement of lithospheric plates. First, I review problems with plate-tectonic models; second, outline elements of an alternative mechanism; third, briefly consider some implications of this mechanism.

One conclusion is worth stating at the outset: to examine alternatives is to recognize that any geological model of the Earth is constrained by our limited knowledge of its early history, or of its interior, and is based largely on a framework of assumptions. In reality, observational data generally can be fitted to a variety of models.

PROBLEMS WITH PLATE TECTONICS: BAFFIN BAY AND LABRADOR SEA

Srivastava and Arthur (1989) stated that drilling results from ODP Leg 105 in Labrador Sea confirm basement age as interpreted from magnetic anomaly identification, and that the crust under the Labrador Sea is oceanic. They concluded that ". . . the crustal age validates the seafloor-spreading model proposed for the Labrador Sea and Baffin Bay."

It is possible to derive quite different conclusions from the Leg 105 drilling data. Basaltic "basement" was reached at only one site (647A, Fig. 1), and the K/Ar ages from that site, which ranged from 35–71 ma, were considered ". . . unreliable for defining the age of basaltic crust" (Roddick, 1989). The age of the basement actually was inferred from the age of the overlying sediments. Moreover, Clarke et al. (1989) noted that they ". . . are not sure whether the cooling units at Hole 647A are flows

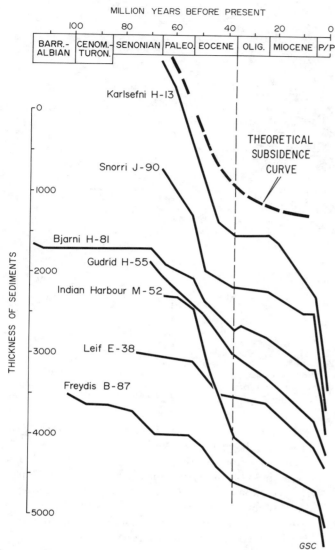

Figure 4. Subsidence curves for seven wells on the Labrador Shelf (After Gradstein and Srivastava, 1980), from Freydis B-87 in the south to Karsefni H-13 in the north (locations in Fig. 1). The dashed line is the theoretical subsidence curve for the Labrador Sea.

or sills." The single channel seismic record from site 647A (e.g., Srivastava and Arthur, 1989, fig. 15) shows faint intrabasement reflectors (sills?), raising the question as to what deeper penetration, multichannel seismic records might show at this site. Clarke et al. (1989) further noted "... there is no geochemical evidence from Hole 647A samples to support or to refute the existence of foundered continental crust in the Labrador Sea" (Kerr, 1967, suggested that Labrador Sea may be underlain by foundered continental crust).

Thus the magnetic lineations in Labrador Sea are not conclusively dated, nor is the underlying crust conclusively proven to be oceanic. The geological and geophysical evidence from Baffin Bay and Labrador Sea indicate only that Cretaceous-Tertiary subsidence and sedimentation occurred in this seaway, accompanied by volcanism.

Additional questions about plate tectonic models for Labrador Sea have been discussed previously (Grant, 1980).

1. Magnetic lineations in Labrador Sea are parallel to gravity anomalies. This suggests that the mass lineation indicated by gravity reflects a more physically substantial source for the magnetic anomalies than mere changes in magnetic polarity.

2. Published interpretations of refraction seismic data from the Labrador continental margin differ regarding the location of the continent-ocean boundary. Such differences are not unusual, and they underline the ambiguity of refraction seismic velocities as an indicator of crustal origin.

3. Subsidence curves for wells along the Labrador shelf (Fig. 4) indicate increasing subsidence from south to north, and renewed subsidence in post Oligocene time, increasingly so to the north, where it equals or even surpasses pre-Oligocene subsidence. The theoretical, sea-floor-spreading subsidence curve for the Labrador Sea (e.g., Sclater et al., 1971) does not account for this late Tertiary subsidence. Moreover, even with this late Tertiary subsidence, the floor of the Labrador Sea is about 1 km shallower than predicted by plate-tectonic theory (Hyndman, 1973).

Late Tertiary subsidence of the Labrador Shelf apparently is matched by late Tertiary uplift of the adjacent landmass; the Labrador highlands have risen about as much as the shelf has subsided, with overall displacement approaching 4 km (Fig. 5).

PROBLEMS WITH PLATE TECTONICS: GENERAL

Labrador Sea is a relatively narrow, "simple," ocean (Fig. 1), and should be a straightforward product of tectonic plate separation as recorded by linear magnetic anomalies. However, the problems outlined above pose questions about the validity of linear magnetic anomalies as a record of sea-floor spreading. The cause of the Earth's magnetic field is not known, and, accordingly, it is not known what causes marine linear magnetic anomalies. The limited information available to date from drilling into oceanic crust does not confirm the "tape-recorder" model (e.g., Becker et al., 1989). As in Labrador Sea, dating of oceanic magnetic anomalies elsewhere commonly is based on the age of sediments overlying basalt.

One remarkable aspect of the magnetic anomaly pattern in Labrador Sea is that the principal fracture zone offsetting linear magnetic anomalies is in line with the inferred offshore projection of the Precambrian

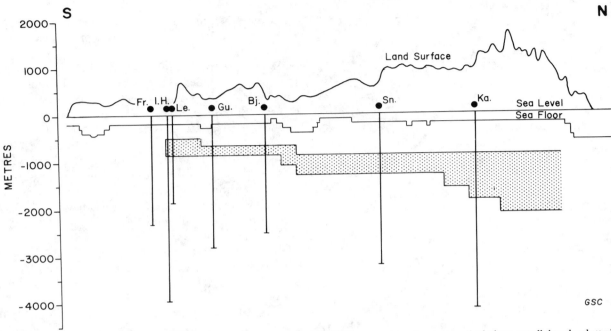

Figure 5. Longitudinal cross section of the Labrador margin (after Grant, 1980), as projected onto a vertical plane parallel to the dotted line in Figure 1. From the top, lines show the profile of the land surface, sea level, and the profile of the sea floor on the continental shelf. The stippled zone indicates the range in subsidence of the inferred late (?) Miocene unconformity beneath the continental shelf.

Grenville Front (Fig. 1). This relationship may indicate a pre-Grenvillian age (approximately 1.0 Ga) for the magnetic anomalies; Meyerhoff and Meyerhoff (1972) reviewed evidence that linear marine magnetic anomalies date from the Archean.

Finally, many Earth scientists will concede that it is not possible to distinguish a plate-tectonic Earth from an expanded Earth. In general, crucial elements of any mechanism depend on models of what cannot be seen; for example, subduction zones, and suture zones, are places where the geologic record is complicated or missing.

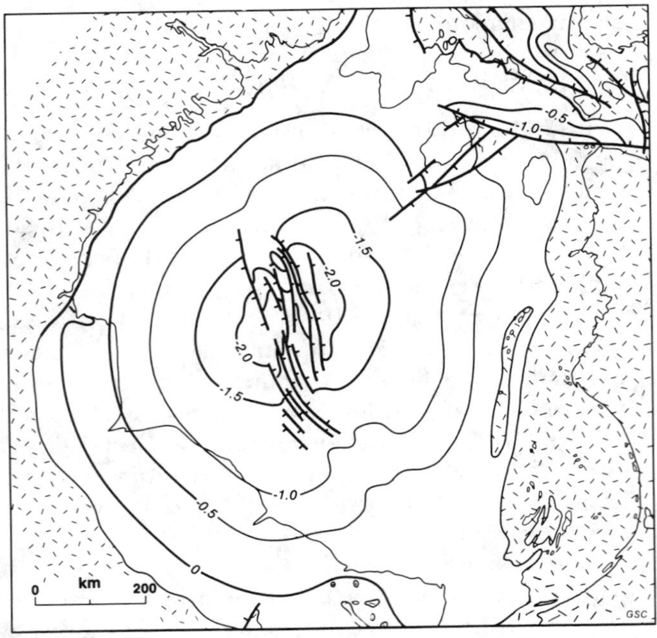

AN ALTERNATIVE MECHANISM

In view of the problems cited above, I provisionally set aside plate-tectonic models for Baffin Bay and Labrador Sea to consider an alternative mechanism. A simple alternative is suggested by the record of vertical displacements and volcanism, which are sufficient to account for the formation of Baffin Bay and Labrador Sea, and satisfy evidence for geologic continuity across Nares Strait (Dawes and Kerr, 1982).

If Baffin Bay-Labrador Sea formed by primarily vertical tectonic displacements, this seaway is the product of intracratonic processes. Because it is now submerged beneath water and sediments, I will examine such processes in a more accessible setting. It appears that aspects of intracratonic tectonism provide relatively

Figure 6. Contours on Precambrian basement (in km), Hudson Bay Basin (after Sanford, 1987).

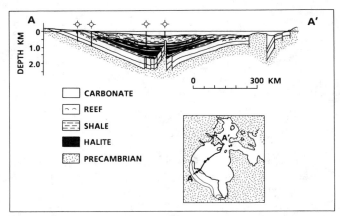

Figure 7. Cross section of the Hudson Bay Basin (after Sanford and Grant, 1990).

unambiguous evidence of the mechanism controlling tectonism.

Hudson Bay Basin is an intracratonic basin on the Canadian Shield (Figs. 2 and 6), with a history of subsidence and uplift similar to that of other Paleozoic basins on the North American Craton (for example, Michigan and Williston basins, Sanford, 1987). The processes that form such basins remain enigmatic in terms of plate-tectonic theory. As summarized by Sloss (1988), these basins appear to evolve, at least in part, independently of activity at the craton margins. I propose that this independent intracratonic behavior is typified by the Hudson Bay Basin, and emphasized by an uplifted block in the center of the basin (Figs. 6 and 7); the saucerlike basin is interrupted by an uplift of about 1 km, approximately 50 km wide and 200 km long. This uplift occurred during the mid-Silurian.

Tectonism in this intracratonic setting (Fig. 7) has occurred at the wavelength of the Hudson Bay Basin itself (hundreds of kilometers), and at the wavelength of the superimposed central uplift (tens of kilometers). The relative time frames of these displacements are respectively in the order of tens of millions of years versus millions of years (Sanford, 1987, fig. 3). Much later, the Hudson Bay region was depressed by Pleistocene ice sheets. Post-glacial rebound from the load of ice, as recorded by raised beaches, amounts to more than 300 m in 8,000 years (Hillaire-Marcel, 1980); this displacement has a time frame of thousands of years. Thus a range of vertical displacements in time and space—in orders of magnitude—is recorded in this intracratonic setting, and these displacements collectively reflect the mechanism that caused them.

Unless there is a physical void beneath the uplift in central Hudson Bay (Fig. 7), there must have been compensation at depth by lateral, fluid movement. A simple mechanism for this uplift involves movement of magmatic fluid, driven by heat from the interior of the Earth. This hypothesized mechanism is illustrated in Figure 8. Loading of the crust by sedimentation causes lateral displacement of subcrustal magmatic fluid, allowing subsidence (8a). In finer detail, fluid emplacement likewise compensates for erosional unloading of the source areas for the sediment. The crust is depressed into a regime of higher temperature and pressure (8b), which causes thermal, chemical, and physical alterations that condition this zone to undergo uplift (8c). This uplift is unstable; there will be erosion of the uplifted basin sediments, retrogression of the subsurface anomaly, and eventual formation of a successor basin.

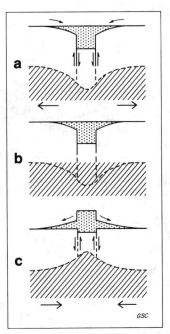

Figure 8. Cartoon illustrating mechanism of diastrophism. Interval between stipple above and crosshatch below corresponds to crust. Explanation in text.

This cycle of recurrent subsidence and uplift constitutes inversion. There are no dimensions on Figure 8, because these processes occur at various scales. This mechanism can accommodate the variety of displacements observed in the Hudson Bay Basin (Fig. 7) and elsewhere, and is attractive because of energy considerations. Conceptually, lateral movements of lithospheric plates (? 100 km thick), with attendant collision, subduction, and obduction, are not energy efficient. In contrast, a mechanism of fluid-impelled vertical displacement, with ancilliary horizontal movement, requires minimal energy expenditure.

The energy efficiency of this mechanism lies in the interaction of external and internal processes, by which loading and unloading of the crust by sedimentation and erosion (Fig. 8) modulate the movement of heat-driven magmatic fluid. The driving mechanism is thus sensitive to atmospheric effects insofar as climate controls erosion and sedimentation. Crustal depression and rebound with waxing and waning ice sheets are a good illustration of this interdependance in a short time frame.

In principle this mechanism is very simple, and not new, but in detail it is infinitely complex because its component processes have been interacting for billions of years. The concept perhaps demanding fresh con-

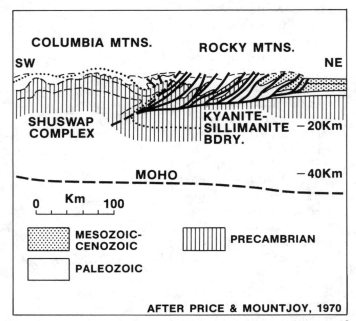

Figure 9. Diagram of large scale inversion in the Cordillera, with formation of mountains and thrust sheets.

sideration is that of energy efficiency; the interaction of internal and external processes reflects a delicate balance in the energy budget of the Earth. This complexity, and energy efficiency, are not recognized in plate-tectonic models.

Other examples

Inversion converts a basinal area into a structural high, or, conversely, a high into a basin. This can occur at lateral scales of several kilometers to several hundred kilometers. The cross section of Hudson Bay (Fig. 7) illustrates gentle inversion processes in an intracratonic setting. The uplifted block in central Hudson Bay is situated in the previous depocenter of the basin.

Inversion processes may be accompanied by magmatic intrusion and extrusion, and in extreme inversion there may be uplift and expulsion of altered sediments, volcanics, and crustal material. The uplift and spreading of this altered mass may cause lateral thin-skinned tectonism. This process was summarized by Price and Mountjoy (1970), who proposed a model in which the thrust sheets of the Rocky Mountains are the result of large-scale bouyant upwelling and lateral spreading of hot mobile rocks from the infrastructure of the Western Cordillera (Fig. 9).

The remnants of the Labrador geosyncline (Fig. 10) provide a Precambrian example of inversion (Dimroth, 1970), structurally similar to that illustrated for the cordillera (Fig. 9). The igneous component is prominent in this example.

Figures 9 and 10 illustrate the mechanism of Figure 8 operating at geosynclinal scale to produce mountains and overthrusts. By this approach, active versus passive continental margins around the world represent different stages of inversion. Continental margins undergo recurrent inversion, because they are primary zones of sedimentation.

Corollaries of the mechanism

This mechanism has operated in the context of various possible evolutionary trends and stages.

1. The Earth cooled from a molten state; a crust formed by separation of sialic and simatic material from the mantle.
2. The crust eventually stabilized, with segregated continental "nuclei" more or less established in their present relative positions; end of continental drift stage.
3. The early crust experienced the phase of meteoritic bombardment recorded prominently on the moon (e.g., 4.6–3.9 Ga; Grieve, 1980).
4. The continental nuclei, particularly the marginal zones, became the loci of tectonic activity, stimulated by erosion and sedimentation; that is, cycles of inversion.
5. Continental areas progressively grew and stabilized by the accretionary effects of inversion.

DISCUSSION

The following discussion expands briefly on questions related to the Earth's magnetic field, aspects of crustal behavior, and new observations that raise further questions about the applicability of plate-tectonic models.

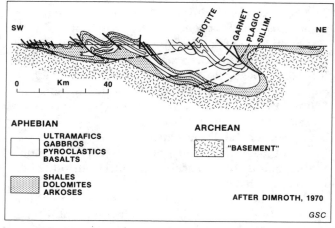

Figure 10. Diagram of large scale inversion in the Precambrian Labrador Geosyncline.

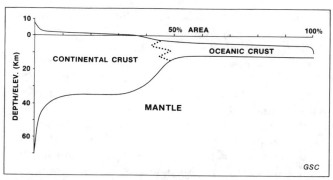

Figure 11. Hypsiographic diagram showing distribution of continental and oceanic elevations relative to sea level, and relative distribution of continental and oceanic crust.

Magnetic anomalies

The primary basis for plate-tectonic reconstructions is the interpretation of marine linear magnetic anomalies as a record of sea-floor spreading. In the alternative mechanism sketched above, this interpretation is not accepted. Rather, these anomalies reflect the long-standing separation of continents and oceans, a dichotomy emphasized and maintained by the mass of seawater. Distribution of the main masses of ocean water and ocean crust are coincident (Fig. 11). The potential of the long-term load of seawater as a tectonic factor may be gauged from the scale of the short-term response of continental crust to waxing and waning ice sheets (water loads), as discussed earlier. Hypsiometrically (Fig. 11), ocean crust lies deeper than continental crust, and geophysical measurements define it as thinner and more dense. Its sensitivity to displacements of underlying fluid is expressed in the depression caused by marginal loads of sediment (especially subduction margins), by the tendency for ocean basins to display a central ridge rift, and in chemical-physical zonation expressed by linear magnetic anomalies. Midocean ridges represent old cracks in the early-formed crust that have been long-standing fissures for egress of heat, conducted by underlying magmatic fluid. Fracture systems developed parallel to midocean fissures (Fig. 12), at spacing related to crustal thickness; effusion of basalt through these fracture grids, at differing geomagnetic field polarity, generated linear magnetic anomalies. Reactivation of these fractures is the oceanic form of inversion. The fluctuations of the Earth's magnetic field suggest it is linked to the primary driving mechanism (heat-driven fluid), and likewise is modulated by external processes such as erosion and sedimentation.

There are phases in geological history, by any scenario (e.g, Sheridan, 1987); the most recent phase of basaltic effusion through the floor of the Atlantic began to abate sometime in the Late Jurassic, and a sediment blanket has developed from the margins toward the mid-Atlantic. Like marine magnetic anomalies, paleomagnetic data from continental rocks also are erroneously interpreted as a record of continental drift. Frequent discrepancies in paleomagnetic interpretations illustrate this fact; for example, McCabe and Elmore (1989) discussed the recent general recognition that many late Paleozoic paleomagnetic poles for North America are unreliable due to remagnetization. They noted that remagnetizations probably also affect orogenic belts other than late Paleozoic ones.

Ancient scars

The persistence of ancient lineaments from continent to ocean (e.g., the Grenville Front to the Cartwright Fracture Zone, Fig. 1) is evidence for early stabilization of continental and oceanic crust. Ancient scars that are arcuate or circular, in various states of preservation, probably record meteorite impacts. Island arcs generally are vestiges of large impact scars that have controlled subsequent patterns of deposition and tectonism. Proximity to a source area is necessary to provide the erosional or volcanic products to register and propagate these arcuate zones, by the processes illustrated in Figure 8. Seamounts may mark sites of smaller impacts, and have been important loci for egress of heat. Because ancient linear scars in the Earth's crust may be important conduits for preferential movement of magmatic fluid, uplift or subsidence may be variable along and across such trends.

The persistence of continent-to-ocean lineaments underlies a basic question about the concept of lithospheric plates: it is not clear how an ocean can form by sea-floor spreading processes, and, for the purpose of plate tectonics, achieve the same lithospheric properties as a continent.

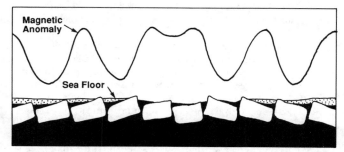

Figure 12. Hypothetical cross section of a midocean ridge showing magnetic anomaly relative to fractures in underlying oceanic crust (white blocks). Stipple represents sediments overlying effusive basalt (black). Discussed in text.

Climatic considerations

Movements of lithospheric plates, particularly north-south displacements, have been used widely to explain global-scale climatic changes. However, with increasing research on climate change, it is increasingly recognized that climatic conditions can alter rapidly relative to plate-tectonic time scales, and irrespective of lateral plate movements. Geologic control on climate may be achieved by vertical crustal displacements altering patterns of atmospheric and oceanic circulation, rather than by shifting landmasses laterally. In this regard, for example, vertical crustal displacements have been considered by Ruddiman and Kutzbach (1989), who demonstrated the importance of uplift as ". . . a key forcing function of late Cenozoic climatic change." Manabe and Broccoli (1990) correlated onset of aridity with uplift of the Rocky Mountains and Tibetian Plateau. Faunal and floral evidence from the Canadian Arctic provide an intriguing example of warm climatic conditions through much of the Tertiary, at latitudes little different from today (Francis and McMillan, 1987).

New data

Crustal studies using deep reflection seismic techniques are revealing complex variability in the seismic character of the crust and Moho. Perhaps the most significant result from this work is the increasing recognition of apparently smooth Moho surfaces beneath regions of complex surface geology (e.g., Brown, 1986). These observations suggest that a displaced Moho may be able to reestablish itself on a relatively short time scale, perhaps on the time scale of inversion processes (Fig. 8). If so, traditional refraction seismic interpretations of continent-ocean crustal relationships, for example, probably are erroneous.

Deep drilling in the oceans (e.g., Becker et al., 1989) is beginning to document the lithologic and structural complexity of oceanic crust. On the continents, results from deep drilling are astonishing; the 12 km hole drilled on the Kola Penninsula (Kozlovsky, 1987) failed to detect the Conrad Discontinuity, encountered much higher temperatures than expected, and found circulating fluids and gasses—including hydrocarbons, throughout. A deep hole started in Germany likewise found much higher temperatures than predicted (Kerr, 1989).

SUMMARY

Nares Strait is a unique place to test the validity of the plate-tectonic theory, and on geological grounds the theory fails the test. Geological and geophysical data in Labrador Sea and Baffin Bay do not support a plate-tectonic interpretation there, and highlight weaknesses in the plate-tectonic theory.

Phanerozoic displacements of the intracratonic Hudson Bay Basin indicate that the primary mechanism of diastrophism is heat-driven magmatic fluid. This mechanism achieves egress of heat from the Earth that is manifested by differential vertical displacements of the brittle crust. Erosion and sedimentation modulate this process, which therefore is ultimately regulated by the Earth's climate; a central theme of the mechanism is that the Earth does not waste energy. The action of this driving mechanism can be described in terms of inversion, the concept of recurrent, reactive crustal response to internal and external processes. The mechanism described in this paper can explain all the phenomena attributed to plate-tectonic processes.

CONCLUSION

By the mechanism proposed in this paper, the crust of Earth is not a passive adjunct of lithospheric plates, but is the dynamic, interactive layer of the Earth. New observations in deep drilling, deep seismic profiling, and climatic modeling, for example, are compelling evidence in support of a dynamic crust. An enormous edifice of interpretations has grown around the assumptions of the plate-tectonic theory; it would be unfortunate if its shadow obscures the exciting implications of these new observations.

Examples in this paper are drawn mainly from Labrador Sea, Baffin Bay, Hudson Bay, and environs. If this region is a representative product of the diastrophic processes that make the Earth work, then the problems that can be identified there with plate-tectonic theory may apply on a global scale.

REFERENCES CITED

Becker, K., Saki, H., et al., 1989, Drilling deep into young ocean crust at hole 504B, Costa Rica Rift: Reviews of Geophysics, v. 27, p. 79-102.

Brown, L.D., 1986, Continents—From accretion to extension: New results from COCORP deep seismic profiling: Geophysics, v. 51, p. 442.

Clarke, D.B., Cameron, B.I., Muecke, G.K. and Bates, J.L., 1989, Petrology and geochemistry of basalts from ODP Leg 105, Hole 647A, Labrador Sea and the Davis Strait area, in Srivastava, S.P., Arthur, M., Clement, B., et al., 1989, Proceedings of the Ocean Drilling Program: Scientific Results, v. 105, p. 683-884.

Dawes, P.R., 1986, The Nares Strait gravity anomaly and its implications for crustal structure: Discussion: Canadian Journal of Earth Sciences, v. 23, p. 2077-2081.

Dawes, P.R., and Kerr, J.W, 1982, Nares Strait and the drift of Greenland: a conflict in plate tectonics: Medelelser om Gronland, Geoscience 8, 392 p.

Dimroth, E., 1970, Evolution of the Labrador Geosyncline: Geological Society of America Bulletin, v. 81, p. 2717-2742.

Francis, J.E. and McMillan, N.J., 1987, Fossil forests in the far north: GEOS, v. 16, p.6-9.

Gradstein, F.M. and Srivastava, S.P., 1980, Aspects of Cenozoic stratigraphy and paleooceanography of the Labrador Sea and Baffin Bay: Paleogeography, Paleoclimatology, Paleoecology, v. 30, p. 261-295.

Grant, A.C., 1980, Problems with plate tectonics: the Labrador Sea: Bulletin of Canadian Petroleum Geology, v. 28, p. 252-278.

Grieve, R.A.F., 1980, Impact bombardment and its role in proto-continental growth on the early Earth: Precambrian Research, v. 10, p. 217-247.

Higgins, A.K. and Soper, N.J., 1989, Short paper: Nares Strait was not a Cenozoic plate boundary: Journal of the Geological Society, London, v. 146, p. 913-916.

Hillaire-Marcel, C., 1980, Multiple component postglacial emergence, eastern Hudson Bay, Canada, in Morner, N.A., ed., Earth rheology, isostasy and eustasy: p. 215-230.

Hood, P.J., Bower, M.E., Hardwick, C.D. and Tesky, D.J., 1985, Direct Geophysical evidence for displacement along Nares Strait (Canada-Greenland) from low-level aeromagnetic data: a progress report, in Current Research, Part A: Geological Survey of Canada, Paper 85-1A, p. 517-522.

Hyndman, R.D., 1973, Evolution of the Labrador Sea: Canadian Journal of Earth Sciences, v. 10, p. 637-644.

Jackson, H.R. and Koppen, L., 1985, The Nares Strait gravity anomaly and its implications for crustal structure: Canadian Journal of Earth Sciences, v. 22, p. 1322-1328.

Kerr, J.W., 1967, A submerged continental remnant beneath the Labrador Sea: Earth and Planetary Science Letters, v. 1, p. 283-289.

Kerr, R.A., 1989, Deep holes yielding geoscience surprises: Science, v. 245, p. 468-470.

Kozlovsky, Ye.A. (editor), 1987, The superdeep well of the Kola Penninsula: Springer-Verlag, 558 p.

Manabe, S. and Broccoli, A.J., 1990, Mountains and arid climates of middle latitudes: Science, v. 247, p. 192-195.

McCabe, C. and Elmore, R.D., 1989, The occurrence and origin of late Paleozoic remagnetization in the sedimentary rocks of North America: Reviews of Geophysics, v. 27, p. 471-494.

Meyerhoff, A.A., and Meyerhoff, H.A., 1972, "The new global tectonics": age of linear magnetic anomalies of ocean basins: American Association of Petroleum Geologists Bulletin, v. 56, p. 337-359.

Price. R.A. and Mountjoy, E.W., 1970, Geologic structure of the Canadian Rocky Mountains between Bow and Athabasca rivers—a progress report, in Wheeler, J.O., ed., Structure of the southern Cordillera: Geological Association of Canada Special Paper No. 6, p. 8-25.

Roddick, J.C., 1989, K-Ar dating of basalts from site 647, ODP Leg 105, in Srivastava, S.P., Arthur, M., Clement, B., et al., 1989, Proceedings of the Ocean Drilling Program: Scientific Results, v. 105, p. 885-887.

Roest, W.R., and Srivastava, S.P., 1989, Sea-floor spreading in the Labrador Sea: a new reconstruction: Geology, v. 17, p. 1000-1003.

Ruddiman, W.F., and Kutzbach, J.E., 1989, Effects of plateau uplift on Late Cenozoic climate: EOS, Transactions American Geophysical Union, v. 70, p.294.

Sanford, B.V., 1987, Paleozoic geology of the Hudson platform, in Beaumont, C., and Tankard, A.J., eds., Sedimentary basins and Basin-forming mechanisms: Canadian Society of Petroleum Geologists, Memoir 12, p. 483-505.

Sanford, B.V. and Grant, A.C., 1990, New findings relating to the stratigraphy and structure of the Hudson Platform, in Current Research, Part B: Geological Survey of Canada, Paper 90-1B, p. 17-30.

Sclater, J.G., Anderson, R.N. and Bell, M.L., 1971, Elevation of ridges and evolution of the central Eastern Pacific: Journal of Geophysical Research, v. 76, p. 7888-7915.

Sheridan, R.E., 1987, Pulsation tectonics as the control of continental breakup: Tectonophysics, v. 143, p. 59-73.

Sloss, L.L., 1988, Conclusions, in Sloss, L.L., ed., Sedimentary Cover—North American Craton: the geology of North America, v. D-2, p. 493-496.

Srivastava, S.P. and Arthur, M.A., 1989, Tectonic evolution of the Labrador Sea and Baffin Bay: constraints imposed by regional geophysics and drilling results from Leg 105, in Srivastava, S.P., Arthur, M. and Clement, B., et al., 1989, Proceedings of the Ocean Drilling Program: Scientific Results, v. 105, p. 989-1009.

ACKNOWLEDGMENTS

I thank K.D. McAlpine, K.M. Storetvedt, J.A. Wade, and an anonymous reviewer for helpful critique of this paper. I take full responsibility for the content. The work of the late H.A. Meyerhoff, and of A.A. Meyerhoff, is a longstanding source of encouragement to persist in trying to figure out how the Earth works.

Geological Survey of Canada Contribution No. 12490

The tectonic framework of Australia

Vadim Anfiloff, Geo Process, P.O. Box 774, Canberra City, ACT 2601, Australia

"My concept of contraction calls for a shrinking globe, with crust fragmented and subsiding differentially" (Landes, 1952).

"Rather does it seem that the primitive (Australian) continental mass has fractured both marginally and internally. One does, in fact, naively wonder what holds the cracked and fractured mass together?" (Hills, 1956).

"Sometimes the adjacent blocks are so dispersed that they carry all of the crustal compression; under these circumstances, a block so freed from lateral compression reacts to the sum of the vertical forces to which it is subjected" (Moody, 1966).

"A system supporting load through stress concentrating blocks is likely to occur in the earth's crust" (Koshtak, 1971).

"The lineaments and fractures... break the crystalline crust into a myriad of interconnected crustal blocks" (Gay, 1973).

"Assume that the Earth has a network of deep channels, or zones of higher penetrability, along which the deep masses bringing heat rise to the surface" (Beloussov, 1990).

ABSTRACT

The tectonic framework of Australia is analyzed using gravity and other geophysical data. It consists of the Cardinal System of vertical crustal fractures, rectilinear systems of compartmented rifts, bifurcating networks of basement ridges, chelogenic uplifts (cratons), and chelogenic downwarps (broad basins). Separately and together, these components imply that fixistic processes have and continue to dominate tectonism in the continent. The coherent orthogonal gravity framework suggests an intact, ancient continent, which has not been shortened, buckled, or stretched, nor would this be expected in a brittle, highly fractured, continental crust.

The mysterious forces that produce pervasive horizontal compression in diverse directions in Australia also may be the forces needed to explain pervasive rifting along the Cardinal Fracture System. Bifurcating basement ridges compartmentalize and envelop rifts, and their main purpose may be to transmit the compression. Such ridges seem to occur in strategic places all over the globe, and may be part of a global load-bearing skeleton. A contracting Earth seems to be the logical source of sustained compression.

It is envisaged that compression is transmitted around the globe through the continuous strong seismogenic upper crust, but is confined to the load-bearing skeleton. When part of this network fails it causes wrenching movements and triggers synchronous global rifting and earthquakes, some of which are associated with vertical movements resulting from isostatic adjustments. If this concept is correct, then maximum earthquake energy is propagated along basement ridges.

A model of intracratonic tectonism is proposed in which three systematic processes and one random process occur. One systematic process is rifting, in which compression wrenches blocks of crust apart, creating narrow deep gaps that translate upwards into much wider rift troughs. Another is "toothpaste tectonism," the movement of molten igneous rocks up such gaps into rift compartments in the seismogenic upper crust. The third is the chanelling of compression in this seismogenic layer, along the load-bearing skeleton, which preferentially includes cratonized igneous bodies in old rifts, which act as stress guides.

The random process is the local chelogenic heat engine driven by large bubbles of hot-mantle material lodging under the lithosphere. In the chelogenic cycle, giant pistons of crust about 1000 km in diameter move up and down, and rifting happens on the downstroke, but is synchronized globally by sudden changes or adjustments in the global compression regime. It is proposed that rifting is a discrete, pervasive process operating in four-dimensional space-time, and perhaps facilitated by crustal thinning during chelogenic subsidence.

When a bubble of hot-mantle material lodges under the lithosphere, it first causes chelogenic subsidence, followed eventually by uplift when the heat is dissipated. As chelogens move up and down, some rifts and ridges stack on top of each other in successive cycles. The "in-phase stacking" (polycyclic rifting) makes the framework more prominent, and underlines the stability and "fixism" of the system. Without this, the framework would be less pronounced, and more complex after two chelogenic cycles. It also suggests the paleostress regime has not changed markedly.

Eventually a chelogen is elevated 1–2 km above sea level and cratonized, and erosion uncovers the roots of rift-mobile belts. The underplated root at the base of the lithosphere, which maintains a craton in an elevated position, may also immobilize fractures by strengthening the lithosphere. The removal of the root during subsidence could explain why rifting appears to be predicated to subsidence—crustal blocks can then be jostled more easily.

INTRODUCTION

This paper mainly deals with steep (subvertical) crustal fractures, rifting, and compression. Global compression can be suspected from various phenomena; the main observation used here is that narrow strips of elevated crust commonly envelop wider areas of lower crust. Some notable examples are the ridge connecting Gibraltar with North Africa, which terminates the Mediterranean Sea; the slender connection between North and South America at Panama; and the two basement ridges connecting the island of Tasmania to the Australian mainland (Fig. 1). Tasmania is supposed to be on the trailing edge of the Australian plate, yet horizontal compression is being propagated between it and the mainland (Denham et al., 1979). Anfiloff (1988a) proposed this compression is transmitted selectively along the basement ridges, and that is their main function.

The role of the Isthmus of Panama is especially interesting. If the huge landmasses of North and South America were being driven by conveyor belts, why would this slender connection persist? It is like connecting a strand of cotton between two bulls and expecting it not to break as they roam over a paddock. Moreover, the flanking trench on the western side of the Isthmus is reminiscent of the basement ridge-rift situation discussed throughout this paper. These slender connections are taken here to represent remnants of continental

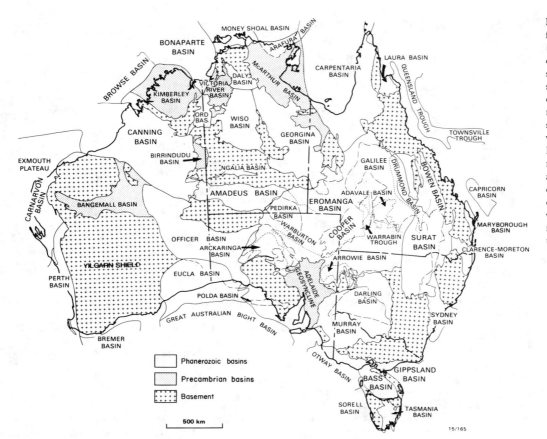

Figure 1. The geological framework of Australia (BMR, 1989a) reveals a checkerboard pattern of subsidence and nonsubsidence during the Phanerozoic. Narrow basement ridges divide many areas of subsidence. The two ridges, which surround the Bass Basin and connect Tasmania with the mainland, are an important example of how ridges envelop pockets of subsidence. Their very presence suggests their role is to transmit horizontal compression.

crust situated in areas of dominant crustal subsidence, and propped up by weak, persistent compression caused by a contracting Earth.

The tectonic framework of Australia comprises the largest structural elements that can be recognized systematically over the whole continent. Australian gravity (Anfiloff et al., 1976; BMR, 1976) reveals rectilinear-orthogonal networks of gravity lows, clusters of parallel anomalies (Fig. 2), and bifurcating networks of gravity highs (Fig. 3). But gravity usually does not reveal the broad epeirogenic movements of crust that topography reveals (Fig. 3).

Most broad depressions fill with sediments and sink more, and some eventually become deep basins with many kilometers of sediments. The problem is to determine the initial outlines of a basin, and this requires mapping the original extent of its sediments. But unfortunately, this is rarely clear, because of later tectonic movements.

One of the main philosophical problems in geology is whether past processes can be guaged from present indicators. Plate tectonics has caused a major schizm by introducing a completely different style of macroprocesses for the Phanerozoic, when plates supposedly began to move. In particular, the concept that sag basins form after failed continental rifting has been a major distraction, and it is argued that although this concept has evolved and diversified, it does not seem to fit the facts.

Hills (1956) was a major pioneer in Australian framework studies, and developed a simple fixistic, and as it turns out, accurate view, without the aid of geophysics. But it seems the effect of massive increases in the geophysical coverage in the next two decades were nullified by the advent of plate tectonics, and the ensuing model-driven approach to analysis. There are several major examples: (1) the accretion interpretation of Wellman (1976), in which the orthogonal gravity framework was discarded in favor of the parallel framework, (2) the detachment rifting concept of Etheridge et al. (1987), where there is a major discrepancy between listric faulting and the lack of curved gravity features, (3) the failed breakup rift model of Gunn (1984) in which massive mantle intrusion into the crust would produce impossibly large gravity highs, and (4) numerous studies of the seismic data in the Eromanga Basin region have failed to recognize vertical faults associated with simple monoclines.

Hills used the term "framed" sedimentary basin to emphasise that broad basins often have polygonal outlines. This does not in itself mean that basins were

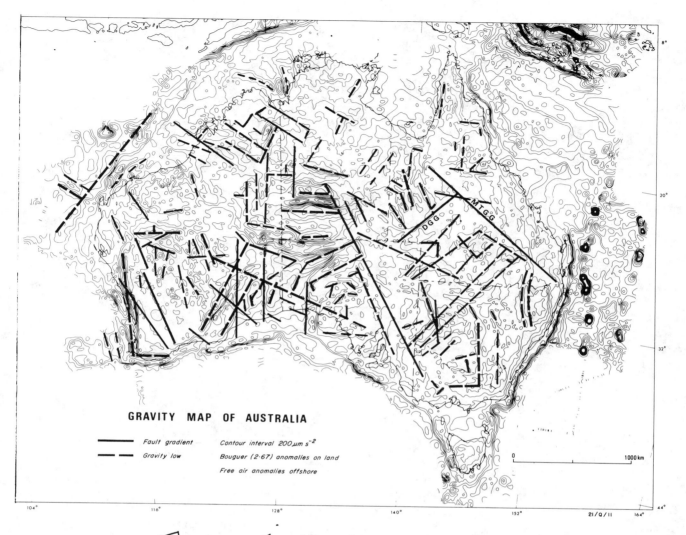

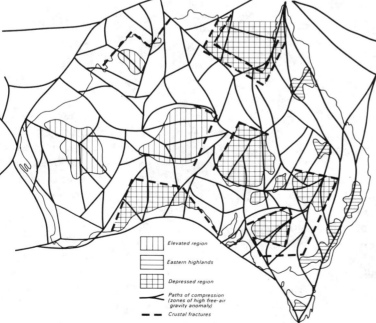

Figure 2. The Gravity Map of Australia reveals a rectilinear pattern of gravity lows and fault anomalies consistent with the Cardinal Fracture System of vertical crustal fractures. The lows suggest rifting has occurred along such fractures.

Figure 3. Polygonal chelogenic uplifts and depressions are revealed by elevation data (digital terrain model) associated with Australian gravity (BMR, 1984) and show the influence of ancient fractures. The Eastern Highlands may be remnants of chelogenic uplift similar to central Australia. Free-air gravity highs form a bifurcating network, which in many cases represents basement ridges cutting through depressions and separating them from each other. The network is consistent with the load-bearing skeleton expected for the transmission of horizontal compression applied radially to the continent.

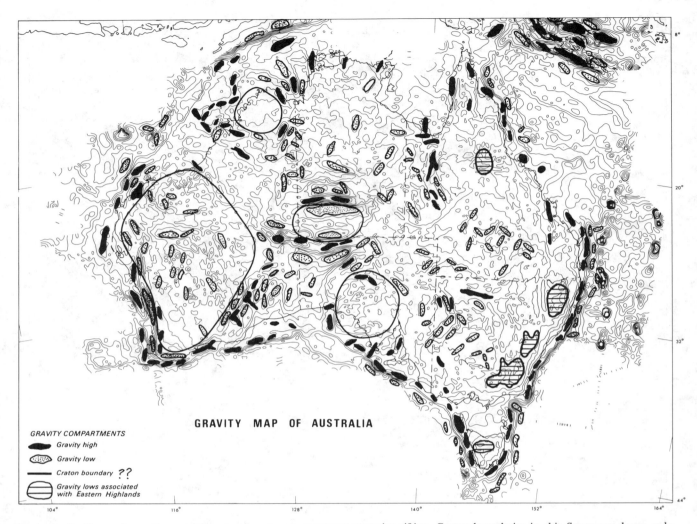

Figure 4. Positive and negative gravity compartments may represent pervasive rifting. Craton boundaries in this figure are drawn only approximately to demonstrate the tendency for exclusion from cratons of positive anomalies associated with mafic rifts. This is relevant to active craton boundaries during the Archaean and Proterozoic, and current boundaries may be different. Negative compartments represent sediments and granites, and have smaller amplitudes in older cratons. The Papunya mafic rift compartments are just north of the central Australian Proterozoic craton. The tendency for mafic compartments to cluster around the edges of cratons, and around the margins of the continent suggests cratons are impervious to lateral penetration of high level underplating fluids originating from adjacent areas of chelogenic subsidence, as shown in Figure 18.

necessarily framed by fractures as they formed, but the fact that most current topographic depressions are framed (Fig. 3), makes this highly probable. The Gulf of Carpentaria (Fig. 1) is distinctly framed (Fig. 3), and even though the Carpentaria Basin extends beyond this frame (Fig. 1), recent sedimentation there probably has occurred within what seems to be a fracture-controlled polygon. Most of the Eucla Basin sediments are located within the framed topographic depression shown in Figure 3, and the Eromanga Basin is also largely framed by the orthogonal gravity system DGG-MTGG (Fig. 2). Hills (1956) demonstrated that the Murray Basin is framed and new sediments are currently accumulating there.

Cratons are represented by broad topographic highs (Figs. 3 and 4), and sometimes, as in the case of Kimberley (Basin) craton (Figs. 1 and 3), they are capped by flat-lying sediments that are remnants of ancient basins. This indicates a large portion of crust has moved down and up. Thus, the formation of broad basins and cratons can be seen as the up and down movement of huge pistons of crust, up to 1000 km in diameter. The broad topographic bulges and depressions produced by this process are referred to in this paper as chelogens, produced by a thermal chelgenic cycle.

Figure 4 shows the approximate outlines of several cratons, and shows how an old craton, such as the Archaean Yilgarn Shield, tends to have smaller gravity anomalies than the Proterozoic craton in central Australia. These relationships also can be observed in elevation and gravity profiles across Australia (Anfiloff, 1982a). The elevation data used in Figure 3 are a by-product of the reconnaissance gravity survey of Australia, and thus, the gravity coverage has supplied the bulk of the data needed to analyze the framework.

This paper also makes use of new high-resolution magnetic pixel maps (Tucker et al., 1985; Anfiloff and Luyendyk, 1986), and seismic data. Appendices A and B contain two framework studies based on geophysical data in the Eromanga Basin, and Murray Basin. Seismic data are indispensible, both for mapping parts of the framework, and to help constrain gravity and magnetics. But it is important to carry out combined analysis of the three data sets only in areas where tectonism has been fairly uncomplicated, otherwise the results can be inconclusive and discouraging.

In the complex central Australian region, for example, early work on major gravity lows attributed them partly to granites (Anfiloff and Shaw, 1973), but subsequent interpretations did not (Mathur, 1976; Wellman, 1978). Many years elapsed without any sign of convergence on a common solution, which may have prompted Lambeck (1984) to conclude that the gravity is "an embarrassment" and "uninteresting and uninformative." Yet gravity is undoubtedly the principal means of mapping the framework, and granites must feature in the explanation of many gravity lows.

Australian gravity reveals a coherent rectilinear-orthogonal network of faults, troughs, and mobile belts (Fig. 2), which suggests the Cardinal Fracture System of Gay, (1973). Many years ago, Hills (1956) found morphological evidence for a similar set of fractures over much of the continent, and concluded that "the Australian block is everywhere underlain by Precambrian basement" and "the concept of the addition of successive folded belts to a proto-Australian neucleus disappears." He was in effect, arguing for an intact continent, based on fractures that could not be proved to have sufficient antiquity. However, the orthogonal gravity framework is produced by structures whose age can be estimated, and it is clear that most of the framework was in place in the Proterozoic, and there is no doubt about the antiquity of the Cardinal Fracture System in Australia.

Granites and sediments are the only confirmed causes of major elongate gravity lows, and produce identical anomalies. Anfiloff (1983) attributed the orthogonal framework of gravity lows to rifting and granite emplacement along ancient fractures in an intact ancient continent. There are also groups of parallel anomalies, which are referred to here as parallel gravity clusters, but attributing these to the accretionary docking of platelets (Wellman 1976, 1984) produces a major conflict—a continent cannot be both accreted and intact, based on the same gravity data, and moreover, one implies fixism, and the other mobilism. The orthogonal and parallel frameworks cannot be equally fundamental, and significantly, Wellman (1988) located many more orthogonal features previously not recognized by Wellman (1976, 1984), and accretion is no longer viable in terms of gravity trends.

The intactness of Australia generally is now being accepted for various reasons (Etheridge et al., 1987), and tectonism can be reduced to two conflicting intracratonic concepts; compressive crustal buckling (Lambeck, 1984), and detachment-rifting produced by small-scale mantle convection (Etheridge et al., 1987). Ironically, these concepts do not acknowledge the vertical crustal fractures that Hills (1956) originally used to postulate the intactness. It seems that most current models are based on the premise that fractures formed late rather than early, but Bevis and Gilbert (1990) concluded that the Cardinal Fracture System "will place major new constraints on models of a wide range of tectonic processes." Anfiloff (1989) noted that vertical crustal fractures and the lack of curved gravity features particularly mitigate against listric faulting as a primary crustal process in detachment models.

The recent finding against large strike-slip movements on the San Andreas Fault (Martin, 1989) also is supported by high-resolution airborne magnetics in Australia. Although some strike-slip movement is essential to produce many types of structures, the largest unambiguous offset detected using magnetics is a movement of about 15 km near Esperance in the southwestern corner of Australia (Tucker et al., 1986a). That is an exceptional occurrence, and in many regions, movements rarely exceed 1–2 km, which implies a high degree of fixism.

The crust seems to be in a state of horizontal compression (Denham et al., 1979), but the compression is not propagated evenly. There are simultaneously diverse directions of principal stress over the continent and the source of this compression is a mystery (Denham, 1988). Such compression favors neither the detachment rifting, nor folding models. Moreover, vertical crustal fractures provide a means of explaining variations in compression directions. Anfiloff (1988a) proposed that rifting could be attributed to wrenching associated with horizontal compression rather than extensional processes. This then explains the question posed by Hills (1956): why a pervasively fractured continent should remain intact, rather than be dispersed.

The Goulburn Graben (Fig. 5) has been analyzed in detail (Bradshaw et al., 1990), and holds great promise

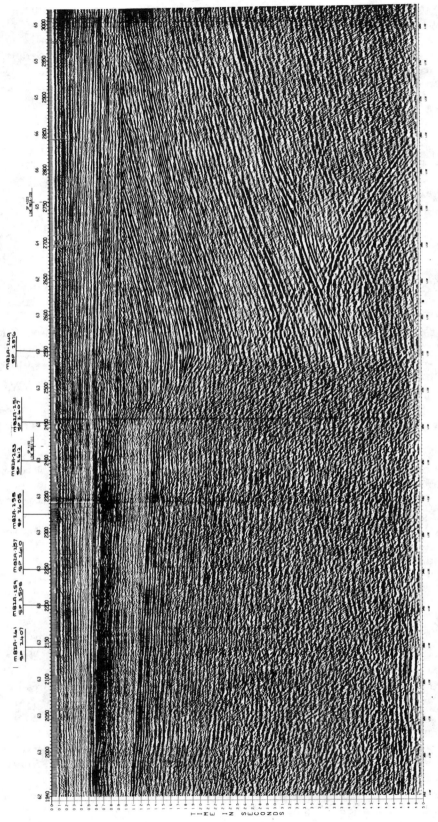

as the missing link in the rifting puzzle. It shows there have been two cycles of chelogenic subsidence separated by uplift and erosion. The rift contains about 10 km of upper Proterozoic platform cover, and is bounded by a near-vertical crustal fracture, which in this case can be traced in the seismic section to a depth of about 13 km. This supports all the main aspects of the new polycyclic rifting model in which rifting utilizes ancient vertical crustal fractures, and follows regional subsidence (Anfiloff, 1988a, fig. 7).

The Proterozoic intracratonic Toko Syncline (Fig. 6) is the same type of structure, but its associated gravity low is much bigger than required to explain the sediments (Harrison et al., 1980). This points to an underlying body in the basement, probably a granite, and is an example of polycyclic rifting—the vertical stacking of a rift directly over an older structure—a process that cannot be explained in terms of detachment rifting.

These two examples also help establish what seems to be a key feature of rifting: that the base of a rift is linear, not curved. This, together with the steep bounding fault, gives the rift a triangular cross section. But

Figure 5. This seismic section to 5.0 seconds shows one of two compartmented parallel troughs associated with the Goulburn Graben (Bradshaw et al., 1990), situated beneath the Arafura Basin (Fig. 1). It is 30 km wide, has a vertical bounding fault, and contains over 10 km of Upper Proterozoic platform cover. This structure is interpreted here as a pristine rift, the type that can explain the orthogonal framework of gravity lows in Australia. It shows that rifting follows regional subsidence and is a passive process in which a whole block of basement subsides on only one side of the fault to form an asymmetrical trough. The asymmetry cannot be explained without invoking an additional process, and it is proposed that a basement ridge once flanked the rift, supporting the crust on that side, but was later eroded. The seismic data indicate two cycles of chelogenic subsidence separated by uplift and erosion.

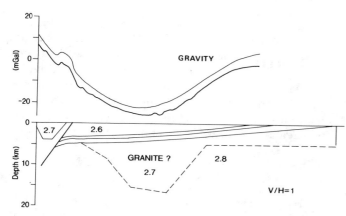

Figure 6. Seismic and gravity data over the Proterozoic Toko Syncline (after Harrison, 1980), situated beneath the Georgina Basin (Fig. 1), show a classic rift with a linear bottom. The main bounding fault is a thrust and has probably been wrenched. The gravity model indicates an underlying deeper trough or granite, consistent with an older rift. This demonstrates the concept of polycyclic rifting in which rifts are stacked on top of each other coaxially to produce major gravity lows.

the steep bounding fault poses the critical problem of why the crust on the other side of it did not also subside. Clearly, that crust is not subjected to the same forces, and here the mystery compression can play a major role. In Australia, rifts and compression both have diverse orientations, and the key question is whether compression acts parallel to, or across the axis of a rift. In this paper, the polycyclic-rifting-wrenching concept of Anfiloff (1988a) is taken to its logical conclusion.

THE CHELOGENIC CYCLE

The chelogenic process is defined as "long-term cycles leading ultimately to the production of shields" (Sutton, 1963). In applying this to Australia on the basis of the cratonization of Archean, Proterozoic, and Phanerozoic rocks over the continent, Rutland (1982) envisaged global thermal events in which whole continents subside, emerge, and all the new sediments formed in that cycle are cratonized at a particular time. The concept is used here on a smaller scale, with local thermal events acting more or less at random causing cycles of subsidence and uplift. Grant (1989) described a similar process in Canada, and Etheridge et al. (1987) and Middleton (1989) recognized the need for local intracratonic thermal cycles in Australia.

The term chelogen is used to denote both the thermal process and the very broad uplifts and downwarps produced in the basement surface by it. Australia is an excellent continent to study the process, because in comparison with most other continents it is currently tectonically flat and inert. The topography reveals various chelogenic depressions ranging in diameter from 500 to 1,000 km (Fig. 3), and presumably, broad interior sag basins developed in the past in the same way. The depressions are roughly circular, but also reveal the influence of crustal fractures that have given them polygonal, framed outlines, and this suggests interior sag basins are fault-bounded.

Etheridge et al. (1987) envisaged upwelling convection causes rifting prior to basin sag, and Middleton (1989) envisaged the opposite, downwelling convection to cause compression during sag. Both link basin sag to the deformation of basement under it, but examples such as the Goulburn Graben cited earlier show that thick platform cover can accumulate before deformation. Consequently, the forces that deform the cover later need to be divorced from this process, and there is a need for a mechanism that produces pure sag.

This is where having a fractured crust is important; it can subside passively in large blocks in response to the most subtle, weak forces arising from changing mantle conditions. Likewise, it could rise again later due to very weak uplifting forces, providing erosion removes the sediments. Erosion and sedimentation play a powerful role because their loading becomes a large proportion of the total driving force, and as noted by Grant (1989) climatic conditions can therefore have a powerful influence.

Thus, the chelogenic engine could be driven by a bubble of hot mantle material lodging under the lithosphere, and the deformation process can be attributed to a variety of processes that produce horizontal compression such as plate collision or a contracting Earth.

Stable shields are uplifted cratons, and often contain evidence of sedimentation that must have preceeded their uplift. If uplift did not follow subsidence, the whole continent would be covered by sediments. The so-called Kimberley Basin (Fig. 1) is a good example of a stable craton covered by flat-lying Proterozoic sediments, which must have been elevated and maintained by a vertical force since the Proterozoic. Why did the region subside initially, why did it rise again, how is it maintained at present, and why has it refrained from subsiding over the past 1,000 m.y.?

It is difficult to establish the perimeters of old cratons and basins, particularly as basins are cut by networks of basement ridges. In principle, basins are defined by the flat–platform cover sediments deposited in them, and cratons are generally areas of high topographic relief, characterized by the absence of Phanerozoic sediments, and by the absence of large gravity anomalies. There are of course many fractures in the crust, and the fractures used on the downstroke of the chelogenic piston are not necessarily the ones used on the upstroke. So it could be difficult to define a particular chelogen that has gone through a complete cycle.

Sutton (1963) recognized four main chelogenic cycles over geological time, but local cycles driven by hot bubbles would have happened more often and at random. In eastern Australia, there have been two main cycles of widespread basin formation during the Phanerozoic (Appendix A), separated by at least one major uplift and erosion episode (Veevers et al., 1984). In the elevated central Australian craton, some 10 km of platform cover preserved in the Amadeus and Ngalia troughs indicates the region was a broad basin in the Late Proterozoic, and some older platform cover identified by Shaw et al. (1984) indicates another earlier depression. In the early to mid-Paleozoic, the region was elevated, cratonized, and eroded to expose a cluster of east-west trending troughs and granites associated with large gravity lows (Fig. 4). Figure 3 shows the craton is close to the depression associated with the Eromanga region, and its eastern part probably has subsided into that depression. It seems that part of an old chelogen is being consumed by a new chelogen, but this has had no effect on the rectilinear pattern of gravity anomalies (Fig. 2). Thus, chelogenic cycles are probably a vertical process, driven by a local heat engine.

The framed, roughly circular Eucla Basin depression (Figs. 1, 3) is an example of what can be considered a genuine chelogen, well isolated from other depressions. But in eastern Australia, the situation is more complex; major basement ridges cross broad depressions, making it difficult to determine the size of chelogens. Figure 7 shows two types of relationships between chelogenic circles of subsidence and basement ridges. In 7A, the circles overlap, indicating each develops independently, and in 7B, circles are always separated by basement ridges, indicating they might be part of a much wider zone of subsidence.

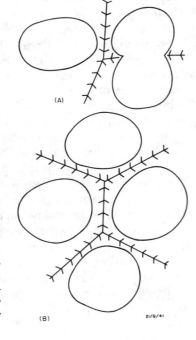

Figure 7. The true extent of chelogenic depressions can be difficult to determine because they are crossed and surrounded by basement ridges. In Case A, the outlines overlap, indicating each is a chelogen. In Case B, they never overlap, thus a much larger chelogen may be cut by active basement ridges into smaller depressions. Compression directed along such ridges would explain the concentrations of seismicity around basins discussed by Dewey (1988).

No examples of case A are observed in Australia, and with only a few examples of case B to analyze in eastern Australia, it is difficult to establish how wide chelogens are there without a detailed knowledge of original sedimenation. It is possible there has been a mega-regional tendency for subsidence during much of the Phanerozoic, culminating in the formation of the Tasman Sea in the Mesozoic by an oceanization process.

PROBLEMS WITH THE RIFT BASIN CONCEPT

There is nothing in Australia to suggest that interior sag basins were produced according to McKenzie's (1978) lithospheric stretching model or variants on that theme. This concept is presumably quite incompatible with vertical crustal fractures, and there are few occurrences of synrift sediments in major troughs. There are no examples of contemporary rifting associated with active thermal domes, and instead, rifting probably is occurring currently in many depressed regions under extensive sediments. Active rifting also may be responsible for the small Lake George drainage basin near Canberra in the passive Eastern Highlands (BMR, 1989b), but this is not a thermal dome situation.

Central Australia is a region dominated by massive gravity anomalies, and continues to be a major enigma. Shaw (1991) discussed various compressional and extensional models for this region, but it is difficult to reconcile crustal extension with the overwhelming evidence for compression there, and this has resulted in highly dubious proposals for oscillating divergence and convergence. The convergence is supposed to buckle and shorten the crust in the center of the continent, but this would require enormous compressive forces. If they originate outside the continent, such forces probably would be dissipated in the weaker oceanic crust before they could be brought to bear on the continent, and the residual forces would be dissipated by vertical faults (Artyushkov, 1973).

There is full agreement with Shaw et al.'s (1984) interpretation of the Papunya Mafic Belt in central Australia (Fig. 4) in terms of a very large upper crustal rift filling with a huge volume of mafic volcanics at about 1,800 m.y. There are two important aspects of the gravity highs associated with this belt: (1) the belt produces two very prominent positive gravity compartments (Fig. 4), and (2) the size of those highs is insufficient to represent a deep axial rift in which a large plug of mantle moves up into a rifted crust, as proposed by Gunn (1984). A viable model of a rifted extensional basin that satisfies gravity has not been developed in Australia, and central Australia is too complex for developing this or any other model.

It is important to recognize the similarities in gravity pattern between central Australia and other parts of the continent, and to develop a lowest-common-denominator approach fo finding a general mechanism. The highly compartmented Fraser Range Mobile Belt, where mafic compartments fringe the southeastern side of the Yilgarn Shield (Fig. 4) is a key area (Anfiloff and Shaw, 1973). There, granites flank the Fraser Complex mafic belt, and granites and sediment troughs therefore must be interchangeable in the gravity "dipole" situation. Moreover, in central Australia, the Ngalia Trough is located on top of granites, and in fact, granites seem to occur at a range of depths under troughs, and may account for broad magnetic highs often being coaxial with gravity lows (Appendices A and B, Anfiloff, 1984). This is evidence that rifting is associated with vertical crustal fractures.

The post-thermal sag model seems to be responsible for the assumption that any major trough containing platform cover could not be a rift—it has to be a crustal fold. Zorin and Lepina (1989) accepted the rift basin concept, but questioned crustal stretching, and proposed that subsidence is sometimes caused by post-thermal densification in the lithosphere. They also acknowledged that major crustal thinning can occur, citing a case where 22 km of subsidence is accompanied by thinning of the crust to a thickness of only 8 km, and this implies dominant vertical movement. Given the fact that the broad basins shown in Figure 1 cannot be distinguished in the gravity trend pattern in Figure 2, it seems the interior sag process does not disrupt the gravity framework, and there seems to be a need to explain this in terms of simple subsidence associated with a deep thermal chelogenic process.

The alternating extension-compression model of Etheridge et al. (1987), involves a local chelogenic-type process within an intact Australian continent. A portion of the continent is domed and extended by upwelling convection, and a group of detachment rifts are formed (resulting in a parallel gravity cluster). Later, the region cools, subsides, and is covered by post-thermal sag sediments, and a compressional phase deforms them. These systematic processes in the upper crust are driven by mantle heat trapped under the lithosphere, but the heat engine would not always behave itself, and when the amount of heat is excessive, this process must progress to some form of oceanization, when continental crust is destroyed and assimilated back into the mantle.

It is extremely difficult to accept models based on alternating extension and compression on a continental or semicontinental scale. The underlying assumption that plates collide is a reasonable explanation for episodic compression, but not for sustained compression to maintain crustal buckling. There must be a major fallacy in the rift basin concept, and it seems to revolve around synrift sediments. The model proposed by Roberts et al. (1990a) demonstrates the problem: synrift sediments are covered by platform cover, and a collisional event then deforms the trough, reversing the angle on the listric fault. The synrift sediments end up at the bottom of the trough, highly deformed, and therein lies the dilemma of the rift basin concept: the evidence for extension and listric faulting is buried and destroyed by the later compression.

The Wernicke (1981) detachment model evolved in the "thick-skinned" Basin and Range Province, where it is difficult to separate the primary framework from the exotic secondary framework. The major low-angle faults there probably can be explained in terms of local compression and tension associated with wrenching and gravity sliding from local uplifted domes, and chelogenic uplifts. The situation in the Eromanga Basin region is much simpler (Appendix A): there are only a few kilometers of relatively undisturbed platform sediments covering a simple, coherent, rectilinear framework of mobile belts in basement. Nevertheless, it is still necessary to separate the primary and secondary frameworks, and without detailed gravity lines, a framework study there, based on seismic lines alone, would be futile.

The geometry of detachment rifting is not reflected in Australian gravity. Etheridge et al. (1988) proposed a 3-D compartmented rift geometry that is not supported by either seismic or gravity data, and one of the main problems is that compartments are not caused by transfer faults, but by basement ridges, which Rosendahl (1987) called accommodation zones. Moreover, there seems to be no evidence for transfer faults in magnetic data, or for much transverse movement generally. Rift systems in east Africa form a rectilinear network, and each main rift hundreds of kilometers long consists of several compartments separated by narrow strips of basement (Rosendahl, 1987). This pattern is clear in Australian gravity, and led to the recognition of pervasive rifting (Anfiloff 1983, 1988a).

Detachment models assume that crust is a homogeneous layer that breaks along large conchoidal listric faults during rifting, but gravity anomalies indicate innumerable bodies in the upper 20 km of crust (Anfiloff, 1982a). If primordial crust were ever a uniform layer, then at that time, the Cardinal Fracture System would have developed in response to vertical forces that produced no horizontal offsets (Gay, 1973). Listric faults can hardly form in crust already cut by ancient fractures, and a recent study by Roberts et al. (1990b) questioned the existence of intracratonic listric faults, stating that they are confined to "the sediment fill of an extended basin," and are likely to be found in situations such as the "unbuttressed sediment pile of a continental margin." Similar concerns were raised by Anfiloff (1988a, 1989).

A new approach is needed to the rifting problem, in which rifting follows regional subsidence. This would explain the many occurrences of platform cover in asymmetrical troughs, and there would be no need for oscillatory opening and closing episodes in continental tectonism. Middleton (1989) recognized the need for an intracratonic mechanism to raise and lower the crust in cycles, and to have horizontal compression during subsidence. His convective downwelling model is based on the need to explain a basin in terms of "a sag which experienced post-depositional reverse faulting." This can be taken several steps further: basement can be jostled and wrenched to produce rifts under platform cover (Anfiloff, 1988a).

THE TRUE CROSS-SECTION OF A RIFT

The Goulburn Graben (Fig. 5) has profound implications: it may hold the key to the rifting mechanism because it clearly is not caused by a listric fault, and therefore can explain the rectilinear network of gravity lows in Australia. It is also consistent with intracratonic structures, such as the Toko Syncline (Fig. 6), whose associated gravity low suggests rift stacking. But primarily these examples show that the cross-sectional shape of a true rift in pristine condition is triangular. Its bottom is fairly linear, and dips down towards the steep bounding fault. This suggests the upper crust acts like a rigid beam and subsides intact when it is undermined, and that rifting is a cold, passive process.

A steep main bounding fault poses the very important problem of why the rift is assymetrical—what supports the crust on the other side of the trough? The answer seems to be that initially, a ridge flanks the trough, but is later removed by erosion. Buck (1986) also described rifting as a passive process, and suggested the flanking ridge is pushed up by local convection created by the rifting process itself. This is a major step in the right direction, away from the popular idea, associated with the rift basin concept, that rifting is driven by grand convection. The need to explain the basement ridge is paramount. In this paper a different explanation is proposed, in which basement ridges are caused by horizontal compression being channelled along them. It identifies basement ridges as a major component in the tectonic framework, and stands or falls on whether basement ridges can be mapped as bifurcating systems (Appendix A).

From the description of the Goulburn Graben given by Bradshaw et al. (1990), the structure would have developed as follows: two parallel, apparently twin rifts, formed after chelogenic subsidence, when over 10 km of upper Proterozoic to mid-Paleozoic platform cover broke cleanly and sagged intact into two asymmetrical troughs. There is no problem postulating a "phantom" basement ridge, because most of the sediments were later eroded prior to new platform cover being laid down during the second cycle of subsidence.

In one trough (Fig. 5), the sediments remain largely undisturbed, whereas the other trough was faulted. This is an important feature, because it suggests two distinct episodes of wrenching. The first would have opened up a narrow gap in the crust, causing the first depression into which sediments sagged undeformed. The second episode wrenched these sediments, and opened up the second gap, forming the second trough, which has remained unwrenched. The same applies to the Warrabin and Quilpie troughs (Appendix A), and these examples suggest parallel rifts developed serially to produce a parallel cluster, in what is a three or four dimensional space-time process.

Other seismic sections across the Goulburn Graben show listric faults that must be exotic structures related to the second wrenching. The Quilpie Trough (Appendix A) and the Amadeus Basin Trough also can be explained as rifts if sediments were bent upwards at the edge as they sagged, masking the steep fault, and giving the appearance of a fold structure. Thus, the simple stratagem that some sediments break cleanly (Fig. 5), and others drape into the depression, can explain all types of troughs, including those usually attributed to folding. This is the basis of a universal rifting solution that would account for a vast number of gravity lows all over the continent.

EFFECTIVENESS OF THE SEISMO-GRAVITY METER

The land gravity meter is an awkward, delicate, mechanical device utilizing intricate levers and springs. It is extremely sensitive, and detects the slightest ground vibration. Some gravity meters originally were designed to be seismographs. Although radical new advances recently have been made in completely redesigning the inner workings to give automatic readout, it is still not possible to have full automation, because the meter has to be stabilized and leveled for every reading. But the accuracy is extremely high, and provides an important margin of safety in survey operations and interpretation.

The gravity meter can provide a direct understanding of the Earth's natural seismicity. During a gravity survey, an earthquake provides welcome relief from tedious gravity measurements, and it is fascinating to watch the energy from a large earthquake rolling by, sometimes for hours, causing the beam in the meter to swing wildly from side to side. The energy reverberates back and forth across the continent, and probably follows certain paths through the crust. If earthquakes were of long enough duration, these paths could be mapped by measuring the amount of swing on the beam in different localities.

The swing varies considerably from place to place, and it can be expected that the resulting map would be most enlightening, but unfortunately, earthquakes are too unpredictable to allow this approach. The alternative is to place thousands of seismographs around the countryside, and record a single earthquake, but this is impossibly expensive. The gravity field does not vary with time, and it is easier to use a single gravity meter to map gravity in whatever detail is required, and then to try to relate the gravity pattern to seismicity. Then a few seismographs can be used to test the relationship in chosen areas, and an understanding of the relationship between crustal structure and seismicity would gradually develop.

The gravity pattern is well established over Australia, and can be correlated with the emerging pattern of compressive stress and seismicity (Denham, 1988). There may well be a relationship between gravity, and stresses and seismicity in the crust. A simple example is that earthquakes often occur along faults, and gravity detects faults, and in a broader sense, major gravity lows probably represent paleotension associated with rifting in the past. The key question is whether the gravity framework is also related to the present stress regime, and it would not be surprising if this were the case, as gravity anomalies represent inhomogeneities in the crust, and these directly influence the stress field (Moody, 1966), and therefore current tectonic processes. Indeed, the framework stacking and polycyclic rifting processes described in this paper suggest that in some regions, the paleostress framework did remain reasonably constant.

There is no doubt therefore, that the humble gravity meter is a powerful tool, and that it carries a major part of the burden of exploring the crust, and pushing back the frontiers of geoscience. The legendary ambiguity of gravity data is far less of a problem in reality than it is in theory, and does not affect the study of the gross patterns representing the tectonic framework of Australia.

GRAVITY FEATURES IN AUSTRALIA

Australian gravity data show four main types of patterns: (1) groups of subparallel trends called parallel gravity clusters (Fig. 2), (2) the orthogonal systems (Fig. 2), (3) bifurcating gravity highs (Fig. 3), and (4) gravity compartments (Fig. 4). Possible isostatic imbalance associated with very long wavelength changes in gravity will be discussed separately.

Parallel gravity clusters

Parallel gravity clusters for many years have been virtually the sole basis for most concepts of tectonism, as they simultaneously cater for crustal buckling models in central Australia (Lambeck, 1984), platelet-collision accretion models (Wellman, 1976, 1984), and failed rifting models (Shaw et al., 1984; Etheridge et al., 1987). They are all oversimplistic two-dimensional concepts, and in central Australia they also fail to address the serious problem of why major granites occur under the Ngalia and southern Amadeus troughs. The failure to attribute gravity lows to granites where they are known to occur seems to stem from the assumption that the upper crust is largely granitic. But it is a fact that granites do produce major gravity lows, and to ignore them completely tends to invalidate key gravity interpretations in central Australia (Mathur, 1976; Wellman, 1978).

Rifting and folding are diametrically opposite interpretations of the same data, and it should be possible to distinguish one from the other. The problem came to a head when Wellman (1988) attributed some parallel gravity clusters to rift reworking, thus undermining the basis for the earlier (Wellman, 1976) accretion-collision concept. The accretionary docking of platelets would imply plate tectonics was active in the Proterozoic, but there is no sign of subduction in Australia (Etheridge et al., 1987). Anfiloff (1983) proposed an intact continent based on the orthogonal systems not recognized by Wellman (1976).

There is another major problem: the assumption that all the structures in a cluster form simultaneously, or at least serially, in one major process. This purely two-dimensional approach is incorporated in the recognition of crustal blocks by drawing boundaries around clusters. But the fact that many different boundaries were produced by Wellman (1976, 1985, 1988) shows that clusters are difficult to define objectively. The problem was analyzed using elevation and gravity profiles across Australia (Anfiloff, 1982a), and it was concluded that crustal blocks (chelogens) cannot be defined accurately using gravity, but are best defined by elevation data, which in some cases show the distinct edges of cratons and broad basins.

Objective trends have now been produced by Geoimage (1989), using a mathematical filter, but there is still the problem that anomalies in cratons are much weaker than elsewhere, and given the finite errors in the reconnaissance gravity survey, trends in cratons still will be unreliable. Nevertheless, this new residual gravity map of the whole of Australia provides an opportunity to study the relationship between gravity trends and chelogens.

The Canning Basin (Fig. 1) is dominated by a northwesterly trend (Fig. 2), which encroaches into the adjacent Kimberley Craton. The same happens at the southeastern edge of the Yilgarn Shield, which includes a major gravity low that is part of the Fraser Range Mobile Belt cluster that extends under the Eucla Basin (see elevation and gravity profiles along 29° latitude, Anfiloff, 1982a). Wellman (1988) explained the gravity low as reworking of the Archaean craton during the docking of the block defined by the Eucla Basin cluster, but this does not explain the fact that it is associated with a granite (Anfiloff and Shaw, 1973), and that there is no sign of subduction here or at any other block junction. The Yilgarn Shield could well have extended much farther east prior to its edge being consumed by the Eucla Basin chelogen.

The Murray Basin region (Fig. 1) contains both a northeast-trending cluster, and a prominent orthogonal framework (Fig. 2 and Appendix B). The new Iona Belt is developing between two older belts, and as discussed in Appendix B, the belts in this cluster probably developed at different times. Clearly, clusters do not imply that crustal blocks formed in one major tectonic event, as required for the accretion model of Wellman (1976). It is simpler to argue that anisotropy caused by igneous activity in the first belt influences the direction of other belts, so that belts develop along a particular grain over a considerable time span (Anfiloff, 1988a, 1988b).

It is important to appreciate that individual gravity anomalies do not have implications for the development of crust, nor its age, nor thickness. In gravity interpretation, the crust is the homogeneous background medium in which anomalous bodies reside. The properties of that medium normally are not reflected in short wavelength anomalies, and its density can be inferred only if the density of the anomalous body is known, which is very rare, and even then the deduced density applies only to the upper crust. In general, the bulk density of crust is not known, and does not need to be known. Longer wavelength anomalies tell even less about the crust, as they can include contributions from various depths. Ambiguity increases with wavelength, and gravity interpretations that impute crustal structure should be viewed with utmost suspicion, including those based on isostatic principles.

The interpretation of trend patterns is a different proposition, and subjected to a different type of ambiguity that applies to all trend analyzes. It depends on the quantity and quality of data, and the strength and clarity of the patterns. The interpretation by Wellman (1988) that clusters establish the relative ages of crustal blocks all over Australia also assumes that accretion can be deduced at the same time. Accretion would be more plausible if all blocks were defined by clusters, and the ages of each block were determined by other means, but there also would have to be evidence of subduction at block junctions.

The ages of crust are not known in most areas, and Wellman's somewhat self-fulfilling interpretation fails to prove accretion, and hence, that plate tectonics was operating in the Proterozoic. If it is not possible to determine relative ages within a cluster from the cluster itself, then that cluster cannot be said to be older or younger than any other cluster, and the Murray Basin cluster described above probably has an enormous spread in age. It might be possible to suggest that weak trends cutting across a strong cluster might represent old structures in a previous craton, because the amplitude of anomalies often decreases with craton age. The possibility that coherent orthogonal frameworks could imply an ancient continent is another reasonable deduction from trends.

It therefore seems that there is no direct relationship between chelogens and clusters. The Murray Basin cluster gradually may have developed as a chelogenic piston moved up and down in the same place. Weak anomalies in a craton are obliterated when it is recycled, especially if new belts develop with a different grain. But, as chelogens consume each other, remnants of old clusters might remain.

In general, chelogenic bulges (cratons and depressions) are not distinguishable in the gravity map of Australia, and there is only one place where the truncation of gravity trends clearly coincides with the boundary between two chelogens. This happens across the Diamantina Gravity Gradient (DGG) (Fig. 2). On the northern side of this line, the Proterozoic Mount Isa Block craton is characterized by a northwest-trending cluster, and on the other side, the large Phanerozoic depression of the Eromanga Basin region is underlain by the Thompson Fold Belt.

The DGG has figured prominently in the concept of Phanerozoic Eastern Australia proposed by Veevers et al. (1984). The concept is based on the supposition that the Thompson Fold Belt is represented by a northeast-trending cluster that represents oceanic crust, produced by the rifting away of part of the continent in the Late Proterozoic. However, the same supposed cluster was used by Wellman (1976, 1984) to indicate accretion, the diametrically opposite process to rifting.

But objective trends show that the Thompson Fold Belt is not dominated by a northeastly trend, and has a near perfect orthogonal framework within the area bounded by DGG and Millungera-Toowomba Gravity Gradient (MTGG) in Figure 2. From the outset,

Australian gravity (BMR, 1976) has presented clear evidence of an orthogonal gravity framework in the Thompson Fold Belt. The region has a coherent, rectilinear and orthogonal framework of gravity lows and fault trends, identical to the pattern over most of the continent, and this was attributed to rifting along the Cardinal Fracture System (Anfiloff, 1983). Therefore, orthogonal systems such as the Tipan and Adavale-Cooladie troughs in the Thompson Fold Belt (Appendix A), probably developed on continental crust.

The abrupt termination of the cluster in the Mount Isa Block at the DGG can be explained in terms of crustal subsidence. The DGG separates two chelogens—a craton representing the end of a cycle, and a broad basin half-way through a cycle. One chelogen may have consumed and reworked the edge of the other, but there is no guarantee that one group of gravity trends formed before the other. There is no reason to doubt that Archaean basement underlies eastern Australia, and some of the differences in crustal properties cited by Veevers et al. (1984) to support "Phanerozoic Eastern Australia" could be attributed to the chelogenic subsidence process.

Orthogonal gravity networks

The orthogonal gravity framework of Australia (Fig. 2) is presumably the direct expression of fundamental basement tectonics, and is one of the main components of the primary tectonic framework. There is no doubt it supports the pioneering work of Gay (1973) who brought attention to pervasive orthogonal vertical fracturing in the Earth's crust and summarized the findings of other workers. He revealed the Cardinal Fracture System, and its various subsets, which in total represent a myriad of presumably near-vertical crustal fractures cutting the crust into a mosaic of small blocks (Fig. 8). There is a need to explain compartmented orthogonal systems by processes operating along these fractures, and schematically this can be done by jostling crustal blocks to produce a general construct (Fig. 9), which can be applied all over the continent.

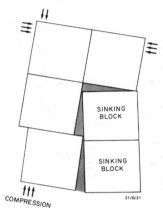

Figure 9. Conjugate and non-conjugate orthogonal rift compartments can in the simplest terms be explained by wrenching movements in a mosaic of rigid blocks of crust. The real situation is more complicated and involves the development of ridges across the pivot points, as shown in Figure 12.

In the Eromanga Basin, Finlayson et al. (1988) explained orthogonal troughs in terms of two separate episodes of crustal shortening. But discrete orthogonal systems such as the Tipan Trough (Appendix A) cannot be explained this way, and shortening would not produce undeformed sediments in asymmetrical downwarps, such as in the Quilpie Trough (Appendix A). A passive rifting mechanism is required that produces rifts along orthogonal vertical fractures without deforming the sediments.

The construct shown in Figure 9 should be viewed as a first approximation to a more complex construct that makes use of the seemingly limitless number of fractures in the crust, which are suggested by the magnetic dykes in south-western Australia (Fig. 10), compiled by Tucker and Boyd (1986). Nonetheless, it is surprisingly accurate in terms of the general gravity expression of compartmented orthogonal rift systems (Figs. 2 and 4).

The recent study by McLennan et al. (1990) provided a detailed examination of an orthogonal system comprising the Goulburn Graben (discussed earlier), and the Calder Graben, which extends from it into the Bonaparte Basin region (Fig. 1). Magnetics shown in that paper reveal a profoundly orthogonal system coherent with the main crustal fractures in that region (Fig. 3). The paper also details the separate development of each arm of the system from the Jurassic to the Tertiary, and there is little doubt that the system was subjected to repetitive wrenching movements. Some

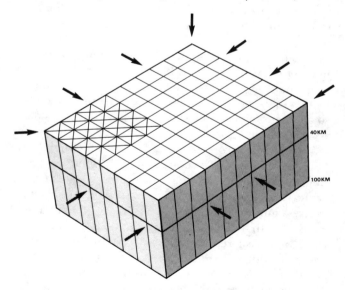

Figure 8. Illustrating the basic concept of a continent cut by pervasive orthogonal fractures of the Cardinal Fracture System, and held together by radial compression. According to Gay (1973), the system consists of at least two sets of orthogonal fractures, resulting in 90° and 45° intersections. Only part of the second set is shown.

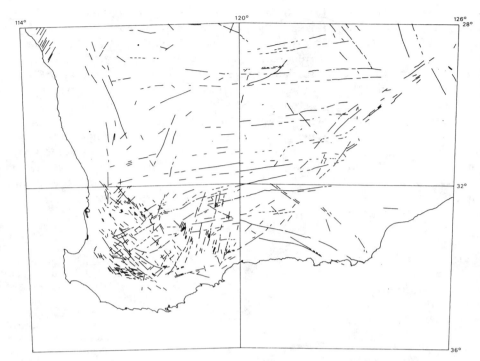

Figure 10. Magnetic dykes compiled by Tucker and Boyd (1986) probably represent the Cardinal Fracture System. Some of the pairs of long dykes are normal and reverse polarised along their whole length, implying a particular method of crustal dilation. Wrench offsets are usually less than 1–2 km.

compartments formed individually, and the wrenching was not constant in time or space. This is, therefore, an example of a nonconjugate orthogonal system, produced by a complex process of wrenching operating in four-dimensional space-time.

Bifurcating gravity ridges

Over the whole of Australia, free-air gravity highs (ridges) tend to form bifurcating systems (Fig. 3), and within broad basins, these often are explained by narrow basement ridges. Such ridges can partition broad basins, and produce compartments in rift systems, but are sometimes very deep and difficult to detect.

Yeates et al. (1984) described the subtle geological effects of the Jones Rise, which separates the Fitzroy and Gregory subbasins (compartments) in the overall Fitzroy Graben in the Canning Basin region.

There are many published examples of bifurcating ridges detected by seismic work, but they are rarely mapped over large distances, and they are not necessarily involved with uplifted basement. Some are produced by wrenching in thick sediments. True basement ridges are rarely detected by the cheapest and most universal method of geophysical mapping, airborne magnetics. Appendices A and B provide examples of detailed, high quality surveys where the relative merits of seismic, gravity, and magnetics for mapping the basement surface can be compared. Seismic data is either patchy, or it can be stymied by problems such as coal seams, and the sediment-basement interface usually is transparent magnetically. But it invariably produces good gravity anomalies, and Appendix B shows an exceptional case where gravity has confirmed with great accuracy and detail the shape of the Iona Ridge as defined by the shallow refraction survey of Odins et al. (1985).

More recent work by Odins et al. (1991) detailed the bifurcation of the Iona Ridge across the northern part of the Murray Basin. Their map of pre-Tertiary basement topography strongly supports the concept of bifurcating ridges, and can be taken to represent the development of a new system of compartmented rifts during the Tertiary subsidence of the Murray Basin chelogen. Moreover, because this process has taken place within the Ivanhoe Block, between the Wentworth and Balranald troughs that are part of an ancient parallel gravity cluster (Appendix B), it seems to demonstrate how the older tectonic grain is constraining the direction of development of new rifts.

When mapping basement ridges using gravity, it is necessary to differentiate between primary basement ridges and superficial structures, such as wrench anticlines in troughs. Fortunately, the latter produce much smaller anomalies and do not appear on the coarse reconnaissance gravity coverage. In some regions, the bifurcating pattern represents topographic highs, and elsewhere, dense bodies contribute to it. Some basement ridges seem to be fortified by large pods of mafic volcanics representing old cratonized rifts that may later act as stress guides. This could occur in the Rankin Trend in the North West Shelf (Anfiloff, 1988a), and in the Eromanga Basin region (Appendix A). In Kansas (USA), basement ridges form a network that envelops troughs, and there is also a rare example of an intact ridge-trough pair associated with the Nemaha Ridge and Forest City Basin (Steeples, 1989, figs. 3 and 5). These examples support the models shown in Figures 11, 12, and 13.

There is reason to be optimistic about a direct correlation between bifurcating ridges and earthquake seismicity.

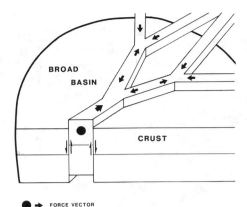

Figure 11. During chelogenic subsidence, basement ridges develop along the load-bearing skeleton network of compressed paths, which by the action and reaction principle must be interlocking. The crust would have to be cut by pervasive vertical fractures to allow such vertical movements to occur freely. No horizontal movements occur while the opposing forces are balanced.

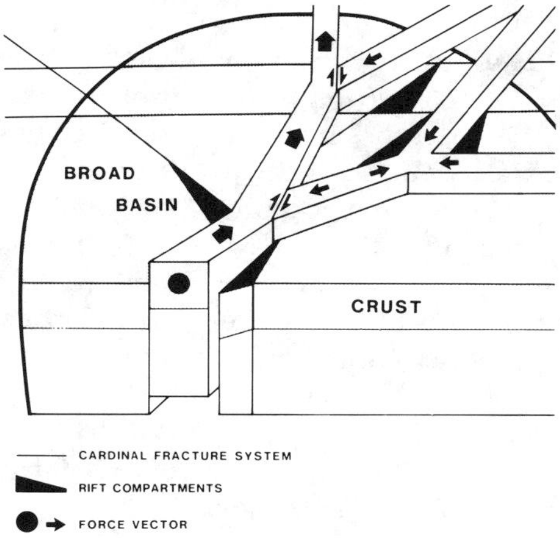

Figure 12. When horizontal forces become unbalanced, a basement ridge will move bodily a small distance as indicated by the large arrows, and narrow gaps will open up in the crust as blocks jam and pivot. The gaps become much wider rift compartments as the lithosphere adjusts to the movements by flowing into the gaps. This combination of bifurcating ridges enveloping a rectilinear framework of rift compartments is the distinctive geometry of the primary tectonic framework (for examples see Appendix Figs. A1 and B3).

Studies of seismicity in the San Fransisco region (Goter, 1988) show that earthquakes are located preferentially along bifurcating topographic highs, and in the New Madrid Seismic Zone (Himes et al., 1988), the distribution also has a bifurcating geometry. This supports the proposal by Anfiloff (1982b, 1988a, 1988b) that basement ridges exist because they transmit horizontal compression. The theoretical basis for this will be discussed later in conjunction with the load-bearing skeleton concept.

In the past, earthquakes usually have been associated with movements along faults, but studies in Kansas show that the whole Nemaha Ridge may have moved, because it is straddled by seismicity (Hildebrand et al., 1988). These data support the possibility of whole ridges being involved in vertical and horizontal movements as shown schematically in Figures 11 and 12. Figure 11 represents the situation where basement ridges form during chelogenic subsidence because horizontal compression is transmitted along them and interferes with their subsidence. When unbalanced, the same forces may cause rifting by bodily moving the whole ridge a small distance, causing crustal blocks to be jostled, as discussed later in relation to Figure 12. However, these constructs require the crust to be fractured extensively by the Cardinal Fracture System.

THE GEOPHYSICAL SIGNATURE OF THE CARDINAL FRACTURE SYSTEM

The Cardinal Fracture System can account for the majority of long faults, lineaments, and straight segments of coastlines in Australia. Geophysically, the system is implied by magnetic dykes, gravity linears, and seismic data. Most of the transcontinental fractures mapped by O'Driscoll (1982, 1990), by enhancing features in elevation and gravity maps, probably represent recent reactivations. The control they exert on crustal subsidence can be seen in the topography of Australia (Fig. 3), and the Gulf of Carpentaria may be a good example of framed chelogenic subsidence utilizing orthogonal fractures. This seems to be a modern analogue of Palaeozoic subsidence in the Eromanga Basin region where the prominent gravity linears DGG and MTGG (Fig. 2) would have framed the subsidence.

Magnetic dykes have been mapped over Australia (Tucker and Boyd, 1986; Tucker et al., 1986b), but mostly are observed in cratons, rather than basins, and this poses the problem of their significance in the

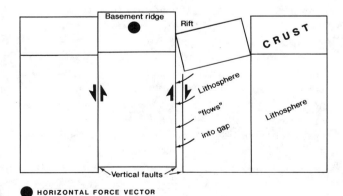

Figure 13. This is the key mechanism in the rifting process. The asymmetry of a rift bounded by a steep fault is explained here in terms of the basement ridge being supported by the same horizontal compression that moves it sideways to open up narrow gaps down through the crust and lithosphere. After the gap develops, the lithosphere will gradually flow towards it, mainly from the side that is not under the ridge, because the ridge is being supported by the compression.

chelogenic cycle (Rutland, 1982). Long Proterozoic dykes in the Archaean Yilgarn Shield (Fig. 10) obviously represent ancient vertical crustal fractures that have tapped basaltic material from great depths. In some cases, long normal and reverse polarized dykes occur side by side, and the polarization does not change along their whole length, indicating that new dykes have formed without greatly dislocating earlier ones. The regularly spaced gaps along dykes probably represent cross-fractures, and it is possible that they cut the dykes into compartments, similar to rift compartments. In general, the behavior of dykes seems to have many portents for the development of rifts.

The orthogonal-rectilinear framework of the major gravity trends over Australia (Fig. 2) inherently implies pervasive vertical fractures, and there is no doubt that an asymmetrical rift filled with sediments will produce a strong linear gravity low alongside the steep main bounding fault. It is therefore clear from the orthogonal Goulburn-Calder Graben system discussed earlier that the gravity framework implies the rifting process, as proposed by (Anfiloff, 1983). The framework commonly displays 90° and 45° intersections (Fig. 2), and is similar to the one mapped by Klassner et al. (1982) for North America using gravity and magnetics.

Gay (1973) concluded that the Cardinal Fracture System represents vertical deformation by a process such as a tidal bulge rather than horizontal forces, and that once formed, the fractures would be reactivated continually and would penetrate upwards into new cover. He noted the lack of offsets in most fracture systems. In Figure 10, horizontal offsets in dykes are usually less than one or two kilometers, and over vast areas of Australia, magnetic data generally show very few major offsets across ancient structures.

Wu (1988) explained the lack of offsets in seismically active areas in terms of repetitive back-and-forth movements cancelling each other. It is also possible some blocks of crust have no internal displacements because they are unfractured, but it is more probable that the opposite is the case: that they are so pervasively fractured that finite movements are rapidly dissipated in many directions.

Whether or not the Cardinal Fracture System exists over whole continents is a very serious issue; it has a direct bearing on the evolution of continental crust. Moreover, the good fit between the matching coastlines of Africa and South America could be explained in terms of crustal subsidence framed by orthogonal fractures, without requiring sea-floor spreading. It could be argued that the fractures observed by Gay and others apply only to areas of exposed basement; in cratons where brittle fracturing of the crust has taken place, after uplift and cratonization. Whether fractures already existed in crust before cratons developed is difficult to tell, but it is probable that most crust has been cratonized and brittle fractured at some stage in its long history, and this is all the more reason to expect pervasive fracturing. There is no doubt, that basement in the Eromanga Basin region is regularly cut by vertical crustal fractures; they can be inferred from monoclinal structures in flat-lying sediments, spaced at regular intervals of 20–30 km (Appendix A).

TOOTHPASTE TECTONISM ALONG THE CARDINAL FRACTURE SYSTEM

The Anabama Granite (Appendix B) is situated west of Broken Hill in southeastern Australia, and produces a distinct elongate gravity low (Tucker, 1983). The granite is located on a rectilinear framework, which is highlighted by the magnetic effects of extrusives, and therefore appears to be a uniquely clear example of igneous activity up a vertical crustal fracture, directly into a rift compartment. Moreover, a detailed gravity profile across the granite shows that the low is asymmetrical, consistent with the expected sloping floor of a rift. This very significant example suggests a granite has been squeezed like toothpaste into a compartment situated on a long crustal fracture, and supports the basic premise of rift development shown in Figure 9 in which rifting is associated with narrow gaps being formed in the crust. The expulsion by an intruding granite of sediments already in the rift compartment would also explain the formation of mountains cored by granites.

This seems to be another example of rifting in four-dimensional space-time; the formation of a discrete compartment along a vertical fracture that is part of a rectilinear framework. Together with the nearby Iona Belt (Appendix B), it suggests that each compartment can develop independently, and that the rectilinear framework of mobile belts gradually builds up along the ancient Cardinal Fracture System by purely vertical processes. Thus, clusters of parallel belts can be attributed to the imposition of local grain on the otherwise pervasive orthogonal system. The grain is caused by the intrusion of granites and volcanics into the first rifts formed, thus setting a preferential direction for other rifts to follow.

RIFTING DRIVEN BY COMPRESSION

There are many reasons to suspect that rifting is driven by compression. A continent pervasively cut by ancient fractures could be dismembered (Hills, 1956) by extensional forces, but it could indefinitely resist compressive forces. The coherent expression of the Cardinal

Fracture System in Australian gravity implies an intact continent, and therefore a continent subjected to persistent compression. The main African rifts are not only compartmented, but they are separated from each other by narrow strips of crust (Rosendahl, 1987), which implies they could not merge together. In an extensional regime, they would have merged, and compression is again implied.

Most of the gravity anomalies in central Australia are compartmented (Fig. 4), including the gravity highs, and this is the basis for the wrenching rift model of Anfiloff (1988a). Rosendahl (1987) explained compartments in terms of wrenching as a continent begins to break up and platelets slide past each other. However, a mechanism that always fails to split the continent is not a viable explanation for pervasive rifting. Moreover, there is no support for large strike-slip movements in magnetic data, and static compression is directly implied in the development of basement ridges.

Wrenching is the means by which local tension can be produced from pervasive compression, but it must be able to produce rifts over 50 km wide. This requires a new perception: a pervasively fractured rigid crust that would not be deformed by squeezing because the forces would be dissipated by many small wrenching movements along fractures. Shortening and overthrusting of the whole crust would not be possible because crustal blocks, being far thicker than they are wide, would slide around each other rather than over each other. Barosh (1986) envisaged such movements associated with dominant north-south compression in North America; but how are forces applied to a continent, and how are they transmitted through a fractured crust?

The principal feature of bifurcating systems is that narrow high zones are interconnected, whereas lower zones are wider and segmented. Meyerhoff et al.'s (1989) surge-tectonics concept explains this in terms of the pressurized movement of mantle fluids along channels in the lithosphere, causing upward forces that lift the ground surface. The pressure is attributed to earth contraction, and the overall concept is therefore similar to the one used here.

In an experiment, Koshtak (1971) applied radial compression to a mosaic of plastic blocks and found that compression was chanelled along a load-bearing skeleton, an interlocking balanced network of narrow paths that divert compression from the rest of the mosaic. He also envisaged this could occur in the Earth's crust. If a subsiding chelogen were subjected to radial compression, this would explain how bifurcating basement ridges form; the ridges would be supported by the compression and would not subside as quickly as the rest of the crust (Fig. 11). Most ridges are quite innoccuous (they remain buried under sediments and are not pushed up to the surface through them), which suggests they are caused by differential subsidence rather than by active uplifting forces. But to attribute their support to a weak upward force resulting from compression at each end of a basement ridge is only feasible if numerous vertical faults already exist for this process to take advantage of. Thus, the Cardinal Fracture System is essential.

Stress data (Denham et al., 1979; Denham, 1988) show that compressive forces are propagated in a variety of directions, and in particular, it appears that compression is being directed around the Murray Basin, and possibly along one of the two ridges connecting with Tasmania. The fact that two ridges connect Tasmania to the mainland (Fig. 3) is important, and constrains development of the Bass Basin that they envelop (Campbell, 1990), as it rules out an extensional origin for the basin.

Stress data support the idea of a load-bearing skeleton for much of eastern Australia, and on a continental scale, the skeleton geometry is clear in the pattern in Figure 3 drawn by connecting zones of positive free-air gravity. The pattern is equivalent to alternating isostatic balance and imbalance; areas under compression would be out of isostatic equilibrium, and if released would move towards equilibrium, which would be down for basement ridges. On average, large regions are presumably statistically in isostatic equilibrium, but as discussed later, there may be significant exceptions.

If compression is produced by a contracting Earth, it would be subjected to the action and reaction principle whereby continuous contiguous crust has to exist if any compression is to be felt anywhere. Earthquakes may therefore be caused by four processes associated with basement ridges: (1) the failure of crust situated on basement ridges as it bears the compression produced by a contracting Earth; (2) the lateral movement of a whole ridge in the direction of its length resulting from gross failure somewhere in the balanced system; (3) the vertical movements associated with the upwards growth of a ridge within a subsiding chelogen; (4) isostatic rebound during prevailing tension.

Koshtak (1971) also showed that while the balanced skeleton was produced, slippages and "hitching of corners" occurred between blocks in the mosaic. In the crust, this would cause blocks to pivot slightly, opening up narrow gaps, perhaps only 1 km wide (Fig. 9), which would translate upwards into much wider troughs as described below.

MODEL FOR THE DEVELOPMENT OF THE PRIMARY TECTONIC FRAMEWORK

The development of the primary tectonic framework can be explained in five stages (Fig. 14).

Stage 1—Broad sediment basins are produced by chelogenic subsidence, and at the same time, balanced compression produces bifurcating networks of ridges by interfering with their subsidence (Figs. 11 and 14, stage 1).

Stage 2—When compression becomes unbalanced, ridges move in the direction of the lesser force, causing some crustal blocks to jam and pivot. This opens narrow deep gaps in the crust and lithosphere with the same rectilinear geometry as rift compartments (Figs. 12 and 14, stage 2), and these gaps become the conduits for igneous activity.

Stage 3—Passive asymmetrical subsidence occurs along the gaps when the crust on one side is supported by horizontal compression transmitted along a basement ridge. This support reduces the load on the lithosphere under it, and the lithosphere flows to fill the gap more from one side than the other. The upper crust on that side then subsides like a rigid beam as the lithosphere flows out from under it, and a wide asymmetrical trough forms at the surface (Figs. 13 and 14, stage 3).

Stage 4—Melting is triggered in the lower crust by the reduction in confining pressure along the gap, and perhaps meteoric water is also introduced. Granite forms, expands, and moves into the zone of tension. It moves up the gap and into the compartment, expelling the sediments, and producing mountains cored by granites (Fig. 14, stage 4).

Stage 5—The second rift cycle takes place in the next chelogenic cycle of subsidence, and stacks rifts on top of each other by repeating the same basic rift mechanism. Granite in one rift becomes basement to a new rift stacking above it (Fig. 14, stage 5). This explains the situation in central Australia where the Ngalia trough overlies granites that were themselves probably intruded into older rifts and cratonized. This results in some 10 km of sediments stacked on top of about the same amount of granite, and accounts for the large 150 mGal gravity lows in terms of low density material extending to a depth of about 20 km (Anfiloff and Shaw, 1973).

DEDUCING PERVASIVE RIFTING FROM GRAVITY COMPARTMENTS

The establishment of the compartmentalized nature of the rifting process by the detailed seismic work in

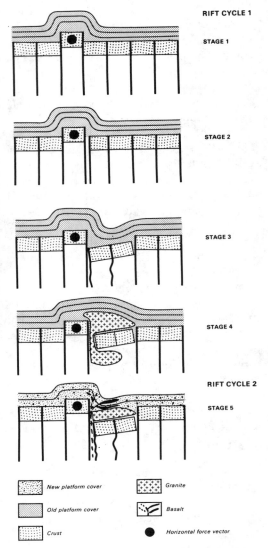

Figure 14. Polycyclic rifting is the repetition, during the next chelogenic cycle, of the process shown in the previous figure. In Stages 1–3, a simple rift develops, and in Stage 4 it is invaded by a granite. The region then rises and is eroded. In the next cycle, rifting is repeated at Stage 5 and a new trough develops on top of the granite, which is now cratonized into the basement. Alternatively, two rifts can stack on top of each other to produce a very thick pile of sediments and a very large gravity low. Mafic volcanics can also fill compartments to produce large gravity highs.

African rift lakes (Rosendahl, 1987) make it possible to interpret gravity compartments as rift compartments. In Australia, gravity compartments (Fig. 4) are elongate gravity highs and lows typically 20–50 km wide, 50–150 km long, with an amplitude of 20–50 mGal. Vertically stacked rifts, and rifts that have been mobilized by acid and mafic igneous activity (Fig. 14), can have bigger anomalies. Appendix B shows one of the best examples of gravity compartments: three colinear gravity lows associated with the Renmark, Tarrara, and Menindee troughs in southeastern Australia. The Bancannia Trough intersects this axis obliquely, and may be a conjugate rift, but it is separated from the others by the major basement ridge called the Scopes Range structure.

A general rifting solution also must explain the occasional mafic belts found in the upper crust. One of the attractions of crustal shortening models is that they provide a means of lifting granulite-grade metamorphic rocks to the surface. However, in central Australia the gravity highs over the Papunya Mafic belt are not large enough to be consistent with uplift of the lower crust to the surface (Anfiloff and Shaw, 1973), and they also have the shape of rift compartments (Fig. 4). In fact, most gravity anomalies in central Australia are compartmentalized. This factor has not been acknowledged in previous studies; they have been essentially two-dimensional, whereas the problem is four-dimensional.

Gravity modeling shows that the mafic bodies must reside in approximately the upper 20 km of the crust (Anfiloff and Shaw, 1973) and, consequently, they are better explained as mafic rifts, produced by toothpaste tectonism, the intrusion of mafic material up dilated vertical crustal fractures to displace sediments from rifts. The granulite-grade metamorphism of the mafic rocks can be attributed to the fact that rifting happens during chelogenic subsidence, under many kilometers of sediments, and in the next cycle, the volcanics can be reburied and metamorphosed again.

Recognizing positive gravity compartments as volcanic rifts completes a general pervasive rifting solution for Australia. The vertical bounding faults of rifts are rarely visible except in exceptional cases (Fig. 5), because most rift troughs are deformed by wrenching or converted to mobile belts by intrusions. The Anabama Granite (Appendix B), seems to be a uniquely explicit example of a granite that neatly has occupied a compartment by expelling sediments, and this demonstrates that granites must be responsible for some types of orogenic uplift and mountain building. Also, the igneous activity in such discrete compartments shows that rifting is not associated with major mantle upheavals, and often there are no intrusions. Rifting represents an opportunity for intrusions to occur if the material is available, and this depends on conditions under the crust. Consequently, rifting can be seen as the "window" to the mantle.

THE DISTRIBUTION OF MAFIC COMPARTMENTS AROUND CRATONS

Throughout Australia, negative gravity compartments can be equated to rifts filled with sediments and granites, and positive ones to rifts filled with mafic volcanics (Fig. 4). When occurring side-by-side, they appear as dipoles, and have been interpreted in terms of the isostatic effects of dense blocks of crust (Wellman, 1978), but Figure 4 shows they are usually offset from each other spatially, and cannot be explained in this manner.

There is a far simpler explanation for dipoles: if rifting is pervasive, then the occasional positive compartment always will be flanked by a negative one. But mafic compartments are usually older than others, and in central Australia, the main Papunya Belt is Early Proterozoic, and was cratonized at about 1800 m.y. (Shaw et al., 1984), whereas the flanking Amadeus and Ngalia troughs are of Late Proterozoic age. This is the basis of the new explanation for gravity dipoles in the polycyclic rifting model (Anfiloff, 1988a, fig. 7). In a variation on the basic model described earlier in Figure 14, the first rift cycle produces mafic compartments, and perhaps this depletes the source of this material, as it does not seem to be repeated. In the next cycle, the cratonized mafic rifts act as stress guides for compression and a basement ridge develops along them. This ridge is then instrumental in the development of flanking rifts, and granites later invade them as shown in Figure 14. This process explains why old mafic compartments are often located on basement ridges.

There is one other interesting property of mafic rifts: their tendency to fringe cratons. The Archaean Yilgarn Shield is characterized by the absence of prominent gravity compartments, because only the remnants of very old rifts remain. This helps accentuate a number of prominent mafic compartments situated around its perimeter (Fig. 4). The mafic material would not have emanated from the cold and inert craton, and more probably, mafic fluids would be generated beyond the craton during chelogenic subsidence around it, while the subsiding crust was being high-level underplated. Not being able to reach the surface, the underplating fluids might move laterally until they reached the edge of the craton, and then obstructed by it, move up into rift compartments. Grant (1989) also noted the exceptional igneous activity around cratons. The activity might be restricted to a particular stage in a chelogenic cycle, and perhaps in the evolution of the crust. This could make the age and distribution of mafic compartments significant for the study of mantle evolution.

Mafic compartments are also distributed around the outer edge of Australia, suggesting the whole continent has behaved like a "supercraton" with respect to the subsidence in the oceanic basins around its edges. Gravity dipoles around the edge of the continent are usually explained as an edge effect, but there are large gaps around the southern edge where there are no dipoles. The dipole explanation also assumes isostatic equilibrium, but this is unlikely. Consequently, the circum-Australian mafic compartments could represent young mafic rifts, with the material moving laterally from the oceanic region. This has similar connotations to Dillon's (1974, fig. 35) ideas about depletion of volcanics from continental lithosphere.

GLOBAL OROGENY AND EARTHQUAKES IN THE UPPER CRUST

In the plate-tectonics concept, episodes of global orogeny are caused by fluctuations in the heat engines that drive plate motion over large distances. Complete synchronization is not possible because some plates move independently of others, and there is, therefore, a problem in explaining simultaneous earthquakes around the globe.

The new approach proposed here is to synchronize orogeny by causing occasional fluctuations in global compression, while small heat engines operate locally and at random to produce chelogenic vertical motion. Artyushkov (1973) casted doubt on small-scale mantle convection, and proposed instead that bubbles of hot mantle material would lodge under the lithosphere. Either way, the trapped heat provides the means to drive the chelogenic engine, while the Earth contracts as a whole, compressing the crust.

As the contracting-Earth type of compression relies on a continuous mantle of rigid crust (the action and equal reaction principle), a mechanical failure anywhere could precipitate an instant global adjustment, triggering earthquakes and episodes of rifting and isostatic adjustment. Fluctuating global compression is also the basis of the new surge-tectonics concept of Meyerhoff et al. (1989).

Denham (1988) noted that most earthquakes in Australia originate in the upper 10 km of crust, and assumed this zone is the weakest. However, the load-bearing skeleton concept requires that the compression that causes earthquakes can only be directed along the strongest paths, and it would therefore seem that the upper 10 km of crust is the strongest part of the entire lithosphere, as proposed by Chen (1988). Most gravity anomalies in Australia originate in the upper crust, and even the largest in central Australia may extend no deeper than 20 km (Anfiloff and Shaw, 1973; Anfiloff, 1982a), and it therefore seems that a stress transmission layer of igneous bodies concentrated in the upper 10 km may be providing much of the crust's resistance to being squeezed.

In the new model proposed here, sudden global orogeny is caused by failure in the strongest seismogenic part of the upper crust, causing "instant" adjustment in the global load-bearing skeleton. Local failures would cause adjustments in the load-bearing skeleton, that would take the form of wrenching movements leading directly to rifting. Major failure would be due to oceanization of crust by a local chelogenic event and would precipitate widespread global tension. All isostatic imbalance caused by horizontal compression would be removed as portions of crust moved back into equilibrium. When the affected crust cooled and regained its strength, global compression would be reestablished, and isostatic imbalances gradually would build up again, and be manifested as differential vertical movements in crust. The new load-bearing skeleton would tend to establish itself along paths that incorporate large igneous bodies, which would act as stress guides in the upper crust.

The stress-transmission concept has two important implications. First, sediments lying on competent basement cannot be deformed directly by horizontal compression, as it cannot be applied to them, and even if it could, they would simply deform at the point of application and not transmit the force. Thus, sediments can be deformed only if the whole basement shortens, or more probably, when a highly fractured basement is jostled under them. Second, the lower crust is weak, and can be expected to "heal" after being deformed, as it would not have the long-term rigidity to preserve deep structures. But the lower crust probably is affected only when rifts stack on top of each other an unusual number of times. In central Australia, Anfiloff and Shaw (1973) proposed all structures there occurred above a depth of roughly 20 km, and that deeper structures had been removed. Although this is still a valid observation, based on the observation that all four major gravity lows bottom at roughly the same depth, the new polycyclic rift model does not require deforming the lower crust at all.

In central Australia, rifts would have stacked downwards to a depth of about 20 km, to produce an extraordinary cluster of the largest gravity anomalies in the continent. To preserve this during the Phanerozoic, central Australia needs to be underplated by a thick root, which helps weld it together and maintain it as an elevated craton as shown later. It may be far from fortuitous that the largest gravity anomalies are clustered in the center of the continent, isolated from the thermal influences of the hotter oceanic environment.

SIGNIFICANCE OF TASMANIA, THE EASTERN HIGHLANDS OF AUSTRALIA, AND THE RANKIN TREND IN THE NORTH-WEST SHELF

The Eastern Highlands, along the eastern edge of Australia, are a narrow chain of passive mountains (Ollier, 1990) and form a continuous bulwark isolating vast regions of Phanerozoic subsidence onshore from the Tasman and Coral Sea basins offshore (Fig. 1). The parts higher than 500 m are confined to several small areas (Fig. 3) and mostly coincide with gravity lows (Fig. 4; Young, 1989) mostly associated with major granites (Wyatt et al., 1980). Most of the highlands are not significantly higher than central Australia, and the whole chain can be seen as the remnants of one or more cratonized chelogenic uplifts.

The highlands can be considered a superbasement ridge in which large granites act as stress guides for horizontal compression. The small Lake George drainage basin, situated in the highlands near Canberra (BMR, 1989b), has the same northerly trend as the highlands, and could represent compression in that direction. It is interpreted here as a modern rift because it has the characteristics of a compartmented rift system flanked by a basement ridge. It suggests that rifting can take place outside the environment of chelogenic subsidence, although it could be inherited from a previous chelogen.

In the scheme of the Australian load-bearing skeleton of compressed paths (Fig. 3), the highlands occupy key locations in the geometry of the balanced system. Without them, Tasmania's position would be untenable, and vice versa. Consequently, although largely in isostatic equilibrium, they might be acting as a basement ridge transmitting compression from Tasmania to New Guinea. Similarly, the Rankin Trend acts as the outer bulwark of the North-West Shelf and is characterized by large gravity anomalies, probably associated with mafic volcanics. It is heavy, and should have subsided, and given the bifurcating geometry of the ridge, it might be supported by horizontal compression; otherwise, the fact that it joins the mainland would be fortuitous (Anfiloff, 1988a).

ISOSTACY AND A KEY CHELOGENIC BOUNDARY

Horizontal compression and isostatic equilibrium are mutually incompatible. Artyushkov (1973, 1974) noted that isostatic equilibrium does not apply over narrow regions when the crust is regularly cut by fractures, and that fractures "strongly complicate distribution of the stresses." Bifurcating basement ridges may be the expression of that complication; the manifestation of local isostatic imbalance.

Wellman (1979) analyzed elevation and gravity data and concluded that isostatic equilibrium prevails over Australia. That analysis was too coarse to detect local imbalances, and failed to detect areas in eastern Australia where significant regional isostatic imbalance is suspected (Shirley, 1979; Anfiloff, 1982). Across the boundary between the Mount Isa Block and the Eromanga Basin region at DGG (Fig. 2), the elevation changes by about 150 m, and Bouguer and free-air gravity changes by about 40 mGal (Figs. 15 and 16). Any change in free-air gravity levels sustained over such a large distance signifies isostatic imbalance, presumably associated with the subsidence in the Eromanga Basin chelogen.

It is generally assumed that the level at which large blocks of crust float reflects their thickness and bulk density. But there also must be a small dynamic component in regional gravity differences, that represents imbalance associated with the thermally-induced forces that move crust up and down. The 40 mGal gravity difference represents a mass deficit in the Eromanga Basin block equivalent to about 5 km of sediments, assuming a 0.2 g/cm^3 contrast with basement (Fig. 16a). However, this long wavelength gravity change cannot be attributed to masses at any particular depth. Sediments alone do not necessarily explain the gravity difference. In Figures 16a and 17, the protruding base of the lithosphere would have to be removed by a process such as subcrustal erosion by a hot mantle bubble. This is an unbalanced situation, implying a dynamic process is operating faster than isostacy can cope, and could explain the present situation under the Eromanga Basin.

Without the erosion, 40 mGal of gravity low is equivalent to only 2.5 km of subsidence and the same amount of protrusion into the mantle (Fig. 16b), and this is also an unbalanced situation that has to be driven by an outside force. To achieve isostatic equilibrium, mantle has to replace part of the lithosphere, making it thinner. Thus, 5 km of sediments could be compensated by 5 km of mantle eroding upwards into the lithosphere, by moving up vertical crustal fractures. If the lithosphere is also high-level underplated by dense mantle material, it becomes denser, subsides more, and more sediments can be accommodated for the same 40 mGal change in gravity (Fig. 16c).

This third case depicts the basification of the lower lithosphere needed to produce isostatic equilibrium by

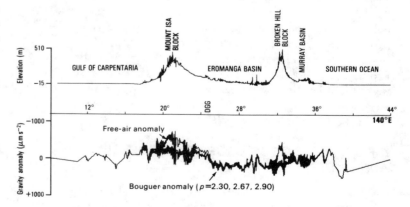

Figure 15. Significant isostatic imbalance is demonstrated by the gravity profiles along 140° longitude, which show regional differences in free-air gravity levels over the Mount Isa Block and Eromanga Basin chelogens over some 6° of arc on each side of the DGG (see Fig. 2).

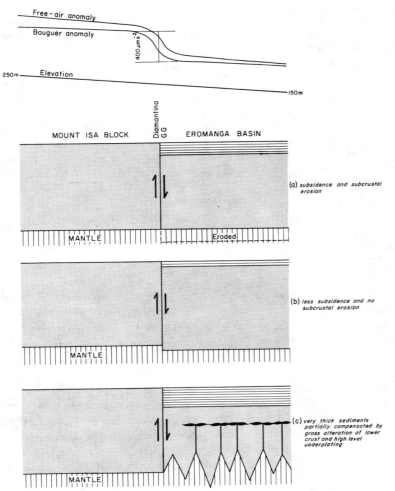

Figure 16. The interpretation of gravity across the edge of the Eromanga Basin at DGG is ambiguous because the long wavelength change cannot be assigned to masses at specific depths. Various combinations of sediments and alterations to the base of the lithosphere can produce the same gravity effect. The regional isostatic imbalance could represent the dynamic process underlying chelogenic subsidence, horizontal compression, or a combination of both.

sediments. The main question is what happens at the start of a thermal event—does it rise due to expansion, or does it subside as its buoyant roots are eroded? In the first case, subsidence would occur at the end of the cycle (the post-thermal sag phase), and in the second case, subsidence would occur at the start of the cycle.

If the base of a block of floating ice were heated, it would sink, not rise, because it would melt underneath and become thinner. This is an analogy for the subcrustal thermal-erosion concept. For crust to rise due to expansion, the heat has to be applied within the lithosphere where there is some rigidity, rather than at its base where it floats. The probable source of heat is hot rising mantle fluids (Beloussov, 1990), and they would erode the base of the lithosphere before they could rise into it to heat it. It seems it would be difficult to heat the upper lithosphere without first heating its base, and subsidence should therefore be the lithosphere's reaction to a heating event.

The problem can also be approached from the viewpoint of what is more prolonged;

balancing sediments with denser mantle in the right proportions. High-level underplating is suggested by sill-like seismic reflectors at the base of the crust in the Eromanga Basin (Mathur, 1983), and these can cause permanent basification. Moreover, if this happens on a large scale, it could act against the emergence of crust at the end of the chelogenic cycle.

THE DYNAMICS OF CHELOGENIC SUBSIDENCE AND UPLIFT

The foregoing discussion covered only the subsidence part of the chelogenic cycle. With cyclic ups and downs, the start of a cycle could be difficult to determine, but clearly, subsidence must be followed by uplift, otherwise continents would be completely covered by thick

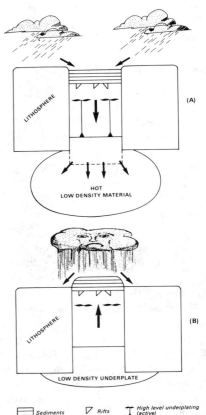

Figure 17. **A**—Chelogenic subsidence of a block of crust is attributed to a hot bubble eroding the base of the lithosphere. Sedimentation helps maintain the downward movement. **B**—When the bubble cools, a low-density underplate must replace the eroded material to push the block upwards, aided by erosion of the sediments. In some cases, a massive pile of sediments will slide off the top. If there is an excess amount of underplate, the region will become a stable uplifted craton.

uplift or subsidence? Uplifted cratons seem to be more enduring than broad basins, which have a relatively briefer existence. The elevation of cold cratons at the end of a chelogenic cycle appears to be an empirical fact, and indicates that uplift is indeed synonymous with cooling, not heating. But for a chelogen to rebound after 10 km of subsidence, it would seem that the material eroded at the base during subsidence has to be restored again by massive underplating when the thermal event wanes (Fig. 17).

In the chelogenic cycle, cratons rise to a height of about 1–2 km and subsidence reaches 10–20 km. This ratio of 1:10 in movement is similar to the ratio of the two operative density contrasts. Thus, the gravity effect of a typical craton with a density of 2.5 g/cm^3 protruding 1 km into the air is roughly the same as for a 10-km pile of sediments in a typical depression for a 0.25 g/cm^3 contrast with basement, suggesting the forces driving uplift and subsidence are roughly equal.

In the equilibrium situation, a craton has to be supported by positive buoyancy in the form of 10 km of low density root, and a basin full of sediments is supported by negative buoyancy in the form of mantle replacing 10 km of lithosphere. After subsidence, uplift would have to be propelled by a large mass of low-density underplate, and this would have to be related to the cooling down of the system. Perhaps acidic material comes out of solution from the mantle as it cools and lodges under the crust providing local buoyancy, and buoyancy might also vary with temperature.

During subsidence and uplift, sedimentation and erosion of sediments plays a key role (Grant, 1989). Subsidence could not continue beyond a certain amount if sediments did not fill the depression as it developed, and after 10 km of sediments accumulated, the uplifted part of the cycle could also not proceed unless sediments are off-loaded by erosion (Fig. 17). This may have happened in the case of the Kimberley (Basin) craton (Fig. 1), where the sediments may have resisted erosion, stabilizing the block before full emergence. Also, sediments can be eroded only if there are depressions alongside, or if rivers can carry them to the edge of the continent and dump them into oceanic basins. Conversely, if rivers flush sediments away from a basin as it is subsiding, it will not continue to subside. This could explain why the Murray Basin has only 0.5 km of sediments; it receives sediments from the Eastern Highlands, but they are not retained, and are flushed offshore by the Murray River. The flanking Bass and Eromanga basins (Fig. 1) have much thicker sediments.

The proposed workings of the chelogenic cycle are shown in Figure 17. A hot bubble lodges at the bottom of the lithosphere and initiates regional subsidence by eroding the base. As soon as subsidence begins, sediments accumulate, facilitating further subsidence. To achieve 10 km of subsidence and a reasonable degree of isostatic equilibrium, 20 km of lithosphere is removed. When the thermal energy of the hot bubble is expended, about 30 km of low density material comes out of solution from the mantle, and underplates the now thinned lithosphere. Twenty kilometers of this replaces the eroded part to propel the block upwards, and another 10 km maintains it indefinitely at a level 1 km above surrounding crust.

Overall, the longevity of cratons, such as the Archaean Yilgarn Shield, suggest that not all parts of the continent are equally susceptible to thermal events. This can be explained in terms of some cratons having deep roots that deflect hot mantle material to neighboring areas, the latter being repeatedly chelogenized while the craton remains intact (Fig. 18).

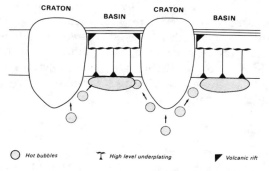

Figure 18. The long-term stability of cratons can be attributed to deep projecting roots deflecting hot bubbles to adjacent areas that are repeatedly reworked (chelogenized) by the heat. One aspect of the reworking involves high-level underplating as mantle fluids penetrate up fractures to the base of the crust and move sideways to the edges of cratons, and then up again to invade rift compartments fringing the cratons. Figure 20 shows this process applied to eastern Australia.

PERMANENT BASIFICATION AND OCEANIZATION

Continental crust would rebound after 10 km of chelogenic subsidence, as long as it was undermined by hot material with a lower density than the lithosphere. If the material is denser and it penetrates up crustal fractures, it could change the bulk composition of the crust sufficiently to prevent rebound. A key indication of crustal composition is the elevation of its surface: dense oceanic crust floats lower than less dense continental crust, and differs from a subsiding chelogen in that it does so without the aid of thick sediments. Consequently, where thick sediments can be demonstrated in an offshore basin, it might indicate underlying continental crust that could rebound later.

If chelogens are caused by thermal reworking of the lithosphere, then in some cases the amount of heat available would be sufficient to completely alter continental crust, or oceanize it (Fig. 19). Bulk crustal composition can be changed by penetration of mafic material, and extrusions at the surface would weigh the crust down further and mask it. Total oceanization would involve assimilation of felsic crust into the mantle, perhaps with the aid of convection. The replacement crust would be oceanic type, and would cool symmetrically inwards towards a central axis, producing symmetrical radiometric ages and magnetic lineations (Fig. 19). The overall process can be seen as the release to the atmosphere of heat trapped under the lithosphere.

The orthogonal gravity framework revealed by Seasat data for the Tasman Sea (BMR, 1989b) is identical to and coherent with that in the Eromanga Basin region (Fig. 2 and Appendix A). This remarkable revelation, together with the fact that the Tasman Sea Basin is framed by lineaments that seem to extend from the Australian mainland, suggests a large block of Australia has simply subsided. The process might have happened without grossly affecting the upper crust so that old rift systems were retained. Alternatively, the orthogonal Cardinal Fracture System was retained, and new rift systems developed in continental crust, and the third option is that the crust was completely oceanized, new orthogonal fractures formed, and new rifts developed along them.

A TRANSECT ACROSS EASTERN AUSTRALIA

Figure 20 shows a schematic interpretation from central Australia to the Tasman Sea. Central Australia is an elevated Proterozoic craton with rifts and granites from several chelogenic cycles stacked on top of each other. Eastern Australia is a region of Phanerozoic subsidence where perhaps one and a half cycles have occurred, but rift stacking has been less pronounced. The Eastern Highlands are remnants of several old cratons similar to central Australia. They are partly in equilibrium and partly maintained by horizontal compression, and flanked by vast

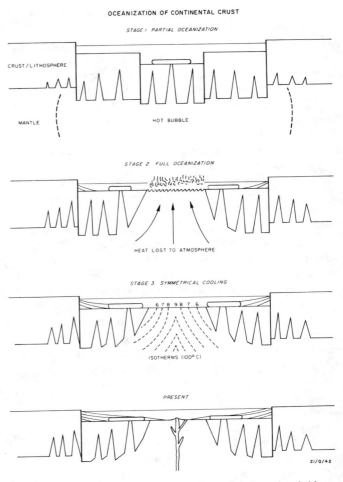

Figure 19. Oceanization is the extreme form of chelogenic subsidence when the crust is penetrated by hot mantle fluids and substantially transformed. When a whole section of continental crust is destroyed by excessive heat at Stage 2, this triggers adjustments in the global compression network, leading to synchronized rifting. When an oceanized region cools symmetrically towards the center, symmetrical belts of magnetic lineations and radiometric ages are produced (Stage 3). The end-product is a volcanic ridge along the central axis (Stage 4).

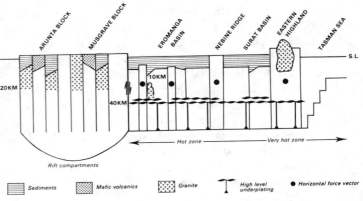

Figure 20. Schematic cross section from central Australia to the Tasman Sea. The large elevated cratonic block in central Australia is maintained by thick underplate, and erosion has exposed a complex system of stacked Proterozoic rifts consisting of sediments, granites, and mafic volcanics. The Eromanga Basin region has been through at least one and a half chelogenic cycles in the Paleozoic. Rifting occurred pervasively, and the crust was probably high-level underplated. Basement ridges cross the region, and the Eastern Highlands are a super-ridge and may represent the remnants of late Proterozoic chelogenic uplifts. All ridges are currently being supported by horizontal compression.

zones of chelogenic subsidence associated with increasing mantle temperature eastwards.

Heat flow measurements show that the basins in eastern Australia are hotter than the rest of Australia (Cull, 1982) and deep seismic profiling shows irregular flat reflectors at the base of the crust under the Eromanga Basin (Mathur, 1983), suggesting high-level underplating during subsidence.

The central Australian craton is elevated (Fig. 3), and would be supported by a thick lithospheric underplate. This strong base might help a craton deflect compression around it, and during subsidence, the weakening of this base by thermal erosion would facilitate strike-slip movements and thereby rifting. There are also factors associated with high-level underplating: perhaps mantle magmas facilitate rifting by lubricating fractures, or decoupling the crust from the lower lithosphere. Few magnetic dykes are evident beneath the Eromanga and Murray basins, whereas they are very prevalent in old cratons, such as the Yilgarn Shield. Perhaps dykes do not penetrate to the surface during subsidence, even though underplating material is accumulating under the crust. Instead, the material might move horizontally to the edge of the nearest craton and rise there (Fig. 18), producing the fringing mafic compartments shown in Figure 4. But the situation is far from clear in Central Australia, and the present craton as defned by high topography (Fig. 3) is wider than the one defined by fringing mafic compartments (Fig. 4).

DISCUSSION

Intracontinental tectonism can be classified into four categories of models. Only the fourth seems capable of explaining all facets of rifting, but it may require a special class of structure, bifurcating basement ridges, to explain the mechanism. The four classes are:

1. Two-dimensional space models, where a parallel gravity cluster develops in one major episode, driven by processes such as plate tectonics. This does not explain orthogonal systems and compartments.

2. Three-dimensional space models, where compartments are caused by wrenching associated with failed plate movements, but the problem with nonconjugate orthogonal systems remains.

3. Three-dimensional space-time models, where a parallel cluster develops gradually by the serial vertical addition of long compartmented belts following an established grain, without conflicting with orthogonal belts just outside the cluster. Dykes seem to fit this category.

4. Four-dimensional space-time models, where a parallel cluster develops gradually by the serial vertical addition of individual compartments following an established grain, and is compatible with all framework geometries and vertical stacking configurations.

Most previous models of tectonism in Australia belong in the first category, and are mobilistic models based on plate tectonics. There are major problems with the detachment rifting model of Etheridge et al. (1987), the accretion model of Wellman (1976), and the crustal buckling model of Lambeck (1984). A common problem has been the failure to separate the primary tectonic framework from exotic structures associated with wrenching in thick sediments. There is also the failure to account for the special problems posed by orthogonal systems and granites. Many gravity analyses have floundered on the granite problem, especially in central Australia. Granites and sediments should be seen as equivalent in terms of gravity lows, and the main problem is to explain how large elongate granites can be located directly under rifts containing undeformed platform cover sediments.

Wrenching models based on failed plate movements (Rosendahl, 1987) belong in the second category, and magnetic dykes in the Yilgarn Shield suggest the third category. But the fourth category is the only one that seems to explain all aspects of discrete rift development, including conjugate and nonconjugate orthogonal systems. It needs to be related to a local intracratonic mechanism to raise and lower the crust in what can be called a chelogenic cycle

Etheridge et al. (1987) and Middleton (1989) proposed diametrically opposite local convection cell concepts for sag basins, but there is no evidence in Australia of classic rift basins, resulting from failed extensional processes. Instead, rifting seems to follow the deposition of platform cover within a chelogenic depression. Middleton's downwelling model links sagging with compression, and is closest to the concept being proposed here, but it is preferable to have pure fixistic sag, because vast amounts of sediment are not deformed during subsidence. Compression should be divorced from the sag mechanism, and crustal fractures should be utilized to produce framed basins. The term sag then becomes a misnomer because a framed basin sinks rather than sags.

Some recent publications demonstrate a growing awareness that continental crust is a highly fractured mosaic of rigid crustal blocks. Pervasive vertical fractures are necessary for the development of bifurcating basement ridges, the ingredient needed to explain rifting in terms of wrenching driven by compression. The geophysical work in Kansas cited in this paper supports key aspects of the ridge-rift model in Figures 11 and 12. Orthogonality is fundamental, and nature of four-dimensional space-time wrenching can be seen in the

Goulburn-Malita Graben orthogonal rift systems cited earlier.

Basement ridges may have the vital role of explaining why crust subsides on only one side of the vertical main bounding fault of a true rift. In producing a rift trough by wrenching, it is envisaged that a narrow intracrustal gap perhaps only 1 km wide, translates upwards into an asymmetrical trough about 5–10 km deep and 30–50 km wide, because the flow of lithosphere into the gap occurs from one side only. In effect, the volume of the narrow gap converts to the volume of the trough.

The process is essentially vertical, and is so discreet that it cannot be linked to plate tectonics. Instead, it can be tied to a combination of global compression and roughly circular zones of chelogenic subsidence and uplift over the continent being driven by a local heat engine associated with large bubbles of hot mantle material lodging under the lithosphere. In essence, the heat causes a large piston of crust to move up and down.

There is strong evidence that after rifting, acid and mafic magmas often penetrate up dilated vertical fractures and occupy discrete rift compartments. This is called toothpaste tectonism. Rifting triggers these extrusions, but only when the material is available under the crust. Rifting therefore provides a window to the lower crust and mantle, and by accepting that it is pervasive (except in cratons), the presence and absence of various intrusives in various rifts provides an insight into processes at great depths. These findings support Gay's (1973) belief in the fundamental nature of the Cardinal Fracture System.

It is envisaged that a continent consisting of a mosaic of rigid crustal blocks is subjected to external compressive forces, while being subjected to random chelogenic cycles of regional subsidence and uplift. Rifting mostly follows regional subsidence; is a cold and passive process; and being driven by external horizontal compression, relies only indirectly on the chelogenic heat engine. It would be synchronized globally by sudden global changes in the normally pervasive and persistent compression regime. Cratons must be buoyed up and supported by thick roots, and this could explain why rifting is not prevalent there. Rifting might be predicated to subsidence because within a chelogenic depression, a thinner, hotter crust might be more susceptible to wrenching by compressive forces.

It is clear that in some parallel gravity clusters, groups of parallel rifts did not all form at the same time; they probably developed over several chelogenic cycles, selectively utilizing fractures with the same trend. Central Australia is the best example where a cluster has developed over several cycles, and rifts have stacked on top of each other. This in-phase overprinting implies fixistic vertical tectonism. The mafic belts between pairs of rifts may represent very old mafic rift compartments, which later acted as stress guides along basement ridges. Mafic compartments seem to be older than others and are often concentrated around the perimeters of ancient cratons. They may derive from high-level underplating in adjacent subsiding chelogens.

The coherence of the Cardinal Fracture System as manifested in rectilinear gravity lows is perhaps extremely profound in terms of indicating an intact continent in which the crust does not buckle or override itself. It responds to horizontal compression by slipping and sliding as blocks are jostled. Perhaps, large offsets cannot occur because movements are rapidly dissipated in many directions. This brings into question the very manner in which horizontal compression is brought to bear on a continent. It is proposed that the forces are applied radially, and are transmitted to the continent from oceanic crust along protrusions such as Tasmania, Cape Yorke, Cape Naturaliste, Fraser Island, and other narrow bathymmetric highs jutting into the oceans.

These forces are necessarily weak because oceanic crust is itself weak, and therefore continental crust cannot be bodily deformed by super-compression, unless it is generated under the continent, and there is no suggestion of that. Across the continent, the forces would be transmitted along the load-bearing skeleton, which is mapped as basement ridges in subsiding areas. In cratons, active compression is difficult to trace, but fossil basement ridges within which old mafic rifts are cratonized, might be represented as gravity highs that envelop gravity lows. The bifurcating gravity framework of Australia (Fig. 3) might therefore directly imply that the strongest parts of the seismogenic upper crust are the many igneous bodies cratonized in it. When earthquakes occur, the skeleton would reverberate as the energy is chanelled along it, and this would not be difficult to test experimentally.

Synchronous episodes of global orogeny can be explained in terms of suspension of global compression when a portion of oceanic crust loses its strength during oceanization. At such brief times, complete isostacy and tension prevail worldwide, and crustal blocks move vertically into equilibrium. When the afflicted crust cools and regains its strength, compression builds up again, and new unbalanced structures develop. Rifting occurs at both stages of adjustment into and out of isostatic equilibrium, when continental crust is jostled by horizontal forces. Basement ridges transmit the compression around the globe and small adjustments in their balanced network causes synchronized global earthquake activity.

CONCLUSIONS

In the past, synrift sediments and listric faults were seen as the main indicators of rifting, in necessarily extensional regimes. But in Australia, neither are important, and the main problem is to explain numerous occurrences of thick platform cover in asymmetrical troughs in a dominantly compressive environment. The first step is to recognize that the crust is highly fractured.

The main emphasis of this paper has been to present a detailed mechanism for producing rift troughs along the Cardinal Fracture System of vertical crustal fractures to explain the rectilinear-orthogonal framework of gravity lows in Australia. In simple terms, it involves wrenching and jostling a mosaic of rigid crustal blocks to produce narrow compartmented gaps in the crust and lithosphere. The substratum adjusts to this by flowing towards the gap, undermining the upper crust, which then sinks, resulting in a much wider trough at the surface. However, the trough is distinctly asymmetrical, and the crust on both sides of the gap could not have been subjected to the same forces. Thus, the basement ridge enters the picture as a key part of the dynamics of the system.

There is clear evidence in some regions where broad basin subsidence has been occurring, that basement ridges form bifurcating systems. Basement ridges trend in many directions, and it is not surprising that horizontal compression also has been found to be pervasive, and to operate in many different directions over the continent. So it is not too presumptuous to suggest that basement ridges exist to transmit compression. Rifts may represent the tension arising from this compression, and therefore the two main components of the gravity framework of Australia can be converted directly to the paleostress framework. The bifurcating (ridge) framework represents compression, and the orthogonal (rift) framework represents tension. Polycyclic rifting and the stacking of basement ridges on top of each other suggests the paleostress framework has been largely constant during the Phanerozoic.

Most of the continental framework consists of alternating ridges and rifts, and it is difficult to establish which is dominant. It is proposed that ridges can exist without rifts, but not vice versa. Ridges can be seen as the expression of pervasive horizontal compression where crust is subsiding, but elsewhere, there is no other structural evidence for compression other than the bifurcating framework of gravity highs. A continent being squeezed by radial compression might be dominated by ridges. To produce major rifts requires a ridge to move bodily a finite distance, so that a narrow gap forms down through the crust and lithosphere. It is envisaged that this can happen when the ridge is under continuous compression, but the opposing forces are not quite balanced (Fig. 12). Thus, on one side of the main bounding fault, the crust constantly is being supported by compression, even as it is being pushed away from the crust on the other side. This is crucial for explaining asymmetrical troughs, and makes the process far more complex than simple jostling and wrenching, although they are a good first approximation.

Attributing pervasive and persistent compression to a contracting Earth, or other sources of radial compression outside the continent, separates basin formation from the processes deforming the platform cover accumulating in basins. Any fixistic mechanism for pure crustal subsidence that produces framed basins is then consistent with Australian gravity, which shows that these movements have not disrupted the ancient Cardinal Fracture System. Having established these basics, it is not important what the mechanism actually is; it is described here as the chelogenic cycle of subsidence and uplift, driven by heat trapped under the lithosphere. This paper has attempted to integrate with this, processes such as high-level underplating, massive mafic volcanism into rifts around cratons, and oceanization, all of which could help understand the chelogenic concept, and processes in the mantle.

A contracting Earth has interesting possibilities for global tectonism. If a global bifurcating load-bearing skeleton can be shown to exist, and indeed, the surge-tectonics concept of Meyerhoff et al. (1989) already suggests it might, then plate tectonics would be in severe jeopardy.

REFERENCES CITED

Anfiloff, V., 1982a, Elevation and gravity profiles across Australia: some implications for tectonism: BMR Journal of Australian Geology and Geophysics, v. 7, p. 47-54.

———, 1982b, Gravity features in the Eromanga Basin: Eromanga Basin Symposium summary papers: GSA/PESA, Adelaide, Abstracts, p. 188.

———, 1983, Gravity evidence for the original fracture pattern of Australia: 5th International Conference on Basement Tectonics, Cairo, October, 1983, Abstracts, p. 12.

———, 1984, Pixel map of gravity with magnetic inserts, Murray Basin region: BMR Journal of Geology and Geophysics, v. 9, n. 4. (front cover)

———, 1988a, Polycyclic rifting—an interpretation of gravity and magnetics in the North West Shelf, in Purcell, P.G. and Purcell, P.R., eds., The North West Shelf, Australia: Proc PESA Symposium, Perth, p. 443-455.

———, 1988b, Rifting and basement ridges in the Murray Basin region, in Brown, C.M., and Evans, W.R., compliers, Murray Basin geology, groundwater and salinity management Conference: BMR Record 1988/7, p. 7-12.

———, 1989, The effect of vertical crustal fractures on the rifting process. Exploration Geophysics, Abstract, v. 20, n. 1/2, p. 175.

Anfiloff, V., and Luyendyk, A., 1986, Production of pixel maps of airbornemagnetic data for Australia, with examples for the

Roper River 1:1000,000 sheet: Exploration Geophysics, v. 17, p. 45-46.

Anfiloff, V., and Shaw, R. D., 1973, The gravity effects of three large, uplifted granulite blocks in separate Australian shield areas, in Mather, R.S., and Angus-Leppan, P.V., eds., Proceedings of symposium on Earth's gravitational field and secular variations in position: Sydney, 1973, p. 273-289.

Anfiloff, V., Barlow, B.C., Murray, A.S., Denham, D., and Sandford, R., 1976, Compilation and production of the 1976 Gravity Map of Australia: BMR Journal of Australian Geology and Geophysics, v. 1, n. 4, p. 273-276.

Artyushkov, E.V., 1973, Stresses in the lithosphere caused by crustal-thickness inhomogeneities: Journal Geophysical Research, v. 78, n. 32, p. 7675-7705.

——, 1974, Can the earth's crust be in a state of isostacy?: Journal Geophysical Research, v. 79, n. 5, p. 741.

BMR, 1976, Bouguer gravity map of Australia, 1:5,000,000 scale: Bureau of Mineral Resources, Canberra.

——, 1984, Bureau of Mineral Resources, Australia, Research Newsletter, n. 1., p. 5.

——, 1989a, Petroleum exploration and development in Australia-activity and results, 1988: Bureau of Mineral Resources, Australia, Record 1989/10.

——, 1989b, Yearbook of the Australian Bureau of Mineral Resources, Geology and Geophysics, July 1988 to June 1989.

Barosh, P.J., 1986, Neotectonic movement, earthquakes and stress state in eastern United States: Tectonophysics, v. 132, p. 117-152.

Beloussov, V.V., 1990, Certain trends in present-day geosciences, in Critical aspects of the plate tectonics theory: Theophrastus Publications, S.A., Athens, v. 1, p. 3-15.

Bevis, M., and Gilbert, L.E., 1990, Lineaments of the southeast and central USA: the case for a regionally organised crustal dislocation fabric, in Critical aspects of the plate tectonics theory: Theophrastus Publications, S.A., Athens, v. 1, p. 237.

Bradshaw, J., Nicoll, R.S., and Bradshaw, M., 1990, The Cambrian to Permo-Triassic Arafura Basin, northern Australia: APEA, v. 30, p. 107-127.

Brown, C.M., Jackson, K.S., Lockwood, K.L., and Passmore, V.L. 1982, Source rock potential and hydrocarbon prospectivity in the Darling Basin, NSW: BMR Journal of Geology and Geophysics, v. 7, p. 23-33.

Buck, W.R., 1986, Small-scale convection induced by passive rifting: the cause of the uplift of rift shoulders: Earth and Planetary Science Letters, v. 77, p. 362-372.

Campbell, I., 1990, The Port Campbell-Netherby north-northwest structural corridor in southeastern Australia: in Le Maitre, R.W., ed., Pathways in Geology, essays in honour of Edwin Sherbon Hills: Blackwell Scientific Publications, p. 280-303.

Chen W., 1988, A brief update on the focal depths of intracontinental earthquakes and their correlations with heat flow and tectonic age: Seismological Research Letters, v. 59, n. 4, p. 263-272.

Cull, J.P., 1982, An appraisal of Australian heat flow data: BMR Journal of Australian Geology and Geophysics, v. 7, n. 11, p. 21.

Denham, D., 1988, Australian seismicity—the puzzle of the not-so-stable continent. Seismological Research Letters, v. 59, no.4, p. 235-240.

Denham, D., Alexander, L.G., and Worotincki, G., 1979, Stresses in the Australian crust, evidence from earthquakes and in-situ stress measurements: BMR Journal of Australian Geology and Geophysics, v. 4, n. 3, p. 289-295.

Dewey, J.W., 1988, Midplate seismicity exterior to former rift basins: Seismological Research Letters, v. 59, no.4, p. 213-218.

Dillon, L.S., 1974, Neovolcanism: A proposed replacement for the concepts of plate tectonics and continental drift: Memoir No. 23, AAPG, p. 167-293.

Etheridge, M.A., Rutland, R.W.R., and Wyborn, L.A.I., 1987, Orogenesis and tectonic processes in the early to middle Proterozoic of northern Australia: American Geophysical Union, p. 131-147.

Etheridge, M.A., Symonds, P.A., and Powell, T.G., 1988, Application of the detachment model for continental extension to hydrocarbon exploration in extensional basins: APEA, v. 18, p. 167-187.

Finlayson, D.M., Leven, J.H., and Etheridge, M.A., 1988, Structural styles and basin evolution in Eromanga region, eastern Australia: Bull AAPG, v. 72, n. 1, p. 33-48.

Gay, S.P., 1973, Pervasive orthogonal fracturing in the earth's continental crust: American Stereo Map Co., Salt Lake City, 124 p.

Geoimage Pty, Ltd., 1989, Residual gravity map of Australia for wavelengths less than less than 250 km.

Goter, S.K., 1988, Map of the seismicity of California, 1808-1987: National Earthquake Information Center, USA.

Grant, A.C, 1989, Intracratonic tectonism-key to the mechanism of diastrophism, in New concepts in global tectonics, Chatterjee, S., and Hotton, N., eds.,: Texas Tech University, Lubbock, Abstracts, p. 13.

Gunn, P., 1984, Recognition of ancient rift systems: Examples from the Proterozoic of South Australia: Bull. ASEG, v. 15, p. 85-97.

Harrison, P.L., 1980, The Toomba Fault and the western margin of the Toko Syncline, Georgina Basin, Queensland and Northern Territory: BMR Journal of Australian Geology and Geophysics, v. 5, p. 201-214.

Hildebrand, G.M., Steeples, D.W., Knapp, R.W., Miller, R.D., and Bennett B.C., 1988, Microearthquakes in Kansas and Nebraska 1977-87: Seismological Research Letters, v. 59, no.4, p. 159-163.

Hills, E.S., 1956, A contribution to the morphotectonics of Australia: Journal of the Geological Society of Australia, v. 3, p. 1-15.

Himes, L., Stauder, W., and Herrmann, R.B., 1988, Indication of active faults in the New Madrid Seismic Zone from precise location of hypocentres: Seismological Research Letters, v. 59, no.4, p. 123-131.

Klassner, J.S., Cannon, W.F., and Van Schmus, W.R., 1982. The pre-Keweenawan tectonic history of southern Canadian Shield and its influence on formation of the Midcontinent Rift: Geological Survey of America Memoir 156, p. 27-46.

Koshtak, C.E., 1971, Models of block systems. Problems of Geomechanics: Yerevan, v. 5, p. 100-112.

Lambeck, K., 1984, Structure and evolution of the Amadeus, Officer and Ngalia basins of central Australia: Australian Journal of Earth Sciences, v. 31, p. 25-48.

Landes, K.K., 1952, Our shrinking globe: Bulletin Geological Society of America, v. 63, p. 225-240

Martin, B.D., 1989, Constraints on major right-lateral movements, San Andreas Fault System, central and northern California, in New concepts in global tectonics, Chatterjee, S., and Hotton, N., eds.: Texas Tech University, Lubbock, Abstracts, p. 20.

Mathur, S.P., 1976, Relationship of Bouguer anomalies to crustal structure in southwestern and central Australia: Australian BMR Journal of Geology and Geophysics, v. 1, p. 177-186.

Mathur, S.P., 1983, Deep crustal reflection results for the central Eromanga Basin, Australia: Tectonophysics, v. 100, p. 163-173.

McKenzie, D., 1978, Some remarks on the development of sedimentary basins: Earth and Planetary Science Letters, v. 40, p. 25-32.

McLennan, J.M., Rasidi, J.S., Holmes, R.L., and Smith, G.C., 1990, The geology and petroleum potential of the Western Arafura Sea: APEA, v. 30, p. 91-106.

Meyerhoff, A., Turner, I., Morris, A.E.L., and Martin, B.D., 1989, Surge Tectonics, *in* New concepts in global tectonics, Chatterjee, S., and Hotton, N., eds: Texas Tech University, Lubbock, Abstracts, p. 25-26.

Middleton, M.F., 1989, A model for the formation of intracratonic sag basins: Geophysical Journal International, v. 99, p. 655-675.

Moody, J.D., 1966, Crustal shear patterns and orogenesis: Tectonophysics v. 3, n. 6, p. 479-522.

Moss, F.J., and Wake-Dyster, K.D., 1983, The Australia central Eromanga Basin project: an introduction: Tectonophysics, v. 100, p. 131-145.

Murray, C.G., Scheibner, E., and Walker, R.N. 1989, Regional interpretation of a digital coloured residual Bouguer gravity image of eastern Australia with a wavelength cut-off of 250 km: Australian Journal of Earth Sciences, v. 36, p. 423-499.

O'Driscoll, E.S.T., 1982, Patterns of discovery—the challenge of innovative thinking: PESA Journal, v. 1, p. 11-31.

———, 1990, Edwin Hills and the lineament-ore relationship, *in* Le Maitre, R.W., ed., Pathways in Geology, essays in honour of Edwin Sherbon Hills: Blackwell Scientific Publications, p. 247-267.

Odins, J.A., Williams, R.M., and O'Neill, D.J., 1985, Use of geophysics for the location of saline groundwater inflow to the Murray River east of Mildura: Exploration Geophysics, v. 16, n. 2/3, p. 256-258.

Odins, J.A., Williams, R.M., O'Neill, D.J., and Lawson, S.J., 1991, Pre-Tertiary basement structure of the central Murray Basin, and its effect on groundwater flow patterns: Exploration Geophysics, v. 22, n. 2, p. 285-290.

Ollier, C.D, 1990, Mountains, *in* Critical aspects of the plate tectonics theory: Theophrastus Publications, S.A., Athens, v. 2, p. 211-236.

Roberts, D.C., Carrol, P.G., and Sayers, J., 1990a, The Kalladeina Formation—A Warburton Basin Cambrian carbonate play: APEA, v. 30, p. 166-183.

Roberts, A., Yeilding, G., and Freeman, B., 1990b, Conference Report: The geometry of normal faults: Journal of the Geological Society of London, v. 147, p. 185-187.

Rosendahl, B.R., 1987, Architecture of continental rifts with special reference to Africa: Annual Review of Earth Planetary Science, v. 15, p. 445- 503.

Rutland, R.W.R., 1982, On the growth and evolution of continental crust: a comparative tectonic approach: Journal of the Proceedings of the Royal Society of NSW, v. 115, p. 33-60.

Shaw, R.D., 1991, The tectonic development of the Amadeus Basin, Central Australia: Bureau of Mineral Resources, Australia, Bulletin 236, p. 429-461.

Shaw, R.D., Stewart, A.J., and Black, L.P. 1984, The Arunta Inlier: a complex ensialic mobile belt in central Australia. Part 2: Tectonic history. Australia: Australian Journal Earth Sciences, v. 31, p. 457-484.

Shirley, J.E., 1979, Crustal structure of north Queensland from gravity anomalies: BMR Journal of Australian Geology and Geophysics, v. 4, n. 4, p. 309-321.

Steeples, D.W., 1989, Structure of the Salina-Forest City interbasin boundary from seismic studies, *in* Steeples, D., ed., Geophysics in Kansas: Kansas Geological Survey, Bulletin 226, p. 31-52.

Sutton, J., 1963, Long term cycles in the evolution of the continents: Nature, n. 4882, p. 731-735.

Tucker, D.H, 1983, The characteristics and interpretation of magnetic and gravity fields in the Broken Hill district. Aus. I.M.M. Conference, Broken Hill, NSW, p. 81-114.

Tucker, D.H., and Boyd, D.M., 1986, Dykes of Australia detected by airborne magnetic surveys: *in* Halls H.C., and Fahrig, W.F., eds., Mafic dyke swarms: Geological Survey of Canada, Special Paper 34.

Tucker, D. H., Anfiloff, V., and Luyendyk, A., 1985, New large area standard format magnetic pixel maps of Australia: Exploration Geophysics, v. 16, n. 2/3, p. 294-299.

Tucker, D.H., Anfiloff, V., and Bagliani, F., 1986a, Albany; total magnetic intensity: Bureau of Mineral Resources, Australia, 1:1,000,000 Aeromagnetic Anomaly Pixel Map Series.

Tucker, D.H., Boyd, D.M., and Anfiloff, V., 1986b, Magnetic dykes of Australia, 1986—A map at 1:5,000,000 scale: Bureau of Mineral Resources, Australia.

Veevers, J.J., and Powell, C.M., 1984, Uluru and Adelaidean Regimes, *in* Veevers, J.J., ed., Phanerozoic earth history of Australia: Oxford, England, Clarendon Press, 270-289.

Wellman, P., 1976, Gravity trends and the growth of Australia: a tentative correlation: Journal Geological Society of Australia, v. 23, p. 11-14.

———, 1978, Gravity evidence for abrupt changes in mean crustal density at the junction of Australian crustal blocks: BMR Journal of Australian Geology and Geophysics, v. 3, n. 2, p. 153-162.

———, 1979, On the isostatic compensation of Australian topography: BMR Journal of Australian Geology and Geophysics, v. 4, p. 373-382.

———, 1985, Block structure of continental crust derived from gravity and magnetic maps, with Australian examples, *in* Hinze, W.J., ed., The utility of regional gravity and magnetic anomaly maps: Society Exploration Geophysicists, p. 102-108.

———, 1988, Development of the Australian Proterozoic crust as inferred from gravity and magnetic anomalies: Precambrian Research, v. 40/41, p. 89-100.

Wernicke, B., 1981, Low-angle normal faults in the Basin and Range Province: Nappe tectonics in an extending orogen: Nature, v. 291, p. 645-648.

Wu, F.T., 1988, Aspects of seismotectonics of eastern China and their implications for eastern U.S.: Seismological Research Letters, v. 59, n. 4, p. 251-261.

Wyatt, B.W., Yeates, A.N., and Tucker, D.H., 1980, A regional review of the geological sources of magnetic and gravity anomaly fields in the Lachlan Fold Belt of NSW: BMR Journal of Australian Geology and Geophysics, v. 5, p. 289-300.

Yeates, A.N., Gibson, D.L., Towner, R.R., and Crowe, R.W.A., 1984, Onshore geology, Canning Basin, *in* Purcell, P.G., ed., The Canning Basin, W.A.: Proceedings GSA/PESA Canning Basin Symposium, Perth, 1984, p. 23-55.

Young, R.W., 1989, Crustal constraints on the evolution of the continental divide of eastern Australia: Geology, v. 17, p. 528-530.

Zorin. Yu.A., and Lepina, S.V., 1989, On the formation mechanism of postrift intracontinental sedimentary basins and the thermal conditions of oil and gas generation: Journal of Geodynamics, v. 11, p. 131-142.

ACKNOWLEDGMENTS

The friendship and support of many colleagues in the Bureau of Mineral Resources and elsewhere is greatfully acknowledged. The job is not done until the paperwork is finished, and the completion of nearly two decades of research was encouraged and assisted by Tony Yeates and Mac Dickins. David Tucker and Tony Yeates reviewed the paper, and Dong Choi presented it at the conference on my behalf.

APPENDIX A— FRAMEWORK STUDY OF THE EROMANGA BASIN REGION

This study is based on a series of detailed seismic and gravity traverses surveyed by the Bureau of Mineral Resources over the central Eromanga Basin (Moss and Wake-Dyster, 1983). Figure A1 shows gravity data and the network of seismic lines, and Figure A2 shows the seismic data along those lines. In Figure A3, gravity has been computed directly from the seismic sections, and reveals other deep structures. Fault lineaments, troughs, and basement ridges are numbered 1–19, although not all are referred to in this paper. The Nebine Ridge (6), Adavale Trough (14), and Canaway Ridge (5) are well-known structures in this area, and form a bifurcating network. The major orthogonal linears MTGG and DGG shown in Figure 2 are numbered 1 and 2, respectively.

The tectonic framework in this region is fairly simple, and there is clear evidence of the Cardinal Fracture System. Modeling the gravity along the seismic lines confirms a good correlation between basement ridges and gravity highs, and gravity lows reveal numerous concealed rift troughs (and possible granites), which seismic data have not revealed because penetration is stopped by coals in Layer 4. In the regions not affected by rifting, there are numerous horsts, flanked by large monoclinal folds that represent deep vertical crustal fractures spaced 20–30 km apart. These simple, innocuous, structures present a far more serious problem for tectonics than is generally appreciated, especially if crustal shortening is not possible. They probably represent lesser basement ridges, but mapping them properly requires more seismic lines.

The Tipan Trough (11) has the classic orthogonal rift geometry shown schematically in Figure 9 where crustal blocks have been pivoted by a wrenching action. The same geometry is evident in the orthogonal Adavale (14) and Cooladdie (16) troughs in Figure A1, but here, it is more explicit that the compartmentalization of these two rifts is caused by ridge 18, the Pleasant Creek Arch, which is part of the main bifurcating ridge system. This is one of the best examples in Australia of how the two primary bifurcating (Fig. 11) and orthogonal (Fig. 9) systems combine to form the total system shown in Figure 12.

Along Traverse 1 (Fig. A4), seismic data reveal the Warrabin and Quilpie troughs where remnants of platform cover from an early Paleozoic chelogenic depression are preserved. Permo-Triassic Layer 4 contains coals that stop seismic penetration, obscuring other deep pockets of Devonian sediments farther west. Gravity reveals these pockets, demonstrating the futility of attempting to understand basement tectonics using seismic data alone. Layers 1–3 are the Mesozoic sediments of the Eromanga Basin, representing the second major cycle of chelogenic subsidence. They have been mildly deformed by Tertiary movements.

In the western half of Traverse 1, there are about seven monoclinal structures that could represent vertical crustal fractures. The broad magnetic high over the Warrabin Trough probably represents a granite directly under it at a depth of 20–30 km. This would account for the broad gravity low centered on the trough, whose western flank is clearly visible. A steep gravity high at the western end coincides with a basement ridge, and forms an elongate gravity high on ridge 4 between T3 and T1 in Figure A1. It represents a shallow mafic body cratonized in the ridge. Likewise, a gravity low situated at the intersection of ridge 5 (Canaway Ridge) and T6 (Fig. A3) represents a known granite, and these examples show how basement ridges seem to preferentially include old igneous bodies in the upper crust.

Along the whole length of Traverse 3 (Fig. A5), coals in Layer 4 have again stopped seismic penetration, and the concealed Warrabin and Tipan troughs are revealed only by gravity lows. A series of monoclinal structures reveal basement ridges, presumably bounded by vertical crustal fractures. Two ridges close together just east of the Tipan Trough may have bifurcated from a common ridge just to the

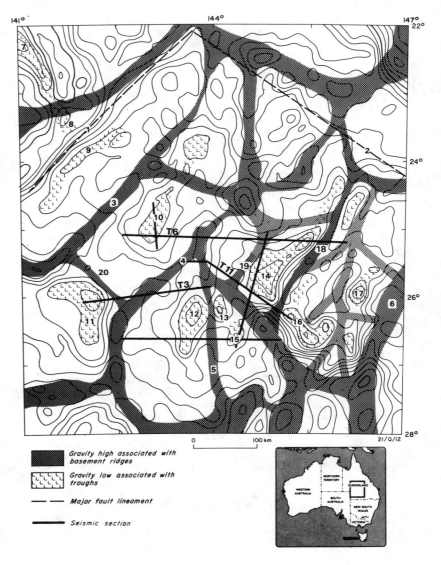

Figure A1. Gravity data in the central Eromanga Basin reveal a coherent orthogonal framework represented by major faults (1 and 2) and rift troughs (11, 14, 16, 17). This implies the Cardinal Fracture System. A bifurcating network of basement ridges envelops the rifts, and the manner in which rift compartments 14 and 16 are separated by ridge 18 is a good example of the main model in Figure 12.

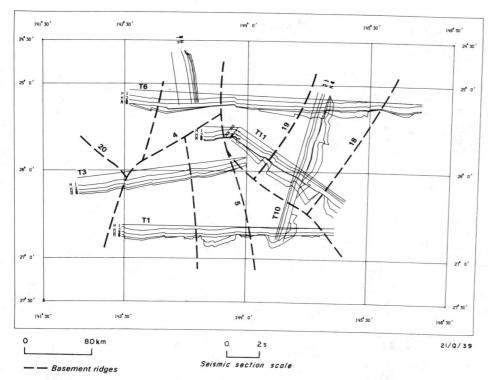

south. The trough is probably an old orthogonal rift (Fig. A6), but the ridge flanking it may represent the most recent Tertiary movements in which a new rift is developing on top of the old one. This may therefore be an example of rift stacking as depicted in Figure 14.

The development of basement ridges 18 and 19 from the Devonian to the present, shown in Figure A7, demonstrates the stacking of the framework on top of itself in successive chelogenic cycles. The ridges were already developed at the time the Warrabin and Quilpie troughs formed, then they were partly eroded during the main uplift. One ridge became active again in the Tertiary, pushing upward into the platform cover of the second chelogenic depression.

Figure A2. Seismic time sections reveal many monoclines (probably representing vertical fractures), basement ridges, and some rift troughs, but Permian coals in the fourth layer prevent deep penetration over much of the area. On Traverse 1, seismic shows two troughs containing the remnants of extensive Devonian platform cover from the first chelogenic cycle. The Eromanga Basin sediments in layers 1–3 represent the second cycle of subsidence.

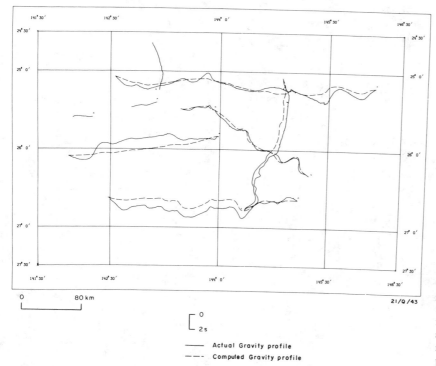

Figure A3. Comparing the gravity computed directly from the seismic sections with the actual gravity confirms the existence of bifurcating basement ridges, although more data would be needed to consolidate the picture. The gravity also reveals many concealed troughs below seismic basement, and seismic data is therefore a poor indicator of the primary tectonic framework.

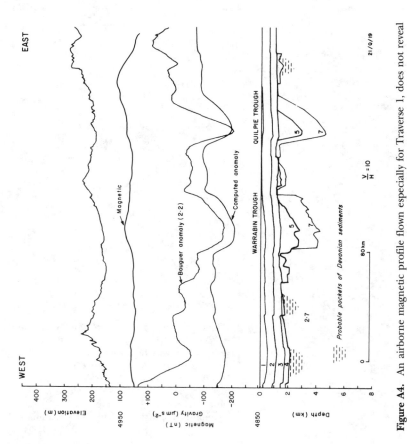

Figure A5. This depth section of Traverse 3 was accurately produced using velscan velocity data. The concealed Tipan and Warrabin troughs are below seismic basement (Layer 4), and represent rifting in the first chelogenic cycle. The basement ridge next to the Tipan Trough would represent the second cycle of rifting in the second cycle of chelogenic subsidence.

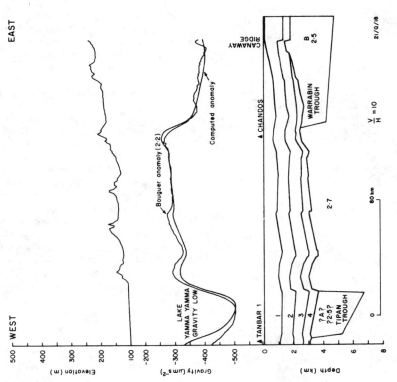

Figure A4. An airborne magnetic profile flown especially for Traverse 1, does not reveal basement topography. The undeformed platform cover sediments in the Quilpie Trough represent passive rifting, and the same sediments in the adjacent Warrabin Trough are faulted, and were probably wrenched. This trough seems to be underlain by a deep granite that produces a broad magnetic high and gravity low, and such deep granites are evidence of the control vertical fractures have on the rifting process.

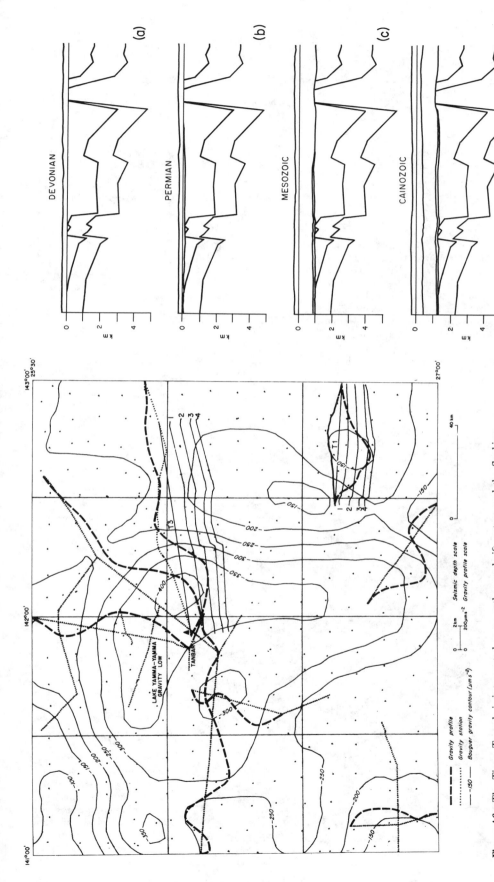

Figure A6. The Tipan Trough is interpreted as an orthogonal rift system and the flanking basement ridge on Traverse 3 connecting the two major ridges 3 and 4 in Figure A1 is an important part of the bifurcating network. The reconnaissance gravity has not revealed compartments in the rift, but one of the detailed gravity lines detected a basement ridge cutting across the east-west trending arm.

Figure A7. A simple palinspastic reconstruction of two chelogenic cycles. Basement ridges 18 (Pleasant Creek Arch on right) and 19 (Cothalow Arch on left) were largely developed at Devonian time, deforming sediments of the first cycle. Much later, platform cover of the next cycle was deformed by the Tertiary reactivation of ridge 18 (this was not repeated on ridge 19, and it is therefore not a compaction process). The reactivation implies a highly fixistic process: the stacking of ridges on top of each other, thus reinforcing the primary framework.

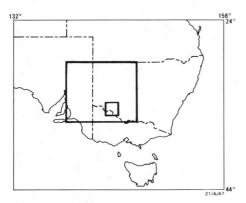

Figure B1. Location diagram showing locations of Figures B2 (large area) and B4 (small area).

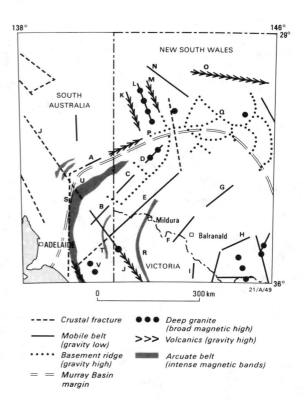

Figure B2. Gravity and magnetics in the Murray Basin-Broken Hill region reveal one of the best examples of the primary tectonic framework in Australia. The region is dominated by a profoundly rectilinear and often orthogonal framework of fractures and rift compartments filled with sediments, granites, and basic volcanics. The nonrectilinear part of the framework includes basement ridges that weave around compartments and bifurcate across the basin, and some arcuate belts. Arcuate features appear to be controlled by the rectilinear system, and are seen as superficial, caused by tectonic sliding of sediments from uplifted zones.

- - - - Crustal fracture
——— Mobile belt (gravity low)
· · · · Basement ridge (gravity high)
= = Murray Basin margin
●●● Deep granite (broad magnetic high)
>>> Volcanics (gravity high)
▓ Arcuate belt (intense magnetic bands)

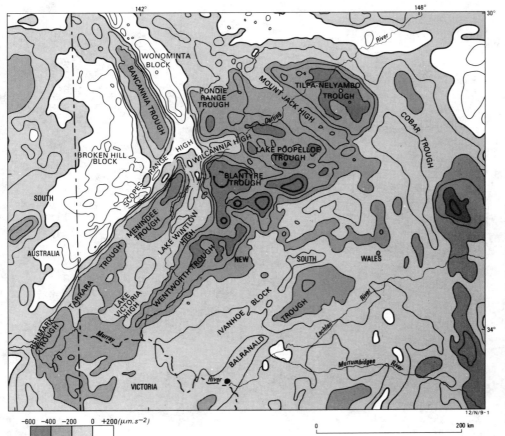

Figure B3. The gravity field demonstrates the primary tectonic framework of rift compartments and basement ridges. The Scopes Range High is a Proterozoic basement ridge, which may have seeded the network of younger ridges to the east. The Anabama granite is responsible for the elongate low half way up the left side of the figure and, being situated at A on the rectilinear system P-A-S in Figure B2, is a good example of toothpaste tectonism—the igneous invasion of a rift, presumably up the same vertical fracture along which the rift developed.

APPENDIX B — FRAMEWORK STUDY OF THE MURRAY BASIN REGION

In the Murray Basin region (Fig. B1), the main elements of the primary tectonic framework (Fig. B2) are revealed with extreme clarity by gravity (Fig. B3) and high resolution magnetic data (Tucker et al., 1985). A thin veneer, about 500 m thick of Tertiary sediments covers Paleozoic basement, and is transparent to gravity and magnetics. A basement ridge emanating from the Scopes Range High at P, is part of a network of bifurcating gravity features. To the south, the basin is separated from the coast by the Padthaway Ridge near V, and this ridge plays a key role in encircling the basin, making it a closed system, except for a small gap where the Murray River carries sediments out to sea. Such ridges suggest broad basins are encircled by compression.

Gravity data (Fig. B3) reveal the primary tectonic framework of rectilinear rift compartments and bifurcating basement ridges clearly, and strongly support the schematic framework shown in Figure 12. The Scopes Range High may be a fossil Proterozoic ridge, which has "seeded" younger basement ridges eastwards across the Murray basin. The Renmark, Tarrara, and Menindie troughs are three en-echelon rift compartments, and the Bancannia Trough is a conjugate compartment, and all contain early Paleozoic sediments (Brown et al., 1982). The Balranald Trough also consists of two compartments.

The Anabama Granite is situated at A on the same axis as volcanics, which form the rectilinear system P-A-S in Figure B2. The granite produces a distinct gravity low (Fig. B3) and is a good example of toothpaste tectonism, because it appears to have invaded a rift compartment situated on a vertical crustal fracture. It suggests single compartments can develop independently and be discreetly invaded by intrusives. A detailed gravity profile shows that the granite probably has the asymmetrical triangular cross section of a pristine rift and would have intruded the rift as shown in stage 4 in Figure 14.

Other rectilinear systems, such as L-D-C-B-J and E-J, also may represent vertical crustal fractures, whereas arcuate features at U, T, and R would represent structures in sediments, probably formed by tectonic sliding. Broad magnetic highs coaxial with troughs suggest granites occur under some rifts. The orthogonal-rectilinear system can be enhanced by combining gravity and magnetics (Anfiloff, 1984), and is very coherent. There are few signs of horizontal offsets in the many basement features visible in the high resolution magnetics of Tucker et al. (1985).

The Iona Belt (Figs. B4 and B5) was detected with gravity and refraction data (Odins et al., 1985), and could be an embryonic Tertiary rift. The ridge-trough combination is 10 km wide, has a vertical relief of 200 m, and appears on three gravity lines (Fig. B4). This linear belt is situated between the much older Balranald and Wentworth troughs (Fig. B3), and the latter have different properties; one is magnetically prominent, the other is magnetically transparent and probably did not form at the same time. These examples indicate that groups of parallel belts should not be attributed to one major episode of tectonism. There are strong indications in this region that rifting is a four-dimensional process in time and space; compartments develop individually over time, utilizing the pervasive Cardinal Fracture System.

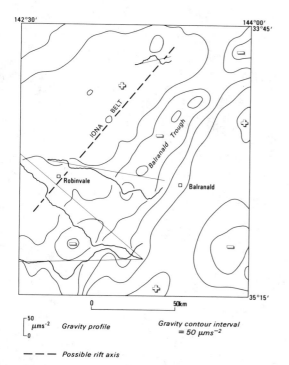

Figure B4. The Iona Belt, situated in the Ivanhoe Block between the Wentworth and Balranald troughs, extends over a distance of 100 km. It appears to be a new Tertiary rift developing in the current chelogenic subsidence of the Murray Basin, and shows that parallel belts can develop independently over a large time span, gradually using the available fractures of the Cardinal Fracture System.

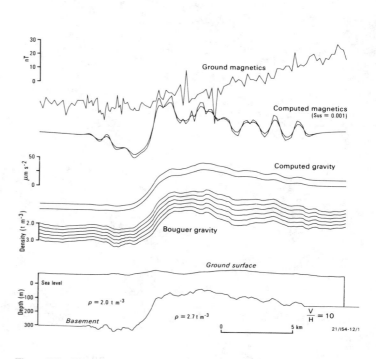

Figure B5. The Iona belt is a ridge-trough combination 10 km wide, with a relief of 200 m, This important structure has a good gravity definition, but is transparent magnetically. It is probably an an embryonic rift.

Conventional plate tectonics and orogenic models

S. E. Cebull and **D. H. Shurbet**, Department of Geosciences, Texas Tech University, Lubbock, Texas 79410 USA

ABSTRACT

The classical plate-tectonic paradigm incorporates two profoundly different explanations for orogeny, noncollisional (Cordilleran) and collisional. The collisional model, in more and more complicated form, increasingly has become favored in the interpretation of ancient orogenic belts, and some workers propose that collision (which includes accretion) may be the sole explanation of orogeny. This emphasis on collision garners part of its strength from weakness in the noncollisional model, which fails to include a trigger that converts continuous subduction into discontinuous orogeny. The appeal of the collision model as a unique explanation of orogeny is diminished partly by its failure to rationalize the nature of contemporary orogenesis along such chains as the Andes and around much of the rest of the Pacific rim.

Indeed, the two models are mutually contradictory, and it is concluded that neither is adequate alone, even in updated and more complex form. Within the framework of plate tectonics, subduction must be viewed as fundamental to a spectrum of orogeneses (excepting intraplate tectonism?) with collision being but an incidental by-product. If parts of the plate-tectonic explanation continue to require patching, nonplate alternatives should be reconsidered.

INTRODUCTION

The conventional plate-tectonic paradigm has been extraordinarily successful in explaining diverse tectonic phenomena, and, accordingly, has acquired a broad measure of acceptance in a relatively short time. By "conventional" we refer specifically to the model presently most widely accepted by geoscientists—that described in most English-language textbooks, at least in the United States—and characterized by plate motion on an Earth of essentially fixed volume and surface area. We exclude less orthodox variations such as that of Carey (1976), which invokes an expanding Earth and rejects the concept of subduction.

Despite the success of this conventional model, questions have arisen among its adherents, at first almost inadvertently and with little fanfare, suggesting that at least parts of the paradigm are not entirely satisfactory. One of these suspect areas is the plate-tectonic explanation of orogeny, which is the focus of this paper.

A DUAL EXPLANATION OF OROGENY

A long-standing and basic feature of the conventional paradigm is the incorporation of two distinctly different explanations for orogeny, as expressed in the terms noncollision (Cordilleran or subduction) and collision. Dewey and Bird (1970) depicted orogeny in their general noncollisional model as being thermally induced by rising masses or thermal fronts that apparently have their origins in the partial melting of a down-going lithospheric slab along B-subduction zones. Orogenic activity in the upper levels of the interior of the orogen was shown as basically extensional, although secondary, and associated with primary vertical stresses. The collision model portrayed orogeny as being mechanically driven, mainly by the impact of colliding continents or island arcs. Interiors of orogenic belts suffered extensive lateral shortening, principally as a product of resulting horizontal compressive stresses.

Subsequent research has led to refinement of these models as well as modifications in the dual-explanation paradigm. For example, greater emphasis has been brought to bear on the observation that convergent (orogenic) boundaries may be compressional or extensional, and, hence, on the thesis that orogenic activity may be a result of a broader spectrum of possibilities than suggested by the Dewey and Bird (e.g., Uyeda, 1982). Some workers added to the paradigm by stressing the possible role of orogen-parallel transcurrent (transpressive) displacements in orogenesis (e.g., Reading, 1980), of accretion (which we consider under the broad heading of collision), and of (postaccretion) intraplate tectonism (Coney, 1987). More or less concurrently, it was advocated that collision, in association with accretion, may be an adequate single explanation for orogeny (e.g., Ben-Avraham et al., 1981). Despite these modifications and additions, and the efforts to both restrict and amplify the plate-tectonic explanation of orogeny, the basic tenets of the dual explanation remain a significant part of current orogenic discussion. These two contrasting views of orogeny are reminiscent of old but long-standing preplate-tectonic arguments as to whether orogenies are primarily horizontal or vertical phenomena.

Perhaps more remarkable than the apparent requirement for two different mechanisms within a single paradigm is that neither the stark contrast between explanations nor the apparent need to more or less continuously amend and expand the explanations have fostered more than minimal overt concern among adherents of the conventional plate-tectonic model, although it appears that at least latent apprehension has been present for some time.

APPREHENSIONS AND RECENT TRENDS

Nearly from the inception of plate-tectonic theory, geologists indirectly expressed unease concerning this

---

Chatterjee, S., and N. Hotton III, eds. *New Concepts in Global Tectonics.* Texas Tech University Press, Lubbock, 1992, xii + 450 pp.

two-fold explanation of orogeny. Expression primarily took the form of an increasing disposition to interpret ancient orogenic belts in terms of the collision model, thereby minimizing utilization of the noncollision alternative. This tendency continued into the early 1980s, chiefly without serious discussion about its implications (exceptions include Gilluly, 1973, and some with antiplate positions), until Nur and Ben-Avraham (West, 1987) and Ben-Avraham et al. (1981) framed their concerns explicitly by questioning not only the necessity of two distinctly different models to explain what appeared to them to be a single phenomenon (mountain building), but, more specifically, the viability of the noncollisional (subduction) model. They wondered why all orogenies could not be explained adequately with a single mechanism, namely collision, and questioned the idea that mountains could form simply because one lithospheric plate dived beneath another, as required in noncollisional orogenies. Some of these questions were addressed subsequently by Dalziel (1984).

The results of the accelerating preference for the collisional alternative are evident. For example, at present most, and perhaps all, past orogenies of the continental United States are interpreted by at least some workers as being primarily the product of collision in one form or another (i.e., between fragments of diverse kinds, in varying tectonic settings, and with differing convergence angles), and this tendency to so interpret ancient orogens is by no means confined to North America. It is possible that the noncollisional model as an explanation of past orogenesis remains in the repertoire of tectonic geologists, even in its present limited role, chiefly because of the presence of the Andes of South America, an orogenic belt that has yet to be interpreted satisfactorily solely in terms of collision tectonics and that lies along a continental margin much of which is Paleozoic to Proterozoic in age (Megard, 1987; Lowman, 1989).

A relatively recent, but well-established by-product of the expanded utilization of the collision model is the increase in number and kind of potentially colliding bodies. Initially, collisions leading to orogenies were thought to take place chiefly between continents and continents, island arcs and continents, or between two arcs. Now, in addition, collisions are perceived to involve virtually any significantly elevated portion of the sea floor, such as plateaus, seamounts, and ridges.

Thus, if such presumably supplementary processes as off-scraping along accretionary wedges and certain orogen-parallel displacements are omitted, the interpretation of ancient orogenic belts in large measure has become a theory of collision (e.g., Coward and Ries, 1986), despite the suggestions for a more diverse explanation. Here, we consider to what degree this emphasis on collision, including accretion, is justified and, if unjustified, what alternatives are available. We begin by reviewing some aspects of orogeny.

OROGENY

"Orogeny" is an ill-defined tectonic concept and subject to widely varying definition and interpretation (e.g., Dennis, 1967; Cebull, 1973; Bates and Jackson, 1980). It is mainly a "fossil" concept; that is, the properties widely attributed to it are chiefly to be found in the rock record of the past, rather than by observation of contemporary activity. In this record, orogenies are characterized by intense internal rock deformation along long linear mountain belts, and at places, by associated regional synkinematic metamorphism. In older belts, mountainous physiography commonly is lacking due to erosion or, where present, may be of postorogenic origin. Orogenies appear to be temporally discontinuous events, commonly diachronous and polyepisodic. Their durations are difficult to determine with accuracy, but are on the order of millions of years to a few tens of millions of years. The specific deformational expression of an orogeny depends partly on the structural level exposed. The deepest of such levels available to direct observation, where deformation commonly is most intense, represents depths of approximately 50 km. Thus, orogenies are observed to occur at relatively shallow depths (although they may occur at greater depths as well).

Although some orogens are relatively well studied, most are not. It is clear that they are complex; they differ greatly in age, depth of erosion, presence or absence of special rock types, such as ophiolites or granitoids, internal structure, degree and kind of metamorphism, nature (or absence) of related fold-thrust belts, space-time progression of deformation, and type of sedimentary (or metasedimentary) rock involved in the deformation (Rodgers, 1987). Given this array of common variables, it is not surprising that they have defied simple classification.

The characteristics of orogeny outlined above, being primarily deformational rather than morphological (that is, mountain building), are ill-suited to the identification of presently active orogenic belts. In such belts evidence of contemporary intense structural deformation commonly is absent (buried?), weak, or obtained only indirectly. Nonetheless, we make the assumption, as do many others, that the tectonically active rim of the Pacific is representative of contemporary orogeny, as evidenced by ongoing seismicity, especially where associated with Benioff zones, volcanism, young mountain ranges, and, to a lesser degree, neotectonism.

SOME CRITICISMS OF THE NONCOLLISIONAL (CORDILLERAN) MODEL

Most of the basic reasons for comparative disuse of the noncollisional model of orogeny are both evident and of some antiquity (e.g., Dennis, 1982). In a sense, the model (Fig. 1) is a theory of orogeny that omits an explanation of why orogeny occurs (to paraphrase crudely Dana's well-known criticism of Hall's explanation of mountain building). First, it requires that an ongoing and comparatively continuous event, subduction (Scotese, 1987, suggested that circum-Pacific subduction has been active for about the last billion years), produces a discontinuous event, orogeny. The model itself contains no mechanism by which this continuous-to-discontinuous transformation is accomplished. Attempts to overcome this difficulty have resulted in a variety of suggestions that include the proposal that orogenies occur as a result of widespread, possibly worldwide, changes of rate or pattern of plate motion (e.g., Schwan, 1980), or because of a sort of discontinuous sticking of the down-going slab, perhaps due to relative rates and motions of adjacent plates. To date, none of these or other similar suggestions have proven demonstrable, perhaps at least partly because of difficulties inherent in the accurate dating of orogenic events, and the model remains devoid of a trigger to set off (or terminate) orogeny.

Second, the model fails to express a means of transmitting the stresses required for orogenic deformation into the core of the orogen. Efforts to depict a mechanism chiefly result in the introduction into the model of some form of rising thermal blob. Dewey and Bird (1970) envisioned a hot, rising mobile core. All such depictions represent attempts to rationalize the essentially vertical transmission of energy from near the top of the down-going slab, at depths of approximately 100 km (to 200 km and possibly deeper) into the upper 50 km of the crust. Thermal energy is developed, probably in the upper part of the down-going slab as a result of partial melting, transmitted upward in the form of rising magma bodies or thermal fronts, and converted into orogenic deformation at shallower depths. It has yet to be demonstrated (perhaps it is not intrinsically demonstratable) that this sort of mechanism is, in fact, operative.

A disturbing implication of the noncollisional model is that a down-going slab, commonly described as cold, can produce a thermally driven orogeny. Another troublesome implication is that such orogenies are fundamentally vertical tectonic events and, hence, the interior and adjacent exterior regions of orogenic belts should be characterized at least partly by structures that are a product of secondary lateral extension (rather than shortening), perhaps associated with gravity spreading, rather than of primary lateral compression.

Because of these and other nagging questions, combined with the inherent appeal of the collision model, it is not surprising that the noncollisional model presently is little used in the interpretation of ancient orogenic belts.

THE COLLISION MODEL

As an explanation of orogeny, the collision model is attractive in its own right, quite apart from its simply being more broadly acceptable than its noncollisional counterpart. It offers an explanation of orogeny that is direct, understandable, and intuitively satisfying. For example, it explains most structures in orogens in terms of lateral shortening through primary horizontal compression, an idea of considerable longevity and broad appeal (although rejected by such workers as Ramberg, 1967; Carey, 1976; and Beloussov, 1980, as well as others), and provides a clear explanation for the compressive stresses (that is, collision). In addition, it explains how subduction can be continuous and orogeny discontinuous, and predicts the occurrence of exotic terrains (which, of course, also may be emplaced by virtue of strike-slip displacement) the presence of which have troubled geologists for decades.

However, some aspects of this model also are disturbing. Perhaps the most important of these is the implication that if the noncollisional orogenic model is deemed inadequate, then collision is the sole cause, or essentially the sole cause, of orogeny. Patterns of contemporary tec-

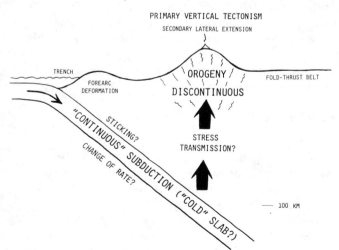

Figure 1. Cartoon illustrating fundamentals of noncollisional (Cordilleran) model of orogeny. Depicts continuous, active subduction and discontinuous orogeny and possible orogenic triggers (sticking and change in subduction rate), vertical transmission of energy from top of slab to shallower crustal depths, and thermally induced vertical-extensional orogeny. See text for further explanation.

Figure 2. Simplified outline map of South America showing principal locations of apparently ongoing collision. Stippled areas are chief zones of active volcanism; seismicity extends along entire western margin of continent (modified from Frutos, 1981, and Cebull and Shurbet, 1990). Size of elevated sea-floor features is exaggerated to emphasize location. Several more detailed portrayals are available (e.g., Ben-Avraham and Nur, 1987, fig. 3).

tonic activity offer some compelling reasons for supposing otherwise. The Andean belt of South America (the "type" for the noncollisional model), as well as much of the rest of the circum-Pacific orogenic system, demonstrates that collision is not the principal cause of ongoing tectonic activity. For example, the Andes display seismicity over their entire length (Gates, 1987), and volcanism over much of it, indicating current, extensive, geographically continuous orogenic activity. Yet, as outlined in Figure 2, the locations of possible ongoing collision are restricted significantly (Cebull and Shurbet, 1990). The chief positive topographic features that may be in collision with the Andean belt are the Nazca Ridge, Carnegie Ridge, Juan Fernandez Islands, San Felix Islands, and Chile Rise (an active spreading center). Each is a long, narrow feature that is oriented approximately perpendicular to the Andean chain, so regions of apparent collision amount to perhaps no more than ten percent of the total length of the orogen. Hence, most of the ongoing orogenic activity along the Andes cannot be due to these contemporary collisions.

Interestingly, Frutos (1981) showed that seismic density is reduced in the area of some (Nazca, San Felix, and San Fernandez ridges) of these collisions, and Ben-Avraham et al. (1981) indicated that gaps in volcanism are promoted by collision, at least by the Nazca and Juan Fernandez ridges. If these observations are correct, it appears that collision may inhibit much of the very activity (seismicity and volcanism) that serves to identify contemporary orogeny.

Nonetheless, some workers have suggested that the Andean belt of the past can be explained in terms of collision tectonics (e.g., Frutos, 1981). Although there is supporting evidence for some past collisions (e.g., Megard, 1987; Ramos, 1988), available data do not support the conclusion that the ancient Andes are simply explained or are exclusively and fundamentally a product of collision. Indeed, Megard (1987, p.203) indicated that "... the Peruvian Andes appear to be the very type of liminal fold belts which are exclusively related to subduction processes."

The inadequate number of contemporary colliding bodies to account for the near-continuous pattern of evident orogenic activity along the Andes also is characteristic of most of the Pacific margin. Indeed, only in a few areas does current tectonic activity clearly appear to be related to collision. It is even more evident that collision also fails to account for the shape and character of the many island arcs of the region.

Still another troubling aspect of the collision model is that it is wedded to the premise of lateral compression and attendant horizontal shortening in orogenic belts. As widely accepted as this premise is, acceptance is not universal. For example, it is rejected by many who oppose plate tectonics generally (e.g., Beloussov, 1980) and by others who reject conventional plate tectonics but otherwise support the concept of plate motion (e.g., Carey, 1976). In addition, and perhaps of greater importance, increasing numbers of adherents of conventional plate tectonics interpret the contemporary presumably orogenic tectonics of most island arc systems as fundamentally extensional (e.g., Hamilton, 1988).

Indeed, primacy of horizontal compression is favorable neither for the injection of the massive synorogenic batholiths that characterize many ancient orogenic belts nor for the volcanism that characterizes present-day island arcs—it is difficult to imagine the emplacement of such magmatic materials into the upper crust under conditions of virtually continual lateral shortening. Accordingly, many models that invoke primary lateral compression call upon some ad hoc modification, such as the assumption of periods of relaxed or extensional stresses, to accommodate intervals of extensive magmatic intrusion.

DISCUSSION

Whatever the merits of the two models individually, in at least one important respect they appear mutually contradictory. The collisional model demands conditions of lithosphere consumption along subduction zones, just as does the noncollisional model, but postulates no orogenic effects until collision. The only function of subduction as regards the origin and development of orogenic belts is to permit collision (hence, orogeny); subduction itself is an orogenically passive phenomenon. In the noncollisional model, subduction is the cause of orogeny (although, as mentioned earlier, the model provides no trigger). Whether subduction by itself causes orogeny or fails to do so, it is unlikely that subduc-

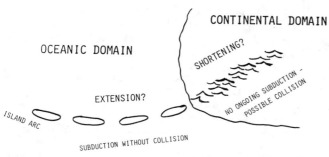

Figure 3. Cartoon of a hypothetical active orogenic belt that extends from oceanic into continental domain, such as the Aleutian arc into Alaska (as this depiction resembles) or the Andaman arc into Burma (view through back side of page). Island arcs in the former domain commonly are considered to be fundamentally extensional; mountain belts that extend into the continental domain commonly are interpreted as compressional by virtue of collision (accretion).

tion that ultimately leads to collision differs fundamentally from that which does not. Nonetheless, the plate-tectonic model suggests that there ought to be some substantive difference between orogenically "active" and "passive" subduction.

This apparent contradiction focuses attention on those places on oceanic crust where island arcs extend into what are essentially continental domains, as illustrated in simplified form in Figure 3. Examples include the Indonesian arc extending into Burma and the Aleutian arc into southern Alaska. If these features are contemporary orogenic belts, they indicate that orogenic activity may proceed without the benefit of collision, as along many island arcs, but that collision with attendant orogeny may also take place, perhaps tens of millions of year earlier, at other locations along the same chain. Thus, orogenic activity that in some respects resembles that depicted in the noncollisional model occurs in essentially extensional (island arc; Mariana type of Uyeda, 1982) regimes, while apparently compressional (continental orogen) regimes prevail elsewhere along the same belt. This common occurrence suggests that orogeny may have many faces and a diverse causality, as emphasized by Rodgers (1987). It might be anticipated legitimately that highly contrasting modes of orogenesis would produce orogens of significantly or, at least perceptibly, different type. However, such distinctive types have not been delineated clearly in the rock record.

Nonetheless, if conventional plate-tectonic models are to explain orogeny, it may be better to view the classical noncollisional and collisional models, probably along with that of transpressive or transform displacement, as part of a continuum of orogenic possibilities, as implied by Uyeda (1982) and as illustrated in Figure 4, than to regard collision as a sole explanation. If this broader perspective is correct, orogenies would be ex-

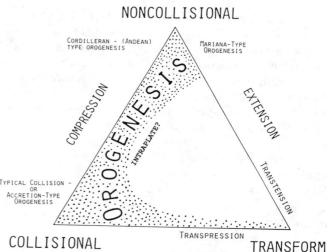

Figure 4. Continuum of possible explanations of orogeny from the current perspective of plate tectonics. Chilean type of Uyeda (1982) is in the approximate position of the Cordilleran type shown here. Complexity and variety of explanations can be interpreted either as progressive modernization or as regressive patching of the classic paradigm.

pected to be of diverse types or combinations of types. They may be collisional, noncollisional, and possibly transpressive (or transcurrent), or a combination thereof, and they may be extensional or compressional, and, perhaps both, even along the same belt and at the same time. Coney (1987) indicated that some may be intraplate. Orogenic histories would be expected to vary appreciably from place to place along such belts. Such disconcertingly complex an explanation has yet to be verified by observations of the admittedly diverse character of ancient orogens.

CONCLUSIONS

We conclude that neither model in the two-fold explanation for orogeny depicted in conventional plate-tectonic theory is sufficiently comprehensive to explain the apparent complexities of orogenies and, moreover, that both together are inadequate. We suspect that collisional regimes have been no more common in the past than they apparently are at present and, hence, that the concept of collision tectonics may be overused in the interpretation of ancient orogens. Inasmuch as the symptoms of contemporary orogeny are associated most closely with subduction (an obvious exception is that of orogenic activity produced by transcurrent displacement), we suggest that in the conventional plate-tectonic model subduction must be fundamental to orogeny (again, with some minor exceptions), and that collisions are better viewed as only occasional and secondary by-products. However, both collisional and noncollisional models may represent part of an broader explanation.

Given the framework of modern plate tectonics, we opt for the more complex model of orogeny, one that accommodates the views of those who support a primarily vertical-tectonic origin of orogeny as well as those who emphasize horizontal shortening, and those who emphasize extensional origins as well as those who favor horizontal compression. Nonetheless, we admit to some discomfort with the resulting complexity and remaining unresolved difficulties. Although the increasing complexity of the plate-tectonic explanation of orogeny commonly is viewed as a product of the refining and modernization of an initially oversimplified paradigm (e.g., Monger and Francheteau, 1987), it may also be interpreted as the result of a series of ad hoc attempts to bolster an inadequate model.

As regards the remaining difficulties in the plate-tectonic explanation, we are especially concerned by the lack of a trigger mechanism for the noncollisional aspects of the broad orogenic model. This lack leaves us, along with others, to wonder why a plate simply diving beneath another would produce an orogeny. Possibly, noncollisional orogenies are triggered by diminished rates or cessation of down-going movement of the subducting plate, which, in turn, may be affiliated with changes in plate motion. Normal subduction rates may result in some lateral compressive stresses across an orogen that are inadequate to produce orogeny, but if subduction ceases or greatly slows, relaxation—representing the trigger—may allow thermally instigated energy to rise and initiate orogenic activity (which implies that resulting orogenies are fundamentally vertical phenomenon). Subduction slowing may be indicated by seismic gaps in Benioff zones, by seismicity and focal mechanisms of shallow and intermediate depth, or by less obvious means. Orogenies during slowed subduction may be affiliated with trench rollback in the fundamentally extensional regimes typical of island arcs and have no important element of lateral compression. However, to the present the matter remains unresolved.

We suggest that an extended reconsideration (even a debate) concerning the adequacy of plate-tectonic explanations for orogeny is both desirable and overdue. If, after reconsideration, plate theory continues to require patching, we suggest that nonplate alternatives be carefully reexamined.

REFERENCES CITED

Bates, R.L., and Jackson, J.A., eds., 1980, Glossary of Geology, (second edition): American Geological Institute, Falls Church, Virginia, 571 p.

Beloussov, V.V., 1980, Geotectonics: Springer-Verlag, New York, 488 p.

Ben-Avraham, Z., and Nur, A., 1987, Effects of collision at trenches on oceanic ridges and passive margins, in Monger, J.W.H. and Francheteau, J., eds., Circum-Pacific Orogenic Belts and Evolution of the Pacific Ocean Basin: American Geophysical Union, Geodynamic Series V. 18, p.9-18.

Ben-Avraham, Z., Nur, A., Jones, D., and Cox, A., 1981, Continental accretion: From oceanic plateaus to allochthonous terrains: Science, v.213, p.47-54.

Carey, S.W., 1976, The Expanding Earth: Elsevier, New York, 488 p.

Cebull, S.E., 1973, Concept of orogeny: Geology, v.1, p.101-102.

Cebull, S.E., and Shurbet, D.H., 1990, Fundamental problems with the plate-tectonic explanation of orogeny, in Augustithis, S.S., ed., Critical Aspects of the Plate Tectonic Theory: Theophrastus Publications S. A. v. 2, p. 435-444.

Coney, P.J., 1987, Circum-Pacific tectogenesis in the North American Cordillera, in Monger, J.W.H. and Francheteau, J., eds., Circum-Pacific Orogenic Belts and Evolution of the Pacific Ocean Basin: American Geophysical Union, Geodynamic Series, V.18, p.59-69.

Coward, M.P., and Ries, A.C., eds., 1986, Collision Tectonics: Blackwell Scientific Publications., Geological Society Special Paper 19, 420 p.

Dalziel, I.W.D., 1984, Circum-Pacific orogenies, in Howell, D.G., Jones, D.L., Cox, A., and Nur, A., eds., Proceedings of the Circum-Pacific Terrane Conference: Stanford University, Stanford, California, p.79.

Dennis, J.G., ed., 1967, International Tectonic Dictionary: American Association Petroleum Geologists, Memoir 7, 196 p.

——, ed., 1982, Orogeny: Benchmark Papers in Geology, v.62: Hutchinson Ross Pub. Co., Stroudsburg, Virginia, 379 p.

Dewey, J.F., and Bird, J.M., 1970, Mountain belts and the new global tectonics: Journal Geophysical Research, v.75, p.2625-2647.

Frutos, J., 1981, Andean tectonics as a consequence of sea-floor spreading: Tectonophysics, v.72, p.T21-T32.

Gates, S.K., compiler, 1987, Global Distribution of Seismicity 1977-1986 (map): National Earthquake Information Center, Washington, D. C.

Gilluly, J., 1973, Steady plate motion and episodic orogeny and magmatism: Geological Society of America Bulletin, v.84, p.499-514.

Hamilton, W.B., 1988, Plate tectonics and island arcs: Geological Society of America Bulletin, v.100, p.1503-1527.

Lowman, P.D., Jr., 1989, Comparative planetology and the origin of continental crust: Precambrian Research, v.44, p.171-195.

Megard, F., 1987, Structure and evolution of the Peruvian Andes, in Schaer, J.-P. and Rodgers, J., eds., The Anatomy of Mountain Ranges: Princeton University Press, Princeton, New Jersey, p.179-210.

Monger, J.W.H., and Francheteau, J., eds., 1987, Circum-Pacific Orogenic Belts and Evolution of the Pacific Ocean Basin: American Geophysical Union, Geodynamic Series V. 18, 165p.

Ramberg, H., 1967, Gravity, Deformation and the Earth's Crust: Academic Press, New York, 214 p.

Ramos, V.A., 1988, The tectonics of the Central Andes; 30° to 33°S latitude: in Burchfiel, B.C. and Suppe, J., eds., Processes in Continental Lithosphere Deformation: Geological Society of America Special Paper 218, p.31-54.

Reading, H.G., 1980, Characteristics and recognition of strike-slip fault systems, in Bullance, P.F. and Reading, H.G., eds, Sedimentation in oblique-slip mobile zones: International Association Sedimentologists, Spec. Pub. no.4, p.7-26.

Rodgers, J., 1987, Differences between mountain ranges, in Schaer, J.-P. and Rodgers, J., eds., The Anatomy of Mountain Ranges: Princeton University Press, Princeton, New Jersey, p.11-17.

Schwan, W., 1980, Geodynamic peaks in Alpinotype orogenies and changes in ocean-floor spreading during Late Jurassic-Late Tertiary time: American Association Petroleum Geologists Bulletin, v.64, p.359-373.

Scotese, C.R., 1987, Development of the Circum-Pacific Panthallassic Ocean during the early Paleozoic, *in* Monger, J.W.H. and Francheteau, J., eds., Circum-Pacific Orogenic Belts and Evolution of the Pacific Ocean Basin: American Geophysical Union, Geodynamic Series V.18, p.49-57.

Uyeda, S., 1982, Subduction zones: An introduction to comparative subductology: Tectonophysics, v.81, p.133-159.

West S., 1982, A patchwork earth: Science 82, v.3, p.46-52.

ACKNOWLEDGMENT

We thank Paul Lowman, Jr. for his constructive criticisms of the manuscript.

Geological evidence for a simple horizontal compression of the crust in the Zagros Crush Zone

Mansour S. Kashfi, 2685 Mum Drive, Richardson, Texas 75082 USA

ABSTRACT

In the popular view, ophiolites are linked with a subduction zone and plate collision. The advocates of plate tectonics interpret the ophiolites of Iran and elsewhere in the Middle East to be remnants of the oceanic crust. They are of the opinion that the Zagros Crush Zone, an alleged suture zone, marks the site of a former continent-continent collision. However, the mere presence of an ophiolite belt is by itself not proof that it was the site of a former subduction zone.

It is argued here that no subduction zone is, or ever was, present in the greater Persian Gulf area. The major proposed subduction zone of the Middle East—the Zagros suture or Zagros Crush Zone—has no associated volcanic rocks, and therefore lacks the lithological suites that typify so-called subduction zones. Moreover, basement-rock exposures and basement reached in deep boreholes indicate that most of the Zagros Fold Belt is underlain by continental, not oceanic crust.

Earthquake-intensity and earthquake-epicenter maps show that shocks of magnitude 6.5 and greater take place in many areas far removed from proposed subduction zones.

Because identical lithofacies of the same ages, Late Precambrian through Late Tertiary, are present on both sides of the alleged subduction zone should be enough to give one cause for thought. Well-known and closely related faunas have been described from Early and Late Paleozoic strata on both sides of the Zagros suture for many decades.

The most compelling evidence that definitely ties Arabia and the Indian subcontinent to Asia is the stratigraphic sequence of the Salt Range. The Salt Range or Punjab Saline series of Late Precambrian-Cambrian age extends southwest from the Himalaya zone and the Indian Shield to Pakistan, Iran, and across the Persian Gulf into Oman, Dhofar, and South Yemen.

Disagreements on the plate tectonics of the Middle East region are many. Some advocate the former presence of two or more plates in this region; others have postulated several microplates; other workers support island-arc interpretations; and a majority favor the existence of at least one suture zone that marks the locus of a continent-continent collision. Nearly all of these hypotheses are mutually exclusive. Most would cease to exist if the field data were honored. These data show that there is nothing in the geologic record to support a past separation of Arabia-Africa from the remainder of the Middle East.

INTRODUCTION

For more than two decades, numerous earth scientists have tried to explain the geology of south-southwestern Iran and part of the Middle East (Fig. 1) within the framework of plate tectonics (Morgan, 1968; Le Pichon, 1968; McKenzie, 1970; Smith, 1971; King, 1972; Nowroozi, 1972; Takin, 1972; Dewey et al., 1973; Haynes and McQuillan, 1974; Welland and Mitchell, 1977; Gealey, 1977; Bird, 1978; Berberian and King, 1981; Berberian, 1983; Jackson and McKenzie, 1984a, 1984b; Sengör, 1983 and 1984; Dewey, et al., 1986). Each of these workers proposed a different model and usually denied the validity of previously proposed models. For example, some interpreted the region in terms of movements of microplates; a few tried to apply island-arc theory to the area; others contended that they identified a suture zone of a former continent-continent collision. The purpose of this paper is to discuss these contradictions and inconsistencies, and to point out some unanswered questions regarding a plate-tectonic interpretation of the Middle East geology.

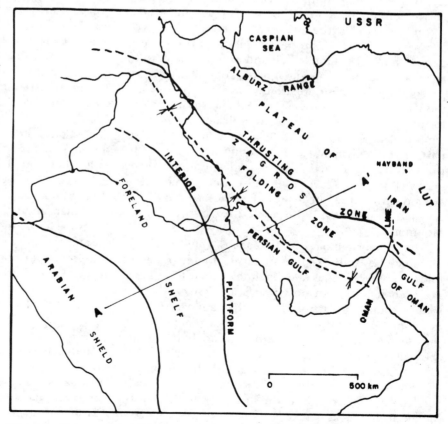

Figure 1. Major structural features of the Middle East.

Chatterjee, S., and N. Hotton III, eds. *New Concepts in Global Tectonics.*
Texas Tech University Press, Lubbock, 1992, xii + 450 pp.

GEOLOGIC SETTING OF THE REGION

The Iranian Plateau formed as a result of tectonic forces that folded and faulted the two bounding ranges (Fig. 1). The Alborz Mountains strike eastward and are folded and thrust toward the sunken foreland of the Caspian Sea; the Zagros Mountains occupy the western part of Iran and portions of Iraq and extend from eastern Turkey to Bandar Abbas at the edge of the Persian Gulf. The Zagros Mountains rise gradually northeastward from an unfolded shelf, extend through a belt of intact anticlines and synclines into a thrust belt and imbricated zone, and merge into an intrusive and metamorphic zone.

Geosynclinal conditions existed here from Cambrian into Cenozoic time and were part of the main Tethys geosynclinal system. At least as early as Jurassic time, two tectonic zones in the Zagros geosyncline were becoming differentiated, as was a much wider platform zone in Arabia. The Zagros, Himalayas, and Alps are the final peripheral products of the vast Tethyan mobile belts of subsidence. Tectonic activity still continues along this line of crustal weakness, as evidenced by frequent earthquakes.

The Zagros Mountains can be divided longitudinally into two zones. The southwestern part is primarily a zone of strong folding that forms the shelf area to the west; it is referred to as the western part of the Zagros geosyncline. The thrust zone occupies the northeastern part of the Zagros mobile belt and extends eastward to an area of metamorphic and igneous rocks. This eastern part of the geosyncline apparently was developed during the Late Cretaceous orogeny, and the western part was formed in later Tertiary time, as indicated by the rock types and more intensive deformations in the eastern part of the geosyncline. Thus, some workers refer to these two zones as the older and younger Zagros geosynclines, respectively. Several periods of compressive stress from the northeast in late Mesozoic and Tertiary time produced the structure of the Zagros Mountains.

The western part of the Zagros geosyncline is dominated by unbroken, asymmetric, doubly plunging folds—large anticlines and synclines similar to those of the French Jura—but on larger scale. There are also some subordinate thrust faults that appear to have developed from simple folds. Anticlines generally have the steeper flank on the southwest and have slightly sheared crests.

The eastern part of the Zagros geosyncline is characterized by northeast-dipping thrust faults. Paleozoic rocks have been brought to the surface by these thrusts. Metamorphic rocks, Paleozoic and Mesozoic in age, occupy much of the northeastern zone of the overthrusting; in places, they are thrust southwestward over younger rocks.

COLORED MELANGE, RADIOLARITES, AND OPHIOLOTES

It is generally popular to link the origin of the common association of colored melange, radiolarites, and ophiolites of the Middle East with a subduction zone and plate collision. Advocates of plate tectonics seem to be of the opinion that a subduction zone must exist wherever ophiolites are present within a major thrust zone, such as the Zagros Thrust Belt (Fig. 1).

Dewey et al. (1973) believed that sea-floor spreading during Late Cretaceous time is indicated by the presence of (1) ophiolitic and metamorphic rock debris with some deformation in southern and southeastern Turkey, (2) ophiolites in the Zagros Crush Zone and in Oman, (3) basalts and andesites in the Alburz Mountains (northern Iran) and southern Caucasus, and (4) ophiolites along the western margin of the Lut Kavir (block) (Fig. 1).

Dewey et al. (1973) wrote that in Late Jurassic time "Additional subduction is also indicated, for the first time, on the southern margin of central Iran by deformation, weak metamorphism, emergence, and granite intrusion in the Sanandaj Zone, and by general deformation and incipient metamorphism throughout central Iran." And further mentioned that "Subduction on the southern margin of Iran is indicated by andesites." They also suggested that flysch deposition and volcanic activity in the Zagros Crush Zone and thrusting in the Zagros Thrust Belt are indications of Early Cretaceous to Late Cretaceous-Paleocene subduction along the southern edge of central Iran. Moreover, ophiolites were obducted onto the northern margin of the Arabian block during Late Cretaceous time.

Furthermore, they pointed out that the presence of Paleocene andesites in the eastern Carpathians, and dacites and tuffs in the Caucasus, suggests the presence of former subduction zones. Dewey et al., concluded that bathyal cherts and carbonates associated with pillow basalts and ophiolites in southern Turkey suggest the development of oceanic crust during the Late Triassic separation of Turkey from Africa.

DISCUSSION

Ophiolites

If the above statements are correct, it seems that many additional fossil subduction and divergent zones exist in various places in the Middle East. For example, in Iran, there are many places, well removed from the Zagros Crush Zone, where Jurassic and Cretaceous ophiolites are exposed (Fig. 2). Ruttner et al. (1968), Ricou (1970),

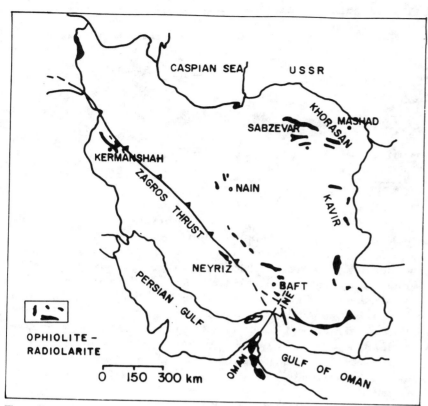

Figure 2. Ophiolite-radiolarites of Iran and Oman.

Stöcklin (1974), and Tehrani (1975) reported ophiolite zones in eastern, central, and northern Iran in association with radiolarites.

In eastern Iran, from Baft in the southeast through Nain to the Kavir and Province of Khorasan, ophiolites are exposed (Fig. 2). These rather large ultramafic bodies were mapped and described by Vialon et al. (1972) and Stöcklin (1974). They consist of basalt associated with spilitic lava, agglomerate, and tuff, with chert of various ages. Do they also indicate the former presence of subduction or divergent zones? Or former oceanic crusts? The same argument was pointed out earlier for the Alps (Hsü and Schlanger, 1971).

The age of the ophiolites in the Middle East has been, and still is, a subject of debate and controversial opinions. Dewey et al. (1973) used Early Jurassic and even Early Cretaceous, for the times of emplacement of ophiolites. Stöcklin (1974) stressed that emplacement of ophiolites took place in Late Cretaceous time and was completed in Campanian or Early Maastrichtian time. The age of emplacement of the ophiolites south of Mashad (Fig. 2) is definitely Early Jurassic (Lias) (Tehrani, 1975).

In central Iran, the ophiolite association were emplaced in Paleocene to Eocene time, and those of the Zagros Crush Zone in Late Campanian time (Vialon et al., 1972; Welland and Mitchell, 1977).

The original rocks of the Sabzevar ophiolite (Fig. 2) consist of green amphibole, plagioclase, and garnet (Tehrani, 1975), which represent the amphibolite facies of regional metamorphism. Ricou (1971) also reported an exceptional case of contact metamorphism around a peridotite-gabbro complex in the Zagros ophiolites. Associated with this complex are bodies of granitic gniess with locally pegmatitic veins. In the Zagros ophiolites, particularly near Neyriz, large bodies of diopside-bearing marbles exist. These marbles were interpreted by Ricou (1971) and Haynes and McQuillan (1974) also to be a result of the contact metamorphism. Advocates of plate tectonics call these marbles metamorphosed parts of the oceanic crust (Dewey et al., 1973; Berberian, 1983). However, they could be fragments of the metamorphosed continental basement because they are present in other parts of Iran.

The ophiolites of Neyriz and Kermanshah along the Zagros Crush Zone (Fig. 2) in Iran are regarded by many as indications of ancient oceanic crust and provided the name for the suture zone along which supposedly the Arabian and Iranian plates, two ancient continents, were welded together (Haynes and McQuillan, 1974; Bird, 1978). However, the existence of ophiolites alone could be interpreted as remnants of the early orogenic phase of the formation of the eugeosynclinal part of the Zagros geosyncline.

The Zagros Mountains rise gradually northeastward from an unfolded shelf, extend through a belt of intact anticlines and synclines into a thrust belt and imbricated zone, and merge into an intrusive and metamorphic zone in central Iran (Fig. 3). This state of affairs is an indication of the structural unity across the greater Persian Gulf area. Geologists who have studied the geology of southwestern Iran recognize that what is called ophiolite in Kermanshah is really a radiolarite (NIOC, 1978), consisting of thin-bedded, varicolored, argillaceous material separated by thin bands of dark grey micritic carbonate with a few chert beds. Radiolarite is not sufficient proof for the former presence of oceanic crust subsequently contracted or compressed.

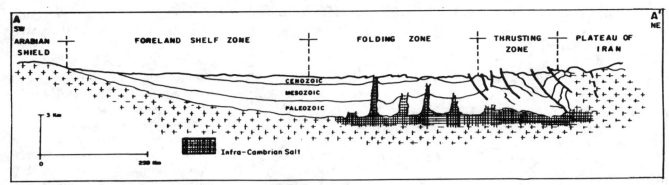

Figure 3. Regional section across the structural zones of the Zagros geosyncline.

Coleman (1984) stated that it is not clear if ophiolites of the Middle East are remnants of back-arc basin crust, midoceanic-ridge crust, or fragments of a small ocean produced by intracontinental rifting and attenuation. He believed that the Oman ophiolite represents oceanic crust formed at a spreading center in the Neo-Tethys during the Late Cretaceous, where its emplacement was nearly synchronous with those ophiolites of the eastern Mediterranean, Syria, and Iran. Berberian and Berberian (1981) disagreed, arguing that the subduction of the Oman oceanic crust may have started much earlier than Cretaceous and that the Oman oceanic crust is older than Cretaceous. Woodcock and Robertson (1984) concluded that the ophiolite terranes in the Tethyan belt are structurally too variable to conform to one emplacement model.

Ricou (1970) interpreted the sheets of sedimentary rock sequences together with ophiolites of the Neyriz area (Fig. 2) as nappe sheets emplaced during Late Cretaceous time. The same mechanism was discussed by Glennie et al. (1973) for the Oman Mountains. The Late Cretaceous Semail ophiolite of Oman is one of the largest ophiolite bodies in the world. However, the seismicity on the Arabian continental margin and field investigations seem to indicate that the Oman Semail ophiolite is a rootless slab lacking communication with the mantle underneath (Gealey, 1977; Welland and Mitchell, 1977; Searle et al., 1980; Searle and Graham, 1982).

Stratigraphic data support the view that the Semail ophiolite is a totally allochthonous body, squeezed between sedimentary sequences. Undisturbed Maastrichtian sediments overlying the ophiolite in the Oman Mountains indicate a minimum of tectonic activity after emplacement. This is in contrast with the Zagros thrust, which continued to be active through Tertiary time well after ophiolite emplacement. Therefore, the emplacement of these ophiolites may be a result of continental basement uplift due to the compressional shortening.

Brinkmann (1972) believed that the Mesozoic radiolarite-ophiolite association in Turkey is the result of crustal tension prior to regional compression, supporting his conclusion on the basis of stratigraphic and paleontologic correlations. Khudoley and Meyerhoff (1971) and Meyerhoff and Meyerhoff (1974) stated that ophiolites are present above continental crust in the entire Caribbean-Mediterranean area. The same conclusion was reached previously for the western Caribbean and Central America by Dengo and Bohnenberger (1969).

Volcanic activities

The supporters of plate tectonics believe that volcanic belts should form about 100 km above the subducting plates (Hamilton, 1988), but Mesozoic and Cenozoic volcanic rocks are unknown in areas adjacent to the alleged subduction zone in Iran (Fig. 4). The Zagros Crush Zone (suture zone) is generally free of volcanic rocks, which should be present if plate tectonics were prevailing phenomenon in this area. Nowroozi (1972) proposed that because the Iranian plate is moving more slowly northward than the Arabian plate, magma was unable to penetrate the overlying sediments. However, two extensive cores of volcanic rocks trending nearly northwest-southeast are located about 200–800 km northeast of the Zagros Thrust Belt (Fig. 4). Here, Nowroozi (1972) explained these cores of volcanic rocks by postulating that the Iranian plate and the Caspian plate were moving northward faster than the Arabian plate. As a consequence, a tensional regime was established north of the alleged subduction zone to make conduits for extrusive rocks. However, the stress system required by this hypothesis contradicts that which Nowroozi (1972) used to explain the absence of volcanic rocks at or near the alleged subduction zone. In central Iran, volcanic activity began in Late Cretaceous time and continued through Early Tertiary time. Volcanism was mainly submarine and geographically scattered.

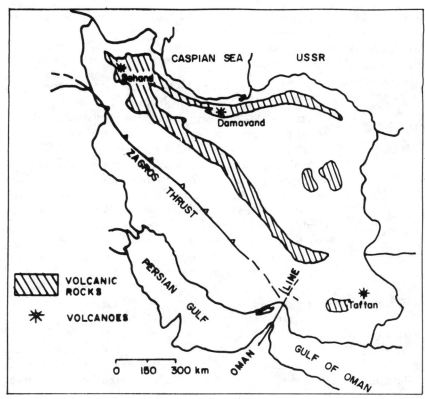

Figure 4. Outcrops of volcanic rocks in Iran.

There are several Late Tertiary-Quaternary, continental, extinct volcanoes in Iran, also randomly scattered. These include Taftan, Damavand, and Sahand (Fig. 4). They show no alignment, and are far from the putative subduction zone.

Basement rocks

The basement characteristics of the Zagros Fold Belt indicate that it is underlain by igneous-metamorphic complexes of nonoceanic origin (Falcon, 1967a; McKenzie, 1972; Berry and Moore, 1975). The strongest evidence is provided by exotic blocks of granite, basalt, ultramafic rocks, gniess, and schist found in the emergent Hormuz Salt that stratigraphically lies directly on the basement. These exotic rocks are actual samples of the basement (Ala, 1974; Kent, 1979; Kashfi, 1976, 1985, 1988). Some outcrops of continental basement rocks (granite, granodiorite-diorite, and gabbros) are scattered throughout the Zagros Thrust Belt and adjacent areas (NIOC, 1978) (Fig. 5), a further indication that the Zagros is underlain by continental crust.

If the basement under the Zagros is continental, its specific gravity and density are too low for it to have been underthrust deeply into the mantle. Thus, a large part of the convergence must be absorbed in the fold belt. Then one has to determine how much crustal shortening has occurred in the fold belt. On the other hand, a northeastward component to reduce the amount of crustal shortening would probably involve large amounts of strike-slip movement in Zagros for which there is no evidence (NIOC, 1978). Sylvester (1988) classified the Zagros Thrust Belt as indent-linked strike-slip faults, apparently in an attempt to solve this problem. Further, the relationship of the Paleo-Tethys compressional events of the Zagros Thrust Belt and Oman to the extensional events in Oman remains unsolved

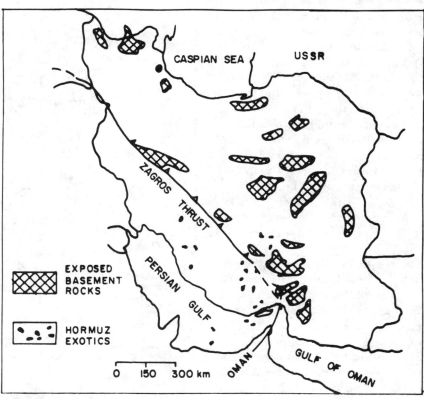

Figure 5. Basement outcrops of Iran.

(Sengör 1984). However, if oceanic crust has been subducted beneath the Iranian plate (due to a change of density at depth, as advocates of plate tectonics claim), then it is difficult to explain the present distribution of the ophiolites (oceanic crust) such as Nyriz, Kermanshah, and ophiolite zones reported in eastern, central, and northern Iran, now exposed at the surface.

A similar situation is observed in the Taurus Fold Belt of Turkey, which is the westward continuation of the Zagros Fold Belt (Yalcin and Görür, 1984). Brinkmann (1972) argued that the Taurus Ranges are underlain by continental crust rather than by oceanic crust. This interpretation and the lateral continuity of the geology eliminate the idea that once-separated blocks are now joined along the Turkish Taurus Mountains (Wolfart, 1967; Meyerhoff and Meyerhoff, 1974).

Gulf of Oman

Advocates of plate tectonics state that the tectonics of the region around the Gulf of Oman (Fig. 1) are characterized by intense deformation resulting from Late Cretaceous plate collision. However, the Gulf of Oman is seismically quiet and earthquakes are generally shallow. The available seismic and bathymetric data do not support the existence of a subduction zone in the Gulf of Oman at any time (Wilson, 1969). Some supporters of plate tectonics state that the Zagros thrust, an alleged subduction zone, continues southeastward into the Oman Mountains (Welland and Mitchell, 1977; Gealey, 1977), but the Zagros thrust does not extend that far south (Kolla and Coumes, 1987) (Fig. 1). The intense folding and dislocation of eugeosynclinal sediments and emplacement of Semail ophiolite in Oman Mountains are probably attributable to gravity sliding.

In any case, the Gulf of Oman is a continuation of the greater Zagros geosyncline, which extends southeastward through it. The northwestern part of this geosyncline was closed, folded, faulted, uplifted, and exposed by Late Tertiary time. In contrast, the eastern-southeastern extension of this regional geosyncline, including the Gulf of Oman, still is in the stage of folding, either because fewer compressive forces have been exerted since Early Tertiary time, or because of absorption of the compressive stress in the Late-Proterozoic Hormuz Salt layer near the base of the sedimentary column.

Earthquakes

The Zagros Fold Belt is characterized by a large number of earthquakes in the magnitude 5 or 5.5 range (Richter scale), and a very small number with magnitude equal to or greater than 6. Most of the earth-quakes occur in the crust (Berberian and Tchalenko, 1975).

Mohajer et al. (1978), on the basis of available isoseismal maps of Iranian major earthquakes, constructed an intensity distribution map of Iran. Figure 6 exhibits observed and calculated intensity distribution of Iranian earthquakes in a modified Mercalli scale, according to which Mohajer et al. (1978) divided Iran into five zones ranging from zero to four. Zone zero includes intensities III and less; zone one includes intensities IV and V, corresponding to Richter magnitude range 4.0–4.7; zone two includes intensities VI and VII, corresponding to Richter magnitude range 5.0–5.9; zone three intensities VIII, IX, and higher, corresponding to Richter magnitude 6.5 and greater; zone four covers areas with earthquakes of unknown instrumental magnitude. A glance at Figure 6 shows that high-intensity earthquakes

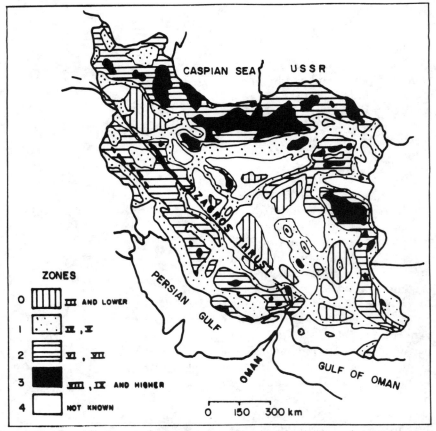

Figure 6. Intensity distribution of major shocks in Iran (after Mohajer et al., 1978).

with Richter magnitude 6.5 and greater take place in areas far from the alleged Zagros subduction zone.

Low seismicity in the southeastern part of Iran and Oman indicates a continuity between the eastern part of the Oman high (Figs. 1 and 6) and eastern Iran. This interpretation is supported by the lack of epicenters in the general area of the Gulf of Oman. Therefore, the concept of a Precambrian Oman-Ural lineament, which was postulated originally by Furon (1941), later by Gansser (1955), and modified by Falcon (1967b), is supported by seismic data.

During Tertiary folding and thrust faulting, shortening of the crust took place. Plate-tectonics advocates believe that the crustal shortening is a result of the convergence of the Afro-Arabian and Eurasian plates (Crawford, 1972; Takin, 1972; Dewey et al., 1973; Haynes and McQuillan, 1974; Rotstein and Kafka, 1982). They promote the idea that seismicity defines positions of the descending slabs that are overridden by advancing upper plates (Hamilton, 1988), and that much of the shortening in the Zagros was accomplished by underthrusting of the Afro-Arabian plate beneath the Iranian plateau. However, current seismicity is concentrated southwest of the Zagros Crush Zone in the Zagros Fold Belt (Nowroozi, 1971; Berberian and Tchalenko, 1975; and GSI seismotectonic map of Iran 1976), not behind surface expressions of a subduction zone northeast of the Zagros (Fig. 7). This observation implies that the shortening of the crust (or plate consumption in plate-tectonics concepts) is essentially under the fold belt and not near the alleged subduction zone as plate tectonics requires.

Salt Basin of the Middle East

The most compelling evidence that ties Arabia and the Indian subcontinent to Asia is the stratigraphic sequence of the Salt Range, northern Pakistan. The Salt Range or Punjab Saline series of Late Precambrian-Cambrian age extends southwestward from the Himalaya zone and the Indian shield to Pakistan, Iran, and across the Persian Gulf into Oman, Dhofar, and South Yemen. An excellent correlation has been established between the Punjab Saline series and Hormuz Salt (latest Precambrian, Figs. 3, 8) (Mina et al., 1967; Wolfart, 1967 and 1969; Stöcklin, 1968; Ahmed, 1969; Lowman and Tiedemann, 1971; Meyerhoff and Meyerhoff, 1974; Kashfi, 1976, 1985, 1988; Jiqing and Bingwei, 1987).

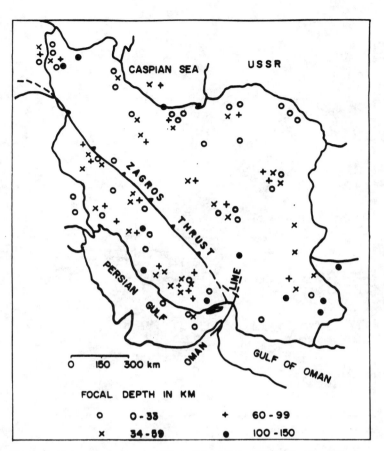

Figure 7. Seismotectonic map of Iran (adapted from GSI Seismotectonic map of Iran, 1976 and USGS).

Stratigraphy—tectonics

The stratigraphic and tectonic continuity in the Middle East provides the most reliable evidence indicating that the Persian Gulf area has long been a single geologic unit (Fig. 3). The facts that prove this are very rarely discussed by the supporters of plate tectonics. An enormous volume of stratigraphic and structural evidence has been compiled through the years by students of Zagros and Middle East geology to indicate that there never has been separation between the Iranian plateau and the Arabian Peninsula. The following discussion provides a few samples from this wealth of evidence.

The purple to red cross-bedded Lalun Sandstone of Cambrian age and the slightly older Hormuz Salt are present both northeast and southwest of the alleged suture zone. These units, or parts of them, extend to Syria, to South Yemen, across Pakistan, and on to the Indian shield south of New Delhi (Falcon, 1967b; Stöcklin, 1974; Meyerhoff and Meyerhoff, 1974; Kashfi, 1976, 1985). The correlation of the Lalun Sandstone is so clear and decisive in both the field and subsurface that

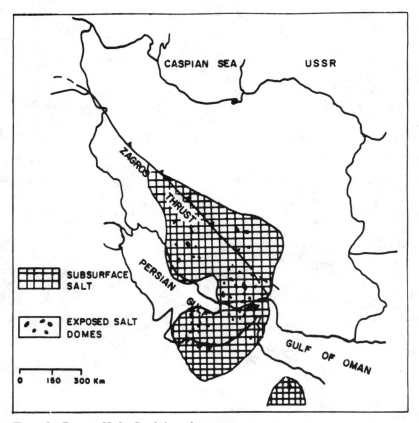

Figure 8. Extent of Infra-Cambrian salt.

there is no doubt concerning the tectonic unity of this region.

Unquestionable correlation of Early Paleozoic trilobites (Redlichia and other faunas) across the Zagros Thrust Belt has been known for years (Stöcklin, 1968; Kashfi, 1976, 1985; Kent, 1979). Moreover, the equatorial and Tethyan nature of the Late Jurassic-Cretaceous fauna of the southern tip of India is well established (Meyerhoff and Meyerhoff, 1974). This equatorial fauna flourished in the tropical Tethys Sea at the time when India was supposed to be adjacent to Antarctica.

The Permian transgressive sea covered a vast area within the greater Persian Gulf region (Kamen-Kaye, 1972). The advancing Permian sea deposited the basinal clastic and shallow epicontinental carbonates. These sediments were deposited unconformably on Devonian in central Iran and gradually overstep the lower Paleozoic in Western Iran; toward the southwest, they overstep the basement rocks of central Arabia (Tschopp, 1967; Ruttner et al., 1968; Stöcklin, 1974). Existence of this regional unconformity is another solid piece of evidence showing the structural continuity of Arabia and the Plateau of Iran.

The Lower Jurassic Neyriz Formation (Lias in coastal Persian Gulf and Lias to Early Dogger in interior Iran in age) of the Zagros Fold Belt is a black marl section associated with dolomite at the base and limestone at the top. Equivalents of this formation are called the Baluti Shale in Iraq, the Marrat Formation (Shale) in Saudi Arabia, and the Musandam Limestone in Oman, all of Lias to Early Dogger age. The fauna of these units is dominated by Pecten species which reasonably can be correlated with the Nayband (Pecten Limestone) Formation of central Iran (Fig. 1).

The Qum Limestone (Oligo-Miocene) in central Iran is equivalent to the Asmari Limestone of south-southwest Iran. The transgressive middle Alpine sea in which the Asmari was deposited in south-southwest Iran invaded the central basin area from the southwest and deposited the Qum. The Qum Formation consists of bioclastic carbonates about 300 m (985 ft) thick with great volumes of organic debris similar to those of Asmari. The same anhydrite beds in the uppermost part of the Asmari carbonates, which provide cap rocks for Asmari oil in the Zagros basins also form the cap rocks for the Qum reservoir in central Iran. These anhydrite beds, which terminate the cycles of marine sedimentation, are time markers.

The regression of late-middle Alpine sea in central Iran produced a lagoonal region of evaporite deposits equivalent to the Gachsaran Formation in south-southwest Iran (Greig, 1958) (Fig. 9). These evaporites were succeeded by the deposition of deltaic river channel sands (Upper Red Formation) in central Iran, equivalent to the Agha-Jari clastics in south-southwest Iran. Stratigraphical evidences regarding the basin unity of the Middle East are also supported by the work of Zahuravleva (1968), Lapparent et al. (1970), Trümpy (1971), Kobayashi (1972), Stöcklin (1974) and Kashfi (1976, 1985, 1988).

Middle East oil fields

The proponents of plate tectonics usually hold that oil and gas accumulations are more likely to be found within back-arc regions (microplates). Bois et al. (1982), in their geotectonic classification of petroleum occurrences, pointed out that foredeeps contain most of the hydrocarbons in the collision areas, such as the Middle East. Their concept is based on the high thermal

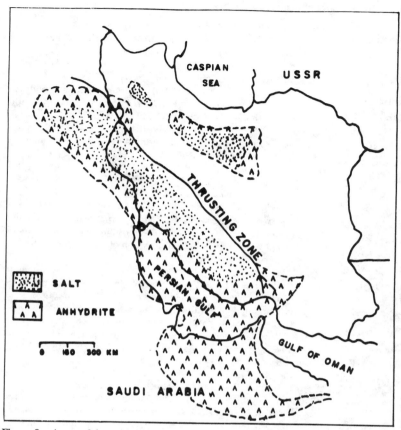

Figure 9. Areas of deposition of the Tertiary evaporites (modified after Greig, 1958 and NIOC, 1978).

gradient required for maturation of petroleum in such areas. However, the supergiant Jurassic and Cretaceous oil fields of Kuwait and Arabia are located on the Arabian shelf, and Cretaceous and Tertiary Iranian oil fields are absolutely confined to the least folded areas adjacent to the Arabian shelf—but definitely far from any foredeep. Furthermore, no unusual heat flow has been reported from the proposed foredeep or from the Oman Semail ophiolite zone (Fig. 10).

Inconsistencies of plate-tectonic models

In general, the advocates of plate tectonics disagree concerning the number of plates involved in the tectonics of the Middle East.

Morgan (1968) recognized five plates that formed the framework for the tectonics of the Middle East. Le Pichon (1968) postulated various small and large plates, which during their rotations and other movements, compressed the plates boundaries. He estimated a northward movement of 4.8 cm/year for the Iranian plate. McKenzie (1970) proposed a number of smaller but faster moving plates between Eurasia and Africa, each in motion with respect to the others. Smith (1971) believed that there have been only African and European plates, but no microplates involved.

Nowroozi (1972) recognized eight plates and claimed that the Indian and Afghanistan plates are moving northeast. He postulated that the Arabian, Persian, and Lut plates moved north, and that the Caspian Sea, Asia Minor, and Black Sea plates moved west. In contrast, Chandra (1981) concluded that the high stresses generated by continental collision caused a rejuvenation of activity along old orogenic belts of the Alburz and the Lut block. Therefore, it seems unnecessary to invoke the concept of a number of small plates to explain the tectonics of the region.

Yet, Dewey et al. (1973) proposed that at all times a combination of compressional, extensional, and transform movements took place between Africa and Europe. Haynes and McQuillan (1974) supported the presence of a suture zone in the Zagros that marks a continent-continent collision.

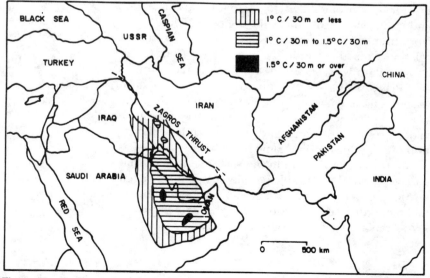

Figure 10. Geothermal gradient of the Middle East oilfields (adapted from NIOC and ARAMCO unpublished data).

Tirrul et al. (1983) envisaged the existence of a separate suture zone in extreme southeastern Iran near the Pakistan border, although there is no evidence for subduction. In contrast, Stöcklin (1974) wrote that all geological evidence suggests that, if any substantial separation existed between Eurasia and Afro-Arabia during Paleozoic time, it must have been in the South Caspian depression (Berberian and Berberian, 1981; Berberian, 1983).

Sengör (1984) referred to central Iran as the "Central Iranian Microcontinent," but he could not establish the age of opening of the oceans that surrounded this Iranian microcontinent. Jackson and McKenzie (1984a) believed that there was a back-arc basin during the Paleo-Tethyan time in central Iran.

In more recent works, Sengör et al. (1985) and Dewey et al. (1986) discussed the triple-junction model and more of microplates for the region. Furthermore, Bond and Kominz (1988) in their recent review of the ancient passive margins encountered a significant problem regarding plate convergence in the Zagros suture. They pointed out that the configuration and setting of the shallow-marine carbonate rocks toward the craton beneath the modern Persian Gulf more likely are indicative of a divergent (passive margin) rather than a convergent or active boundary between the two Asian and African plates.

CONCLUSION

There is nothing in the known geological record in the Middle East to suggest a former separation between Arabia-Africa and the Middle East portion of Eurasia. The Iranian plateau and southwestern Iran and Arabia have been a single geologic zone since the beginning of Proterozoic time, as shown by the following (1) the Late Proterozoic through Tertiary stratigraphic correlations and continuity across Iran and the entire Middle East, from India to Yemen and Jordan; (2) the biozonal correlation from the Middle East to central Asia; (3) the structural unity across the greater Persian Gulf area; (4) the Precambrian-Cambrian salt connection along western India, Pakistan, Iran, Persian Gulf, and Arabia; (5) the existence of strong seismicity away from the alleged subduction zone; and (6) the random distribution of ophiolites and volcanic rocks in the Middle East.

There is no indication in the surface and subsurface geology, or in the available geophysical data, to suggest anything other than simple and homogeneous horizontal (tangential) compression of the crust between the Afro-Arabian block and southwestern Asia.

REFERENCES CITED

Ahmed, S.S., 1969, Tertiary geology of part of south Makran, Baluchistan, West Pakistan: American Association of Petroleum Geologists Bulletin, v. 53, p. 1480-1499.

Ala, M.A., 1974, Salt diapirism in southern Iran: American Association of Petroleum Geologists Bulletin, v. 58, p. 1758-1770.

Berberian, F., and Berberian, M., 1981, Tectonic-plutonic episodes in Iran, in Gupta, H.K., and Delany, F.M., eds., Zagros, Hindu Kush, Himalaya: Geodynamics Series, v. 3, p. 5-32.

Berberian, M., 1983, The Southern Caspian: a compressional depression floored by a trapped, modified oceanic crust: Canadian Journal of Earth Sciences, v. 20, p. 163-183.

Berberian, M., and King, G.C.P., 1981, Towards a paleogeography and tectonic evolution of Iran, Canadian Journal of Earth Sciences, v. 18, p. 210-265.

Berberian, M., and Tchalenko, J.S., 1975, On the tectonics and seismicity of the Zagros Folded Belt: International Geodynamics project proceedings of Tehran symposium on the geodynamics of southwest Asia. Special publication of the Geological Survey of Iran, Tehran, Iran, p. 53-69.

Berry, R.H., and Moore, D., 1975, Evidence for two different basement types beneath the Zagros Range, Iran: International Geodynamics project proceedings of Tehran symposium on the geodynamics of southwest Asia, Special publication of the Geological Survey of Iran, Tehran, Iran, p. 70-88.

Bird, P., 1978, Finite element modeling of lithosphere deformation: The Zagros collision orogeny: Tectonophysics, v. 50, p. 307-336.

Bois, C., Bouche, P, and Pelet, R., 1982, Global geologic history and distribution of hydrocarbon reserves: American Association of Petroleum Geologists Bulletin, v. 66, p. 1248-1270.

Bond, G.C., and Kominz, M.A., 1988, Evolution of thought on passive continental margins from the origin of geosynclinal theory(~1860) to the present: Geological Society of America Bulletin, v. 100, p. 1909-1933.

Brinkmann, R., 1972, Mesozoic troughs and crustal structure in Anatolia: Geological Society of America Bulletin, v. 83, p. 819-826.

Chandra, U., 1981, Focal mechanism solutions and their tectonic implications for the Eastern Alpine-Himalaya Region, in Gupta, H.K., and Delany, F.M., eds., Zagros, Hindu Kush, Himalaya Geodynamic Evolution: Geodynamics Series, v. 3, p. 243-271.

Coleman, R.G., 1984, Ophiolites and the tectonic evolution of the Arabian Peninsula, in Gass, I.G., Lippard S.J., and Shelton, A.W., eds., Ophiolites and oceanic lithosphere: Geological Society of London Special Publication 17, p. 359-366

Crawford, A.R., 1972, Iran: Continental drift and plate tectonics: 24th International Geological Congress Proceedings, Section 3, p. 106-112.

Dengo, G., and Bohnenberger, O., 1969, Structural development of northern Central America: American Association of Petroleum Geologists Memoir 11, p. 203-220.

Dewey, J.F., Hempton, M. R., Kidd, W. S. F., Saroglu, F., and Sengör, A. M. C., 1986, Shortening of continental lithosphere: The neotectonics of Eastern Anatolia—a young collision zone, in Coward, M. P. and Ries, A. C. eds., Collision tectonics: Geological Society of London Special Publication 19, p. 3-36.

Dewey, J.F., Pitman, W.C. III, Ryan, W.B.F., and Bonnin, J., 1973, Plate tectonics and the evolution of the Alpine System: Geological Society of America Bulletin, v. 84, p. 3137-3180.

Falcon, N.L., 1967a, The geology of the north-east margin of the Arabian Basement Shield: British Association for the Advancement of Science, v. 24, p. 31-42.

———, 1967b, Equal areas of Gondwana and Laurasia: Nature, v. 213, p. 580-581.

Furon, R., 1941, Geological du Plateau Iranian (Perse-Afghanistan-Baloutchistan). Memoir, Museum of Natural History, Paris, n.s., 7, pt. 2, p. 177-414.

Gansser, A., 1955, New aspects of the geology in central Iran, 4th World Petroleum Congress, Rome, Proceedings, Section 1, (a), 5, p. 280-300.

Gealey, W. K., 1977, Ophiolite obduction and geologic evolution of the Oman Mountains and adjacent areas: Geological Society of America Bulletin, v. 88, p. 1183-1191.

Geological Survey of Iran, 1976, Seismotectonic map of Iran, Scale 1:2,500,000, Tehran, Iran.

Glennie, K.W., Boeuf, M.G.A., Hughas Clarke, M.W., Moody-Stuart, M., Pilaar, W.F.H., and Reinhardt, B.M., 1973, Late Cretaceous nappes in Oman Mountains and their geological evolution: American Association of Petroleum Geologists Bulletin, v. 57, p. 5-27.

Greig, D.A., 1958, Oil horizons in the Middle East, in Weeks, L.G. ed., The habitat of oil: American Association of Petroleum Geologists p. 1182-1193.

Hamilton, W.B., 1988, Plate tectonics and island arcs: Geological Society of America Bulletin, v. 100, p. 1503-1527.

Haynes, S.J., and McQuillan, H., 1974, Evolution of the Zagros suture zone, southern Iran: Geological Society of America Bulletin, v. 85, p. 739-744.

Hsü, K.J., and Schlanger, S.O., 1971, Ultrahelvetic flysch sedimentation and deformation related to plate tectonics: Geological Society of America Bulletin, v., 82, p. 1206-1218.

Jackson, J.A., and McKenzie, D., 1984a, Rotational mechanisms of active deformation in Greece and Iran: Geological Society of London Special Publication 14, p. 743-754.

———, 1984b, Active tectonics of the Alpine-Himalayan belt between western Turkey and Pakistan: Royal Astronomical Society Geophysical Journal, v. 77, p. 185-264.

Jiqing, H., and Bingwei, C., 1987, The evolution of the Tethys in China and adjacent regions: Beijing, Geological Publishing House, 109 p.

Kamen-Kaye, M. 1972, Permian Tethys and Indian Ocean: American Association of Petroleum Geologists Bulletin, v. 56, p. 1984-1999.

Kashfi, M.S., 1976, Plate tectonics and structural evolution of the Zagros geosyncline, southwestern Iran: Geological Society of America Bulletin, v. 87, p. 1486-1490.

———, 1985, Pre-Zagros integrity of the Iranian platform: Journal of Petroleum Geology, v. 8, p. 353-360.

———, 1988, Evidence for non-collision geology in the Middle East. Journal of Petroleum Geology, v. 11, p. 443-460.

Kent, P.E., 1979, The emergent Hormuz salt plugs of southern Iran: Journal of Petroleum Geology, v. 2, p. 117-144.

Khudoley, K.M., and Meyerhoff, A.A., 1971, Paleogeography and geologic history of Greater Antilles: Geological Society of America Memoir 129, 199 p.

King, L.C., 1972, An improved reconstruction of Gondwanaland, in NATO Symposium on Continental Drift, Newcastle upon Tyne, April 1972: Scottish Academic Press.

Kobayashi, T., 1972, Three faunal provinces in the Early Cambrian: Proceedings Japan Academy of Science, v. 48, p. 242-247.

Kolla, V., and Coumes, F., 1987, Morphology, internal structure, seismic stratigraphy, and sedimentation of Indus Fan: American Association of Petroleum Geologists Bulletin, v. 71, p. 650-677.

Lapparent, A.F., Termier, H., and Termier, G., 1970, Sur la stratigraphic et al. Paleobiologia de la serie Permo-Carbonifere due Dacht-e-Nawar (Afghanistan): Bulletin Societe de Geologique de France, ser, 7, t. 12, no. 3, p. 565-572.

Le Pichon, X., 1968, Sea floor spreading and continental drift: Journal of Geophysics Research 73, p. 3661-3697.

Lowman, P.D., and Tiedemann, H.A., 1971, Terrain photography from Gemini spacecraft—final geologic report: Greenbelt, MD., Goddard Space Flight Center, Report X-644-71015, 75 p.

McKenzie, D.P., 1970, Plate tectonics of the Mediterranean region: Nature, v. 226, p. 239-243.

———, 1972, Active tectonics of the Mediterranean region: Royal Astronomical Society Geophysical Journal, v. 30, p. 109-185.

Meyerhoff, A.A., and Meyerhoff, H.A., 1974, Tests of plate tectonics, in Kahle, C.F., ed., plate tectonics—assessments and reassessments: American Association of Petroleum Geologists Memoir 23, p. 43-145.

Mina, P. Razagnia, M.T., and Paran, Y., 1967, Geological and geophysical studies and exploratory drilling of the Iranian continental shelf Persian Gulf: 7th World Petroleum Congress, Mexico, Proceedings, v. 2, p. 870-903.

Mohajer, A., Ashjai, M., and Nowroozi, A.A., 1978, Observed and probable intensity zoning of Iran: Tectonophysics, 49, p. 149-160.

Morgan, W.J., 1968, Rises, trenches, great faults and crustal blocks: Journal of Geophysical Research 73, p. 1959-1982.

National Iranian Oil Company, 1978, Geological maps and sections of Iran, sheet no. 1, north-west Iran: scale 1:1,000,000. Tehran, Iran.

Nowroozi, A.A., 1971, Seismo-tectonics of the Persian Plateau, eastern Turkey, Caucasus, and Hindu Kush region: Seismological Society of America Bulletin, v. 61, p. 317-341.

———, 1972, Focal mechanism of earthquakes in Persia, Turkey, West Pakistan, and Afghanistan and plate tectonics of the Middle East: Seismological Society of America Bulletin, v. 62, p. 823-850.

Ricou, L.E., 1970, Comments on radiolarites and ophiolites nappes in the Iranian Zagros Mountains: Geological Magazine, v. 107, p. 479-480.

———, 1971, Le metamorphisme au contact des peridotites de Neyriz (Zagros interne, Iran): development de skarus a pyroxene. Bull. Socéité Gãologique de France, ser. 7, v. XIII, p. 146-155.

Rotstein, Y., and Kafka, A.L., 1982, Seismotectonics of the southern boundary of Anatolia, eastern Mediterranean region: Subduction, collision, and arc jumping: Journal of Geophysical Research, v. 87, p. 7694-7706.

Ruttner, A., Nabavi, M.H., and Hajian, J., 1968, Geology of the Shirgesht area (Tabas, East Iran): Geological Survey of Iran Rept. 4, p.133.

Searle, M.P., and Graham, G.M., 1982, "Oman Exotics"—Oceanic carbonate built-ups associated with the early stages of continental rifting: Geology, v. 10, p. 43-49.

Searle, M.P., Lippard, S.J., Smewing, J.D., and Rex, D.C., 1980, Volcanic rocks beneath the Semail ophiolite nappe in the northern Oman mountains and their significance in the Mesozoic evolution of Tethys: Journal of Geological Society of London, v. 137, p. 589-604.

Sengör, A.M.C., 1983, Gondwana and "Gondwanaland": A discussion: Geologische Rundschau, v. 72, p. 397-400.
——, 1984, The Cimmeride orogenic system and the tectonics of Eurasia: Geological Society of America Special Paper 195.
Sengör, A.M.C., Görür, N., and Saroglu, F., 1985, Strike-slip faulting and related basin formation in zones of tectonic escape: Turkey as a case study, *in* Biddle, K.T., and Christie-Blick, N., eds., Strike-slip deformation, basin formation and sedimentation: Society of Economic Paleontologists and Mineralogists Special Publication 37, p. 227-264.
Smith, A.G., 1971, Alpine deformation and the oceanic areas of the Tethys, Mediterranean, and Atlantic: Geological Society of America Bulletin, v. 82, p. 2039-2070.
Stöcklin, J., 1968, Salt deposits of the Middle East, *in* Mattox. R.B., ed., Saline deposits: Geological Society of America Special Papers 88, p. 157-181.
——, 1974, Possible ancient continental margins in Iran, *in* Burk, C.A., and Drake, C.L., eds., The geology of continental margins: New York-Heidelberg-Berlin, Springer Verlag, p. 873-887.
Sylvester, A.G., 1988, Strike-slip faults: Geological Society of America Bulletin, v. 100, p. 1666-1703.
Takin, M., 1972, Iranian geology and continental drift in the Middle East: Nature, v. 235, p. 147-150.
Tehrani, N.A., 1975, On the metamosphism in the ophiolitic rocks in the Sabzevar region, northeast Iran: International Geodynamics project proceedings of Tehran symposium of the geodynamics of southwest Asia, Special publication of the Geological Survey of Iran, Tehran, Iran, p. 25-52.
Tirrul, R., Bell, I.R., Griffis, R.J., and Camp, V.E., 1983, The Sistan suture zone of eastern Iran: Geological Society of America Bulletin, v. 94, p. 134-150.
Trümpy, R., 1971, Stratigraphy in mountain belts: Geological Society of London Quarterly Journal, v. 126, pt. 3, no. 503, p. 293-318.
Tshopp, R.H., 1967, The general geology of Oman: 7th World Petroleum Congress, Mexico, Proceedings, v.2, p. 231-241.
Vialon, P. Houchmand-Zedeh, A. and Sabzehi, 1972, Propositions d'un modele de l évolution pétro-structurale de quelques montagnes iraniennes, comme une consequence de la tectonique de plaques: 24th International Geological Congress, Montreal, 1972, section 3, p. 196-208.
Welland, M.J.P. and Mitchell, A.H.G., 1977, Emplacement of the Oman ophiolite: A mechanism related to subduction and collision: Geological Society of America Bulletin, v. 88, p. 1081-1088.
Wilson, H.H., 1969, Late Cretaceous eugeosynclinal sedimentation, gravity tectonics, and ophiolite emplacement in Oman Mountains, southeast Arabia: American Association of Petroleum Geologists Bulletin, v. 53, p. 626-671.
Wolfart, R., 1967, Zur Entwicklung der paläozoischen Tethys in Vorderasien: Erdöl und Kohle-Erdgas-Petrochemie, Jaharb., no. 3, p. 168-180.
——, 1969, Die Kambro-Ordovizische schichtenfolge von Surkh bum bei Panjaw in Östlichen Zentral-Afganistan: Geologische Jahrabucher, no. 87, p. 541-550.
Woodcock, N.H., and Robertson, A.H.F., 1984, The structural variety in Tethyan ophiolite terrains, *in* Gass, I.G., Lippard S.J., and Shelton, A.W., eds., Ophiolites and oceanic lithosphere: Geological Society of London Special Publication 17, p. 321-330.
Yalcin, M.N., and Görür, N., 1984, Sedimentological evolution of the adana Basin, *in* Tekeli, O., and Goncoglu, M.C., eds., Geology of the Taurus Belt: Ankara, Mineral Research and Exploration Institute, p. 165-172.
Zhuravleva, I.T., 1968, Early Cambrian biogeography and geochronology to the Archaeocyathi: 23th International Geological Congress, Prague, Proceedings I PU, p. 361-373.

Constraints to major right-lateral movements, San Andreas fault system, central and northern California, USA

B. D. Martin, Bruce Martin Associates, Inc., P.O. Box 234, Leonardtown, Maryland 20650-0234 USA

ABSTRACT

Constraints to major right-lateral movements of the San Andreas fault (SAF) and/or the San Andreas fault system (SAFS) have been described in the literature for over 30 years; additional field work supplies details of other constraints. Major constraints occur in three regions. These are: (1) northern California in the vicinity of Cape Mendocino together with the eastern Mendocino fault zone in the Pacific Ocean basin; (2) central California's Monterey Bay region both on land and in the sea; and (3) southern California in the Transverse Ranges north of Los Angeles. Since the Early Cenozoic, movements of the SAFS have been more nearly vertical than horizontal (lateral), which suggests strongly that most stratigraphic evidence of vertical movements has been erosionally or tectonically destroyed because of high relief, whereas stratigraphic evidence for the less important horizontal (lateral) movements has been better preserved, albeit incompletely, because of both low relief and protection by adjacent fault blocks. To most investigators, such an imbalance of preserved evidence versus missing or eroded evidence has favored greatly, but incorrectly, hypotheses of major right-lateral movements rather than hypotheses of vertical movements. Missing or eroded evidence can be determined from detailed paleogeologic and paleogeomorphologic studies.

Because major right-lateral, long-range horizontal movement is either mainly incorrect or impossible in some cases by the described constraints, inference and comparison make northwest rotation of the Pacific plate along this part of the SAF impossible to justify, and rotation itself becomes moot. Whatever faulting occurred or presently occurs between Cape Mendocino and the Transverse Ranges, lateral movements did not and do not continue into the Pacific Ocean basin at the northern end and do not continue into southern California through the Transverse Ranges. Granitoid rocks (granites to granodiorites) of Salinia were not transported laterally from batholithic rocks at the southern end of the Sierra Nevada or from other locales in southern California, but conversely, these Salinian intrusive rocks may not have been emplaced in situ in the Coast Ranges. This conundrum may be better addressed or actually solved by data presented in a recently announced hypothesis of geodynamics called surge tectonics.

Horizontal or lateral crustal movements between separate fault blocks from Cape Mendocino to the Transverse Ranges have occurred, as shown by geodetic investigations of the past 10 years conducted by the National Air and Space Administration NASA (Crustal Dynamics Project), but vertical fault-block movements during the past 70 million years (Cenozoic) overall are far more evident as shown by field work. Changed force and movement vectors amounting to almost 90 degrees will be of importance to professional engineers, especially to civil engineers, in their reconsideration and introduction of revised safety factors in present and future plans for manmade structures to prevent or to ameliorate the effect of earthquakes on such structures. Vertical force directions as proved by the collapse of the San Francisco Bay Bridge, the destruction of Interstate 880, and as recorded on and in other collapsed or damaged structures as a result of the Loma Prieta earthquake of 17 October 1989, certainly show dramatically the need for reconsideration and redesign of safety factors, among other engineering aspects, in older, new, and proposed structures including landfill and compacted soils (if such occur) on which such structures are presently sited or will be constructed.

INTRODUCTION

In need of reexamination in the light of constraints to be presented in this study, is the well-held and well-liked hypothesis (at least by most geologists and geophysicists) of long-range right-lateral movements during the Cenozoic of up to 350 miles (560 km) or more of the west block (Pacific plate) in relationship to the east block (North American plate) along the major fracture zone here termed the San Andreas fault (SAF) and its system (SAFS) of related, generally parallel, faults. Constraints are major and can be verified in the field or by well-presented sub-surface sections based on geophysical methods and by well logs. For use in discussion, some plate tectonics terms are used such as Pacific plate and North American plate, but as will be developed and discussed, these terms may not be valid in the overall presentation of constraints to major right-lateral movements.

This paper presents a number of constraints to major right-lateral movements between Cape Mendocino on the north and the Transverse Ranges on the south—basically northern and central California (Fig. 1). For brevity, the main fracture, the San Andreas fault itself where distinctly recognized will be termed SAF; related

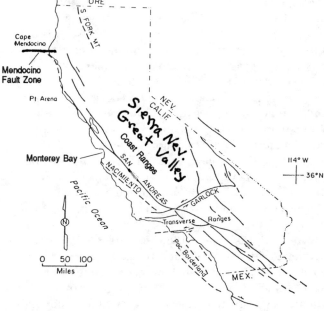

Figure 1. Location map of northern and central California including part of the northeast Pacific Ocean basin (modified from Hill, 1971).

Chatterjee, S., and N. Hotton III, eds. *New Concepts in Global Tectonics.* Texas Tech University Press, Lubbock, 1992, xii + 450 pp.

parallel faults including the SAF if necessary will be termed the San Andreas fault system (SAFS). Where and when it is necessary to identify individual faults in the SAFS, complete names will be used. Part of the earlier California work was presented in units of the British system of measurement, and most recent work in units of the metric system, both systems will be presented where possible.

Following the San Francisco earthquake of 1906, the SAF and SAFS have become one of the most geologically and geomorphologically studied fracture zones on Earth, owing mainly to accessibility, to the location of major academic institutions within this seismically unstable region, and because of effects on large numbers of people and on manmade structures by earthquakes and related events. These investigations increased dramatically since the early 1960s with the introduction of the concept of plate tectonics and with the determination of its supporters to use this concept to further the ideas that they already held as valid with regard to the tectonics of California. The proponents of major right-lateral movements base their concepts on many correlations of near-surface and surface units on opposite sides of the SAF, or in some cases, of the total SAFS. When units quite distant (in many cases) from one another seem alike in mineralogy, paleontology, sedimentology, gravimetrics, seismics, or aerial and space photography, or in any combination of these, the supporters of major right-lateral movements or the concept of plate tectonics or both, have concluded that their tectonic concepts are correct. This concept has become so pervasive in both geologic-geophysics literature and in the popular public press that little thought is given to alternative views including vertical-fault block movements; such is quite apparent in the scientific and popular presentation of events occurring during the 17 October 1989, Loma Prieta earthquake in northern California.

In most cases, no consideration is given to the probability that many rock units have been uplifted and subsequently eroded between areas of supposed correlation matches for proof of major right-lateral movements. Regardless of the many methods used to prove right-lateral movements, there is no consensus as to total distances; each investigator usually arrives at an amount of lateral movement at variance with distances determined by other investigators.

Since approximately 1980, seismologic records and new field evidence have demonstrated to this author that most of the SAF and SAFS have had or continue to have more nearly vertical movements within 5 to 10 kilometers of ground surface rather than major right-lateral movements, and that these vertical planes appear to flatten and become thrust planes below this depth (Lowman, 1980; Zoback et al., 1987; Namson and Davis, 1988; Hauksson et al., 1989; Stein and Yeats, 1989). Martin (1964) and Martin and Emery (1967, 1982) showed vertical SAFS fault planes, but no deeper thrust planes, because of lack of instrumentation at that time to investigate deeper tectonics. Thus, right-lateral movements on the SAF/SAFS in most cases appear to be secondary effects of rebound after primary vertical or near-vertical movement on the steeply dipping near-surface fault planes. The result is a major constraint, and one that affects adversely the amount of right-lateral movement between Cape Mendocino and the Transverse Ranges (Fig. 1).

Lateral movements measured and reported by the use of instruments on the land surface and from space in the Crustal Dynamics Project of NASA (Geodynamics Branch, Code 621) will be discussed to attempt an understanding between the constraints evidence presented herein (mainly field evidence) and the lateral movement evidence (mainly by instrumental and remote sensing evidence). Additionally, these constraints make moot the prevailing concept of right-lateral movement of the granitic-granodioritic (granitoid) rocks of Salinia from the south end of the Sierra Nevada batholith to their present major outcrop and subcrop localities (Gabilan Range, Santa Lucia Range, Santa Cruz Mountains, Monterey Peninsula, Farallon Islands, and isolated locations to the north of San Francisco Bay); this constraint to major movement of Salinia, however, should not be construed to mean that the intrusive igneous rocks of Salinia therefore formed in situ. Other concepts have been proposed to solve this seeming geologic Gordian Knot.

CONSTRAINTS AT CAPE MENDOCINO AND MENDOCINO FRACTURE ZONE

Arguments for major right-lateral movement on the SAF and SAFS are weakest at Cape Mendocino (Fig. 1) precisely where they should be strongest to prove maximum tectonic activity, because the SAF/SAFS is mapped as taking an almost right-angle change of trend (Jennings, 1975) from north-south to east-west, parallel to the Mendocino fracture zone, and mapped also as becoming a part of the Mendocino fracture zone. As noted by the intense tectonics mapped (Jennings, 1975; Namson and Davis, 1988) in the SAF/SAFS trend change in the western Transverse Ranges where the trend changes from the southern Coast Ranges to the Transverse Ranges (Tehachapi Mountains), there should be ample evidence at Cape Mendocino of similar present and past tectonic activity, especially thrusting. Griscom (1973) presented an acceptable argument from gravity and magnetic data that the SAF/SAFS just did not reach to the Mendocino fracture, but in later papers, for example, Griscom and Jachen (1989), the hypothesis of plate tectonics is used to avoid the problem of curvature from the SAFS trend to the Mendocino fracture trend. These problems become more acute and difficult the steeper the SAFS fault plane or planes become. In light of major constraints that occur farther south in the Coast Ranges and Transverse Ranges, it appears moot even to consider that major right-lateral movement has or had occurred in the Cape Mendocino area.

Especially revealing in this context is the major offshore constraint provided by the presence of stable ocean-floor sedimentary structures at the base of the north facing slope of the Mendocino escarpment. Figure 2 is a photograph of these ocean floor sediments (Anonymous, 1985) made during one of the first uses of the GLORIA scanning sonar system at the base of the north-facing slope taken about 70 km west of the coastline. In these sediments, between 125.2° and 125.6° degrees W longitude at a depth approximately 3000 m (depth from topographic map by Menard, 1964), there are developed highly complex and convoluted meander patterns apparently caused by ocean currents flowing parallel to the escarpment (east-west). Figure 2 shows a longitudinal distance at the latitude given of approximately 21 statute miles (33.6 km) determined for the author by C. D. Meekins and Associates, Inc., registered engineers and land surveyors, Annapolis, Maryland, from the differences of the longitudes—0.4 degree. Daniel J. Stanley, senior oceanographer, Smithsonian Institution (pers. comms., 1985, 1989) stated that these highly sinuous current-eroded and deposited channels with associated sedimentary structures are very delicate and can be destroyed easily by even minor tectonic activity. Based on his experience, especially with similar structures in the Mediterranean Sea, these Cape Mendocino ocean-floor sedimentary features probably are no more than 1 million years old, having formed during the Late Pleistocene.

If major right-lateral movements had occurred during the past one million years, or if such movements are presently occurring, and if the SAF or SAFS is rotated almost 90 degrees into the trend or the plane of the Mendocino fault, then tectonic activity associated with rotation should have occurred, which in turn should have destroyed these delicate sedimentary features at the base of the Mendocino escarpment. As these sedimentary features appear pristine and not destroyed, as shown in Figure 2, little or no tectonic activity has occurred west of Cape Mendocino within the longitudes given for at least the past one million years. This conclusion runs completely counter to the geodynamic and kinetic solutions of tectonics in this area proposed by plate tectonics.

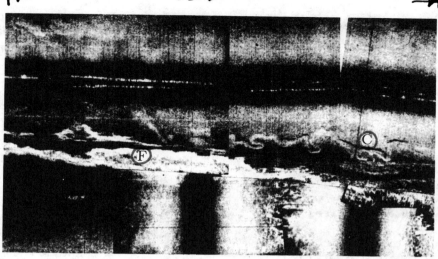

Figure 2. Sea-floor mosaic taken with GLORIA scanning sonar system. The formation and preservation of such channels depend on lack of major tectonic activity during their existence. Channel age is probably Pleistocene to Recent and formed during the last one million years (published with permission of Geotimes).

F = Mendocino Fracture
C = Current Channels

The author believes that engineering principles play an important part in ideas concerning the mechanics of rotating a solid mass of rock (west block of the SAF/SAFS) through an arc of approximately 90 degrees to enter and to parallel or merge with the fault plane of the Mendocino fault. The apparent lack of stress features especially in the west block (compression-tension features) seems most unlikely, and is not reasonable from an engineering standpoint. (See the discussion about changes of SAF/SAFS trend from the southern Coast Ranges to the western part of the Transverse Ranges—here a great amount of tectonism is noted, especially thrusting both to the north and to the south of the fault (Martin, field work, 1956; Namson and Davis, 1988).

An example of no SAFS movement, described farther south near Point Delgada, may be compared with Griscom's work of 1973. McLaughlin et al. (1985) showed that the SAF has not moved since the Middle Miocene based on K-Ar dating of adularia (feldspar) zones, which cross this investigated part of the SAF/SAFS. Their explanation of this constraint to any movement, yet to permit major right-lateral SAF/SAFS movement still, is that although this part of the SAFS is and has been inactive since the Middle Miocene, SAFS faults to the west of the inactive fault are actively moving.

These authors ignore the evidence that no movement on any or all of the SAF/SAFS has occurred since the Middle Miocene.

Between Punta Gorda (next major cape south of Cape Mendocino) and Point Arena (Fig. 1), Perry and Pryor (1986) stated that "... there is no good evidence from the sea floor morphology or the seismic records that the San Andreas fault continues offshore between P. Delgada and Punta Gorda...." (P. Delgada is the next cape south of Punta Gorda.) There is no solid field evidence to justify a solution of tectonics by major right-lateral movement of the extension of the SAF or SAFS to the north of Point Arena (R. B. Perry, NOAA, pers. comm., 1989).

CONSTRAINTS IN THE COAST RANGES
Background

(Pajaro Misspelled in report - Not Parajo (underlined))

From Point Arena southward to the Transverse Ranges (Fig. 1), two details are obvious from inspection of geologic maps (e.g., Page, 1966) and from published papers dealing with local areas: (1) deposits have been stripped from uplifted areas, especially Salinia (Fig. 3), and (2) fault planes are mainly vertical on near-surface expressions (Gribi, 1963; Martin, 1964; Martin and Emery, 1967, 1982; Hill, 1984; and Zoback et al., 1987). Stripped sedimentary deposits and other rocks range from Late Cretaceous to Pleistocene; deposits on either side of the uplifted Salinian block (containing the Santa Lucia and Santa Cruz mountains where granitoid rocks are widely exposed) received some of the stripped debris. From the first investigations, it has been obvious on the one hand and a geologic riddle (still unsolved mainly) on the other, that Franciscan rocks do not occur on Salinia between the SAF and the San Gregorio–Carmel Canyon–Nacimiento faults (one and the same trend with different local names before the continuity was recognized and established by Martin and Emery (1967) via the connecting Carmel Canyon fault and published in addition by Jennings (1975). Furthermore, granitoid rocks, although exposed in the Santa Cruz Mountains, Santa Lucia Mountains, Gabilan Range, and other Salinian-block localities, do not occur east of the SAF nor are exposed on the block immediately west of San Gregorio–Carmel Canyon–Nacimiento fault. Farther west, in another northwest-trending fault block situated offshore, granitoid rocks again are exposed, especially on the Farallon Islands to the west of San Francisco Bay. Granitoid rocks do not occur along the west side of the Great Valley of California either in outcrop or at depth; the closest occurrences are those in wells along the east side of the Great Valley as well as in Sierran exposures.

Constraints to major right-lateral movements occur as follows: at Pinnacles National Monument; along the buried Gabilan Escarpment (Fig. 3); along Pliocene to Pleistocene seaways and surface water drainage from the Great Valley, and by similar considerations of the Eocene Vallecitos seaway; by vertical fault movements in the Monterey graben under Monterey Bay (Fig. 3); and by the depositional age of basal sediments in the Monterey deep-sea fan. Although each constraint will be discussed separately, and although constraints may not be the same in each cited case, the geologic history, including overlaps of tectonism, define and describe a total concept in the Coast Ranges that disproves major right-lateral movements during the Cenozoic.

Pinnacles National Monument

A. A. Meyerhoff (1987) made new field investigations in Pinnacles National Monument to determine whether major right-lateral movement had occurred along this portion of the SAF/SAFS. He stated (pers. comm., 1989) that he could find no evidence of right-lateral movement in the Tertiary intrusive-extrusive igneous rocks exposed on both sides of the bisecting SAF/SAFS.

Gabilan escarpment (King City hinge line-Salinas flexure-Davenport escarpment)

Martin (1964) and Martin and Emery (1967, 1982) demonstrated that a distinct southwest-facing fault escarpment, later erosionally modified, is a single escarpment (Fig. 3) with buried relief in the Santa Cruz Mountains adjacent to Pigeon Point of over 9,000 feet (2,790 m) and about 3,000 feet (930 m) near its southern end near King City; the buried relief increases from south to north to the maximum cited. Distance between the two end points is about 150 miles (240 km). Total uplift was probably greater, as east of Monterey Bay, subaerial erosion is considered to have created the Elkhorn erosion surface (Fig. 3) with subsequent down-dropping to permit eventual deposition of marine Miocene and Plio-Pleistocene sediments atop the erosion surface. Additional evidence for subaerial erosion of the Elkhorn erosion surface is Parajo Gorge, which was cut into the uplifted block with a minimum of relief below the block's surface of approximately 7,000 feet (2170 m). In Parajo Gorge, a variety of marine and nonmarine sediments was deposited that show lithologic relationship to adjacent nearshore land deposits especially red-weathering Oligocene volcanics of the nearby Santa Cruz Mountains (Martin and Emery, 1967, 1982, p. 2296). This major erosional feature proves additionally that uplift of at least 7,000 feet (2,170 m) occurred to permit this geomorphological expression, does not

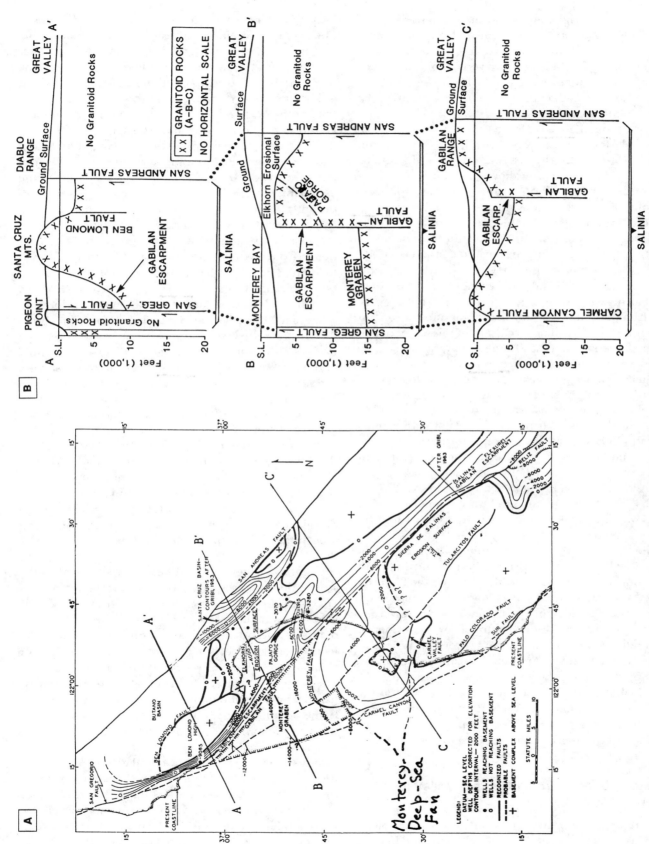

Figure 3. A—Structure under Monterey Bay and in adjacent land onshore (Martin and Emery, 1967 and 1982). B—"xs" indicate surface exposures of granitoid (basement) rocks. No granitoid rocks dredged on Carmel Canyon's west wall nor on north wall of Monterey Canyon.

allow for major right-lateral movement to occur simultaneously. To account for present geologic and geomorphologic conditions even when buried, if such a combination of right-lateral and vertical movements occurred, later down-dropping of the block containing the Parajo Gorge would have had to move downward and in a left-lateral direction.

Even though Parajo Gorge was not considered an ancestral Monterey Canyon (Martin, 1964; Martin and Emery, 1967, 1982), the headward reaches of Monterey Canyon today are directly above the sedimentary fill in the buried Parajo Gorge suggesting that (for many reasons discussed by Martin and Emery) the Parajo Gorge influenced the later headward location of Monterey Canyon. Any major lateral movements on faults within the three-dimensional aspects of the Monterey Bay area seemingly would have negated this unique location of headward reaches of a major, open submarine canyon above a buried and filled gorge, which in part, may have been a submarine canyon also in seaward reaches. The discussion in this paper of land drainage debouching at Monterey Bay, at or near the headward reaches of both Parajo Gorge and of the Monterey submarine canyon, adds support to evidence for little or no right-lateral movement on the SAF/SAFS.

Gabilan escarpment was named not only to replace the rather confusing tripartate names of earlier usage (Martin, 1964), but also to demonstrate that a major fault existed, and exists today, at the base of the escarpment. This demonstrated basal fault appears to be a northwest extension of the Reliz fault (Martin and Emery, 1967 and 1982, Figs. 3A, and 6), and similarly was named Gabilan fault, which merges with the San Gregorio fault in the western Santa Cruz Mountains.

Movement on the Gabilan fault (and others in the Monterey Bay area) is explained best by vertical movement during the Middle-Late Miocene and the Pliocene. These movement vectors negate any lateral movement force vectors at this time and elsewhere where vertical movement vectors are demonstrated by the evidence. Such relief features cannot occur without vertical uplift. Major right-lateral block movements cannot account for physiographic elevations needed to produce the erosional effects. Comment will also be made on this need for topographic relief in the discussion of Monterey graben, including the need for a depression to contain sediments introduced to the marine environment from off the adjacent land.

Drainage patterns from the Great Valley to Monterey Bay

Patterns of major drainages off the ancestral Sierra Nevada across the Great Valley of California and into the Pacific Ocean basin during the Cenozoic add evidence of constraints to major right-lateral movements on the SAF and SAFS. Naturally, configuration of the coast was not the same as today, and especially during the Miocene many points of freshwater debouchment occurred, owing to extensive marine transgressions into the San Joaquin basin, but from the Late Miocene to the mid-Pleistocene's violent orogenic events, the only point of debouchment of rivers from both the San Joaquin basin and the Sacramento basin was at Monterey Bay (Martin and Emery, 1967, 1982). Additionally, this same debouchment point at Monterey Bay appears likely as the only place where rivers of the Sacramento basin earlier, from Oligocene to Late Miocene time, entered the marine environment. Both the earlier and later flows (from the Sacramento basin and from the San Joaquin basin) then joined in the Late Miocene to flow as one via the ancestral Parajo River through Elkhorn Slough to the east of Monterey Bay. Visual inspection today of the fluvially eroded slough should be enough to demonstrate the vast quantities of water and contained loads of sediment that cut the greatly oversized channel of the present miniscule Parajo River within the slough. These sediments were carried into the marine environment at Monterey Bay and are recorded in the extensive thickness and the expanse of the Monterey deep-sea fan (and no doubt, the Delgada fan also because of northward littoral drift at times). Menard (1964) stated the fans' sediment volumes are 1.6×10^5 km^3. Wilde et al. (1982) said that the earliest sediments at the Monterey fan's base are Oligocene, and these rest on an Oligocene basement. This correlates well with the river-flow data of Martin and Emery (1967, 1982) indicating a beginning of major river debouchment at Monterey Bay.

Emphasis is being placed here on the aspects of the stream flows and sediment loads with resultant erosion effects to show the vast hydrologic and hydrogeologic conditions in central and northern California that did occur to demonstrate the requirement that the hydrogeologic aspects could not have existed if major right-lateral SAF and SAFS movements at any time in the interval from Oligocene to the mid-Pleistocene orogeny had disrupted and dammed continental river drainage to the ocean, especially in the Sacramento basin and in the San Joaquin basin from the Late Miocene to the mid-Pleistocene.

A major east-west cross-structural divide between the two basins in the Great Valley near Stockton was shown by Safonov (1962), and was investigated by this author also by electric log correlations when with Shell Oil Company. The Stockton Arch, which became active in the Late Cretaceous was covered finally with Great Valley alluvial-pluvial-fluvial deposits during and after the mid-Pleistocene. This deposition, and damming of earlier drainages (Jenkins, 1973) changed the course of the ancestral Kern River (the major river system then and now in the San Joaquin basin) and caused it to join with the Sacramento basin rivers to jointly flow to the ocean via the newly opened debouchment point through the Golden Gate (Taliaferro, 1951; Martin and Emery, 1967, 1982). Once the ancestral Kern River and other San Joaquin basin rivers had changed course, and had cut through the easily eroded Great Valley sediments above the Stockton Arch, the major earlier drainage (as discussed next) through Priest Valley (from Coalinga northwestward toward Monterey Bay) was never reestablished because of both increased topographic elevations across the former drainage (Jenkins, 1973) and the fact that the new course of the San Joaquin basin drainage (mainly the Kern River) was established on a lower and more easily maintained gradient.

The final phase of disruption of drainage channels in central California was discussed by Jenkins (1973) relative to the creation of the short-lived, but comparatively large, Lake San Benito. This lake was formed by the damming of earlier river channels during mid-Pleistocene tectonic events. At its greatest extent, Lake San Benito was some 30 miles (48 km) long, and it lay almost entirely to the east of the SAF, which certainly moved vertically and perhaps laterally to create this lake's dam. The lake was drained when down-cutting through the Parajo River drainage (including Elkhorn Slough) lowered the base level through the orogenic dam. By this time, however, the new course of the Kern River and other San Joaquin basin rivers to meet the Sacramento River had been firmly established and previous drainage conditions from the San Joaquin basin through Priest Valley were not restored. Martin and Emery (1967, 1982) showed part of this before and after drainage sequence in their figure 8, diagrams V and VI.

Priest Valley seaway and drainage

Besides the aforementioned river systems, another constraint related to the Priest Valley seaway and subsequent drainage can be cited. The only channel through which marine Pliocene waters entered the San Joaquin basin was through Priest Valley, which had a northward trend from Coalinga on the south to just east of Monterey Bay. Marine Pliocene deposits in Priest Valley are approximately 6,000 feet (1,860 m) thick along its trend which is parallel to that of the SAF/SAFS. These deposits and those with similar thicknesses were described by Gribi (1963) in the Hollister Trough to the northeast of Priest Valley. Faunas at Kettleman Hills near Coalinga, furthermore, were described and shown to have normal upward-grading facies from marine, through brackish, to freshwater, and then to continental (Woodring et al., 1940). These faunas, from the Pliocene and the Early Pleistocene up to the mid-Pleistocene orogeny, reflect what is expected (before the concepts of plate tectonics confused such issues) when marine waters retreat from the continents and when their faunas are replaced by regional lacustrine faunas, then by local freshwater faunas in streams, and finally by continental faunas both aqueous and terrestrial. The large lake that formed in the San Joaquin basin as a result of the closing of the Priest Valley seaway was the ancestral Lake Tulare. The remnant that remains today is small compared to the Plio-Pleistocene lake and varies in size relative to overflows and surplus from irrigation farming. That ancestral lake filled most of the San Joaquin basin to discharge into the marine waters at Monterey Bay via the Priest Valley channel as marine waters changed to fresh. This connection, whether for marine water inflow or later freshwater outflow, was not closed until uplift caused by mid-Pleistocene orogeny.

Thus, for at least 10 million years, and perhaps more, the Priest Valley channel remained open both to marine water incursions and later to freshwater discharge to the Pacific Ocean via Elkhorn Slough. This sequence of geologic and geomorphic events constitutes a major constraint to right-lateral movements on the SAF and SAFS during this time because of the narrow width of the Priest Valley channel, and whereby even a little right-lateral movement would have been enough to completely close the channel. Conversely, slow vertical movement could have occurred as long as erosional processes (currents in the seaway or river scour) could have kept the channel open; or in other words, as long as the channel's axial trend moved in the vertical direction, water movements would not have been constrained.

Monterey Bay submarine canyons and the Monterey deep-sea fan

Physiographic and geomorphologic characteristics in these two contiguous provinces can provide additional data to prove constraints to major right-lateral movements of the SAFS (faults to the west of the main San Andreas fault). Martin (1964) and Martin and Emery (1967, 1982) carried out the first work in the central Coast Ranges to correlate continental geology with that found by dredgings and echo soundings under the waters of Monterey Bay out to the seaward limits of the Monterey deep-sea fan. Greene (1979) added to this knowledge, but his work was limited mainly to the marine geology of Monterey Bay and geology of nearshore land areas. By 1989, physiographic data mapped by the National Oceanographic and Atmospheric Administration

(NOAA) using SeaBeam technology (developed by the General Instrument Corporation) had been declassified. These first charts detail the Monterey Bay submarine canyons, nearshore marine topography, and bottom extension to the limits of the deep-sea fan. This author was given an opportunity on 15 August 1989 to inspect some of these SeaBeam charts, especially to compare them with the earlier work of Martin and Emery (1967, 1982). The main interest for such inspection, besides intense personal interest, was to determine whether constraints in the marine environment to major right-lateral SAFS movements were changed or were themselves constrained by the new SeaBeam data. They are not changed, because these SeaBeam data support completely Martin's previously held concepts of constraints related to the San Gregorio-Carmel Canyon–Nacimiento faults.

On first inspection, and concepts held when Martin's Monterey Bay research began in 1962, was that because of the size of the Monterey submarine canyon, the total Monterey deep-sea fan accumulated as a result of most sediments having been channeled down the canyon. Martin (1964) later gave reasonable and accepted proof that the Monterey Canyon was not cut or initially eroded until beginning in the Late Pliocene or the Early Pleistocene. The fan's sediment volume (1.6×10^5 km^3 calculated by Menard, 1964) and its size similarly shown by Menard (topographic map, 1964), and in a conversation with Menard, 1962, strongly suggested that sediments were being transported to the deep-sea environment much earlier than Late Pliocene or Early Pleistocene even though Monterey Canyon now functions as the only major conduit feeding nearshore sediments to the extremes of the Monterey fan. To Martin (1964) this demonstrated that sediments were being transported seaward by an older, open submarine canyon in the vicinity of the Monterey canyon yet to be eroded (Parajo Gorge was possibly too small to intercept littorally drifted sediments or to be the needed conduit prior to the initial erosion of Monterey Canyon).

The most likely candidate for a major, earlier sediment conduit was Carmel Canyon, but today this canyon appears to have had its axial trend and sediment load captured by Monterey Canyon at a present depth of approximately 6,300 feet (1,953 m) after fault movement on the Carmel Canyon fault (Fig. 3A) changed the axial trend from east-west in headward reaches to northwest in lower canyon reaches to its intersection with Monterey Canyon. This problem is not yet resolved using present topography, for it means that the modern Carmel Canyon could not have been eroded before the beginning of the erosion of Monterey Canyon (Late Pliocene), but if Carmel Canyon could have been open and eroding before the initiation of Monterey Canyon submarine erosion, Carmel Canyon could have been the missing or unrecognized sediment conduit in use prior to the opening of the major sediment conduit of Monterey Canyon.

Proof that Carmel Canyon was the earlier or ancestral submarine canyon, perhaps open and transporting sediments along its axial trend to the Monterey deep-sea fan since the Oligocene or the Early Miocene, is based on (1) need for a sediment conduit to the Monterey deep-sea fan earlier than the initial erosion of Monterey Canyon to account for the great volume of sediments comprising the fan (calculations were made to estimate the present fan's volume to eroded material from Monterey Canyon itself and estimates of sediment carried down the canyon from Pliocene to present; there is still an excess of sediment in the fan requiring another source of sediments), and (2) need of a conduit to bypass the sediment trap caused by the creation of the Monterey graben (see next section). Without a separate, but nearby source of sediments, to feed the deep-sea fan from Oligocene to present there is an imbalance of fan volume compared to sediments available from nearer to shore from the Pliocene to present.

Topographic proof comes also from the existence of an abandoned east-west tributary to the west of the present east-west axial headward trend of Carmel Canyon. This decapitated abandoned tributary (Fig. 3A, shown on the west wall by a dashed line where cross-section C crosses Carmel Canyon) begins about 2 miles (3.2 km) west of the point where the modern Carmel Canyon makes an abrupt axial change to the northwest. This abandoned tributary based on PDR echo soundings was mapped by Martin and Emery (1976, 1982, fig. 2) to extend further westward and seaward to intersect the modern Monterey Canyon at a depth of 7,800 feet (2,418 m), which is 1,500 feet (465 m) deeper that the present intersection depth of Carmel Canyon and Monterey Canyon at 6,300 feet (1,953 m).

Martin and Emery believed that this abandoned tributary represents the extension of Carmel Canyon prior to uplift of its west wall, which in turn changed the canyon's axial trend to the present northwest direction. The importance of this extension for the purpose of this paper in proving major right-lateral constraints, is that the decapitated extension is a direct east-west continuation of the headward (east-west) reaches of Carmel Canyon: there is no right-lateral offset across the Carmel Canyon fault. This in turn proves that no right-lateral movement has occurred on the San Gregorio or the Nacimiento fault of which the Carmel Canyon fault is a part of the total fracture.

Inspection of the SeaBeam contours of the bottom of Monterey Bay and especially of the area of the decapitated tributary shows that the earlier contours of Martin (1964) and Martin and Emery (1967, 1982) are correct. The only difference is in enhanced detail of the SeaBeam data. A definite abandoned or decapitated channel, west of Carmel Canyon from that small portion shown (Fig. 3A) complete to its intersection at the depths given with Monterey Canyon as mapped and described 25 years ago, must be a seaward part of Carmel Canyon prior to its west wall uplift and subsequent sediment axial sediment load capture by Monterey Canyon. The outward gradients shown by the SeaBeam contours are evidence for a major tributary; it is difficult to explain the existence of such a well-defined tributary so far from shore without its reason for being there. In the interval on ocean bottom between the decapitated head of the ancestral canyon and the present open Carmel Canyon of 2 miles (3.2 km), smooth bottom contours may be explained by (1) submarine erosion on the high west wall, (2) deposition by marine pelagic sediments after the abandoned canyon channel stopped being actively eroded, or (3) a combination of both. (Keep in mind also that since the abandonment of the decapitated tributary, ocean bottom sedimentation has been active, no doubt, to smooth the bottom contours and to fill in the abandoned channel, but even today, the tributary is very evident suggesting strongly that the channel when open and being eroded was much more impressive in terms of relief between the tops of the canyon walls and the axis.) Critics might observe that many "decapitated" channels start at some distance from the shore in the area of Monterey Bay (e.g., Soquel Canyon and Ascension Canyon), but the heads of these canyons start at the shelf break. The decapitated head of Carmel Canyon starts across a deep, active submarine canyon. Where are the sediments and energy sources to keep such a decapitated tributary active and eroding today unless the tributary was initially cut when it was connected to a submarine canyon that had headward reaches closer to shore?

Finally, uplift on the Carmel Canyon fault of the west wall or west block that changed the canyon axis to northwest occurred post-Middle Miocene based on west wall dredgings that obtained indurated siltstones in which a Middle Miocene microfauna was identified. The point again for emphasis is that there was vertical uplift on a major SAFS fault (Carmel Canyon fault) and that no right-lateral fault movement can be demonstrated or shown on that same fault. The microfauna also show that west wall or west block uplift was post-Middle Miocene, which demonstrates that the now abandoned extension of Carmel Canyon was open and actively being eroded to transport sediments in its axis before Middle Miocene time; evidence to be presented later found the basal Monterey deep-sea fan shows sediments to have been dated as Oligocene. Thus, Carmel Canyon was transporting sediments to the Monterey deep-sea fan along its now abandoned axis since Oligocene time. The conclusion is that lateral movement on the Carmel Canyon fault (and San Gregorio and Nacimiento in the vicinity of Monterey Bay) has not occurred since the Oligocene.

Monterey graben

The Monterey graben with contained marine Upper Miocene and Pliocene sediments (Fig. 3B) illustrates additional vertical fault movements on the SAFS to include: the San Gregorio and Carmel Canyon faults; extension of the Tularcitos fault in the Santa Lucia Mountains; the Gabilan fault at the base of the Gabilan escarpment; and the Monterey fault, which follows the axial trend of the canyon to account for the juxtaposition of granodiorites on the south wall to sediments on the north wall (Fig. 3A). In order to account for the thickness and the volume of the marine sediments recovered by dredgings on the north wall of Monterey Canyon, a sediment trap must have occurred during the Late Miocene and the Pliocene under the present north wall of the canyon. The only solution, because of extremely limited seismic data in 1962–1963 concerning the structure underlying Monterey Bay, was to assume that a major, deep graben actually did occur, an assumption that was later acceptable to the geophysicists of a major oil company. Gary Greene, USGS (pers. comm., circa 1985) who did later work for his dissertation on Monterey Bay, stated that he also agreed with the concept of a Monterey Graben and the concept of graben depths resulting from basement down-dropping, however, Greene's concept was based on later seismic data. Greene also sent this author a copy of his dissertation (1975) and accompanying uncut maps showing the graben by basement rock contours—quite similar to those contours of Martin (1964) and Fig 3A.

The top of the Monterey graben lies almost 17,000 feet (5,720 m) below sea level (Fig. 3A). This graben depth again proves the need for a submarine canyon, as was the ancestral Carmel Canyon, to feed sediments to the Monterey deep-sea fan prior to the Pliocene, because vast quantities of nearshore sediment, which might have been transported to the fan were trapped in the graben as later exposed by subsequent downcutting of the Monterey Canyon.

The existence and the fault relationships of the Monterey graben prove that vertical fault movements are dominant, and that right-lateral fault movements are either minimal or nonexistent. Physiographic relief is absolutely necessary to produce (mainly by gravity, erosion, or both) such features as the Parajo Gorge, the Gabilan escarpment, and the Monterey graben. Such features and others described in this present paper can form only when a difference in elevation occurs between adjacent fault blocks; only vertical movement on adjacent fault blocks can produce the required elevation difference or relief. Lateral movement on adjacent fault blocks that produce little or no relief cannot produce physiography needed to result in the geology as noted by field evidence. The outstanding example is the marine sedimentary fill (1,000 km^3 by first approximation) of a thickness of over 17,000 feet (5,720 m) that was deposited in the Monterey graben during the Late Miocene and the Pliocene.

Monterey deep-sea fan

For those who believe strongly in the concepts of plate tectonics to explain the geology of California and the northeastern Pacific Ocean basin adjacent to California, an additional series of investigations of the age of fan sediments at the base of this huge marine deposit would be revealing, especially with the relationship to supposed subduction of the Pacific plate below the North American plate at the SAF/SAFS boundary fracture. If subduction did and is occurring, fan sediments near the contact with underlying basement rocks (pre-fan ocean floor) should be Late Cenozoic (Pliocene to Recent) because earlier fan sediments should have been subducted and now should be missing. Wilde et al. (1982) stated, however, that the earliest age of the basal fan sediments is Oligocene, and that the age of the basement-ocean floor on which these fan sediments are deposited is Oligocene.

Proponents of a plate-tectonics solution to present geologic conditions in the offshore waters of central California in the Pacific Ocean basin are thus faced with a double conundrum and problem to explain a marine plate-tectonics solution to the continental plate-tectonics solution: they are forced to insist on a major right-lateral movement of the Monterey fan (Delgada fan also) because of deposition on the northwest rotating Pacific plate in addition to a general eastward movement of the fan(s) to be eventually subducted beneath the North American plate. This should be an extremely challenging problem to those who are actively working on matters dealing with the Monterey and Delgada deep-sea fans.

It is hoped and suggested that by judicious use of the new SeaBeam contour data, and with realization of the proof that most major fault movements on the SAF and SAFS in central and northern California (and probably in southern California, but not a part of this paper's discussion) are vertical, these submarine data may help to understand better the relationship of deep-sea fans to more nearshore marine geology and to geology on land.

CONSTRAINTS IN THE TRANSVERSE RANGES

Structural constraints to major right-lateral SAF and SAFS movements are quite evident in the Transverse Ranges (Fig. 1), and recent studies using geochemical methods indicate constraints that are additional to structural constraints. Because of these two diverse constraints, a major connection between the SAF/SAFS of the Coast Ranges and the faults of the Imperial Valley (San Jacinto and Imperial faults) via the Transverse Ranges seems unreasonable (Fig. 1). Allen (1957) discussed problems of mapping a recognizable continuation or connection of the SAF/SAFS in the San Gorgonio Pass. Baird (1974) and Baird and Miesch (1984) on the basis of their geochemical investigations and data concluded ". . . no large amount of displacement has occurred along this particular strand of the fault since the emplacement of the batholithic rocks." Such batholithic explacement took place during the Middle Mesozoic according to most workers.

Namson and Davis (1988), and Figure 4B, show major vertical and thrust movements of the SAF in the western Transverse Ranges (Tehachapi Mountains) at the south end of the San Joaquin basin. Thrusts both to the north and to the south from the vertical San Andreas fault plane are exactly what should be expected as the SAF/SAFS trend changes from that in the Coast Ranges to a more easterly direction in the Transverse Ranges even if right-lateral movements are not involved. Even with vertical fault movements there must be structural compensation of compression-tension in such a drastic change of trend of any major fracture (SAF) or fracture zone (SAFS), and in this area, such effects appear in mapping.

Just to the north (Fig. 4A), Zoback et al. (1987) depicted similar vertical SAF movements accompanied by thrusts orthogonal to the fracture trend. The Carrizo Plain is famous for aerial photographs showing right-lateral surface drainage offsets along the SAF/SAFS, but these are considered rebound effects secondary to major vertical primary movements as presented by Zoback et al. (1987) in their cross-section, and as additionally presented in the various constraints given in this paper.

Field evidence by this author in the San Emigdio Canyon about 10 miles (16 km) west of Grapevine Canyon (where Interstate 5 crosses the Tehachapi Mountains) concerns an exposed synclinal structure showing intense northward thrusting from the SAF along the same general line, section B, of Namson and Davis (1988), and Fig. 4B. This syncline, with a relief on the vertical south limb of approximately 2,000 feet (620 m)

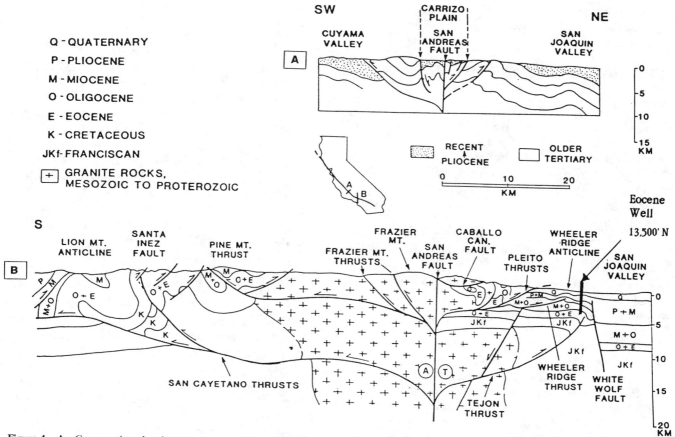

Figure 4. **A**—Cross-section showing structure and thrust fault tectonism that is normal (orthogonal) to the San Andreas fault (adapted from Zoback et al., 1987). **B**—Note syncline in Eocene (E) rocks to the right of the San Andreas fault and below Cabello Canyon fault; this is the same syncline investigated and described by Martin (1964). Also note the location of the well to Eocene rocks as described (adapted from Namson and Davis, 1988).

contains tightly folded Eocene (Tejon formation) and Oligocene beds; it is one of the most magnificent structural features that this author has ever seen within such a relatively small area. The axial plunge is unusually steep eastward. This structure is not recognized in the Grapevine Canyon nor on the west wall of the San Emigdio Canyon, owing to granodiorites and rocks younger than Eocene and Oligocene having been thrust over the syncline by the Caballo Canyon fault (Namson and Davis, 1988, Fig. 4). The Eocene Tejon formation, exposed in the syncline at an elevation of about 2,000 to 4,000 feet (620 to 1,840 m) above sea level, is again found subsurface at 12,700 feet (3,968 m) below sea level in an oil well approximately 20 miles (38 km) to the north (Fig. 4-B) at the south end of the San Joaquin basin. This correlation shows additionally the tremendous tectonic forces acting in part from the south to down-drop the Tejon approximately 17,000 feet (5,270 m) in the relatively short horizontal distance between the well and the exposed syncline.

The distinct tectonic stress patterns shown by examples cited above make the lack of similar patterns at Cape Mendocino suspect as a region where major right-lateral fault movement or block rotation had been supposed to occur (Griscom and Jachens, 1989, and others cited by them). Richard B. Perry (NOAA, pers., comm., 1989) stated that if any movement on the SAF/SAFS occurred or occurs north of Point Arena (Fig. 1) to Cape Mendocino, such movement is vertical and not right-lateral. As discussed and also depicted at the juncture of the fracture zones from the southern Coast Ranges to the western Transverse Ranges (Tehachapi Mountains), vertical fault movements did and do result in intense and widespread thrust faulting, which should have occurred many times at the junction of the SAF/SAFS with the Mendocino fracture zone (Fig. 1) if vertical faulting had occurred there also. Finally, a recent test well to prove the SAF in the Cajon Pass area (Fritz, 1987), was shut down because of funding problems (as stated in the article), but word along the "geologic grapevine" indicated that the SAF in this area did not have the proper stress conditions or stress directions to be both a major fault, as the SAF is supposed to be, and to show major right-lateral movements.

CONSTRAINTS RELATIVE TO CORRELATIONS OF EOCENE ROCKS

Discussion and evidence previously had shown constraints to major right-lateral SAF and SAFS movement during the latter part of the Cenozoic. Similar constraints now need to be shown for the earlier part of the Cenozoic, as Eocene rock correlations made by Nilsen (1984) and others cited by him, to prove 305 to 320 km (183 to 192 miles) of right-lateral movement on the SAF and SAFS need further study to support their long-range movement thesis. Based on evidence and methods used previously in the present paper, constraints in the earlier Cenozoic seem reasonable, but the evidence may not be as conclusive as that for the later Cenozoic owing mainly to more early Cenozoic deposits on uplifted areas possibly having been eroded and dispersed than during the later Cenozoic.

Nilsen (1984) correlated the outcrops of Eocene rocks in two localities: the San Emigdio Canyon exposures on the east side of the SAF in the southern San Joaquin basin and on the west side of the SAF (or SAFS) in the San Juan Bautista area to the east of Monterey Bay (Fig. 1). Much field evidence is cited by Nilsen to relate the similarities of Eocene rocks in these two end localities, and how such similarities prove major right-lateral fault movement on the SAF and SAFS during the last 10 million years. The flaw in Nilsen's paper (1984) and in other papers investigated previously lies in paleogeography and paleogeology. Eocene and other early Cenozoic rocks may have been deposited extensively atop Salinia between the two end localities (San Juan Bautista and the San Emigdio mountains), but these deposits could have been removed by erosion after the Eocene to have been deposited elsewhere as detritus. Therefore, the conclusion by Nilsen that the Eocene rocks in the two end localities were later separated by major right-lateral fault movement appears tenuous and incorrect, because these present end-locality deposits may represent in situ deposits at the ends of a once continuous deposit of Eocene sediments along the whole length of Salinia and the San Andreas fault. (The section on Eocene seaways will consider this pattern of deposition further.) Nilsen (1984, fig. 1) depicted other lower Tertiary outcrops between the two end localities described above within the boundaries of Salinia. These deposits (Eocene?) need to be considered additionally as part of a more widely distributed Eocene deposit atop Salinia, but now eroded. The thesis of Nilsen (1984) cannot be considered complete until he and others study and report on the effects of vertical block movements, and how Eocene paleogeology and paleogeography almost totally negate right-lateral movement of Salinia.

Hoots et al. (1954, and their figures 3 through 6) discussed and depicted the boundaries of early Cenozoic seas, and the related deposits of their sediments in the San Joaquin basin, the southern Sacramento basin, and across the Transverse Ranges (Tehachapi Mountains). These data are still valid in general presentation even though later field data may have refined the marine limits somewhat. From Paleocene to Oligocene time (Early Cenozoic), present deposits show that marine waters in the San Joaquin basin were limited on their west by the San Andreas fault, which indicate uplift of the block to the west (Salinia). In light of the previous discussion, however, Eocene marine water may have covered all or part of this west block, and deposits subsequently may have been eroded from atop the block (except in protected areas (Nilsen, 1984)). The study of Hoots et al. (1954) seemed to tip the scales toward vertical uplift to allow the erosion of early Cenozoic deposits (Paleocene and Eocene) from most areas of the Coast Ranges to the south of Monterey Bay.

A major seaway during the Eocene, the Vallecitos, permitted marine water to enter the San Joaquin basin as did other possible other seaways at times from the Sacramento basin across the Stockton Arch and from the south across the Transverse Ranges. Regardless of the last two, the Vallecitos remained open until the end of the Eocene. This seaway lay generally in the same vicinity and along the same trend as the Pliocene Priest Valley channel. As argued previously for the Priest Valley channel, it is impressive that the Vallecitos remained open similarly as a major seaway for at least 20 million years. Major right-lateral fault movements along the SAF/SAFS hardly could have occurred without closing the seaway. It seems more reasonable to consider that the Vallecitos seaway was closed at or near the end of the Eocene because of vertical uplift on the Salinian block; this uplift through the rest of the Cenozoic in turn allowed the erosion, all or in part, of previously deposited Eocene sediments. This concept also relates well to the consideration that the Eocene deposits described as transported laterally after deposition by Nilsen (1984) were in reality laid down in place and are remnants of a far more extensive area of marine deposition. The distinct boundary of Eocene seas shown by Hoots et al. (1954) also relates to the post-Eocene uplift that created the sharp boundary between Eocene deposits in the San Joaquin basin and Salinia itself along the SAF/SAFS.

Thus, two major seaways into the San Joaquin basin in early and in late Cenozoic time both paralleled and crossed the SAF and SAFS, yet for an approximate total at a minimum of 30 million years during the Cenozoic, these seaways remained open to permit marine waters to enter this basin. Once again, this is a major constraint to major right-lateral movements on the SAF and SAFS.

Vallecitos stratigraphy and structure are well known to the author and not because of literature review only. The first well drilled into the structure within the generally synclinal Vallecitos was drilled by Shell Oil Company; this author sat on this well and described the cores as taken.

SALINIA: THE GORDIAN KNOT OF CALIFORNIA GEOLOGY

From the previously discussed constraints to major right-lateral fault movements presented in this paper, Salinia could not have been transported laterally to its present locale, either by fault-block movement from the south or southeast or by rotation of the Pacific plate to the northwest according to the precepts of plate tectonics. On the other hand, the evidence for in situ emplacement of the igneous rocks of Salinia is not proved by default by the lateral movement constraints, for it is difficult to understand how the granitoid rocks were intruded only into the Salinian block, and not across the plane of the SAF on the east side. Even if the San Andreas fault were extant when and if such magmatic intrusions did take place, the fault plane at the depths needed for the intrusions would seem unlikely to constitute much of a barrier, or any barrier, to prevent the magmatic intrusions from also intruding into Late Mesozoic rocks along the west side of the San Joaquin basin; but as no granitoid rocks occur in outcrop or at depth along the west side of the basin that lies to the east of the SAF, a new concept must be introduced to explain this riddle. A reverse problem, in a sense, is that no outcrops of Franciscan rocks occur in or near Salinia, but a discussion of this problem is beyond the scope of this paper.

Thus, the term seems appropriate: the Salinian Gordian Knot. Constraints to major right-lateral transport or movement of Salinia are strong, as demonstrated in this present paper by the major vertical fault movements, yet confinement of the deep intrusion of granitic to granodioritic magmas to Salinia alone also seems a constraint to in situ development of Salinian igneous rocks.

A third contribution that might help solve the Salinian problem recently has been proposed by Meyerhoff et al. (1989 and this volume) as part of an alternative approach to global dynamics. This new approach is called surge tectonics. This third concept to explain Salinia, by surge tectonics methods, must be given due consideration. The first two concepts mentioned suffer from serious to unalterable constraints.

CONSTRAINTS RELATED TO PRESENT TECTONISM AND INVESTIGATIONS

Some considerations of major vertical movements.

Acceptance or even realization of the proofs of major vertical movements both up and down on the SAF and SAFS means a drastic and dramatic shift both in thinking and in presentation of data by geologists to their colleagues and to the general public. Geologists can argue in their own groups and publications the data concerning constraints and the changed direction vectors of faulting (from right-lateral to vertical), but a direct and necessary application of these vectorial direction changes of movement belong to the engineering professions, and especially to that of civil engineering as related to construction and safety factors designed in manmade structures to prevent or ameliorate damage from earthquakes, especially those above 5.0 M (on the moment-magnitude scale now being used to replace the Richter scale). The failure to consider vertical motion with respect to design of the San Francisco Bay Bridge was discussed by a civil engineer (Cheskin, 1990) relative to the Loma Prieta earthquake and the collapse of the bridge's span.

As a geologist with engineering affiliations, and a full-member of the American Society of Civil Engineers, this author places great importance on adequate and safe engineering design in manmade structures. A 90 degree change of a force vector during earthquakes, and resultant fault movement or movements must be reevalutated in terms of design stresses and in terms of factors related to vertical changes.

Examples of constraints with attendant vertical movements in this paper have dealt mainly with central and northern California, but Lowman (1980) proved that vertical fault movement was the correct direction on the Elsinore fault in southern California, and other examples either are now known or may come to be recognized. Thus, the Cenozoic tectonism of all California must be reevaluated in the light of primary vertical fault movement and of secondary effects (rebound) resulting from those movements.

Analyses of the effects of vertical fault movements, except in historically reported and analyzed earthquakes, are difficult to reconstruct and to present. Vertical forces because of the movement direction either are ameliorated or dissipated at the contact between the land surface and the atmosphere, or the stratigraphic or structural records or both, in uplifted blocks are later destroyed by erosion and gravity effects. Only in a presentation such as discussed previously in the Monterey Bay area and as preserved in buried conditions

(Fig. 3) can the importance of vertical movements be recognized. On the other hand, rocks involved in lateral movements can be expected to be preserved better or completely because such rocks are not subject to as much erosion or mass wasting by processes or agents associated with gravity. Thus, unless the rocks involved in vertical movements are preserved in some way (e.g., by burial under and to the east of Monterey Bay), rocks involved in lateral or horizontal movements are preserved better mainly because of the lack of topographic relief. This provides evidence even to the most astute field geologist that right-lateral movement is dominant, especially when evidence for vertical uplift has been eroded and the detritus dispersed to be redeposited elsewhere.

Since about 1980, field investigators have been studying and reporting on thrust faults that are orthogonal to the SAF and SAFS (Fig. 4B; Zoback et al., 1987; Namson and Davis, 1988). Although these thrusts have been studied for years prior to 1980, mainly by oil companies by seismic and electric log correlations, recognition of relationships of these orthogonal thrusts to vertical fault planes is comparatively new. When these data are coupled with the realization of vertical movements on the SAF and SAFS as discussed herein, the results of such investigations will aid greatly in a revised understanding of tectonics in central and northern California, and perhaps of tectonics in southern California.

Fault movements: evidence by instruments versus field studies

One of the major problems evident in this study is the need to understand crustal movements as determined by geodetic measurements to those determined by field geologic investigations. It is obvious that geodetic measurements favor right-lateral movements (Sauber, 1989), whereas the thesis presented here favors vertical movements in central and northern California.

The basic point of contention deals with the matter of time: the NASA database collection for the western United States began in 1979; the geology for field investigations began some 70 million years ago. That is, 10 years of geodesy versus 70 million years of geology. It seems obvious that it is almost impossible to establish a meaningful comparison by the time ratio of 10 to 70 million or one to seven million years. It is known and recognized during this 70 million year period (Cenozoic) that right-lateral movements did occur and it may be that the measurements being made now by NASA are mainly during one of these periods of lateral movement, but even if such a position is considered, the information released by Robert Hamilton (1989) after the Loma Prieta earthquake of 17 October 1989, showed that vertical fault movement was just as important as right-lateral movement (further discussion follows).

Geodetic data as presented by Sauber (1989) certainly are valid and well presented, but field evidence as presented here, shows that many constraints to major right-lateral movements occur in central and northern California, especially at Cape Mendocino and vicinity, and at the Transverse Ranges (Fig. 1). Thus, any right-lateral crustal movements in central and northern California are restricted to the region between Cape Mendocino and the Transverse Ranges.

Future investigations and discussions must attempt to rectify this major point of disagreement between crustal dynamics as presented by Sauber (1989) and lateral movement constraints as discussed herein. However, again with reference to the vast difference in time to develop evidence for each viewpoint (geodesy versus field geology), such discussions may only emphasize that the on-going Crustal Dynamics Project of NASA is only a small part of the total time and, even with presentation of instrumental data and calculations (Sauber, 1989), these great amounts of data representing many man-hours of work cannot compare well with the geologic field evidence. Scientific research favors investigations of long duration: one year of geodesy to seven million years of geology does not seem a favorable comparison.

Perhaps this is another example of the present not being a key to the past, as has been recognized in trying to understand, when compared to present conditions on and in the Earth, such enigmas as the vast outpourings of Columbia River basalts in the Tertiary, the thickness and widespread distribution of Paleozoic salt formations in southeastern United States, or, even so close in time to the Recent, the reasons for the vast and thick accumulations of Pleistocene ice sheets in the Northern Hemisphere. Again, this author finds it difficult to accept that 10 years of geodetic measurements, and data preparation and publishing, can or should be directly compared or related to the field geologic evidence developed for the past 70 million years during the Cenozoic.

Loma Prieta Earthquake (17 October 1989)

After a major earthquake, such as the Loma Prieta, much is soon written in both popular and scientific journals to discuss the initial event and later findings, however, this earthquake and its related geology and geophysics confounded the experts (e.g., an article published after the seismic event was titled "Bay Area quake fails to fit textbook model" [Monastersky, 1989]). From other articles also, it is obvious that engineers, geologists, geophysicists, and others involved in earthquake investigations were by various degrees confused and they stated, as did Hamilton (1989), "The upward component of the shift is unusual for fault movement in the central California section of the San Andreas fault." Hamilton, U.S. Geological Survey, made this statement along with more written information about the earthquake at a meeting of the Geologic Society of Washington (DC).

In reviewing articles about the Loma Prieta earthquake both in geologic and in engineering journals, it is evident that no one could explain the geologic phenomena and the structural damage to manmade works based on lateral movement only on the San Andreas fault. Personal observations of television and photographs reenforced the idea that lateral movement alone could not account for the types of damage and destruction that occurred. Finally, eight days later, the vertical component of the movement was announced; this was 5 feet (1.5 m) upward along a fault plane dipping about 70 degrees to the southwest (Hamilton, 1989). According to Hamilton, the shock measured 7.1 Richter (the moment-magnitude was not given) at a focal depth of 11 miles (17.6 km). Surprisingly, no lateral movement was mentioned, however, 5 feet (1.5 m) of horizontal slip was stated elsewhere (Monastersky, 1989). This vertical movement probably was a major factor in the failure of the San Francisco Bay Bridge (Cheskin, 1990) and the I-880 freeway (both failures discussed with Thomas L. Brown, P.E., geotechnical engineer, pers. comm., 1989—see following section). Overall, the compression-tension forces related to vertical upward movement during the Loma Prieta earthquake seem more apparent from the observed damage than those related to lateral (horizontal) movements, however, both vectorial forces when combined in the total seismic event certainly resulted in the final damage effects noted.

As the quake's focus occurred on a fault plane dipping about 70 degrees to the southwest (Hamilton, 1989) in the Santa Cruz area, a likely candidate for a buried fault in this area having such characteristics is the Gabilan fault (Fig. 3).

This brief discussion of the Loma Prieta earthquake is given mainly to make scientific plea to other geoscientists and engineers to give due consideration now to the recognition of vertical fault movements on the San Andreas and related faults in its system. The Cenozoic geologic and geomorphologic evidence cited to prove vertical movements on these faults hopefully should spur geoscientists and engineers, and those concerned with public safety when apprised of this present paper's data and conclusions, to reevaluate their thinking to prevent or ameliorate seismic effects during future earthquakes. These persons should not take an attitude that Loma Prieta was one-of-a-kind because of the vertical movements recognized. The previous geologic evidence detailed herein, obtained mainly by field work, proves that this earthquake was not unique with respect to fault movement—it was unique only because many types of manmade structures designed and built since the 1906 earthquake were not designed to withstand vertical fault movements.

Certain engineering considerations related to the Loma Prieta earthquake

This author and his company have been associated with a number of consulting engineers and their firms. One in particular has been T.L.B. Associates, Inc., Crofton, Maryland, and Washington D.C., owned by Thomas L. Brown, professional engineer (civil). On 18 October 1989, at 2300 hours, Brown called to obtain a geologic opinion as to what could have caused the damage to the San Francisco Bay Bridge and to Interstate 880; in his geotechnical experience, the collapse of either structure should not have occurred because of design safety factors (I-880 was later shown to be underdesigned for safety as discussed by Fehr and Gorney, 1989). The answer was, and was reinforced by later data, that a component of vertical fault movement had to have been involved to account for the noticeable collapse of the bridge span, where it was attached to a supposed stable tower, according to Brown, and in an article by Lawson and Ichniowski, (1989). Attachment bolts were sheared when the vertical force component was added to the force component in the horizontal direction resulting in design safety factors being exceeded for the horizontal stress alone (T. L. Brown, pers. comm., 1989; and later by Cheskin, 1990).

Collapse of I-880 was shown later to be primarily a result of construction on manmade fill causing the stability of the soils to fail and to transmit wave-like motions of the shock through the fill. This created the undulations to cause the collapse of the various sections of the highway (an eye-witness reported this wave-like effect as she raced ahead of the shock wave in a vehicle on the upper deck). It was apparent to the author and to Brown, however, that vertical movement during the primary shock had to have been involved also, as in the Bay Bridge collapse. Each damaged section of I-880 showed no lateral offset from adjacent roadway sections. It is suggested that the damaged upper sections of the double highway were lifted vertically by the vertical component of the earthquake to a height that the immense weight of the mass (of each section combined) overcame any strength or safety factors (albeit inadequate at best) as the road crashed down on the upright upper highway supports. It was proved later (Fehr and Gorney, 1989), that no connecting steel rods or rebar were emplaced in the original construction between the upright support and the lower roadway! The upper roadway was not constrained in any way (except by actual mass) from rising upwards and then falling down onto the supports and the lower roadway. In any final report, it is hoped that discussion will be made relative to the lack of lateral movement noted between damaged sections of the upper roadway to show the amount of vertical movement caused by the earthquake and what appears to be little or no lateral fault movement between sections.

CONCLUSIONS

At least eight examples are detailed by field evidence for severe constraints to major long-range right-lateral movement on the San Andreas fault and related faults within the San Andreas fault system. During the Cenozoic, vertical fault-block movements on the SAF and SAFS were by far more important than right-lateral movements, and such vertical directions of movement are occurring now in a way to make suspect many older engineered manmade structures for safety: this was proven rather dramatically by the Loma Prieta earthquake of 17 October 1989.

The purpose of this paper is to demonstrate by field evidence mainly that vertical movement has occurred many times during the Cenozoic in central and northern California. No hypotheses of plate tectonics based on major right-lateral movements on faults or terranes can be valid until these constraints are removed or proved invalid for each case. However, because the individual constraint is a part of the total hypothesis of major vertical fault movement in this area of California, the total hypothesis must be disproved by opponents.

Although the recently announced hypothesis of surge tectonics (Meyerhoff et al., 1989 and this volume) has not been promulgated yet to enough geologists and geophysicists for them to understand the principles of surge tectonics as related to tectonic investigations in California, it is strongly recommended that present and future publications based on major right-lateral movements on the SAF and SAFS be reconsidered in light of the data from surge tectonics and in light of principles involved. This hypothesis is based on over 500 references of data previously published, but reevaluated in light of Laws of Physics (Newton's three Laws of Motion, the Law of Gravity, and related laws) and present and past well-considered geologic and geophysics investigations.

Authors of recently published papers (e.g., Yeats et al., 1989) should review their data to determine if a factor of vertical faulting has been ignored, and if so, can a paper such as this with much effort and time involved for preparation be revised to ameliorate such a strong statement as theirs about right-lateral movements. Overall, authors planning to publish papers that continue the emphasis on right-lateral movements, and on plate-tectonic motions as proved incorrect or suspect in the present paper, should review their data to determine if a vertical movement factor has either been ignored or not recognized in the preparation of data for publication.

Eventually it will be necessary to have a determination made of differences in data collection and preparation between geodetic studies and evidence presented by field investigators to understand more about crustal dynamics in central and northern California. Field evidence for major vertical fault movements or constraints to major right-lateral fault movements are strong and evident in these regions. The Loma Prieta earthquake of 17 October 1989, provided an opportunity to update evidence from previous field investigations.

REFERENCES CITED

Allen, C.R., 1957, San Andreas fault zone in San Gorgonia Pass, southern California: Geological Society of America Bulletin, v. 68, p. 316-350.

Anonymous, 1985, EEZ-SCAN program maps sea-floor: Geotimes, v. 30, p. 13-15.

Baird, A.K., Morton, D.M., Woodford, A.O., and Baird, K.W., 1974, Transverse Ranges province: A unique structural-petrochemical belt across the San Andreas fault system: Geological Society of America Bulletin, v. 85, p. 163-174.

Baird, A.K., and Miesch, A.T., 1984, Batholithic rocks of southern California—a model for the petrochemical nature of their source materials: U.S. Geological Survey Professional Paper 1284, Washington, p. 1-42.

Cheskin, David B., 1990, Look at vertical motion: ENR Letters, January 18, 1990, p. 9.

Fehr, S.C., and Gorney, C., 1989, Freeway design may have been flawed; Engineer blames placement of expansion joints in support columns: The Washington Post, October 20, p. A-1.

Fritz, M., 1987, Test well probing San Andreas fault: American Association of Petroleum Geologists Explorer, June, p. 14-15.

Greene, H.G., 1975, Geology of the Monterey Bay region: PhD Dissertation, University Microfilms International, Ann Arbor, Michigan, original on file at Stanford University.

Greene, H.G., and Clark, J.C., 1979, Neogene paleogeography of the Monterey Bay area, California, in "Cenozoic Paleogeography of Western U. S." Pacific Coast Paleogeography, Symposium no. 3, 20 p.

Gribi, E.A., Jr., 1963, Hollister trough oil province: Guidebook, Geological Society of Sacramento, Annual Field Trip, p. 77-80.

Griscom, A., 1973, Tectonics at the junction of the San Andreas fault and Mendocino fracture zone from gravity and magnetic data, in Proceedings of the Conference on Tectonics Problems of the San Andreas Fault System, Stanford University Publications, Geological Sciences, v. XIII, Stanford University, September.

Griscom, A., and Jachen, R.C, 1989, Tectonic history of the north portion of the San Andreas fault system, California, inferred from gravity and magnetic anomalies: Journal of Geophysical Research, v. 94 (B3), p. 3089-3099.

Hamilton, R., 1989, October 17, 1989, Loma Prieta earthquake: U. S. Geological Survey, Reston, Virginia, handout prepared for distribution at the October 25, 1989, meeting of the Geological Society of Washington (DC), 1 data page, 8 maps and/or charts.

Hauksson, E., Jones, L., Davis, T.L., Hutton, L.K., Brady, A.G., Reasenberg, P.A., Michael, A.J., Yerkes, R.F., Williams, P., Reagor, G., Stover, C.W., Bent, A.L. Shakal, A.K., Etheredge, E., Porcella, R.L., Bufe, C.G., Johnstron, M.J.S., Cranswick, E., 1988, The 1987 Whittier Narrows earthquake in the Los Angeles Metropolitan Area, California: Science, v. 239, p. 1409-1412.

Hill, M.L., 1971, A test of new global tectonics: comparisons of Northeast Pacific and California structures: American Association of Petroleum Geologists Bulletin, v. 55 (1), p. 3-9.

———, 1984, Earthquakes and folding, Coalinga, California: Geology, v. 12, p. 711-712, December.

Hoots, G.W., Bear, T.L., and Kleinpell, W.D., 1954, Geological summary of the San Joaquin Valley, California, in Geology of southern California: California Division of Mines and Geology, Bulletin 170, Chapter II, sec. 8, p. 113-129.

Howard, A.D., 1951, Development of the landscape of the San Francisco Bay counties, in Geologic guide to the San Francisco Bay counties: California Division of Mines and Geology, Bulletin 140, 75 p.

Jenkins, O.P., 1973, Pleistocene Lake San Benito: California Geology, California Division of Mines and Geology, v. 26 (7), p. 151-163, July.

Jennings, C.W., 1975, Fault map of California: California Division of Mines and Geology.

Lawson, M. and Ichniowski, T., November 2, 1989, Contractors jack Bay Bridge truss; Washington jacks up quake funding: ENR News, The McGraw-Hill Construction Weekly, New York, N.Y., p.11-13.

Lowman, P.D., Jr., 1980, Vertical displacement on the Elsinore fault of southern California: Evidence from orbital photographs: Journal of Geology, v. 88, p. 415-432.

Martin, B.D., 1964, Monterey submarine canyon, California: genesis and relationship to continental geology: unpublished PhD dissertation, University of Southern California, 249 p.

Martin, B.D., and Emery, K.O., 1967, Geology of Monterey canyon, California: American Association of Petroleum Geologists Bulletin, v. 51 (11), p. 2281-2304.

———, 1982, Geology of Monterey canyon, California, in Deep water canyons, fans and facies: models for stratigraphic trap exploration: American Association of Petroleum Geologists Reprint Series No. 26, p. 401-424.

McLaughlin, R.J., Sorg, D.H., Morton, J.L., Theodore, T.G., Meyer, C.E., 1985, Paragenesis and tectonic significance of base and precious metal occurrences along the San Andreas fault at Point Delgada, California: Economic Geology, v. 80, p. 344-359.

Menard, H.W., 1964, Marine geology of the Pacific: McGraw-Hill Book Company, New York, 271 p.

Meyerhoff, A.A., Taner, I., Morris, A.E.L., Martin, B.D., Agocs, W.B., and Meyerhoff, H., 1989, Surge tectonics, in New Concepts in Global Tectonics, Abstracts Volume, S. Chatterjee and N. Hotton III, eds., Lubbock, Texas, Texas Tech University.

Monastersky, R., 1989, Bay area quake fails to fit textbook model: Science News, v. 136 (18), p. 277, October 28.

Namson, J. and Davis, T., 1988, Structural transect of the western Transverse Ranges, California: Implications for lithospheric kinematics and seismic rise evaluation: Geology, v. 16 (8), p. 675-679.

Nilsen, T.H., 1984, Offset along the San Andreas fault of Eocene strata from the San Juan Bautista area and western San Emigdio Mountains, California: Geological Society of America Bulletin, v. 95, p. 599-609.

Page, B.M., 1966, Geology of Northern California: geology of the Coast Ranges of California: Bulletin 190, California Division of Mines and Geology, Sacramento, p. 255-275.

Perry, R.B., and Pryor, D.E., 1986, Bathymetric evidence for the San Andreas fault north of Point Arena: abstract Eos, v. 67 (44), p. 1215, Nov. 4.

Salinas Valley Guidebook, 1963, Guidebook to the Geology of Salinas Valley and the San Andreas Fault, Annual Spring Field Trip, E. Gribi, Jr., and R. R. Thorup, SEPM, leaders, 168 p.; other authors.

Safanov, A., 1962, Tectonic diagram of the Great Valley of California, Redding to Wheeler Ridge, in Geologic guide to the gas and oil fields of northern California: California Division of Mines and Geology, Bulletin 181, p. 78.

Sauber, J., 1989, Geodetic measurement of deformation in California: NASA Technical Memorandum #100732, Geodynamics Branch (Code 621), Laboratory for Terrestrial Physics, NASA/Goddard Space Flight Center, Greenbelt, Maryland, 211 p.

Stein, R.S., and Yeats, R.S., 1989, Hidden earthquakes: Scientific American, June, p. 48-57.

Taliaferro, N.L., 1951, Geology of the San Francisco Bay counties, California: California Division of Mines and Geology Bulletin 154, p. 117-150.

Wilde, P., Normark, W.R., and Chase, T.E., 1982, Channel sands and petroleum potential of Monterey deep-sea fan, California, in Deep water canyons, fans and facies: models for stratigraphic trap exploration: American Association of Petroleum Geologists Reprint Series no. 26.

Woodring, W.P., Stewart, R.B., and Richards, R.W., 1940, Geology of the Kettleman Hills oil field, California, stratigraphy, paleontology, and structure: U.S. Geological Survey Professional Paper 195, 170 p.

Yeats, R.S., Calhoun, J.A., Nevins, B.B., Schwing, H.F., and Spitz, H.M., 1989, Russell fault: Early strike-slip fault of California Coast Ranges: American Association of Petroleum Geologists Bulletin, v. 73 (9), p. 1089-1102, September.

Zoback, M.D., Zoback, M.L., Mount, V.S., Suppe, J., Eaton, J.P., Healy, J.H., Oppenheimer, D., Reasenberg, P., Jones, L., Raleigh, C.B., Wong, I.G., Scotti, O., Wentworth, C., 1987, New evidence on the state of stress of the San Andreas fault system: Science, v. 238 (4830), p. 1105-1111.

ACKNOWLEDGMENTS

Special appreciation is given to Jane Riddle, Research Librarian, National Aeronautics and Space Administration, Goddard Space Flight Center, Greenbelt, Maryland, for her fine effort especially with providing computerized reference data of central and northern California. Paul D. Lowman, Jr., Crustal Dynamics Branch, Code 921, at NASA-GSFC provided both written information and moral support; Patrick T. Taylor, Crustal Dynamics Branch, Goddard, also is thanked for supplying information; and Jeanne Sauber, Geodynamics Branch, although onto the scene late, helped with understanding the geodetic aspects. Maurice Kamen-Kaye, Cambridge, Massachusetts, kindly read and commented on both a preliminary version of the report and a final one—his comments and suggestions are most appreciated. Thanks especially are given to Arthur A. Meyerhoff, Tulsa, Oklahoma, for his comments and early report review. Early in the paper considerations, the late A. K. Baird, Pomona College, California, provided much needed literature, references, and support. Advice and consent were available and given at all times by Daniel J. Stanley, Smithsonian Institution, especially for the interpretation of the 1985 GLORIA image along the Mendocino Fault Scarp; he is additionally acknowledged for his support and professional comments made during some phases of the Monterey Canyon re-survey at the Smithsonian from 1982 to 1986. A stimulating conversation with Richard B. Perry, and other staff, National Oceanic and Atmospheric Administration, Rockville,

Maryland, was helpful along with their providing an investigation of the SeaBeam charts. Special appreciation is given to Kay Meyerhoff for the fine redrafting of part of Figure 3. Irfan Taner and W. B. Agocs provided additional written and verbal support of the constraints concept. Thomas L. Brown, TLB Associates, Inc., Consulting Engineers, Crofton, Maryland, reviewed and approved sections dealing with engineering aspects and such is most appreciated. C. D. Meekins and Associates, Inc., Annapolis, MD, Engineers & Surveyors, are thanked for their help. Finally, K. O. Emery, Woods Hole Oceanographic Institution, reviewed an earlier version of this paper and made comment in a letter which showed that he is appreciative of knowing about the concepts of SAFS constraints and he wishes to know more as evidence continues to be obtained and analyzed.

All these acknowledged may not agree at all or only in part to the findings and conclusions in this paper, but as true scientists and researchers, they allow free thinking of new and exciting concepts.

Tectonics of the Ocean Basins

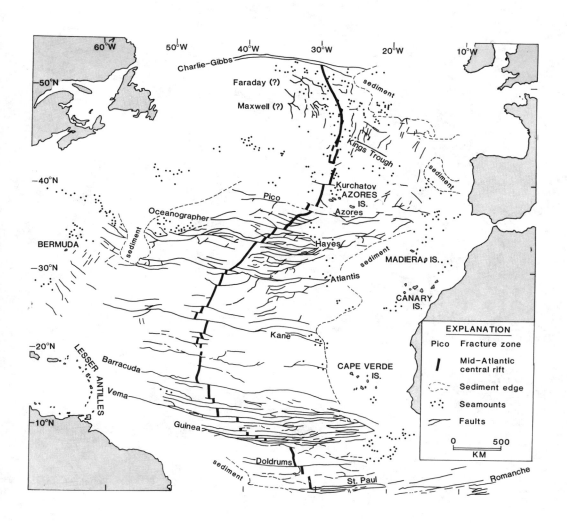

Origin of midocean ridges

Arthur A. Meyerhoff, P. O. Box 4602, Tulsa, Oklahoma 74159 USA

William B. Agocs, 968 Belford Road, Allentown, Pennsylvania 18103 USA

Irfan Taner, 3336 East 32nd Street, Suite 210, Tulsa, Oklahoma 74135 USA

Anthony E. L. Morris, P. O. Box 64610, Los Angeles, California 90064 USA

Bruce D. Martin, P. O. Box 234, Leonardtown, Maryland 20650 USA

ABSTRACT

In plate tectonics, the 65,000-km long chain of midocean ridges is regarded as a zone of sea-floor generation, with sea-floor spreading orthogonal to the ridges. High-resolution side-scanning sonar imagery developed largely since 1970 provided abundant new structural data that show that plate motions orthogonal to the ridges do not occur. GLORIA Mark I, GLORIA Mark II, SeaMARC I, and SeaMARC II sonographs of the midocean ridges reveal the ubiquitous presence on them—from their crests to the margins of the adjacent abyssal plains—of linear faults, fractures, and fissures that extend for thousands of kilometers parallel with the ridges. The linear fault, fracture, and fissure systems resemble those of ice streams, glaciers, lava tunnels, and lava tubes. This indicates that the ridge-parallel structures, as those in ice streams, glaciers, lava tunnels, and lava tubes, are a response to Stokes's Law, which is one expression of Newton's Second Law of Motion, and that the principal sense of movement beneath the midocean ridges is ridge-parallel. The linearity of ridge-parallel faults, fractures, and fissures indicates that flow along the base of the midocean ridge crust is laminar.

Other structural features on the midocean ridges that demonstrate ridge-parallel flow include pinch-and-swell structure that segments the ridges, en echelon fracture patterns, a family of eddylike features (called overlapping spreading centers in plate tectonics), and ridge-transverse fracture zones (transform faults of plate tectonics). Genesis of these ridge-transverse fractures and related structures is due to the pinch-and-swell segmentation.

New reflection-seismic (CDP) studies of the Mid-Atlantic Ridge have shown the presence adjacent to the abyssal plain of a 300–400-km-wide zone of steep thrust faults. These studies suggest that midocean ridges are gigantic bivergent structural fans that originated in much the same way as continental foldbelts. In the parlance of Meyerhoff et al. (this volume), midocean ridges develop into kobergens.

The demonstration here that the most fundamental tenet of plate tectonics—that flow beneath midocean ridges is at right angles to them—is wrong has many implications. One is the assumption that linear magnetic anomalies are produced by magnetic field reversals. A second assumption is the requirement that the age of the ocean floor increases with increasing distance from the ridge crest. A detailed review of the Deep Sea Drilling Project results shows that the age of the ocean floor does not increase systematically from the crests. Moreover, the linear magnetic anomalies are explained better by magnetic susceptibility contrasts than by magnetic field reversals.

Midocean ridges are unlikely to be edifices sustained by convective mantle upwelling. Instead, we interpret them to be the loci of shallow magma channels attached to the asthenosphere below. In a companion paper in this volume, we show that midocean ridges form part of a worldwide network of shallow magma channels (which we call surge channels) that underlie both continents and ocean basins and are the primary cause of all forms of tectogenesis.

INTRODUCTION

Although the origin of midocean ridges has been a subject for speculation for well over 100 years (Heezen et al., 1959), adequate volumes of data on which to base a sound hypothesis were not available until the late 1970s. Such data include not only the information gleaned from geophysical studies (magnetics, gravity, seismic-refraction, heat flow), but also cores from deep-sea drilling and data from submersibles (e.g., Ballard and Moore, 1977). Most important, however, was the development of side-scanning sonar that produced images of the sea floor (Rusby and Somers, 1977; Somers et al., 1978). The structural detail revealed by the GLORIA Mark I and II side-scanning sonar systems was a giant forward stride in deep-sea research. Successor systems, such as SeaMARC I and SeaMARC II, provide even greater structural detail (e.g., Crane, 1987).

Side-scanning imagery revealed a structural style beneath the midocean ridges that was quite different from what had been anticipated. Plate tectonics predicts that the structural style will be dominated by normal tension structures (Sears et al., 1974). Instead the structures observed are of the type produced by viscous drag in laminar flow (Sears et al., 1974). Study of the side-scanning sonar imagery made it possible for us to demonstrate that midocean ridges are not the sites of sea-floor spreading. Hence the purpose of this paper, beyond showing that midocean ridges are not sea-floor spreading sites, is to lay the foundation for a wholly new hypothesis of Earth dynamics.

PREVIOUS HYPOTHESES OF ORIGIN

General—Although the existence of a median ridge in parts of the ocean basins was known by the late 1800s

(Murray, 1913), almost nothing was known of the rocks underlying it. The first deliberate sampling of a median ridge to determine its composition appears to have been done during the French *Talisman* expedition in 1883 (Furon, 1949). The worldwide extent of the midocean-ridge system was not proved until the late 1950s (Ewing and Heezen, 1956, Heezen, 1957), but its continuity through the Atlantic and Indian oceans was known by 1927 (Kober, 1928). As early as 1925, the worldwide extent of the ridge system had been confidently predicted (Kober, 1925, 1928), partly on the basis of the 1925–1927 *Meteor* expedition results (Kober, 1928; Stocks, 1938; Stock and Wüst, 1938). Although Kober (1928) was largely wrong in his projection of the midocean ridge through the Pacific, his reasoning was far advanced for his time. His predictions were borne out by Ewing and Heezen (1956; see Heezen, 1957). The central rift, although it appears on some of the *Meteor* depth profiles, was not identified until 1954 (Hill, 1956), and its extent not dreamed of until 1957 (Heezen, 1957). Heezen (1960) published a full account of the rift system as it was known 30 years ago.

Because so little was known about midocean-ridge structure before 1977, most of the hypotheses of its origin that were published were based more on personal perceptions than on sound knowledge. Another feature common to these hypotheses was the conviction that, whatever system of forces produced the midocean ridge, it acted at right angles to the ridge. Finally, most of the published hypotheses were based on the Mid-Atlantic Ridge, because so little was known of the midocean ridges of the Arctic, Indian, and Pacific oceans.

Compression—The published hypotheses can be divided somewhat arbitrarily into four groups—compression, tension, a combination, and passive. In the first group, compression can result from tangential stress in the lithosphere that is generated by Earth contraction, or cooling; compression can be generated in the convergence zone between two oppositely moving limbs of a convection cell; and compression can be generated in a plate-tectonic subduction zone. Haug (1900), a supporter of Earth contraction, viewed the Mid-Atlantic Ridge as a median welt rising in the middle of the Atlantic basin, which he regarded as a geosyncline. Similar views were held by many geologists, including Kober (1921, 1925, 1928), Washington (1930), Willis (1932), Hans Cloos (1937, 1939a, 1939b), Stille (1939), Gutenberg and Richter (1949, 1954), and many others. Washington (1930), noting the presence of metamorphosed peridotites on the Mid-Atlantic Ridge, interpreted these as proof of extreme compression and upward squeezing.

Vening Meinesz (1942), Heiskanen and Vening Meinesz (1958), and Keith (1972) suggested that mantle convection currents rise beneath the continents. Therefore, they reasoned, the downward currents are the loci of midocean ridges, which are squeezed between the limbs of adjacent convection cells.

Extension—Extension as the cause of midocean ridges also was held by many distinguished earth scientists. Taylor (1910) and Sonder (1939) interpreted the ridges as horst-and-graben complexes. Holmes (1931), the father of some of the fundamental concepts of plate tectonics, suggested that upward-moving limbs of adjacent convection cells beneath a continent would split that continent, carrying its two parts in opposite directions, leaving behind a sialic strip to form a midocean ridge (see Hess, 1962; Meyerhoff, 1968).

Menard (1958, 1965), Ewing and Ewing (1959), and many others interpreted the circum-Pacific deep-sea trenches as deeps formed between the downward-going limbs of adjacent convection cells. The corresponding upward-moving limbs produced the midocean ridges out of oceanic crust, rather than sialic crust. Hess (1954, 1955), a few years before his pioneer sea-floor spreading paper, suggested that the upward-moving limbs beneath midocean ridges carried with them strips of peridotite that became serpentinized when raised to the sea floor. The resultant swelling, in Hess's opinion, produced midocean ridges.

Carey (1958) and Heezen (1962), advocates of Earth expansion, thought that midocean ridges were surface manifestations of deep tension cracks; the ridges formed from the mafic and ultramafic lavas that poured through the cracks. Wegener (1920) proposed that midocean ridges are the loci of mafic and ultramafic rocks that rose like diapirs between two continents as they began to move away from one another. Molengraaff (1928) held a similar view.

Van Bemmelen, who originally thought that midocean ridges were created by sial differentiated from the mantle below them (van Bemmelen, 1933), later interpreted them to be elongate geotumors whose axes became midocean ridges as the continent above them split in two and slid off to either side (van Bemmelen, 1976).

Alternate compression and extension—A combination of compression and tension for origin of the ridges was proposed by Gutenberg and Richter (1949, 1954). They suggested that the ridges are compressional foldbelts (primarily because of their great size) that later became the sites of block-fault and rift development (because the hypocenters are shallow, as in continental rifts). Tolstoy and Ewing (1949) and Tolstoy (1951), on the other hand, wrote that the ridges could be either of compressional or tensional origin.

Passive hypotheses—In the category of passive hypotheses for the origin of midocean ridges, we include a variety of proposals. Van Bemmelen (1933), Umbgrove (1947), and Kuenen (1950), for example, stated that midocean ridges are strips of sial, but did not specify how they might have been created. Betz and Hess (1942) proposed that they are piles of vesicular basalt, several thousand meters thick, that oozed forth from convection-current-related giant cracks in the ocean floor. A similar origin was postulated by Scheidegger (1958).

Beloussov (1980, 1989), on the basis of ocean drilling results, agreed with the plate tectonicists that the ridge is young, Middle Jurassic and younger. Beloussov wrote that before Jurassic time the present sites of the Atlantic and Indian oceans were occupied by continental (granitic) platforms. Noting that midocean ridges are seismically active, he wrote (1980) that these ridges are to the deep ocean basins flanking them what active continental rifts and geosynclines are to the adjacent ancient continental platforms. He implied that oceanic and continental rifts are genetically related. Carey (1981, p. 390–391) emphasized the close genetic relation between midocean ridges and foldbelts, writing that "The essential difference between an orogenic belt and an oceanic spreading-ridge is the presence of large volumes of added terrigenous sediments in the one, but not in the other."

Plate tectonics, in contrast, emphasizes the great differences between orogenic belts and midocean ridges. The former, according to plate tectonics, is a site of intense compression; the latter, where new oceanic crust is created, is a site of tension (Hess, 1962; Morgan, 1968; Wyllie, 1971).

Beloussov's (1980) and Carey's (1981) conclusions are essentially correct in our opinion, as they draw attention to the close genetic relations between foldbelts and midocean ridges. Meyerhoff et al. (this volume) document this statement in full. One of the reasons why the close genetic relations were missed in plate tectonics is that the postulate was developed with almost no knowledge of the midocean-ridge structure that was to be recognized much later.

PHYSICAL BASIS FOR A NEW HYPOTHESIS

To provide a proper understanding of this paper, the writers review briefly Newton's Second Law of Motion, and one of its major expressions, Stokes's Law. (We summarize all laws used in this paper and our companion paper [this volume] in the appendix of the latter paper.) The second law of motion states that the acceleration of an object is equal to the resultant of all external forces acting upon it, divided by the mass of the object, and acting in the same direction as the resultant force. Stokes's application of this law of motion is expressed in various ways. One commonly known form is: as a homogeneous sphere moves through an ideal (Newtonian) fluid of zero viscosity (or, conversely, as such a fluid moves around a homogeneous and stationary sphere), the flow lines or streamlines in the fluid form a perfectly symmetrical pattern around the sphere and parallel the movement direction of the sphere. However, if the fluid is viscous, there will be a viscous drag upon the sphere. The force required to move a sphere through a given viscous fluid at a low uniform viscosity is directly proportional to the velocity and radius of the sphere, and acts in the same direction.

To this point, homogeneity of both the fluid and the spherical surface has been assumed. If the surface of the sphere consists of alternate parallel strips of rough and smooth texture—with the boundaries of the strips parallel with the direction of flow—a viscous fluid flowing over that surface will flow more slowly over the rough strips than over the smooth. This is the phenomenon of viscous drag. As a consequence of such drag, a flow-parallel laminar (streamline or Poiseuille) flow pattern develops, with flow-parallel shear between the strips of different velocities. Such a pattern illustrates Stokes's Law.

NATURAL EXAMPLES OF STOKES'S LAW

Familiar examples of laminar flow, and therefore of Stokes's Law, abound. They include ice streams, glaciers, and patterns created by flowing lava, whether in vents, lava tubes, lava tunnels, or flows.

Ice streams and glaciers—Figures 1 and 2 are satellite photographs of parts of Antarctica. Figure 1 shows a part of the upper Lambert Glacier. Figure 2 shows part of the Byrd Glacier and ice stream. Each exhibits a well-developed pattern of laminar flow, a pattern produced by viscous drag of the ice on the underlying surface of bedrock and sediments. The pattern is most pronounced where the ice stream narrows between bedrock walls.

Lava tubes—Figure 3 is a schematic drawing of a lava tube (Yamagishi, 1985). Three features are especially noteworthy. First is the feature referred to as a longitudinal spreading crack. These consist of tube-parallel cracks, fractures, minigrabens, and minihorsts that parallel the direction of flow in the tube. Second are the constrictions of the lava tube that form at right angles to it. Third and last are transverse spreading cracks. These are confined to the deepest parts of the constrictions, indicating a cause-and-effect relationship. The overall pinch-and-swell geometry of the tube is a universal phenomenon in lava tubes and tunnels, and its forma-

Figure 1. Upper Lambert Glacier (ice stream), Antarctica. Flow is from bottom to top. The flowlines show that the flow is laminar. This is an example of Stokes's Law. This satellite photographic was obtained through the courtesy of Baerbel K. Lucchitta and Harold Masursky, U.S. Geological Survey, Flagstaff, Arizona.

tion is related closely to lava surges in the tube as the tube is extended. Pinch-and-swell geometry is also a major tectonic feature of the midocean ridges.

Vent flow, Mauna Ulu eruption, Kilauea Volcano—An unusually clear illustration of Stokes's Law was published some 19 years ago by Duffield (1972). He described the structural fabrics and patterns created during the Mauna Ulu eruptions of Kilauea that began on 24 May 1969. These patterns are so strikingly like those predicted by the plate-tectonic model that they merited the careful description and prompt publication they received. It should be kept in mind that the structures described by Duffield (1972), and predicted by the plate-tectonic model, were published two to three years before sonographs of the midocean ridges generally became available, although the first crude sonographs from the GLORIA I system were taken in the United Kingdom during 1969 (Searle and Laughton, 1977).

Duffield's (1972) principal findings can be observed on Figures 4 and 5. These are: (1) lava rises vertically in its vent to the surface where it wells up, forming a low ridge several meters high. (This ridge is equivalent to the midocean ridges.) (2) A central rift forms at the ridge crest. (This is equivalent to the axial valley of the midocean ridge.) (3) The central rift has a decided zigzag shape, with no important straight segments. (4) On either side of the central rift, the lava flows away from it at approximately right angles to the trend of the rift. The lava exhibits laminar flow (Duffield, 1972). (5) Lava flowing away from the central rift cools almost instantly, forming a dark crust perhaps 2 cm thick. (6) The laminar flow of the lava beneath the crust exhibits viscous drag, presumably a response to the irregular lower surface of the crust. That laminar flow produces pronounced linear features (many are thin fractures) that are orthogonal to the central rift.

Thus, this natural plate-tectonic model from the Mauna Ulu vent at Kalauea contains all of the tectonic features predicted by the plate-tectonic model. Therefore, the important question is: what actually is observed in sonographs of the midocean ridges?

SONOGRAPHS OF MIDOCEAN RIDGES

Sonograph images are available from many areas of midocean ridges. They provide a scattered but thoroughly consistent picture of midocean-ridge structure, whether in the Atlantic, Indian, or Pacific oceans.

Figure 6 is a SeaMARC II image of the Juan de Fuca Ridge at 47° N latitude, 129°10′ W longitude in the northeastern Pacific Ocean, approximately 550 km due west of Ocean City, Washington. The width of the image is 10 km, and the width of the axial depression ranges from 2.5 to 4 km. In the axial depression, the abundance and close spacing of the fractures, and their continuity along strike, are apparent. All fractures parallel the ridge.

Figure 7 is a tracing of the fracture pattern along the Juan de Fuca Ridge at 48° N latitude as interpreted by Tivey and Johnson (1987). Figure 8, a tracing of an area on the East Pacific Rise at 11°55′ N latitude shows a nearly identical pattern as interpreted by Crane (1987). A most important feature illustrated on Figure 8 is the en echelon relationship between fracture "Bundle 1" and fracture "Bundle 2." This en echelon pattern can be produced only by ridge-parallel flow.

Figure 2. Byrd Glacier (ice stream), Antarctica. Flow is from bottom to top. The flowlines show that the flow is laminar. This is an example of Stokes's Law. This satellite photograph was obtained through the courtesy of Baerbel K. Lucchitta and Harold Masursky, U.S. Geological Survey, Flagstaff, Arizona.

Figure 9, from Searle and Laughton (1977) and from Laughton and Searle (1979), is a tracing of GLORIA Mark II sonographs of the Mid-Atlantic Ridge between 36° and 38° N latitude. Figure 10, from Laughton and Searle (1979), shows a part of the Reykjanes Ridge between 58° and 61° N latitude just southwest of Iceland. Figure 11, from Schäfer (1972), shows that the fractures extend across Iceland. Once again, all fractures parallel the ridge.

Figure 12 presents two SeaMARC sonographs from the Juan de Fuca Ridge and East Pacific Rise respectively, together with a satellite photo image of the Upper Lambert Glacier, Antarctica. The SeaMARC images are from Kappel and Normark (1987). The close similarity among the three images suggests that a common cause is much more than just a possibility.

These figures show that the fissure patterns displayed on them are a consequence of Stokes's Law. All motions beneath midocean ridges—except at intersections with transform faults—have been and are parallel with the strike of the ridges. We acknowledge that some types of lineations produced in nature do not parallel the stresses that produce them (M.L. Keith, pers. comm., 1989). However, the linear features that are the topic of this paper are fissures and faults, which are actual breaks in the rocks, tens to hundreds of kilometers long, and are not rotated crystals or similar phenomena. The dense concentrations of fissures, fractures, and faults that are ubiquitous on midocean ridges and extend along strike for thousands of kilometers can be produced only by motions that parallel them (N. Sylvester, pers. comm., 1989).

Figure 13 shows the zigzag structure that would characterize the crest of midocean ridges if the movements beneath it were orthogonal and tensile. Note how closely the fracture pattern on Figure 13 matches the zigzag patterns at the ridge crest of Figures 4 and

Figure 3. Idealized sketch of a lava tube, or tunnel. Flow is from right to left. The major structures shown are the longitudinal spreading cracks, transverse spreading cracks at contrictions in the tube, a tension crack, and the ropy wrinkles where a new pillow is forming (on the left). The longitudinal spreading cracks form as a result of laminar flow of the lava accompanied by swelling of the tube. This is another example of Stokes's Law. Note the similarity in morphology to the midocean ridge (Figs. 14, 15), in which the longitudinal spreading cracks are analogues of the ridge-parallel fissures, fractures, and faults; the transverse spreading cracks and constrictions are analogues of the transform faults (ridge-transverse fracture zones). Both the lava tube shown here and the midocean ridges exhibit pinch-and-swell structure. From Yamagishi (1985). Published with the permission of Geological Society of America.

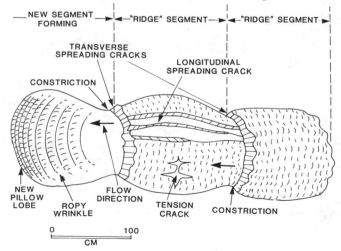

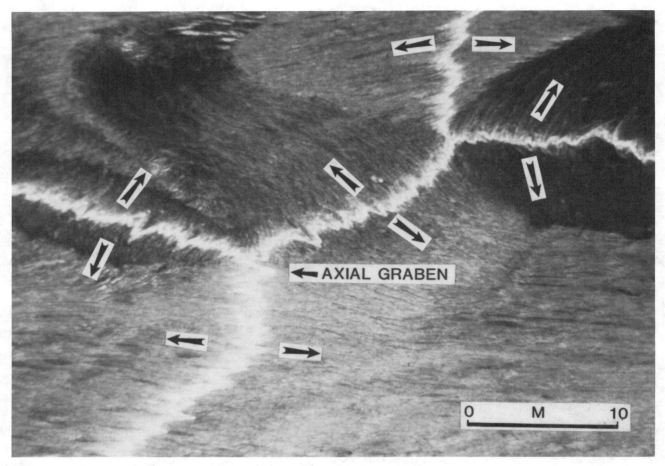

Figure 4. Oblique view of lava ridges in a vent, 1969 Mauna Ulu eruption, Kilauea Volcano, Hawaii. A bright zigzag crestal graben runs along the axis of each lava ridge. Compare this zigzag pattern with that shown on Figure 13. The following is a description of the events and features illustrated on this figure: Uncooled lava rises beneath the lava ridges and is exposed along the ridge axis where a graben forms. On either side of the crestal grabens, the lava flows down the ridge flanks in the directions indicated by the arrows. As the lava flows away from the grabens, a thin crust 1-2 cm thick forms. Viscous drag on the base of this produces lineations parallel with the flow and orthogonal to the crestal graben. This is the lineation pattern predicted by plate tectonics, with flow away from the central graben. From Duffield (1972). The original photographs were loaned to the writers by Duffield.

5. Thus the fault pattern predicted in plate tectonics is totally different from that actually observed.

ADDITIONAL EVIDENCE FOR RIDGE-PARALLEL MOTIONS

The writers have mentioned two structures from midocean ridges that demonstrate that ridge-parallel flow characterizes the midocean ridges, specifically (1) ridge-parallel fissures, fractures, and faults (Figs. 6–12), and (2) the en echelon bundles of fissures and faults that have been observed on the ridges (Fig. 8). Three additional major structures on midocean ridges demonstrate that ridge-parallel flow takes place. These are (3) pinch-and-swell structures, which segment midocean ridges everywhere, (4) the so-called overlapping spreading centers, and (5) transform faults.

Figure 14, from White (1989), shows a north-south refraction-seismic line along a segment of the Mid-Atlantic Ridge. Wherever a bathymetric deep is present, a ridge-transverse (transform) fracture zone is present, and the Mohorovicic discontinuity rises to within 1 or 2 km of the sea floor. Thus, the ridge-transverse fracture zones are major partitions that segment the midocean ridges and impart to them a pinch-and-swell geometry.

The upper part of Figure 15 shows a similar segmentation along the East Pacific Rise (Macdonald et al., 1988b). A familiar small-scale analogue of a midocean ridge is a lava tube, as illustrated on Figure 3 and in the lower part of Figure 15. Both the lava tube and the midocean ridge have these analogous along-strike structures (Figs. 3, 14, 15): ridge-transverse fracture zone (transverse spreading cracks of Figs. 3, 15), and an associated bathymetric deep (constriction) that separates long and bathymetrically much shallower ridge segments (the lobes of Figs. 3, 15).

The analogy between midocean ridges and lava tubes-lava tunnels is an ideal illustration that physical laws are expressed at all scales in nature. In the case of the lava tube or tunnel, lava flow paralleling the tube moves in pulses, or surges. The same type of pulsating, or surging flow must take place beneath the midocean ridges, only on a far grander scale.

In our explanation of the segmentation of midocean ridges, we noted that ridge-transverse fracture zones (transform faults) commonly lie in bathymetrically depressed zones separating the different ridge segments (Figs. 3, 14, 15). The so-called overlapping spreading centers described by Macdonald and colleagues (1988b), among others, also occur in depressions that separate ridge segments, and to date are described only from the East Pacific Rise (e.g., Macdonald et al., 1988b). It seems logical to conclude that the overlapping spreading centers and the ridge-transverse fracture zones have a common origin.

Because of their shape, we refer to the overlapping spreading centers as eddylike features, or eddylike structures. Figure 16 illustrates a typical eddylike structure along the East Pacific Rise (Macdonald et al., 1984). An eddy is a current of air or water moving in a swirling or circular manner within a relatively limited area; one part of the eddy current is moving contrary to the main current. Eddies form where some obstruction (e.g., a constriction, as illustrated on Fig. 15) impedes normal laminar flow. For example, Figure 17 shows the flow pattern within the Gulf Stream off the east coast of the United States (J.O. Blanton, pers. comm., 1988). The pattern is that of a series of eddies produced by friction between the waters of the northeast-moving current and the sub-

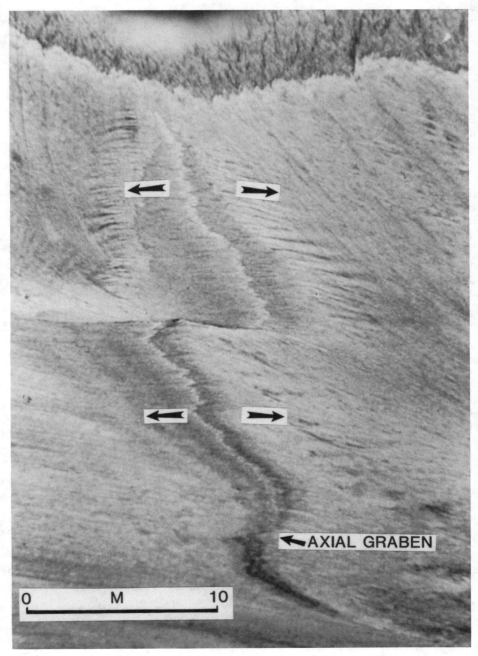

Figure 5. Lava in a vent formed during the 1969 Mauna Ulu eruption of Kilauea Volcano, Hawaii, is flowing toward the reader. This oblique view shows the crestal graben of the lava ridge. Hot lava rises beneath the graben and spreads laterally at right angles to the graben. Arrows show the flow directions. A strike-slip fault offsets the graben midway across the photograph. From Duffield (1972), who loaned the photograph to the writers.

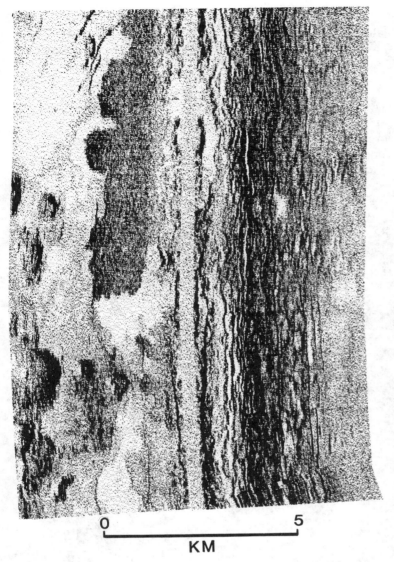

Figure 6. Sonograph of the axis of the Juan de Fuca Ridge at 47° N latitude The light-colored straight bar down the center of the sonograph is the ship's track. The lineations to the right of the bar are fissures, fractures, and faults in the axial depression. The straightness and continuity of this fissure and fracture system show that the underlying flow is laminar and that it parallels the axis of the ridge. This too is an example of Stokes's Law. The sonograph was provided to us by Mark L. Holmes, U. S. Geological Survey, Seattle, WA.

jacent surface of the continental shelf. Overlapping spreading centers (Fig. 16) closely resemble the eddies of Figure 17. For example, the region between the overlapping and curved ridge segments is invariably a bathymetric deep, commonly the deepest part of the depression that separates adjacent ridge segments (Fig. 16). The central position of this deep between the two overlapping ridge segments is typical of all vortical structures (e.g., whirlpools), as is demonstrated in the laboratory (Tritton, 1988) and in hundreds of vortical structures that have been mapped geologically (Meyerhoff et al., this volume).

Lonsdale (1982, 1983), Macdonald and Fox (1983), and Macdonald et al. (1984, 1987, 1988a, 1988b) also recognized that flow beneath midocean ridge segments is ridge parallel. Macdonald et al. (1987, p. 995) wrote that, "Sometime after freezing, the [propagating] ridge segment in question receives another pulse of magma. The axial chamber will swell, and magma will migrate along strike."

F. G. Koch (pers. comm., 1989) suggested another possible origin for these overlapping structures. He brought to our attention the Olson and Pollard (1989) paper describing en echelon crack propagation in rocks, and especially the origin of the curvature in the overlapping parts of en echelon on cracks. Figure 18, from Olson and Pollard (1989), shows that the curvature of overlapping parts of propagating cracks is in the same sense as that observed in the eddylike structure of Figure 16. Olson and Pollard (1989) showed that (1) the presence and amount of curvature are related directly to the magnitude of stress that produces the cracks, and (2) the axis of the stress producing the cracks parallels them. Thus—although nowhere do Olson and Pollard specifically state this—these authors showed how Hooke's Law (this last is to certain solids what the Stokes's Law is to liquids) applies to crack propagation. (Hooke's Law states that the stress within an elastic solid, up to the elastic limit, is proportional to the strain producing it; see Meyerhoff et al., this volume, their appendix).

A laboratory study by Emmons (1969) independently reached the same conclusions as Olson and Pollard (1989). Experimenting with sand layers and compressing them parallel with their bedding, Emmons (1969) produced overlapping cracks that match those of Figure 18 in every detail. Regardless of whether one uses Stokes's Law of fluids or Hooke's Law of solids, the principal motion (and stress) producing the eddylike structures on Figure 16 parallels the strike of the midocean ridge.

WHY DO EDDYLIKE STRUCTURES AND RIDGE-TRANSVERSE FRACTURE ZONES FORM?

We noted that both types of structures—the eddylike structures and the ridge-transverse fracture zones—form where the midocean ridge is constricted, that is,

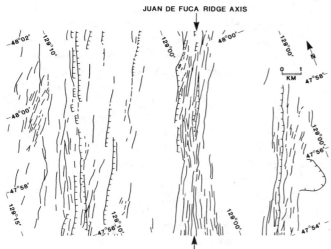

Figure 7. This is a tracing of ridge-parallel fissures, fractures, and faults along the Juan de Fuca Ridge at 48° N latitude This is another example of laminar flow and of Stokes's Law. From Tivey and Johnson (1987).

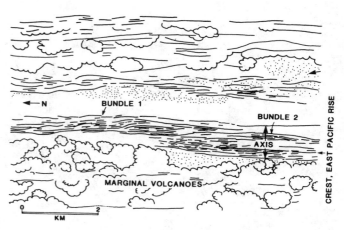

Figure 8. This is another example of ridge-parallel fissures, fractures, and faults along the crest of a midocean ridge, in this case, the East Pacific Rise at 11°55′ N latitude. The ridge-parallel structures are a consequence of ridge-parallel laminar flow, and illustrate Stokes's Law. *Note:* The fractures in the axial depression form en echelon, overlapping "bundles." The en echelon overlap proves that the flow which produced them is ridge-parallel. Such en echelon structures originate presumably in much the same manner as the eddylike structures that have been called "overlapping spreading centers." From Crane (1987). Reproduced with permission of Elsevier Science Publishers B. V.

where a bathymetric deep is present. These deeps are fairly regularly spaced, approximately 100 to 1,000 km, and most commonly 300 to 500 km (Macdonald et al., 1988b). This regularity of segmentation is another property that midocean ridges share with lava tubes-lava tunnels, and constitutes another proof demonstrating the great importance of ridge-parallel flow beneath midocean ridges.

Because the lava beneath midocean ridges, like the lava in a tube or tunnel, moves in pulses or surges and thereby produces segmentation, it follows that eddylike structures and ridge-transverse fracture zones are younger than the segments. We believe that the presence of bathymetric deeps or constrictions leads directly to the formation of eddylike and ridge-transverse structures in the following manner.

The introduction along a magma-filled channel of a bathymetric deep, or constriction, has several physical consequences as is known from Pascal's and Poiseuille's laws (defined in A. Meyerhoff et al., this volume, appendix). For example, the magma in the channel, whether static or in motion, exerts a pressure transverse to the direction of dominant flow, and perpendicular to the channel walls. The pressure is greater for dynamic flow than for static magma. Thus, wherever the channel diameter is greatly reduced, magma forces are most likely to exceed the shear strength of the midocean ridge lithosphere (which can support a stress of about 10^8 Nm^{-2}, or 1 kbar), and faulting orthogonal to the channel will take place. The fault zones thus formed are fairly wide (5–20 km, in a few cases up to 100 km). The magma in the channel turns approximately 90°, as shown in Figure 19. This figure shows that some fractures and

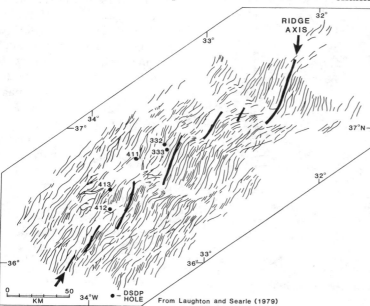

Figure 9. Ridge-parallel, curvilinear, slightly anastomosing fissures, fractures, and faults of the Mid-Atlantic Ridge at 37° N latitude (the FAMOUS area). Note the numerous offsets similar to those along ridge-transverse fracture zones (transform faults). This is another example of Stokes's Law, with flow paralleling the ridge. The area shown is 125 km wide, demonstrating that physical laws operate at all scales.

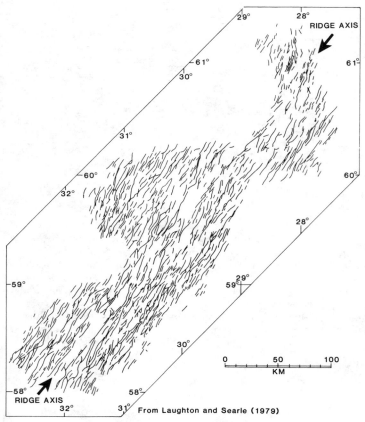

Figure 10. This is another example of ridge-parallel fissures, fractures, and faults prouced by laminar, ridge-parallel flow (Stokes's Law). This example, from the Reykjanes Ridge at 58°-61° N latitude, spans a band up to 150 km across and 400 km long; this is an area larger than the state of Florida (USA) and half the size of the United Kingdom.

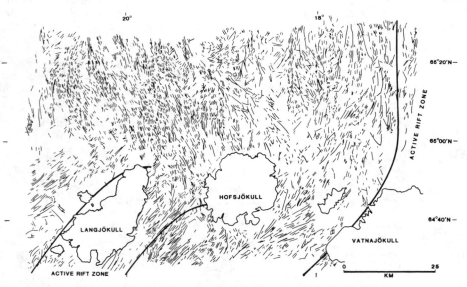

Figure 11. The ridge-parallel fractures of the Mid-Atlantic-Reykjanes Ridge continue across all of Iceland. This illustration shows part of central-western Iceland. The black lines are borders of active volcanic zones. The white areas (jökull) are glaciers. From Schäfer (1972). Published with permission of *Geologische Rundschau*.

fissures in the ridge-transverse fracture zones are direct continuations of fractures or fissures on the ridge and that they are continuous from one ridge segment to the next. (This fact shows that Macdonald et al.s' [1988b] model of bilateral ridge-parallel flow beneath each ridge segments is wrong.) The transverse fracture zones thus formed probably propagate laterally outward on both ridge flanks, but the total length of a given fracture zone must be controlled partly by the pressure gradient. The great extents of some fracture zones, particularly on the East Pacific Rise, are probably a consequence of long-period stresses that are exerted episodically, producing shear motions along the fracture zones.

We interpret the sonograms of the midocean ridges as proof that an evolutionary series of structures exists. At one end of the spectrum is laminar flow parallel with the ridge. This grades into the beginning of vortical laminar flow (Fig. 16), which in turn grades into ridge-transverse laminar flow that creates the ridge-transverse fracture zones (Fig. 19). As Meyerhoff et al. (this volume) demonstrate, the other end of the spectrum is dominated by vortex structures characterized by vortical, but still laminar, flow (Tritton, 1988).

GEOMETRY OF THE MAGMA CHANNELS OF MIDOCEAN RIDGES

Little is known about the sizes and shapes of the magma channels beneath the midocean ridges. Vogt (1974), who also recognized the existence of long-distance, ridge-parallel magma flow, made the first attempt that we know of to describe midocean-ridge magma channels. He wrote (p. 116, 118), "In the model I assume there is a pipe-like region below the spreading axis, extending subhorizontally away from a plume such as Iceland . . . . This mid-oceanic pipe extends from the base of the axial lithosphere, about 5 or 10 km deep, down to maximum depths (30 to 50 km?) from which basalt melts segregate and rise. Tholeiitic fluids would be released

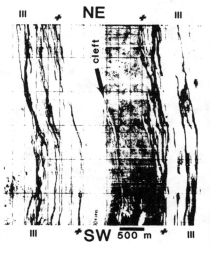

Figure 12. This figure compares sonographs of the Juan de Fuca Ridge at 44°45′ N (Kappel and Normark, 1987) and of the East Pacific Rise at 10°30′ N (Eisen et al., 1987) with the laminar flow pattern of the Upper Lambert Glacier (Antarctica). These images show that Stokes's Law applies as well to the mid-ocean ridges as to glaciers, ice streams, and lava tubes or tunnels. The three images were supplied very kindly, from left to right, by Ellen Kappel and William Normark; by P. J. Fox; and by Baerbel K. Lucchitta and Harold Masursky.

a. JUAN DE FUCA RIDGE b. EAST PACIFIC RISE c. LAMBERT GLACIER, ANTARCTICA

from the entire pipe; origin depths of 23 km for the Mid-Atlantic Ridge and 16 km for the East Pacific Rise ... would approximate depth to the center of the pipe." Continuing, he wrote, "The ultrabasic mush in this pipe is assumed to be flowing away from the hot spot at a rate determined by pipe diameter, viscosity, and horizontal pressure gradient.... The roof of the pipe is probably formed by the plate bottoms.... [The] flow is confined either to a circular pipe or, in the other limit, to a sheet-like region between two parallel plates. Seismic data ... suggest that the pipe is more nearly elliptical in cross-section, with the region of extensive fusion greater in width than in thickness. The actual pipe will therefore be intermediate between a circular pipe and a sheet flow."

We, on the other hand, assume that the magma channels (which Meyerhoff et al. this volume, call surge channels) are contained within an envelope of what has been called anomalous upper mantle. This envelope has an elliptical or lens shape. It is characterized by P-wave velocities of 7.0–7.8 km/s (Ewing and Ewing, 1959) and contains within it one or more low-velocity zones (Pavlenkova, 1989). The approximate dimensions of such a lens were determined by Talwani et al., (1965) from refraction-seismic and gravity data. Their results appear on Figure 20. The magma channel and its enclosing lens (Fig. 20) underlie the Mid-Atlantic Ridge between 27° and 47° N latitude. A nearly identical Mid-Atlantic Ridge cross section was obtained at 11° S latitude by Zverev et al. (1988) and by Pavlenkova (1989). Channel thickness ranges from 20–28 km, well within Vogt's (1974) estimate of 22 to about 38 km. It

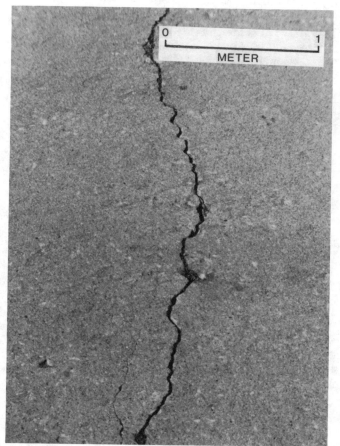

Figure 13. This is a photograph of a tension crack in pavement. Compare Figures 4 and 5. Note the zigzag fracture trace. Tension acted at right angles to the trend of the crack. If the midocean ridges were produced by tensile stress acting at right angles to the trend of the ridge, this is the type of fault that should be observed on the sonographs of the ridges. The fact that such fractures are not observed in sonographs of the ridges indicates that the plate-tectonic model of the midocean ridges is wrong.

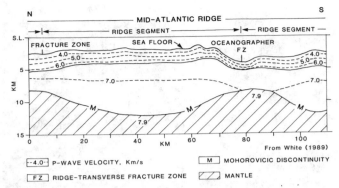

Figure 14. North-south refraction-seismic section of the Mid-Atlantic Ridge from 37° to 40° N latitude This figure shows the pinch-and-swell structure of lava tubes (compare with Figs. 3, 15). Figure also shows the velocity structure of the Mid-Atlantic Ridge parallel with its strike. Published with permission of the Geological Society (London).

contains about 15 km of 7.7-km/s (P-wave velocity) material, and includes at least one low-velocity zone, with the slowest velocities (approximately 5.4 km/s) between 13 and 21 km below the ridge crest (Zverev et al., 1988).

The volume of potential magma in the channel is unknown, but it must be very large for reasons discussed by Meyerhoff et al. (this volume). The magma apparently is generated from an ultramafic parent in the asthenosphere (Basalt Volcanism Study Project, 1981), where it is gravitationally unstable. Following the law of gravity—or, more specifically, the Peach-Köhler climb force

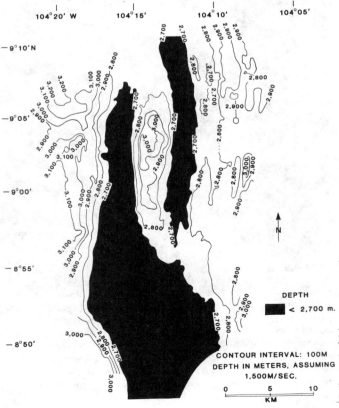

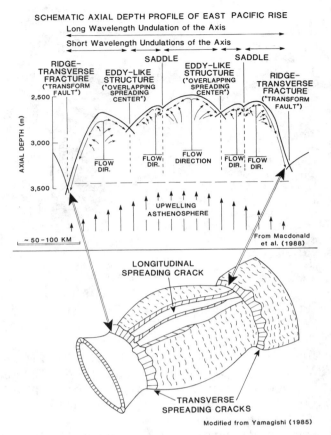

Figure 15. Comparison of a depth profile (schematic) of the East Pacific Rise with a lava tube or tunnel. As explained in the caption for Figure 3, the bathymetric lows along the midocean ridge where ridge-transverse fracture zones (transform faults) cross the ridge correspond to the transverse spreading cracks (constrictions) of the lava tube: the ridge-parallel fissures, fractures, and faults of the midocean ridge correspond to the longitudinal spreading cracks. From Macdonald et al. (1988b) and Yamagishi (1985). The lower diagram is published with permission of Geological Society of America.

(Weertman and Weertman, 1964; Gretener, 1969; Weertman, 1971)—the basaltic magma rises beneath the midocean ridge until it is in equilibrium with its surroundings (that is, where the density of the basalt equals the density of the surrounding lithosphere), as outlined in detail by G. K. Gilbert (1877; see Corry, 1988, for an excellent review). At the level where the magma is in equilibrium with its surroundings—Walker (1989) called it the level of neutral bouyancy—the magma spreads laterally, forming a giant sill- or laccolithlike body (Krylov et al., 1979; Beloussov, 1980; Sychev, 1985). Figure 21, from Beloussov (1980), relates

Figure 16. Eddylike structure from the East Pacific Rise at 9° N latitude. This structure, according to Macdonald et al. (1987, p. 995), is formed by flow parallel with the axis of the East Pacific Rise, and was called an "overlapping spreading center" by Macdonald and Fox (1983). The writers believe that this is a classic eddy structure. Compare this figure with Figure 17. From Macdonald et al. (1984).

Figure 17. Eddy patterns formed by the Gulf Stream along the eastern coast of North America. Illustration and interpretation were supplied by J. O. Blanton of the Skidaway Institute of Oceanography. The point is that these are structures that parallel the direction of flow, in this case, flow from southwest to northeast.

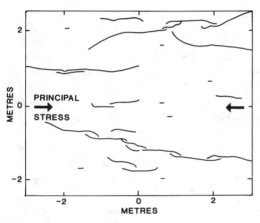

Figure 18. This figure shows crack paths generated by a principal compressive stress that parallels the crack paths. This is an alternate explanation for the overlapping spreading center illustrated on Figure 16. From Olson and Pollard (1989). Published with permission of Geological Society of America.

the magma-channel geometry directly to the Talwani et al. (1965) models of the Mid-Atlantic Ridge.

In the East Pacific Rise, one or more smaller magma channels overlie the principal magma channel. According to several recent studies (e.g., Detrick et al., 1987; Caress et al., 1989), these smaller channels are of the order of 4–7 km wide, are 3–4 km high, and lie 0.8–2 km below the sea bed at the crest of the midocean ridge. The smaller channels probably contain differentiates from the underlying principal magma chamber. The existence of these smaller chambers accounts for occurrences of titanium-enriched basalt, andesite, dacite, rhyodacite, and many other peculiarities in rocks collected along the midocean ridges (Fornari et al., 1983; Perfit and Fornari, 1983; Langmuir et al., 1986; Thompson et al., 1989).

POSITION AND CONFIGURATION OF MIDOCEAN RIDGES

Three additional features of midocean ridges that are important in considering their origin are their positions in each ocean basin, sense of offset along the ridge-transverse fracture zones, and links with various continental massifs.

Positions of midocean ridges—The approximate median position of the midocean ridges in the Atlantic, Indian, and Southern oceans has been known for many years (Menard, 1958, 1965). The absence of a median ridge in much of the Pacific also has been known for some time (Menard, 1965). The presence of an Arctic midocean ridge was suspected as a result of the work by Gutenberg and Richter (1949, 1954), and was confirmed by Heezen and Ewing (1961).

Menard (1958) postulated that some (unspecified) process centers midocean ridges in their respective basins. He wrote that this process has the following characteristics: (1) it is sensitive to ocean-basin margins; (2) it can act across distances of up to 5,000 km, or nearly 50° of arc; (3) it acts on both sides of the ocean basin at the same time; and (4) the process is either ephemeral or intermittent. Menard (1965) modified his earlier position, and presented evidence that the positions of the

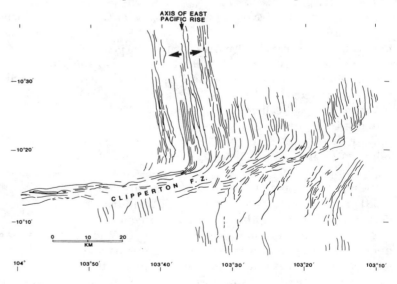

Figure 19. Tracing of a sonograph at the intersection of the East Pacific Rise with Clipperton fracture zone. Note that some fissures, fractures, and faults continue unbroken through the fracture zone. See the text for the physical explanation. Sonograph is based on work by Eisen et al. (1987) and was kindly supplied to the writers by P. J. Fox.

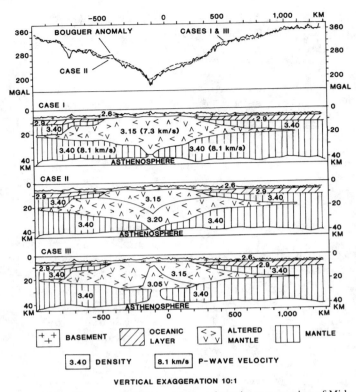

Figure 20. West-east refraction-seismic and gravity cross section of Mid-Atlantic Ridge (average between 27 and 47° N lat.). Three possible models that fit the gravity profiles are illustrated. In each a great lens of low-density, low-velocity anomalous mantle is required just below the surface of the midocean ridge. The writers interpret this lens to be the source of the lavas at the crest of the midocean ridge and to be the container for the magma chambers that underlie the midocean ridge system. The P-wave velocity of the walls of the lens is 7.0 to 7.7 km/s, and is equivalent to the 7.0+ km/s layer shown on Figure 14. From Talwani et al. (1965).

midocean ridges are related more closely to the positions of early Precambrian continental nuclei than to the median lines of the existing ocean basins. The writers believe that Menard's (1958) earlier concepts are closer to the mark, specifically, that the position of the ridges with respect to the median line is more important than their position with respect to Precambrian nuclei. In support of this opinion, the writers present the following hypothesis.

Meyerhoff et al. (this volume), following Stacey (1981) and Lyttleton (1982), noted that Earth has undergone a long cooling history. As a consequence of this history, the lithosphere has already cooled and is under constant compression in all directions parallel with its surface (that is, the lithosphere is in a state of equiplanar tangential compression, tangent to the Earth's surface). However, the oceanic lithosphere is thinner (weaker) than the continental lithosphere (Miyashiro et al., 1982). Therefore, equiplanar compression is likely to cause buckling of the oceanic lithosphere, because it is weaker than the adjacent continental lithosphere. Provided that the oceanic lithosphere is of overall uniform thickness, buckling ideally should occur in a zone equidistant from the adjacent continental massifs. Thus the positions of midocean ridges are a logical result of a cooling Earth.

The only place where a midocean ridge is not present is in the Pacific basin north and northeast of the Eltanin fracture zone (53°30′ S latitude, 135° W longitude). South of this fracture zone, the Pacific basin is 4,200 km wide or less (between the continental Chatham Rise on the northwest and the base of the Antarctic Peninsula on the southeast). Northeast of the Eltanin fracture zone, the Pacific basin is more than 9,000 km across (nearly 95° of arc and more). We believe that a midocean ridge cannot be buckled up in a midoceanic position through an arc of curvature greater than about 60° (Menard, 1958), not because compressive stress is not transmitted through the Pacific lithosphere, but because the arc of curvature is too great for uniform transmittal of stress.

One reason why the East Pacific Rise does not occupy a median position north and east of the Eltanin fracture zone is that the arc of curvature across this part of the Pacific basin exceeds the ability of the lithosphere to transmit stress uniformly. However, there is a much more important reason for the position of the East Pacific Rise north and east of the Eltanin fracture zone. The position is largely a response to the Earth's rotation, and is explained in the companion paper in this volume by Meyerhoff et al.

Sense of offset along the ridge-transverse fracture zones—Menard (1958) wrote that the process that produces midocean ridges is sensitive to ocean-basin margins. Figure 22 shows just how sensitive that process is. Each major offset of the continental-shelf edge in North and South America, and in Africa, is mirrored by

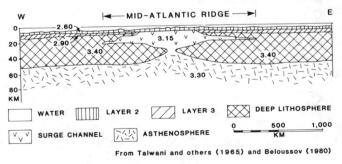

Figure 21. Same as Figure 20, except that the manner in which the anomalous mantle pod connects to the ultimate magma source, the asthenosphere, is shown as interpreted by Beloussov (1980). The point of this illustration is to show the relation between the low-velocity zone close to the surface below the midocean ridge and its magma source, the asthenosphere.

a shift in position of the axial rift of the midocean ridge. For each major shift in the continental margin of the Americas, Africa, or both, there is a corresponding shift in the midocean rift of the adjacent oceanic block (e.g., blocks A through J, Fig. 22). Note that each oceanic block is bounded by bathymetrically very deep and prominent ridge-transverse fracture zones. In general, the fracture zones between the block-bounding fractures are bathymetrically shallower and much less prominent. They also tend to be discontinuous. Similar relations between continental margins and midocean rifts are observed in the Arctic, Indian, and Southern oceans.

Such geometric relations imply that the continental nuclei were in their present positions when midocean ridges formed. Seismic-velocity and heat-flow studies (e.g., MacDonald, 1963; Jordan, 1975, 1978) show that the asthenosphere is very thin or absent beneath Precambrian nuclei, and that these continental nuclei have very deep roots, of the order of 300 to 400 km, a result that is supported by neodymium and strontium isotope studies (Wasserburg and DePaolo, 1979), as well as by strontium isotope ratios in young volcanic rocks (Brooks et al., 1976). The absence, or near-absence, of a low-velocity asthenosphere below the Precambrian nuclei led Lowman (1985, 1986) to propose that the continents are fixed and, if sea-floor spreading takes place, it is limited to oceanic regions.

Age of the midocean ridges and rifts—How long have the existing ridges and rifts been in their present positions? The question is answerable only in part because so few data are available. Some of these data are discussed by Meyerhoff et al. (this volume). However, bathymetric data also provide clues to the age of the midocean-ridge system.

Figure 22 shows that the Mid-Atlantic rift lies very close to the center of the Atlantic basin within the area included in the figure. One notable exception is that part of the Atlantic east of the Caribbean Sea (blocks C and D, Fig. 22). Here, the rift is about 3,400 km west of the African continental margin, but only about 1,700 km east of the Lesser Antilles. This suggests that, if the midocean rift originally was midway between the continents (as is the case north of block C and south of block D), the western, or American margin of the Atlantic must have been close to eastern Cuba. Although it may be no more than a coincidence, it is worth noting that the easternmost Precambrian outcrops known in the northern Caribbean are in Villa Clara Province, central Cuba, where radiometric dating has identified a metamorphic event in the age range 952 to 903 Ma (Somin and Millán, 1977; Hatten et al., 1988; Renne et al., 1989). Subsequently—but no later than about Carboniferous time—active island arcs appeared along both the northern and southern margins of the Caribbean (Joyce and Aronson, 1983, 1987; Benjamini et al., 1988; Morris et al., 1990). We conclude that the midocean-rift system is older than the Permian; it probably is much older as Meyerhoff et al., (this volume) explain.

Continental extensions—Active segments of the midocean-ridge system and its branches extend into the continents at several localities. The principal places are (1) the Lena River delta, northeastern Siberia, where the Mid-Atlantic-Gakkel Ridge enters the northern end of the Verkhoyansk Range; (2) the Red Sea-Gulf of Aden area, where the Carlsberg Ridge strikes southeast into the Indian Ocean; and (3) the Gulf of California, where the East Pacific Rise enters the Great Basin (Heezen, 1960). An extension of the East Pacific Rise, the Juan de

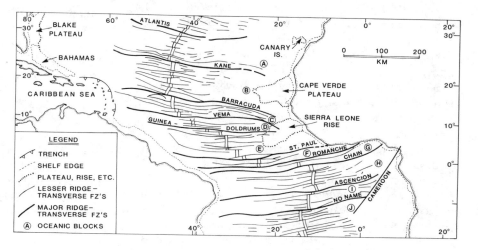

Figure 22. Map of the Mid-Atlantic Ridge between 19° S latitude and 32° N latitude The ridge-transverse fracture zones are from Heirtzler (1985) and Haxby (1987). Two types of ridge-transverse fractures characterize this segment of the Mid-Atlantic Ridge. The first fracture type is long, continuous, and bathymetrically deep. This type forms the boundaries between adjacent oceanic blocks A through J. The 7.0-7.8 km/s layer of anomalous upper mantle is cut out locally along these fracture zones as shown on Figure 19 at the Oceanographer f.z. Figure 22 shows also that the positions of the axial valley of the ridge are extremely sensitive to shift in the position of the adjacent continental-shelf edge (see text). The second type of fracture zone is discontinuous and not so deep as the first. The 7.0-7.7 km/s layer is not cut out along it. Relations between the axial depression and the continental margins are not nearly so sensitive as they are with the first type of ridge-transverse fracture zone. A ridge-transverse fracture zone of the second type is visible at the north end of Figure 19.

Fuca Ridge, enters the Coast Range of Canada just north of Vancouver Island.

Branches of the midocean-ridge system are joined to continental areas in several additional places. In the eastern Pacific, the Galapagos rift zone enters the northern Andes. Father south, the Chile Rise connects the East Pacific Rise with the southern Andes-Scotia arc complex. The complex in turn connects with the Mid-Atlantic Ridge via the America-Antarctic Ridge. In the North Atlantic, the Mid-Atlantic Ridge connects with the western Mediterranean via the Azores-Gorringe Bank. Other, less prominent ridge-to-continent connections (e.g., Murray Ridge in the northwestern Indian Ocean) are known.

INDEPENDENT OBSERVATIONS OF RIDGE-PARALLEL MOTIONS

Since 1970, several workers have invoked a component of asthenosphere flow parallel with midocean ridges to explain certain morphological features (e.g., Vogt, 1971, 1974). Since Vogt's early papers, an increasing number of authors proposed—for a variety of reasons—that a component of ridge-parallel flow exists. Although most of the reasons for invoking ridge-parallel motions have been valid, no one until Nicolas (1987) recognized the significance of the ridge-parallel linear fissures, fractures, and faults that appear on GLORIA and SeaMARC sonograph images of midocean ridges. Nicolas (1987, p. 54) recognized the significance of these linear fractures in various ultramafic complexes where they are exposed in the Red Sea, the Mediterranean region, and the Western Cordillera of North America, and concluded that motion beneath the ridges "... has an important component parallel with the rift."

Vogt (1971, 1974), and Vogt and Johnson (1975) proposed ridge-parallel movements beneath the Reykjanes Ridge and its continuation north of Iceland to explain certain morphological characteristics of the midocean ridge. Gorshkov and Lukashevich (1989) extended Vogt and Johnson's (1975) observations. They proposed that magma generated in hot spots beneath Iceland and Antarctica flowed the full length of the Mid-Atlantic Ridge.

In the Pacific basin, Herron (1972) postulated northward propagation of the East Pacific Rise from the Eltanin fracture zone at 55°S latitude although nowhere did she specifically suggest ridge-parallel flow of magma. Ridge-parallel rift propagation has been proposed by several workers to explain certain bathymetric characteristics of the Galápagos rift zone (e.g., Vogt and Johnson, 1975; Hey, 1977; Hey and Vogt, 1977; Hey et al., 1980). After the discovery of eddylike structures (overlapping spreading centers) along the crest of the East Pacific Rise (Fig. 16) by Lonsdale (1982, 1983), K. C. Macdonald, P. J. Fox, and others, explained them by the migration of magma along the strike of the ridge (Macdonald et al., 1987, p. 995). Schouten and Klitgord (1982) and White (1989) invoked ridge-parallel flow to explain several structural features of midocean ridges, although the ridge-parallel motions that they postulated were believed to be on a limited scale. Similar ideas were published by Ceuleneer et al. (1988), Nicolas et al., (1988a, 1988b), and Nicolas (1989).

The prevalent view of ridge-parallel movements is that they are associated with rows of upwelling mantle diapirs that lie on strike beneath the midocean ridge crest, with one diapir postulated beneath each ridge segment (Macdonald et al., 1988b). At the crest of each diapir, radial horizontal flow takes place such that a significant part of the flow parallels the strike of the ridge, but in the opposite direction from the crest of each diapir. In this way, ridge segmentation also is explained (Figs. 14, 15). The crest of each mantle diapir underlies the highest part of each midocean-ridge segment, and ridge-transverse fracture zones (transform faults) are where the flow from adjacent diapirs meets. Winterer et al. (1989, p. 271) explained it thus: "... ridge crests seem to be naturally segmented, the ends of the segments being marked by transform faults, propagaing rifts, overlapping spreading centers, or small misalignments called devals.... This segmentation appears to reflect the structure of the upwelling asthenosphere beneath the ridge axes. The disjunctures all tend to occupy deep spots along the ridge-crest profile and to accompany abrupt changes in chemical signature, suggesting that they delimit individual upwelling cells.... Magma is postulated to upwell primarily at the centers of the segments, and to flow sideways to the ends."

Except for Nicolas, Ceuleneer, and their colleagues, very few workers recognized the significance of the ubiquitous ridge-parallel fissures, fractures, and faults (Nicolas, 1987, 1989; Ceuleneer et al., 1988; Nicolas et al., 1988a, 1988b). Those fissures, fractures, and faults (Figs. 6–12) form a clear pattern. They are present across the full width of the ridge, overlapped by the sediments of the flanking abyssal-plain basins. They parallel, or subparallel, the ridge axis (Figs. 6–8); and they turn nearly 90° where they intersect ridge-transverse fracture zones (Fig. 19). In places, individual fissures can be traced without interruption from one ridge segment to the next (Fig. 19). If the ridge segments had been produced by upwelling mantle diapirs, the pattern of fissures, fractures, and faults would be interrupted near the center of each diapir, where a pattern of radial fissures, fractures, and faults would (unfailingly) develop. No radial (or even annular) pattern is observed on any ridge segment for which sonographs are available. Moreover, in the depressions between adjacent ridge segments, pressure ridges and similar compressive structures should be clearly visible; they are not.

PLATE-TECTONIC TENETS

Plate tectonics was founded on the assumption of ridge-orthogonal flow beneath the ocean basins (Hess, 1962; Morgan, 1968). Several corollary assumptions follow from the primary assumption of ridge-orthogonal flow. The principal of these are: (1) a state of tension characterizes all parts of midocean ridges (Hess, 1962); (2) new crust forms at midocean ridge crests (Hess, 1962); (3) this new crust records the magnetic field at the time of its formation; (4) the crustal record of the magnetic field shows that the new crust originated at alternate times of normal and reversed magnetization (Vine and Matthews, 1963; Morley and Larochelle, 1964); therefore, rocks at the ridge crests are the youngest and those at the basin margins are the oldest (Hess, 1962); (5) the ridge-transverse fracture zones (transform faults) are continuous and approximately parallel across the full width of each ridge; (6) midocean ridge crust exhibits seismic anisotropy, with the fast direction at right angles to the ridge, providing additional evidence that flow beneath midocean ridges is orthogonal to their strikes (Silver and Chan, 1988; Nicolas, 1989); (7) geochemical studies along the ridges show important, nonsystematic chemical changes, suggesting that ridge-parallel flow is not possible (Batiza et al., 1988); and (8) even if ridge-parallel flow were possible, it would be halted by the ridge-transverse fracture zones which are, in effect, partitions or walls that provide barriers to ridge-parallel flow (Vogt, 1974; Vogt and Johnson, 1975). If the primary assumption of plate tectonics is wrong—that of ridge-orthogonal flow—then some or all of its corollary assumptions either must be wrong, or—if some are proved by additional data to be correct—those assumptions that are correct must support some other interpretation.

A state of tension characterizes all parts of midocean ridges—A necessary corollary of plate tectonics is that stresses everywhere in midocean ridges must be tensile and act at right angles to the ridge axes. Hence, the principal structures should resemble those shown on Figures 4, 5, and 13. We demonstrated, however, that the structures actually observed are the long, linear fractures seen on Figures 6–12, and 19. If this were not enough to show that the stress state differs greatly from that which is predicted, the recent discovery of a broad belt of thrust faults in a zone along the flank of the Mid-Atlantic Ridge should dispel any lingering doubts.

Antipov et al. (1990) during an extensive geophysical study of the western flank of the Mid-Atlantic Ridge between 10° and 20° N latitude discovered a broad band of reverse and thrust faults in the lower part of the ridge at the edge of the abyssal plain (Fig. 23). This zone, 300–400 km wide, lies at the foot of the western flank of the Mid-Atlantic Ridge. Antipov and his colleagues (1990) made no attempt to trace the zone everywhere along the western flank of the ridge, but found that it does have a north-south extent of not less than 1,000 km; that is, it extends north-south throughout the area they investigated. The zone shows up on single-channel reflection-seismic data obtained by Rabinowitz et al. (1978), but only in the bathymetry. The records are not good enough to discern structural detail. A CDP line published by Rona (1980) showed the zone as a deformed belt characterized by abundant diffractions. Similar bathymetry on the eastern flank of the ridge (Rabinowitz et al., 1978) provides good evidence that a second band of reverse and thrust faults is present there. If these observations are confirmed by future studies, it will mean that midocean ridges are characterized by a bivergent fan structure, with a 300–400-km-wide band of thrusts on either side, those on the west being west-vergent and those on the east being east-vergent. Between the two deformed belts is a 900–1,000-km-wide region of block faults and linear fracture zones (Fig. 23). Thus, midocean ridges in surge tectonics (Meyerhoff et al., this volume) are surge channels that have been deformed into bivergent fan structures, or kobergens.

Age of oceanic crust increases away from midocean ridges—The Deep Sea Drilling Project (DSDP) scientists, through numerous publications, gave the impression that their drilling had adequately defined the basement of the ocean basins. We examined all the data in the 106 volumes and 42 supplemental volumes published for Legs 1 through 96 (a total of 99,564 pages). As of this writing only the first 54 legs have been examined in detail (58,588 pages). These legs include Sites

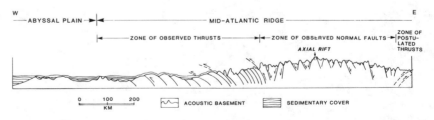

Figure 23. This is a west-to-east schematic structural interpretation of the Mid-Atlantic Ridge between 13° and 14° N latitude The interpretation is from Antipov et al. (1990), and is based on CDP reflection-seismic data.

1 through 429, of the total of 624 sites drilled during Legs 1–96.

Table 1 presents some revealing facts concerning the basement studies that were conducted from 1968 through 1980. This was the period of time during which most major tenets of plate tectonics were said to have been documented and confirmed. Of the first 429 sites that were drilled, only 165 (38%) reached basalt. All but 12 of them were called basement, including 19 sites where the contact between sediment and basalt was baked. Although the practice of referring to baked contacts as basement is disturbing, it is not nearly so disturbing as the claim that 101 contacts that were never recovered in cores—therefore never actually seen—were treated as normal depositional contacts. A third disturbing practice was the use of the term basement in sequences that may contain thick sections of lava and sills. Kamen-Kaye (1970) pointed out that in such sections, true basement may lie thousands of meters below the highest flow or sill designated as basement.

The DSDP practice of identifying baked contacts as basement was explained during Leg 9 in the Pacific Ocean (see Figs. 24–26). The purpose of this leg was to confirm the ages of magnetic anomalies in the eastern Pacific. At Site 79 on a magnetic anomaly believed to be 21 Ma, a baked contact was penetrated beneath the sedimentary column. The ship then went to a second location on the same anomaly, Site 80, where a baked contact was reached beneath sediments of the same age as at Site 79. The Shipboard Scientific Party (1972, p. 401) concluded that, "The baked sediment...overlying the basalt is included in the same foraminiferal and nannofossil zones as those at the base of Site 79 [where a baked contact also was found]. Because these two sites are located on the same [longitude], such age equivalence is in accord with the concepts of sea floor spreading and although the basalt encountered at Site 80 is a sill, suggests that the silling process does not invalidate the basal sediment age as an indicator of basement age."

Figures 24 and 25 show for the Atlantic and Pacific basins respectively plots of age of basement versus distance from the ridge crest. The basement ages are taken exclusively from DSDP data as corrected by Bolli (1980a, 1980b). The fact that several basalts called basement by the shipboard scientists later turned out to have radiometric dates younger than the overlying sedi-

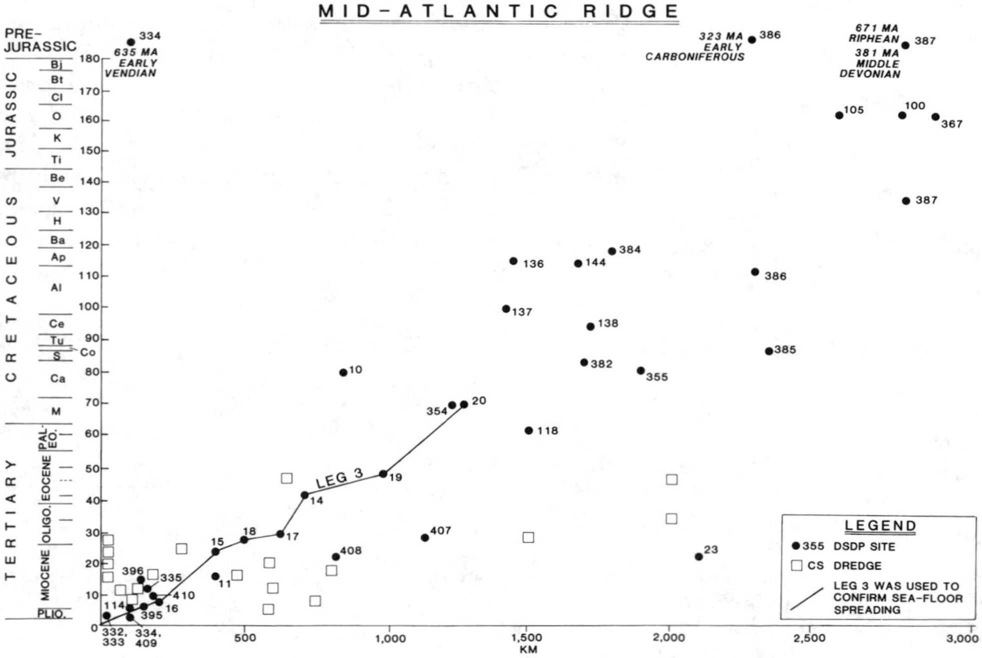

Table 1. Basement penetration summary, Legs 1–54 (Sites 1–429), Deep Sea Drilling Project, August 1968–June 1977. Source: DSDP volumes 1-54, 1969-1980.

	Total Sites Drilled	As a % of 429	As a % of 165
TOTAL SITES DRILLED	429	100	–
Basalt not penetrated	264	62	–
Basalt penetrated	165	38	100
BASALT DATA			
Total number of sediment-basalt contracts *not seen*	101	24	61
Sediment-basalt contact data recovered	64	15	39
Contact is depositional	33	8	20
Contact is baked	31	7	19
Total number of baked contacts that were called "basement"	19	4	12

Figure 24. This is a plot of rock age versus distance from the crest of the Mid-Atlantic Ridge, and is based mainly on DSDP core holes from Legs 1–54 (1969–1980). Pre-Jurassic dates are from Reynolds and Clay (1977) and Houghton et al. (1979). Leg 3, on which Atlantic sea-floor spreading is based, is shown. Radiometrically dated Cenozoic basalts have been added to show that the linear relation alleged between Leg 3 and distance from the ridge axis is illusory. Dredge-haul dates are from Carr and Kulp (1953), Miller (1964), Saito et al. (1966), Melson et al. (1972), and Meyerhoff and Meyerhoff (1974).

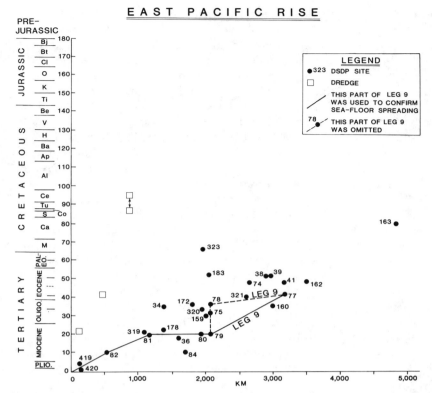

Figure 25. This is a plot of rock age versus distance from the crest of the East Pacific Rise, and is based mainly on DSDP cores from Legs 1–54 (1969–1980). Leg 9, on which sea-floor spreading in the Pacific is based, is shown. Radiometrically dated Cenozoic and Cretaceous basalt dredges have been added to show that the linear relation alleged between Leg 9 and the distance from the ridge crest is illusory. In addition, the Shipboard Scientific Party (1972) omitted Site 78 which does not fit. Dredge dates are from Budinger and Enbysk (1967) and Ozima et al. (1968).

ments was not taken into account.

Figures 24–26 do not support the DSDP scientists' claim that the age of the basement becomes older away from the ridge crests. The data (and dates) exhibit a very large scatter. (The fact should be noted that Leg 3, on Fig. 24 and Leg 9, on Fig. 25, were the only two legs used by DSDP scientists to prove the linear relationship that they claimed between age and increasing distance from the ridge crests.) The addition to each figure of several critical dredges from midocean ridges increases the scatter. It is true that, overall, the DSDP sites show a general trend from younger rocks at the ridge crests to older rocks away from the ridge crests, but on some anomalies, the age scatter is tens of millions of years. (In one seamount just west of the crest of the East Pacific Rise, the radiometric dates range from 2.4 to 96 Ma; Ozima et al., 1968.) Actually, an overall "younging" should be expected toward the axis, which is the highest and most mobile part of the ridge. We conclude that the risks of extrapolating geological interpretation across an area as great as the combined Atlantic and Pacific oceans (24,766,000 km^2, excluding the marginal seas) are too large, particularly in view of the facts that 19 of the contacts with basement were baked, and 101 were not seen.

Mid-Atlantic Ridge geology north of 37° N latitude—More dredge collections have been made along the northern Mid-Atlantic Ridge than in any other part of the midocean-ridge system. Only a small part of these dredges has been studied in detail, but even so, a respectable literature exists on North Atlantic dredge hauls. This literature suggests that North Atlantic geology is far more complex than generally realized.

For example, in the area bounded by 37° and 45°30′ N latitude and by 19° and 35° W longitude an area greater than 800,000 km^2, thousands of outcrops of Proterozoic and Paleozoic rocks appear to be present. This conclusion is based on the fact that nearly every dredge haul from uplifted blocks on the Mid-Atlantic Ridge contains old rocks, many of them granitic rocks (Aumento and Loncarevic, 1969). Aumento and Loncarevic (1969, p. 13) wrote that dredge hauls in the Bald Mountain region, just west of the Mid-Atlantic Ridge crest at 48° N latitude, consistently recovered 74% silicic rocks. They commented that this is ". . . a remarkable phenomenon. . . ." The casual dismissal of such hauls as evidence of "glacial erratics" or "ship ballast" (Heezen et al., 1959) can no longer be given credence. Moreover, the number of pre-Jurassic rock localities is now so great that such localities can no longer be ignored.

The Bald Mountain locality, mentioned above, is centered at 45°13′ N latitude, 28°52′ W longitude. The whole mountain—some 13 km long—has an estimated volume of 80 km^3. It consists of granitic and silicic metamorphic rocks that have yielded radiometric dates ranging from 1,550 to 1,690 Ma (Wanless et al., 1968), and is intruded by mafic rocks yielding a 785-Ma date. Such an outcrop is not likely to be either a glacial erratic or a discarded piece of ship's ballast.

Along the eastern flank of the Mid-Atlantic Ridge north of the Azores, along the northern flank of the Azores Ridge (Azores-Gorringe Bank area), and along the scarps facing the King's Trough (44° N, 22° W), fossiliferous Paleozoic sedimentary rocks have been recovered during dredging operations, and even on fishing nets, since 1883 or earlier (Furon, 1949; Meyerhoff,

1981). The *Talisman* (1883) collections were reported by Furon (1949). The more recent collections from fishing nets are in the Sedgwick Museum, Cambridge University. C.P. Hughes (pers. comms., 1978–1980) identified many of them. One of his collections from the southern escarpment of the King's Trough contains the trilobite *Triarthus* aff. *T. spinosus* and the graptolite *Climacograptus typicalis* (Meyerhoff, 1981), both North American taxa. Sparker profiles of the sea bottom suggest that many of the specimens are from outcrops in low-lying cuestas (C.P. Hughes, pers. comm., 1980). The faunas in every collection to date include only ostracods, trilobites, graptolites, worm trails, and similar faunal remains that would be expected on a Paleozoic ocean floor. Shallow-water taxa so far have not been recovered.

Proterozoic gabbroid rocks were recovered at Site 334, Leg 37 (37°02.13′ N, 34°24.87′ W). The gabbroid rock appears to be part of a mélange (Shipboard Scientific Party, 1977) in a middle Miocene nannofossil ooze (Bukry, 1977). Brecciated and serpentinized zones within the complex, and a wide scatter of magnetic inclinations, together indicate the mélange, or rubblelike, nature of the deposit. It is overlain by flat-lying late Miocene and younger deposits. The fact that the boulders are buried beneath undisturbed late Miocene and younger strata indicates that the deposit is not an assemblage of glacial erratics, or fragments of ship ballast. Reynolds and Clay (1977) determined the age of the gabbroid boulders to be 635 ± 102 Ma. They wrote that the age had to be wrong, because the ocean floor at this location could not be older than about 10 Ma. They attributed the ancient date to excess ^{40}Ar gas. They admitted that the excess-gas explanation was fraught with "difficulty" (Reynolds and Clay, 1977, p. 629).

These are but a few of the Paleozoic and older localities known from the Atlantic basin. They and several others are shown on Figure 26. In addition to the older rocks, Figure 26 shows a sufficient number of younger localities to demonstrate that the linear relation suggested by Leg 3 is largely illusory. As Meyerhoff and Meyerhoff (1974, p. 420) pointed out, the drilling results from the eight sites used to prove sea-floor spreading on Leg 3 most decidedly did not prove sea-floor spreading, as we document with the following observations. (1) The Site 15 basalt-sediment contact probably is depositional, but the basalt contains sedimentary clasts, suggesting the presence of older sediments below. (2) Basalt-sediment contacts were not seen at Sites 16 and 17, yet both basalts

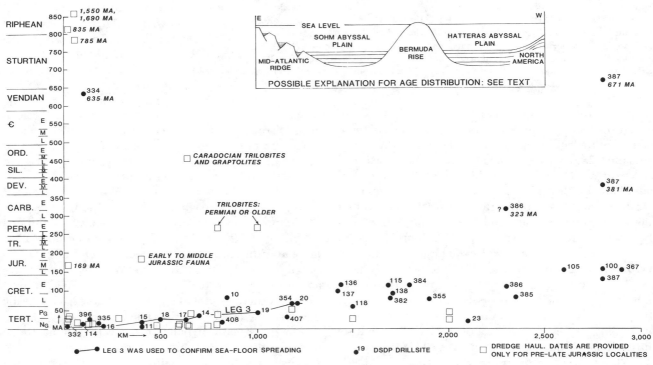

Figure 26. This is a plot of rock age vs. distance from the crest of the Mid-Atlantic Ridge designed to show rocks of *all* ages to scale, not just the post-Middle Jurassic rocks, as in Figure 24. A cross section of the western North Atlantic basin is inserted to show the writers' explanation for the distribution of pre-Late Jurassic rocks. See the text for details. DSDP radiometric dates, including those for Paleozoic and Proterozoic cores, are from Reynolds and Clay (1977), Houghton et al., (1979), and Ozima et al., (1979). Dredge-haul dates are from Furon (1949), Carr and Kulp (1953), Miller (1964), Saito et al., (1966), Wanless et al., (1968), Melson et al., (1972), Meyerhoff and Meyerhoff (1974), Honnorez et al., (1975), Ozima et al., (1976), and C. P. Hughes of the Sedgwick Museum, Cambridge University, as reported in Meyerhoff (1981).

were called basement. (3) Site 18 provided a baked contact, yet the basalts were called basement. (4) Site 19 basalt, even though the contact is depositional, contains clasts of older sediments. (5) Site 20 displayed a depositional contact. (6) Site 21 did not reach basalt, but was claimed as further support for sea-floor spreading, because the sedimentary rocks at total depth were older than those at Site 20 (Shipboard Scientific Party, 1970, p. 466).

The cross section on Figure 26 presents our explanation for the distribution of the oldest rocks in the Atlantic basin. All older rocks in the North Atlantic are either on the Mid-Atlantic Ridge or on the Bermuda Rise. We interpret them to be pre-Late Jurassic edifices that subsequently were overlapped by the overlap assemblage (successor basin) of the Sohm and Hatteras Abyssal Plain basins. The cross section has the merit of explaining all the facts so far known from the Atlantic basin, something no other hypothesis does.

Linear magnetic anomalies—Magnetic field reversal during generation of new oceanic crust is the only hypothesis that has been used to explain the alternate positive and negative magnetic anomalies that characterize much (but not all) of the world's midocean-ridge system. Agocs et al. (this volume) presented an alternative interpretation, specifically, that the anomalies are a consequence of magnetic susceptibility contrasts. In their scheme, the anomalies are not alternate strips of positive and negative magnetism, but of more positive and less positive (or, conversely, less negative and more negative) magnetism. The hypothesis that the magnetic stripes result from susceptibility contrasts was proposed by Luyendyk and Melson (1967). These authors, as well as Opdyke and Hekinian (1967), reported magnetization values from oceanic rocks that are more than adequate to produce the observed anomalies. The linearity of the anomalies in some parts of the ridge system is explained as a natural consequence of ridge-parallel laminar and linear flow beneath the midocean ridges. A more detailed explanation appears in the Agocs et al. (this volume) paper.

Seismic anisotropy—Raitt et al. (1969), on the basis of refraction-seismic studies in a very limited part of the Pacific basin, determined that the seismic velocities in an east-west direction were faster than those in the north-south direction, thereby confirming in a general way the results of an earlier study by Hess (1964). Hess (1964) interpreted the difference in velocity to reflect the preferred optical orientation of olivine crystals, which he believed to be stretched parallel with the direction of flow in the asthenosphere. Subsequent workers saw these results as confirmation of the fundamental tenet of plate tectonics, namely, that flow beneath the midocean ridges is orthogonal to them. Silver and Chan (1988) reviewed the history of this hypothesis.

The most recent studies of anisotropy, demonstrate that the subject is far more complex than either Hess (1964) or Raitt et al. (1969) realized, and the results of those recent studies are quite different from anything predicted by plate tectonics. For example, Stephen (1985) studied the Costa Rica rift some 550 km south-southwest of Panama in the Pacific basin, and found that the fast direction parallels the rift. Kuo et al. (1987) obtained the same result along part of the Mid-Atlantic ridge, that is, the fast direction parallels the strike of the ridge. Burnett and Orcutt (1989) in a study of the northern part of the East Pacific Rise found that the fast direction parallels the trend of the Rise. In all continental rift studies of which we know, the fast direction parallels the trend of the rift system (e.g., Rhine graben: Fuchs, 1983; Salton trough: Hearn, 1987; etc.).

Studies of anisotropy, however, have scarcely begun on a global scale. It will be many years before sufficient data are available to evaluate the full significance of anisotropy.

Geochemical variations parallel with the midocean ridge—Batiza et al. (1988) studied the chemical compositions of axial basalts on the Mid-Atlantic Ridge at 26° S latitude. They found numerous significant along-strike variations in composition, and from this fact concluded that ridge-parallel magma migration does not take place. The authors also inferred that large, well-mixed magma chambers are not present below the Mid-Atlantic Ridge at 26° S latitude.

Thompson et al. (1989) came to the opposite conclusion on the basis of a study along the East Pacific Rise at 10°–12° N latitude. They concluded that a deep common magma chamber is present, above which smaller individual and intermittent magma chambers develop, each with its own chemistry.

These opposing view points probably are related to the belief that shallow magma chambers are absent along the Mid-Atlantic Ridge but are widely developed beneath the East Pacific Rise (Gass, 1989). Although this may be the situation today, it may not have been true in the past. Moreover, seismic studies of the Mid-Atlantic Ridge are still in their infancy, and are still too few to permit many generalizations.

Ridge-transverse fracture zones act as dams to ridge-parallel flow—This concept appears to have originated with Vogt (1974), and was expounded most recently by Gorshkov and Lukashevich (1989). Data are insufficient to determine whether the ridge-transverse zones can prevent some lateral flow, but almost certainly material flows from ridge segment to ridge segment, as two lines of evidence—one of them direct, the other indirect—show.

The direct evidence is seen on Figure 19. Some fissures on the ridge segments continue uninterrupted through the fracture zones. Following Stokes's Law, this constitutes proof that, at one time at least, movement between adjacent ridge segments took place.

Earthquake epicenters provide the indirect evidence. Lines of epicenters are known to connect ridge segments with one another in all parts of the midocean-ridge system. Some of the earthquakes are tectonic, but Ouchi et al. (1982) discovered that many of the earthquakes are more like those associated with an active magma chamber. Young, fresh basalt flows also are visible on sonographs of the transverse faults. Thus, epicenter and outcrop data, as well as sonograph images, demonstrate that the magma chambers of the individual ridge segments are joined to one another beneath the ridge-transverse fracture zones.

Ridge-transverse fractures are continuous and approximately parallel—The continuity and general parallelism of the ridge-transverse (transform) fractures have been basic tenets of plate tectonics almost since the concept's inception (Morgan, 1968). These tents have been perpetuated and refined to the present (e.g., Klitgord and Schouten, 1986). This simplistic view of the geometry of the ridge-transverse fracture zones was disproved by the United States Naval Oceanographic chart series published in 1974 (Bergantino, 1974), but the charts were ignored. Instead, the idealized bathymetric diagrams of Bruce Heezen and Marie Tharp (e.g., Heezen and Tharp, 1968, 1977) were adopted by advocates of plate tectonics. These diagrams were effectively engraved in stone by the National Geographic Society's ocean-floor chart series (e.g., Grosvenor, 1968). Subsequently the highly sophisticated ocean-relief maps of Heirtzler (1985) were published. These were soon followed by the Haxby (1987) gravity map of the world's oceans. The Heirtzler (1985) and Haxby (1987) maps show unequivocally that the Bergantino (1974) charts are, in the main, correct. Subsequently, Smoot (1989) published a map of the principal faults of the North Atlantic Ocean, basing his mapped faults on detailed bathymetry. His map further confirms the published results of Bergantino (1974), Heirtzler (1985), and Haxby (1987). Tucholke (1990) tried to discredit Smoot's (1989) map, but the attempt was unsuccessful. As Smoot (1990, p. 912) wrote, "I cannot change real bathymetry to agree with the proliferating hypothetical models. Maybe it is time to change the inbred model of the evolution of the Atlantic basin."

Figure 27 is a reproduction of the Smoot (1989) map. The map shows that: the ridge-transverse fracture zones are segmented and discontinuous; they do not exist in large areas of the North Atlantic; where the ridge-transverse fracture zones are few or absent, seamount chains are present; several postulated major fracture zones do not exist; many do not cross the midocean ridge; poles of rotation cannot be inferred from the fracture zones (see Morgan, 1968); and many major fracture zones that strike diagonally across the North Atlantic have never been discussed in plate-tectonic literature. Smoot (1990, p. 914) concluded his various observations by stating, "Clearly, the study of the Atlantic

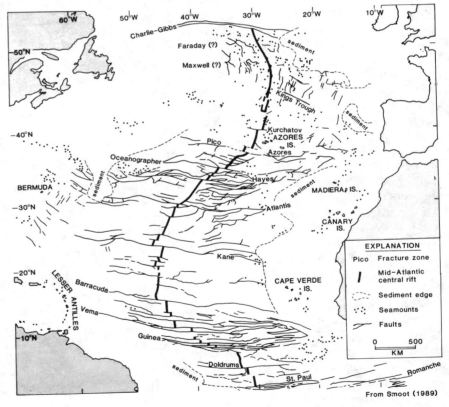

Figure 27. Map of part of the North Atlantic basin. Shown are fracture zones and seamounts, together with areas of heaviest sediment fill. The fracture zones are drawn from actual bathymetry, and are not based on theoretical models or geophysical data. As Smoot (1989) noted, the fracture pattern predicted by theoretical plate-tectonic models and the fracture pattern that actually exists are vastly different. Smoot wrote further that an entirely new tectonic model must be devised to explain the fault pattern of the North Atlantic basin. Published with permission of Geological Society of America.

basin is barely in its infancy as far as hard data are concerned."

CONCLUSIONS

1. Midocean ridges form in response to equiplanar compression, tangent to the Earth's surface, in the lithosphere.

2. Magma rises from the asthenosphere beneath these ridges in response to the law of gravity, and is emplaced in the manner described by Gilbert (1877) for the formation of sills (Peach-Köhler force of Weertman and Weertman, 1964).

3. The channel system beneath midocean ridges is interconnected, and flow within them is parallel with the ridges, as shown by the ridge-parallel fissure, fracture, and fault systems (and, because of the linear fractures, required by Stokes's Law).

4. Ridge-parallel flow when it occurs is laminar adjacent to the magma-channel walls. Flow is likely a response to pressure gradients produced by fluctuations in the compressive stress in the lithosphere (Meyerhoff et al., this volume). Flow, just as in lava tubes and tunnels, takes place in surges that produce an overall pinch-and-swell structure.

5. Where a magma channel is pinched, nonlaminar flow may occur, creating a variety of vortex-related structures that range from eddylike features (overlapping spreading centers) to ridge-transverse fracture zones (transform faults). These vortex-related structures divided midocean ridges into 300–500-km-long segments (Macdonald et al., 1988b; White, 1989).

6. The midocean-ridge system is probably as old as the lithosphere, almost certainly older than the Paleozoic. The Mid-Atlantic Ridge is at least as old as early Paleozoic, and probably is much older (Meyerhoff et al., this volume).

Meyerhoff et al. (this volume) show that midocean-ridge magma channels are part of a worldwide network that underlies parts of continents and ocean basins everywhere. In surge tectonics (Meyerhoff et al., this volume), these channels are called surge channels and provide the primary mechanism for every known type of tectogenesis.

REFERENCES CITED

Agocs, W.B., Meyerhoff, A.A., and Kis, K., 1992, Reykjanes Ridge: quantitative determinations from magnetic anomalies: This volume.

Antipov, M.P., Zharkov, S.M., Kozhenov, V.Ya., and Pospelov, I.I., 1990, Structure of the Mid-Atlantic Ridge and adjacent parts of the abyssal plain at latitude 13° N: International Geology Review, v. 32, no. 5, p. 468-478.

Aumento, F., and Loncarevic, B.D., 1969, The Mid-Atlantic Ridge near 45° N., III. Bald Mountain: Canadian Journal of Earth Sciences, v. 6, no. 1, p. 11-23.

Ballard, R.D., and Moore, J.G., 1977, Photographic atlas of the Mid-Atlantic Ridge rift valley: New York, Springer-Verlag, 114 p.

Basalt Volcanism Study Project, 1981, Basaltic volcanism on the terrestrial planets: New York, Pergamon Press, Inc., 1286 p.

Batiza, R., Melson, W.G., and O'Hearn, T., 1988, Simple magma supply geometry inferred beneath a segment of the Mid-Atlantic Ridge: Nature, v. 335, no. 6189, p. 428-431.

Beloussov, V.V., 1980, Geotectonics: Moscow, Mir Publishers and New York, Springer-Verlag, 330 p.

——, 1989, Osnovy geotektoniki: Moscow, Nedra, 382 p.

Benjamini, C., Shagam, R., and Menendez-V., A., 1987, (Late?) Paleozoic age for the "Cretaceous" Tocutunemo Formation, northern Venezuela: Geology, v. 15, no. 10, p. 922-926.

Bergantino, R.N., 1974, Physiographic diagram of the North Atlantic: United States Naval Oceanographic Office Charts NA-6, NA-7, NA-9, NA-10, and NA-11.

Betz, F., and Hess, H.H., 1942, The flow of the North Pacific Ocean: Geographical Review, v. 32, no. 1, p. 99-116.

Bolli, H.M., 1980a, Ages of sediments recovered from the Deep Sea Drilling Project Pacific Legs 5 through 9, 16 through 21, and 29 through 35, in Rosendahl, B.R., Hekinian, R., et al., Initial reports of the Deep Sea Drilling Project, v. 54: Washington, D.C., U.S. Government Printing Office, p. 881-886.

——, 1980b, The ages of sediments recovered from DSDP Legs 1-4, 10-15, and 36-53 (Atlantic, Gulf of Mexico, Caribbean, Mediterranean, and Black Sea), in Rosendahl, B.R., Hekinian, R., et al., Initial reports of the Deep Sea Drilling Project, v. 54: Washington, D.C., U.S. Government Printing Office, P. 887-895.

Brooks, C., James, D.E., and Hart, S.R., 1976, Ancient lithosphere: its role in young continental volcanism: Science, v. 193, no. 4258, p. 1086-1094.

Budinger, T.F., and Enbysk, B.J., 1967, Late Tertiary date from the East Pacific Rise: Journal of Geophysical Research, v. 72, no. 8, p. 2271-2274.

Bukry, D., 1977, Coccolith and silicoflagellate stratigraphy, central North Atlantic Ocean, Deep Sea Drilling Project Leg 37, in Aumento, F., Melson, W.G., et al., Initial reports of the Deep Sea Drilling Project, v. 37: Washington, D.C., U.S. Government Printing Office, p. 917-927.

Burnett, M.S., and Orcutt, J.A., 1989, Behavior of amplitudes and travel times of P-waves propagating about the magma chamber at 12°50′ N on the East Pacific Rise (Abstract): EOS, v. 70, no. 15, p. 455.

Caress, D.W., Burnett, M.S., and Orcutt, J.A., 1989, Tomographic imaging of the magma chamber at 12°50′ N on the East Pacific Rise (Abstract): EOS, v.70, no. 15, p. 455-456.

Carey, S.W., 1958, The tectonic approach to continental drift, in Carey, S.W., convener, Continental drift. A symposium: Hobart (Tasmania), University of Tasmania, Geology Department, p. 177-355.

——, 1981, The necessity for earth expansion, in Carey, S.W., ed., The expanding earth. A symposium: Hobart, University of Tasmania, Symposia of the Geology Department, p. 377-396.

Carr, D.R., and Kulp, J.L., 1953, Age of a Mid-Atlantic Ridge basalt boulder: Geological Society of America Bulletin, v. 64, no. 2, p. 253-254.

Ceuleneer, G., Nicolas, A., and Boudier, F., 1988, Mantle flow patterns at an oceanic spreading centre: the Oman peridotites record: Tectonophysics, v. 151, no. 1-4, p. 1-26.

Cloos, H., 1937, Zur Grosztektonik Hochafrikas und seiner Umgebung: eine Fragestellung: Geologische Rundschau, Bd. 28, Hft. 3-4, p. 333-348.

——, 1939a, Hebung-Spaltung-Vulkanismus: Elemente einer geometrischen Analyse irdischer Grossformen: Geologische Rundschau, Bd. 30, Hft. 4A, p. 405-527.

——, 1939b, Hebung-Spaltung-Vulkanismus, II: Geologische Rundschau, Bd. 30, Hft. 6, p. 637-640.

Corry, C.E., 1988, Laccoliths; mechanics of emplacement and growth: Geological Society of America Special Publication 220, 110 p.

Crane, K., 1987, Structural evolution of the East Pacific Rise axis from 13°10′ N to 10°35′ N: interpretations from SeaMARC I data: Tectonophysics, v. 136, no. 1/2, p. 65-124.

Detrick, R.S., Buhl, P., Vera, E., Mutter, J., Orcutt, J., Madsen, J., and Brocher, T., 1987, Multi-channel seismic imaging of a crustal magma chamber along the East Pacific Rise: Nature, v. 326, no. 6107, p. 35-41.

Duffield, W.A., 1972, A naturally occurring model of global plate tectonics: Journal of Geophysical Research, v. 77, no. 14, p. 2543-2555.

Eisen, M.F., Fox, P.J., and Macdonald, K.C., 1987, SeaMARC II studies of the Clipperton fracture zone and the Orozco transform fault (Abstract): EOS, v. 69, no. 44, p. 1505.

Emmons, R.C., 1969, Strike-slip rupture patterns in sand models: Tectonophysics, v. 7, no. 1, p. 71-87.

Ewing, J., and Ewing, M., 1959, Seismic-refraction measurements in the Atlantic Ocean basins, in the Mediterranean Sea, on the Mid-Atlantic Ridge, and in the Norwegian Sea: Geological Society of America Bulletin, v. 70, no. 3, p. 291-317.

Ewing, M., and Heezen, B.C., 1956, Some problems of Antarctic submarine geology: American Geophysical Union, Geophysical Monograph 1, p. 75-81.

Fornari, D.J., Perfit, M.R., Malahoff, A., and Embley, R., 1983, Geochemical studies of abyssal lavas recovered by DSRV *Alvin* from eastern Galapagos rift, Inca transform, and Ecuador rift, 1. Major element variations in natural glasses and spacial [sic] distribution of lavas: Journal of Geophysical Research, v. 88, no. B12, p. 10519-10529.

Fuchs, K., 1983, Recently formed elastic anisotropy and petrological models for the continental subcrustal lithosphere in southern Germany: Physics of the Earth and Planetary Interiors, v. 31, p. 93-118.

Furon, R., 1949, Sur les trilobites draguées à 4255 m de profondeur par le *Talisman* (1883): Paris, Académie des Sciences, Compte Rendu, v. 258, no. 19, p. 1509-1510.

Gass, I.G., 1989, Magmatic processes at and near constructive plate margins as deduced from the Troodos (Cyprus) and Semail nappe (N Oman) ophiolites, in Saunders, A.D., and Norry, M.J., eds., Magmatism in the ocean basins: Geological Society of London Special Publication no. 42, p. 1-15.

Gilbert, G.K., 1877, Geology of the Henry Mountains, Utah: U. S. Geographical and Geological Survey of the Rocky Mountain region, 170 p.

Gorshkov, A.G., and Lukashevich, I.P., 1989, Computing of magma chamber temperatures in rift zones of the world ocean: Tectonophysics, v. 159, no. 3/4, p. 337-346.

Gretener, P.E., 1969, On the mechanics of the intrusion of sills: Canadian Journal of Earth Sciences, v. 6, no. 6, p. 1415-1419.

Grosvenor, M.B., ed.-in-chief, 1968, Atlantic Ocean floor: National Geographic Society, 1 sheet, scale 1:30,412,800.

Gutenberg, B., and Richter, C.F., 1949, Seismicity of the earth and associated phenomena: Princeton University Press, 273 p.

——, 1954, Seismicity of the earth and associated phenomena, 2nd edition: Princeton University Press, 310 p.

Hatten, C.W., Somin, M., Millán, G., Renne, P., Kistler, R.W., and Mattinson, J.M., 1988, Tectonostratigraphic units of central Cuba, in Barker, L., and Gordon, J., eds., Transactions of the Eleventh Caribbean Geological Congress, 20-26 July, 1986, Bridgetown: Bridgetown, Barbados, Ministry of Finance, Energy and Natural Resources Division, p. 35-1 - 35-13.

Haug, E., 1900, Les geosynclinaux et les aires continentales: Société Géologique de France Bulletin, v. 28, p. 617-711.

Haxby, W.F., 1987, Gravity field of the world's oceans: National Geophysical Data Center, National Oceanic and Atmospheric Administration, scale, 1: 40,000,000.

Hearn, T.A., 1987, Crustal structure and tectonics in southern California: United States Geological Survey Circular 956, p. 56-57.

Heezen, B.C., 1957, Deep-sea physiographic provinces and crustal structure (Abstract): American Geophysical Union Transactions, v. 38, no. 3, p. 394.

——, 1960, The rift in the ocean floor: Scientific American, v. 203, no. 4, p. 98-110.

——, 1962, The deep-sea floor, in Runcorn, S.K., ed., Continental drift: New York, Academic Press, p. 235-288.

Heezen, B.C., and Ewing, M., 1961, The mid-oceanic ridge and its extension through the Arctic basin, in Raasch, G.O., ed., Geology of the Arctic, v. I: University of Toronto Press, p. 622-642.

Heezen, B.C., and Tharp, M., 1968, Physiographic diagram of the North Atlantic Ocean (revised): Geological Society of America, 1 sheet, scale, 1:5,700,000.

——, and Tharp, M., 1977, World ocean floor: United States Navy, Office of Naval Research, 1 sheet, scale, 1:46,460,000.

Heezen, B.C., Tharp, M., and Ewing, M., 1959, The floors of the oceans, I. The North Atlantic: Geological Society of America Special Paper 65, 122 p.

Heirtzler, J.R., ed., 1985, Relief of the surface of the earth: National Geophysical Data Center, National Oceanic and Atmospheric Administration, 3 sheets, scale, 1:39,000,000.

Heiskanen, W.A., and Vening Meinesz, F.A., 1958, The earth and its gravity field: New York, McGraw-Hill Book Company, Inc., 470 p.

Herron, E.M., 1972, Sea-floor spreading and the Cenozoic history of the east-central Pacific: Geological Society of America Bulletin, v. 83, no. 6, p. 1671-1691.

Hess, H.H., 1954, Geological hypotheses and the earth's crust under the oceans: Royal Society of London Philosophical Transactions, ser. A, v. 222, p. 341-348.

——, 1955, Serpentines, orogeny, and epeirogeny, in Poldervaart, A., ed., Crust of the earth (a symposium): Geological Society of America Special Paper 62, p. 391-407.

——, 1962, History of ocean basins, in Engel, A.E.J., James, H.L., and Leonard, B.F., eds., Petrologic studies: a volume in honor of A.F. Buddington: Geological Society of America, p. 599-620.

——, 1964, Seismic anisotropy of the uppermost mantle under oceans: Nature, v. 203, no. 4945, p. 629-630.

Hey, R.N., 1977, A new class of pseudofaults and their bearing on plate tectonics: a propagating rift model: Earth and Planetary Science Letters, v. 37, p. 321-325.

Hey, R.N., and Vogt, P.R., 1977, Spreading center jumps and sub-axial asthenosphere flow near the Galapagos hotspot: Tectonophysics, v. 37, no. 1-3, p. 41-52.

Hey, R.N., Buennebier, F.K., and Morgan, W.J., 1980, Propagating rifts on midocean ridges: Journal of Geophysical Research, v. 85, no. B7, p. 3647-3658.

Hill, M.N., 1956, Notes on the bathymetric chart of the northeast Atlantic: Deep-Sea Research, v. 3, p. 229-231.

Holmes, A., 1931, Radioactivity and earth movements: Geological Society of Glasgow Transactions, v. 18, pt. 3, p. 559-606.

Honnorez, J., Bonatti, E., Emiliani, C., Brönnimann, P., Furrer, M.A., and Meyerhoff, A.A., 1975, Mesozoic limestone from the Vema offset zone, Mid-Atlantic Ridge: Earth and Planetary Science Letters, v. 26, p. 8-12.

Houghton, R.L., Thomas, J.E., Jr., Diecchio, R.J., and Tagliacozzo, A., 1979, Radiometric ages of basalts from DSDP Leg 43, Sites 382 and 385 (New England Seamounts), 384 (J-anomaly), 386 and 387 (central and western Bermuda Rise), in Tucholke, B.E., Vogt, P.R., et al., Initial reports of the Deep Sea Drilling Project, v. 43: Washington, D.C., U.S. Government Printing Office, p. 739-753.

Jordan, T.H., 1975, The continental tectosphere: Reviews of Geophysics and Space Physics, v. 13, no. 1, p. 1-12.

———, 1978, Composition and development of the continental tectosphere: Nature, v. 274, no. 5671, p. 544-548.

Joyce, J., and Aronson, J., 1983, K-Ar dates for blueschist metamorphism on the Samaná Peninsula, Dominican Republic: Tenth Caribbean Geological Conference, 14-19 August, 1983, Cartagena: Bogotá, Ingeominas, preprint, 8 p.

———, 1987, K-Ar dates for blueschist metamorphism on the Samaná Peninsula, Dominican Republic, in Duque-Caro, H., coordinator, Tenth Caribbean Geological Conference, Cartagena, Colombia, 1983, Transactions: Bogotá (Colombia), INGEOMINAS, p. 454-458.

Kamen-Kaye, M., 1970, Age of the basins: Geotimes, v. 15, no. 7, p. 6, 8.

Kappel, E.S., and Normark, W.R., 1987, Morphometric variability within the axial zone of the southern Juan de Fuca Ridge. Interpretation from Sea MARC II, Sea MARC I, and deep-sea photography: Journal of Geophysical Research, v. 92, no. B11, p. 11291-11301.

Keith, M.L., 1972, Ocean-floor convergence: a contrary view of earth tectonics: Journal of Geology, v. 80, no. 3, p. 249-276.

Klitgord, K.D., and Schouten, H., 1986, Plate kinematics of the central Atlantic, in Vogt, P.R., and Tucholke, B.E., eds., The geology of North America, v. M, The western North Atlantic region: Geological Society of America, Decade of North American Geology, p. 351-378.

Kober, L., 1921, Der Bau der Erde: eine Einführung in die Geotektonik: Berlin, Gebrüder Borntraeger, 324 p.

———, 1925, Gestaltungsgeschichte der Erde: Berlin, Gebrüder Borntraeger, 200 p.

———, 1928, Der Bau der Erde: eine Einführung in die Geotektonik, 2nd edition: Berlin, Gebrüder Borntraeger, 499 p.

Krylov, S.V., Mishen'kin, B.P., Petrik, G.V., and Seleznev, V.S., 1979, O seysmicheskoy mideli verkhov mantii v Baykal'skoy riftovoy zone: Novosibirsk, Geologiya i Geofizika, no. 5, p. 117-129.

Kuenen, P.H., 1950, Marine geology: New York, John Wiley and Sons, Inc., 568 p.

Kuo, B.-Y., Forsyth, D.W., and Wysession, M., 1987, Lateral heterogeneity and azimuthal anisotropy in the North Atlantic determined from SS-S differential travel times: Journal of Geophyscial Research, v. 92, no. B7, p. 6421-6436.

Langmuir, C.H., Bender, J.F., and Batiza, R., 1986, Petrological and tectonic segmentation of the East Pacific Rise, 5°30′–14°30′ N: Nature, v. 322, no. 6075, p. 422-429.

Laughton, A.S., and Searle, R.C., 1979, Tectonic processes on slow spreading ridges, in Talwani, M., Harrison, C.G.A., and Hayes, D.E., eds., Deep drilling results in the Atlantic Ocean: ocean crust: American Geophysical Union Maurice Ewing Series 2, p. 15-32.

Lonsdale, P., 1982, Small offsets of the Pacific-Nazca and Pacific-Cocos spreading axes (Abstract): EOS, v. 65, no. 45, p. 1108.

———, 1983, Overlapping rift zones at the 5.5° S offset of the East Pacific Rise: Journal of Geophysical Research, v. 88, no. B11, p. 9393-9406.

Lowman, P.D., Jr., 1985, Plate tectonics with fixed continents: a testable hypothesis—I: Journal of Petroleum Geology, v. 8, no. 4, p. 373-378.

———, 1986, Plate tectonics with fixed continents: a testable hypothesis—II: Journal of Petroleum Geology, v. 9, no. 1, p. 71-87.

Luyendyk, B.P., and Melson, W.G., 1967, Magnetic properties and petrology of rocks near the crest of the Mid-Atlantic Ridge: Nature, v. 215, no. 5097, p. 147-149.

Lyttleton, R.A., 1982, The earth and its mountains: New York, John Wiley and Sons, 206 p.

MacDonald, G.J.F., 1963, The deep structure of continents: Review of Geophsyics, v. 1, no. 4, p. 587-665.

Macdonald, K.C., and Fox, P.J., 1983, Overlapping spreading centers: new accretion geometry on the East Pacific Rise: Nature, v. 302, no. 5903, p. 55-57.

Macdonald, K.C., Sempere, J.-C., and Fox, P.J., 1984, East Pacific Rise from Siquieros to Orozco fracture zones: along-strike continuity of axial neovolcanic zone and structure and evolution of overlapping spreading centers: Journal of Geophysical Research, v. 89, no. B7, p. 6049-6069, 6301-6305.

Macdonald, K.C., Sempere, J.-C., Fox, P.J., and Tyce, R., 1987, Tectonic evolution of ridge-axis discontinuities by the meeting, linking, or self-decapitation of neighboring ridge segments: Geology, v. 15, no. 11, p. 993-997.

Macdonald, K.C., Haymon, R.M., Miller, S.P., Sempere, J.-C., and Fox, P.J., 1988a, Deep-tow and Sea Beam studies of duelling propagating ridges on the East Pacific Rise near 20°40′ S: Journal of Geophysical Research, v. 94, no. B4, p. 2875-2898, 3514-3515.

Macdonald, K.C., Fox, P.J., Perram, L.J., Eisen, M.F., Haymon, R.M., Miller, S.P., Carbotte, S.M., Cormier, M.-H., and Shor, A.N., 1988b, A new view of the mid-ocean ridge from the behaviour of ridge-axis discontinuities: Nature, v. 335, no. 6187, p. 217-225.

Melson, W.B., Hart, S.R., and Thompson, G., 1972, St. Paul's Rocks, equatorial Atlantic: petrogenesis, radiometric ages, and implications on sea-floor spreading, in Shagam, R., Hargraves, R.B., Morgan, W.J., Van Houten, F.B., Burk, C.A., Holland, H.D., and Hollister, L.C., eds., Studies in earth and space sciences: Geological Society of America Memoir 132, p. 241-272.

Menard, H.W., 1958, Development of median elevations in ocean basins: Geological Society of America Bulletin, v. 69, no. 9, p. 1179-1185.

———, 1965, Sea floor relief and mantle convection, in Ahrens, L.H., Press, F., Runcorn, S.K., and Urey, H.C., eds., Physics and chemistry of the earth, v. 6: London, Pergamon Press, p. 315-364.

Meyerhoff, A.A., 1968, Arthur Holmes: originator of spreading ocean floor hypothesis: Journal of Geophysical Research, v. 73, no. 20, p. 6563-6565.

———, 1981, Ordovician (Caradocian) trilobites and graptolites near the North Atlantic Ridge: International Stop Continental Drift Society Newsletter, v. 3, no. 3, p. 1.

Meyerhoff, A.A., and Meyerhoff, H.A., 1974, Ocean magnetic anomalies and their relation to continents, in Kahle, C.F., ed., Plate tectonics—Assessments and reassessments: American Association of Petroleum Geologists Memoir 23, p. 411-422.

Meyerhoff, A.A., Taner, I., Morris, A.E.L., Martin, B.D., Agocs, W.B., and Meyerhoff, H.A., 1992, Surge tectonics: a new hypothesis of earth dynamics: this volume.

Miller, J.A., 1964, Age determinations made on samples of basalt from the Tristan da Cunha group and other parts of the Mid-Atlantic

Ridge, *in* The volcanological report of The Royal Society Expedition to Tristan da Cunha, 1962: Royal Society of London Philosophical Transactions, series A, v. 256, no. 1075, p. 565-569.

Miyashiro, A., Aki, K., and Şengör, A.M.S., 1982, Orogeny: New York, John Wiley and Sons, 242 p.

Molengraaff, G.A.F., 1928, Wegener's continental drift, *in* Theory of continental drift. A symposium on the origin and movement of land masses both inter-continental and intra-continental, as proposed by Alfred Wegener: American Association of Petroleum Geologists, p. 90-92.

Morgan, W.J., 1968, Rises, trenches, great faults, and crustal blocks: Journal of Geophysical Research, v. 73, no. 6, p. 1959-1982.

Morley, L.W., and Larochelle, A., 1964, Palaeomagnetism as a means of dating geological events, *in* Osborne, F.F., ed., Geochronology in Canada: Royal Society of Canada Special Publication no. 8, p. 39-51.

Morris, A.E.L., Taner, I., Meyerhoff, H.A., and Meyerhoff, A.A., 1989, Tectonic evolution of the Caribbean region, *in* Dengo, G., and Case, J.E., eds., The geology of North America, v. H, The caribbean region: Geological Society of America, Decade of North American Geology, p. 433-457.

Murray, Sir John, 1913, The ocean. A general account of the science of the sea: London, Williams and Norgate, 256 p.

Nicolas, A., 1987, Asthenosphere structure and anisotropy beneath rifts, *in* Noller, J.S., Kirby, S.H., and Nielson-Pike, J.E., eds., Geophysics and petrology of the deep crust and upper mantle— a workshop sponsored by the U.S. Geological Survey and Stanford University: U.S. Geological Survey Circular 956, p. 53-54.

Nicolas, A., 1989, Structures of ophiolites and dynamics of ocean lithosphere: Dordrecht, Kluwer Academic Publishers, 367 p.

Nicolas, A., Reuber, I., and Benn, K., 1988a, A new magma chamber model based on structural studies in the Oman ophiolite: Tectonophysics, v. 151, no. 1-4, p. 87-105.

Nicolas, A., Ceuleneer, G., Boudier, F., and Misseri, M., 1988b, Structural mapping in the Oman ophiolites: mantle diapirism along an oceanic ridge: Tectonophysics, v. 151, no. 1-4, p. 27-56.

Olson, J., and Pollard, D.D., 1989, Inferring paleostresses from natural fracture patterns: a new method: Geology, v. 17, no. 4, p. 345-348.

Opdyke, N.D., and Hekinian, R., 1967, Magnetic properties of some igneous rocks from the Mid-Atlantic Ridge: Journal of Geophysical Research, v. 72, no. 8, p. 2257-2260.

Ouchi, T., Ibrahim, A.-B.K., and Latham, G.V., 1982, Seismicity and crustal structure in the Orozco fracture zone: Project ROSE Phase II: Journal of Geophysical Research, v. 87, no. B10, p. 8501-8507.

Ozima, M., Kaneoka, I., and Yanagisawa, M., 1979, ^{40}Ar-^{39}Ar geochemical studies of drilled basalts from Leg 51 and Leg 52, *in* Donnelly, T., Franchetau, J., et al., Initial reports of the Deep Sea Drilling Project, v. 51, 52, 53, part 2: Washington, D.C. U.S. Government Printing Office, p. 1127-1128.

Ozima, M., Ozima, M., and Kaneoka, I., 1968, Potassium-argon ages and some magnetic properties of some dredged submarine basalts and their geophysical implications: Journal of Geophysical Research, v. 73, no. 2, p. 711-723.

Ozima, M., Saito, K., Matsuda, J., Zashu, S., Aramaki, S., and Shido, F., 1976, Additional evidence of existence of ancient rocks in the Mid-Atlantic Ridge and the age of the opening of the Atlantic: Tectonophysics, v. 31, no. 1/2, p. 59-71.

Pavlenkova, N.I., 1989, Struktura zemnoy kory i verkhney mantii i tektonika plit, *in* Belousov, V.V., ed., Tektosfera yeye stroyeniye i razvitiye: Akademii Nauk SSSR, Mezhduvedomstvenniy Geofizicheskiy Komitet, Geodinamicheskiye Issledovaniya no. 13, p. 36-45.

Perfit, M.R., and Fornari, D.J., 1983, Geochemical studies of abyssal lavas recovered by DSRV *Alvin* from eastern Galapagos rift, Inca transform, and Ecuador rift, 2. Phase chemistry and crystallization history: Journal of Geophysical Research, v. 88, no. B12, p. 10530-10550.

Rabinowitz, P.D., Heirtzler, J.R., Aitken, T.D., and Purdy, G.M., 1978, Underway geophysical measurements: *Glomar Challenger* Legs 45 and 46, *in* Melson, W.B., Rabinowitz, P.D., et al., Initial reports of the Deep Sea Drilling Project, v. 45: Washington, D.C., U.S. Government Printing Office, p. 55-118.

Raitt, R.W., Shor, G.G., Jr., Francis, T.J.G., and Morris, G.B., 1969, Anisotropy of the Pacific upper mantle: Journal of Geophysical Research, v. 74, no. 12, p. 3095-3109.

Renne, P.R., Mattinson, J.M., Hatten, C.W., Somin, M., Millán-Trujillo, G., and Linares-Cala, E., 1989, Confirmation of late Proterozoic age for the Socorro Complex of north-central Cuba from ^{40}Ar/^{39}Ar and U-Pb dating (Abstract), *in* Geología '89, Primer Congreso Cubano de Geología, Programa y Resúmenes: La Habana, Sociedad Cubana de Geología, p. 118.

Reynolds, P.H., and Clay, W., 1977, Leg 37 basalts and gabbro: K-Ar and ^{40}Ar/^{39}Ar dating, *in* Aumento, F., Melson, W.G., et al., Initial reports of the Deep Sea Drilling Project, v. 37: Washington, D.C., U.S. Government Printing Office, p. 629-630.

Rona, P.A., 1980, The central North Atlantic Ocean basin and continental margins: geology, geophysics, geochemistry, and resources, including the Trans-Atlantic Geotraverse (TAG): United States Department of Commerce, National Oceanic and Atmospheric Administration, NOAA Atlas 3, 99 sheets.

Rusby, J.S.M., and Somers, M.L., 1977, The development of the 'Gloria' sonar system from 1970-75, *in* Angel, M.V., ed., Voyage of discovery: Deep-Sea Research, v. 24 (Supplement), p. 611-625.

Saito, T., Ewing, M., and Burckle, L.H., 1966, Tertiary sediment from the Mid-Atlantic Ridge: Science, v. 151, no. 3714, p. 1075-1079.

Schäfer, K., 1972, Transform faults in Island: Geologische Rundschau, Bd. 61, Hft. 3, p. 942-960.

Scheidegger, A.E., 1958, Physics of marine orogenesis: Alberta Society of Petroleum Geologists Bulletin, v. 6, no. 11, p. 266-291.

Schouten, H., and Klitgord, K.D., 1982, The memory of the accreting plate boundary and the continuity of fracture zones: Earth and Planetary Science Letters, v. 59, p. 255-256.

Searle, R.C., and Laughton, A.S., 1977, Sonar studies of the Mid-Atlantic Ridge and Kurchatov fracture zone: Journal of Geophysical Research, v. 82, no. 33, p. 5313-5328.

Sears, F.W., Zemansky, M.W., and Young, H.D., 1974, College physics, 4th edition: Reading, Massachusetts, Addison-Wesley Publishing Company, 751 p.

Shipboard Scientific Party, 1970, Summary and conclusions, *in* Maxwell, E.A., Von Herzen, R.P., et al., Initial reports of the Deep Sea Drilling Project, v. 3: Washington, D. C., U. S. Government Printing Office, p. 441-467.

Shipboard Scientific Party, 1972, Site 80, *in* Hays, J.D., Cook H.E., III, et al., Initial reports of the Deep Sea Drilling Project, v. 9: Washington, D. C., U. S. Government Printing Office, p. 401-447.

Shipboard Scientific Party, 1977, Site 334, *in* Aumento, F., Melson, W.B., et al., Initial reports of the Deep Sea Drilling Project, v. 37: Washington, D. C., U. S., Government Printing Office, p. 239-287.

Silver, P.G., and Chan, W.W., 1988, Implications for continental structure and evolution from seismic anisotropy: Nature, v. 335, no. 6185, p. 34-39.

Smoot, N.C., 1989, North Atlantic fracture-zone distribution and patterns shown by multibeam sonar: Geology, v. 17, no. 12, p. 1119-1122.

———, 1990, North Atlantic fracture-zone distribution and patterns shown by multibeam sonar: reply: Geology, v. 18, no. 9, p. 912-914.

Somers, M.L., Carson, R.M., Revie, J.A., Edge, R.H., Barrow, B.J., and Andrews, A.G., 1978, GLORIA II—an improved long range side-scan sonar, in Oceanology International 1978, IEEE/IERE sub-conference on offshore instrumentation and communications, Technical Session J: London, BPS Publications Ltd., p. 16-24.

Somin, M.L., and Millán, C., 1977, Sobre la edad de rocas metamórficas cubanas: Academia de Ciencias de Cuba, Instituto de Geología y Paleontología, Informe Científico-Técnico no. 2, 11 p.

Sonder, R.A., 1939, Zur Tektonik des Atlantischen Ozeans: Geologische Rundschau, Bd. 30, Hft. 1, p. 28-51.

Stacey, F.D., 1981, Cooling of the earth—a constraint on paleotectonic hypotheses, in O'Connell, R.J., And Fyfe, W.S., eds., Evolution of the earth: America Geophysical Union and Geological Society of America, Geodynamic Series, v. 5, p. 272-276.

Stephen, R.A., 1985, Seismic anisotropy in the upper oceanic crust: Journal of Geophysical Research, v. 90, no. B13, p. 11, 383-11, 396.

Stille, H., 1939, Kordillerisch-atlantische Wechselbeziehungen: Geologische Rundschau, Bd. 30, Hft. 3-4, p. 315-342.

Stocks, Th., 1938, Morphologie des atlantischen Ozeans: Berlin, Deutsche Atlantische Expedition auf dem Forschungs- und Vermessungsschiff "Meteor" (1925-1927), Wissenschaftliche Ergebnisse, Bd. 3, t. 1, 2 Lief, p. 35-151.

Stocks, Th., and Wüst, G., 1938, Die Tiefverhältnisse des offenen Atlantischen Ozeans: Berlin, Deutsche Atlantische Expedition auf dem Forschungs- und Vermessungsschiff "Meteor" (1925-1927), Wissenschaftliche Ergebnisse, Bd. 3, T. 1, 1 Lief, p. 1-31.

Swift, S.A., and Stephen, R.A., 1989, Lateral heterogeneity in the seismic structure of upper oceanic crust, western North Atlantic: Journal of Geophysical Research, v. 94, no. B7, p. 9303-9322.

Sychev, P.M., 1985, Anomal'nyye zony v verkhney mantii, mekhanism ikh obrazovaniya i rol' v razvitii struktur zemnoy kory: Akademiya Nauk SSSR, Tikhookeanskaya Geologiya, no. 6, p. 25-35.

Talwani, M., Le Pichon, X., and Ewing, M., 1965, Crustal structure of the midocean ridges, 2. Computed model from gravity and seismic refraction data: Journal of Geophysical Research, v. 70, no. 2, p. 341-352.

Taylor, F.B., 1910, Bearing of the Tertiary mountain belt on the origin of the earth's plan: Geological Society of America Bulletin, v. 21, no. 2, p. 179-226.

Thompson, G., Bryan, W.B., and Humphris, S.E., 1989, Axial volcanism on the East Pacific Rise, 10-12°, in Saunders, A.D., and Norry, M.J., eds., Magmatism in the ocean basins: Geological Society of London Special Publications no. 42, p. 181-200.

Tivey, M.A., and Johnson, H.P., 1987, The Central Anomaly magnetic high: the implications for ocean crust construction and evolution: Journal of Geophysical Research, v. 92, no. B12, p. 12685-12694.

Tolstoy, I., 1951, Submarine topography in the North Atlantic: Geological Society of America Bulletin, v. 62, no. 5, p. 441-450.

Tolstoy, I., and Ewing, M., 1949, North Atlantic hydrography and the Mid-Atlantic Ridge: Geological Society of America Bulletin, v. 60, no. 10, p. 1527-1540.

Tritton, D.J., 1988, Physical fluid dynamics, 2nd ed.: Oxford, Clarendon Press, 519 p.

Tucholke, B.E., 1990, North Atlantic fracture-zone distribution and patterns shown by multibeam sonar: comment: Geology, v. 18, no. 9, p. 911-913.

Umbgrove, J.H.F., 1947, The pulse of the earth: The Hague, Martinus Nijhoff, 358 p.

van Bemmelen, R.W., 1933, The undation theory of the development of the earth's crust: 16th International Geological Congress, Washington 1933, Repts. v. 2, p. 965-982.

———, 1976, Plate tectonics and the undation model: a comparison: Tectonophysics, v. 32, no. 3/4, p. 145-182.

Vening Meinesz, F.A., 1942, Topography and gravity in the North Atlantic Ocean: Koninklijke Nederlandsche Akademie van Wettenschappen Proceedings, v. 45, no. 2, p. 120-125.

Vine, F.J., and Matthews, D.H., 1963, Magnetic anomalies over oceanic ridges: Nature, v. 199, no. 4897, p. 947-949.

Vogt, P.R., 1971, Asthenosphere motion recorded by the ocean floor south of Iceland: Earth and Planetary Science Letters, v. 13, p. 153-160.

———, 1974, The Iceland phenomenon: imprints of a hot spot on the ocean crust, and implications for flow beneath the plates, in Kristjansson, L., ed., Geodynamics of Iceland and North Atlantic area: Dordrecht, D. Reidel Publishing Company, p. 105-126.

Vogt, P.R., and Johnson, G.L., 1975, Transform faults and longitudinal flow below the midoceanic ridge: Journal of Geophysical Research, v. 80, no. 11, p. 1399-1428.

Walker, G.P.L., 1989, Gravitational (density) controls on volcanism, magma chambers and intrusions: Australian Journal of Earth Sciences, v. 36, no. 2, p. 149-165.

Wanless, R.K., Stevens, R.D., Lachance, G.R., and Edmonds, C.M., 1968, Age determinations and geological studies. K-Ar isotopic ages, report 8: Geological Survey of Canada Paper 67-2, pt. A, p. 140-141.

Washington, H.S., 1930, The origin of the Mid-Atlantic Ridge: Maryland Academy of Sciences Journal, v. 1, no. 1, p. 20-29.

Wasserburg, G.J., and DePaolo, D.J., 1979, Models of earth structure inferred from neodymium and strontium isotopic abundances: Washington, National Academy of Sciences Proceedings, v. 76, p. 3591-3598.

Weertman, J., 1971, Theory of water-filled crevasses in glaciers applied to vertical magma transport beneath oceanic ridges: Journal of Geophysical Research, v. 76, no. 5, p. 1171-1183.

Weertman, J., and Weertman, J.R., 1964, Elementary dislocation theory: New York, Macmillan and Company, 213 p.

Wegener, A., 1920, Die Entstehung der Kontinente und Ozeane, 2nd edition: Braunschweig, Friedrich Vieweg & Sohn, 135 p.

White, R.S., 1989, Asthenospheric control on magmatism in the ocean basins, in Saunders, A.D., and Norry, M.J., eds., Magmatism in the ocean basins: Geological Society of London Special Publication no. 42, p. 17-27.

Willis, B., 1932, Isthmian links: Geological Society of America Bulletin, v. 43, no. 4, p. 917-952.

Winterer, E.L., Atwater, T.M., and Decker, R.W., 1989, The northeast Pacific Ocean and Hawaii, in Bally, A.W., and Palmer, A.R., eds., The geology of North America, v. A, The geology of North America; an overview: Geological Society of America, Decade of North American Geology, p. 265-297.

Wyllie, P.J., 1971, The dynamic Earth: New York, John Wiley and sons, Inc., 416 p.

Yamagishi, H., 1985, Growth of pillow lobes—evidence from pillow lavas of Hokkaido, Japan, and North Island, New Zealand: Geology, v. 13, no. 7, p. 499-502.

Zverev, S.M., Yaroshevskaya, G.A., Tulina, Yu.V., and Pavlenkova, N.A., 1988, Skorostnaya struktura litosfery no vostochnoy chasti Angolo-Brazil'skogo geotraversa, in Zverev, S.M., and Boldyrev, S.A., eds., Geofizicheskiye polya Atlanticheskogo okeana: Moscow, Akademiya Nauk SSSR, Mezhduvedomstvennyy Geofizicheskiy Komitet, p. 7-23.

ACKNOWLEDGMENTS

We thank, first of all, J. M. Dickins whose idea it was to hold the Smithsonian Conference on "New Concepts in Global Tectonics." We thank him, Sankar Chatterjee, Nicholas Hotton III, and Arthur J. Boucot for inviting us to present at this conference the results of our research. We are grateful to Curt Teichert who was also a key figure in the initiation of the conference. J. O. Blanton (Skidaway Institute of Oceanography) and N. Sylvester (Texas A&M University) helped us throughout the fluid mechanics. Paul D. Lowman, Jr., guided us through the vast Washington (D.C.) labyrinth of government publications. Other valuable publications were provided by V. V. Beloussov, Kathleen Crane, and M. L. Keith. Key illustrations were provided us by W. A. Duffield, P. J. Fox, Mark D. Holmes, Ellen S. Kappel, Baerbel K. Lucchitta, and W. R. Normark. Mid-Atlantic Ridge collections of Paleozoic fossils were made available to us by C. P. Holmes (Sedgwick Museum). Kathryn Meyerhoff did the drafting; Ernestine R. Voyles typed the final manuscript. Valuable critiques and suggestions came from Donald L. Baars, M. I. Bhat, Stanley E. Cebull, Ashok K. Dubey, C. W. Hatten, Jean M. S. Jenness, Stuart E. Jenness, M. Kamen-Kaye, M. L. Keith, F. G. Koch, Konrad B. Krauskopf, Donna K. Meyerhoff, James C. Meyerhoff, Curt Teichert, and Walter L. Youngquist. However, the paper would never have been written had it not been for the enthusiastic help given to us at all times by Donald L. Baars, M. I. Bhat, Arthur J. Boucot, Stanley E. Cebull, Dong R. Choi, Ashok K. Dubey, Charles W. Hatten, Jean and Stuart E. Jenness, Maurice Kamen-Kaye, T. T. Khoo, Franklyn G. Koch, Paul D. Lowman, Jr., Wallace D. Lowry, Donna K. Meyerhoff, James C. Meyerhoff, Kathryn Meyerhoff, William Stannage, Bock K. Tan, Sukran Taner, Curt Teichert, Ernestine R. Voyles, and Walter L. Youngquist. Jean and Stuart Jenness's and Donna and James Meyerhoff's suggestions and help in organizing the paper improved it several-fold.

Finally, we point out that the views presented here are our own; those who helped us are in no way responsible for our interpretations.

All research was financed by the writers.

Paleoland, crustal structure, and composition under the northwestern Pacific Ocean

D. R. Choi, 6 Mann Place, Higgins, A.C.T. 2615, Australia

B. I. Vasil'yev, Pacific Oceanology Institute, USSR Academy of Sciences, Radio 7 Street, Vladivostok 690032 USSR

M. I. Bhat, Wadia Institute of Himalayan Geology, 33 General Mahadeo Singh Road, Dehradun 248001, India

ABSTRACT

Crust beneath the NW Pacific Ocean is not composed of oceanic crust, but is formed of Archean and Proterozoic continental crust. The interpretation is suggested by the reevaluation of deep-penetration seismic profiles combined with data from dredgings and paleogeographic reconsiderations.

The provenance of widely distributed Proterozoic orthoquartzite clasts in the Japanese Phanerozoic sediments is sought in the present Pacific Ocean and the present Japan Sea areas. This scenario is supported by interpretation of seismic profiles across the Japan Trench and the Nankai Trough. These profiles reveal a distinct sediment progradation toward island arcs during Paleozoic to Early Mesozoic time.

Seismic profiles show Precambrian crust to be composed of two layers: a mound-forming lower unit (Arhcean?) and a depression-filling, well-layered upper unit (Proterozoic?). The upper unit is considered to have been the source of orthoquartzite clasts to the Tethys Sea. Some of the dredged Precambrian rocks off Kuril-Kamchatka Trench may have been derived from the lower unit.

The structure of the Precambrian crust is characterized by tensional tectonics, represented by the development of horsts and grabens, which are crossed perpendicularly or diagonally by deep fractures. Mapped magnetic anomalies appear to coincide with major fault zones that accompanied volcanic activity. These Precambrian structures are overprinted by younger compressional arc tectonics that developed subsequent to the Paleozoic and are characterized by the Wadati-Benioff zone.

Most of the present NW Pacific Ocean appears to have been either subaerially exposed, forming paleoland (the Great Oyashio Paleoland), or was partly very shallow sea during the Paleozoic to Early Mesozoic times, becoming deep sea towards the end of the Jurassic.

These interpretations require the fundamental revision of plate-tectonic models for the geological development of the island arcs and trenches in the NW Pacific region.

INTRODUCTION

Numerous tectonic models have been proposed for the origin of the island arc and trench systems of the northwest Pacific (Fig. 1). Most models (e.g., Dewey, 1980; Uyeda, 1982) are based on the plate-tectonics hypothesis.

In plate-tectonics models, the trench-island arc systems in the northwest Pacific are considered to be regions of convergent plate boundaries, where Pacific plates originating at distant ridges accreted at, or were subducted along, the trench system associated with the Wadati-Benioff zone. One fundamental assumption of such models is that the plate is composed of oceanic crust, some of it carrying presumably allochthonous continental fragments. However, analyses of field-geological evidence from the Japanese islands, and of marine geological and geophysical data, together with dredging data collected chiefly by Soviet scientists, from the northwestern Pacific, lead us to the view that the plate-tectonics hypothesis is not relevant to this region, and to a new insight into the crustal composition and structure in the region. Here we review the data and recently published new interpretations.

PROVENANCE OF ORTHOQUARTZITE CLASTS AND USUGINU-TYPE CONGLOMERATES

Numerous occurrences of orthoquartzite clasts in sedimentary sequences ranging in age from Paleozoic to Cenozoic (but mainly from Triassic to Paleogene) have been documented throughout Japan (Fig. 2; Tokuoka, 1967; Adachi, 1971; Kano, 1971 and 1974; Konishi et al., 1973; Okami, 1973; Harata and Tokuoka, 1978; Shibata, 1979; Tokuoka and Okami, 1982; Okami and Kano, 1983). K-Ar dating of these orthoquartzite clasts reveal three age groups: Precambrian (778 m.y.), Early Paleozoic (470–550 m.y.), and Late Paleozoic (260–310 m.y.), with a concentration in the Precambrian and early Paleozoic. Because these ages are considered to represent the times of metamorphism or alteration of the orthoquartzite, Shibata (1979) concluded that all the clasts originated in the Precambrian.

Orthoquartzite clasts have also been found in the Usuginu-type conglomerates (Tokuoka and Okami, 1982), which are characterized by granodiorite clasts as well as clasts of aplites, gabbro, tuff, limestone, and other lithologies. The provenance of the Usuginu-type conglomerates recently has been reviewed by one of us. Choi (1984b) demonstrated that paleolands of subcontinental size had once been present on both sides of the Japanese islands (Fig. 3). In the southern Japan,

---

Chatterjee, S., and N. Hotton III, eds. *New Concepts in Global Tectonics.*
Texas Tech University Press, Lubbock, 1992, xii + 450 pp.

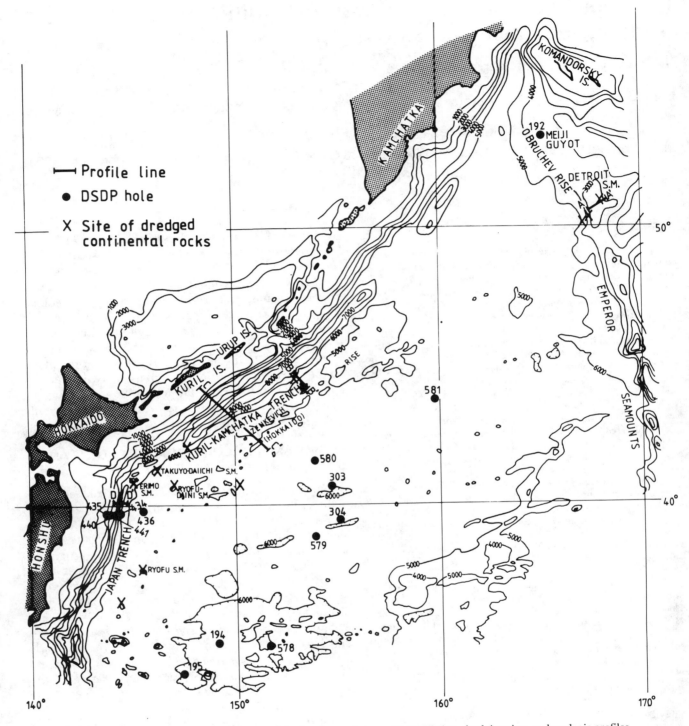

Figure 1. Generalized bathymetry of northwestern Pacific showing study area with DSDP sites, dredging sites, and geologic profiles.

paleocurrent data (Fig. 4) suggest that the Kuroshio paleoland was the provenance of the above-mentioned orthoquartzite clasts during Cretaceous to Paleogene times (Kano, 1974; Harata et al., 1978; Harata and Tokuoka, 1978; Tokuoka and Okami, 1982).

The proposed paleolands were evidently largely emergent during Paleozoic-Mesozoic-Paleogene times, but were totally submerged during Paleogene to Miocene times. The subsidence, which apparently was related to the orogenic upheaval of the Honshu geosyncline, is

Figure 2. The occurrence of orthoquartzite clasts in Japan.

confirmed by structural relics detected in seismic profiles.

SEISMIC STRATIGRAPHY AND CRUSTAL STRUCTURES BELOW THE JAPAN TRENCH

Seismic stratigraphy has been examined in detail in the Japan Trench (Fig. 1), where deep-penetration seismic reflection profiles are available (Fig. 5). The basement (Unit I) is 4 to 5 km thick, and consists of two major units: a basal, massive, mound-forming unit, and an overlying depression-filling unit of layered sediments. The sedimentary units are separated into five units (Units II–VI). Sediments in Units II–IV prograde toward the present land, implying a provenance in the present northwestern Pacific (Fig. 6). Comparison of the seismic stratigraphy in the trench area with well-established stratigraphy of northern Honshu (Minato et al., 1979) makes it tempting to speculate that these prograding units correlate to lithologies of Cambrian-Ordovician, Silurian-Carboniferous, and Permo-Triassic age, respectively (Choi, 1984a).

In the Japan Trench area, the oceanic crust consists of (1) 4- to 5-km-thick basement that includes a massive mound-forming base unit (Unit Ia; Archean?) and layered units (Units Ib–Id; Archean-Proterozoic?), (2) thin middle units (Unit II or III; Lower to Middle Paleozoic?) that are well layered and tectonically little disturbed, and (3) veneering top units (Units V–VI;

Jurassic? to Quaternary), the bottom of which includes the horizon of radiolarian cherts recovered at the bottom of one of the Deep Sea Drilling Program (DSDP) holes (site 436; von Huene et al., 1980). Similar seismic stratigraphy as the above Japan Trench was also established in the Aleutian Trench (Fig. 7) from analysis of preexisting seismic data published by McCarthy and Scholl (1985). The crust under the abyssal plain is composed of massive mound-forming Units I and II, and the overlying well-layered, flat, tectonically undisturbed Unit III. Unit III onlaps toward the south where a basement high exists. These units are overlain thinly by Units V–VI under the abyssal plains. However, in the landward trench slope, thick wedges of sedimentary rock, Units VI and V, develop. Both units show northward sediment progradation (toward the present Aleutian arc).

The above observation and interpretation in both the Japan and Aleutian trenches contradict plate-tectonic interpretation, which assumes that the landward prograding reflectors indicate accretionary wedges formed in relation to subducting Pacific plates. However, their interpretation is negated by the fact that major composition of the accretionary wedges as revealed in many DSDP holes (sites 434–436, 440–441) is not oceanic sediments but terrigenous clastics that obviously derived from the western source or present land. The westward sediment progradation, as described and discussed above, is indicative of a provenance that existed in the present abyssal plains adjacent to the trenches. This provenance is likely to have supplied sediments including Proterozoic orthoquartzite clasts into the former Tethys Sea.

Sediment filling of present trenches, as well as acoustic units under the abyssal plain are relatively undisturbed. This fact has been pointed out and was puzzled by many previous workers who studied the trenches in the northwestern Pacific (e.g., Ludwig et al., 1966; von Huene and Shor, 1969; Scholl and Marlow, 1974; Gnibidenko et al., 1978). Scholl and Marlow (1974) once stated "evidence for subduction or off-scraping of trench deposits is not glaringly apparent" (p. 268).

As is clear in seismic profiles, the geological structure under the trench-deep abyssal area is characterized by a combination of normal- (tensional) and thrust-fault (compressional) systems. Block-faults that form horst and graben structures are conspicuous, largely dislocating the lower crust as well as the Cenozoic sedimentary cover beneath the abyssal plain. The large faults are undoubtedly deep-seated, perhaps rooted in the upper mantle. Several large, west-dipping thrusts (30° to 35°) are present at the base of the trench slope. From the interpreted dislocation produced by these thrusts, these

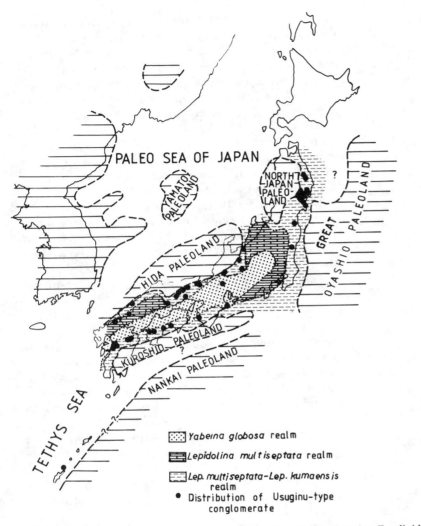

Figure 3. Paleogeography around the Japanese islands in the Late Permian. Fusulinid realms (Choi, 1976) and Usuginu-type conglomerates are also indicated.

movements appear to have ended before the deposition of Layer IV, or Permian-Triassic. An exception is the thrust at the very foot of the slope, which is still active and disturbs the sea-bottom topography. This eastward migration of the thrust-zone with time coincides exactly with the present Wadati-Benioff zone (Fig. 8), implying that the thrust-zone may have been more or less continuously active since Early Paleozoic time, the activities of which may have been related to repeated orogenic movements that occurred in northern Honshu (Minato et al., 1979).

DREDGING RESULTS

Solid information of the composition of the oceanic crust (Fig. 1) has been presented by Russian scientists who made numerous dredgings in the northern and northwestern Pacific. Their localities are (1) the Obruchev Rise located south of the western end of the Aleutian Trench (Vasil'yev, 1982), (2) east of the Kuril-Kamchatka Trench (Smirnov, 1982; Vasil'yev and Evlanov, 1982; Sergeev et al., 1983; Vasil'yev et al., 1986a, 1986b), and (3) east of the Japan Trench (Vasil'yev, 1986).

Precambrian metamorphic rocks have been dredged from the Detroit Seamount (Vasil'yev, 1982), about 300 km southeast of the Meiji Guyot on which DSDP site 192 was located (Creager et al., 1973). Vasil'yev (1982) reported the recovery of banded biotite gneiss, amphibole, and biotite-bearing granitic gneiss and other metamorphic rocks from what had been interpreted as acoustic basement and assumed to be basaltic oceanic crust. At least some of the recovered rocks were judged in situ based on various evidence. These continental basement rocks are overlain by thick geosynclinal Late Cretaceous acid-volcanic and sedimentary rocks, on which younger (Eocene to Pliocene) subaerial volcanic effusives flowed (Fig. 9). Vasil'yev (1982) noted the lithologic similarity of the dredged rocks with those in the Kamchatka Peninsula, and concluded that the rise had been part of a continent until Pliocene time, after which rapid submergence took place in the Pleistocene to form the present deep trench. The fact that a Cretaceous geosyncline developed in the area of Precambrian basement in the present Obruchev Rise area could imply long-lasting subaerial exposure of a Precambrian basement before the development of the geosynclinal stage in the Cretaceous.

Further evidence for the existence of continental crust under the northwestern Pacific comes from the dredgings made at two other sites; one on the seaward slope of the Kuril-Kamchatka Trench and the other on the Zenkevich Rise. Both of these sites are southeast of Urup Island and lie at water depths ranging from 8,500 to 9,000 m, and about 5,400 m respectively (Fig. 1; Vasil'yev et al., 1986a, 1986b). The recovered rock types include crystalline garnet-bearing apoterrigenous (slatey) rocks, and at one site with associated pyroxenite from the basement section of the trench. These rocks were considered to be in situ based on the freshly broken surfaces. The dredged crystalline schists represent a part of what has previously been thought of as oceanic layer 3.

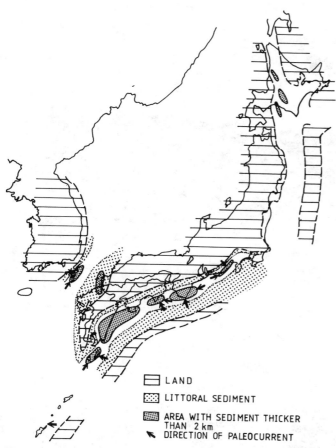

Figure 4. Paleogeographic map of the Japanese islands during Paleogene time. Compiled mainly from Minato et al. (1965) and Harata and Tokuoka (1978).

This basement complex is mostly now covered by thick (2–3 km) Late Jurassic to Early Cretaceous pillow lavas (tholeiitic basalt) intruded by hyperabyssal dolerite, gabbro-dolerite and gabbro-anorthosite of Late Cretaceous age (Fig. 10). These facts led Vasil'yev et al. (1986a, 1986b) to reconstruct the following history of the region.

1. Paleozoic-Early Cretaceous: sedimentation, alternating with periods of volcanism, and plutonism. At the end of this stage, elevation and erosion.
2. Late Jurassic-Early Cretaceous: areal outpouring of pillow lavas in relatively shallow-water environments and intrusion into the pile of hypabyssal bodies with basic composition.
3. Paleogene: uplift and fracturing.
4. Miocene: slow subsidence, sporadic accumulation of tuffaceous and organic-clay sediments with further terrestrial pyroclastic volcanism (basic-intermediate composition) in places.
5. Pliocene: compression, and formation of thrusts in the basement of the present arc-trench system.
6. Late Pliocene-Pleistocene: sharp subsidence with formation of faults and tectonic scarps in the trench slope as a result of extension. Formation of trench in its present form and commencement of deposition of its currently undisturbed sediments.

A number of intrusive rocks represented by granite and granodiorite (Middle to Late Cretaceous in K-Ar age) together with younger volcanic rocks (including Eocene to Oligocene dacites and basalts) have been obtained by Sergeev et al. (1983) from the south of Hokkaido Rise. Freshly broken surfaces of these dredged rock fragments indicate that they are in situ constituents of the acoustic basement. They have similar mineralogical and petrochemical affinities. Judging from the absolute ages of the dredged rocks (Sergeev et al., 1983), they concluded that an orogenic movement took place in the Middle Cretaceous, followed by eruption of subalkaline basaltic flows in the Early Tertiary.

Another dredging and seismic cruise by Russian scientists obtained continental rocks from Takuyo-Daiichi, Ryofu-Daini, and Erimo seamounts located southeast of Hokkaido Island (Fig. 1; Vasil'yev and Evlanov, 1982). There, fragments of granitic rocks, granodiorites, and porphyritic granites together with biotite-amphibole and plagioclase-bearing crystalline slates having freshly broken surfaces were dredged along with other in situ basalts of both submarine and subaerial origin. Vasil'yev and Evlanov (1982) considered that the granitoids and crystalline rocks formed the acoustic basement in this area.

In addition, gneisses and granitic rocks (including granites, granite porphyries, dioritic porphyries) have been known to occur at several submarine rises and seamounts in the east of the Japan Trench (Fig. 1; Vasil'yev, 1986). Despite some difficulty in determining whether all the dredged samples were in situ or exotic, Vasil'yev (1986) concluded that continental crust is likely to form the acoustic basement in this region.

Besides the above-mentioned continental rocks dredged from the seaward slope of the trenches and from deep abyssal plains, several other dredged rocks of continental affinity and of possible Paleozoic-Mesozoic age have been obtained from the landward slope of the trenches, suggesting a basement complex there.

Granodiorite clasts have been dredged at the deep sea slope off north Honshu. The clasts originally had been considered to be in situ by scientists of the University of Tokyo, but were later thought to be ice-rafted (for details see Choi, 1987). Based on reinterpreted seismic profiles in the dredge station, however, Choi (1987) considered that the clasts were most likely to have been in situ, possibly originating from the Upper Paleozoic

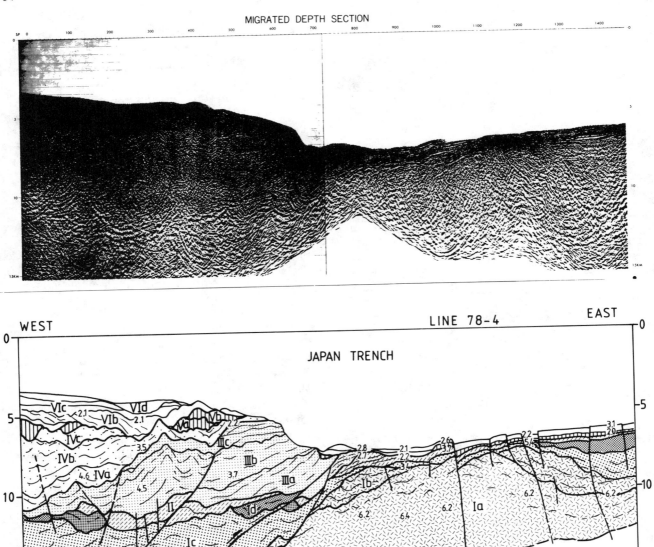

Figure 5. Seismic profile (top) and interpretation (bottom) across the Japan Trench (D-D' in Fig. 1).

conglomerates that crop out near the dredge site. Granodiorite is the most characteristic clast in the Upper-Permian Usuginu-type conglomerates in northern Honshu (Minato et al., 1979), some 150 km away from the dredge site.

A similar situation is known from the Aleutian Trench slope. Grow (1973) originally regarded large blocks of fresh amphibolite schist and subangular boulders of hard greywacke as in situ, but reserved the conclusion whether they are in situ or not on the basis of gravity data. Choi (1990) concluded on the basis of seismic profile information that the blocks would be in situ, and possibly derived from Paleozoic or Mesozoic formations cropping out along a fault scarp in the lower slope.

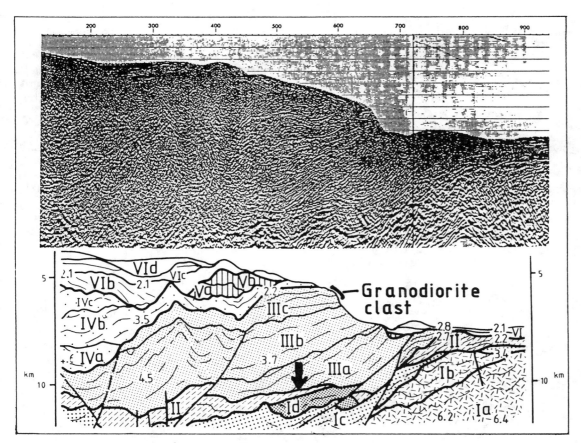

Figure 6. Blow-up of a seismic profile across the Japan Trench (top) and its interpetation (bottom) to show the landward progradation of sediments in Layers II, III, and IV.

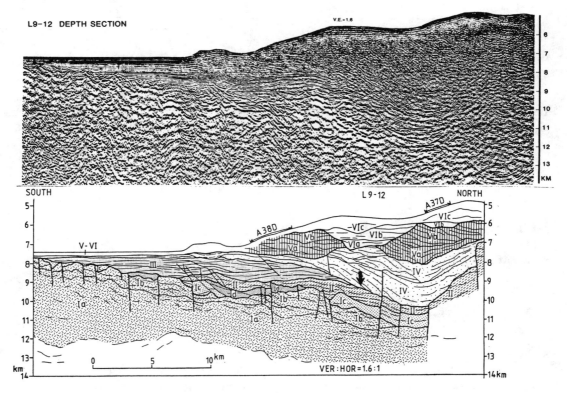

Figure 7. Seismic profile (top; McCarthy and Scholl, 1985) and its interpretation (bottom; Choi, 1990) across the Aleutian Trench. Arrow shows prograding sediments. Location, south of Amila Island, Aleutian Islands.

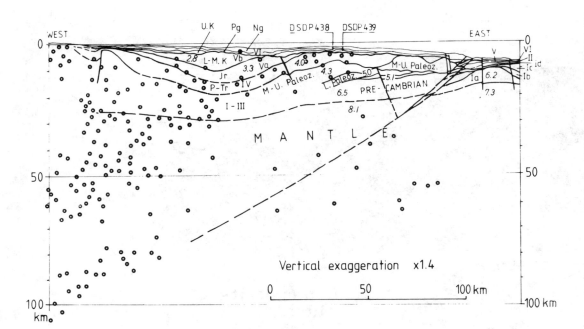

Figure 8. Interpreted geologic section across the continental shelf–fore-arc slope–Japan Trench. Interpreted from Line JNOC-1 (right half), and unpublished seismic profile (left half). Earthquake foci (open circles; modified from Rodnikov et al., 1982) superimposed.

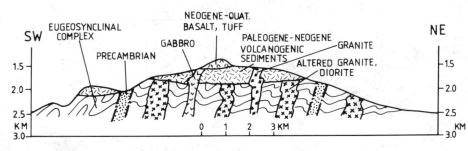

Figure 9. Schematic profile across the Detroit Seamount on the Obruchev Rise, compiled from dredging and seismic studies (Vasil'yev, 1982; Line A-A', Fig. 1).

DRILLING RESULTS

Despite numerous DSDP drillings made in abyssal plains of the northwestern Pacific (Fig. 1), no holes have penetrated oceanic layer 3. The oldest sediments recovered in the southeast of the Japan Trench are Jurassic-Cretaceous cherts—nannofossil chalks (sites 195, 197) that overlie tholeiitic basalt (Heezen et al., 1973). However, the basalt had chilled fine-grained margins at its top and base. It therefore could have been a sill.

Several other DSDP holes (sites 303 and 304; Larson et al., 1975: sites 580 and 581, Heath et al., 1985) drilled on submerged topographic highs southeast of the Kuril-Kamchatka Trench also bottomed in basalt after penetrating Early Cretaceous chert and nannofossil ooze. However, we speculate that these mounds of basalts are lavas that overlie the so-called oceanic layer 3, based on the relationship established from nearby

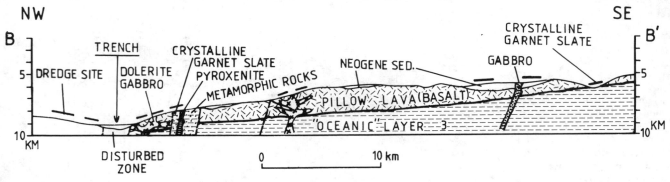

Figure 10. Simplified cross section across the Kuril-Kamchatka Trench-Zenkevich Rise (modified from Vasil'yev, 1986a, 1986b; Line B-B', Fig. 1). Crystalline garnet slate was dredged from two sites in this profile. Therefore, the oceanic layer 3 is considered to include, at least partly, this rock.

dredging and seismic surveying in the Zenkevich Rise off Urup Island as described earlier (Fig. 10).

Therefore, all the basalts recovered by DSDP drilling in the northwestern Pacific are considered to be sills or lavas that are not necessarily indicative of real oceanic crust. Similar conclusions have also been reached by several authors (Kamen-Kaye, 1970; Meyerhoff, 1974; Meyerhoff and Meyerhoff, 1974a).

DISCUSSION

As summarized above on the basis of paleogeographic, seismic stratigraphic, and dredging data, there is strong evidence for the presence of continental rocks that formed paleoland (the Great Oyashio Paleoland, Choi et al., 1990) during Paleozoic and Mesozoic times under the deep abyssal plains of the present northwestern Pacific. These conclusions argue strongly against the application of oversimplified plate-tectonics models in the study area, and suggest that a reevaluation of the whole plate-tectonics hypothesis is necessary.

The standard oceanic crust has been classified into Layer 1 (2.0 km/sec), Layer 2 (5.12 km/sec), and Layer 3 (6.7 km/sec) in descending order. Layer 1 is inferred to be sedimentary, Layer 2 alternate sedimentary and basaltic, and Layer 3 gabbroic (Worzel, 1974; Christensen and Salisbury, 1975; Garkalenko and Ushakov, 1980; Peive and Pushcharovskii, 1982). This classification is basically applicable to the study area: Layers Ia–Ic (Archean to Lower Proterozoic) under the Japan Trench correspond to Layer 3 in standard sections; Layers Id (Upper Proterozoic) and II (Cambro-Ordovician) to Layer 2; and Layers V and VI (Jurassic to Quaternary) to Layer 1.

The continental origin of the present oceanic crust has been advocated by many authors. Beloussov (1979, 1981, 1989) and Beloussov and Ruditch (1961) regarded the present oceanic crust as a product of basification of continental crust. The combination of the modified version of basification by Beloussov and surge tectonics by Meyerhoff et al. (1989) may explain the foundering of the paleolands and subsequent tectonic devel-opment of the present-day oceans, although their detailed processes need to be elaborated. On the other hand, Peive and Markov (1973), after examination of island-arc basement rocks in the western Pacific, concluded that the ultramafic-gabbro-amphibolite association could correlate with the present oceanic crust.

Possible presence of continental crust under the ocean has been postulated by Bullin (1980) and Orlenok (1983). They stated the idea that "the oceanic crust is thin and graniteless" is a mistake. The mistake is due to (1) lack of detailed seismic research in the ocean; and (2) the difficulty of distinguishing the boundary between the upper mantle and lower oceanic crust in areas where the high- and low-velocity layers alternate. This view was shared by Shilo and Tuezov (1985), who reviewed geological and geophysical evidence in the Pacific-Asia transition zone (Fig. 11).

The presence of Proterozoic rocks under the world ocean floor has been strongly advocated by Meyerhoff (1974), Meyerhoff and Meyerhoff (1972, 1974a, 1974b) and Meyerhoff et al. (1972). This series of publications contains an extensive review of evidence that led these workers to argue against the alleged young age (Jurassic to Recent) and basaltic composition of the oceanic basement, which the plate-tectonics hypothesis demands. On the basis of the sedimentation rate of the deep-ocean sediments, as well as dredging results supported by paleontological and radiometric evidence, they concluded that the oceanic basement must include widespread Proterozoic rocks. Meyerhoff and his associates' conclusions are supported by our studies in the East Asia and the northwestern Pacific.

Further support comes from the super-deep drill hole in the Kola Peninsula, USSR (Kozlovsky, 1984). This hole reached a depth of 12 km in 1984, and the drilling

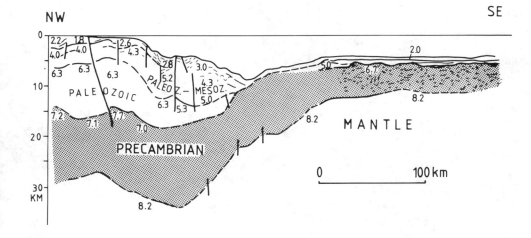

Figure 11. Interpreted geologic profile of the crust across the Kuril-Kamchatka Trench (C-C' in Fig. 1). Modified from Vasil'yev et al. (1983) and Shilo and Tuezov (1985). Note the extension of the oceanic crust (dotted area) under the island arc without subduction.

is still in progress toward the target depth of 15 km. The results have shown that the postulated basaltic layer beneath the granitic layer appears to be nonexistent. Fractured Archean silicic metamorphic rocks filled with mineral-rich fluids were intersected instead. Although the drilling has penetrated an Archean Shield, the implication from the results to date is that caution should be adopted when considering the composition of unpenetrated oceanic crust.

The presence of continental crust under the northwestern Pacific is harmoneous with its existence under the Sea of Japan (Choi, 1984b), which was a reasonable conclusion from analyses of regional geology and deep crustal structures of the surrounding countries. The timing of the deepening or foundering of the Japan Sea, which started in the Jurassic after long-lasting subaerial exposure during the Triassic (Choi, 1984b) also coincides with that of the northwestern Pacific.

Another aspect to emerge from our new interpretation concerns the age of the present ocean. The evidence presented in this and previous studies (Choi, 1987; Choi et al., 1990) suggests that (1) a broad area of the present northwestern Pacific was land that contributed sediments to the Tethys Sea during Paleozoic to Early Mesozoic times, and (2) this area first became deep in Jurassic to Early Cretaceous time (Choi et al., 1990). This conclusion concurs with that of Hoshino (1976), Timofeyev et al. (1983), Timofeyev and Kholodov (1984), Orlenok (1986), and Ruditch (1990) who found that the present deep ocean was formed in Mesozoic to Cenozoic times.

The presence of continental crust in the northwestern Pacific casts doubt over the validity of the use of magnetic anomalies for determination of spreading age and rate. The magnetic anomalies in the study area have been studied by Larson and Chase (1972). These anomalies are located within the area of continental crust (Fig. 12). They appear to coincide with the major fracture patterns accompanied with intrusives, or possibly with structural trends developed in Precambrian or Paleozoic or both continental crust under the ocean. The coincidence between magnetic anomalies and Precambrian structures has been pointed out by Meyerhoff et al. (1972) and Meyerhoff and Meyerhoff (1974b) on the basis of the concentric pattern of magnetic anomalies around Archean shields of the world. Therefore, a fundamental reevaluation of the magnetic anomaly interpretation from that currently operative is necessary (see also Pratch, 1986; Wezel, 1988).

Finally, this study focuses on the validity of the assumed subduction and accretion of Pacific plates. The observations (summarized above) across the Wadachi-Benioff zone, and the crustal structure across the Japan Trench, indicate that neither accretion nor subduction of the Pacific plates has taken place along the present Japan Trench. The Wadachi-Benioff zone is considered to coincide with a deep-rooted thrust zone, or boundary of density contrast in the upper mantle formed in relation to the movement of the asthenosphere but not the plane of subduction. This view has been repeatedly discussed by Tarakanov and Leviy (1968), Meyerhoff et al. (1972), Sychev (1973), Krebs (1975), Carey (1976),

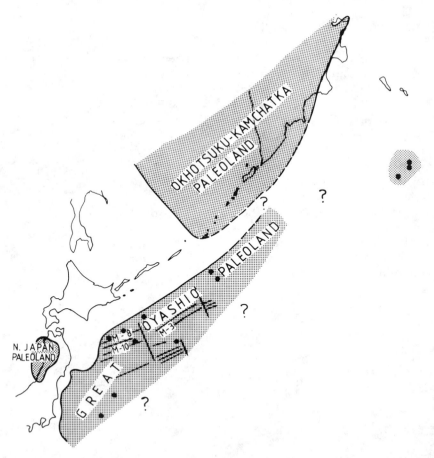

Figure 12. Distribution of paleolands and seas during the Late Permian. North Honshu area from Choi (1987), and Okhotsuk-Kamchatka area from Likharev et al. (1966). Dredge sites of continental crust and magnetic anomalies and their ages identified by Larson and Chase (1972) are superimposed.

Meyerhoff and Meyerhoff (1977), Shiki and Misawa (1982), Ciric (1983), Shilo and Tuezov (1985), Beloussov (1986), Pratsch (1986), Wezel (1986), Rodnikov (1987), Meyerhoff et al. (1989), and Krayushkin (1990).

REFERENCES CITED

Adachi, M., 1971, Permian intraformational conglomerate at Kamiaso, Gifu Prefecture, central Japan: Journal of Geological Society Japan, v. 77, p. 471-282.

Beloussov, V.V., 1960, Development of the earth and its tectonogenesis: Sovetskaya Geologiya, no. 7, p. 3-27.

———, 1979, Why do I not accept plate tectonics?: EOS, v. 60, p. 207-210.

———, 1981, Some problems of structure and development conditions in continent-ocean transition zones: Geotectonics, v. 15, p. 201-213.

———, 1986, Structure and evolution of transitional zones between continents and ocean, in Beloussov, V.V., Artemiev, M.E., and Rodnikov, A.G., eds., Structure and dynamics of transition zones from the continent to the ocean; Results of researches on the international geophysical projects: Academy of Sciences, USSR, Moscow (in Russian with English abstact).

———, 1989, Fundamentals of Geotectonics: Moscow, NEDRA, 380p. (in Russian)

Beloussov, V.V., and Ruditch, E.M., 1961, Island arcs in the development of the earth's structure (especially in the regions of Japan and the Sea of Okhotsk): Journal of Geology, v. 69, p. 647-658.

Bullin, W.K., 1980, On the grounds of a new seismic model of the earth's oceans' crust, in Abyssal structure of lithosphere of the Far East region, p. 87-101, Vladivostok, Far East Sci. Centre, Acad. Sci. USSR.

Carey, S.W., 1976, The expanding earth: Development in Geotectonics: Elesevier Sci. Pub. Co. 488 pp.

Choi, D.R., 1972, Discovery of Uralian fusulinids from the Upper Permian conglomerates in the southern Kitakami Mountains, Japan: Hokkaido University Faculty of Science Journal, ser. IV, Geology and Mineralogy, v. 15, p. 479-492.

———, 1976, Distribution of the Upper Permian fusulinids with relation to limestone lithofacies in the southern Kitakami mountains, N.E. Japan: Journal of Geological Society of Japan, v. 82, p. 113-125.

———, 1984a, The Japan Basin—a tectonic trough: Journal of Petroleum Geology, v. 7, p. 437-450.

———, 1984b, Late Permian-Early Triassic paleogeography of northern Japan: Did microplates accrete to Japan?: Geology, v. 12, p. 728-731.

———, 1987, Continental crust under the northwestern Pacific: Journ. Petrol Geology, v. 10, p. 425-440.

———, 1990, Plate subduction in the Aleutian Trench questioned: a new interpretation of seismic profiles: Tikhookenskaya Geologiya (Pacific Geology) no. 5, p. 23-33.

Choi, D.R., Vasil'yev, B.I., and Tuezov, I.K., 1990, The Great Oyashio Paleoland: A Paleozoic-Mesozoic landmass in the northwestern Pacific, in Critical aspects of the plate tectonics theory, v. 1 (Criticism on the plate tectonics theory): Theophrastus Publications, S.A., Athens, Greece, p.197-213.

Christensen, N.J., and Salisbury, M.N., 1975, Structure and constitution of the lower oceanic crust: Review of Geophysics and Space Physics, v. 13, p. 728-731.

Ciric, B.M., 1983, Is subduction a real phenomenon? in Carey, S.W., ed., Expanding Earth Symposium: Sydney, 1981, p. 247- 257.

Creager, J.S., Scholl, D.W., Boyce, R.E., Echols, R.J.,Fullam, T.J., Grow, J.A., Koizumi, I., Lee, H.J., Ling, H.Y., Stewart, R.J., Supko, P.R., and Worsley, T.R., 1973, Initial Reports of DSDP: v. 19, Washington (U.S. Government Printing Office).

Dewey, J.F., 1980, Episodicity, sequence, and style at convergent plate boundaries,in Strangway, D.W., ed., The continental crust and its mineral deposits: p. 553-573. Geological Association of Canada Special Paper 20.

Garkalenko, I.A., and Ushakov, S.A., 1980, The earth's crustin the Kuril region: Sovetskaya Geologiya, no. 11, p. 46-69 (in Russian).

Gnibidenko, H.S., Krasny, M.L., and Popov, A.A., 1978, Tectonics of the Kuril-Kamchatka deep-sea trench: EOS (Transactions, Amer. Geophys. Union), v. 59, no. 12, p. 1184.

Grow, J.A., 1972, A geophysical study of the central Aleutian arc [Ph.D. dissert.]: Scripps Institution of Oceanography, University of California-San Diego, California, 132p.

Harata, T., and Tokuoka, T., 1978, A consideration on the Paleogene paleogeography in southwestern Japan, in Fujita, K. et al., eds., Cenozoic geology of Japan: Professor Nobuo Ikebe memorial volume, p. 1-12.

Harata, T., Hisatomi, K., Kumon, F., Nakazawa, K., Tateishi, M., Suzuki, H., and Tokuoka, T., 1978, Shimanto geosyncline and Kuroshio paleolands: Journal of Physics of the Earth, v. 26, suppliment, p. 357-366.

Heath, G.R., Burckle, L.H., D'Agostino, A.E., Bleil, U., Horai, K., Jacobi, R.D., Janecek, T.R., Koizumi, I.,Krissek, L.A., Monechi, S., Lenotre, N., Morley, J.J.,Shultheiss, P., and Wright, A.A., 1985, Initial Reports of the DSDP: v. 86, Washington (U.S. Government Printing Office).

Heezen, B.C., MacGregor, I.D., Foreman, H.P., Forristall, G., Hekel, H., Hesse, R., Hoskins, R.H., Jones, J.W., Kaneps, A.G., Krasheninnikov, V.A., Okada, H., and Ruef, M.H., 1973, Initial Reports of the DSDP: v. 20, Washington (U.S. Government Printing Office).

Hoshino, M., 1975, Eustacy in relation to orogenic stage: Tokai University Press, Tokyo, 397p.

Kamen-Kaye, M., 1970, Age of the basins: Geotimes, v. 115, p. 6-8.

Kano, H., 1971, Studies on the Usuginu conglomerates in theKitakami Mountains: Journal of Geological Society of Japan, v. 77, p. 415-440 (in Japanese with English abstract).

———, 1974, Review of the basement geology of the Japanese Islands from the standpoint of conglomerate: Kaiyo Kagaku (Marine Science Monthly), v. 6, p. 617-622 (in Japanese with English abstract).

Konishi, K., Ishibashi, T., and Tsuruyama, K., 1973, Find of nummulites and orthoquartzite pebbles from the Eoceneturbidites in Shimajiri Belt, Okinawa: Kanazawa University Science Reports, v. 18, p. 43-s53.

Kozlovsky, V.A., 1984, The world's deepest well: Scientific American, v. 251, p. 106-112.

Krayushkin, V.A., 1990, About imporant geological events incompatible with some fundamentals of plate tectonics, in Critical aspects of the plate tectonics theory, v. 1 (Criticism on the plate tectonics theory): Theophrastus Publications, S.A., Athens, Greece, p. 49-71.

Krebs, W., 1975, Formation of southwest Pacific island arc-trench and mountain systems: Plate or global-vertical tectonics?: American Association of Petroleum Geologists Bulletin, v. 59, p. 1639-1666.

Larson, R.L., and Chase, C.G., 1972, Late Mesozoic evolution of the western Pacific Ocean: Geological Society of America Bulletin, v. 83, p. 3627-3644.

Larson, R.L., Moberly et al., 1975, Initial Reports of the DSDP. v. 32, Washington (U.S. Government Printing Office).

Likharev, B.K., Miklukho-Maklai, A.D., Miklukho-Maklai, K.V., Stepanov, D.L., Forsh, N.N., and Schvedov, N.A., 1966, Permian system: Stratigraphy of USSR, v. 14, 536 pp. (in Russian).

Ludwig, W.J., Ewing, J.I., Ewing, M., Murauchi, S., Den, N., Asano, S., Hotta, H., Hayasaka, M., Asanuma, T., Ichikawa, K., and Noguchi, I., 1966, Sediments and structure of the Japan trench: Journal of Geophysical Research, v. 71, p. 2121-2137.

McCarthy, J., and Scholl, D.W., 1985, Mechanism of subduction accretion along the central Aleutian Trench: Geological Society of America Bulletin, v. 96, p. 691-701.

Meyerhoff, A.A., 1974, Crustal structure of northern Atlantic ocean—review, in Kahle, C.F., ed., Plate tectonics—assessments and reassessments: American Association of Petroleum Geologists Memoir 23, p. 43-145.

Meyerhoff, A.A., and Meyerhoff, H.A., 1972, The new global tectonics: age of linear magnetic anomalies of ocean basins: American Association of Petroleum Geologists Bulletin, v. 56, p. 337-359.

———, 1974a, Tests of plate tectonics, in Kahle, C.F., ed., Plate tectonics—assessment and reassessment: American Association of Petroleum Geologists Memoir 23, p. 43-145.

———, 1974b, Ocean magnetic anomalies and their relation to continent, in Kahle, C.F. ed., Plate tectonics—assessments and reassessments: American Association of Petroleum Geologists Memoir 23, p. 411-422.

Meyerhoff, A.A., Meyerhoff, H.A., and Briggs, R.S. Jr., 1972, Continental drift, V: Proposed hypothesis of earth tectonics: Journal of Geology, v. 80, p. 663-692.

Meyerhoff, A.A., Taner, I., Morris, A.E.L., and Martin, B.D., 1989, Surge tectonics (abstract): "New Concepts in global tectonics," Washington, D.C., p. 25-26.

Meyerhoff, H.A., and Meyerhoff, A.A., 1977, Genesis of island arcs, in International Symposium on Geodynamics in south-west Pacific: Noumea, p. 357-370.

Minato, M., Gorai, M., and Hunahashi, M., eds., 1965, The geologic development of the Japanese Islands: Tsukiji Shokan, Tokyo, 442p.

Minato, M., Hunahashi, M., Watanabe, J., and Kato, M., eds., 1979, The Abean orogeny: Tokai University Press, Tokyo, 427p.

Nasu, N., von Huene, R., Ishiwada, Y, Langseth, M., Bruns, T., and Honza, E., 1980, Interpretation of multichannel reflection data, Legs 56 and 57, Japan Trench transect, in Initial Reports of the DSDP, vol. 56 and 57, Washington, D.C. (U.S. Government Printing Office).

Okami, K., 1973, The Sarukubo conglomerate: Journal of Geological Society of Japan, v. 70, p. 145-156.

Okami, K., and Kano, H., 1983, The provenance of the orthoquartzite pebbles which occurred in the eastern terrain of the Abukuma Plateau, NE Japan: Geological Society of Japan Memoir no. 21, p. 231-243.

Orlenok, V.V., 1983, Paleogeography of the world ocean in the Late Phanerozoic: Tikhookeanskaya Geologiya (Pacific Geology), v. 4, p. 88-100 (in Russian).

———, 1986, The evolution of ocean basins during Cenozoic time: Journal of Petroleum Geology, v. 9, p. 207-216.

Peive, A.V., and Pushcharovskii, Y.M., 1982, Oceanic geology, in Exploration and problems (Science in the USSR): p.40-47, NAUKA, Moscow.

Pratch, J.C., 1986, Petroleum geologist's view of oceanic crust age: Oil and Gas Journal, July 14, p. 114-116.

Rodnikov, A.G., 1988, Correlation between the asthenosphere and the structure of the earth's crust in active margins of the Pacific Ocean: Tectonophysics, v. 146, p. 279-289.

Ruditch, E.A., 1990, The world ocean without spreading, in Critical aspects of the plate tectonics theory, v. 1 (Criticism on the plate tectonics theory): Theophrastus Publications, S.A., Athens, Greece, p. 343-395.

Schevaldin, Y.V., 1974a, Heat flow and some problems of tectonics in Japan sea region, in Problems of geology and geophysics of marginal sea in the northwestern part of the Pacific Ocean: p. 162-167, Vladivostok (in Russian).

———, 1974b, Magnetic study in central Japan Sea, in Problems of Geology and geophysics of marginal sea in the western part of the Pacific Ocean: p. 168-174, Vladivostok (in Russian).

Scholl, D.W., and Marlow, M., 1974, Global tectonics and the sediments of modern and ancient trenches: some different interpretation: American Association of Petroleum Geologists Memoir 23, p. 255-271.

Sergeev, K.F., and Krasny, M. L., 1984, New data of structure of marginal oceanic Hokkaido Rise: Tikhookeyanskaya Geologiya, (Pacific Geology) no. 3, p. 100-103 (in Russian).

Sergeev, K.F., Argentov, V.V., and Bikkenina, S.K., 1983, Earth's crust seismic model of the southern Okhotsk Sea region and some results of its geological interpretation: Tikhookeanskaya Geologiya (Pacific Geology), no. 6, p. 3-12 (in Russian).

Sergeev, K.F., Krasny, M.L., Neverov, Y.L., and Ostapenko, V.F., 1983, Substance of crystalline basement of the Zenkevich rampart southeast flank: Tikhookeyanskaya Geologiya (Pacific Geology), no. 2, p. 3-8 (in Russian).

Shibata, K., 1979, Geochronology of pre-Silurian rocks in the Japanese Islands, with special reference to age detremination of orthoquartzite clasts, in The basement of the Japanese Islands: Prof. Hiroshi Kano Memorial volume, p. 625-639. Akita Univ., Japan.

Shiki, T., and Misawa, Y., 1982, Forearc geological structure of the Japanese Islands, in Legett, J.K, ed., Trench forearc geology: Sedimentation and tectonics in modern and ancient active plate margins: Blackwell Scientific Pub., Oxford, p. 63-73.

Shilo, N.A., and Tuezov, I.K., 1985, Tectonics and geological nature of Asian-Pacific transition zone: Tikhookeanskaya Geologiya (Pacific Geology), no. 3, p. 3-15 (in Russian).

Smirnov, A.M., 1982, "Continental" rocks of the Pacific: Tikhookenanskaya Gologiya (Pacific Geology), no. 4 (in Russian).

Sychev, P.M., 1973, Upper mantle structure and nature of deep processes in island arcs and trench systems: Tectonophysics, v. 19, p. 343-359.

Tarakanov, R.Z., and Leviy, N.V., 1968, A model for the upper mantle with several channels of low velocity and strength, in Knopoff, L., Drake, L., and Hart, P.J., eds., The crust and upper mantle of the Pacific area: American Geophysical Union Monograph, no. 12, p. 43-50.

Timofeyev, P.P., and Kholodov, V.N., 1984, The problem of existense of oceans in geologic history: Doklady Academii Nauk, USSR, v. 276, p. 689-692.

Timofeyev, P.P., and Kholodov, V.N., and Khvorova, I.V., 1983, Evolution of sedimentation processes on continents and in the oceans: Lithology and Mineral Resources, no. 5, p. 3-23 (in Russian).

Tokuoka, T., 1967, The shimanto terrain in the Kii Peninsula, southwest Japan—with special reference to its geologic development viewed from coarser clastic sediments: Memoir of Faculty of Science, Kyoto University, Series Geology & Mineralogy, v. 34, p. 35-74.

Tokuoka, T., and Okami, K., 1982, Orthoquatzite rocks as Precambrian basements of the Japanese Islands: Geological Society of Japan Memoir, no. 21, p. 283-295.

Tuezov, I.K., ed., 1978, The basic feature of geological structure of the Japan Sea floor: NAUKA, 246 pp., Moscow (in Russian).

Uyeda, S., 1982, Subduction zones: an introduction to comparative subductology: Tectonophysics, v. 81, p. 133-159.

Vasil'yev, B.I., 1982, Preliminary data on the dredged results of the Obruchev submarine rise (the Pacific Ocean): Tikhookeanskaya Geologiya (Pacific Geology), no. 5, p. 96-99 (in Russian).

——, 1986, The results of dredging of some submarine mountains in Japan marginal oceanic rampart: Tikhookeyanskaya Geologiya (Pacific Geology), no. 5, p. 35-42 (in Russian).

Vasil'yev, B.I., and Evlanov, Y.B., 1982, Geologic structure of submarine mountains in the region near Kuril-Kamchatka and Japan Trenches: Tikhookeanskaya Geologiya (Pacific Geology), no. 4, p. 37-44 (in Russian).

Vasil'yev, B.I., Tararin, I.A., Govarov, I.N., and Konovalov, Yu.I., 1986a, New data on the structure of the Kuril-Kamchatka Trench: Tikhookeanskaya Geologiya (Pacific Geology), no. 3, p. 64-73 (in Russian).

——, 1986b, New data on geological structure of Zenkevitch's Rampart: Tikhookeanskaya Geologiya (Pacific Geology), no. 4, p. 99-103 (in Russian).

von Huene, R., Langseth, M., Nasu, N., and Okada, H., 1980, Summary, Japan Trench transect, in Initial Reports of the DSDP: vol. 56 and 57, Washington, D.C. (U.S. Government Printing Office), p. 473-488.

von Huene, R., and Shor, G.G. Jr., 1969, The structure and tectonic history of the eastern Aleutian trench: Geological Society of America Bulletin, v. 80, p. 1889-1902.

Wezel, F.-C., 1986, The Pacific Island Arcs, in Wezel, F.-C., ed., the origin of Arcs: Development in Geotectonics 21, p. 529-567. Elsevier.

——, 1988, A young Jura-type fold belt within the central Indian Ocean. Bollettino Oceanologia ed Applicata, v. 6, p. 75-90.

Worzel, J.L., 1974, Standard oceanic and continental structure, in Burk, C.A., and Drake, C.L., eds., The geology of continental margins: p. 59-66, Springer-Verlag, Berlin.

Past distribution of oceans and continents

J. M. Dickins, Bureau of Mineral Resources, G.P.O.Box 378, Canberra, A.C.T., 2601, Australia

D. R. Choi, 6 Mann Place, Higgins, A.C.T., 2615, Australia

A. N. Yeates, Bureau of Mineral, Resources, G.P.O.Box 378, Canberra, A.C.T., 2601, Australia

ABSTRACT

This paper examines evidence that indicates that considerable areas of the present oceans were formerly land or relatively shallow oceans, or both, before the Neogene and especially before the mid-Cretaceous. The evidence comprises ocean-floor sampling and drilling, seismic data, information from paleocurrents and provenance of sediments, and the environment of ocean-bed floras and faunas. An alternative explanation of what has purported to have been subduction, is indicated.

Our conclusion, that substantial areas of the present oceans were once land and areas with shallow water, calls for a reassessment of current theories on the development of the Earth. It also suggests that present ocean depths are a relatively recently developed feature. During the Palaeozoic and at least the Early Mesozoic, seas would have been largely epicontinental, with oceanic depths (in the modern sense) being apparently restricted to geosynclines.

INTRODUCTION

From earliest geology, the oceans and lands basically as they are now, generally have been presumed to have existed from the time there was sufficient water to form seas. This has been ingrained in geological thinking and incorporated into theories such as the formation of continents by accretion from the oceans, and in hypotheses such as plate tectonics. This is manifest, for example, in the idea that present continental margins were also margins in the past, outside of which was the ocean. In reconstructions the continents generally are thought to have moved about as a whole or in pieces that together make up the present continents.

Despite the covering of large parts of the continents from time to time by seas, and the formation of deep sea in the geosynclines within continents, the evidence from geosynclines has been taken on faith or even manipulated to indicate that geosynclines are formed at the margins of plates at the boundaries between continents and oceans.

The evidence summarized here suggests that, until the beginning of the Jurassic and perhaps even later, the present oceans had considerable land area and the oceans actually may have been represented in large part by epicontinental seas and the geosynclines within the present continents.

One of us has published extensively on the presence of land and shallow water in the northwestern Pacific before the Neogene (Choi, 1984, 1987, 1990; Choi et al., 1990; see also Choi, this volume). Along with this, we draw attention to data from other areas, and conclude that the observations collectively can no longer be ignored.

GEOPHYSICAL EVIDENCE

Oceanic plateaus are high parts of the sea floor that show from seismic information crustal characteristics not consistent with oceanic crust. Ben-Avraham et al. (1981) show the distribution of more than 100 oceanic plateaus (Fig. 1). They are distributed in all oceans and in some places, such as the western Pacific, make up a significant part of the area of the present sea floor. The oceanic plateaus share common characteristics of morphology and may rise above sea level. They have crustal thicknesses of 20 to 40 km; seismic compressional wave velocities of 6.0 to 6.3 km per second, more characteristic of continental rocks than oceanic basaltic crust; some contain granitic rocks or other features considered indicative of continental crust, as for example the Seychelles, and they may be topped by rocks in which fauna, flora, or both indicate shallow-water deposition. Details of some of these features are discussed later in this paper.

Ben-Avraham et al. (1981) concluded that although some are volcanic, many represent continental fragments. They considered that the plateaus could be the sources for some allochthonous terranes. Their information and interpretation, however, seem to have been little recognized by advocates of plate tectonics, presumably because of the embarrassing implications.

The presence of continental-type crust in the oceans where oceanic crust might be expected has been recognized from seismic information by a number of authors. In some cases this identification has followed from mapping of land areas. In seismic profile, Choi (1987) traced Precambrian continental crust, consisting partly of Proterozoic orthoquartzite, from Japan under the Japan Trench and Nankai Trough. Choi (1990) applied his interpretation from the Japan Trench and Nankai Trough to the recognition of continental crust in the Aleutian Trench. In the landward trench slopes of these areas, sediments show a strong landward progradation, indicating the presence of provenance in the present northwestern Pacific during Paleozoic to Mesozoic.

Chatterjee, S., and N. Hotton III, eds. *New Concepts in Global Tectonics.* Texas Tech University Press, Lubbock, 1992, xii + 450 pp.

Figure 1. World oceans showing oceanic plateaus (shaded) including features mentioned in text (from Ben-Avraham et al., 1981).

Choi et al. (1990) called this paleoland "the Great Oyashio Paleoland." They regard the Wadachi-Benioff Zone not as a subduction zone but a zone of density change along which stress adjustment is focused (see also Choi, 1987), thus offering an alternative explanation of this feature.

On the basis of geophysical and drilling information, Udintsev and Koreneva (1982) postulated that the Broken and Ninetyeast ridges of the Indian Ocean were part of a Tertiary continent, Lemuria, before sinking to their present depth. Wezel (1988) recognized the probability of Alpine-Himalayan-type folding associated with a foundered continental basement, the Lemurian foreland basement, in the northwestern Indian Ocean (Fig 2). This folding has much in common with the Cretaceous geosyncline postulated for the present Obruchev Rise in the northwestern Pacific by Vasil'yev (1982) (see also Choi et al., 1990).

SEDIMENTARY FEATURES

Much data has been accumulating on the shallow nature of the present oceans during the Palaeozoic and Mesozoic. Timofeev et al. (1983) reviewed data from the Deep Sea Drilling Project indicating that before the Jurassic, sediments from all the oceans indicate shallow water. Orlenok (1986) found that in 149 of the first 493 *Glomar Challenger* boreholes, shallow-water deposits, or igneous rocks with subaerial weathering profiles, were present. From his examination of the drilling materials he concluded that the oceans progressively deepened from the Jurassic onwards. Furthermore, Orlenek considered that the amount of water has increased rapidly since the Jurassic, and that the view that there are two types of crust, continental and oceanic, had no factual basis.

Kaz'min et al. (1987) reported basaltic volcanic rocks regarded as Lower Cretaceous, with shallow water and subaerial erosion features from the Magellan Seamounts in the central Pacific, now at depths of 1,400 m since post-Early Cretaceous sinking.

Much information is available from continents on the source of sediments from present oceanic areas, including from the areas of the present deep trenches. Eardley (1947) concluded that the Permian trough west of the present Rocky Mountains from California to Alaska contained volcanic material, mostly andesitic, which had come from the west, and that a volcanic archipelago flanked the Permian trough on the west where the Pacific Ocean is now found (Fig. 3). Miller (1970), on the basis of structural trends of pre-Mesozoic orogens, concluded a former sialic (continental) crust, which has now disappeared was present west of the present coast of Chile. Clemmy et al. (1983) described a Lower Cretaceous prox-imal molasse in the northern Andes derived from a south-west andesitic and granodioritic (sialic) source. Presumably the pre-Cambrian outcropping along the western side of South America in the Coast Range and pre-Cordillera was part of this sialic crustal region. Isaacson (1975) described a land source west of the Central Andes in the Devonian based on the thickening and coarsening of the sediments (Fig. 4).

Kamp (1980) concluded that sediments of the Canterbury Suite east of the Alpine Fault in New Zealand were derived from a sialic source in the Pacific. He suggested also that similar quartzo-feldspathic sediments in California were derived from a similar source.

Banks and Clarke (1987) indicated a source for ice west of Tasmania in the late Carboniferous or early Permian and both a western and eastern source for sediments in the Permian from present ocean areas. The isopachs indicate the Permian structures continue into existing oceanic areas. The studies of one us (J.M.D) on the Permian indicate that the counterparts of these structures, faunas, and sequences are not found in any existing land areas.

Mollan et al. (1970) reported a thick Upper Triassic clastic sequence in Ashmore Reef No. 1 Bore on the northwest margin of the northwest Australian Shelf, facing Sumba and Timor. They suggested a nonmarine and paralic depositional environment for this sequence and showed it shedding from a northwest landmass in the direction of the present Java Trench and Timor Trough.

Stauffer and Mantajit (1981) suggested a western Indian Ocean source for the Late Palaeozoic pebbly mudstone found in eastern Burma and peninsular Thailand and Malaya.

Choi (1984, 1987) and Choi et al. (1990) tabulated the evidence for an eastern sialic source, including the deep region of the present Japan Trench, for sediments in the Japanese Geosyncline. These data are extensive and are not repeated here. The references are given earlier in this report. Yamakita (1988) described in detail, Jurassic and earliest Cretaceous olistrosomes and pile nappes in the Japanese geosyncline, not only from the north but also from a land area to the south now occupied by the Pacific Ocean. This explanation fits conditions of very strong tensional faulting and deep crustal foundering rather than plate collision and accretion for the Jurassic and Lower Cretaceous of Japan.

FAUNAL AND FLORAL DATA

In 1956, Hamilton reported shallow-water Albian-Cenomanian sediments on flat-topped guyots of the sunken mid-Pacific Mountains, now at depths of 1,400-1,800 m. These sediments contained shallow-water

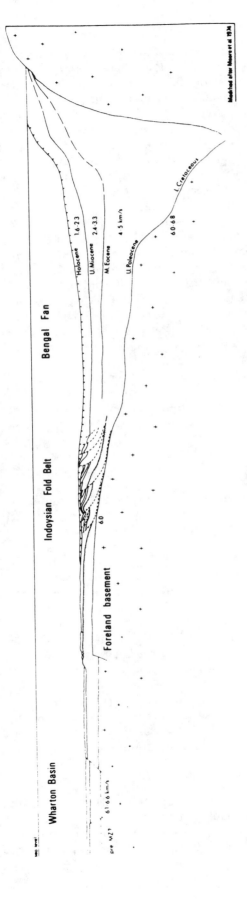

Figure 2. Diagrammatic interpretation of Alpine-Himalayan-type folding and thrust folding associated with foundered basement in the northwestern Indian Ocean. Note 6.0–6.8 p-wave seismic velocity layer (after Wezel, 1988).

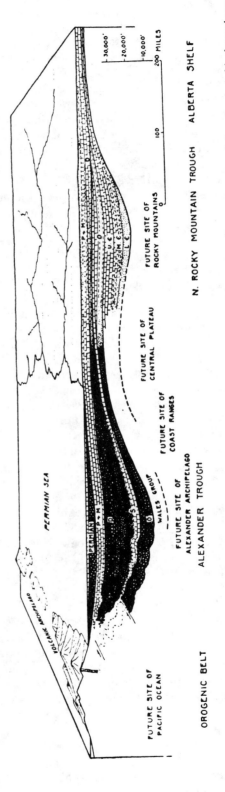

Figure 3. The Cordilleran geosyncline in southeastern Alaska and British Columbia at the close of the Permian showing the Permian sea and the volcanic archipelago where the Pacific Ocean is now found (from Eardley, 1947).

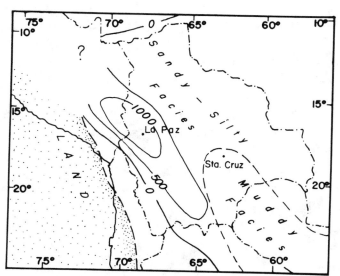

Figure 4. Postulated sialic landmass including Archaean of western South America and areas now occupied by the Pacific Ocean. Interpretation based on direction of thickening and coarsening of Devonian sediments in Bolivia (after Isaacson, 1975).

faunas such as rudistids. The significance of this discovery has been much overlooked and ignored, although subsequent information has shown that strong sinking since the Lower Cretaceous, is a widespread feature of the present oceans.

Particularly interesting and significant are data from the Indian Ocean synthesized in Wezel (1988). Drilling has shown the presence of shallow-water and subaerial deposits until at least the Oligocene. Udintsev and Koreneva (1982) argued that palynological evidence indicated an extensive ancient landmass associated with what is now the Broken and Ninetyeast ridges. Continental basement is also thought to underlie the Scott, Exmouth, and Naturaliste plateaus west of Australia (Yeates et al., 1987). Yeates et al. regarded the Westralian Superbasin, marginal to the present continent, as a tongue extending south from Tethys with land on the east and west.

East of the Mid-Atlantic Ridge, Lower to Middle Paleozoic *Trilobites* are known to occur (Schneck, 1974; Kidd et al., 1982). However, dredged sandstones and limestones containing brachiopods, trilobites, and graptolites have been ascribed to ice-rafting, and little attention has been paid to their occurrence, despite their obvious indications of local origin.

TIME OF FORMATION OF THE OCEANS

From data considered above, no evidence is apparent for deep oceans in the form of the present oceans before the Jurassic to mid-Cretaceous. The configuration of sea and land was apparently quite different from the present up until the end of the Hunter-Bowen Indosinian Orogenic Phase at the end of the Triassic (Dickins, 1988). The seas (oceans) were represented, partly, by epicontinental seas and geosynclines, which in places were deep water, and partly by nonland areas within present oceans. Conversely, present oceans contained considerable areas of land, presumably also with epicontinental seas and geosynclines. In the Upper Palaeozoic and Lower Mesozoic, for example, Tethys may have encircled the world, although this is a conjecture. Such a configuration of land and sea would have strongly affected the migration possibilities of plants and animals and had a profound effect on climate.

From the beginning of the Jurassic a large-scale foundering of the Earth commenced that formed a framework for present oceans (Beloussov, 1962), and appears to reflect some form of Earth expansion. The present ocean basins are characterized by the large-scale outpouring of basalt. The continental basins also have extensive basaltic extrusion and intrusion of extensive dolerite sheets under conditions of tension. At no time during the Phanerozoic was there such basaltic intrusive and extrusive activity as in the Jurassic and Lower Cretaceous. The tectonism of the Jurassic and Lower Cretaceous is marked by block faulting, in most if not in all, parts of the world (Dickins, 1988a; 1988b; in press). The beginning of this block faulting usually is explained by plate tectonics as the beginning of the breakup of Gondwana and/or Pangaea and similar movements that are later than the beginning of the Jurassic are also explained as breakup. These explanations generate a plethora of conclusions on the time of breakup, each based on reasoning that does not include examination of all data available, and does not account for inconsistencies among the different explanations. This period of Jurassic to Lower Cretaceous block faulting comes to an end at the end of the Lower Cretaceous with the beginning of the Alpine-Himalayan Folding Phase. During this phase, which lasts to the present, the oceans developed into their present form.

The deep trenches and other deeper parts of the oceans appear to be recent features formed only in the Neogene. This is shown particularly in the works of Carey (1976), Choi (1987), Choi et al. (1990), and Wezel (1988). Further indirect support for the presence of shallow seas during the Paleozoic-Mesozoic comes from the study on ostracoda (Mackenzie, 1987) in the Tethys; true oceanic ostracoda first appeared after the mid-Cretaceous.

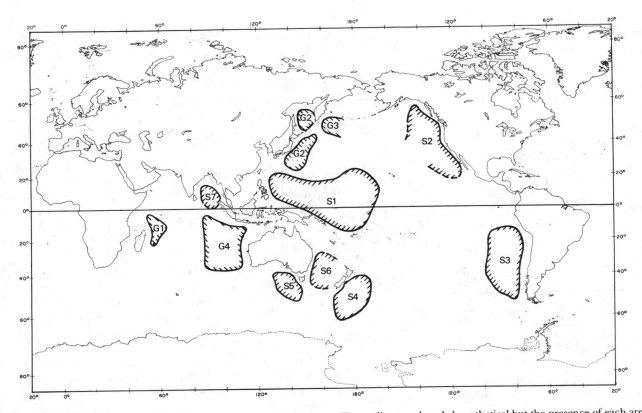

Figure 5. Previous land areas in the present oceans identified in the text. The outlines are largely hypothetical but the presence of each area is based on data regarded as reliable. Not all of these areas were necessarily present as land at any particular time. **G1**—Seychelles; **G2**—Great Oyashi Paleoland; **G3**—Obruchev Rise; **G4**—Lemuria; **S1**—area including Ontong-Java Plateau, Magellan Sea Mounts, and mid-Pacific Mountains; **S2**—Northeast Pacific; **S3**—Southwest Pacific including Chatham Rise and Campbell Plateau; **S5**—area including South Tasman Rise; **S6**—East Tasman Rise and Lord Howe Rise; **S7**—Northeast Indian Ocean.

CONCLUSIONS

Although this paper only summarizes some of the available data, what we have tabulated (Fig. 5) gives us confidence that before the Jurassic, much land and shallow water was to be found in present ocean areas. The oceans became deeper during the Mesozoic and Early Tertiary but the great depths of the modern oceans were formed only during the Neogene. Much sialic material appears to be present beneath the oceans and we remain skeptical as to the distinction between what is designated continental and oceanic crust. We are surprised and concerned for the objectivity and honesty of science that such data can be overlooked or ignored.

So far we are just at the beginning of understanding the surface of the Earth, and particularly what is under the modern oceans. What we know of the interior of the Earth is very limited. Any theory that claims to give an all-embracing explanation of the tectonic development of the Earth is likely to be premature and in danger of becoming dogma. There is a vast need for future Ocean Drilling Program intiatives to drill below the base of the basaltic ocean floor crust to confirm the real composition of what is currently designated oceanic crust.

REFERENCES CITED

Banks, M.R., 1987, Changes in the geography of the Tasmania Basin in the Late Palaeozoic, *in* McKenzie, G.D., ed., Gondwana Six: Stratigraphy, Sedimentology, and Palaeontology: American Geophysical Union, Geophysical Monograph, v. 41, p. 1-14.

Beloussov, V.V., 1962, The principal problems of geotectonics: Gosgeoltechpress, Moscow.

Ben-Avraham, Z., Nur, A., Jones, D., and Cox, A., 1981, Continental accretion: From oceanic plateaus to allochthonous terranes: Science, v. 213, p. 47-54.

Carey, S.W., 1976, The expanding Earth: Development in Geotectonics: Elsevier Scientific Publishing Company, 488 p.

Choi, D.R., 1984, Late Permian-Early Triassic paleogeography of northern Japan: Did Pacific microplate accrete to Japan?: Geology, v. 12, p. 728-731.

——, 1987, Continental crust under the northwestern Pacific Ocean: Journal of Petroleum Geology, v. 10, p. 425-440.

——, 1990, Plate subduction in the Aleutian Trench questioned: a new interpretation of seismic profiles: Tikhookeanskaya Geologiya, 1990 (5): p. 23-33 (in Russian).

Choi, D.R., Vasil'yev, B.I., and Tuezov, I.K., 1989, The Great Oyashio Paleoland: A Paleozoic-Mesozoic landmass in the northwestern Pacific, *in* Barto-Kyriakidis, A., eds., Critical aspects of the plate tectonics theory. Theophrastus Publications S.A., Athens, Greece, v. 1, p. 197-213.

Clemmy, H., Flint, S., and Turner, P., 1983, Cretaceous molasse of northern Andes. Evidence for subducted sial? (abstract): Geological Society, Newsletter, v. 12, p. 30.

Dickins, J.M., 1988a, The world significance of the Hunter/Bowen (Indosinian) mid-Permian to Triassic folding phase: Memorie Societa Geologica Italiana, v. 34, p. 345-352.
———, 1988b, Jurassic-Lower Cretaceous tectonic style: Deep structure of the Pacific Ocean and its continental surroundings: USSR Blagoveshchensk August 17-24, 1988, Abstracts, p. 13-14.
———, in press, Major sea level changes, tectonism and extinctions: Compte Rendu, 11th International Congress of Carboniferous Stratigraphy and Geology.
Eardley, A.J., 1947, Paleozoic Cordilleran geosyncline and related orogeny: Journal of Geology, v. 55, p. 309-342.
Hamilton, E.L., 1956, Sunken islands of the mid-Pacific mountains: Geological Society of America, Memoir 64, 97p.
Isaacson, P.E., 1975, Evidence for a western extracontinental land source during the Devonian Period in the Central Andes: Geological Society of America, Bulletin, v. 86, p. 39-46.
Kamp, P.J.J., 1980, Pacifica and New Zealand: proposed eastern elements in Gondwanaland's history: Nature, v. 288, p. 659-664.
Kaz'min, V.G., Matveyenkov, V.V., Raznitsin, Yu.N., Rudnik, G.B., and Skolotnev, S.G., 1987, New data on rocks of the Magellan seamounts, western Pacific: Doklady Academy of Sciences, USSR, v. 296, p. 942-946.
Kidd, R., Searle, R.C., Ramsay, A.T.S., Prichard, H., and Mitchell, J., 1982, The geology and formation of King's Trough, northeast Atlantic ocean: Marine Geology, v. 48, p. 1-30.
McKenzie, K.G., 1987, Tethys and her progeny, in McKenzie, K.G., ed., Shallow Tethys 2: A.A. Balkema Publishers, p. 501-523.
Miller, H., 1970, Das Problem des hypothetischen "Pazifischen Kontinentes" gesehen von der chilenishen Pazifikkuste: Geologische Rundschau 1970, v. 59, p. 927-938.
Mollan, R.G., Craig, R.W., and Lofting, M.J.W., 1970, Geologic frame work of Continental Shelf off Northwest Australia: American Association of Petroleum Geologists, Bulletin, v. 54, p. 583-600.
Orlenk, V.V., 1986, The evolution of ocean basins during Cenozoic time: Journal of Petroleum Geology, v. 9, p. 207-216.
Schneck, M.C., 1974, Mid-Atlantic trilobites: Geotimes, v. 19 (4), p. 16.
Stauffer, P.H., and Mantajit, N., 1981, Late Permian tilloids of Malaya, Thailand and Burma, in Hambrey, M.J., and Harland, W.B, ed., Earth's pre-Pleistocene glacial record: Cambridge University Press, p. 331-335.
Timofeev, P.P., Kholodov, V.N., and Khvorova, I.V., 1983, Evolution of sedimentation processes on continents and in the ocean: Lithology and Mineral Resources, no. 5, p. 3-23 (in Russian).
Udintsev, G.B., and Koreneva, E.V., 1982, The origin of aseismic ridges of the eastern Indian Ocean, in Scrutton, R.A. and Talwani, M., eds., The Ocean floor: John Wiley and Sons, p. 203-209.
Vasil'yev, B.I., 1982, Preliminary data on the dredged results of the Obruchev submarine rise (the Pacific Ocean): Tikhookenskaya Geologiya, no. 5, p. 96-99 (in Russian).
Wezel, F.-C., 1988, A young Jura-type fold belt within the central Indian Ocean?: Bollettino. di Oceanologia Teorica ed Applicata, v. 6, p. 75-90.
Yamakita, S., 1988, Jurassic-earliest Cretaceous allochthonous complexes related to gravitational slidings in the Chichibu Terrane in eastern and central Shikoku, Southwest Japan: Journal of the Faculty of Science, University of Tokyo, Sec. II, v. 21, p. 467-514.
Yeates, A.N., Bradshaw, M.T., Dickins, J.M., Brakel, A.T., Exon, N.F., Langford R.P., Mulholland, S.M., Totterdell, J.M., and Yeung, M., 1987, The Westralian Superbasin: an Australian link with Tethys, in McKenzie, K.G., ed., Shallow Tethys 2: A.A. Balkema Publishers, p. 199-213.

ACKNOWLEDGMENTS

Figure 1 is reproduced by permission of Z. Ben-Avraham and *Science* from Volume 213, page 48, 1981, copyright 1981 by the AAAS. Figure 2 by permission of F.-C. Wezel, Figure 3 by permission of the University of Chicago from *Journal of Geology*, Volume 55, page 342, 1947. Figure 4 by permission of P.E. Isaacson and the Geological Society of America from *Bulletin*, Volume 86, page 42, 1975.

Paleomagnetism

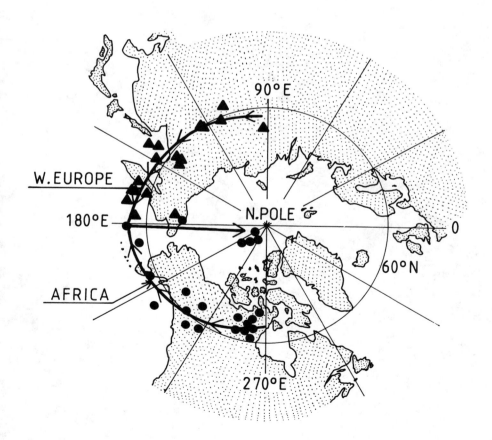

Rotating plates: new concept of global tectonics

K. M. Storetvedt, Institute of Solid Earth Physics—Geomagnetism, University of Bergen, Allégt. 70, N-5007 Bergen, Norway

> Other dominant theories have been toppled: surely this could not happen to the new global tectonics—or could it?
>
> P. J. Wyllie

ABSTRACT

Reconsideration of Meso-Cenozoic paleomagnetic data has uncovered a novel mobilistic principle of the lithosphere accountable for both oceanic and continental structural features. The new theory, which establishes a kinematic system modulated by the Earth's axial spin, consists of a hierarchal order of plates in relative intermittent rotation. The theory, which can be extended easily to pre-Mesozoic time, seems to have the necessary explanatory power required of a general global tectonic model. Sea-floor spreading is discounted as an important mechanism of oceanic crustal evolution. By removing the constraints posed by conventional plate tectonics, a large number of paradoxes accumulated in the wake of that model, disappears.

The apparent polar-wander curves seem strongly dominated by true polar wander, that is, a systematic but jerky meridional reorientation of the body of Earth relative to its axis of spin. The latter effect is related to redistribution of mass and associated changes in the Earth's moments of inertia caused by mantle diapirism and oceanization processes. The present wide distribution of thin oceanic crust is apparently a relatively recent geological phenomenon, having developed principally during Alpine (Meso-Cenozoic) time. The midoceanic ridge system is interpreted as a new class of Alpine foldbelts. Rotation of plates at various hierarchal levels provided a transpressive tectonic regime in the oceanic domain, including a transoceanic system of en-echelon shear zones (so-called transform faults) as well as a "central" belt of crustal-lithospheric thickening. The thickened crustal belt was isostatically uplifted, giving rise to the general ridge-parallel tectonotopographic grain. A wide variety of hitherto unexplained geological and geophysical observations (e.g., the pattern of orogenic belts, deep continental detachment structures, widespread intracontinental remagnetization, fundamental differences in Benioff zone characteristics, etc.) seems to be accounted for by the new global tectonic model.

INTRODUCTION

The orthodox theory of global tectonics holds that the outer shell of Earth, the lithosphere, constitutes a series of rigid plates that perform relative motions on a "plastic" asthenosphere. Slow convective currents in the mantle are thought to play a leading role in this dynamic system, which in a geological perspective, is thought to have caused significant changes of paleogeography. It is hypothesized that the world-encircling system of midoceanic ridges is the surface expression of mantle currents ascending toward the Earth's surface, giving rise to sea-floor spreading. The fact that the observed heat flow over these ridges is both highly variable and generally of an intensity much below that anticipated for crustal spreading models was explained by invoking extensive hydrothermal circulation in the crust. The thermal problem is well demonstrated in the Indian Ocean where all parts of the ocean, except the Central Indian basin of alleged within-plate setting, display a cold and featureless heat flow picture (Weissel et al., 1980; Geller et al., 1983). According to the theory of plate tectonics, rigid lithospheric plates deform at their boundaries (due to interaction with other plates) but not internally. Therefore, the fact that the Central Indian Ocean is characterized by high seismicity, intense tectonic deformation, and unexpected high heat flow (Stein and Okal, 1978; Geller et al., 1983; Neprochnov et al., 1988) is problematic in the context of conventional plate tectonics.

The trench-island arc systems of the Pacific (associated with the Benioff zones) commonly are regarded as convergent plate boundaries (i.e., regions in which the oceanic crust is consumed into the mantle). However, there is a remarkable difference in stress regime between the trenches of the Pacific: the Mariana-type structure (W Pacific) is characterized by a steeply dipping Benioff zone and low stress tensional or neutral tectonic features, whereas the Chilean-type margin (E. Pacific) displays a shallow dipping Benioff zone with high compressive stress (Uyeda and Kanamori, 1979). This basic tectonic discrepancy between the eastern and western Pacific margins has been related to differences in mechanical coupling between subducting and overriding plates (Dewey, 1980). Although explanations of this kind might be accepted to account for smaller scale differences, the contrasting characteristics of the Benioff zones on the two opposite sides of the Pacific are likely to have a much more direct and fundamental geodynamic cause (see below).

Another problem has been posed by the difficulty of finding the sheeted-dyke complex assumed to constitute a vital rock component of the oceanic crust. After two decades of deep-sea drilling a sheeted-dyke system has been found only in one drilling location (Deep Sea Drilling Project site 504, Legs 69, 70, and 83), yet numerous discoveries of more deep-seated constituents of so-called ophiolite complexes (thought to be the characteristic feature of oceanic crust formed by sea-floor spreading) like gabbros and serpentinites-peridotites, have been made at surface level. Furthermore, the assumed remanence source to account for the marine magnetic anomalies (in terms of changing geomagnetic field polarity) has remained a mystery.

Chatterjee, S., and N. Hotton III, eds. *New Concepts in Global Tectonics*. Texas Tech University Press, Lubbock, 1992, xii + 450 pp.

Most of the geological record is contained by the continents, so we must turn to them to unravel the long-term dynamic processes of the Earth. Some of the outstanding structural features are posed by the trans-Eurasian orogenic belts, which display systematically increasing ages northward, and by the circum-Pacific mountain belts that apparently have been the sites of repeated orogenic activity during Phanerozoic time. These and other prominent intracontinental features, such as the many deep detachment structures that cut into the upper mantle, are enigmatic within the context of conventional plate tectonics.

The inadequacies of the current model to explain predominant structural features of our globe, the undue complexity of geophysical processes in the wake of the model, and despite two decades of ocean drilling, an adequate verification of the theory appears more remote than ever raise the question of whether orthodox plate tectonics represents a realistic approach to "crustal" kinematics. Is it just another fallible theory?

To get out of the present deadlock it is clear that we are in need of stronger pluralism in global tectonic theorizing. To contribute to this important task this paper takes a fresh look at the global tectonic picture. It reviews Meso-Cenozoic paleomagnetic data and arrives at a mobilistic geodynamic system that is radically divergent from the model presently in vogue. The new dynamic pattern proposed tends to have great explanatory power in a wider geological perspective and accommodates a significant amount of geological evidence that has remained problematic within the current tectonic framework.

MESO-CENOZOIC PALEOMAGNETISM AND CRUSTAL KINEMATICS

Relative motion between Africa and Eurasia: Tethys and the Alpine-Himalayan orogenic belt

Reconfigurations of the Atlantic bordering continents (e.g., Bullard et al., 1965) have led to the conclusion that, prior to the assumed Meso-Cenozoic fragmentation of the ancient hypothetical landmass of Pangea, the Tethys Sea formed a major eastward-widening oceanic embayment (Fig. 1). The closure of this seaway, principally by counterclockwise rotation of Africa, provided the plate tectonic causal mechanism for the origin of the Alpine orogenic system (e.g., Dewey et al., 1973). However, a variety of paleo-climatic–geographic data and faunistic evidence are difficult to reconcile with this model (e.g., Meyerhoff and Meyerhoff, 1972; Sonnenfeld, 1981). First, the Tethyan faunal endemism in Paleozoic and Mesozoic time, (Sonnenfeld, 1981) seems incompatible with a wide oceanic Tethys. Furthermore, the Indian dinosaurs have a close affinity and generic identity with those of the northern continents (Chatterjee and Hotton, 1986), and their lack of endemism contradicts the plate tectonic conception of an independent island continent that drifted northward prior to collision with Asia in Upper Cretaceous-Lower Tertiary time (e.g., Dewey and Bird, 1970; Crawford, 1974). It is obvious that the actual configuration of the Tethys (either the traditional view of a long, narrow epicontinental sea, or the plate tectonics requirement of a wide oceanic seaway) has utmost consequences for global tectonic modeling. Figures 2 and 3 display Meso-Cenozoic paleomagnetic polar patterns that are relevant for the Tethyan issue as well as for

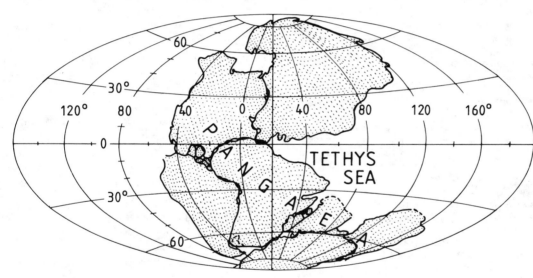

Figure 1. The commonly suggested landmass of Pangea some 200 my ago, simplified from Dietz and Holden (1970). In this popular paleogeographic model thought to have been in existence prior to the hypothesized Meso-Cenozoic continental dispersal, the Tethys Sea becomes a major eastward-widening oceanic embayment. This picture of the Tethys is, however, problematical with respect to both sedimentological, paleontological and paleoclimatological evidence.

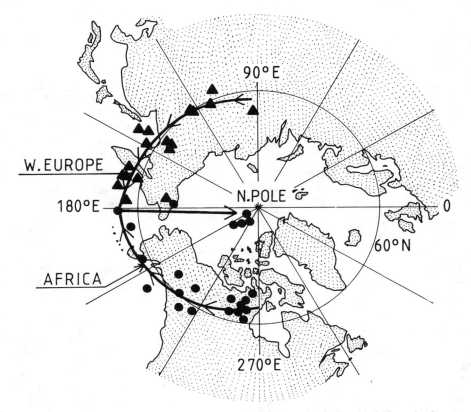

Figure 2. Meso-Cenozoic paleomagnetic poles for Africa (circles) including results from the Canary and Cape Verde islands (see Stortvedt et al., 1990a, for references to total database) in comparison with relevant data for W. Europe (triangles). The European data, which include younger (i.e., Meso-Cenozoic) igneous rocks or inferred recent overprints on older rocks, are gathered from the following sources: Girdler (1968), Halvorsen (1970, 1972), Lövlie et al. (1972), Bylund (1974), Lövlie and Kvingedal (1975), Storetvedt (1978, 1990), Storetvedt et al. (1978a, b), Storetvedt and Carmichael (1979), Thorning and Abrahamsen (1980), Lövlie and Mitchell (1982), Sturt and Torsvik (1987), Torsvik et al. (1987, 1988). The symmetrical and oppositely trending APW paths signify that the two plates have rotated in opposite directions (see also Fig. 5), their common cusp at approximately 180°E, 55°N, apparently corresponding to the termination of relative plate motion. There is evidence that most of the latitudinal APW branches are of Lower Tertiary age (Storetvedt, 1990; Storetvedt et al., 1990a). The subsequent joint near-meridional APW segment formed at around the Eocene-Oligocene boundary. See text for further discussion.

the Afro-Eurasian plate tectonic relationship. Paleomagnetic data in the wider perspective of the Alpine-Himalayan orogenic belt have been discussed in some detail elsewhere (Storetvedt, 1990), so in the present context only main overall characteristics are considered. Figure 2 depicts the relevant African branches of apparent polar wander (APW) with those inferred for West Europe (data mainly from southern France, Scotland, and Scandinavia; see Fig. 2). The credibility of the APW structure defined by West European results is enhanced by a significant number of poles from the northern flank of the Alpine chain in Central and East Europe (South Germany, South Poland, Czechoslovakia, and southern USSR) as demonstrated by Figure 3. European poles along the latitudinal APW branch east of approximately 120°E seem to be of Lower Tertiary age (Storetvedt et al., 1990a), implying that significant remagnetization along the Alpine belt, as well as along major and minor tectonic lineaments across the continent, occurred in the Lower Tertiary (Storetvedt, 1990, and see below). The essence of the paleomagnetic information presented in Figure 2 may be summarized as follows.

1. The African and European latitudinal APW paths are symmetrically arranged, forming age progressive trends that end in a common cusp at around 180°E, 55°N (i.e., the paleomagnetic poles are inferred to increase in age with increasing angular distance from the cusp).

2. In terms of plate motion the oppositely trending APW paths signify clockwise rotation of Europe and counterclockwise rotation of Africa. The cusp corresponds to the termination of this relative plate movement, which was of the order of 25 degrees per plate.

3. During the early phase of latitudinal polar progression (caused by plate rotation) there is approximately 15 degrees of polar movement away from the present geographic pole (PGP), the most distant position being reached in the Eocene (Storetvedt, 1990; Storetvedt et al., 1990a).

4. Around the Eocene-Oligocene boundary there was a rapid meridional shift of relative pole position towards the PGP, giving rise to 30–35 degrees of angular change of the paleoequatorial bulge. This latter segment of relative polar motion is interpreted as an event of true polar wander (i.e., the Earth changed its orientation in space relative to the axis of rotation).

The paleomagnetic-geographic implications summarized above (points 3 and 4) are consistent with paleoclimatic evidence. Thus, Figure 4 compares the Eocene paleolatitude system for the northern hemisphere (the pole being defined by the joint cusp of the Eurasian and African polar wander curves) with occurrences of rich fauna and flora of Lower Tertiary age (see

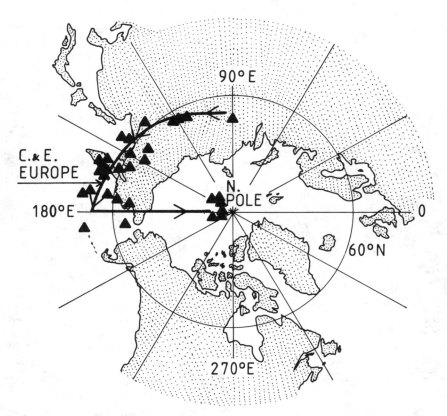

Figure 3. Paleomagnetic data from Mesozoic-Lower Tertiary rock formations from the northern flank of the Alpine orogenic belt (Central Europe-Southern European USSR), adding to the W European database depicted in Figure 2. Data are selected from Piper (1988).

the Paleocene shows a gradual temperature increase, consistent with the inferred lowering of paleolatitude at that time. The temperature reached its peak in the Eocene during which the paleolatitudes had an interim period of relatively low values, and, finally, a drastic cooling took place around the Eocene-Oligocene boundary, consistent with the major relative change of the axis of rotation at that time. The latter axial shift brought the poles of rotation into their approximate present geographic positions, which were well suited for the growing of polar ice caps. This onset of the Cenozoic ice age is also demonstrated by an abrupt change in oxygen isotope values of sea water (e.g., Poore and Matthews, 1984).

The properties of the latitudinal APW paths (see points 1 and 2 above) are vitally important for the question of Tethyan paleogeography and the lithospheric mechanism that caused the Alpine orogeny. Thus, the symmetry of the two APW curves is clearly at variance with the closing of a wide oceanic Tethys as required by the plate tectonic model, but, on the contrary, the data are supportive of the more traditional geological view of a

Fig. 4). Whereas the Siberian and Alaskan sites are dominated by abundant fossil deciduous forests, areas like Ellesmere Island, Greenland, and Svalbard were characterized by a warmer flora and fauna (including broad-leaved evergreens and reptiles). The subtropical latitudes for Central Europe inferred from paleomagnetic evidence (Fig. 4) are fully consistent with the paleoclimatic evidence (e.g., Pomerol, 1982). Furthermore, isotope paleotemperature data from the North Sea region (Buchardt, 1978) are in all respects conformable with the paleomagnetic evidence:

Figure 4. In addition to the Upper Cretaceous-Lower Tertiary plate rotations leading to the pronounced longitudinal spreads of paleomagnetic poles as envisaged by Figure 2, the Lower Tertiary experienced significant meridional changes of true polar wander (see text). In Upper Paleocene-Eocene time the relative geographic axis was located in the North Pacific, approximately 35 degrees of arc from the present axis of rotation. Diagram displays Northern Hemisphere latitudes with respect to this paleopole (approximately 180°E, 55°N) drawn on present geography. Numbers 1–7 refer to locations of rich fossil floras and vertebrate faunas in present day Arctic regions: **1**—Svalbard (Manum, 1962; Schweitzer, 1980); **2**—Disco Island (Heer, 1880); **3**—Nugssuaq (Koch, 1963); **4**—Ellesmere Island (Christie, 1964); **5**—Peel River (Rouse and Srivastava, 1972); **6**—Mackenzie Delta (Staplin, 1976); **7**—Yakutia (Tomskaya, 1976).

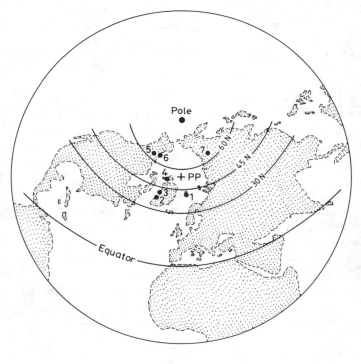

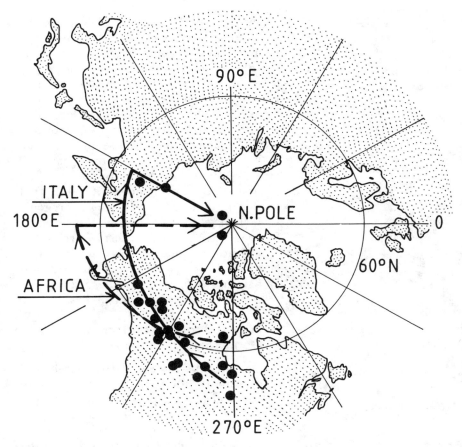

Figure 5. Paleomagnetic poles from Mesozoic-Lower Tertiary rocks of Italy (circles) with suggested eastward age progression (solid line), in association with the corresponding African APW path (broken line) as defined in Figure 2. The Italian data describe an eastward polar pattern suggestive of an African plate tectonic relationship, the observed angular divergence versus the African path being accounted for by a relative microplate rotation of approximately 10–15 degrees in the clockwise sense. The Italian data used are selected from the pole listings of Piper (1988), but the majority of available results, notably for rocks of Upper Cretaceous-Lower Tertiary age (i.e., pelagic limestones Southern Alps, Gargano limestones, Vincentinian limestones (combined result), Gargano Peninsula limestones, various data from the Scaglia Rossa limestones and the Colli Euganei volcanics, etc.), have been included.

narrow epicontinental Tethys. The symmetry and similarity in total polar spread for the two latitudinal APW paths suggest simple plate rotations consistent with the Coriolis effect (see below); the two plates display closely similar marginal velocities. This mobilistic mechanism turns the common boundary zone, the Alpine orogenic belt, into a megaband of an overall transpressive deformation. The new kinematic model accounts for the general paleomagnetic declination discrepancy between Africa and Eurasia without disrupting the observed geological continuity between the plates. The latter evidence is a cornerstone in any sensible tectonic model (e.g., Kent, 1969; Sander, 1970; Meyerhoff and Meyerhoff, 1972; Bonini et al., 1973; and many others). However, due to probable variations in the instantaneous speed of rotation for the two plates, "local" transtentional regimes as well as shear dislocations on widely differing scales would easily develop within their common boundary terrain. The model permits complex rotational behavior of micro-plates within the orogenic belt. It has been shown recently that Iberia performed a two-way tectonic rotation (counterclockwise followed by clockwise), terminating close to the K-T boundary (Storetvedt et al., 1990a). The net effect of these two rotations is approximately 30 degrees in the clockwise sense, probably providing the main impetus for the strong tectonic deformation in the Bay of Biscay. The paleomagnetic evidence suggests that the Afro-Eurasia plate boundary runs through the Pyrenees and the main Alpine chain (Storetvedt, 1990), and this observation has important implications for western Mediterranean microplate issues. For example, the African affinity of Italian paleomagnetic data is demonstrated by the eastward age progression of corresponding paleomagnetic poles (see Fig. 5), the minor angular difference between the African and Italian APW curves accounted for by 10–15 degrees of clockwise microplate rotation. The Alpine tectonic framework presented here would give rise to both compressional and extensional regimes on a varying scale, providing the physical basis for important tectonic processes such as nappe transport, and the development of ophiolites as short-lived intrusive events within the Tethyan realm. The tectonic model outlined here appears capable of accomodating all important geological features of the Alpine orogenic belt.

Figure 6 presents a schematic picture of the suggested Afro-Eurasia tectonic relationship. As will be discussed more fully these two plates are only constituents of two considerably larger plates (hemispherical megaplates) that rotated in opposite senses. During their rotation, the two megaplates suffered tectonic disruptions that, notably for the southern megaplate, led to the development of a complex system of smaller plates in relative motion. For Upper Cretaceous-Lower Tertiary time there is evidence of transpressive reactivation along

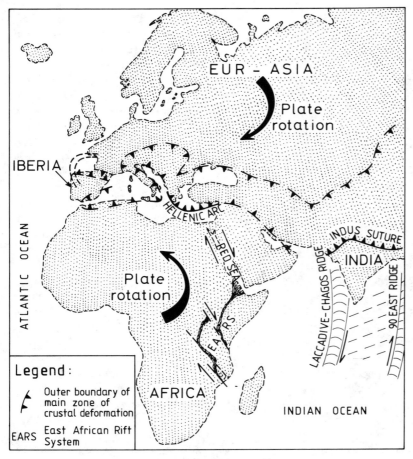

Figure 6. Schematic diagram depicting the Afro-Eurasia kinematic system in Upper Cretaceous-Lower Tertiary time as inferred from paleomagnetic data (Fig. 2). The model explains the Alpine-Himalayan orogenic belt in terms of transpressive deformation, the two plates in question displaying similar overall marginal velocities. The popular notion of the Tethys originally constituting a wide oceanic embayment is not in evidence. During rotation the Southern Megaplate (see text) broke up into a multiplicity of smaller plates creating shear-type plate boundaries such as the East Africa-Red Sea Rift System, and the Central Indian Fracture Zone (bounded by the Laccadive-Chagos and Ninetyeast ridges). See text for further discussion.

structures such as the East African-Red Sea Rift System, the Laccadive-Chagos Ridge and the Ninetyeast Ridge; the elevated topography associated with these lineaments being explained by processes of crustal-lithospheric thickening (Storetvedt, 1990). With respect to active plate boundaries, India was in a similar tectonically unstable position as Iberia, and at the K-T boundary just after the Deccan Trap volcanism the peninsula rotated approximately 135 degrees clockwise toward its present azimuthal orientation. The combined paleomagnetic, geological, and geophysical evidence are (1) consistent with a close paleogeographic relationship between India and Asia, in agreement with the paleontological inferences referred to above; and (2) add further substance to the view (Wezel, 1988; Storetvedt, 1990) that the Central Indian Ocean constitutes a huge shear zone or transpressive orogenic belt (as indeed evidenced by its high seismicity, relatively high heat flow, and structural complexity). The tectonic rotation model for India, in terms of explaining the presently anomalous orientation of the Deccan Trap paleomagnetic axis, was already discussed by Clegg et al. (1956), but at that time the concept of a Gondwana continental assembly was gaining recognition so the authors naturally chose to relate their results to that framework. In conclusion, it is suggested that the entire Alpine-Himalayan orogenic belt developed by overall transpressive processes associated with plate rotations, which imply that the preexisting Tethys Sea, along which the orogenic belt developed, must have been a long and relatively narrow epicontinental sea.

Paleomagnetism and the Gondwanaland issue

The paleomagnetic and paleontological arguments for placing India in close paleogeographic proximity to Asia, and the associated evidence for an Alpine foldbelt crossing the Indian Ocean (probably extending southward along the SW Indian Ridge) have devastating effects for the orthodox Gondwanaland concept. However, if the Indian Ocean is in fact subdivided by a sinistral shear zone as depicted in Figure 6 (with Africa to the west of this tectonic belt, and Antarctica and Australia to the east of it), it should be possible to find corroborative evidence for this proposition through adequate paleomagnetic comparison across the suggested tectonic belt. Both Antartica and Australia are surrounded by thin oceanic crust as well as by major systems of crustal dislocations. In the new global tectonic frame being developed here, these continents should be particularly vulnerable to rotational instability. Figure 7 shows that a complete match of the mean Jurassic paleomagnetic poles for the two continents can be obtained by appropriate in situ rotations (around their centroids).

The Jurassic is chosen for a comparison because that period is the only one for which there are reliable paleomagnetic data for Antarctica. The fact that the Australian and Antarctic poles can be brought into complete agreement in a position that is to the northeast of the corresponding African pole, gives further substance to the view that the Indian Ocean is transected by an

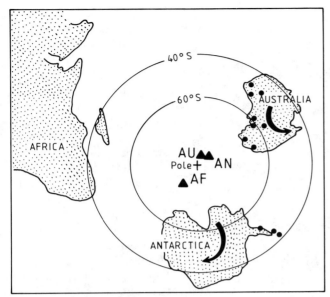

Figure 7. Mid-Mesozoic reconfiguration of the continents around the Indian Ocean. Triangles are mean poles of assumed Jurassic age relative to Africa (AF), Antarctica (AN), and Australia (AU) (Tarling, 1983). Arrows indicate sense of subsequent Alpine-age continental rotations, leading to present-day geographic orientation of Antarctica and Australia. Circles indicate occurrences of mild temperature Cretaceous-Lower Tertiary biota according to Axelrod (1984).

Alpine foldbelt of sinistral transpressive nature. Figure 7 gives a tentative first-order approximation of Jurassic paleolatitudes for the Indian Ocean and adjacent continents. During the Upper Cretaceous and Lower Tertiary, the overall relative south pole of the higher hierarchy plate frame (including Africa) is inferred to have defined a westward path (with decreasing time), that is, trending parallel but opposite to the corresponding African north pole path depicted in Figure 2. It is reasonable to assume that the individual rotations of Australia and Antarctica (within the framework of the southern megaplate), from their pre-Alpine relative orientation delineated in Figure 7 towards their present position, were basically simultaneous with those of Africa and Eurasia, which apparently terminated around the Eocene-Oligocene boundary. Compared with the Mesozoic-Lower Tertiary APW path for Africa, which runs roughly along the 60°S parallel and has a longitudinal spread from approximately 0°E to approximately 80°E (see Fig. 2 for northern pole path), the corresponding Australian data show much more easterly longitudes, varying between approximately 100°E and approximately 180°E (Piper, 1988, for listing of data). This longitudinal paleomagnetic discrepancy apparently can be accounted for by the inferred relative counterclockwise rotation of Australia depicted in Figure 7. However, if the cusp in the Australian APW path is several tens of degrees east of that for Africa, as indicated by available data, then the completion of the Australian tectonic rotation must have gone beyond the Eocene-Oligocene boundary (see later discussion).

The reconfiguration of Figure 7 also gains support from paleoclimatological considerations. As discussed above, the Earth underwent approximately 15 degrees of meridional reorientation in the Lower Tertiary (relative to the axis of rotation "fixed" in space), which lead to a geographic change that would bring areas like South Australia and West Antarctica into somewhat lower latitudes than those displayed by Figure 7. This consideration implies that in the Upper Cretaceous-Lower Tertiary, at a time when these continents became subjected to tectonic rotations, southern Australia and western Antarctica experienced climatic conditions characteristic for middle latitudes. Therefore, the mild temperate Cretaceous-Lower Tertiary fossil occurrences in these areas (see Figure 7) readily can be accounted for by their paleolatitudes as inferred from paleomagnetic data. Within the framework of orthodox plate tectonics (e.g., Le Pichon and Heirtzler, 1968; Smith and Hallam, 1970) the flora and fauna that populated these locations would have experienced anomalously high latitudes, and special astronomical and biological mechanisms have been invoked to explain these biological-latitudinal inconsistencies (e.g., Axelrod, 1984). However, it seems likely that this paradox has been a consequence of forcing biological data into a wrong paleogeographic framework.

In conclusion, the conception of Gondwanaland as a confined continental assembly (prior to its alleged breakup and dispersal in Meso-Cenozoic time) does not seem to rest on a reasonable scientific foundation.

Paleomagnetic and tectonic setting of the Atlantic

It is well established that the European and North American APW paths, and notably their Carboniferous-Triassic parts, display a systematic disagreement of some 50 degrees of longitude (e.g., Tarling, 1983). In tectonic terms, such systematic polar offsets can be explained in at least two entirely different ways: (1) there may have been continental separation of the Wegenerian type between the two continents (as adopted by the conventional plate tectonic model), or (2) the continents have been in their present positions but have been subjected to minor relative rotation. Figure 8, which compares the Mesozoic-Lower Tertiary polar pattern for cratonic North America with that for Europe north of the Alpine chain, shows two important features.

1. The sense of relative polar progression with decreasing time is apparently the same for both continents, but (as for Europe) there is evidence of remagnetization,

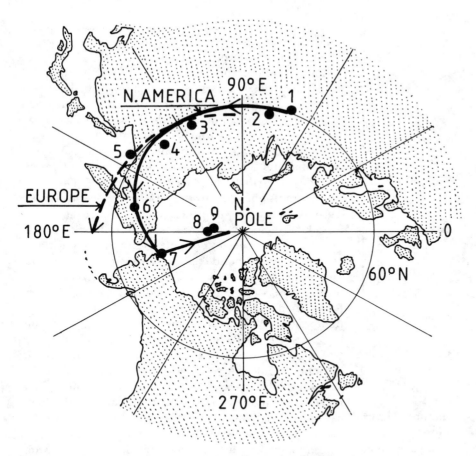

Figure 8. Comparison of Mesozoic-Lower Tertiary paleomagnetic polar trends for Europe (north of the Alpine belt) and cratonic North America. The North American data (numbered circles) are those compiled by May et al. (1989): 1—Reeve, 1975; 2—Steiner and Helsley, 1974; 3—May et al., 1986; 4—Kluth et al., 1982; 5 and 6—Steiner and Helsley, 1975; 7—Globerman and Irving, 1988; 8 and 9—Diehl et al., 1983. The age-paleopole relationship should not be taken too literally as magnetization and physical rock ages may easily have been disconnected by remagnetization caused by the major phase of Alpine plate rotation in the Lower Tertiary.

for example, Mesozoic rocks that are expected to have poles in the eastern end of the latitudinal polar branch have magnetizations that conform to Lower Tertiary ages (see discussion of APW calibration and remagnetization for the Afro-European paleomagnetic system).

2. There is an angular discordance between the APW curves that (apart from the difference in the sense of relative polar movement) is similar to the Italy-Africa relative kinematic picture (Fig. 5).

The combined evidence posed by the North American data is consistent with the tectonic model based on Afro-Eurasian relative rotation. A mobilistic tectonic system of the type considered here would be prone to create torsional stress throughout the crust and upper mantle causing jointing, reactivation of older fault zones, triggering the release and ascent of fluids from deeper levels, and so forth. The paleomagnetic consequences of such processes would be widespread remagnetization.

In the discussion on the Tethyan and Gondwanaland issues there is no positive evidence of sea-floor spreading. In fact, we removed a number of observational paradoxes that developed in the wake of that model. If we can further bolster the preliminary paleomagnetic indication that spreading processes may be absent in the Atlantic too, then the impact on the orthodox plate tectonic model would be serious. The Central Atlantic has a series of paleomagnetic data (from basalts of Layer 2) that are highly pertinent to this delicate question. Figure 9 shows the distribution of a relatively large number of DSDP sites from which considerable paleomagnetic inclination data are available. The drill sites are located close to the crestal zone, and according to conventional interpretation of marine magnetic lineations, the age of the crust at these sites should be late Miocene and younger. For the latitude range concerned the present dipole field inclinations vary between approximately 45 degrees for the southern sites and approximately 56 degrees for the northern locations. However, with very few exceptions the various basement sections studied display characteristic inclinations that are anomalously shallow when compared to those predicted by the present axial dipole field (PADF) model. The most detailed series of results come from sites 332A and B (Leg 37) (Hall and Ryall, 1977) and 396B (Leg 46) (Petersen, 1979) (see Fig. 9 for site locations). The upper 65 m of basalt in site 332A has inclination values that agree with the PADF, but all the remaining basaltic stratigraphy of sites 332A and B (approximately 440 m) is entirely dominated by inclinations in the range of approximately ±10–30 degrees. Apart from a 16 m thick sequence (77 to 93 m below top Layer 2), site 396B, which represents the southernmost part of the surveyed region covering altogether 244 m of basalts, has very shallow inclinations. Also, an important observation is that the percentage of practically horizontal inclinations (i.e., below 10 degrees) is much higher here than in the more northern DSDP locations.

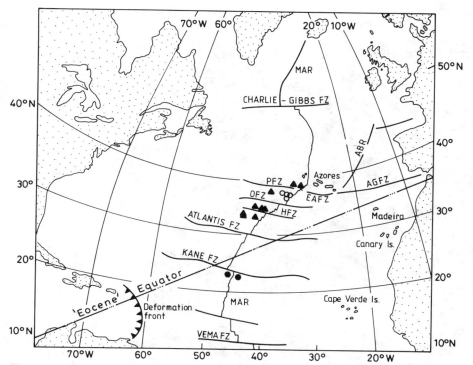

Figure 9. Sketch map of the Central Atlantic showing: (i) the location of the Mid-Atlantic Ridge (MAR) crest; (ii) prominent members of the transverse fault system, and (iii) deep sea drilling sites with particular emphasis on basement rocks. The DSDP sites are as follows: open circles, leg 37 (sites 332-335); closed circles, leg 45 (sites 395 and 396); closed triangles, leg 82 (sites 556-564). **ABR**—Azores-Biscay Ridge; **AGFZ**—Azores-Gibraltar Fracture Zone; **EAFZ**—East Azores Fracture Zone; **PFZ**—Pico Fracture Zone; **OFZ**—Oceanographer Fracture Zone; **HFZ**—Hayes Fracture Zone.

Unspecified tectonic tilting is suggested repeatedly as the cause of the Central Atlantic inclination anomaly, but no conceivable tectonic model can explain the observations (Storetvedt, 1990). However, the entire problem evaporates if one views the observed inclinations in the context of the Alpine-age paleomagnetic field frames, ignoring the model of sea-floor spreading. In the Eocene paleogeographic setting, the equator runs across the Central Atlantic as depicted in Figure 9 (see also Fig. 4). The Upper Mesozoic equator (not shown in diagram) has a slightly more southerly position, passing between the Canary and Cape Verde islands, intersecting the Eocene equator at around 90°E. These paleogeographic frames account for the observed inclination pattern; the generally shallow dip and tendency to northward steepening of the inclination are apparently concordant with the paleogeographic setting of the Central Atlantic as envisaged by Figure 9. These results demonstrate again how paradoxes disappear when the constraints posed by the accepted paradigm are removed. There seems to be no alternative but to conclude that the crust of the Central Atlantic Ridge was already in existence in Upper Cretaceous-Lower Tertiary time. In fact, there is no apparent reason to invoke sea-floor spreading as a general process of ocean crust evolution. Therefore, a series of intriguing geological and geophysical observations, including unexpectedly strong vertical oscillations of the crust, polyphase metamorphic processes, "awkward" rock types, anomalously high rock ages, seismic crustal anisotropy, and so forth (see Storetvedt, 1990), are explained within the context of a long history of oceanic crustal development during which thinning processes, mantle diapirism, and tectonic deformation must have featured repeatedly (see below).

Within a dynamic system of lithospheric block rotations the paleomagnetic puzzle of South America also may find its solution. Paleomagnetically South America represents a special case in that the poles determined from Permian and younger rocks form a somewhat elongated cluster across the present South Pole region (e.g., McElhinny, 1973; Tarling, 1983). Conceivably this phenomenon is directly related to the development of the North Brazilian Ridge, a gravimetrically compensated structure with low P-wave velocities (below that for oceanic Layer 2), running off and along the north coast of Brazil (Fig. 10) for a distance of 1300 km (Hayes and Ewing, 1970). The ridge is now submerged in deep water, but in the mid-Miocene at least parts of it experienced shallow-water conditions (Bader et al., 1970). The North Brazilian Ridge may be considered a compressional feature, constituting a belt of fractured and thickened crust (Storetvedt, 1990). Due to the relatively narrow width and special shape of the equatorial Atlantic, South America appears to have been hooked up in the relative motion of Africa imposing transpressive tectonic forces along the North Brazilian margin, compelling South America to swing clockwise. A clockwise rotation of some 20 degrees in Alpine time (probably in the Eocene) around a Euler pole in NE Brazil is likely responsible for the offset of the South American APW path to the southwest of that for Africa, bringing the Permian-Mesozoic poles into positions in Antarctica. Around the Eocene-Oligocene boundary the Earth's axis of rotation changed by coincidence to

Figure 10. The paleomagnetically inferred rotational pattern in Upper Cretaceous-Lower Tertiary time (i.e., during the main Alpine diastrophic event) for the Atlantic bordering continents plus India. The kinematic picture gives the overall impression of two hemispherical megaplates moving in opposite senses, suggesting that the Coriolis effect maybe of fundamental importance in global tectonics. In the new global tectonic concept, the oceanic regions are viewed as broad tectonized regions of thinned-assimilated continental crust, constituting complex plate boundaries. The term plate is confined to the continental masses (bounded by the continental margins). Note the consistent tectonotopographic features, convex to the southeast, of the South Atlantic and Indian oceans. The ocean floor structure is according to Heezen and Tharp (1977) (reproduced with permission).

the same general position. Thus, the unusually stationary South American paleopole for Permian-Recent times appears to be the fluky corollary of a tectonic rotation and an event of true polar wander in the Lower Tertiary.

CONCEPTION OF A NEW DYNAMIC MODEL

Hierarchal principle of rotating plates

Figures 10 and 11 delineate the paleomagnetically inferred rotational pattern for the Earth's major continents. Relatively small continental masses, surrounded by thin oceanic crust, are quite unstable, and their tectonic behavior is unpredictable. We must turn to the Afro-Eurasia block, the largest continental assembly on Earth, to gain deeper insight into the Meso-Cenozoic kinematic system. The picture gives the intuitive impression of two megaplates of hemispherical size rotating in opposite senses, their common boundary corresponding roughly to the paleoequator. This pattern suggests that the Earth's rotation plays a significant role in global tectonics. For such a dynamic system the Euler axis and the Earth's spin axis would be coincident, so that this first-order plate motion cannot be measured directly by paleomagnetic techniques. However, paleomagnetic data for North America fall into the framework of a northern megaplate. Adjusting for the independent rotations of Australia and Antarctica, their inferred pre-Alpine reconfigurations define a relative mid-Mesozoic South Pole in a position that is axially misaligned with respect to corresponding poles for the northern megaplate, located in NE Asia. These results give further substance to the notion of two hemispherical megaplates. In fact, the entire South Atlantic and Indian oceans seem to have memorized the anticlockwise rotation of the southern megaplate by a large number of tectonotopographic lineaments, convex to the southeast (e.g., Fig. 10). Such "flow" structures are expected to develop in thin oceanic crust as a result of the rotation of the southern hemisphere. It is submitted, therefore, that there is reasonable evidence for the model of two hemispherical megaplates. The overall impression is that some outer layer of the Earth experienced some kind of westward drift. The fact that this mobile outer layer, the "tectonosphere," consisted of two hemispherical plates suggests

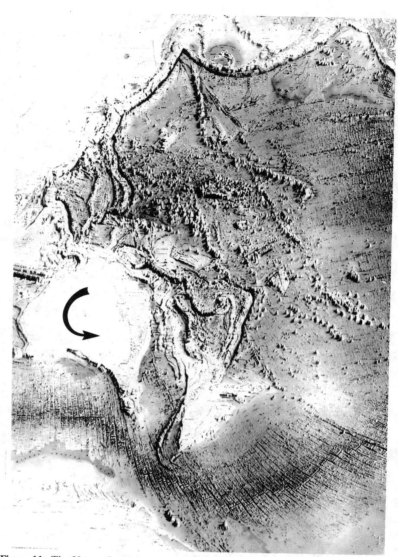

Figure 11. The Upper Cretaceous-Lower Tertiary movement of Australia appears to have been that of a simple rotation, approximately 70 degrees counterclockwise (Storetvedt, 1990). This motion is consistent with the major dextral transcurrent fault zone that runs (in a NNE-SSW direction) across New Zealand, and may explain the westward convexity of the Pacific ocean structures east of Australia.

either the onset of a simple dipolar convective system in the mantle, or that major events of tectonic activity are related to periods of faster Earth rotation. This leads to the interesting conclusion that tectonic activity on Earth appears more strongly linked to the Earth's rotation than previously thought. The inhomogeneity of the outer mobile layer, principally due to the weaker oceanic substratum, caused the hemispherical plates to break up into a multiplicity of smaller plates at various plate levels. The end product is a hierarchal system of plates in relative rotation. As a general rule, large plates are much more stable tectonically than small plates. The greatest tectonic instability is found in the southern hemisphere (i.e., Australia, Antarctica—due to the large proportion of oceanic crust there) and along the principal plate boundary that became the locus of the Alpine-Himalayan orogenic belt (India, Italy, Iberia, etc.). It is the higher orders of plates with which we are principally concerned in paleomagnetism and which produce the observed polar differences between continental blocks.

A new picture of the oceanic crust

In the mobilistic system developed here, sea-floor spreading is eliminated as a general ocean-forming process. Therefore, the geological history of the oceanic crust must be viewed in a perspective very different from that imposed by orthodox plate tectonics. Without the help of paleomagnetic data, the new kinematic framework of the globe would be difficult to unfold and the situation might look more fixed than it actually is. It is difficult to escape from the conclusion that in the outset, after the original cooling of the Earth, the entire surface was covered by continental crust ("granitic scum"). A major part of this original crust was, throughout the course of geological time, transformed into present oceanic structure through rhythmic processes of attenuation and tectonic deformation. Evidence of earlier continental nature and gradual transformation into oceanic crust can be seen in many parts of the oceanic domain. Continental fragments like the Seychelles-Mascarene Plateau (Indian Ocean), Jan Mayen Ridge (Norwegian-Greenland Sea) and the seamount of Bald Mountain (North Atlantic) reflect different stages of a continental crust destruction process that reached a more advanced stage of development in the thinner oceanic crust surrounding such oceanic topographic features. Apart from possible azimuthal changes, depending on the actual tectonic setting, such continental fragments are in place with respect to adjacent continental and oceanic crust. However, to delineate older geological structures from one continental block to another, and in particular if the crossing of a prominent tectonic boundary is involved, it is necessary to account for relative plate rotations to avoid artificial cusps in matching geological trends. Nevertheless, in our understanding of the development of the oceanic crust the pendulum seems to swing back to oceanization (e.g., Beloussov, 1962; Van Bemmelen, 1972).

Evidence for crustal thinning—During the past decade or so numerous examples of oceanward-dipping subacoustic basement reflectors were delineated from seismic studies of many continental margins (e.g., Hinz, 1981). Such structures are frequently associated with block rotation and listric faulting indicative of an extensional tectonic regime. However, restoration of the upper crustal layering has proved entirely inadequate to explain the associated crustal thinning; assuming that the thinning relates to stretching processes in the context of incipient rifting and early stage of sea-floor spreading (e.g., de Charpal et al., 1978). Also, recent deep-seismic reflection profiling across the Parentis Basin (the inner marginal basin of the Bay of Biscay) provides strong evidence of some kind of subcrustal erosion-assimilation by the upper mantle (Pinet et al., 1987). It is also pertinent that along the northern Biscay margin the listric faulting as well as the overprinted Alpine tectonic structures continue onto the abyssal plain, beyond the alleged (but dubiously defined) ocean-continent boundary (see discussion by Storetvedt, 1990). The results are consistent with the view that continental margins constitute transition zones with gradually increasing attenuation seaward. Crustal thinning caused by mantle processes create the necessary extensional regime in the upper crustal layering to account for the observed listric faulting. Therefore, a constant headache of the plate-tectonics paradigm, to define the required ocean-continent boundary, is removed.

More indirect evidence for the thesis that the oceanic crust was formed through gradual attenuation and chemical alteration processes can be acquired from a spectrum of geophysical and geological data. If the assimilation theory is correct, then the markedly larger proportion of oceanic crust in the Southern Hemisphere compared to that in the Northern Hemisphere should be reflected in different ways. First of all, one would anticipate that the mantle beneath the globe-encircling southern oceans would have the greater chance of showing signs of any continental crust isotope enrichment. Indeed, the Southern Hemisphere mantle, between the equator and 60°S, displays a unique case of anomalous Sr, Nd, and Pb isotopic characteristics (Dupré and Allègre, 1983; Hart, 1984), consistent with a continental crust contaminant (McKenzie and O'Nions, 1983).

Following the same thread of hemispherical dissimilarity, a more extensive assimilation of continental crust in the south would be expected to give rise to changes in the Earth's moments of inertia. Apart from smaller continuous changes the relative pole will have a fairly stationary position as long as the major axis of inertia stays near the spin axis, but in the case of a reorganization of matter, such as during an oceanization process, the Earth will be expected to readjust its position relative to the axis of spin. Such an event of true polar wander can occur quickly, limited only by the relaxation time of the equatorial bulge (Goldreich and Toomre, 1969). For the larger continental blocks (for which the paleomagnetic pole positions are not too affected by plate rotation) the APW paths delineate spasmodic and near-meridional patterns towards the present geographic poles. It seems likely, therefore, the major cause of the general meridional polar trends is the Earth's spatial adjustment to changes in the moments of inertia following the difference in ocean crustal development between the two hemispheres. The apparently rapid meridional changes in the APW curves can be considered as events of true polar wander, related to events of crustal loss to the mantle.

Other geophysical arguments for oceanization are based on studies of the length of the day, as determined by growth rings in corals, and from past tidal evidence. The major factor that affects past rates of the Earth's spin angular momentum is lunar tidal torque, which apparently had a general decelerating effect on the angular velocity throughout Phanerozoic time (Pannella et al., 1968). However, the paleontological evidence for the length of day during Phanerozoic time (e.g., Berry and Barker, 1975; Mohr, 1975; Pannella, 1975) suggests much slower deceleration in the past than at present. The most likely cause of this discrepancy is that general oceanic bathymetry did not allow significant tidal magnification, that is, the oceans of the geological past must have been much shallower than they are today (Tarling, 1975). Also, there is ample evidence for extensive distribution of marine deposits on the continents for most of the Phanerozoic. Apparently significant development of deep oceanic basins did not take place before Alpine time. Despite the loss of continental crust to the mantle over a long period of geological time, there is more dry land on the Earth now than there has ever been before. In conclusion, the time is ripe for readvocating oceanization models for the oceanic crust; that is a rhythmic development through processes of attenuation and chemical alteration of an original continental crust.

Origin of midocean ridges—Within the plate-tectonic scheme outlined above a gradually developing oceanic crust will become increasingly vulnerable to tectonic deformation. The weakening of the crust that follows advancing attenuation leads to fragmentation of the hemispherical megaplates as well as tectonic deformation of the oceanic crust and upper mantle. Plate rotations will provide the basis for a general transpressive tectonic regime, and it is suggested that the so-called transform faults are nothing else than a system of en-echelon shear zones. These faults have a long history of development. The seismic anisotropy configurations that have been found repeatedly in the crust and upper mantle of the oceans also would be a natural consequence of a transpressive tectonic regime.

In the Lower Tertiary, when Alpine plate rotations were at their most intense, the transpressive deformation appears to have been centered along median belts, causing thickening of the crust and possibly parts of the upper mantle also. By termination of plate rotation in the late Eocene and associated release of the compressive forces, the thickened layer became isostatically uplifted. The uplift must have created a tensional regime in the upper crust producing magmatic activity, the central extensional rift (the median valley), as well as the predominant ridge-parallel tectonotopographic lineament. Metamorphic processes during the transpressive regime, as well as subsequent low temperature alteration along the major ridge-parallel fault system, are likely to have developed significant contrasts in magnetic susceptibility across the ridge. This hypothetical two-dimensional susceptibility pattern may be the most important cause of the ridge-parallel magnetic anomalies, which basically requires a shallow source. It is pertinent to reiterate that the remanence properties of Layer 2 have so far proved entirely inadequate to explain the anomalies. Also, Agocs et al. (this volume) found a generally poor correlation coefficient both along strike and across the ridge for the Reykjanes Ridge anomalies; the best correlation obtained with bathymetry. These results are consistent with the view that the magnetic lineations are caused by a two-dimensional susceptibility variation in the upper part of oceanic Layer 2.

In a tectonic setting like the one considered here, the narrowest parts of an ocean will be affected most significantly by the orogenic effects of plate rotation. In the Atlantic regions, areas of particular concern are the northern North Atlantic and the equational Central Atlantic. The Lower Tertiary West-Spitsbergen orogeny and the major tectonic uplift of western Scandinavia at that time fall in line with expectations. Also, the evidence for substantial vertical movements in the equatorial Atlantic (Bonatti and Honnorez, 1971; Honnorez et al., 1975) is explained readily by tectonic processes implied by the present model.

Global events of relative plate rotations may dynamically affect wide areas of the oceanic crust, and already elevated topographic features may be subjected to further uplift, renewed magmatism, and subsequent subsidence. For example, Miocene carbonate banks of the equatorial Atlantic (Bonatti et al., 1977; Bonatti and Chermak, 1981), found on summits that may have been above sea level at that time, are seen here as the result of a Miocene tectonic uplift, following a mid-Miocene Alpine event of plate rotations and orogenesis. The absence of mid-Miocene sediments in the DSDP Leg 3 sites of the South Atlantic (Maxwell et al., 1970) also can be explained by the same mechanism.

Considering the overall relative intensity of Alpine orogenic events, the suggested mid-Miocene phase of plate rotations must have been much less spectacular than the one that operated in the Lower Tertiary. However, apart from tectonic effects exemplified above the mid-Miocene event clearly reactivated oceanic crust over a wide area, giving rise to volcanism in regions like Cape Verde and the Canary Islands, Madeira, and Iceland.

A subpopulation of green turtle lives on the Brazilian coast but breeds and nests 2,000 km offshore NE Brazil, on Ascension Island in the equatorial Atlantic (Carr and Coleman, 1974). How did this migration habit develop? This species is characterized by a herbivorous feeding habit, which would restrict migration routes to a geographical system of protected, shallow-water pasture grounds. Such natural conditions repeatedly have characterized the equatorial Atlantic in the geological past; the various events of plate rotation in Alpine time apparently led to crustal uplift that created chains of islands across the equatorial Atlantic. Hence, the green-turtle puzzle seems to have a simple explanation within the tectonic model presented here.

Oceanic trenches and Benioff zones—Having discounted the sea-floor spreading mechanism, oceanic trenches and associated Benioff zones require a different explanation from that of spreading-related subduction. The suggested "stationary" position of the oceanic regions, and gradual attenuation of the oceanic crust, bring us back to the old idea that the circum-Pacific Benioff zones developed as contraction fractures during the Earth's original cooling. If so, modifications of the original structural pattern would occur as a consequence of later tectonospheric movements. For example, in the context of the tectonic model advocated here, one expects significant structural modification of the trench system in the SW Pacific caused by the inferred counterclockwise rotation of Australia. As can be seen in Figure 11 the tectonotopographic features in this region of the Pacific are indeed very complex, and their general pattern, convex to the east, may be associated with the suggested rotation of Australia.

Evidence of a certain westward drift of the tectonosphere is consistent with the present tectonic picture: the Mariana-type structures (W Pacific) display a low stress or neutral regime, whereas the Chilean-type margin (E Pacific) has high compressive stress (Uyeda and Kanamori, 1979). Also, the marked difference in dip of the Benioff zones along the two opposite margins of the Pacific: steep for the Mariana-type and shallow for the Chilean-type (Uyeda and Kanamori, 1979) is consistent with the new mobilistic system. Other detachment-trench structures of smaller scale (Banda Arc, Lesser Antilles Arc, etc.) are understood as underthrusting of oceanic crust beneath thicker and lighter continental crust, caused by relative rotation and associated tectonic interaction, between bordering plates.

Orogenic belts

In the new tectonic model orogenic belts are zones of transpressive deformation. These fall into three main categories.

1. A globe-encircling system of foldbelts, running roughly along the paleoequator at the time of a major event of crustal mobility, developed along the common boundary of two hemispherical megaplates. The Alpine-Himalayan belt formed in this way in the Lower Tertiary. The fact that the Alpine, Hercynian, and Caledonian belts across Eurasia tend to follow consecutively, with increasing ages northward, is consistent with the temporal change of the paleoequator as determined from paleomagnetism, gives further substance to the notion of equatorial orogenic belts. Major parts of these diastrophic zones were destroyed, overprinted, or both by subsequent oceanization processes and oceanic ridge formation.

2. A special situation arises in the circum-Pacific region. The old detachment structures (the Benioff zones) undergo intermittent tectonic reactivation, in concert with the events of plate motion. The orogenic effect is developed particularly along the Eastern Pacific margins where the Americas had an over-riding effect on the Pacific crust. The fact that the various orogenic events along this margin, from Paleozoic to Tertiary times, are either superimposed or display perhaps a slight westward shift with decreasing age (e.g., Hervé et al., 1987; Megard, 1987) is consistent with the new model.

3. The gradual development of the oceanic crust, reaching its present state in late Mesozoic-Tertiary time, ultimately changed the tectonic significance of these domains completely. The existence of a substantial amount of thin and tectonically fragile oceanic crust in Alpine time induced a more diversified plate kinematic system than there was during previous periods of major crustal unrest. Thus, it seems that the Alpine plate rotations turned the oceanic regions into huge transpressive belts. The degree of tectonic deformation is likely variable, but at least the thickened part of the oceanic crust, now outlined by the midoceanic ridge system, may be considered a new class of Alpine foldbelts.

Other consequences of plate rotation

The extensive jointing seen throughout the continental crust (e.g., Hast, 1973; Scheidegger, 1985), frequently delineating systematic fracture patterns over large areas, is most likely the result of torsional deformation produced by plate rotations. Furthermore, the deep detachment structures frequently found in continental crust, sometimes cutting below the Moho, are likely another consequence of plate rotation. Reactivation of tectonic structures easily can release hydrothermal activity, which in turn may cause widespread remagnetization. In all, to judge from the present evaluation of Meso-Cenozoic paleomagnetic data, remagnetization triggered by plate-tectonic processes appears far more important than hitherto envisaged.

CONCLUSIONS

The considerations above lead to the following main conclusions.

1. The primeval crust of Earth was probably thick and buoyant, capped by a granitic layer. Cooling led to the development of deep contraction dislocations, for example, the Benioff zones circumscribing the Pacific, as well as to the onset of crustal attenuation processes.

2. Throughout geological time some outer shell of Earth repeatedly was in a state of mobility relative to the underlying mantle. The cause of this lithospheric mobility is likely thermal convection in the mantle, which was controlled by the Earth's axial spin. Associated with the events of mantle convection, diapiric offshoots have again and again ascended towards the surface where processes of attenuation and chemical transformation have developed the present oceanic crustal structure.

3. As can be judged from the total distribution of oceans and continents, the Southern Hemisphere must have been affected most strongly by oceanization processes. This explains the unique globe-encircling geochemical anomalies of the upper mantle beneath the southern oceans as contaminants from assimilated continental crust. The hemispherical dissimilarity in crustal structure implies the development of progressive differences in mass distribution that in turn led to systematic changes in the Earth's moments of inertia. The jerky meridional shifts of the relative paleomagnetic axis of the larger continental blocks, at least since mid-Paleozoic time, are consistent with an oceanization model, reflecting reorientations of Earth relative to the celestial axis.

4. During stages of mobilism (corresponding to the orogenic periods) Earth's outer shell, the tectonosphere, split into two hemispherical plates. These plates, which apparently had similar angular velocities, are unlikely to have undergone significant motion relative to the underlying mantle, but a certain overall westward drift is inferred. This latter behavior of the tectonosphere is consistent with the marked differences in angle of dip and tectonic regime of the Benioff zones along the opposite margins of the Pacific. The common boundary of the two hemispherical plates formed major transpressive orogenic belts along the corresponding paleoequators. Thus, the relative position of lower-Paleozoic and younger orogenic belts across Eurasia is explained by the model.

5. During Alpine time the oceanization processes reached an advanced stage in many regions of the world. This weakening of the tectonosphere beneath the oceans led to the break-up and tectonic fragmentation of the hemispherical plates, and the Southern Hemisphere in particular developed a diversified plate system (due to its high proportion of weak oceanic crust). The final tectonic system was a hierarchal order of plates in relative rotation. Plate rotations turned the oceanic regions into transpressive tectonic regimes.

6. The transpressive deformation of oceanic areas thickened parts of their crust and shallow mantle, and these regions subsequently became subjected to isostatic uplift. The midoceanic ridges are regarded as a system of Alpine foldbelts, and the so-called transform faults are interpreted as en-echelon shear zones. Smaller arc-trench structures (Banda Arc, Scotia Arc, and Lesser Antilles Arc) have a tectonic significance different from those of Benioff zones, constituting structural fronts within the system of interacting rotating plates.

7. A system of rotating plates imposes torsional stress throughout the crust and the development of joints therefore can be accounted for.

By dropping the constraints of the conventional plate tectonic model, awkard observations such as deep continental detachment structures, intracontinental remagnetization, the heat-flow problem of the oceans, the general absence of sheeted dykes in the oceanic crust, occurrences of granitic rocks (or very old rocks) along midoceanic ridges, enigmatic structural deformation and anomalous vertical tectonic movements of the oceanic crust, and so forth, can be explained. Ophiolitic complexes developed in local transtensional regimes along plate boundaries; therefore, the chances of such structures being recovered in oceanic areas would be slim. The many geophysical and geological paradoxes that have accumulated during the past two or three decades are apparently the consequence of forcing observational data into an inadequate tectonic model.

This is not the end. It is not even the beginning of the end. But it is, perhaps, the end of the beginning.

Winston Churchill

REFERENCES CITED

Axelrod, D.I., 1984, An interpretation of Cretaceous and Tertiary biota in polar regions: Paleogeography, Paleoclimatology, Paleoecology, v.45, p.105-147.

Bader, R.G., Gerard, R.D., et al., 1970, Initial Reports of the Deep Sea Drilling Project: Washington, U.S. Government Printing Office, v.115.

Beloussov, V.V., 1962, Basic Problems in Geotectonics: McGraw-Hill, New York.

Berry, W.B.N., and Barker, R.M., 1975, Growth increments in fossil and modern bivalves, in Rosenberg, G.D., and Runcorn, S.K., eds., Growth rythms and the history of the Earth's rotation: John Wiley and Sons, London, p.9-24.

Bonatti, E., and Honnorez, J., 1971, Nonspreading crustal blocks at the Mid-Atlantic Ridge: Science, v.174, p.1329-1331.

Bonatti, E., Sarntheim, M., Boersma, A., Gorini, M., and Honnorez, J., 1977, Neogene crustal emission and subsidence at the Romanche F.Z., equatorial Atlantic: Earth and Planetary Science Letters, v.35, p.369-383.

Bonatti, E., and Chermak, A., 1981, Formerly emerging crustal blocks in the equatorial Atlantic: Tectonophysics, v.72, p.165-180.

Bonini, W.E., Loomis, T.P., and Robertson, J.D., 1973, Gravity anomalies, ultramafic intrusions, and the tectonics of the region around the Strait of Gibraltar: Journal of Geophysical Research, v.78, p.1372-1382.

Buchardt, B., 1978, Oxygen isotope paleotemperatures from the Tertiary period in the North Sea area: Nature, v.275, p.121-123.

Bullard, E.C., Everett, J.E., and Smith, A.G., 1965, The fit of the continents around the Atlantic: Philosophical Transactions of the Royal Society, v.258A, p.41-51.

Bylund, G., 1974, Paleomagnetism of dykes along the southern margin of the Baltic Shield: Geologiska Föreningens i Stockholm Förhandlingar, v.96, p.231-235.

Carr, A., and Coleman, P.J., 1974, Sea-floor spreading theory and the odyssey of the green turtle: Nature, v.249, p.128-130.

Chatterjee, S., and Hotton III, N., 1986, The paleoposition of India: Journal of Southeast Asian Earth Sciences, v.1, p.145-189.

Christie, R.L., 1964, Geological reconnaissance of northeastern Ellesmere Island, District of Franklin, Canada: Memoir of the Canadian Geological Survey, v.331, p.1-79.

Clegg, J.A., Deutsch, E.R., and Griffiths, D.H., 1956, Rock magnetism in India: Philosophical Magazine, Ser.8, v.1, p.419-431.

Crawford, A.R., 1974, The Indus Suture Line, the Himalaya, Tibet and Gondwanaland: Geological Magazine, v.111, p.369-383.

De Charpal, O., Guennoc, P., Montadert, L., and Roberts, D.G., 1978, Rifting, crustal attenuation and subsidence in the Bay of Biscay: Nature, v.275, p.706-711.

Dewey, J.F., and Bird, J.M., 1970, Mountain belts and the new global tectonics: Journal of Geophysical Research, v.75, p.2625-2647.

Dewey, J.F., Pitman, W.C., Ryan, W.B.F., and Bonnin, J., 1973, Plate tectonics and the evolution of the Alpine system: Geological Society of America Bulletin, v.84, p.3137-3180.

Dewey, J.F., 1980, Episodicity, sequence and style at convergent plate boundaries, in Strangway, D., ed.: Geological Association of Canada, Special Paper, v.20, p.553-574.

Diehl, J.F., Beck, M.E., Beske-Diehl, S., Jacobsen, D., and Hearn, B.C., 1983, Paleomagnetism of the late Cretaceous and early Tertiary north central Montana alkalic province: Journal of Geophysical Research, v.88, p.10.593-10.602.

Dietz, R.S., and Holden, S.C., 1970, The breakup of Pangaea: Scientific American, v.223, p.30-41.

Dupré, B., and Allègre, C.J., 1983, Pb-Sr isotope variation in Indian Ocean basalts and mixing phenomena: Nature, v.303, p.142-146.

Geller, C.A., Weissel, J.K., and Anderson, R.N., 1983, Heat transfer and intraplate deformation in the Central Indian Ocean: Journal of Geophysical Research, v.88, p.1018-1032.

Girdler, R.W., 1968, A paleomagnetic investigation of some late Triassic and early Jurassic volcanic rocks of the North Pyrenees: Annales de Géophysique, v.24, p.1-24.

Globerman, B.R., and Irving, E., 1988, Mid-Cretaceous paleomagnetic reference field for North America: re-study of 100Ma intrusive rocks from Arkansas: Journal of Geophysical Research, v.93, p.11.721-11.733.

Goldreich, P., and Toomre, A., 1969, Some remarks on polar wandering: Journal of Geophysical Research, v.74, p.2555-2567.

Hall, J.M., and Ryall, P.J.C., 1977, Paleomagnetism of basement rocks, leg 37, in Aumento, F., Melson, W.G., et al.: Initial Reports of the Deep Sea Drilling Project, Washington, U.S. Government Printing Office, v.37, p.425-448.

Halvorsen, E., 1970, Paleomagnetism and the age of the younger diabases in the Ny-Hellesund area, S.Norway: Norsk Geologisk Tidsskrift, v.50, p.157-166.

Halvorsen, E., 1972, On the paleomagnetism of the Arendal diabases: Norsk Geologisk Tidsskrift, v.52, p.217-228.

Hart, S.R., 1984, A large-scale isotope anomaly in the Southern Hemisphere mantle: Nature, v.309 p.753-757.

Hast, N., 1973, Global measurements of absolute stress: Philosophical Transactions of the Royal Society London, A, v.274, p.409-419.

Hayes, D.E., and Ewing, M., 1970, North Brazilian Ridge and adjacent continental margin: American Association of Petrolium Geologists Bulletin, v.54, p.2120-2150.

Heer, O., 1880, Nachträge zur fossilien flora Grönlands: Kungliga Svenska Vetenskaps-Akademins Handlingar, v.18, p.1-17.

Heezen, B.C., and Tharp, M., 1977, World Ocean Floor: U.S. Navy, Office of Naval Research, Washington D.C.

Hervé, F., Godoy, E., Parada, M.A., Ramos, V., Rapela, C., Mpodozis, C., and Davidson, J., 1987, A general view on the Chilean-Argentine Andes, with emphasis on their early history, in Monger, J.W.H., and Franchteau, J., eds., Circum-Pacific orogenic belts and evolution of the Pacific Ocean basin: Geodynamics Series American Geophysical Union, v.18, p.97-113.

Hinz, K., 1981, A hypothesis on terrestrial catastrophes; wedges of very thick oceanward dipping layers beneath passive continental margins: Geologisches Jahrbuch, v.E22, p.3-28.

Honnorez, J., Bonatti, E., Emiliani, C., Brönnimann, P., Furrer, M.A., and Meyerhoff, A.A., 1975, Mesozoic limestone from the Vema offset zone, Mid-Atlantic Ridge: Earth and Planetary Science Letters, v.26, p.8-12.

Kent, P.E., 1969, the geological framework of petroleum exploration in Europe and North Africa and the implications of continental drift, in The exploration for petroleum in Europe and North Africa, Inst. of Petroleum, London, pp. 3-17.

Kluth, C.F., Butler, R.F., Harding, L.E., Shafiqullah, M., and Damon, P.E., 1982, Paleomagnetism of late Jurassic rocks in the northern Canelo Hills, southeastern Arizona: Journal of Geophysical Research, v.87, p.7079-7086.

Koch, E.B., 1963, Fossil plants from the Lower Paleocene of the Agatdalen (Angmartussut) area, Central Nugssuaq Peninsula, NW.Greenland: Meddelelser om Grönland, v.152, p.1-119.

Le Pichon, X., and Heirtzler, J., 1968, Magnetic anomalies in the Indian Ocean and sea-floor spreading: Journal of Geophysical Research, v.73, p.2101-2117.

Lövlie, R., Gidskehaug, A., and Storetvedt, K.M., 1972, On the magnetization history of the Northern Irish Basalts: Geophysical Journal of the Royal Astronomical Society, v.27, p.487-498.

Lövlie, R., and Kvingedal, M., 1975, A paleomagnetic discordance between a lava sequence and an associated interbasaltic horizon from the Faeroe Islands: Geophysical Journal of the Royal Astronomical Society, v.40, p.45-54.

Lövlie, R., 1981, Paleomagnetism of coast-parallel alkaline dykes from western Norway; ages of magmatism and evidence for crustal uplift and collapse: Geophysical Journal of the Royal Astronomical Society, v.66, p.417-426.

Lövlie, R., and Mitchell, J.G., 1982, Complete remagnetization of some Permian dykes from western Norway induced during burial/uplift: Physics of the Earth and Planetary Interiors, v.30, p.415-421.

Manum, S., 1962, Studies in the Tertiary flora of Spitsbergen, with notes of Tertiary floras of Ellesmere Island, Greenland and Iceland: Norsk Polarinstitutts Skrifter, v.125, p.1-127.

Maxwell, A.E. et al., 1970, Initial Reports of the Deep Sea Drilling Project, Washington (U.S. Government Printing Office), v. 3, p. 1-806.

May, S.R., Beck, M.E., and Butler, R.F., 1989, North American Apparent Polar Wander, plate motion, and left-oblique convergence: late Jurassic-early Cretaceous orogenic consequences: Tectonics, v.8, p.443-451.

May, S.R., Butler, R.F., Shafiqulla, M., and Damon, P.E., 1986, Paleomagnetism of Jurassic volcanic rocks in the Patagonia Mountains, southeastern Arizona: implications for the North American 170Ma reference pole: Journal of Geophysical Research, v.91, p.11.545-11.555.

McElhinny, M.W., 1973, Paleomagnetism and plate tectonics: Cambridge University Press.

McKenzie, D., and O'Nions, R.K., 1983, Mantle reservoirs and ocean island basalts: Nature, v.301, p.229-231.

Megard, F., 1987, Cordilleran Andes and Marginal Andes: A review of Andean geology north of the Arica elbow (18° S), in Monger, J.W.H., and Francheteau, J., eds., Circum-Pacific orogenic belts and evolution of the Pacific Ocean basin: Geodynamics Series American Geophysical Union, v.18, p.71-95.

Meyerhoff, A.A., and Meyerhoff, H.A., 1972, The new global tectonics: major inconsistencies: Bulletin of the American Association of Petroleum Geologists, v.56, p.269-336.

Mohr, R.E., 1975, Measured periodicities of the Biwabik (Precambrian) stromatolites and their geophysical significance, in Rosenberg, G.D., and Runcorn, S.K., eds., Growth rhythms and the history of the Earth's rotation: John Wiley & Sons, London, p.43-55.

Neprochnov, Y.P., Levchenko, O.V., Merklin, L.R., and Sedov, V.V., 1988, The structure and tectonics of the intraplate deformation area in the Indian Ocean: Tectonophysics, v.156, p.89-106.

Pannella, G., MacClintock, C., and Thompson, M.N., 1968, Paleoontological evidence of variation in length of synodic month since late Cambrian: Science, v.162, p.792-796.

Pannella, G., 1975, Paleontological clocks and the history of the Earth's rotation, in Rosenberg, G.D., and Runcorn, S.K., eds., Growth rhythms and the history of the Earth's rotation: John Wiley & Sons, London, p.253-284.

Petersen, N., 1979, Rock-and paleomagnetism of basalts from site 396B, in Dimitriev, L., Heirtzler, J., et al.: Initial Reports of the Deep Sea Drilling Project: Washington, U.S. Government Printing Office, v.46, p.357-362.

Pinet, B., Montadert, L., et al., 1987, Crustal thinning on the Aquitaine shelf, Bay of Biscay, from deep seismic data: Nature, v.325, p.513-516.

Piper, J.D.A., 1988, Paleomagnetic database: Open University Press, Milton Keynes.

Pomerol, C., 1982, The Cenozoic Era: Ellis Horwood Limited, Chichester.

Poore, R.Z. and Matthews, R.K., 1984, Late Eocene-Oligocene oxygen- and carbon-isotope record from the South Atlantic Ocean, DSDP site 522, in Hsu, K.J., LaBrecque, J.L., et al. Initial Reports of the Deep Sea Drilling Project: Washington, U.S. Government Printing Office, v.73, p.725-736.

Reeve, S.C., 1975, Paleomagnetic studies of sedimentary rocks of Cambrian and Triassic age: Ph.D. thesis, University of Texas at Dallas, p.1-426.

Rouse, G.E., and Srivastava, S.K., 1972, Palynological zonation of the Cretaceous and early Tertiary rocks of the Bonnet Plume Formation, northeastern Yukon, Canada: Canadian Journal of Earth Sciences, v.9, p.1163-1179.

Sander, N.J., 1970, Structural evolution of the Mediterranian region during the Mesozoic era, in Alvarez, W., and Gohrbandt, K.H.A., eds., Geology and History of Sicily: Petroleum Exploration Society of Libya, Annual field conference no.12, p.43-132.

Scheidegger, A.E., 1985, The significance of surface joints: Geophysical Surveys, v.70, p.259-271.

Schweitzer, H.-J., 1980, Environment and climate in the early Tertiary of Spitsbergen: Paleogeography, Paleoclimatology, Paleoecology, v.30, p.297-311.

Smith, A.G., and Hallam, A., 1970, The fit of the southern continents: Nature, v.225, p.139-146.

Sonnenfeld, P., 1981, Phanerozoic Tethys Sea, in Sonnenfeld, P., ed., Tethys, the Ancestral mediterranean: Hutchinson Ross Publication Company, Stroudsburg, p.18-53.

Staplin, F.L., 1976, Tertiary biostratigraphy, MacKenzie Delta region: Bulletin Canadian Petroleum Geology, v.24, p.117-134.

Stein, S., and Okal, E.A., 1978, Seismicity and tectonics of the Ninetyeast ridge area: evidence for internal deformation of Indian plate: Journal of Geophysical Research, v.83, p.2233-2245.

Steiner, M.B., and Helsley, C.E., 1974, Magnetic polarity sequence of the Upper Triassic Kayenta Formation: Geology v. 2, p. 191-194.

Steiner, M.B., and Helsby, C.E., 1975, Reversal pattern and apparent polar wander for the Jurassic: Bulletin of the Geological Society of America, v. 36, p.1537-1543.

Storetvedt, K.M., 1978, Structure of remanent magnetization in some Skye lavas, NW.Scotland: Physics of the Earth and Planetary Interiors, v.16, p.45-58.

Storetvedt, K.M., 1990, The Tethys Sea and the Alpine-Himalayan orogenic belt; mega-elements in a new global tectonic system: Physics of the Earth and Planetary Interiors, v.62, p.141-184..

Storetvedt, K.M., Carmichael, C.M., Hayatsu, A., and Palmer, H.C., 1978a, Paleomagnetism and K/Ar results from the Dunscanby volcanic neck, NE.Scotland:superimposed magnetizations, age of igneous activity, and tectonic implications: Physics of the Earth and Planetary Interiors, v.16, p.379-392.

Storetvedt, K.M., Pedersen, S., Lövlie, R., and Halvorsen, E., 1978b, Paleomagnetism in the Oslo rift zone: in, Ramberg, I.B., and Neumann, E.R., eds., Tectonics and Geophysics of Continental Rifts: D. Reidel Publishing Company, Dordrecht, p.289-296.

Storetvedt, K.M., and Carmichael, C.M., 1979, Resolution of superimposed magnetization in the Devonian John O'Groats Sandstone, N.Scotland: Geophysical Journal of the Royal Astronomical Society, v.58, p.769-784.

Storetvedt, K.M., Mitchell, J.G., Abranches, M.C., and Oftedahl, S., 1990a, A new kinematic model for Iberia; further paleomagnetic and isotopic age evidence: Physics of the Earth and Planetary Interiors, v.62, p.109-125.

Storetvedt, K.M., Tveit, E., Deutsch, E.R., and Murthy, G.S., 1990b, Multicomponent magnetization in the Foyers Old Red Sandstone (N.Scotland) and their bearing on lateral displacements along the Great Glen Fault: Geophysical Journal International, v.102, p.151-163.

Sturt, B.A., and Torsvik, T.H., 1987, A late Carboniferous paleomagnetic pole recorded from a syenite sill, Stabben, Central Norway: Physics of the Earth and Planetary Interiors, v.49, p.350-359.

Tarling, D.H., 1975, Geological proccesses and the Earth's rotation in the past, in Growth Rhythms and The History of The Earth's Rotation, eds. Rosenberg, G.G. and Runcorn, S.K.: John Wiley & Sons, London, p.397-412.

Tarling, D.H., 1983, Paleomagnetism: Chapman and Hall, London.

Thorning, L., and Abrahamsen, N., 1980, Paleomagnetism of Permian-multiple intrusion dykes, *in* Bohuslan, SW.Sweden: Geophysical Journal of the Royal Astronomical Society, v.60, p.163-185.

Tomskaya, A.I., 1976, Results of palynological studies of Cenozoic deposits in Yakutia, *in* Palynology in USSR: Papers of the Soviet palynologists to the IV International Palynological Congress, Lucknow, Nauka, Moscow, p.109-112.

Torsvik, T.H., Sturt, B.A., Ramsay, D.M., Bering, D., and Fluge, P.R., 1988, Paleomagnetism, magnetic fabrics and the structural style of the Hornelen Old Red Sandstone, Western Norway: Journal of the Geological Society, London, v.145, p.413-430.

Torsvik, T.H., Sturt, B.A., Ramsay, D.M., and Vetti, V., 1987, The tectonomagnetic signature of the Old Red Sandstone and Pre-Devonian strata in the Håsteinen area, Western Norway, and implications for the later stages of the Caledonian orogeny: Tectonics, v.6, p.305-322.

Uyeda, S., and Kanamori, H., 1979, Back-arc opening and the mode of subduction: Journal of Geophysical Researh, v.84, p.1049-1061.

Van Bemmelen, R.W., 1972, Geodynamic models, an evaluation and a synthesis: Elsevier, New York.

Weissel, J.K., Anderson, R.N., and Geller,C.A., 1980, Deformation of the Indo-Australian plate: Nature, v.287, p.284-291.

Wezel, F.-C., 1988, A young Jura-type fold belt within the Central Indian Ocean?: Bolletino di Oceanologia Teorica ed Applicata, v.6, p.75-90.

ACKNOWLEDGMENTS

The present contribution is an extended version of two papers presented at the discussion meeting on "New Concepts in Global Tectonics" held in Washington D.C., 20–21 July 1989. The present work basically was triggered by a paper by S. Chatterjee and N. Hotton III on the paleoposition of India. Various aspects of the study benefited from discussions with S. Chatterjee, P. Davies, N. Hotton III, A. A. Meyerhoff, H. G. Owen, J. H. Parry, S. K. Runcorn, D. Tozer and F.-C. Wezel. I acknowledge with gratitude the assistance of K. Breyholtz for drafting and E. J. Eriksen for typing. Critical reading and comments on the manuscript by A. C. Grant is greatly appreciated.

Reykjanes Ridge: quantitative determinations from magnetic anomalies

William B. Agocs, 968 Belford Road, Allentown, Pennsylvania 18103

Arthur A. Meyerhoff, P. O. Box 4602, Tulsa, Oklahoma 74159

Karoly Kis, Geophysics Department, Lorand Eötvös University, Ludovika tér 2 1083 Budapest, Hungary

ABSTRACT

Correlation of the linear magnetic anomalies of ocean basins has been largely qualitative, with limited effort to test correlations quantitatively. Consequently, the well-studied anomaly pattern of the Reykjanes Ridge southwest of Iceland was chosen for a quantitative study. Fifteen magnetic profiles across the ridge were digitized and correlation coefficients were determined along strike and across the ridge crest. The coefficient value along strike averaged 0.31, and that across the ridge 0.17, with limits of +1 to –1. It is worth noting that correlation between the anomalies and the bathymetry was superior, and averaged 0.42.

To correlate the magnetic anomalies of the Reykjanes Ridge with anomalies elsewhere in the world, it is necessary to transform all magnetic anomalies to a common magnetic datum. This may be done by transformation of all profiles to the pole (that is, recalculating each profile to a common latitude). In general, such a procedure has not been followed, which makes most correlations claimed in the literature highly suspect.

The depths to the tops and bottoms of the sources generating the magnetic anomalies were determined by using spectral energy bands. The efficacy of this procedure was checked against the available seismic-refraction control. The upper surface of the sources lies 2–4 km deep, and the lower surface is 4–7 km deep. Average thickness of the source is about 2 km.

We determined that the linear magnetic anomalies of the Reykjanes Ridge could be caused just as easily by differences in magnetic susceptibility as by changes in polarity, as required by the Morley-Vine-Matthews hypothesis. Although the latter is a simple explanation, it is not consistent with the complex geology of the Reykjanes Ridge and its continuation, the Mid-Atlantic Ridge. On the latter, numerous outcrops of Proterozoic rocks have been found (1,690–785 Ma), together with large volumes of silicic igneous and metamorphic rocks. Clearly the geological model that produces the linear magnetic anomalies is far more complex than the currently accepted models.

INTRODUCTION

Analysis of the linear magnetic anomalies of ocean basins might be published more properly in a geophysical journal, but because so much global geological interpretation hinges on magnetic-anomaly interpretations, this paper is presented for the benefit of a predominantly geological audience.

For more than 25 years, the linear magnetic anomalies of ocean basins have been assumed to represent times of alternate normal and reversed polarity. The assumption has scarcely been questioned in print for more than 15 years. Yet during the past quarter century, we repeatedly have encountered situations where geological data are not amenable to the assumptions derived from magnetic data. Accordingly we have studied the quantitative bases for the assumptions.

Search of the literature failed to reveal examples of in-depth, detailed, magnetic modeling that must have been done to support the assumption that linear magnetic anomalies on midocean ridges represent polarity reversals. Numerous papers have been published to develop procedures for evaluating reduction methods, but none treats the problem of pole reversals and its effects in detail. Therefore, the present study offers a partial remedy for what appears to be a gap in the literature.

The Reykjanes Ridge was selected for a detailed study of the linear anomalies for several reasons. First, geophysical data are abundant here. Second, as a "type locality" of the world's midocean ridge system, the Reykjanes Ridge is familiar to most Earth scientists. Finally, the midocean ridge supposedly is exposed on Iceland at the northern end of the Reykjanes Ridge. If this is true, Iceland might be expected to display evidence for the hypothesis of sea-floor spreading (Dietz, 1961; Hess, 1962; Vine and Matthews, 1963; Morley and Larochelle, 1964). However, Iceland has been less than successful as a source of information supportive of the sea-floor spreading model, as shown by many workers (e.g., Serson et al., 1968).

This lack of correlation between Iceland and the Reykjanes Ridge is not limited to magnetic data. It is also true of gravity data (Tr. Einarsson, 1954, 1965), seismic-refraction data (Zverev et al., 1975; Bott, 1983), and field geology (Tr. Einarsson, 1950, 1960, 1965, 1967, 1968; Th. Einarsson, 1967; Einarsson and Meyerhoff, 1973). Meyerhoff spent three field seasons (1971–1973) in Iceland attempting unsuccessfully to reconcile the observed field data with accepted geophysical models.

Chatterjee, S., and N. Hotton III, eds. *New Concepts in Global Tectonics.* Texas Tech University Press, Lubbock, 1992, xii + 450 pp.

HISTORICAL BACKGROUND

Mason (1958) published the pioneer discovery paper on the linear magnetic anomalies of the northeastern Pacific Ocean. Shortly afterward, the great extent of the anomalies became apparent (Mason and Raff, 1961; Raff and Mason, 1961) and gave rise to an extensive literature. Possible causes of the anomalies were proposed, among them (1) contrasts in magnetic susceptibility, with and without polarity reversals (Mason, 1958; Mason and Raff, 1961; Raff and Mason, 1961; Drake and Girdler, 1964; Tr. Einarsson, 1967; Gudmundsson, 1967; Luyendyk and Melson, 1967; Harrison, 1968; van Andel, 1968); and (2) reversals of polarity in the Earth's magnetic field, coupled with sea-floor spreading (Vine and Matthews, 1963; Morley and Larochelle, 1964). The former hypothesis is known as the Vine-Matthews hypothesis, despite Morley's clear priority (F.F. Osborne, in a footnote to the Morley and Larochelle paper, p. 50; see P.J. Wyllie, 1971).

As plate tectonics became accepted, alternative hypotheses to explain the linear magnetic anomalies were abandoned, and research to find either alternatives to or confirmation of the Morley-Vine-Matthews postulate ended. One of the models proposed to explain the linear magnetic anomalies was by Luyendyk and Melson (1967, p. 148–49). They wrote that, "Inspection of the map of magnetic intensity shows that the magnetic anomaly pattern in the [Mid-Atlantic Ridge] area . . . may be related mainly to zones of fresh basalts, alternating with hydrothermally altered zones which perhaps surround fracture zones. This offers an alternative hypothesis in contrast to recent models of sea floor spreading of adjacent . . . intrusive bodies with alternate normal and reversed magnetization." The sources of the magnetic anomalies in this hypothesis are magnetic susceptibility contrasts. This hypothesis was never developed despite the fact that Luyendyk and Melson (1967) also reported magnetization values from oceanic rocks that showed their magnetic susceptibility contrast model to be quite capable of explaining the observed anomalies. Likewise, Opdyke and Hekinian (1967) published magnetization values from oceanic rocks with magnetic susceptibility contrasts greater than five orders of magnitude, again fully capable of producing the observed anomalies without recourse to polarity change.

Among the models that differed from the Morley-Vine-Matthews postulate were two published by van Andel (1968). One involved sea-floor spreading; the other did not. Both required the injection and lateral spreading of mafic magmas during alternate epochs of normal and reversed polarity. However, van Andel, like Luyendyk and Melson, did not pursue his models and ultimately accepted the Morley-Vine-Matthews postulate. Van Andel (1968) made several observations that are refreshing and most cogent. He wrote (p. 154) that ". . . magnetic symmetry of rifted rises has been claimed . . . but not convincingly demonstrated" in response to the widely believed claim that the magnetic anomalies show excellent symmetry from one side of the midocean ridge to the other. On a cautionary note, he stated (p. 144) that the reasoning presented in his paper ". . . suggests that there is need for continuing examination of the reality of the magnetic models and for rigorous proof of the validity of magnetic correlations." He stated (p. 158) that "model testing is necessary to determine whether [the proposed models] are also consistent with the geomagnetic patterns. If accepted as reasonable, they imply that the geomagnetic correlations from non-rifted to rifted rises and the geomagnetic time-scale extrapolations require further examination and more conclusive proof than is available at present." This rigorous proof still is lacking, although many claim that sea-floor spreading is indicated by so many types of data that van Andel's rigorous proof is no longer required. Such a proof would involve several phases, all of them possible with today's technology. The following data are essential.

1. A complete section of oriented volcanic rocks for each interval of geologic time must be sampled on land, dated radiometrically, and studied for polarity. All epochs of normal and reversed polarity would have to be recorded and their absolute ages determined. Much of this has been done for the late Cenozoic, but detailed studies for the pre-Miocene are scarce.

2. An equivalent section of oriented and dated volcanic rocks must be sampled in the ocean basins. This essential step has never been done. On the only Deep Sea Drilling Project (DSDP) leg where this type of work was attempted (Leg 37), the directions of magnetization that were found generally were different from plate-tectonic predictions (Hall and Ryall, 1977).

3. If the samples from (1) and (2) have the same magnetic polarity for the same time intervals, then the Morley-Vine-Matthews postulate will be proved.

REGIONAL GEOLOGIC INFORMATION

The plate-tectonic model for the Reykjanes Ridge (Talwani et al., 1971) predicts the presence at the ridge crest of an axial graben containing the youngest igneous rocks—presumably dikes and flows—on the ridge. According to this model, the age of the rocks flanking the graben must increase progressively away from the ridge crest. Iceland, as a northern continuation of the Reykjanes Ridge, should—at least in the southern part—exhibit this same overall symmetry.

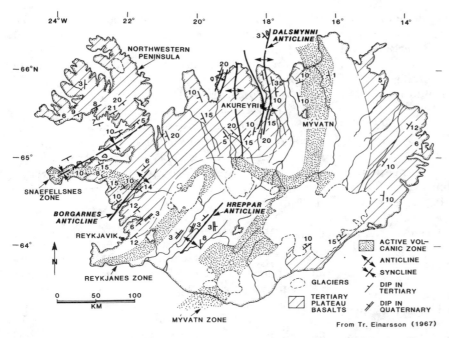

Figure 1. Map of Iceland showing (1) anticlines and synclines, (2) extent of Tertiary Plateau Basalt (9–13 Ma), and (3) the three neovolcanic zones.

Such is not the case; the geology and geophysics of Iceland are different from those predicted by the plate-tectonic model. The same statement applies to the entire North Atlantic region surrounding the Reykjanes Ridge and Iceland. Some of the contradictory data are enumerated.

1. Contrary to the popular picture of Iceland's geology, the island is cut by three—not one—northeast-to north-striking neovolcanic zones (Fig. 1). From west to east, these are the Snaefellsnes, Reykjanes, and Mývatn (Eastern) volcanic zones (Tr. Einarsson, 1950, 1965, 1968; Einarsson and Meyerhoff, 1973). The axis of the Reykjanes Ridge is continuous with the central, or Reykjanes, volcanic zone where a short axial graben is present at Thingvellir (Fig. 1). However, the only volcanic zone that traverses the full 300 km of the island from south to north is the easternmost, or Mývatn zone (Fig. 1).

2. According to the worldwide magnetic-anomaly numbering scheme (Nunns, 1983), the Mývatn and Snaefellsnes volcanic zones are equal to Anomaly 5, or are about 8 Ma old. Yet, because the Mývatn zone traverses the island, plate tectonics assigns to it the role of the central axial zone! For this to be so, an east-west fault system would have to separate the neovolcanic zones onshore from the Reykjanes Ridge. Intensive field work shows that no such fault system exists.

3. The quality of the magnetic data deteriorates from the Reykjanes Ridge northward, and no recognizable marine magnetic anomaly can be mapped from north to south through Iceland (Serson et al., 1968), although repeated attempts have been made (e.g., Nunns, 1983). The Serson et al. (1968) and Nunns (1983) studies indicated that important structural changes take place between the Reykjanes Ridge and Iceland (see also, Heirtzler et al., 1966). Moreover, on Iceland the sources of the anomalies are buried deeply below tuff and volcanic ash.

4. The only axial graben is the Thingvellir graben in the Reykjanes volcanic zone. Even so, the graben is only 30–50 km long, less than 16% of the length of the island.

5. The three active, or recently active, volcanic zones lie in synclinal depressions (Tr. Einarsson, 1950, 1967, 1968). Between them from the Snaefellsnes zone on the west to the Mývatn zone on the east, the volcanic layers have been cast into 60–100-km-long anticlines and synclines (Fig. 1). These fold trends traverse the 300-km length of the island, and have dips that range from minima of 2° to 9°, to maxima in the 25°–35° range, locally as great as 45° (Thoroddsen, 1907; Tr. Einarsson, 1960, 1965; Th. Einarsson, 1967; Einarsson and Meyerhoff, 1973). This aspect of Iceland's geology—that of a north-south folded lava plateau—certainly has not been stressed in the literature outside of Iceland.

6. The oldest stratigraphic unit, the Tertiary Plateau Basalt, underlies the entire island, 300 km from north to south and 500 km from west to east (Fig. 1), and radiometric dates are consistently in the 9–13 Ma range (Pálmason and Saemundsson, 1973; J. B. Otto et al., in Meyerhoff and Meyerhoff, 1974; Saemundsson et al., 1980).

7. South of the Reykjanes Ridge, on the Mid-Atlantic Ridge, between 37°02′ and 45°30′ N latitude, and 19°20′ W longitude, a region with an 800,000-km^2 area contains large areas of Paleozoic and Proterozoic rocks. The number of such localities is now so great that they no longer can be ignored. One locality, submarine Bald Mountain (45°13′ N latitude, 28°51′–28°54′ W longitude), is a whole mountain of largely granitic rock with a volume of 80 km^3, and radiometric dates in the 1,550–1,690 Ma range (Wanless et al., 1968; Aumento and Loncarevic, 1969). The granitic rocks are intruded by 785-Ma mafic rocks. Aumento and Loncarevic (1969, p. 13) wrote that dredge hauls in the Bald Mountain region—from three widely separated areas—consis-

tently recovered 74% silicic rocks; they commented that this is "... a remarkable phenomenon...."

Along the eastern flank of the Mid-Atlantic Ridge and the northern flank of the Azores-Gibraltar Ridge, Paleozoic sedimentary rocks have been recovered repeatedly for more than 130 years, mainly by Spanish and Portuguese fishermen (Furon, 1949; C.P. Hughes, pers. comm., 1980). One collection contains the trilobite *Triarthus* aff. *T. spinosus* and the graptolite *Climacograptus typicalis* (Meyerhoff, 1981). These are Caradocian (Late Ordovician) taxa from North America. This collection apparently came from a small cuesta identified on sparker profilers (C.P. Hughes, pers. comms., 1978–1981). Some are on display in the Sedgwick Museum at Cambridge University. Neither these nor the Bald Mountain locality are likely to be glacial erratics or ballast as has been claimed for most collections.

8. Within the same 800,000-km² area, a late Proterozoic gabbro boulder (635 ± 102 Ma) was identified in flat-lying chalky middle Miocene ooze, and below younger sediments and basalts at DSDP Site 334, Leg 37 (Reynolds and Clay, 1977), atop the Mid-Atlantic Ridge. The radiometric date, dismissed as possibly a result of excess argon gas, was not considered sufficiently important to mention in the cruise synthesis report (Robinson et al., 1977).

9. Seismic-refraction studies of the Greenland-Faeroe Ridge that passes beneath Iceland show that the crust is exceptionally thick. Bott (1983) reported the crustal thickness to be 29 to 32 km, but detailed studies by Zverev et al. (1975) show the range to be 16–60 km: 16–18 km beneath Faeroe-Shetlands Channel, and 40–60 km beneath Iceland. Meyerhoff, while in Moscow, was invited to inspect the refraction data, and computed independently a minimum crustal thickness beneath Iceland of 50 km above the 7.8–8.2-km/s layer.

The preceding facts indicate that the geology of this region is indeed more complex than had been envisioned on the basis of the early Reykjanes Ridge magnetic surveys (Baron et al., 1965; Heirtzler et al., 1966; Godby et al., 1968). This complexity provided us with an even greater incentive for studying the magnetics of this region.

DATA FOR THE PRESENT STUDY

Baron et al. (1965) made an extended, systematic, airborne magnetometer survey that includes 24,000 km of magnetic profiles (Fig. 2); Heirtzler et al. (1966) published a synopsis of these data with some analysis. Godby et al. (1968) described six aeromagnetic profiles extending from southern Greenland across the Reykjanes Ridge to the European continental shelf (Fig. 2). Talwani et al. (1971) presented a series of variously oriented sea-level magnetic profiles, with several at right angles to the ridge strike. The coverage of these surveys is outlined on Figure 2. Other valuable geophysical surveys pertinent to this study have been published. These include papers by Vogt et al. (1970, 1980, 1981), Johnson and Vogt (1973), Fleischer (1974), Vogt and Avery (1974), Johnson (1975), Johnson et al. (1975), Kristjánsson (1976), Talwani and Eldholm (1977), Voppel and Rudloff (1980), and many more. However, the oldest studies contain most of the types of data that are basic to the present study.

We selected 12 sea-level profiles from Talwani et al. (1971) and three aeromagnetic profiles from Baron et al. (1965). The locations of these 15 profiles are shown on Figure 3. They were chosen because they are geographically close to one another, are oriented at right angles to the Reykjanes Ridge, and are in profile from (and thus yield maximum magnetic data upon digitizing). We preferred the sea-level profiles to the aeromagnetic profiles because they are closer to the magnetic anomaly sources. The three aeromagnetic profiles were included to permit a comparison

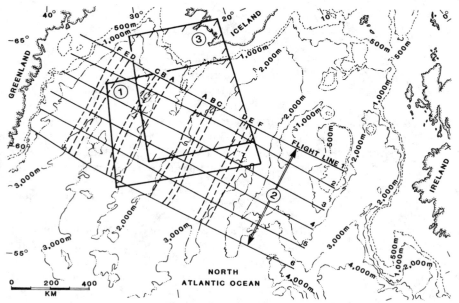

Figure 2. Map showing outlines of the three magnetic surveys made over the Reykjanes Ridge: area 1 = Baron et al. (1965) and Heirtzler et al. (1966); area 2 = Godby et al. (1968); area 3 = Talwani et al. (1971).

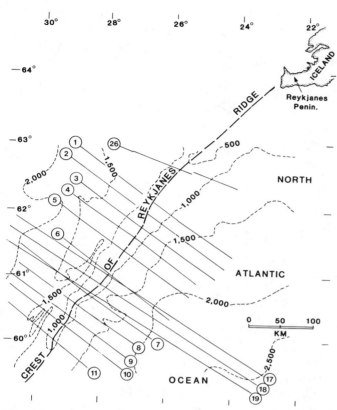

Figure 3. Index map to the 12 seaborne (Talwani et al., 1971) and 3 airborne (Baron et al., 1965) magnetic profiles used in this study.

with sea-level data. All 15 profiles (4340 km of profile) were digitized.

TECHNIQUES UTILIZED

As is well known, the succession of positive (that is, relatively positive) anomalies from the midocean-ridge crests to the adjacent abyssal plains has been numbered. The central anomaly is no. 1, and those in successive order from the ridge crests are numbered 2, 3, 4, ... Only the laterally more persistent anomalies are numbered. It is tacitly assumed that the character of each anomaly, both along the ridge flank and across the ridge crest, is sufficiently similar that a correlation is possible, not just on the basis of position but also of character. Unfortunately, these assumptions never were tested with an objective technique. Such a test is essential, because several of the numbered anomalies may be absent in a given area, and the correlations made are entirely subjective.

Because of this subjectivity and because the correlations have been extrapolated throughout the world's oceans, we elected to use the statistician's correlation coefficient to see whether the correlations that have been made could stand close scrutiny. We found no similar quantitative studies in the literature. Almost no correlation claimed in the literature considers the fact that magnetic-anomaly character changes substantially from one place to another on the Earth's surface. This is true because of changes in (1) magnetic latitude, (2) strike and configuration of the anomaly sources, (3) orientation of the magnetic profiles with respect to magnetic latitude and strike of the source bodies, and (4) depths to the sources. Moreover, our literature review showed that few authors understand the conditions for the permanent magnetization of minerals. Therefore, we review these conditions, and include references to explanatory publications.

Magnetite grains cannot become permanently magnetized unless they pass through the Curie point during cooling and unless the magnetite crystals are single domain with grain sizes less than 9 to 10 microns. Thus samples should be examined petrologically with great care to see whether they can be magnetized reversely. The theory behind this has been developed by Kittel (1949), Néel (1955), and Moorish (1965). In addition, fine-grained magnetite in a dolerite has low to zero susceptibility. Coarse-grained magnetite has greater susceptibility but cannot be permanently magnetized. Thus it is possible to have normal and reversed polarity within a single basaltic sill or flow. Nagata (1961) has explained why coarse-grained magnetite cannot be magnetized permanently.

We therefore used the correlation coefficient, which permits an objective correlation of magnetic anomalies both along strike on the ridge flanks, and across the ridge crest. The data were used also to determine the spectral energy variations along the profiles. Depth-to-source determinations were made from magnetic anomaly variations, using the second derivative maximum distance. We also calculated spectral energy depths and autocorrelation factor depths. The spectral energy depths were contoured on two horizons that coincide remarkably well with two refraction-seismic horizons that have P-wave velocities of 4 and 6 km/s respectively (Fleischer, 1974). The refraction-seismic horizons of 4 and 6 km/s tie the spectral energy depths with deviations of less than 0.2 km. The second derivative depths tie the 6-km/s horizon with a deviation of less than 0.1 km.

PROBLEMS IN INTERPRETING MAGNETIC PROFILES

The interpretation of magnetic profiles can be difficult, and should not be made without a sizable input of geology. Moreover, to make a geological interpretation, the factors that produce an anomaly must be understood.

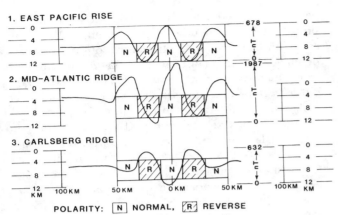

Figure 4. Magnetic profiles from East Pacific Rise, Mid-Atlantic Ridge, and Carlsberg Ridge, as published by Vine and Matthews (1963). Anomaly sources are alternate blocks of normal and reversed magnetism.

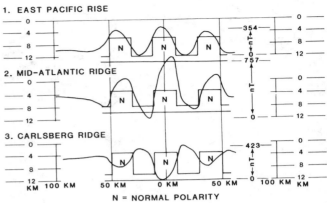

Figure 5. This figure shows the same profiles as Figure 4, except that the profiles have been recomputed for normally polarized magnetism. Cause of the anomalies is differences in magnetic susceptibility. Note the close similarity with Figure 4. This similarity means that polarity reversals have no important effect on the character or magnitude of the anomalies. (Adapted from Vine and Matthews, 1963)

Percentage of reversely magnetized oceanic crust

A logical prediction of the Morley-Vine-Matthews postulate is that about 50% of the basalt of the ocean floors is normally magnetized, with a like amount reversely magnetized. This assumption finds some support from studies of lava flows and related rocks in central France (Roche, 1951), Iceland (Einarsson, 1962), and elsewhere. However, the problem is much more complicated. Blackett (1956), for example, wrote that the Earth's magnetic field was reversed about half of the time from the Carboniferous to the Cretaceous, but more recent work shows that the greatest frequency of reversal was during Cenozoic time, with fewer reversals as one goes back in time (Irving, 1964). Clearly, a great deal of work must be done to justify the 50% assumption. Such work will involve collecting oriented and dated samples from both the continents and ocean basins.

Latitude

Figures 4 and 5 illustrate the effects of magnetic latitude. Vine and Matthews (1963) showed the computed magnetic anomalies for the East Pacific Rise, the Mid-Atlantic Ridge, and the Carlsberg Ridge (Fig. 4). The considerable change with latitude is evident. For example, the curve of the Carlsberg Ridge (centered at 5° N latitude) is almost the opposite of the curves from the higher magnetic latitudes.

Figure 5 shows the effect on the same three curves of a normally magnetized crust. Comparison of Figures 4 and 5 show that the effect of introducing reversed polarity is nil. Thus the Vine and Matthews (1963) profiles provide no clue to the polarity of the anomaly sources.

Shape and attitude of the source bodies

Figure 6 shows the magnetic anomalies for two bodies, one normally magnetized and the other reversely magnetized, at the latitude of the Reykjanes Ridge (approximately 60° N latitude); the profile strikes at 125° (at about right angles to the ridge). Each body is 2 km thick and lies between 1 and 3 km depth. Such widely spaced anomalies have a distinctive character or signature.

Figure 7, with the same latitude and azimuth, has a very different appearance. The source bodies are close to one another and have a bottom near the Curie isotherm at about 20 km. Thus the depth of the base of the source body is important, as is the proximity of adjacent bodies.

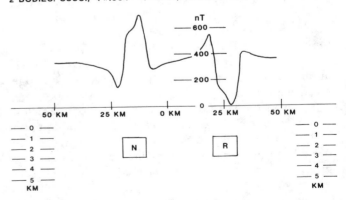

Figure 6. The figure shows magnetic anomalies for two bodies, one normally (susceptibility, $+\pi\,0.004$), the other reversely magnetized (susc., $-4\pi\,0.004$) at the latitude of the Reykjanes Ridge (approximately 60° N lat.). Each body is 2 km thick and lies between the depths of 1 and 3 km.

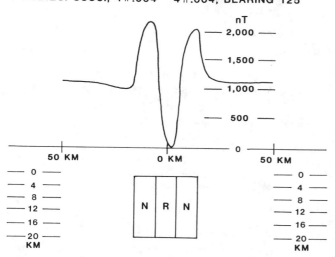

Figure 7. The figure shows magnetic profiles produced by source bodies that are close together. As in Figure 6, latitude is approximately 60° N. Base of source bodies is at 20 km, close to the Curie isotherm. Figure 7 shows the importance of (1) depth and (2) proximity of source bodies in creating anomaly character. Susceptibility values are same as for Figure 6.

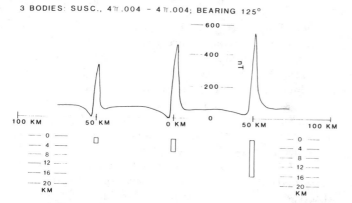

Figure 9. The figure shows the effect on magnetic anomalies of both the depth to the base of the source body and the width of the source body. Source is only 2 km wide. Thus, (1) depth to the base of an anomaly source and (2) width also are important in creating anomaly character.

Figures 8 and 9 show the importance of width, as well as the importance of depth. In Figure 8, we show the anomalies produced by changing the depth of the base of the source body. The body is 1 km deep and 10 km wide, and we show how a change of depth progressively from 2 km to 15 km alters the anomaly appearance. Figure 9 shows the same effect, except that the width of the source is only 2 km wide.

Reykjanes Ridge

Figure 10 shows the magnetic anomalies of the Reykjanes Ridge from northwest to southeast (Fig. 3). All are

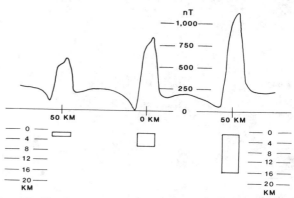

Figure 8. The figure shows magnetic anomalies produced by changing the depth of the base of the source body. Body is 1 km deep and 10 km wide.

sea-level profiles except for 17 through 19, which are airborne profiles from Baron et al. (1965).

The central anomaly of profiles 26, 1, 4, 5, 6, 18, 7, 9, and 11 has a shape that indicates the presence beneath it of a source body with its base well above the Curie isotherm; that is, the source body must have an appearance similar to that of the body in Figure 6. The central anomaly source on the profiles listed (above) and the source for the anomaly of Figure 6 are tilted, as shown by the fact that one side of the central anomaly

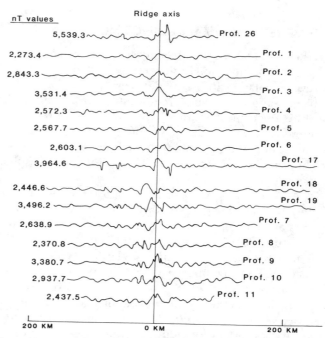

Figure 10. Magnetic profiles across the Reykjanes Ridge. All profiles except 17 through 19 are from the Talwani et al.'s (1971) seaborne magnetometer survey; profiles 17–19 are from the Baron et al. (1965) airborne magnetometer study.

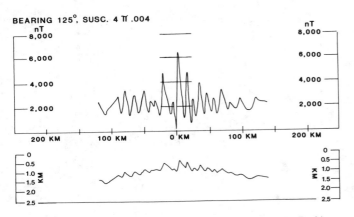

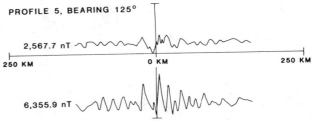

Figure 11. The lower profile shows the bathymetry across the Reykjanes Ridge (as recorded along profile 5). The upper profile shows the magnetic profile computed from the bathymetry, with an assumed constant magnetization of 51.5 Am^{-1}.

Figure 12. Upper profile is the recorded magnetic-anomaly pattern along profile 5, Reykjanes Ridge. Lower profile is the magnetic profile computed from the bathymetry. The correlation coefficient between these two profiles is 0.42 at the 99% confidence level. This value is higher than those for magnetic anomalies that supposedly are equivalent, both along the strike of and across the Reykjanes Ridge!

registers higher than the other (that is, the source is not symmetrical). The source for the central anomaly of the other profiles is more symmetrical and less tilted (e.g., profiles 2, 3, 17, 19, 8, 10), and has a deeper base than the equivalent anomaly on the other profiles.

The anomalies that flank the central anomaly generally have a smaller amplitude than the central anomaly, and also are more symmetrical. The low amplitudes and their general uniformity on each profile indicate that the anomaly sources are narrow with limited vertical dimensions.

MAGNETIC ANOMALY SOURCE MODELS

The model most widely accepted is one of central vertical dike injection in alternate periods of normal and reverse polarity (Vine and Matthews, 1963). One of the biggest and still unanswered questions of this model is: why is the central anomaly larger than any of the other anomalies? Even if one accepts the Morley-Vine-Matthews model and combines matching anomalies from opposite flanks of the ridge, the central anomaly still is larger than any on the ridge.

A weakness of many of the proposed models is that they involve constant depth, constant thickness, and constant magnetization, features that are rarely observed in nature. Matthews and Bath (1967) presented a variation of this usual model, with a two-layered crustal model, each layer having different properties. Schouten (1970) proposed that the various bodies producing the anomalies had different inclinations to produce the varied shapes of the anomaly curves. Almost every model involving sea-floor spreading and polarity reversals raises problems that have not been adequately tested and resolved.

The same statement is not true of models that do not require sea-floor spreading. We have mentioned the Luyendyk and Melson (1967) model, and one nonspreading model proposed by van Andel (1968). A. Mesko (pers. comm., 1989), however, suggested that we reexamine the effects of bathymetry, despite the fact that bathymetry as the prime source of the magnetic anomalies has been rejected (e.g., Heirtzler and Le Pichon, 1965).

Accordingly, we prepared Figures 11 and 12. Figure 11 shows the bathymetry recorded along profile 5 (Fig. 3). The magnetic profile (we assume a constant magnetization of 51.5 Am^{-1}) was computed from the bathymetry. Its resemblance to the recorded magnetic profile is at once apparent. The positive anomalies are due to elevations and the negative elevations to valleys; normal and reverse magnetizations are not involved.

Figure 12 shows the recorded magnetic pattern along profile 5 (top), compared with that calculated from the bathymetry (bottom). A quantitative comparison of the observed and computed magnetic profiles yields a coefficient of correlation of 0.42 at the 99% confidence level for the 240 stations recorded. (In the next section, we explain the coefficient of correlation.) It is clear that bathymetry alone could account for much of the anomaly configuration, despite opinions to the contrary (e.g., Heirtzler and Le Pichon, 1965).

QUANTITATIVE CORRELATION OF MAGNETIC ANOMALIES

The coefficient of correlation permits one to make objective correlations of magnetic anomalies along strike on one flank of a midocean ridge, and from one side of a ridge to the other. Its main purpose is to reduce drastically, or even avoid, subjectivity in anomaly correlations. The coefficient of correlation compares like data sets. The coefficient ranges in value from +1 to 0 to -1, indicating a range from perfect correlation to zero correlation to reversal or wholly out-of-phase correlation. The correlation coefficient r between n quantities, $x_0, x_1, \ldots x_n$ and n other quantities, $y_0, y_1, \ldots y_n$ (Green and Margerison, 1978) is defined by

$$r = \frac{n\Sigma(x_i y_i) - (\Sigma x_i)(\Sigma y_i)}{[(n\Sigma x_i^2 - (\Sigma x_i)^2)(n\Sigma y_i^2 - (\Sigma y_i)^2)]} \quad (1)$$

Tables 1 and 2 show the results. Table 1 gives profile-to-profile correlation factors along one flank of the Reykjanes Ridge; Table 2 shows the anomaly correlations across the Reykjanes Ridge axis. The proportion of all values favoring a correlation factor of r of n samples for a confidence limit of 95% is

$$\text{Prof} = r \pm 1.96 \left(\frac{r(1-r)}{n}\right)^{1/2}. \quad (2)$$

The mean interprofile correlation coefficient (Tables 1 and 2) is 0.31. The standard deviation is 0.26. The mean transridge correlation coefficient is 0.17, with a standard deviation of 0.23. Both the interprofile and transridge correlation coefficients are low and by objective methods cannot be regarded as reliable except within the imposed limits of the correlation factor.

Data-alteration procedures were considered, tried, and rejected. One of these procedures is the mapping function of Martinson et al. (1982). Use of this function alters the original data to improve correlations and fits. Although the coefficient of correlation is improved, it then becomes impossible to evaluate the original observed data.

Table 1. Profile-to-profile correlation factors along one flank of the Reykjanes Ridge. All correlation factors are within the 95 to 99% confidence limit.

Profile no. (Fig. 3)	No. of samples	Coefficient from Equation (1)
26-1	300	0.0
1-2	300	0.67
2-3	260	0.75
3-4	260	0.42
4-5	260	0.29
5-6	240	0.15
6-7	280	0.0
7-8	240	0.43
8-9	240	0.51
9-10	240	0.50
10-11	180	0.45
6-17	340	0.0
17-18	340	0.0
18-19	360	0.42
19-7	280	0.0

Table 2. Anomaly-to-anomaly correlation on single profiles, from one side of the Reykjanes Ridge axis to the other. All correlation factors are within the 95 to 99% confidence limit.

Profile no. (Fig. 3)	No. of samples	Coefficient from Equation (1)
26	100	0.0
1	140	0.0
2	140	0.45
3	120	0.63
4	120	0.02
5	120	0.0
6	100	-0.25
7	140	0.0
8	120	0.25
9	120	0.17
10	120	0.40
11	80	0.31
17	120	0.18
18	150	0.0
19	120	-0.27

ANOMALY CORRELATIONS FROM HIGH TO LOW LATITUDES AND POLE TRANSFER

The low values of the correlation coefficients in a small compact area such as the Reykjanes Ridge suggest that correlations of anomalies from high latitudes with those of low latitudes are likely to be incorrect unless a transformation is made to a common magnetic latitude. The great changes that do take place from high to low latitudes are illustrated dramatically on Figures 4 and 5. On both figures, the anomaly at high latitudes directly overlies its source body (Mid-Atlantic Ridge), but at low latitudes the anomaly is entirely out of phase with the source body (Carlsberg Ridge).

Figure 13 shows the magnetic anomalies produced at the magnetic equator or profiles with different bearings. The anomaly for a profile with zero bearing is pronounced (top profile, Fig. 13). However, as the bearing increases to 90°, the anomaly decreases to zero and vanishes. Comparison of Figure 13 with Figures 4 and 5 illustrates the importance of this point.

To make direct comparisons possible among anomalies recorded at different latitudes, all profiles must be recalculated (that is, transformed) to a common latitude. Procedures for such transformations to the pole have been published by Baranov (1957), Blakely and Cox (1972), Le Mouel et al. (1972, 1974), Nabighian (1972), Cande and Kent (1976), and Kis (1981, 1990). We have, in the past, used extensively the procedures developed by Le Mouel et al. (1972) and Nabighian (1972), and do so in this paper. Improvements intro-

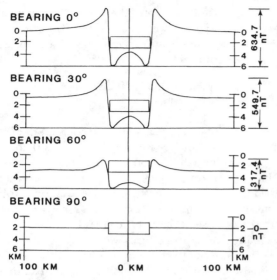

Figure 13. This figure compares the magnetic anomaly that results at the magnetic equator for profiles with different azimuths (bearings).

duced by Kis (1990) eventually should supplant other procedures.

To illustrate the effects of polar transforms, we prepared Figures 14 and 15. Figure 14 shows a magnetic anomaly caused by a single body that has a magnetization inclination of 15°, and an azimuth of 30°. We placed the polar transform anomaly below it. The radical change in anomaly character is evident.

Figure 15 is a more complicated case, in which the inclination is 0°, and the azimuth of the three bodies producing the anomaly also is 0°. We have placed the polar transform anomaly below it. Again, a complete change in character is evident.

We found very few such examples of transformations to the pole in the literature. Apparently the procedure is used only rarely in studies of the linear anomalies of the ocean basins. In view of this and the results displayed on Figures 4, 5, 14, and 15, our concerns about the many correlations claimed in the literature—but unsupported by quantitative analyses such as those presented here—are justified.

DEPTH DETERMINATIONS

Spectral energy

Heirtzler and Le Pichon (1965) and Harrison (1976) made limited spectral-energy depth determinations. Heirtzler and Le Pichon (1965) obtained and showed a wavelength position-isoamplitude plot that shows the magnetic anomalies along the ridge axis and parallel with it out to 600–650 km east and west of the Mid-Atlantic Ridge. Two spectral-energy density curves were obtained—one for the central anomaly and one for the flanking anomalies. Harrison's (1976) analyses dealt with the variation of the (1) power spectrum as a function of depth, (2) source thickness, and (3) reversal rate. He showed that (1) the reversal rate must be greater than that measured, (2) the depth to the source of the anomalies is greater than that of the basement surface, and (3) the thickness of the magnetic sources must be greater than 0.5 km.

The spectral energy density used by us was obtained to identify characteristics that could be related to the central anomaly source and to the flank anomaly sources. We present three profiles (Figs. 16, 17, 18). Each shows the magnetic anomaly curve and the sectional spectral energy across a 32-km sampling section, advanced 20 km from section to section. The spectral energy covers the range from 0 to 0.5 cycles/km; the spectral energy is in the range of $\ln(X) = -10$ to $\ln(X) = +10$. The spectral energy curve is placed directly below the associated magnetic anomaly curve.

Figure 16 (profile 26) is from Talwani et al.'s (1971) sea-surface magnetic reconnaissance of the Reykjanes Ridge. Figure 17 (profile 1), south of profile 26, is from the same survey. Figure 18 (profile 18) shows an aeromagnetic example from Baron et al. (1965) and Heirtzler et al. (1966).

The central anomaly is quite clear on Figures 16 and 18, but is masked slightly on Figure 17. The variation in the size and

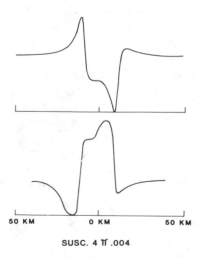

Figure 14. Above is the model magnetic anomaly for a body with an inclination of 15° and an azimuth of 30° (top). Below it is the polar transform of the same anomaly.

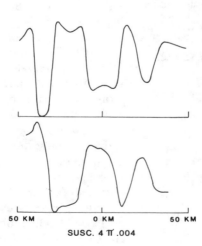

Figure 15. Above are three bodies producing magnetic anomalies. Inclination is 0° and the bearing of the three bodies also is 0°. The polar transform anomalies are below.

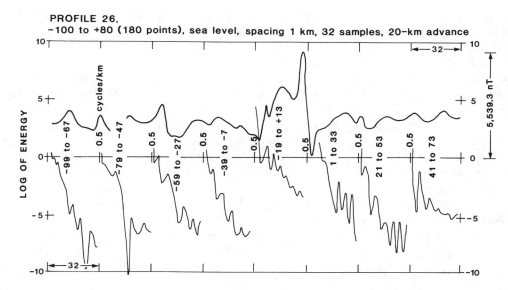

Figure 16. Profile 26 across the Reykjanes Ridge, the observed magnetic anomaly is above. Below we show the spectral energy density obtained using 32-km sample intervals and a 20-km advance from sample interval to sample interval. Profile 26 is from Talwani et al. (1971).

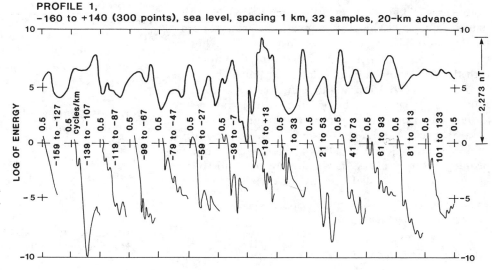

Figure 17. Same as Figure 16, except that this is profile 1 from Talwani et al. (1971).

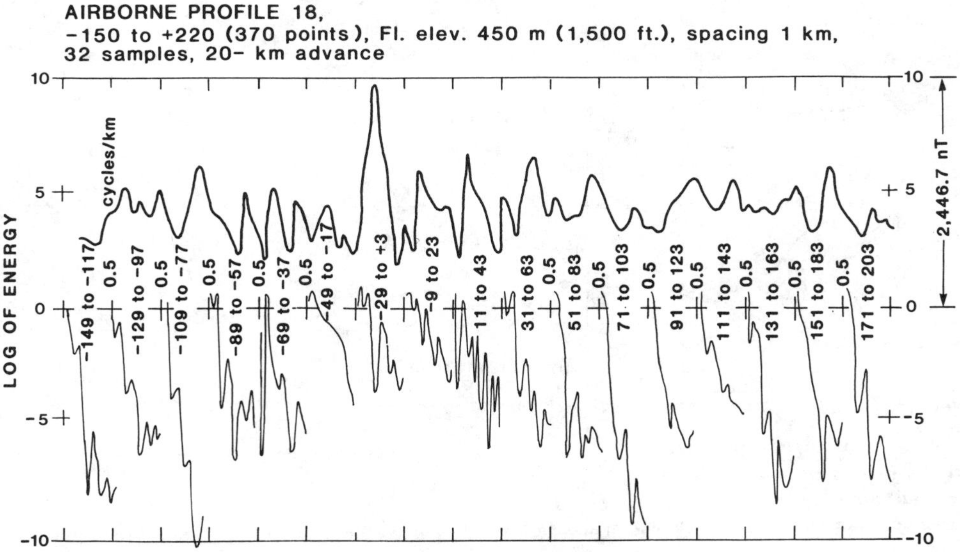

Figure 18. Same as Figure 16, except that this is an airborne profile (no. 18) from Baron et al. (1965) and Heirtzler et al. (1966).

shape of the central anomaly is reflected in the variations of the spectral energy for this anomaly. Thus, on profile 26, the spectral energy decreases in amplitude as it approaches the limiting frequency of 0.5 cycles/km. On profile 1, the DC low-energy zone is succeeded by a decreasing-energy, high-frequency zone. Profile 18 (Fig. 18) shows a low-frequency break at 0.09 cycles/km, and decreasing amplitudes as the 0.5-cycles/km frequency is approached. As expected, the flank anomalies differ from one profile to the next. Therefore, spectral energy distribution cannot be used to correlate events among profiles.

Depth determination to magnetic anomaly sources

Vine and Matthews (1963) did not determine depths to possible sources. Instead, they assumed that the base of the sources is the Curie isotherm, and that this isotherm is 20 km deep beneath ocean basins and ridge flanks, rising to 11 km depth beneath the crestal anomaly. Vine and Wilson (1965) arbitrarily chose the higher of these two values (11 km) so that direct comparison with the results of Vine and Matthews (1963) could be made. Hess (1965) implied that the magnetic anomaly sources are in layer 3 of the oceanic crust, just above the Moho, which he believed to be the boundary between basaltic rocks (above) and serpentinized peridotite (below). Heirtzler et al. (1966) examined the source problem more carefully, stating that the top could be the ocean floor, but was no deeper than 1 km below the floor; thickness, they reasoned, depending on the position of the Curie isotherm, was not likely to be less than 2 km or greater than 9 km; and the base of the source rocks should lie between 4 and 10 km.

Talwani et al. (1971) tied the magnetic source depths to the position of oceanic crustal layer 2 with a velocity of 4.76 km/sec (standard deviation, 0.63 km/s) and a thickness of 1.96 km (standard deviation, 0.64 km). Godby et al. (1968) used an assumed depth of 2 km.

Methods for determining the depth to the sources of magnetic anomalies have been in use for many years. An early paper by Peters (1949) discussed the techniques of computing slope-length depths and second-derivative zero-maximum characteristic depths. The slightly later monograph by Vacquier et al. (1951) is so well known that comment is unnecessary. Werner (1953) developed a procedure for obtaining source depths and positions from magnetic anomalies; this technique was adapted for computer solution. Spector and Grant (1970), Treitel et al. (1971), Green (1972), and Mesko and Kis (1978) developed the spectral-energy density method of depth determination. Phillips (1979) used autocorrelation procedures for depth determination. Mohan et al. (1982) showed the use of the Hilbert transform for depth determinations. Hospers and Rathore (1984) illustrated the results obtained by using slope depths and the second derivative-maximum depth method. Mohan et al. (1986) developed a use of the Mellin transform for gravity interpretation. This technique can be modified for calculating the depths to the magnetic sources.

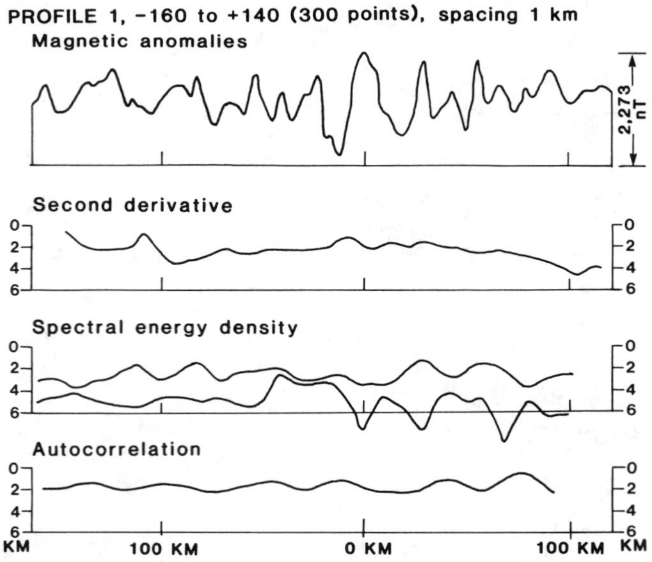

Figure 19. Profile 1 across Reykjanes Ridge; at top is the magnetic anomaly curve. Below is are (consecutively) (1) depths obtained from the second derivative; (2) depths to top and bottom of the anomaly source layer obtained from the spectral layer determined by (3) the autocorrelation method.

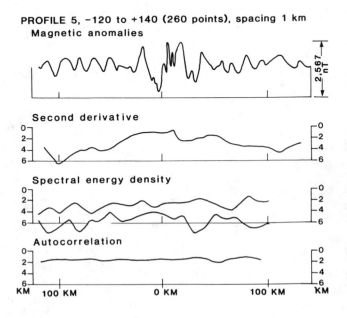

Figure 20. Profile 5 across Reykjanes Ridge: same as for Figure 19.

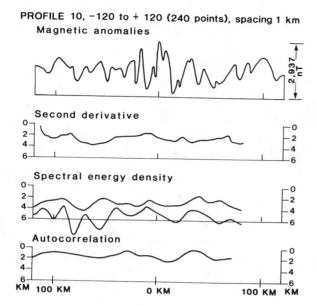

Figure 21. Profile 10 across Reykjanes Ridge: same as for Figure 19.

Our depth determinations on the Reykjanes Ridge were made using the second derivative zero-maximum, and the spectral energy density using a sampling of 32 km with a 20-km advance between adjacent depth zones. We also used the autocorrelation procedure using 30 sampling values (in km) with three lags.

Figures 19, 20, and 21 show profiles 1, 5, and 10 respectively. Each figure shows the magnetic anomaly curve plus (1) depths obtained for the second derivative, (2) spectral energy depths, and (3) depths attained by the autocorrelation method. Use of all of these procedures and presentation on a single figure permit a comparison of the magnetic anomaly with the depth section, as well as the results of depth computations by different methods.

On the three profiles, the general average depth to the top of the magnetic source is 2 km subsea. The second derivative results show a broad high with its apex close to the central anomaly. The spectral energy density indicates a common level at about (average) 2 km subsea, with a deeper horizon (base of anomaly source?) about 2 km below the upper horizon. The autocorrelation results yield three curves that are so close to one another that they could not be shown separately. They yield a general horizon for all three lags at average depth slightly less than 2 km. These results suggest that the anomaly sources lie in layer 2, whose seismically determined depths are of the same order (2.4 km, standard deviation 0.63 km). The thickness of layer 2 by our calculations (approximately 2 km) is close to that indicated for layer 2 by refraction seismology (1.96 km, standard deviation 0.64 km; Talwani et al., 1971).

Spectral energy-derived structure contour maps

The spectral energy depths on profiles 1 through 11 were smoothed at two horizons, one shallow and one deep. Fourier smoothing was used to reduce the sharp and abrupt variations of depth produced by the high frequencies. The results are two structural contour maps (Figs. 22, 23).

Figure 22 shows a high at or very close to the axis of the Reykjanes Ridge. The high is closed locally, but the control indicates that the high extends northeastward and southwestward beyond the limits of our control. The high axis is flanked both on the northwest and southeast by narrow, elongate synclines. Another high is suggested southeast of, and subparallel with, the high above the ridge axis. A low reentrant at the northwestern ends of profiles 7 through 9 may indicate the presence

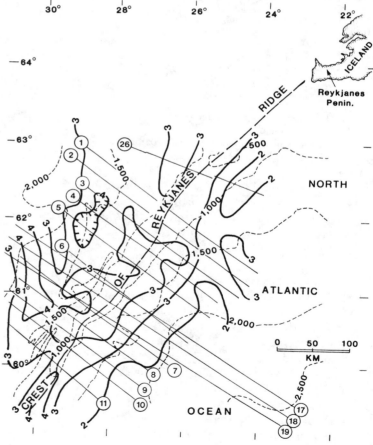

Figure 22. Structural contour map of the Reykjanes Ridge. Structural datum is the top of the anomaly source layer. Depths were obtained from the spectral energy curves. Values are in kilometers. (Dashed lines are water depths in meters.)

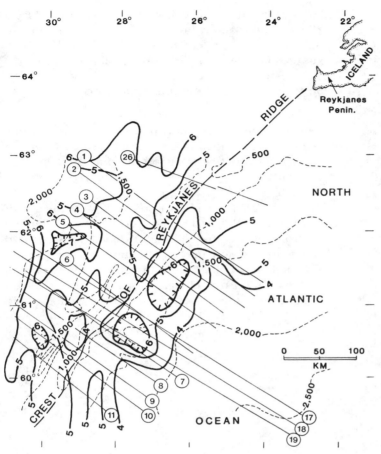

Figure 23. Structural contour map of the Reykjanes Ridge. Structural datum is the base of the anomaly source layer. Depths were obtained from the spectral energy curves. Values are in kilometers. (Dashed lines are water depths in meters.)

there of a ridge-transverse (transform) fault. A similar low trend extends from the northwestern end of profile 5, across the high axis, and eastward to the southeastern end of profile 1. This trend also could indicate the presence of a fault zone.

The deeper horizon is shown on Figure 23. The structure is similar to that of the upper horizon. A high axis coincides with the crest of the Reykjanes Ridge. This high is flanked on the northwest and southeast by lows. A cross-trending low extending southeastward from the western end of profiles 5 and 6 to the low at the eastern ends of profiles 1 through 3 may be the same as the second transverse trend mentioned for Figure 22, and probably indicates the presence here of a ridge-transverse fault.

Thus our study shows the top of the magnetic anomaly source to lie generally 2 to 4 km below sea level; the base is about 4 to 7 km below sea level; the thickness averages slightly more than 2 km. Overall, the maps suggest that layer 2 (if this is what it is) deepens slightly toward the northwest, with the highest parts of both horizons 75–100 km southeast of the ridge crest. More control is needed to confirm this. It is apparent from Figures 22 and 23 that, had maps such as these been prepared for the DSDP, some interesting tests of the magnetic source layer could have been drilled. However, to the best of our knowledge, such maps were not constructed and investigations of layer 2 were based on seismic data (e.g., Leg 27).

Our work indicates that layer 2 is essentially continuous across the axial part of the Reykjanes Ridge. It may be broken by vertical faults that parallel the ridge, but there is little if any horizontal separation. Our work further indicates that whatever processes are at work beneath the ridge, those processes produce a horizontally layered ridge structure, not one that would be anticipated if vertical dikes are the principal building blocks. Our work further suggests that lateral heterogeneity is a characteristic of midocean ridge structure. The drilling results from the DSDP's Leg 37 support this conclusion.

LEG 37 AND LEG 43 RESULTS

During the course of Leg 37, four sites were drilled, two of them close to anomaly 5 (alleged to be 8–10 Ma). All sites are west of the ridge crest in an area bracketed by the following coordinates, 36°51′–37°18′ N and 33°47′–35°12′ W. This area is 2,600 km south of the Reykjanes Ridge. The principle purpose of Leg 37 was to determine the petrological and geophysical characteristics of layer 2 (Robinson et al., 1977). Of the sites drilled (332–335), three penetrated more than 500 m of mafic rock, with the deepest penetration being 721.5 m at site 332. Poorly oriented cores were recovered so that meaningful magnetic studies could be carried out. (Before Leg 37, oriented cores had not been taken.)

Hall and Ryall (1977, p. 425) assessing the magnetic results obtained from Leg 37 wrote that "In terms of remanent intensity and induced magnetization, the basalts fit the requirements of the Vine-Matthews hypothesis. However, in terms of vertical polarity layering, and abundant very shallow nondipole NRM inclinations at Sites 332 and 333, they do not fit the hypothesis."

Before Leg 37, layer 2 was believed to be the source of the linear magnetic anomalies. Layer 2, as it turned out, was one surprise after the other (Robinson et al., 1977). This layer has (1) a great range in magnetization values, from 1 Am^{-1} to 10 Am^{-1}; (2) a dearth of rocks capable of producing the observed anomalies; (3) a very wide range in magnetic inclinations, from +86° through horizontal to -73°; (4) many alternate layers of normally and reversely magnetized rock; and (5) rocks of many ages within single sites. For example, a huge gabbro block in core 22, site 334, Leg 37, yielded a $^{40}Ar/^{39}Ar$ date of 635 ± 102 Ma (Reynolds and Clay, 1977). This block, possibly part of a mélange, lies in a middle Miocene deep-sea ooze and is overlain by flat-lying Miocene and younger sediments and basalt flows. The great size of the block—several meters across—requires a nearby source, a fact that was not addressed in the cruise synthesis (Robinson et al., 1977).

Two Paleozoic (323, 381 Ma; Carboniferous, Devonian) and one Proterozoic (671 Ma) dates were obtained at Sites 386 and 387, Leg 43, on the Bermuda Rise. Two of the three may involve an Ar retention problem, but the 381-Ma date cannot be explained easily (Houghton et al., 1979).

The findings of Legs 36 and 43—and of subsequent legs in which mafic rock drilling was conducted—strengthen our conclusions that (1) great lithologic diversity characterizes the midocean ridges, (2) the rocks present are of many ages, and (3) the measured magnetic susceptibility values from ocean-floor rocks are more than adequate to produce the linear magnetic anomalies. Like Hall and Ryall (1977), we found many facts that militate against the Morley-Vine-Matthews hypothesis.

CONCLUSIONS

Our principal conclusions follow.

1. An objective technique using a correlation coefficient to test magnetic anomaly correlations, both along strike and across the crest of the Reykjanes Ridge, demonstrates that the good correlations claimed (e.g., Heirtzler et al., 1966) are at best poor. The coefficient for ridge-parallel correlations is 0.31 (out of 1.00); that for transridge correlations is 0.17 (also out of 1.00).

2. Correlations claimed in the literature have been entirely subjective with no attempt to quantify them and render them more objective. Most of the claims for good worldwide anomaly correlations do not take into account (a) changes in magnetic latitude, (b) orientation and shape of the anomaly source, (c) orientation of the magnetic profile, and (d) variable depths to the sources. Our results show that the numerous undocumented claims of good correlations worldwide are questionable.

3. Magnetic susceptibility values of normally magnetized rocks on the ocean floor are more than sufficient to produce the linear magnetic anomalies. The Morley-Vine-Matthews hypothesis does not explain the observed anomalies nearly as well as the magnetic susceptibility-contrast hypothesis documented here.

4. The source of the anomalies is a flat-lying body, slightly more than 2 km thick, whose top lies at 2 to 4 km. The vertical dike model accepted by most workers is unlikely.

5. Studies similar to the one presented here should be conducted for all parts of the midocean ridge system. Such studies should include transforms to the pole as a routine procedure.

6. Moreover, rigorous testing of the Morley-Vine-Matthews hypothesis must be conducted. This hypothesis can be proved or disproved by the steps outlined in this paper.

As a general observation, we note that some workers are seeking new tectonic hypotheses and concepts (Wood, 1985; Meyerhoff et al., 1988; Salvador, 1988; Morris et al., 1990). This suggests that problems with currently popular hypotheses are multiplying, perhaps to the point that a new hypothesis is needed. Certainly some new ideas would be welcome. In recent months, at least one totally new hypothesis has been proposed (Meyerhoff et al., 1988, and this volume), specifically, that of surge tectonics. Whatever the final answer, any new hypothesis must be consistent with the findings reported here, for it is clear that the geology that produces the linear magnetic anomalies is far more complex than any model currently in vogue.

REFERENCES CITED

Aumento, F. and Loncarevic, B.D., 1969, The Mid-Atlantic Ridge near 45° N., III. Bald Mountain: Canadian Journal of Earth Sciences, v. 6, no. 1, p. 11-23.

Baranov, V., 1957, A new method for interpretation of aeromagnetic maps: pseudo-gravimetric anomalies: Geophysics, v. 22, no. 2, p. 359-383.

Baron, J.G., Heirtzler, J.R., and Lorentzen, G.R., 1965, An airborne geomagnetic survey of the Reykjanes Ridge, 1963: United States Naval Oceanographic Office, Information Report H-3-65, 23 p.

Blackett, P.M.S., 1956, Lectures on rock magnetism: Jerusalem, The Weizmann Science Press of Israel, 131 p.

Blakely, R., and Cox, A., 1972, Identification of short polarity events by transforming marine magnetic profiles to the pole: Journal of Geophysical Research, v. 77, no. 23, p. 4339-4349.

Bott, M.H.B., 1983, The crust beneath the Iceland-Faeroe Ridge, in Bott, M.H.B., Saxov, W., Talwani, M., and Thiede, J., eds., Structure and development of the Greenland-Scotland Ridge: New York, Plenum Press, p. 63-75.

Cande, S.C., and Kent, D.V., 1976, Constraints imposed by the shape of marine magnetic anomalies on the magnetic source: Journal of Geophysical Research, v. 81, no. 23, p. 4157-4162.

Dietz, R.S., 1961, Continent and ocean basin evolution by spreading of the sea floor: Nature, v. 190, no. 4779, p. 854-857.

Drake, C.L., and Girdler, R.W., 1964, A geophysical study of the Red Sea: Royal Astronomical Society, Geophysical Journal, v. 8, no. 5, p. 473-495.

Einarsson, Th., 1967, The extent of the Tertiary basalt formation and the structure of Iceland, in Björnsson, S., ed., Iceland and mid-ocean ridges: Societas Scientiarum Islandica, v. 38, p. 170-178.

Einarsson, Tr., 1950, The eruption of Hekla 1947-1948, IV, 5. The basic mechanism of volcanic eruptions and the ultimate causes of volcanism: Societas Scientiarum Islandica, 30 p.

———, 1954, A survey of gravity in Iceland: Societas Scientiarum Islandica, v. 30, p. 1-22.

———, 1960, The plateau basalt areas in Iceland, in Askelsson, Jr., Bödvarsson, G., Einarsson, Tr., Kjartansson, G., and Thorarinsson, S., eds., On the geology and geophysics of Iceland: 21st International Geological Congress, Norden 1960, Guide to Excursion No. H-2, p. 5-20.

———, 1962, Upper Tertiary and Pleistocene rocks in Iceland: Societas Scientiarum Islandica, v. 36, 197 p.

———, 1965, Remarks on crustal structure in Iceland: Royal Astronomical Society, Geophysical Journal, v. 10, no. 3, p. 283-288.

———, 1967, The Icelandic fracture system and the inferred causal stress field, in Björnsson, S., ed., Iceland and mid-ocean ridges: Societas Scientiarum Islandica, v. 38, p. 128-139.

———, 1968, Submarine ridges as an effect of stress fields: Journal of Geophysical Research, v. 73, no. 24, p. 7561-7576.

Einarsson, Tr., and Meyerhoff, A.A., 1973, Continental drift, VI: tectonic features of Iceland: unpubl. ms., Iceland Science Institute Archives, 43 p. + 21 figures.

Fleischer, U., 1974, The Reykjanes Ridge—a summary of geophysical data, in Kristjánsson, L., ed., Geodynamics of Iceland and the North Atlantic area; Dordrecht-Holland, D. Reidel Publishing Company, p. 17-31.

Furon, R., 1949, Sur les trilobites draguées à 4255 m de profondeur par le Talisman (1883): Paris, Académie des Sciences, Compte Rendu, v. 228, no. 19, p. 1509-1510.

Godby, E.A., Hood, P.J., and Bower, M.E., 1968, Aeromagnetic profiles across the Reykjanes Ridge southwest of Iceland: Journal of Geophysical Research, v. 73, no. 24, p. 7637-7649.

Green, A.F., 1972, Magnetic profile analysis: Royal Astronomical Society, Geophysical Journal, v. 30, p. 393-403.

Green, J.R., and Margerison, D., 1978, Statistical treatment of experimental data: Amsterdam, Elsevier Scientific Publishing Company, 383 p.

Gudmundsson, G., 1967, Magnetic anomalies, in Björnsson, S., ed., Iceland and mid-ocean ridges: Societas Scientiarum Islandica, v. 38, p. 97-105.

Hall, J.M., and Ryall, P.J.C., 1977, Paleomagnetism of basement rocks, Leg 37, in Aumento, F., Melson, W.B., et al., 1977, Initial reports of the Deep Sea Drilling Project, Leg 37: Washington, D.C., United States Government Printing Office, p. 425-448.

Harrison, C.G.A., 1968, Formation of magnetic anomaly patterns by dyke injection: Journal of Geophysical Research, v. 73, no. 6, p. 2137-2142.

———, 1976, Magnetisation of the oceanic crust: Royal Astronomical Society, Geophysical Journal, v. 47, p. 257-283.

Heirtzler, J.R., and Le Pichon, X., 1965, Crustal structure of the mid-ocean ridges, 3. Magnetic anomalies over the Mid-Atlantic Ridge: Journal of Geophysical Research, v. 70, no. 16, p. 4013-4033.

Heirtzler, J.R., Le Pichon, X., and Baron, J.G., 1966, Magnetic anomalies over the Reykjanes Ridge: Deep-Sea Research, v. 13, p. 427-443.

Hess, H.H., 1962, History of ocean basins, in Engel, A.E.J., James, H.L., and Leonard, N.F., eds., Petrologic studies: a volume in honor of A.F. Buddington: Geological Society of America, p. 599-620.

———, 1965, Mid-ocean ridges and tectonics of the sea-floor, in Whittard, W.F., and Bradshaw, R., eds., Submarine geology in geophysics. Proceedings of the 17th Symposium of the Colston Research Society: London, Butterworths, p. 317-332.

Hospers, J., and Rathore, J.S., 1984, Interpretation of aeromagnetic data from the Norwegian section of the North Sea: Geophysical Exploration, v. 32, no. 5, p. 929-942.

Houghton, R.L., Thomas, J.E., Jr., Diecchio, R.J., and Tagliacozzo, A., 1979, Radiometric ages of basalts from DSDP Leg 43: Sites 382 and 385 (New England Seamounts), 384 (J-anomaly), 386 and 387 (central and western Bermuda Rise), in Tucholke, B.E., Vogt, P.R., et al., Initial reports of the Deep Sea Drilling Project, v. 43: Washington, D.C., United States Government Printing Office, p. 739-753.

Irving, E., 1964, Paleomagnetism and its application to geological and geophysical problems: New York, John Wiley and Sons, Inc., 399 p.

Johnson, G.L., 1975, The Jan Mayen Ridge, in Yorath, C.J., Parker, E.R., and Glass, D.J., eds., Canada's continental margins and offshore petroleum exploration: Canadian Society of Petroleum Geologists, Memoir 4, p. 224-233.

Johnson, G.L., and Vogt, P.R., 1973, Mid-Atlantic Ridge from 47° to 51° North: Geological Society of America Bulletin, v. 84, no. 10, p. 3443-3461.

Johnson, G.L., Sommerhoff, G., and Egloff, J., 1975, Structure and morphology of the West Reykjanes basin and the southeast Greenland continental margin: Marine Geology, v. 18, no. 3, p. 175-196.

Kis, K., 1981, Transfer properties of reduction of the magnetic anomalies to the magnetic pole and the magnetic equator: Universitatis Scientiarum Budapestinensis de Rolando Eötvös Nominatae Annales, Sectio Geologica, t. 23, p. 75-88.

———, 1990, Transfer properties of the reduction of magnetic anomalies to the pole and to the equator: Geophysics, v. 55, no. 9, p. 1141-1147.

Kittel, C., 1949, Physical theory of magnetic domains: Reviews of Modern Physics, v. 21, p. 451-483.

Kristjánsson, L., 1976, A marine magnetic survey off southern Iceland: Marine Geophysical Researches, v. 2, p. 315-326.

Le Mouel, J.L., Galdeano, A., and Le Pichon, X., 1972, Remanent magnetization vector direction and the statistical properties of magnetic anomalies: Royal Astronomical Society, Geophysical Journal, v. 30, p. 353-371.

Le Mouel, J.L., Courtillot, V.E., and Galdeano, A., 1974, A simple formalism for the study of transformed aeromagnetic profiles and source location problems: Journal of Geophysical Research, v. 79, no. 2, p. 324-331.

Luyendyk, B.P., and Melson, W.G., 1967, Magnetic properties and petrology of rocks near the crest of the Mid-Atlantic Ridge: Nature, v. 215, no. 5097, p. 147-149.

Martinson, D.G., Menke, W., and Stoffa, P., 1982, An inverse approach to signal correlation: Journal of Geophysical Research, v. 87, no. B6, p. 4807-4818.

Mason, R.G., 1958, A magnetic survey off the west coast of the United States between latitudes 32° and 36° N, longitudes 121° and 128° W: Royal Astronomical Society, Geophysical Journal, v. 1, no. 4, p. 320-329.

Mason, R.G., and Raff, A.D., 1961, Magnetic survey off the west coast of North America, 32° N latitude to 42° N latitude: Geological Society of America Bulletin, v. 72, no. 8, p. 1259-1265.

Matthews, D.H., and Bath, J., 1967, Formation of magnetic anomaly pattern of Mid-Atlantic Ridge: Royal Astronomical Society, Geophysical Journal, v. 13, p. 349-357.

Mesko, A., and Kis, K., 1978, Interpretation of magnetic anomalies by power spectrum: Universitatis Scientiarum Budapestinensis de

Rolando Eötvös Nominatae Annales, Sectio Geologica, t. 20, p. 103-126.

Meyerhoff, A.A., 1981, Ordovician (Caradocian) trilobites and graptolites at 43°54′61″ N. latitude, 29°03′22″ W. longitude.: International Stop Continental Drift Society Newsletter, v. 3, no. 3, p. 1.

Meyerhoff, A.A., and Meyerhoff, H.A., 1974, Tests of plate tectonics, in Kahle, C.F., ed., Plate tectonics—assessments and reassessments: American Association of Petroleum Geologists, Memoir 23, p. 43-145.

Meyerhoff, A.A., Taner, I., Morris, A.E.L., Meyerhoff, H.A., and Martin, B.D., 1988, Surge tectonics: alternative hypothesis of Earth dynamics: Paper presented before Department of Geosciences, Texas Tech University, Lubbock, Texas, April 20, 1988.

Meyerhoff, A.A., Taner, I., Morris, A.E.L., Martin, B.D., Agocs, W.B., and Meyerhoff, H.A., 1991, Surge tectonics: a new hypothesis of Earth dynamics: this volume.

Mohan, N.L., Sundararajan, N., and Seshagiri Rao, S.V., 1982, Interpretation of some two-dimensional magnetic bodies using Hilbert transforms: Geophysics, v. 47, no. 3, p. 376-387.

Mohan, N.L., Anandababu, L., and Seshagiri Rao, S.V., 1986, Gravity interpretation using the Mellin transform: Geophysics, v. 51, no. 1, p. 114-122.

Moorish, P.H., 1965, The physical principles of magnetism: New York, John Wiley and Sons, Inc., 643 p.

Morley, L.W., and Larochelle, A., 1964, Palaeomagnetism as a means of dating geological events, in Osborne, F.F., ed., Geochronology in Canada: Royal Society of Canada, Special Publication no. 8, p. 39-51.

Morris, A.E.L., Taner, I., Meyerhoff, H.A., and Meyerhoff, A.A., 1990, Tectonic evolution of the Caribbean: alternative hypothesis: Geological Society of America, The Geology of North America, Gulf of Mexico-Caribbean Region, v. H. p. 433-457.

Nabighian, M.N., 1972, The analytic signal of two-dimensional magnetic bodies with polygonal cross-sections: its properties and use for automated interpretation: Geophysics, v. 37, no. 3, p. 507-517.

Nagata, T., 1961, Rock magnetism, revised edition: Tokyo, Maruzen Company Ltd., 350 p.

Néel, L., 1949, Théorie du trainage magnétique des ferromagnétiques au grains fins avec applications aux terres cuites: Annales de Géophysique, t. 7, p. 99-136.

———, 1955, Some theoretical aspects of rock magnetism: Philosophical Magazine, Supplement, Advances in Physics, v. 4, p. 191-243.

Nunns, A.G., 1983, Plate tectonic evolution of the Greenland-Scotland Ridge and surrounding regions, in Bott, M.H.P., Saxov, S., Talwani, M., and Thiede, J., eds., Structure and development of the Greenland-Scotland Ridge: New York, Plenum Press, p. 11-30.

Opdyke, N.D., and Hekinian, R., 1967, Magnetic properties of some igneous rocks from the Mid-Atlantic Ridge: Journal of Geophysical Research, v. 72, no. 8, p. 2257-2260.

Pálmason, G., and Saemundsson, K., 1973, Iceland in relation to the Mid-Atlantic Ridge: Reykjavik, Orkustofnun Jardnitadeild OSJHD 7309, 60 p.

Peters, L.J., 1949, The direct approach to magnetic interpretation and its practical application: Geophysics, v. 14, no. 3, p. 290-320.

Phillips, J.D., 1979, ADEPT: a program to estimate depth to magnetic basement from sampled magnetic profiles: United States Geological Survey Open-File Report 79-367, 35 p.

Raff, A.D., and Mason, R.G., 1961, Magnetic survey off the west coast of North America, 40° N. latitude to 52° N. latitude: Geological Society of America Bulletin, v. 72, no. 8, p. 1267-1270.

Reynolds, P.H., and Clay, W., 1977, Leg 37 basalts and gabbro: K-Ar and ^{40}Ar-^{39}Ar dating, in Aumento, F., Melson, W.G., et al., 1977, Initial reports of the Deep Sea Drilling Project, v. 37: Washington, D.C., United States Government Printing Office, p. 629-630.

Robinson, P.T., Hall, J.M., Aumento, F., Melson, W.G., Bougault, H., Dmitriev, L., Fischer, J.F., Flower, M., Howe, R.C., Hyndman, R.D., Miles, G.A., and Wright, T.L., 1977, Leg 37 cruise synthesis: the lithology, structure, petrology and magnetic history of Layer 2, in Aumento, F., Melson, W.G., et al., 1977, Initial reports of the Deep Sea Drilling Project, v. 37: Washington, D.C., United States Government Printing Office, p. 987-997.

Roche, A., 1951, Sur les inversions de l'aimantation remanente des roches volcaniques dans les monts d'Auvergne: Paris, Académie des Sciences, Compte Rendu, v. 233, no. 19, p. 1132-1134.

Saemundsson, K., Kristjánsson, L., McDougall, I., and Watkins, N.D., 1980, K-Ar dating, geological and paleomagnetic study of a 5-km lava succession in northern Iceland: Journal of Geophysical Research, v. 85, no. B7, p. 3628-3646.

Salvador, A., 1988, Caribbean tectonics—facts and fantasy (Abstract): Houston Geological Society Bulletin, v. 30, no. 8, p. 11.

Schouten, J.A., 1970, A fundamental analysis of magnetic anomalies over oceanic ridges: Marine Geophysical Researches, v. 1, p. 111-144.

Serson, P.H., Hannaford, W., and Haines, G.V., 1968, Magnetic anomalies over Iceland: Science, v. 162, no. 3851, p. 355-357.

Spector, A., and Grant, F.S., 1970, Statistical models for interpreting aeromagnetic data: Geophysics, v. 35, no. 3, p. 293-302.

Strangway, D.W., 1970, History of the Earth's magnetic field: New York, McGraw-Hill Book Company, Inc., 168 p.

Talwani, M., and Eldholm, O., 1977, Evolution of the Norwegian-Greenland Sea: Geological Society of America Bulletin, v. 88, no. 7, p. 969-999.

Talwani, M., Windisch, C.C., and Langseth, M.D., Jr., 1971, Reykjanes Ridge crust: a detailed geophysical study: Journal of Geophysical Research, v. 76, no. 2, p. 473-517.

Thoroddsen, Th., 1907, Island, Grundriss der Geographie and Geologie: Petermanns Mitteilungen, Heft 152-153, 358 p.

Treitel, S., Clement, W.C., and Kaul, R.K., 1971, The spectral determination of depths to buried magnetic basement rocks: Royal Astronomical Society, Geophysical Journal, v. 24, p. 415-428.

van Andel, Tj., 1968, The structure and development of rifted midoceanic rises: Journal of Marine Research, v. 26, no. 2, p. 144-161.

Vacquier, V., Steenland, N.C., Henderson, R.G., and Zietz, I., 1951, Interpretation of aeromagnetic maps: Geological Society of America Memoir 47, 151 p.

Vine, F.J., and Matthews, D.H., 1963, Magnetic anomalies over oceanic ridges: Nature, v. 199, no. 4897, p. 947-949.

Vine, F.J., and Wilson, J.T., 1965, Magnetic anomalies over a young oceanic ridge off Vancouver Island, Science, v. 150, no. 3695, p. 485-489.

Vogt, P.R., and Avery, O.E., 1974, Detailed magnetic surveys in the northwest Atlantic and Labrador Sea: Journal of Geophysical Research, v. 79, no. 2, p. 363-389.

Vogt, P.R., Anderson, C.R., Bracey, D.R., and Schneider, E.D., 1970, North Atlantic magnetic smooth zones: Journal of Geophysical Research, v. 75, no. 20, p. 3955-3968.

Vogt, P.R., Johnson, G.L., and Kristjánsson, L., 1980, Morphology and magnetic anomalies north of Iceland: Journal of Geophysics, v. 47, p. 67-80.

Vogt, P.R., Perry, R.K., Feden, R.H., Fleming, H.S., and Cherkis, N.Z., 1981, The Greenland-Norwegian Sea and Iceland environment: geology and geophysics, in Nairn, A.E.M., Churkin, M., Jr., and

Stehli, F.G., eds., The ocean basins and margins, v. 5. The Arctic Ocean: New York, Plenum Press, p. 493-598.

Voppel, D., and Rudloff, R., 1980, On the evolution of the Reykjanes Ridge south of 60° N between 40 and 12 million years before present: Journal of Geophysics, v. 47, p. 61-66.

Wanless, R.K., Stevens, R.D., Lachance, G.R., and Edmonds, C.M., 1968, Age determinations and geological studies, K-Ar isotopic ages, Report 8: Geological Survey of Canada, Paper 67-2, Part A, 141 p.

Werner, S., 1953, Interpretation of magnetic anomalies as sheet-like bodies: Sveriges Geologiska Undersökning, Årsbok 43, no. 6, series C, no. 508, 130 p.

Wood, B.G.M., 1985, The mechanics of progressive deformation in crustal plates—a working model for southeast Asia: Geological Society of Malaysia Bulletin, no. 19, p. 55-99.

Wyllie, P.J., 1971, The dynamic Earth: New York, John Wiley and Sons, 416 p.

Zverev, S.M., Kosminskaya, I.P., Krasil'shchikova, G.A., and Mikhota, G.G., 1975 (1977), Deep structure of Iceland and the Iceland-Faeroe-Shetlan region based on seismic studies (NASP-72): International Geology Review, v. 19, no. 1, p. 11-24.

ACKNOWLEDGMENTS

We thank Evelyn Rivas for her invaluable help with the literature; Ernestine R. Voyles for typing; and Kathryn L. Meyerhoff for drafting. Encouragement, suggestions, and/or reviews were made of earlier drafts by Donald L. Baars, Charles Ducloz, Peter Gretener, Maurice Kamen-Kaye, Konrad B. Krauskopf, Paul D. Lowman, Jr., Bruce D. Martin, A. Mesko, Danilo Rigassi, William Stannage, Nelson C. Steenland, Irfan Taner, Patrick T. Taylor, and George A. Thompson. The research was done at the writers' own expense.

Paleobiogeography

Paleofloras, faunas, and continental drift: some problem areas

Charles J. Smiley, Department of Geology and Geological Engineering, College of Mines and Earth Resources, University of Idaho, Moscow, Idaho 83843 USA

ABSTRACT

Certain facts and factors relating to the later Paleozoic, Mesozoic, and Cenozoic distributions of terrestrial plants and vertebrate tetrapods, and of contemporaneous marine faunal realms, do not seem to support conventional ideas of plate tectonics—especially as they relate to significant continental movements (continental drift) and polar wandering. The evidence that ranges from the Paleozoic to Recent comes from numerous fossil sites that are widely distributed over present continental areas. Major concerns are: (1) latitudinal zonations of biologic and sedimentary provinces, (2) temporal replacements of floras and faunas of one region by a different flora or fauna from an adjacent area, and (3) similar biologic history on currently adjacent land areas.

These data seem best explained by continental and polar stability rather than by classical drift since later Paleozoic times: (1) the Australia-New Guinea block at or near its present location, (2) India at or near its present position, (3) North America near its present position, which would provide land routes of exchange for terrestrial organisms across both the North Pacific and North Atlantic avenues of interchange, and (4) past latitudinal zonations of floras and faunas, and of zonal boundaries that approximate modern isotherm lines. Conventional geophysical theories that require significant past movements of major continental masses and of rotational poles create certain biogeographic problems that otherwise would not exist.

INTRODUCTION

This report attempts to bring together certain facts and factors of historical biogeography as they relate to modern concepts of plate tectonics, continental drift, and polar wandering. Primary emphasis is placed on the historical records of land plants and of their latitudinal zonations across continents in present positions. The floral data are compared to documented records of (1) terrestrial vertebrates, (2) marine faunas, and (3) temperature-controlled sedimentary deposits.

The global history of vegetation and factors that seem to control the latitudinal distributions of continental floras through time, are well-documented in the literature (e.g., Bailey and Sinnott, 1916; Seward, 1933; Chaney, 1940, 1947; Just, 1952; Krishnan, 1954; Wagner, 1962; Axelrod, 1963; Vakhrameev, 1964, 1986; Asama, 1966; Surange, 1966; Smiley, 1967, 1974, 1976, 1989; Zimina, 1967; Lacey, 1975; Mi et al., 1984; Duan, 1987; Wolfe and Upchurch, 1987). Comparisons between phytogeography and the recorded past distributions of vertebrates are based in large part on the recent analysis of Indian tetrapod genera by Chatterjee and Hotton (1986) who emphasized continental interchanges for the later Paleozoic, Mesozoic, and Cenozoic, and the analysis of North American Cenozoic mammals by Simpson (1947) who dealt with Eurasia-North America interchanges.

Comparisons between terrestrial organisms and marine faunas are based on syntheses of marine faunal-sedimentary realms as they relate to latitudinal zonations and interregional exchanges, such as: Durham, 1950, 1952 (Cenozoic molluscan faunas); Minato et al., 1965 (geology and paleobiology of the Japanese Islands); Khudoley, 1974, 1988 (Mesozoic ammonoids and associated carbonate rocks); Teichert, 1974 (Gondwana marine faunas and sediments); C.R. Ross, 1976 (Paleozoic fusulinid faunas); J.R.P. Ross, 1976 (Paleozoic bryozoan faunas); Jin, 1981 (faunal-sediment relations in China); Wen, 1981 (Paleozoic marine faunas of Tibet and contiguous areas); Yang et al., 1983 (marine faunal realms of China); Xi (Ed.), 1985 (paleogeography of China); Waterhouse and Gupta, 1979, and Dickins and Shah, 1989 (India-Eurasia fossil relations); and Taylor et al., 1973 (data from the southeast Asia-Indonesia region).

At this point, it is important to emphasize that the present report represents a compilation of documented facts and a global integration of diverse data and, further, that specific studies when taken out of global context may lead to quite different interpretations of global tectonics.

SOME RELEVANT FACTORS

Definitions

For present considerations, historical biogeography refers to past distributions of land animals and plants over continental surfaces; to marine faunas of coastal and epeiric seas; to latitudinal zonations of cooler and warmer floras, faunas, and depositional settings; to past ecotonal sites of taxonomic admixing, comparable to transition zones that exist between contiguous ecosystems of the present day; to routes of biologic interchange that may be inferred between terrestrial or oceanic habitats; and to documented changes in plant and animal distributions through time.

In the present concept of plate tectonics, continental drift is used here in reference to putative dislocations of large continental masses across many degrees of latitude or longitude. Polar wandering refers to poles of rotation

---

Chatterjee, S., and N. Hotton III, eds. *New Concepts in Global Tectonics.* Texas Tech University Press, Lubbock, 1992, xii + 450 pp.

rather than to geomagnetic poles. The concept of paleolatitudes is used here in reference to concentric lines or zonations that circumscribe a pole of rotation. The common assumption of a close historical coincidence between rotational and magnetic poles seems unwarranted, as the two poles are now known to be widely divergent on different planets of the Solar System (e.g., Earth, Neptune, Uranus). Of importance for present considerations is the fact that global isotherm lines, and the latitudinal zonations of plants and animals, generally conform to rotational rather than to geomagnetic points of reference.

Plant and animal paleogeography

Important biogeographic factors, ones that have become recognized by botanists and zoologists for more than a century, have been summarized by me before (1967, 1974).

1. Fluid environments above the solid crust of the Earth (hydrosphere and atmosphere) are influenced by the energy relations that exist between sun and planet. This Sun-Earth relationship dictates the presence of a warmer equatorial belt and a progressive poleward cooling for the planet as a whole. Such latitudinal zonations seem always to have prevailed, as widely recognized from the distributions of fossil organisms and of temperature-controlled sediments throughout Phanerozoic time. One could expect, therefore, that climatic zonations and resultant paleogeographic distributions of plants and animals will serve as one of the scientific truisms to be considered.

2. The cooler global climates of the Quaternary differ from the typically warmer climates of earlier geologic time. The Quaternary can, however, serve as a general guide for the interpretation of a similarly cool-climate episode such as that of the Permo-Carboniferous: (a) during intervals of cooler global climates as at present, the tropical belt constricted, and temperate zones shifted equatorward; and (b) during times of warmer global climates the tropical zone expanded, and temperate zones shifted poleward.

3. Continental climates are more extreme (that is, they can exert greater temperature-stress and water-stress on terrestrial organisms) than those that persist under the moderating influences of large bodies of surface water. For terrestrial organisms, critical climatic factors are (a) occurrences of freezing temperatures and (b) a continual supply of potable water. For shallow marine and planktonic organisms, temperature stresses are the most important climatic factors.

4. Fossil records of terrestrial organisms represent a body of interrelated biologic-ecologic factors that correspond to a mosaic of physical conditions that prevailed between: (a) the relative positions of continents, oceans and rotational poles; (b) continental margins versus interiors of landmasses; and (c) orographic or rain-shadow results of regional tectonic and climatic history.

5. Well-preserved megafossil remains provide the best tangible evidence that a past organism actually lived in a particular place at a particular time. In contrast, many microstructures may be subject to transport over considerable distances through the agencies of wind and water currents, or to a greater degree of epigenetic erosion and reworking, than is possible for intact fossils of megastructures. Plant and animal megafossils are considered more likely to represent those organisms that lived near the depositional site at the time of burial.

6. Fossil assemblages, coupled with evidence of taphonomy and sedimentology, are considered best for paleoecologic interpretations (that is, an emphasis on paleosynecology rather than on paleoautecology, see Ager, 1963).

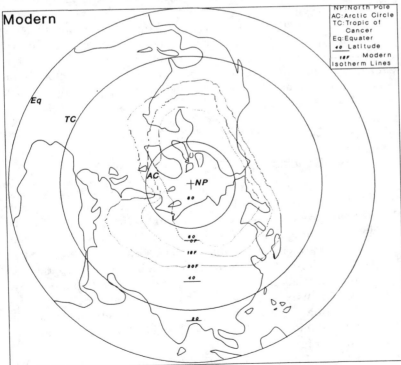

Figure 1. Base map on a north-polar projection. Modern positions of landmasses, oceans, and the North rotational pole. Present isotherm lines are midwinter (after Reed, 1941, p. 666, fig. 1).

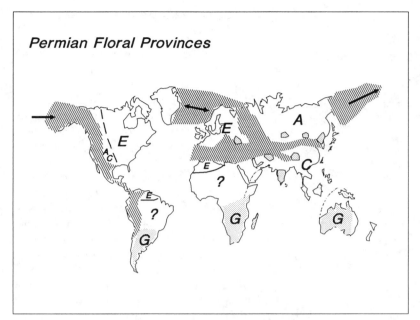

Figure 2. Later Paleozoic floral provinces (after Smiley, 1974, 1976). Note the south and southeast "Laurasian" records of typical "Gondwana" floras characterized by species of *Glossopteris* and *Gangamopteris*.

7. Ecotones containing mixed floras or faunas now occur in transition zones between adjacent biologic-ecologic systems. The paleogeographic plotting of past ecotonal floras and faunas thus can provide clues to the geographic closeness of the different paleoprovinces that provided the taxa of admixing. Ecotonal areas, when based on well-preserved plant and animal megafossil records, are considered here to represent the most critical evidence for interpreting the relative proximity of past terrestrial realms or the occurrence of past avenues of intercontinental interchange.

Plant criteria used for paleoclimatic inferences

Paleoclimatic, especially latitudinal, interpretations of past floral provinces are based on factors long known to reflect the relative temperateness and the relative tropicality of vegetation. When such factors are coupled with a paleogeographic plotting of cooler and warmer floras, certain past global patterns become evident. For paleoclimatic analyses of past floras, particularly for the Mesozoic and Cenozoic, the following criteria apply.

Relative temperateness (cooler climates)—(1) ferns of known high-latitude distribution; (2) a low proportion of Mesozoic Cycadophytes, with *Nilssonia* species common; (3) a high proportion of Ginkgophyte species; (4) a high proportion of needled conifer species; and (5) after angiosperms became dominant in later Mesozoic floras, a relatively high proportion of toothed-margin and thin-texture "deciduous" dicot leaves and a relatively low proportion of entire-margin and thick-texture "evergreen" dicot leaves.

Relative tropicality (warmer climates)—(1) ferns of known lower latitude distribution; (2) a high proportion of Cycadophyte species, with *Otozamites* species common; (3) species of Ginkgophytes are few or absent; (4) species of needled conifers are rare or absent, although scale-leaved conifers may be present; and (5) a relatively high proportion of "evergreen" dicot species compared to "deciduous" species.

Ecotonal floras of the Carboniferous and Permian

Ecotonal areas of taxonomic admixing are well-documented in paleobotanical literature for later Paleozoic time (e.g., Just, 1952; Krishnan, 1954; Wagner, 1962; Zimina, 1967; Smiley, 1974, 1976; Lacey, 1975). These include: (1) western New Guinea (a Permian flora with Gondwana and Cathaysia taxa); (2) Hazro flora of eastern Turkey (a Permian assemblage with Gondwana, Euramerica, Angara, and Cathaysia taxa); (3) eastern and northeastern Africa (Permian floras with Gondwana and Euramerica taxa); (4) South America (Permian floras with Gondwana and Euramerica taxa); (5) Himalaya, Tibet, and regions to the north and northeast (Permian floras with Gondwana, Angara, and Cathaysia taxa); (6) the Beringia-North Pacific region (Permian, Mesozoic, and Cenozoic floras with east Asia, Siberia, and North America taxa); and (7) the circum-North Atlantic region (Carboniferous to Recent floras with North America and west Eurasia taxa).

For the later Paleozoic, on the basis of ecotonal floras, it would appear that continental India was near the Angara and Cathaysia land floral provinces of Eurasia, and the Australia–New Guinea continental block was near the Cathaysia province of southeastern Eurasia. Positions of these Gondwana (India and Australia) landmasses at high southern latitudes in a classical Pangaea reconstruction would result in oceanic separation that seems impossibly wide for interchange of viable floral structures to have occurred. Furthermore, land floras of eastern North America show affinities with ones of western Eurasia (Euramerica), whereas land floras of western North America show affinities with ones of eastern Eurasia (Angara and Cathaysia). Avenues of interchange, both eastward and westward from the North America continent, would seem to be required for such floral relationships to have existed in late Paleozoic time.

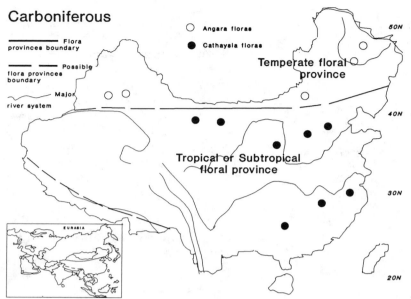

Figure 3. Later Carboniferous floras and floral provinces of China. Compare with Figures 2 and 4.

Evidence for latitudinal zonations of land floras

A plotting of locations of numerous fossil floras across Northern Hemisphere continents shows latitudinal zonations of contemporaneous and sequential floral regions during later Paleozoic, Mesozoic, and Cenozoic times (see Vakhrameev, 1964, 1987; Smiley, 1974, 1976). Prior published maps (equatorial projection) depicting the locations of fossil floras, and of latitudinal zonations of floral provinces, are replotted here on a north-polar projection to reduce cartographic distortion at middle and higher latitudes of the Northern Hemisphere. Added to these paleogeographic maps are modern isotherm lines, for comparisons with interprovince boundaries that have been interpreted between past floral regions.

Global vegetation tends to conform most closely to long-term factors of temperature (coupled with humidity) minima. The midwinter isotherm lines result primarily from relations that exist between poles of rotation and the relative positions of continents and oceans. That is, cold-season isotherm lines now extend into higher latitudes along more equable continental margins, and into lower latitudes across continental interiors, whereas the reverse generally is true for warm-season isotherm lines (Reed, 1941, and Figs. 1 and 2). Of interest is the fact that such modern cold-season isotherm lines generally coincide with past boundary lines that have been inferred between cold and warm floral provinces of the Mesozoic and Cenozoic eras (see later discussions and paleogeographic maps).

Terrestrial vertebrate (tetrapod) evidence

The dispersal of some organs of certain land plants (vegetative or reproductive) across narrow bodies of water has been inferred to result from wind-dissemination, water currents, or rafting; such inferences may be true for certain taxa under specific circumstances, and may be problematical. For larger terrestrial vertebrates, however, a land avenue of interchange would seem a necessity. It is important, therefore to compare the fossil records of these two groups of organisms (terrestrial plants and animals) to determine past links of interchange between ancient land areas.

Chatterjee and Hotton (1986, tables 3 through 11) summarized the following tetrapod relationships between India and other land areas during later Paleozoic, Mesozoic, and Cenozoic times. For the various ages indicated below, the number of genera from India listed (plus Indian endemics) is noted first, followed by the number of Indian genera reported from deposits on other continents (in parentheses):

Permian—9 Indian genera (1 endemic genus): Europe (7); Africa (5); North America (3); East Asia (1) and South America (1); no Indian genus was reported from Australia or Antarctica.

Early Triassic—9 Indian genera (2 endemic genera): Africa (7); Europe (5); east Asia, Australia and Antarctica (4 each); no Indian genus was reported from North or South America.

Middle Triassic—8 Indian genera (no endemic genus): Africa (8); Europe (7); East Asia (5); North and South America (3 each); Australia (2); Antarctica (0).

Later Triassic—10 Indian genera (no endemic genus): North America (9); Europe (8); Africa (6); East Asia and South America (5 each); no Indian genus was reported from Australia or Antarctica.

Lower Jurassic—4 Indian genera (no endemic genus): Europe (4); Africa and South America (2 genera each); North America and Australia (1 genus each); Antarctica (0).

Upper Cretaceous—8 Indian genera (2 endemic genera): North America (6); East Asia (5); Europe (3); Africa and South America (2 genera each); no Indian genus was reported from Australia or Antarctica.

Cenozoic mammals—Chatterjee and Hotton (1986, p. 166–168) noted that recent discoveries of Cenozoic mammals from India "...clearly indicates that mammals were already present in India before its alleged union with Asia . . ." and that affinities of Cenozoic mammals of India always have been with Eurasia and North America. They noted also that at least periodic isolation of Australia, Antarctica, and South America seems to have occurred during Cenozoic time.

Land avenues of intercontinental interchange

Certain land avenues of biologic interchange seem apparent from: (1) documented locations of ecotonal floras and faunas; (2) presence of similar fossil taxa and fossil assemblages on different land areas; (3) fossil records in or near likely interchange routes; (4) stratigraphic documentations of similar biologic histories on presently adjacent land areas; and (5) complementary evidence of different land organisms (e.g., land floras and terrestrial vertebrates).

Indian tetrapod records documented by Chatterjee and Hotton (1986) were interpreted as follows (p. 168):

> No endemic tetrapod family is known from the Lower Permian to Pleistocene deposits of India. From Lower Permian to Cretaceous . . . Indian fauna is overwhelmingly 'Northern.' The highest degree of faunal similarity between India and North America during the Upper Triassic (Carnian) and Upper Cretaceous is striking. The presence of *Lystrosaurus* and *Cynognathus* faunas in China could only be explained if India occupied the position between Africa and Asia, and the Tethys was narrow . . . . The Upper Triassic-Lower Jurassic dinosaurs of India and China are identical, indicating proximity of these two landmasses. High correlations of faunal similarities between India and Africa can be seen from Upper Permian to Mid Triassic indicating a land connection. Indian tetrapods show least similarities with Australia and Antarctica fossils.

Simpson (1947) observed that mammalian interchanges between North America and Eurasia occurred periodically during Cenozoic time, and that (p. 686): "All the faunal evidence is consistent with a single land route, the Bering bridge between Alaska and Siberia, as the sole means of mammalian interchange between Eurasia and North America." The Bering bridge seems to have been a major route of terrestrial vertebrate interchange, although perhaps not the sole Eurasia-America route envisioned by Simpson. Major mammalian interchanges between North and South America did not take place until late in Cenozoic time when the Panamanian land bridge was formed by orogenic activity.

On the basis of such fossil evidence of terrestrial organisms, land routes of interchange from later Paleozoic time would seem to have occurred between (1) India and east Eurasia, (2) India and Africa, (3) the Australia-New Guinea landmass and southeast Eurasia, (4) east Eurasia and west North America, via the Bering bridge, and (5) east North America and west Eurasia, via the North Atlantic route. In contrast, periodic and partial isolation seems evident for Gondwana continents such as South America, Australia, and Antarctica through much of the Mesozoic and Cenozoic.

Numerous studies of tectonic and petrologic data suggest that the ancient circum-Arctic routes of interchange have been floored by continental rather than by oceanic rocks from probably Precambrian time (see compilations of data by Meyerhoff, 1974). Furthermore, an analysis of sea-floor data relating to geophysics, basalts, and sediments of the North Atlantic Ocean area can be ascribed as readily to vertical crustal dynamics as to conventional continental drift (Storetvedt, 1987; Khudoley, 1988). Additionally, minor vertical tectonic movements (or sea-level changes) on the order of approximately 100–200 meters will either emerge or submerge these circum-Arctic routes of intercontinental interchange for terrestrial organisms.

Marine faunal evidence

Marine faunas of shallow continental shelves or of epeiric seas provide additional data on paleogeographic distributions of temperature-controlled biotic associations, especially those that occur on or near continental masses, as follows:

India-South Eurasia—The paleogeographic distributions of marine faunal assemblages conform in general

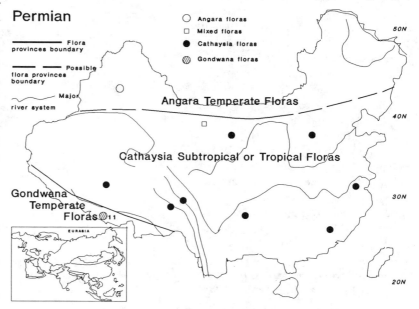

Figure 4. Permian floras and floral realms of China. Floral boundaries conform to marine faunal boundaries of the age, and to modern isotherm lines based on present locations of continents and poles.

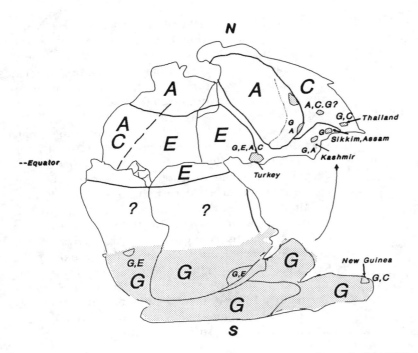

Figure 5. Permian floral provinces on a Pangaea reconstruction (From Smiley, 1974, 1976). Compare with Figures 2 through 4.

to the fossil evidence of terrestrial organisms. In the brief summary of Dickins and Shah (1989), they noted that the earliest Permian marine fauna of India was the cold-water "*Eurydesma*" fauna, overlain by sediments containing temperate-water faunas of Gondwana aspect, and finally by warm-water faunas of east Tethyan relationships. They noted also (p. 10): "The biota of both land and sea indicate India was not close to other Gondwana land masses during the Permian but was part of the southern Asian Tethyan region." Such observations closely parallel climatic trends and interregional relationships that have been recognized for land floras in this part of the world (Krishnan, 1954; Vakhrameev, 1964, 1986; Surange, 1966; Smiley, 1974, 1976), as well as for intercontinental interchanges of terrestrial vertebrates (Simpson, 1947; Chatterjee and Hotton, 1986).

China—Major tectonic provinces of China, including geosynclines and marginal shelves, were recognized in a large volume of paleogeographic maps and accompanying text edited by Xi (1985): (1) a northern Siberia-Mongolia platform; (2) a north China-Tarim platform; (3) a south China domain including the Yangtze platform and the Qiangtang Massif; (4) a southern (Gondwana) domain including the Himalaya and Grandise Massifs; and (5) an east China (circum-Pacific) domain.

Yang et al. (1983) recognized three marine faunal realms in continental eastern Eurasia, north of the Himalayas: (1) a Siberia-Sino/Korea-Tarim region of northern China; (2) the Yangtze-Pearl rivers area of southern China; and (3) the Tibet Plateau and contiguous areas of southwestern China. Concerning marine faunal relations, they concluded (p. 37): (1) the northern marine province has later Paleozoic faunas similar to ones in Siberia and North America with, on the west, an admixture of taxa from central Asia and Europe; a marine regression in later Carboniferous time cut the marine faunal communication between eastern Asia and North America. (2) The southern China marine province contained coral-brachiopod faunas that were at first provincial and, later in the Carboniferous, showed affinities with east and southeast Asia, Europe and North America. (3) The southwestern Xizang marine province (Tibet and vicinity) had coral-brachiopod faunas that showed relationships with both the northern and southern provinces of China.

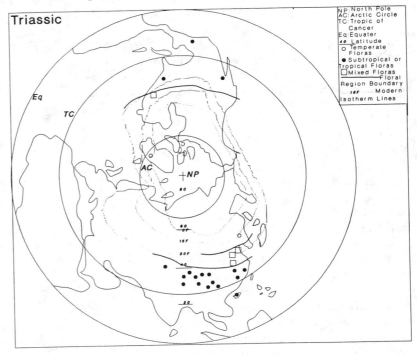

Figure 6. Northern Hemisphere distribution of "cooler" and "warmer" floras of Triassic age (after Smiley, 1974, 1976; Duan, 1987; and Mi et al., 1984). Compare with Figure 7.

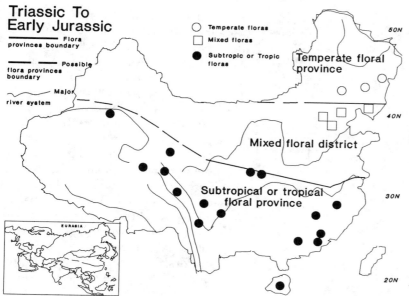

Figure 7. Later Triassic–Early Jurassic floras and floral zonations of China (mainly after Duan, 1987, and Mi et al., 1984). Compare with Figure 6.

Jin (1981), considering Permian faunal realms of the Xizang (Tibet) region, emphasized that faunal composition and lateral distributions appear to be related as much to depositional environments in a geosynclinal setting (onshore-offshore) as to paleoclimatic factors stating (p. 175): ". . . apart from climatic conditions, distribution of marine animals is also dependent on other ecological factors." Wen (1981, and fig. 1) traced the boundaries between the three Permian marine provinces of China westward to Europe. Except for a southeast extension of the warm Cathaysia faunas into equatorial areas of Indonesia (see also Yancey, 1976, fig. 4), he showed the faunal boundaries to approximate modern lines of latitude from eastern to western Eurasia. Wen also showed the Gondwana faunal realm to comprise the region of southern Tibet southward across the Himalaya area to India and Australia, and that it extended westward to encompass regions south of the present Mediterranean Sea. Wen further noted that the later Paleozoic marine faunal zones of eastern Eurasia conform to the Gondwana, Angara, and Cathaysia floral provinces on adjacent landmasses.

In further reference to late Paleozoic geology and paleontology of the Xizang (Tibet) region, Yang et al. (1983) reported a thick sequence of clastics with marine faunal assemblages like ones of Gondwana areas: for example, brachiopod faunas are similar to ones from both the Tianshan area on the north and Australia on the south. They noted, in addition, that the later Paleozoic Gondwana faunas of the Tibet region are present in stratigraphic sections that contain fluvioglacial (diamictite) deposits as occur also in Gondwana regions. Waterhouse and Gupta (1979) noted that Permian marine faunas of southern Tibet resemble those of India. And J.R.P. Ross (1976) reported that late Paleozoic bryozoan faunas of east Tethyan type extend from mainland Eurasia southward to western Australia.

Such marine biologic-paleoecologic relations closely conform to the recorded evidence of terrestrial organisms for the same interval of time: (1) the vertebrate interchanges as summarized by Chatterjee and Hotton (1986), and (2) the distribution of characteristic Gondwana land plant taxa (e.g., *Glossopteris* and *Gangamopteris*) across south and southeast Eurasia (Krishnan, 1954; Surange, 1966; Zimina, 1967; Li, 1981).

Japanese islands—Minato et al. (1965) summarized the geologic development of the Japanese islands and observed that Silurian, Devonian, and early Car-

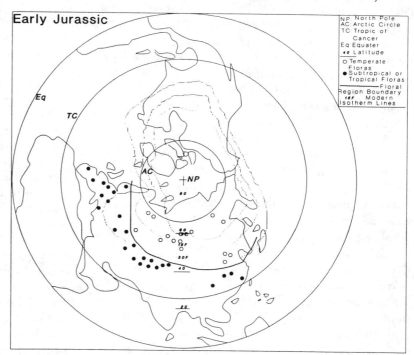

Figure 8. Early Jurassic floras and floral provinces of the Northern Hemisphere (mainly after Vakhrameev, 1964). Compare with Figures 6 and 7. Note the conformance between past province boundaries and modern isotherm lines.

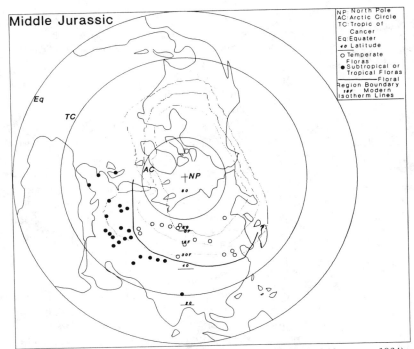

Figure 9. Middle Jurassic floras and floral provinces (mainly after Vakhrameev, 1964). Compare with Figures 6, 7, and 8, and with modern isotherm lines.

boniferous marine faunas were related to ones of similar age that extend southward from the mainland of eastern Eurasia to eastern Australia. Upper Carboniferous and Permian faunas in marine limestones (notably corals and fusulinid forams) also are latitudinally zoned: those from northern Japan show affinities with northeastern Eurasia and North America; ones from southern Japan are related to faunas of the east Tethyan realm that extends from southern China to eastern Australia.

Mesozoic marine faunas of Japan were shown by Minato et al. (1965) to have the following relationships: (1) Triassic faunas are comparable to ones of the east Eurasia mainland and especially to the east Tethyan realm, displaying some affinities also to those of similar age in North America; (2) Jurassic faunas are composed largely of taxa of cosmopolitan distribution, coupled with ones showing closest affinities with the east Tethys-South Pacific region (a few taxa occur in the Boreal-North Pacific province that includes Siberia, Alaska, and Greenland); and (3) Cretaceous marine faunas can be correlated with those of Europe via India and the Mediterranean region, as well as showing affinities with faunas of North America, the Soviet Far East, the Himalayas, Timor, New Zealand, and Australia.

Of particular note for the present analysis is the fact that later Paleozoic to Recent plant fossil assemblages show that the Japanese islands were populated by land plants identical to those found in contemporaneous fossil floras on the mainland to the west: Late Devonian (Kimura et al., 1986); Permian (Asama, 1967); Mesozoic (Oishi, 1940; Kimura, 1988); Tertiary (Tanai, 1963, 1972); and Quaternary (Miki, 1941). At present, Japanese vegetation shows latitudinal changes from cool-climate forests on the north to subtropical forests on the south. Similarly, the latitudinal zonation of modern marine planktonic associations across the Pacific Ocean (Loeblich and Tappan, 1964, p. 126, and fig. 76) shows a subarctic fauna in seas marginal to northern Japan and a tropical fauna on the south.

Such floral and faunal data, present and past, coupled with sedimentary environments and geophysical relations, indicate that the Japanese islands area has served as the eastern margin of the Eurasia continent during much of Phanerozoic time, apparently retaining latitudinal zonations of plants and animals comparable to zonations of the present day.

Southeast Asia-Indonesia—South of the Himalayas and east of India occurs the southeastern extension of the eastern (warm water) Tethyan realm of Phanerozoic age (see also Yancey, 1976, fig. 4; and Wen, 1981, fig. 1). Since the later Paleozoic, this region seems to have been related biologically to eastern Asia on the north and to the Australia-New Guinea-New Zealand area on the south: Gobbett (1973) and Waterhouse (1973) showed that later Carboniferous fusulinid-brachiopod-coral faunas are correlative with those of the Russian Platform; and Permian fusulinid limestones are correlative with ones of the Salt Range of Pakistan, continental east Asia, and southern Japan.

Tectonically, Hamilton (1973) noted that New Guinea, now located along the southeastern border of this region, has been connected at least partly to the Australia landmass since Precambrian time: that is, the southern part of New Guinea was a stable continental shelf of northern Australia until the middle Cenozoic. In reference to accepted plate tectonic theory, Hamilton observed that the northern part of New Guinea was a volcanic island arc during late Mesozoic and early Cenozoic time, representing a tectonic belt that moved southward when the New Guinea-Australia block was moving northward. In theory, an eventual Miocene collision melded the historically different parts to form the large New Guinea island of today. Despite such geophysical interpretations based on plate tectonic theory, it is difficult to explain a Permian ecotonal (megafossil) flora in western New Guinea (Cathaysia taxa derived from southeast Eurasia on the north, admixed with Gondwana taxa derived from Australia on the south), when such floral source areas would have been separated by thousands of kilometers of open ocean according to conventional drift (plate tectonic) assumptions.

Mesozoic ammonoid zonations—In his global analysis of Mesozoic ammonoids and associated sediments emphasizing the circum-Pacific region, Khudoley (1974, figs. 1 and 7) showed these data to be latitudinally zoned: (1) a cooler ammonoid-clastic sediment realm ("Boreal") of circum-polar distribution centering around the present rotational (not geomagnetic) pole; (2) a warmer ("Tethyan") realm of circum-equatorial distribution; and (3) some evidence for a cool realm in high southern latitudes ("Antiboreal"). Throughout the Mesozoic, these temperature-controlled faunal-sediment realms remained distinctive and constant, although the boundaries between them seem to have shifted latitudinally as global climates changed. Varying degrees of endemism were noted at times, serving to delineate faunal provincialism and also the migration routes of taxa from one province to another. Khudoley concluded: (1) that ammonoids and associated sediments were latitudinally zoned on a global basis; (2) that such zonal belts do not conform to conventional models of continental drift or polar wandering; (3) that the data show a high degree of intermigration between regions, conforming to present geographic proximity and present oceanic currents; (4) that they show some degree of regional provincialism that can be inferred from isolation or temperature factors; and (5) that such provincialism suggests a distinctive province in middle and higher latitudes of South America, and also an eastern Tethys province that encompasses the region southward from the Japanese islands, through southern China and India, to Australia and New Zealand (see also Yancey, 1976, fig. 4; Wen, 1981, fig. 1).

The more recent analysis of Mesozoic ammonoids of the circum-Atlantic region by Khudoley (1988) resulted in conclusions that are virtually the same as those of his earlier circum-Pacific study: (1) latitudinal zonations of faunas, (2) intermigrations of some taxa and isolation of others, (3) an apparent requirement of direct oceanic connections from higher to lower latitudes as occurs today along the existing Atlantic Ocean, and (4) sedimentary and fossil evidence (outcrop and drill-core) suggesting the presence of continental shorelines near their present locations on both sides of the Atlantic Ocean. Such integrations of data led Khudoley (1988, p. 632) to conclude that: (1) "The foregoing analysis of the distribution and dispersal of ammonoids shows that throughout the Jurassic, between Western Europe and the adjoining regions of the Mediterranean on the one hand, and the Caribbean and adjacent waters off the American continents and the Boreal ocean on the other, *there was a marine basin inhabited by a community of ammonites* (italics mine)," and (2) "The dispersal of ammonoids in various directions can be explained by global ocean currents, without recourse to continental drift. The character and direction of the marine paleocurrents were close to those of the present day . . . ."

Cenozoic marine faunas—Durham (1950, 1952) noted (1) that Cenozoic marine faunas along coastal areas of the circum-Pacific region have been latitudinally distributed throughout the Cenozoic, and (2) that their zonal boundaries have shifted in conformance with global climatic changes as previously indicated by Chaney (1940) for the Cenozoic records of land floras. Durham observed also (1952, p. 339–340) that Paleogene shallow marine faunas on both sides of the present Atlantic Ocean exhibit low taxonomic relationships and little evidence of transoceanic interchange, and that distributions in the circum-Pacific region conform most closely with oceans, continents, and rotational poles of the present. He noted, in addition, that Oligocene faunas of India show a 55 percent faunal similarity with ones of the same age in Europe, concluding that India was then a part of the Tethyan realm and probably was close to its present geographic position.

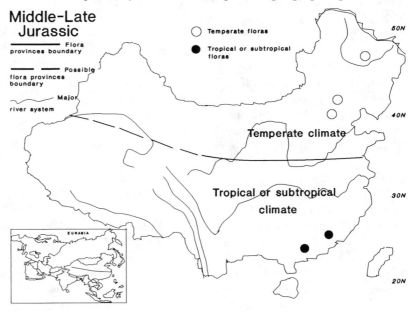

Figure 10. Middle to Late Jurassic floras and floral zonations of China (mainly after Chen et al., 1980-1984). Compare with Figures 9 and 11.

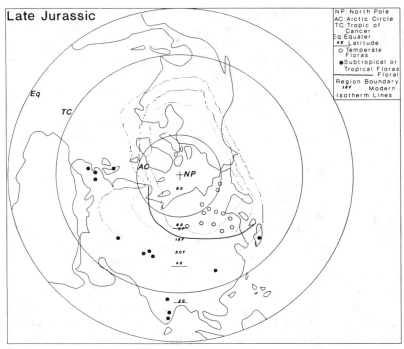

Figure 11. Later Jurassic floras and floral zonations of the Northern Hemisphere (mainly after Vakhrameev, 1964). Compare with Figures 6–9, and modern isotherm lines.

Depositional environments

A global analysis of latitudinal zonations of sedimentary deposits undertaken by Meyerhoff and Meyerhoff (1974), included records of evaporites, desert deposits, coals, tillites, and marine carbonates. For paleoclimatic inferences, evaporites and desert deposits represent conditions of low atmospheric humidity, generally at lower to middle latitudes. Peat and coal represent continental areas of high atmospheric humidity, with stagnant surface water in bottomland sites where plant debris can accumulate in anoxic aquatic environments (regardless of latitude). Tillites (or diamictites) represent areas usually marginal to cold and humid (commonly upland) areas where ice-field concentrations exceed approximately 200 meters. Organic limestones with invertebrate megafossils are most typical of warm, shallow seas such as those of epeiric and shelf environments of lower latitudes.

Desert and evaporite (low humidity) records—"In all rocks younger than middle Proterozoic, at least 95 percent of all evaporite deposits, by area and by volume, is in those areas under the influence of *today's* dry-wind belts." (Meyerhoff and Meyerhoff, 1974, p. 49, and figs. 3 through 7). Khudoley (1974, p. 323) noted that Mesozoic evaporite and carbonate rocks are "widespread along the eastern and western margins of the Pacific Ocean" and that they form an equatorial belt that is "offset slightly toward the north, similar to the offset of today's meteorological equator." Khudoley (1988) continued his studies across the circum-Atlantic area, reaching similar conclusions.

The latitudinal zonations of desert and evaporite deposition appear to have remained constant since Precambrian times, with the widths of dry-climate zones fluctuating throughout Earth history (Meyerhoff and Meyerhoff, 1974, and fig. 5). A historical example of such latitudinal zonation is the Jurassic Sundance Sea of western North America, which shows a change from a lower latitude (warmer-dry) climate to a higher latitude (cooler-wet) climate (Smiley, 1974, p. 346-350, and figs. 7 and 8); this epicontinental sea extended southward from the Arctic Ocean to terminate in the lee (rain shadow) of the Nevadan orogenic belt (Kummel, 1970, fig. 9-4). Near the southern margin of the seaway occur the spectacular "frozen sanddunes" of Navajo age that are associated locally with evaporite (gypsum) deposits; but northward in southwestern Canada are rich deposits of coal indicating a more humid climatic and depositional setting.

Marine carbonates—Meyerhoff and Meyerhoff (1974) summarized a large body of published data on marine carbonate distribution, and noted (p. 53): "Phanerozoic carbonate rocks occupy a broad belt that is approximately symmetrical with the present *thermal* equator."

The later Carboniferous and Permian distribution of fusulinid limestones (warm-water, shallow marine) was plotted by C. A. Ross (1976) who stated (p. 219–220): "In general, fusulinaceans occur in normal marine limestones that are associated with bioherms and banks in which coral, algal, and echinodermal fragments are common." Ross (p. 221 and fig. 8) showed that such limestones occur largely in the Northern Hemisphere, mainly at middle latitudes of North America and Eurasia. For the eastern Tethys, Ross noted also a southward extension into the Malay-Indonesia domain similar to evidence documented by Yancey (1976) and Wen (1981). Regarding Gondwana continents, Ross reported fusulinid limestones from the northern part of South America, northern Africa, and New Zealand, and noted that they are widespread in the Himalaya region—a region that is known to have been biologically transitional from northern India to continental Eurasia. No fusulinid limestones were reported by Ross from higher latitudes of the Northern or Southern hemispheres, apparently representing marine waters too cold for fusulinids (or extensive marine limestones) to prevail.

Ammonoid-bearing limestones of Mesozoic age had a similar equatorial zonation (Khudoley, 1974, fig. 1; 1988). Khudoley (1974) observed that during the Mesozoic (p. 296–298): "... there were two zones—one in the Northern Hemisphere and one in the Southern Hemisphere—where terrigenous and volcanic deposition predominated and an intermediate zone where carbonate and evaporite deposition occurred with the terrigenous clastic and volcanic deposition."

Coal deposition—The accumulation of terrestrial plant debris in sufficient quantity and purity to result in coal deposits of commercial value involves a variety of factors: (1) paleoenvironments of aquatic continental conditions; (2) stagnant-water (anoxic, acidic, toxic) settings; (3) subsidence of nonmarine basins or swampy coastal plains; and (4) subsequent capping of peaty material by clays of low permeability (various reports in Ross and Ross, 1984).

For uniformitarian considerations, the conditions of peat accumulation are apparent from Quaternary evidence: modern accumulations of peat are known to occur in regions ranging from (1) the high Arctic, commonly in tundra areas on top of permafrost; to (2) humid, cool-temperate areas such as marginal to the North Sea (coastal bogs); to (3) humid, warm-temperate areas such as southeastern United States (coastal swamps); to (4) humid, tropical lowlands of southeast Eurasia (plant debris washed down from adjacent slopes, in areas of tropical rain forests). A humid environment with a luxuriant source vegetation is a constant factor; temperature seems to be a variable.

A global survey of coal deposits by Meyerhoff and Meyerhoff (1974) showed the following paleogeographic distribution: (1) major Paleozoic coal regions are known within present latitudinal belts that range from approximately 70° N to approximately 75° S; (2) major Mesozoic coal regions are known from approximately 70° N to approximately 70° S; (3) major Tertiary coals extend from approximately 55° N to approximately 40° S; and (4) Quaternary peat accumulations are known from high-latitude tundra to humid equatorial areas.

For Gondwana and southeast Eurasia regions, White (1925), Krishnan (1954), Kummel (1970), Meyerhoff and Meyerhoff (1974), and Yang et al. (1983) reported that later Paleozoic coals occur in stratigraphic sections that also contain glacial deposits (in India, Australia, Antarctica, southern Africa, southern South America, and in Himalayan areas such as Kashmir and Tibet). As Meyerhoff and Meyerhoff stated (1974, p. 53): "Since Devonian time, coal deposits have formed in two latitudinal belts which are broadly symmetrical with respect to the thermal equator and are poleward from, and parallel with, the evaporite belts..." (see also Krishnan, 1954, fig. 2; Kummel, 1970, fig. 11–9; Meyerhoff and Meyerhoff, 1974, fig. 9).

Glacial deposits—The two most recent glacial episodes of major import occurred during late Paleozoic and late Cenozoic times. Continental glaciation now occurs on Greenland (approximately 60°–80° N) and Antarctica (approximately 70°–90° S). Present upland ice fields with lateral glaciers that terminate at or near sea-level occur in areas such as southern Alaska (approximately 50°–60° N) and the South Island of New Zealand (approximately 45° S). Based on decades of research on the glacial fields of southeastern Alaska, Miller (1985) noted that much of the deposition of clastic debris derived from upland glacial processes occurs at lower elevations far removed from the upland ice-fields, and that thick glacially derived deposits occur also in offshore marine sites adjacent to the montane glacial regions. Furthermore, the temperature of oceanic water is significantly reduced in areas marginal to the glaciated uplands.

When comparing later Paleozoic glaciation of various Gondwana areas, Krishnan (1954, p. 4–5) noted: "In India, these Upper Carboniferous glacial beds are known as the Talchir tillites, the equivalents being the

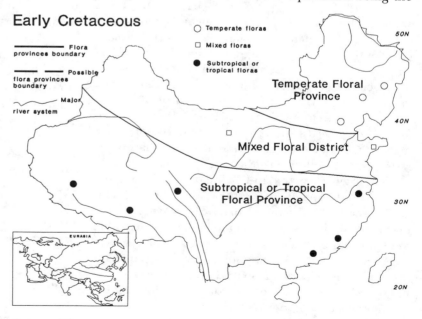

Figure 12. Early Cretaceous floras and floral boundaries of China. Compare with Figures 11 and 13.

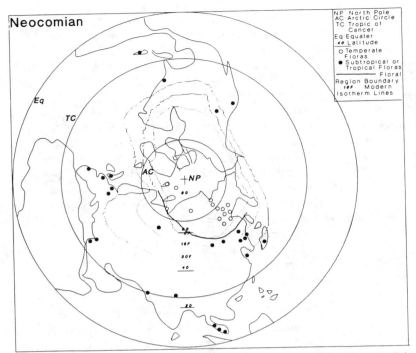

Figure 13. Early Cretaceous (Neocomian) floras of the Northern Hemisphere (mainly after Vakhrameev, 1964). Compare with Figures 11 and 12, and with modern isotherm lines.

Dwyka tillites of South Africa, the Tubarao series of Brazil, the lower part of Patquia in Argentina and the Kutting glacials of Australia." Such paleogeographic centers of glacial origin also were plotted by Kummel (1970, fig. 11-20) on a Pangaea basis, and by Meyerhoff and Meyerhoff (1974, fig. 11) on the basis of continents at present locations. Whereas the paleogeographic map of Kummel showed major glacial areas centering around the "South Pole" with glacial movements generally outward from Antarctica, Meyerhoff and Meyerhoff showed the glacial centers to be scattered, and more restricted in area, with glacial movements outward from each center of glacial accumulation; the latter authors, furthermore, show glacial centers to be distributed widely across high latitudes of northern Eurasia, rather than being restricted to Gondwana landmasses.

In reference to the India-Himalaya-Tibet region, Krishnan (1954) noted that the clasts of the Talchir boulder beds of India appear to have been derived from the ancient mountain range of the Eastern Ghats within the landmass of India, rather than from some source external to India. Furthermore, glacial deposits (or diamictites) are reported to extend northward from India through the Himalayas to the southern Tibet region during the later Carboniferous-Permian interval, where they are associated with cool-water marine faunas and with cool-climate land floras (Krishnan, 1954; Surange, 1966; Zimina, 1967; Meyerhoff and Teichert, 1971; Meyerhoff and Meyerhoff, 1974; Smiley, 1974, 1976, 1989; Waterhouse and Gupta, 1979; Jin, 1981; Li and Yao, 1981; Wen, 1981; Waterhouse, 1982; Yang et al., 1983; Chatterjee and Hotton, 1986).

When one considers such paleo-ecologic-paleo-geographic information relating to India-south Eurasia geologic history, an explanation that could accommodate virtually all facts and factors would be: (1) the present location of India as an upland center of glaciation; (2) marginal sites of glacigene deposition, with peripheral cooling of adjacent marine waters; and (3) paleoclimatic influence on land vegetation, especially at lower elevations and at higher latitudes toward the paleo-North Pole.

PHYTOGEOGRAPHIC EVIDENCE

The following series of maps plots important floral localities of the Northern Hemisphere, spanning an interval of time from the later Paleozoic to the later Mesozoic. Locations and regions of cooler and warmer floras are depicted, as are the approximate boundaries between the temperate and tropical vegetation based on floral assemblage analyses. These floral data are placed on maps of North-Polar projection that show continents, oceans, and rotational poles of the present. Midwinter isotherm lines that largely delimit the global distribution of temperature-controlled vegetation, faunas, and sediments also are presented; their geographic patterns across present continental areas reflect the climatic interrelations that exist between large landmasses and adjacent oceans for specific regions and latitudes (Reed, 1941).

For the present day, latitudes are based on the rotational axis (and poles), rather than on geomagnetic data. For the past, paleolatitudes commonly are based on inferred geomagnetic (rather than on rotational) geographic coordinates; this latter is unfortunate because (as noted earlier) planetary variations between geomagnetic and rotational poles differ by significant degrees of latitude (NASA recently noted a differential of approximately 50°–55° for Neptune and Uranus). The attention of the reader is directed to the degree of conformance that exists between the boundaries of past vegetation zones and the geographic patterns of present isotherm lines. Any significant past deviation in continental and polar positions—and in their historical relations as can be illustrated on a temporal sequence of paleogeographic maps—should become readily apparent.

Later Paleozoic—Figures 2 through 5 document later Paleozoic floras and floral provinces that have long been recognized. Such provinces (or regions) have been referred to by various paleobotanists and paleozoologists. Figure 2 shows the distribution of Gondwana floras (characterized by presence of *Glossopteris* and *Gangamopteris* leaf forms) on the basis of continents in present positions. Figure 5 shows similar data plotted on a classical Pangaea map (after Dietz and Holden, 1970). Figures 3 and 4 depict floras and floral provinces recently reported for China, which show latitudinal floral zonations that conform more to Figure 2 than to Figure 5, and to latitudinal zonations of the present than to inferred past latitudes based largely on geomagnetic data. Documented distributions of Gondwana plant taxa across southern Laurasia would seem to require Gondwana source areas that are in close proximity to Eurasia, rather than at high southern latitudes of an isolated floral region as is required by a classical reconstruction of an original Pangaea supercontinent after Dietz and Holden (1970) or by most later paleogeographic reconstructions.

Triassic to Early Jurassic—Figures 6, 7, and 8 depict Northern Hemisphere floras and floral zonations of the later Triassic to Early Jurassic interval. For latitudinal comparisons of later Triassic floras (Fig. 6), eastern North American (including Greenland) floras show an equatorward warming, from the East Greenland Scorseby Sound flora, to the northern Newark basins, to the southern Newark basins (Fontain, 1883; Newberry, 1888; Harris, 1926; Axelrod, 1963; Smiley, 1974, 1976). For western North America, Axelrod (1963) compared the ecological requirements of the later Triassic floras of Arizona (Daugherty, 1941) to the requirements of a Triassic flora from southern Mexico, noting a similar equatorward warming trend. Such latitudinal floral zonations on both sides of the continent suggest that North America experienced little or no rotation or other movement during this interval of time.

For southeastern Eurasia (China), numerous Triassic floras (Fig. 7) show a latitudinal zonation similar to that noted for North America, with inferred floral boundaries that conform to both the records of North America and to the patterns of modern isotherm lines as shown on Figure 6 (later Triassic), Figure 8 (Early Jurassic) and Figure 1 (present day). Such later Triassic and Early Jurassic floral documentations can be seen to conform to isotherm lines of the present, rather than to inferred isotherm deviations that would be required by significant dislocations of landmasses or of the rotational axis of the planet.

Early to Late Jurassic—Figures 8 through 11 depict Northern Hemisphere floras and floral zonations, largely from records documented by Vakhrameev (1964). On all paleogeographic maps, boundary lines between cooler and warmer vegetational regions conform to isotherm lines of the present. Again, there is little paleoclimatic (paleofloral-paleoecologic) evidence that polar and continental positions have shifted appreciably.

Late Jurassic to Early Cretaceous (Neocomian)—Figures 11, 12, and 13 show the distribution of floras and floral regions of the Northern Hemisphere, largely from Eurasia after Vakhrameev (1964), and from Chen et al. (1980, 1981, 1982, 1983, 1984) for China. As for previous intervals of time, the later Jurassic-Neocomian floral records and vegetational boundaries coincide with the patterns of modern isotherm lines for this part of the world.

Cretaceous—Figures 13 and 14 show Early Cretaceous (Neocomian) and medial Cretaceous (Aptian-Albian) floras of the Northern Hemisphere, with data from

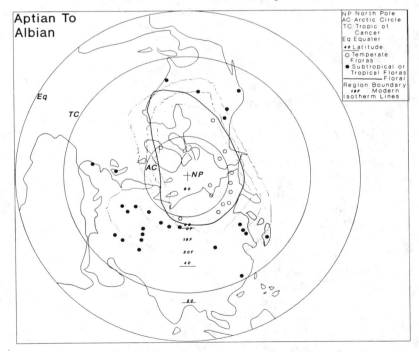

Figure 14. Medial Cretaceous (Aptian–Albian) floras of the Northern Hemisphere (mainly after Vakhrameev, 1964, and Smiley, 1974, 1976). Compare with Figures 12 and 13, and with modern isotherm lines.

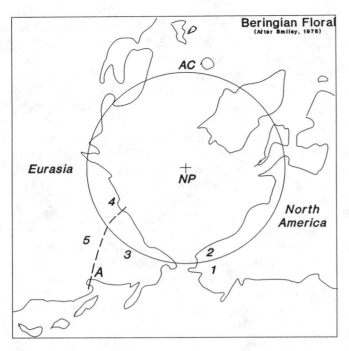

Figure 15. Locations of earlier to later Cretaceous floral successions in the Beringian region (after Smiley, 1974, 1976). Compare with Figure 16.

Eurasia largely after Vakhrameev (1964). Data for North America are based on summations of Smiley (1974, 1976) for early to late Cretaceous time, and from Wolfe and Upchurch (1987) for angiosperm (di-cotyledon) floras of late Cretaceous time. From Early to medial Cretaceous, paleofloral boundaries of Eurasia and North America conform to isotherm lines of the present, as was also the case for floral zonations of previous Mesozoic ages (Figs. 6 through 12).

Figures 15 and 16 from Smiley (1976) show later Mesozoic floral distributions across the Beringian intercontinental route of interchange. For this entire region, floral similarities and differences show only what one might expect on the basis of relative distance from a single point of reference (northern Alaska). Oceanic separations such as are commonly inferred in relation to the present Verkhoyansk foldbelt area (see line A) are not supported by comparative floral data on the east and west side of this Verkhoyansk area of assumed later continental collision (see Churkin, 1972). Later Cretaceous and Cenozoic floral zonations and interchanges conform to isotherm lines and continental positions of the present; they do not conform to paleogeographic conditions that are virtually dictated by later concepts or by recent applications of plate tectonics, continental drift, or polar wandering.

CONCLUSIONS

No single body of data, point of perspective, or concept can serve as a sole viewpoint on which to interpret the complexity of geologic and paleobiologic history of Earth. Attempts have been made here to integrate certain records of past land floras and faunal-sedimentary data, especially for the later Paleozoic-Mesozoic interval, to determine paleozonations of global temperature belts. For determining routes of continental interchange, comparisons are made between the records of land floras and of terrestrial vertebrates (tetrapods). For determining paleolatitudes, reference is made to patterns of modern isotherm lines across continental areas in comparison to past floral, faunal, and sedimentary provinces, rather than to geomagnetic poles that are now known to differ markedly from rotational poles on various planets.

The combined data are best explained by continental and polar stability, rather than by significant past dislocations, as follows.

1. Paleozoic marine carbonates containing fusulinid-coral-bryozoan-brachiopod faunas are distributed in lower latitude belts that conform globally to present geographic coordinates of rotational poles and equatorial zonations.

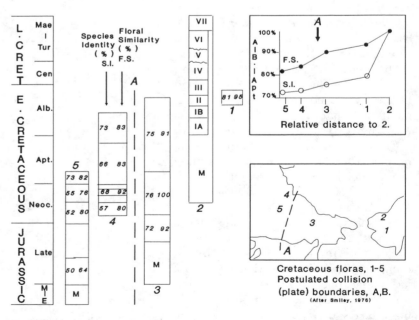

Figure 16. Later Jurassic to medial Cretaceous floral sequences of the Beringian region (after Smiley, 1976). Biostratigraphic correlations and comparative taxonomic affinities show no evidence of oceanic separation of east Asia–North America as required by present continental drift inferences.

2. Mesozoic marine carbonates with ammonoid faunas show a latitudinal zonation similar to that known for the later Paleozoic.

3. Cenozoic marine faunas, both planktonic and benthonic, seem to have remained latitudinally zoned, conforming to modern geographic coordinates.

4. Later Paleozoic, Mesozoic, and Cenozoic records of terrestrial vertebrates (tetrapods) indicate at least periodic land routes of interchange between India and Eurasia, and between Eurasia and North America—suggesting proximity of land areas, and locations of land interchange routes, similar to those of the Quaternary.

5. Ecotonal admixtures of plants from the later Paleozoic through the Cenozoic suggest a past proximity of land areas that conforms with inferences that can be made from records of terrestrial vertebrates (amphibians, reptiles, and mammals).

6. Later Paleozoic to Recent zonations of land vegetation conform lati-tudinally to the global records of marine faunas and sediments, as well as to patterns of modern isotherm lines across continents.

7. The combined evidence most conforms to continents, oceans, rotational poles, and equatorial belts of the present.

The biosphere of Earth occurs, in essence, on top of the planet's solid crust. It is controlled in large measure by principles of solar system and planetary physics, as is the underlying crust. Major physical factors that control the biosphere seem to be those that control temperatures of the hydrosphere and the atmosphere, and consequently the constant or periodic availability of liquid water within a temperature range of approximately 0°– approximately 100° C. Geophysical theories that have been proposed to explain the documented facts of physical geology (that is, of the planet's solid body as currently exemplified) would seem to be suspect, especially when such theories appear to be in nonconformance with other known laws of physics that govern the evolution and distribution of organisms on the planet's surface. Whenever any theory of planetary physics neglects to take into account the known distributions of temperature-controlled floras, faunas, and chemical sediments— especially when postulating relative locations of continents, of warm-climate or of cold-climate belts, and of poles of rotation—that basic theory is, in my opinion, subject to question. No single concept of planetary geophysics can be legitimately defended unless it also takes into account the physics of the biosphere—particularly as factually documented by the fossil-sedimentary records of the geologic past.

The only reasonable conclusion that can be made, it seems to me, is that present concepts of plate tectonics-continental drift-polar wandering must be re-evaluated, revised, or rejected. The global facts of the paleobiological world and of the paleoecological world simply do not fit the hypotheses that now prevail regarding the geophysics of the planet. Something has to give! In the opinion of this author, documented scientific fact must take precedence over widely held scientific theory— there is no alternative.

REFERENCES CITED

Ager, D.V., 1963, Principles of Palaeoecology: McGraw Hill Book Company, New York, 371 p.

Asama, K., 1966, Permian plants from Phetchabun, Thailand and problems of floral migration from Gondwanaland: Tokyo Natural Science Museum Bulletin, v. 9, p. 171-211.

———, 1967, Permian plants from Maiya, Japan: Tokyo Natural Science Museum Bulletin, v. 10, p. 139-153.

Axelrod, D.I., 1963, Fossil floras suggest stable, not drifting, continents: Journal of Geophysical Research, v. 68, p. 3257-3263.

Bailey, I.W., and Sinnot, E.W., 1916, The climatic distribution of certain types of Angiosperm leaves: American Journal of Botany, v. 3, p. 24-39.

Chaney, R.W., 1940, Bearing of forests on theory of continental drift: Science Monthly, v. 51, p. 489-499.

———, 1947, Tertiary centers and migration routes: Ecological Monographs, v. 17, p. 139-148.

Chatterjee, S., and Hotton, N. III, 1986, The paleoposition of India: Journal of Southeast Asian Earth Science, v. 1, p. 145-189.

Chen, F., and Yang, G.X., 1980, The Jurassic Mentougou-Yudaishan flora from western Yanshan, North China: Acta Paleontologica Sinica, v. 19, p. 423-432 (in Chinese, with English Abstract).

———, 1982, Lower Cretaceous plants from Pingquan, Hebei Province and Beijing, China: Acta Botanica Sinica, v. 24, p. 575-580 (in Chinese).

———, 1983, Early Cretaceous plants in Shiquanhe area, Xizang (Tibet), China: Earth Science, v. 19, p. 129-136 (in Chinese).

Chen, F., Yang, G.X., and Chow, H.Q., 1981, Lower Cretaceous flora in Fuxin basin Liaoning Province, China: Earth Science, v. 15, p. 39-51 (in Chinese).

Chen, F., Dou, Y.W., and Huang, Q.S., 1984, The Jurassic flora of West Hills, Beijing (Peking): Geology Publishing House, Beijing, 136 p. (in Chinese).

Churkin, M. Jr., 1972, Western boundary of the North American plate in Asia: Geological Society of America Bulletin, v. 83, p. 1027-1036.

Daugherty, L.H., 1941, The Upper Triassic Flora of Arizona: Carnegie Institution of Washington, Publication 526, 108 p. and 34 Plates.

Dietz, R.S., and Holden, J.C., 1970, The breakup of Pangaea: Scientific American, v. 223, p. 30-41.

Dickins, J.M., and Shah, S.C., 1989, Permian flora and fauna and the position of India: in S. Chatterjee and N. Hotton III (Eds.), New Concepts in Global Tectonics, Abstracts Volume, Texas Tech University, p. 27.

Duan, S.Y., 1987, A comparison between the Upper Triassic floras of China and the Rhaeto-Liassic floras of Europe and East Greenland: Lethaia, v. 20, p. 177-184.

Durham, J.W., 1950, Cenozoic marine climates of the Pacific Coast: Geological Society of America Bulletin, v. 61, p. 1234-1264.

———, 1952, Early Tertiary marine faunas and continental drift: American Journal of Science, v. 250, p. 321-343.

Fontaine, W.M., 1883, Contributions to the knowledge of the Mesozoic flora in Virginia: United States Geological Survey, Monograph 6, 144 p.

Gobbett, D.J., 1973, Carboniferous and Permian correlation in Southeast Asia: Geological Society of Malaysia Bulletin, v. 6, p. 131-142.

Hamilton, W., 1973, Tectonics of the Indonesian region: Geological Society of Malaysia Bulletin, v. 6, p. 3-10.

Harris, T.M., 1926, The Rhaetic flora of Scoresby Sound, east Greenland: Medd om Gronland, v. 68, p. 45-147.

Jin, Y-G., 1981, On the paleoecological relation between Gondwana and Tethys faunas in the Permian of Xizang: Symposium on Qinghai-Xizang (Tibet) Plateau, Geological and Ecological Studies, Beijing, Proceeding Volumes, v. 1, p. 171-178.

Just, T.K., 1952, Fossil floras of the Southern Hemisphere and their phytogeographical significance: American Museum of Natural History Bulletin, v. 99, p. 189-202.

Khudoley, K.M., 1974, Circum-Pacific ammonoid distribution: relation to hypotheses of continental drift, polar wandering, and Earth expansion, in C. F. Kahle (Ed.), Plate Tectonics—Assessments and Reassessments, American Association of Petroleum Geologists, Memoir 23, p. 295-330.

———, 1988, The paleobiogeography of the Atlantic Ocean in the Jurassic period: International Geology Review, v. 30, p. 623-634 (English translation).

Kimura, T., 1986, Discovery of late Devonian plants from the "Yuzuruha" Formation, Kyushu, southwest Japan: Geological Society of Japan, v. 11, p. 813-816.

———, 1988, Jurassic macrofloras in Japan and palaeogeography in East Asia: Tokyo Gakugai University Bulletin, v. 40, p. 147-164.

Krishnan, M.S., 1954, History of the Gondwana era in relation to the distribution and development of flora: Sir Albert Charles Seward Memorial Lectures, B. Sahni Institute of Palaeobotany, Lucknow, India, p. 1-15.

Kummell, B., 1970, History of the Earth (Second Edition): W. H. Freeman and Company, San Francisco, 707 p.

Lacey, W.S., 1975, Some problems of mixed floras in the Permian of Gondwanaland: In K. S. W. Campbell (Ed.), Gondwana Geology, Australian National University Press, p. 125-136.

Li, X-X., and Yao, Z-Q., 1981, Discovery of Cathaysia flora in the Qinghai-Xizang Plateau with reference to its Permian phytogeographical provinces: Symposium on Qinghai-Xizang (Tibet) Plateau, Geological and Ecological Studies, Beijing, Proceeding Volumes, v. 1, p. 171-178.

Loeblich, A.R., Jr. and Tappan, H., 1964, Treatise on Invertebrate Paleontology, Part C, Protists: Geological Society of America, 510 p.

Meyerhoff, A.A., 1974, Crustal structure of northern North Atlantic Ocean—a review: in C.F. Kahle (Ed.), Plate Tectonics—Assessments and Reassessments, American Association of Petroleum Geologists, Memoir 23, p. 423-433.

Meyerhoff, A.A., and Meyerhoff, H.A., 1974, Tests of plate tectonics: in C.F. Kahle (Ed.), Plate Tectonics—Assessments and Reassessments, American Association of Petroleum Geologists, Memoir 23, p. 43-145.

Meyerhoff, A.A., and Teichert, C., 1971, Continental drift III: late Paleozoic glacial centers, and Devonian-Eocene coal distribution: Journal of Geology, v. 79, p. 285-421.

Mi, J., Chuanbo, Z., Maoqing, L., Chunbin, S., and Gueichang, L., 1984, On the problem of the division of Late Triassic paleobotanic provinces in the north of China: Changchun College of Geology, Jilin, China, p. 1-16.

Miki, S., 1941, On the change of flora in eastern Asia since Tertiary Period: Japanese Journal of Botany, v. 11, p. 237-303.

Miller, M.M., 1985, Recent climatic variations, their causes and Neogene perspectives: in C.J. Smiley (Ed.), Late Cenozoic History of the Pacific Northwest, American Association for the Advancement of Sciences (Pacific Division), San Francisco, p. 357-414.

Minato, M., Gorai, M., and Hunahashi, M. (Eds.), 1965, The Geologic Development of the Japanese Islands: Tsukiji Shokan Company, Tokyo, 442 p.

Newberry, J.S., 1888, Fossil fishes and fossil plants of the Triassic rocks of New Jersey and the Connecticut Valley: United States Geological Survey, Memoir 14, 95 p.

Oishi, S., 1940, The Mesozoic Floras of Japan: Journal of the faculty of Science, Hokkaido Imperial University, Series IV, v. 5, p. 123-480.

Reed, W.W., 1941, Climates of the World: Climate and Man, Yearbook of Agriculture, United States Department of Agriculture, Washington, D.C., p. 665-684.

Ross, C.A., 1976, Evolution of Fusulinacea (Protozoa) in late Paleozoic space and time: in J. Gray and A.J. Boucot (Eds.), Historical Biogeography, Plate Tectonics, & the Changing Environment, Oregon State University Press, p. 215—226.

Ross, C.A., and J.R.P. Ross, 1984, The Geology of Coal: Benchmark Papers in Geology 77, Hutchinson and Ross Publishing Company, Stroudsburg, PA, 349 p.

Ross, J.R.P., 1976, Permian Ectoprocts in space and time: in J. Gray and A.J. Boucot (Eds.), Historical Biogeography, Plate Tectonics, & the Changing Environment, Oregon State University Press, p. 259-276.

Seward, A.C., 1933, Plant Life Through the Ages: Cambridge University Press, Cambridge, England, 603 p.

Simpson, G.G., 1947, Holarctic mammalian faunas and continental relationships during the Cenozoic: Geological Society of America Bulletin, v. 58, p. 613-687.

Smiley, C.J., 1967, Paleoclimatic interpretations of some Mesozoic floral sequences: American Association of Petroleum Geologists Bulletin, v. 51, p. 849-863.

———, 1974, Analysis of crustal relative stability from some late Paleozoic and Mesozoic floral records: in C. F. Kahle (Ed.), Plate Tectonics—Assessments and Reassessments, American Association of Petroleum Geologists, Memoir 23, p. 331-360.

———, 1976, Pre-Tertiary phytogeography and continental drift—some apparent discrepancies: in J. Gray and A.J. Boucot (Eds.), Historical Biogeography, Plate Tectonics, & the Changing Environment, Oregon State University Press, p. 311-319.

———, 1989, Continental floras and continental drift: some problem areas: in S. Chatterjee and N. Hotton III (Eds.), New Concepts in Global Tectonics, Abstract Volume, Texas Tech University, p. 29.

Storetvedt, K.M., 1987, Evidence for ocean-continent crust boundary beneath the abyssal plain of the East Central Atlantic: Physics of the Earth and Planetary Interiors, v. 48, p. 115-129.

Surange, K.R., 1966, Distribution of *Glossopteris* flora in the Lower Gondwana formations of India: Symposium on Floristics and Stratigraphy of Gondwanaland, B. Sahni Institute of Palaeobotany, Lucknow, India, p. 55-68.

Tanai, T. (Ed.), 1963, Tertiary Floras of Japan I: Geological Survey of Japan, 80th. Anniversary Commemoration Volume, 262 p.

——— (Ed.), 1972, Tertiary Floras of Japan II: Japan Association of Palaeobotanical Research, 452 p.

Taylor, D. (Ed.), 1973, Regional Conference on the Geology of Southeast Asia: Proceedings Volume: Geological Society of Malaysia Bulletin, v. 6, 334 p.

Teichert, C., 1974, Marine sedimentary environments and their faunas in Gondwana area: in C.F. Kahle (Ed.), Plate Tectonics—Assessments and Reassessments, American Association of Petroleum Geologists, Memoir 23, p. 361-394.

Vakhrameev, V.A., 1964, Jurassic and Early Cretaceous floras of Eurasia and the paleofloristic provinces of this period: Academy of Sciences, USSR, Transactions, v. 102, 263 p. (in Russian).

———, 1986, Climates and the distribution of some Gymnosperms in Asia during the Jurassic and Cretaceous: Review of Palaeobotany and Palynology, v. 51, p. 205-212.

Wagner, R.H., 1962, On a mixed Cathaysia and Gondwana flora from S. E. Anatolia (Turkey): 4th. Congress of Stratigraphic Geology of the Carboniferous, Heerlin, 1958, v. 3, p. 745-752.

Waterhouse, J.B., 1973, Permian Brachiopod correlations for South-East Asia: Geological Society of Malaysia Bulletin, v. 6, p. 187-210.

———, 1982, An early Permian cool-water fauna from pebbly mudstones in south Thailand: Geological Magazine, v. 119, p. 337-354.

Waterhouse, J.B., and Gupta, V. T., 1979, Early Permian fossils from southern Tibet, like faunas from peninsular India and Lesser Himalayas of Garhwal: Journal of the Geological Society of India, v. 20, p. 461-464.

Wen, S-X., 1981, Palaeobiogeography of Qinghai-Xizang Plateau, evidence for continental drift: Symposium on Qinghai-Xizang (Tibet) Plateau, Geological and Ecological Studies, Beijing, Proceedings Volumes, v. 1, p. 149-157.

White, D., 1925, Environmental conditions of deposition of coal: American Institute of Mining, Metallurgical and Petroleum Engineers, Transactions, v. 71, p. 3-34.

Wolfe, J.A., and Upchurch, G.R., Jr., 1987, North American nonmarine climates and vegetation during the Late Cretaceous: Palaeogeography, Palaeoclimatology, Palaeoecology, v. 61, p. 33-77.

Xi, Y. (Ed.), 1985, Atlas of the Palaeogeography of China: Cartographic Publishing House, Beijing, 143 p. of maps and map captions, 139 p. of text (in Chinese and English).

Yancey, T.E., 1976, Permian positions of the Northern Hemisphere continents as determined from marine biotic provinces: *in* J. Gray and A.J. Boucot (Eds.), Historical Biogeography, Plate Tectonics, & the Changing Environment, Oregon State University Press, p. 239-247.

Yang, S., Hou, H., Gao, L., Wang, Z., and Wu, X., 1983, The Carboniferous System in China: IX Congres International du Stratigraphie et Geologie, Official Reports, v. 1, p. 31-43.

Zimina, V.G., 1967, *Glossopteris* and *Gangamopteris* from the Permian deposits of the South Maritime Territory: Paleontology Journal (USSR), v. 2, p. 98-106.

ACKNOWLEDGMENTS

Thanks are extended to S. Chatterjee and N. Hotton III for an invitation to participate in the Smithsonian conference on *New Concepts in Global Tectonics*, July, 1989. Appreciation is given to the China University of Geosciences, Wuhan and Beijing, and to Changchun Geology University, for support of scientific exchange programs in 1987 and 1990. And special thanks are noted for Yang Hong, geology graduate student at the University of Idaho, for translations of Chinese literature, and for computer drafting of text figures.

Paleozoogeographic relationships of Australian Mesozoic tetrapods

R. E. Molnar, Queensland Museum, P. O. Box 300, South Brisbane, Queensland 4101, Australia

ABSTRACT

The Australian Early Cretaceous terrestrial tetrapod fauna differs more from those of Africa and South America, at the generic level, than they do from each other. The Australian fauna seems more closely related at the familial level to that of South America (Simpson's similarity index value of 25) than to that of Africa (9). These values are far below those for the Early Triassic.

Australia was faunally isolated from South America and Africa. Two lines of evidence for isolation are that some species exhibit functionally related complexes sharply differing from those of their nearest non-Australian relatives, whereas others are relicts. The factors that caused their extinction elsewhere did not affect Australia.

The Jurassic and Late Cretaceous faunas of Australia are too poorly known for comparison. The Australian Early Triassic fauna also differs from those of the other Gondwanan continents.

INTRODUCTION

What are the zoogeographical relationships of the Australian Cretaceous terrestrial tetrapod fauna to those of the other southern continents? The terrestrial tetrapods of these lands constituted a distinct fauna, different from that of the Laurasian continents (Bonaparte, 1986). The well-known arrangement of Gondwanaland continents during the Cretaceous (e.g., Howarth, 1981) suggests the terrestrial tetrapods of Australia would be expected to be similar to those of Antarctica, South America, Africa, and India. Before the Early Cretaceous the closest similarity should be to the faunas of Antarctica and India (for purposes of this study, the proposal of Chatterjee and Hotton (1986) that India was not part of Gondwanaland is not considered). During the Early Cretaceous, the spreading of the Indian Ocean altered the continental arrangement such that the Australian fauna should be most like those of Antarctica and South America. This scenario also applies to the Late Cretaceous.

There is little evidence for this expected similarity of the Australian terrestrial tetrapod fauna to those of the other southern continents (Molnar, 1980a, 1989, 1990). Thus it is worth considering in detail the zoogeographical relationships of the Australian Cretaceous fauna. The Australian terrestrial fossil record is poor compared to those of the Northern Hemisphere. Thirteen species representing 11 families are known from the Early Cretaceous. For South America nine species representing eight families, and for Africa 21 species representing at least 16 families have been reported. Because Early Cretaceous tetrapods are not known from India, only the fossil records from Australia, Africa, and South America are comparable.

In addition to general comparisons of the faunas, examination of the differences of each Australian taxon from its nearest overseas relatives is useful. The Australian taxa differing from their overseas relatives fall into two categories. First, some Australian species differ in functionally linked suites of features not found in their common ancestor. These species have, presumably, adapted to new or unusual environments in Australia. Because these species were unable to migrate out from Australia, a barrier that isolated Australia from the other southern continents is suggested. Either no suitable migration route existed, suitable habitat did not exist outside Australia, or any suitable habitat was occupied by species that resisted displacement. Those Australian species with single or few marked differences less clearly indicate isolation, and those without major differences from their overseas relatives provide no evidence for isolation.

Second, some Australian Mesozoic species are relicts. These species indicate that the factor causing their extinction (or evolution into daughter species) elsewhere did not apply in Australia. They provide a minimum date for the separation of the Australian lineage from the overseas lineage, for they could not have arrived in Australia after they became extinct (or diverged) elsewhere.

ASSUMPTIONS AND LIMITATIONS OF THIS STUDY

This study focuses on terrestrial tetrapods, excluding pterosaurs and birds. Members of these groups could fly and hence, potentially, disperse over seas. In view of Walker's (1981) doubts regarding the flying ability of enantiornithines, this group is considered. Marine tetrapods, including the hesperornithines, were not considered because of their potential for dispersal by sea. However, freshwater forms such as pholidosaurids, which apparently did not disperse across seas, are considered. The term continental vertebrates (or tetrapods) is used to include both terrestrial and freshwater taxa.

Although comments are made about other periods of the Mesozoic, this study focuses on the Early Cretaceous. This period is the only one (other than the Early Triassic—studied by Thulborn, 1986) for which more than

---

Chatterjee, S., and N. Hotton III, eds. *New Concepts in Global Tectonics.* Texas Tech University Press, Lubbock, 1992, xii + 450 pp.

two species of Australian terrestrial tetrapods are known. The Early Cretaceous Australian fossil record is limited largely to the Aptian and Albian (Molnar, 1980, 1990, in press; Rich and Rich, 1989); there is also a little material from the Cenomanian (Coombs and Molnar, 1981; Molnar, 1990; see Thulborn and Wade 1984 for taxa represented only by tracks). Aptian and Albian tetrapods are reasonably well known in Africa (de Lapparent, 1960) and South America. Unfortunately the South American deposits, principally those of Ceara in Brazil, are strongly biased toward the preservation of pterosaurs with almost no land-dwelling tetrapods represented (Mabesoone and Tinoco, 1973; Campos and Kellner, 1985). Similarly, whereas Cenomanian reptiles are known from Africa (Stromer, 1936), they are poorly known in South America (only those of Rusconi, 1933). Excepting the comparison of the Aptian-Albian faunas of Australia with those of Africa and South America, the best that can be done is to lump all the Early Cretaceous terrestrial tetrapods for each continent and compare them. Necessarily, this gives a more vague and less exact comparison than one at a lower taxonomic level.

The only reasonably well-known Mesozoic terrestrial tetrapod fauna from Antarctica is Early Triassic, and has been compared with that of Australia by Thulborn (1986). Although potentially interesting, comparison with later faunas of Antarctica cannot be carried out because they are so poorly known (see Molnar, 1989).

Because the Australian Cretaceous terrestrial tetrapod fauna appears to be more different from those of other southern continents than expected, a conservative course regarding uncertain identifications of nonendemic Australian families will be adopted here. For example, *Austrosaurus*, which may belong to either the Brachiosauridae or Cetiosauridae (Coombs and Molnar, 1981), is regarded here as a brachiosaurid because brachiosaurids are also known from Early Cretaceous Africa and South America. This course overemphasizes the similarity of the Australian fauna to those of the other southern continents. Because the intent is to determine if the Australian fauna is different from those overseas, any error will lead to underestimating this difference.

One genus, but no species, of Australian Cretaceous continental tetrapods is found on another continent (Molnar et al., 1981, 1985). Because there are so few genera in common between Australia and the other southern continents, comparison will be made at the familial and the superfamilial or subordinal levels.

Bonaparte (1986) recognized three components of the Cretaceous Gondwanan terrestrial tetrapod fauna. These differ from the two used here. Bonaparte's three components are a) endemic lineages that diverged in Gondwanaland after its separation from Laurasia; b) Pangaean lineages that persisted in Gondwanaland, but evolved endemic genera; and c) lineages that invaded South America and Africa late in the Cretaceous after connections were established with North America and Europe respectively. Component c has no relevance to Australia because none of the Laurasian lands were approached until well after the Mesozoic (in the Pliocene) and its Late Cretaceous terrestrial tetrapod fauna is unknown. The relicts and Australian taxa apparently not differing greatly from their overseas relatives (e.g., the hypsilophodontians), would belong in Bonaparte's component b. The forms differing from their overseas relatives may belong to either component a or b.

Finally, in spite of recent discoveries of Cretaceous tetrapods in the Southern Hemisphere, it must be remembered that the faunas are still poorly known. The occurrence of a pachycephalosaurid in Madagascar (Sues and Taquet, 1979) and an ankylosaurid in Antarctica, both represented by single specimens widely separated from their nearest relatives in the Laurasian continents, suggests that other such forms may have lived in Gondwanaland.

COMPARISON OF EARLY CRETACEOUS FAUNAS

During the Early Cretaceous, 11 families of continental tetrapods are known from Australia, eight from South America, at least 17 from Africa, and none from India. These are given in the Appendix. Australia shares two families with Africa (Brachiosauridae and Iguanodontidae) and one with South America (Brachiosauridae). It seems likely that both these families are shared with South America, as tracks attributed to iguanodontids have been described from that continent (Casamiquela and Fasola 1968; Leonardi, 1984).

The African and South American faunas share four families (Araripemyidae, Dicraeosauridae, Pholidosauridae, and Uruguaysuchidae). Simpson's (1947) index of faunal similarity (S) represents the ratio of taxa common to the two faunas under comparison (C) to the number of taxa in the smaller of the faunas (N). Specifically $(C/N) \times 100 = S$. Comparing Australia to Africa gives an S value of 9, to South America of 25 (or 37 including the iguanodontids, based on tracks), and comparing Africa to South America, 50. This does not suggest a close relationship between the Australian Early Cretaceous continental tetrapod fauna and those of Africa and South America. None of these values is as great as those found by Thulborn (1986) for the Early Triassic, which ranged from 71 to 100. Thus by the Early Cretaceous, the Australian tetrapod fauna seemingly diverged considerably from those of the other southern continents. The index values suggest as proposed by

Buffetaut (1981) that African and South American faunas are related. In calculating these index values, the classification of *Rapator ornitholestoides* as an abelisaurid based on the unusual form of its first metacarpal (Molnar, 1980a) has been assumed correct. The resultant values are maximal, and minimize the difference of the Australian from the other faunas.

Consideration of only Aptian-Albian taxa does not alter substanially these results. The index values are unchanged for the comparison of Australia and South America and of Australia and Africa. The index value for Africa and South America is raised to 66. Aptian and Albian ages are lumped to allow comparison of the largely Albian Australian fauna with the largely Aptian South American fauna. Without lumping, the sample sizes would be too small for confidence in the results.

Thulborn (1986) proposed a modification of the Simpson index taking into account the proportional representation of taxa present in a fauna. The method involves adding shared percentiles of taxa making up the fauna, rather than adding numbers of shared taxa. Comparing Australia to Africa with Thulborn's modified index gives 13%, to South America 16%, and Africa to South America 24%. Using this measure the Australian terrestrial tetrapod fauna is still less similar to those of Africa and South America than they are to each other, but the disparity is not so striking. The significance of this difference between the two indices is unclear. But bearing in mind that only half as many genera are known from Early Cretaceous South America as from Early Cretaceous Africa, it seems likely that further discoveries in South America will bring the results of the two indices in concordance.

Of the 11 families present in the Australian Early Cretaceous, two (those represented by *"Crocodylus" selaslophensis* and *Kakuru kujani*) are too poorly known to allocate to any of Bonaparte's three components. Five (Allosauridae, Brachiosauridae, Hypsilophodontidae, Iguanodontidae, and Nodosauridae) belong to component a, the remainder to component b. This is between South America (with six a and two b) and Africa (seven a and ten b). Three of Australian families in component a (Allosauridae, Brachiosauridae, and Iguanodontidae) are reported from South America, suggesting that the remaining two families may be discovered there.

Interestingly, Australia has the greatest proportion, 33%, of endemic families (Chelycarapookidae, Enantiornithidae, and Ornithorhynchidae?) in the Early Cretaceous. South America has only 12.5% (one, the Vincelestidae) and Africa 18% (Libycosuchidae, Simoliopheidae?, and Spinosauridae). (Note that following Molnar (1990) this paper does not accept as demonstrated the synonymy of the Spinosauridae and Baryonychidae.) South America has the greatest proportion of families endemic to Gondwanaland (75%).

At the superfamilial-subordinal level, ten taxa are found in Australia, Africa, or South America. Of these, only three—ornithopods, sauropods, and theropods—are common to Australia, Africa, and South America. Because these suborders each include large, easily preserved, and easily recognized animals, it is likely that the occurance of only three shared taxa reflects the rather poor quality of the known fossil records of Australia and South America, rather than any regional differentiation of the faunas.

The relationships of individual Australian taxa are somewhat more informative. Those Australian taxa that differ from their nearest overseas relatives in functional suites of characters include *Minmi, Muttaburrasaurus,* and *Steropodon*, and, perhaps, *"Crocodylus" selaslophensis* and *Nanantius. Minmi* and (maybe) *Muttaburrasaurus* pertain to Bonaparte's (1986) component b. *Steropodon* and *Nanantius* pertain to his component a. *"Crocodylus" selaslophensis* is too poorly know to allocate. These forms will be briefly discussed in the above order. *Minmi paravertebra* is probably a nodosaurid ankylosaur (Molnar, 1980c; Molnar and Frey, 1987). It differs from all other nodosaurids, indeed from all other dinosaurs, in the possession of paravertebrae. These are ossified tendon-aponeurosis structures associated with the epaxial vertebral musculature. They are similar in form to the tendons and aponeuroses of those muscles in crocodilians (Molnar and Frey, 1987). Molnar and Frey considered these structures were associated with locomotion. Nodosaurids are not otherwise known from the Southern Hemisphere (Molnar and Frey, 1987), indeed there is only a single other Gondwanan ankylosaurian, an ankylosaurid from Antarctica (Gasparini et al., 1987).

Muttaburrasaurus langdoni is a large ornithopod with an unusual skull (Bartholomai and Molnar, 1981). The snout is dorsally expanded into an inflated chamber, all teeth are erupted to the same degree, the laterotemporal fenestra is reduced and placed high on skull, there is a large foramen between the quadratojugal and quadrate, an ascending process on the surangular just anterior (and lateral) to the craniomandibular contact, and a fibular-astragalar contact. Not all of these unique features are associated with a single functional system. However, it is interesting that most are associated with the skull, and most of those with the teeth and jaws, suggesting that they are related to the trophic system. *Muttaburrasaurus* is most similar to the camptosaurs (Bartholomai and Molnar, 1981), but the opinion has been expressed (Weishampel, pers. comm., 1987) that it might represent an endemic family. Camptosaurs are well known from Laurasia, but tracks have been reported from South America (Casamiquela and Fasola, 1968; Leonardi, 1984).

Steropodon galmani is a monotreme (Archer et al., 1985), a group unknown outside Australia. Monotremes differ from other mammals in the structure and function of the reproductive system.

Two other species may also belong in my first category. *Nanantius eos* is the oldest enantiornithine (Molnar, 1986; but Cracraft, 1986, doubts that this is a real group). It is the only Early Cretaceous enantiornithine. If the enantiornithines are a valid group, and if the poor Early Cretaceous fossil record for birds is reliable in that enantiornithines lived only in Australia at this time, then *Nanantius* belongs in this category. Late Cretaceous enantiornithines are more widespread, occurring in Asia and North America, but most diverse in Argentina (Walker, 1981). They seem to be a group that originated in Gondwanaland and later spread to the northern continents.

The second species is "*Crocodylus*" *selaslophensis* (Etheridge, 1917; Molnar, 1980b). The type is a dentary fragment that is unusual in having isodont teeth with the alveolar partitions set below the level of the dorsal margin, in an alveolar groove (Molnar, 1980b). The toothrow is centrally placed on the jaw, rather than along the lateral margin, as in most other crocodilians. These features are found among no other Australian crocodilians. The single referred caudal has the neurocentral suture extending well onto the transverse process, which also has a posteriorly-opening cavity. It is not certain that only a single species is represented, but if so, no other known crocodilian has this combination of features (Molnar, 1980b). However, the jaws of the type of *Eocaiman cavernensis*, from the Eocene of Argentina, have the posterior portion of the toothrow placed centrally on the jaw (Simpson, 1933). Examination of the specimen shows that the partitions of the posteriormost five alveoli are set in a groove, as in "*C.*" *selaslophensis*. The caudals of *Eocaiman* are unknown, but this genus is the most similar to "*C.*" *selaslophensis* yet seen.

The evolution of these species in Australia, and the absence of closely related species overseas, suggests that the Australian continental fauna was, to some degree, isolated during the Early Cretaceous. The discovery of *Muttaburrasaurus*-like ornithopods, *Minmi*-like ankylosaurs, enantiornithines, or monotremes in Lower Cretaceous rocks outside Australia would weaken or invalidate this evidence of isolation.

A second line of evidence suggests isolation of the Australian continental tetrapods during the Early Cretaceous. The Jurassic Australian temnospondyl *Siderops kehli* (Warren and Hutchinson, 1983) can no longer be considered a significant relict with the discovery of Middle Jurassic temnospondyls in China (Dong, 1985) and Soviet Central Asia (Nessov, 1988). But an Early Cretaceous temnospondyl from Victoria (Jupp and Warren, 1986; Milner, 1989; Rich et al., in press) indicates that this group lived on in Australia well after it became extinct elsewhere. The existence of *Allosaurus* (or a very similar taxon—Molnar et al., 1981, 1985) in Australia at least five million years after its disappearance elsewhere provides another example of a Mesozoic relict tetrapod.

The sauropod *Austrosaurus* also may be interpreted as a relict (Molnar, 1989). Unlike other Cretaceous sauropods, the dorsal pleurocoels are small and simple in form, and the caudal neural spines and transverse processes are simple in structure (Coombs and Molnar, 1981). Although it has other unusual features (a humerus distally only about half as wide as proximally, ulnar alae extending approximately 80% ulnar length, and an elongate metacarpus but without elongate humerus or antebrachium), *Austrosaurus* is unusual mainly in that it is not a titanosaurid. Its vertebral structure is more reminiscent of Jurassic than Cretaceous sauropods (Coombs and Molnar, 1981).

Relict taxa in Australia indicate that whatever factors led to their extinction overseas did not affect Australia. If these factors were biological (e.g., over-predation), isolation is suggested, in that these factors (e.g., predators) could (or at least did) not reach Australia. The ancestors of the Mesozoic relicts did not all enter Australia at the same time. Xenacanths and temnospondyls presumably entered Australia in the Permo-Carboniferous, and the ancestors of *Allosaurus* and *Austrosaurus* in the Middle or Late Jurassic.

The remainder of the Early Cretaceous tetrapods (see Appendix) are fragmentary. Some—the hypsilophodontids (Molnar and Galton, 1986; Rich and Rich, 1989)—are similar to contemporaneous animals from Laurasia. Although unknown, members of this family presumably lived in Jurassic or Early Cretaceous Africa or South America.

Other species, although fragmentary, seem different from those found elsewhere. The theropod *Kakuru kujani* shows a form of the distal tibia, and presumably astragalus (Molnar and Pledge, 1980), found in only one other known theropod (described but not named by Taquet, 1985). Its tibia is matched in slenderness only by those of *Avimimus portentosus* and *Borogovia gracilicrus*, among dinosaurs. The turtle *Chelycarapookus arcuatus* also shows unusual morphology (Warren, 1969). The incompleteness of the fossils, however, renders uncertain whether or not these morphological differences reflect any distinct functional complex.

No Cretaceous continental tetrapods are known in Australia that must have entered after the Jurassic, unless *Rapator ornitholestoides* proves to be an abelisaurid. Camptosaurids, nodosaurids, hypsilophodontians, sauropods with vertebral structure similar to that of *Austrosaurus* and allosaurids are all known from the Jurassic.

COMPARISONS OF FAUNAS OF OTHER PERIODS

The Late Cretaceous terrestrial tetrapod faunas of South America, Africa, and India are better known, with 46, 15, and eight genera respectively. Unfortunately only a single genus, the sauropod *Austrosaurus* (Coombs and Molnar, 1981), is known from Australia and it belongs to a family not represented elsewhere at this time.

At the superfamilial-subordinal level three suborders are known from Australia, two of these, Ornithopoda and Theropoda, from tracks (Thulborn and Wade, 1984). These suborders are known also from Africa and South America. As these suborders all include large, easily recognized animals it is not clear that the similarity has much significance. There is no indication of ornithopods in Late Cretaceous India. If this is not an artifact of the record, the Indian Late Cretaceous fauna was quite unusual.

A comparison of Triassic continental tetrapod faunas of Australia and the other southern continents was made by Thulborn (1986). He concluded that the Australian fauna was different from the others not in composition, but in proportional representation of the taxa present. Only two continental tetrapods are known from the Australian Jurassic, the primitive sauropod *Rhoetosaurus brownei* (Longman, 1926) and the temnospondyl *Siderops kehli* (Warren and Hutchinson, 1983). These are Early or Middle Jurassic in age, and it is interesting that the Middle Jurassic at Zigong, Sichuan also has yielded primitive sauropods (e.g., *Shunosaurus lii*) and a temnospondyl (*Sinobrachyops placenticephalus*) confamilial with that from Australia. Primitive sauropods are also known from the Middle Jurassic of South America (Casamiquela, 1963) and Africa (de Lapparent, 1955). From the sparse Australian data no firm conclusions (beyond those of Thulborn) may be drawn.

SUMMARY AND CONCLUSIONS

Thulborn (1986) showed that the Early Triassic tetrapod fauna of Australia differs from those of other Gondwanan continents not in composition of the fauna, but in proportions of its components. Later Triassic and Jurassic terrestrial tetrapod faunas of Australia are known too poorly for profitable comparison.

By the Early Cretaceous, the Australian terrestrial fauna is more different from those of Africa and South America than they are from each other. This is less obvious at the superfamilial-subordinal level than at the familial or generic levels. The Australian fauna seems more closely related at the familial level to that of South America (Simpson similarity index value of 25) than to that of Africa (9). This accords with expectations derived from the arrangement of the continents during the Early Cretaceous. Australia seems to have been isolated faunally to some degree from South America and Africa. Unfortunately, Early Cretaceous terrestrial tetrapod faunas of India and Antarctica remain unknown. There are two lines of evidence for isolation of Australia at this time. First, some species exhibit functionally related complexes that sharply differ from those known in their nearest non-Australian relatives. Second, some species are relicts—whatever factors brought about their extinction elsewhere did not occur in Australia.

Cretaceous Australia was at the extreme tip of Gondwanaland, connected via Antarctica and the narrow horn of South America. It was perhaps the most geographically isolated landmass of the time. This position is consistent with the presence of distinctive and of relict taxa, and with the apparently high proportion of endemic families. Bonaparte (1986) argued that the distinct Gondwanan Cretaceous tetrapods added significantly to our understanding of the faunal diversity of the time, and of their adaptations. This is especially true of the Australian fauna, which seems unusual even by Gondwanan standards.

REFERENCES CITED

Archer, M., Flannery, T.F., Ritchie, A., and Molnar, R.E., 1985, First Mesozoic mammal from Australia: an Early Cretaceous monotreme: Nature, v. 318, p. 363-366.

Bartholomai, A., and Molnar, R.E., 1981, *Muttaburrasaurus*, a new iguanodontid (Ornithischia: Ornithopoda) dinosaur from the Lower Cretaceous of Queensland: Memoirs of the Queensland Museum, v. 20, p. 319-349.

Bonaparte, J.F., 1984, I dinosauri dell'Argentina, in Ligabue, G., ed., Sulle Orme dei Dinosauri: Venice: Erizzo Editrice, pp. 125-143.

———, 1985, A horned Cretaceous carnosaur from Patagonia: National Geographic Research, v. 1, p. 149-152.

———, 1986, History of the terrestrial Cretaceous vertebrates of Gondwana: IV Congresso Argentino Paleontologia y Bioestratigrafia, v. 2, p. 63-95.

de Broin, F., 1980, Les tortues de Gadoufaoua (Aptien du Niger), aperçu sur le paléobiogéographie des Pelomedusidae (Pleurodira): Mémoires de la Societe géologique de France, (ns) v. 139, p. 39-46.

Buffetaut, E., 1976, Der Land-Krokodilien *Libycosuchus* Stromer und die Familie Libycosuchidae (Crocodilia, Mesosuchia) aus der Kreide Afrikas. Mitleilungen der Bayerischen Staatssammlung für Paläontologie und Historische Geologie, v. 16, p. 17-28.

———, 1981, Die biogeographische Geschichte der Krokodilier, mit Beschreibung einer neuen Art, *Araripesuchus wegeneri*: Geologische Rundschau, v. 70 p. 611-624.

Buffetaut, E., and Taquet, P., 1977, The giant crocodilian *Sarcosuchus* in the Early Cretaceous of Brazil and Niger: Palaeontology, v. 20, p. 203-208.

Campos, P. de A., and Kellner, A.W.A., 1985, Panorama of the flying reptiles study in Brazil and South America: Anais da Academia brasileira de Ciencias, v. 57, p. 453-466.

Casamiquela, R.M., 1963, Consideraciones acerca de *Amygdalodon* Cabrera (Sauropoda, Cetiosauridae) del Jurásico medio de la Patagonia: Ameghiniana, v. 3, p. 79-95.

Casamiquela, R.M., and Fasola, A., 1968, Sobre pisadas de dinosaurios del Cretacico inferior de Colchagua (Chile): Univerisidad de Chile, Departamento de Geologia, Publicaciones, 30.

Chatterjee, S., and Hotton, N.,1986. The paleoposition of India: Journal of Southeast Asian Earth Sciences, v. 1, p. 145-189.

Coombs, W.P., Jr., and Molnar, R.E., 1981, Sauropoda (Reptilia, Saurischia) from the Cretaceous of Queensland: Memoirs of the Queensland Museum, v. 20 p. 351-373.

Cooper, M.R., 1985, A revision of the ornithischian dinosaur *Kangnasaurus coetzeei* Haughton, with a classification of the Ornithischia: Annals of the South African Museum, v. 95, p. 281-317.

del Corro, G., 1975, Un Nuevo sauropodo del Cretácico superior *Chubutisaurus insignis* gen. et sp. nov. (Saurischia - Chubutisauridae nov.) del Cretácico Superior (Cubutiano), Chubut, Argentina: I Congreso Argentino Paleontologia y Bioestratigrafia, Actas, v. 1, p. 229-240.

Cracraft, J., 1986, The origin and early diversification of birds: Paleobiology, v. 12, p. 383-399.

Dong Z., 1985, A Middle Jurassic labyrinthodont (*Sinobrachyops placenticephalus* gen. et sp. nov.) from Dashanpu, Zigong, Sichuan province: Vertebrata PalAsiatica, v. 23, p. 301-306.

Etheridge, R., Jr., 1917, Reptilian notes: *Megalania prisca*, Owen, and *Notiosaurus dentatus*, Owen; lacertilian dermal armour; opalized remains from Lightning Ridge: Proceedings of the Royal Society of Victoria, v. 29, p. 127-133.

Galton, P.M., and Coombs, W.P., Jr., 1981, *Paranthodon africanus* (Broom) a stegosaurian dinosaur from the Lower Cretaceous of South Africa: Géobios, v. 14, p. 299-309.

Galton, P.M., and Taquet, P., 1982, *Valdosaurus*, a hypsilophodontid dinosaur from the Lower Cretaceous of Europe and Africa: Géobios, v. 15, p. 147-159.

Gasparini, Z., Olivero, E., Scasso, R. and Rinaldi, C., 1987, Un ankylosaurio (Reptilia, Ornithischia) Campaniano en el continente Antartico: Anais do Congresso brasileiro de Paleontologa, v. p. 131-141.

Hoffstetter, R., 1960, Un serpent terrestre dans le Crétacé inférieur du Sahara: Bulletin de la Societe géologique de France, v. 7, p. 877-902.

Howarth, M.K., 1981, Palaeogeography of the Mesozoic, *in* Cocks, L.R.M., ed., The Evolving Earth: Cambridge, Cambridge University Press, p. 197-220.

von Huene, F., 1932, Die Saurischia, ihre Entwicklung und Geschichte: Monographien zur Geologie und Paläontologie, v. 1, p. 1-361.

Jupp, R., and Warren, A.A., 1986. The mandibles of the Triassic temnospondyl amphibians: Alcheringa, v. 10, p. 99-124.

Kellner, A.W.A., 1987, Ocorrência de um novo crocodiliano no Cretáceo inferior da bacia do Araripe, nordeste do Brasil: Anais do Academia Brasileira de Ciencias, v. 59 p. 219-232.

de Lapparent, A.F., 1955. Étude paléontologique des vertébrés du Jurassique d'El Mers (Moyen Atlas): Notes et Mémoires, Service géologique du Maroc, v.124, p. 3-36.

———, 1960, Les dinosauriens du "Continental intercalaire" du Sahara central: Mémoires de la Societe géologique de France, v. 88A, p. 1-57.

Leonardi, G., 1984, Le impronte fossili di dinosauri, *in* G. Ligabue, ed., Sulle Orme dei Dinosauri: Venice, Erizzo Editrice, pp. 163-186.

Longman, H.A., 1926, A giant dinosaur from Durham Downs, Queensland: Memoirs of the Queensland Museum, v. 8, p. 183-194.

———, 1933, A new dinosaur from the Queensland Cretaceous: Memoirs of the Queensland Museum, v. 10, p. 131-144.

Mabesoone, J.M., and Tinoco, I.M., 1973, Palaeoecology of the Aptian Santana Formation (north-eastern Brazil): Palaeogeography, Palaeoecology, Palaeoclimatology, v. 14, p. 97-118.

Milner, A., 1989, Late extinctions of amphibians: Nature, v. 338, p. 117.

Molnar, R.E., 1980a, Australian late Mesozoic terrestrial tetrapods: some implications: Memoirs de la Societe géologique de France, v. 139 p. 131-143.

———, 1980b, Procoelous crocodile from Lower Cretaceous of Lightning Ridge, N.S.W.: Mémoirs of the Queensland Museum, v. 20, p. 65-75.

———, 1980c, An ankylosaur (Ornithischia: Reptilia) from the Lower Cretaceous of southern Queensland: Memoirs of the Queensland Museum, v. 20, p. 77-87.

———, 1986, An enantiornithine bird from the Lower Cretaceous of Queensland, Australia: Nature, v. 322 p. 736-738.

———, 1989. Terrestrial tetrapods in Cretaceous Antarctica, *in* Crame, J.A., ed., Origins and Evolution of the Antarctic Biota: London, The Geological Society, pp. 131-140.

———, 1990, Problematic Theropoda: "Carmosaurs," *in* Weishampel, D., Dodson, P. and Osmolska, H., eds., Dinosaur Biology: Berkeley, University of California Press pp. 306-317.

———, in press, Palaeozoic and Mesozoic reptiles and amphibians from Australia, *in* Clayton, G., Hand, S. and Archer, M., eds., Vertebrate Zoogeography and Evolution in Australasia: Carlisle, Hesperian Press.

Molnar, R.E., and Frey, E., 1987, The paravertebral elements of the Australian ankylosaur *Minmi*: Neues Jahrbuch für Geologie und Paläontologie, Abhandlungen, v. 175, p. 19-37.

Molnar, R.E., and Galton, P.M., 1986, Hypsilophodontid dinosaurs from Lightning Ridge, New South Wales, Australia: Géobios, v. 19, p. 231-239.

Molnar, R.E., and Pledge, N.S., 1980, A new theropod dinosaur from South Australia: Alcheringa, v. 4, p. 281-287.

Molnar, R.E., Flannery, T.F., and Rich, T.H.V., 1981, An allosaurid theropod dinosaur from the early Cretaceous of Victoria, Australia: Alcheringa, v. 5, p. 141-146.

———, 1985, Aussie *Allosaurus* after all: Journal of Paleontology, v. 69, p. 1513-1515.

Mones, A., 1980, Nuevos elementos de la Paleoherpetofauna el Uruguay (Crocodilia y Dinosauria): Actas í Congreso Argentino Paleontología y Bioestratigrafía y I Congreso Latinoamericano de Paleontología, v. 1, p. 265-274.

Nessov, L., 1988, Late Mesozoic amphibians and lizards of Soviet Middle Asia: Acta Zoologica Crakoviensia, v. 31, p. 478-486.

Price, L. I., 1959, Sobre um crocodilídeo notossúquio do Cretáceo brasilero: Diviso de Geologia e Mineralogia, Boletim, v. 188, p. 1-55.

———, 1973, Quelónio Amphichelydia no Cretáceo Inferior do Nordeste do Brasil: Revista Brasileira Geociencias, v. 3, p. 84-96.

Rich, T.H.V. and Rich, P.V., 1989, Polar dinosaurs and biotas of the Early Cretaceous of southeastern Australia. National Geographic Research, v. 5, p. 15-33.

Rich, T.H., Rich, P.V., Wagstaff, B.E., Mason, J.R.C. McE., Flannery, T.F., Archer, M., Molnar, R.E. and Long, J.A., in press, Two possible chronological anomalies in the early Cretaceous tetrapod assemblage of southeastern Australia, *in* Chen P., and Mateer, N., eds., Aspects of Nonmarine Cretaceous Geology: Beijing, China Ocean Press.

Rusconi, C., 1933, Sobre reptiles Cretáceos del Uruguay (*Uruguaysuchus Aznaresi*, n. g. n. sp.) y sus relaciones con los notosuquidos de Patagónia: Boletin, Instituto de Geologia y Perforaciones, v. 19 p. 3-64.

Simpson, G.G., 1933, A new crocodilian from the Notostylops beds of Patagonia: American Museum Novitates, v. 623, p. 1-9.

———, 1947, Holarctic mammalian faunas and continental relationships during the Cenozoic: Bulletin of the Geological Society of America, v. 58, p. 613-688.

Stromer, E., 1915, Ergebnisse der Forschungsreisen Prof. E. Stromers in den Wüsten Ägyptens. 1. Wirbeltier-Reste der Baharije-Stufe (unterstes Cenoman). 3. Das Original des Theropoden *Spinosaurus aegyptiacus* nov. gen., nov. spec. Abhandlungen der Königlich Bayerischen Akademie der Wissenschaften: Mathematische-physicalische Klasse, v. 28, p. 1-32.

———, 1936. Ergebnisse der Forschungsreisen Prof. E. Stromers in den Wüsten Ägyptens. VII. Baharije-Kessel und -Stufe mit deren Fauna und Flora: Eine erganzende Zusammenfassung: Abhandlungen der Bayerischen Akademie der Wisenschaften, Mathematische-naturwisssschaftliche Abteilung, v. 33, p.

Sues, H.-D., and Taquet, P., 1979, A pachycephalosaurid dinosaur from Madagascar and a Laurasia-Gondwanaland connection in the Cretaceous: Nature, v. 279, p. 633-635.

Taquet, P., 1976, Géologie et paléontologie du gisement de Gadouafaoua (Aptien du Niger): Cahiers de Paléontologie, 191 pp.

———, 1984, Une curieuse spécialisation du crâne de certaine Dinosaures carnivores du Crétacé: le museau long et étroit des Spinosauridés: Comptes Rendus de l'Academie des Sciences, Paris, v. 299, p. 217-222.

———, 1985, Two new Jurassic specimens of coelurosaurs (Dinosauria), *in* Hecht, M. K., Ostrom, J. H., Viohl, G., and Wellnhofer, P., eds., The Beginnings of Birds: Eichstätt: Freunde des Juramuseums Eichstätt, pp. 229-232.

Thulborn, R.A., 1986, Early Triassic tetrapod faunas of southeastern Gondwana: Alcheringa, v. 10, p. 297-313.

Thulborn, R.A., and Wade, M., 1984, Dinosaur trackways in the Winton Formation (mid-Cretaceous) of Queensland: Memoirs of the Queensland Museum, v. 21, p. 413-517.

Walker, C., 1981, New subclass of birds from the Cretaceous of South America: Nature, v. 292, p. 51-53.

Warren, A.A., and Hutchinson, M.N., 1983, The last labyrinthodont? A new brachyopid (Amphibia, Temnospondyli) from the Early Jurassic Evergreen Formation of Queensland, Australia: Philosophical Transactions of the Royal Society of London, B, v. 303, p. 1-62.

Warren, J.W., 1969, A fossil chelonian of probably Lower Cretaceous age from Victoria: Memoirs of the National Museum of Victoria: v. 29, p. 23-28.

ACKNOWLEDGMENTS

Don Baird first drew my attention to the unusual proportions of the tibia of *Kakuru kujani* several years ago. José F. Bonaparte, Dale A. Russell and Tom H. V. Rich supplied helpful comments during the evolution of this paper.

APPENDIX—EARLY CRETACEOUS GONDWANAN TERRESTRIAL TETRAPODS

References are cited, as far as possible, for the most recent review of locality, stratigraphy and associated fauna. Aptian and Albian taxa marked with asterisk.

AUSTRALIA

ORNITHOPODA
Hypsilophodontidae
* *Atlascopcosaurus loadsi* (Rich and Rich, 1989)
* *Fulgurotherium australe* (Molnar and Galton, 1986; Rich and Rich, 1989)
* *Leaellynosaura amicagraphica* (Rich and Rich, 1989)
Iguanodontidae
* *Muttaburrasaurus langdoni* (Bartholomai and Molnar, 1981)
ANKYLOSAURIA
Nodosauridae
* *Minmi paravertebra* (Molnar and Frey, 1987)
SAUROPODA
Brachiosauridae or Cetiosauridae:
* *Austrosaurus mckillopi* (Longman, 1933; Coombs and Molner, 1981)
THEROPODA
Allosauridae
* *Allosaurus* sp. (Molnar et al., 1981; 1985)
Abelisauridae?
* *Rapator ornitholestoides* (von Huene, 1932)
Family *incertae sedis*
* *Kakuru kujani* (Molnar and Pledge, 1980)
ENANTIORNITHES
Enantiornithidae
* *Nanantius eos* (Molnar, 1986)
CROCODILIA *incertae sedis*
family *incertae sedis*
* "*Crocodylus*" *selaslophensis* (Molnar, 1980b)
MONOTREMATA
Ornithorhynchidae?
* *Steropodon galmani* (Archer et al., 1985)
TESTUDINES *incertae sedis*
Chelycarapookidae
* *Chelycarapookus arcuatus* (Warren, 1969)

SOUTH AMERICA

SAUROPODA
Brachiosauridae
* *Chubutisaurus insignis* (del Corro, 1975)
Dicraeosauridae
Amargasaurus groeberi (Bonaparte, 1984)
THEROPODA
Abelisauridae
* *Carnotaurus sastrei* (Bonaparte, 1985)
METASUCHIA
Pholidosauridae
Meridiosaurus vallisparadisi (Mones, 1980)
* *Sarcosuchus hartii* (Buffetaut and Taquet, 1977)
Trematochampsidae
* *Caririsuchus camposi* (Kellner, 1987)
Uruguaysuchidae
* *Araripesuchus gomesi* (Price, 1959)
EUPANTOTHERIA
Vincelestidae
Vincelestes neuquenianus (Bonaparte, 1986)

TESTUDINES *incertae sedis*
  Araripemyidae
  * *Araripemys barretoi* (Price, 1973)

AFRICA
  ORNITHOPODA
  Dryosauridae
    Kangnasaurus coetzeei (Cooper, 1985)
  * *Valdosaurus nigeriensis* (Galton and Taquet, 1982)
  Iguanodontidae
  * *Ouranosaurus nigeriensis* (Taquet, 1976)
  STEGOSAURIA
  Stegosauridae
    Paranthodon africanus (Galton and Coombs, 1981)
  SAUROPODA
  Brachiosauridae
  * *Rebbachisaurus garasbae* (de Lapparent, 1960)
  * *Rebbachisaurus tamesnensis* (de Lapparent, 1960)
  Camarasauridae?
    Algoasaurus bauri (von Huene, 1932)
* Dicraeosauridae (Taquet, 1976)
  Titanosauridae
  * *Aegyptosaurus baharijensis* (de Lapparent, 1960)
  THEROPODA
* Baryonychidae (Taquet, 1984)
  Ornithomimidae(?)
  * *Elaphrosaurus gautieri* (de Lapparent, 1960)
  * *Elaphrosaurus iguidiensis* (de Lapparent, 1960)
  Spinosauridae
  * *Spinosaurus* sp. (Stromer, 1915)
  family *incertae sedis*
  * *Bahariasaurus ingens* (de Lapparent, 1960)
  * *Carcharodontosaurus saharicus* (de Lapparent, 1960)
  * *Inosaurus tedreftensis* (de Lapparent, 1960)
  METASUCHIA
  Pholidosauridae
  * *Sarcosuchus imperator* (Taquet, 1976)
  Libycosuchidae
  * *Libycosuchus* sp. (Buffetaut, 1976)
  Uruguaysuchidae
  * *Araripesuchus wegeneri* (Buffetaut, 1981)
  SIMOLIOPHEOIDEA
  Simoliopheidae
  * *Lapparentophis defrennei* (Hoffstetter, 1960)
  PLEURODIRA
  Pelomedusidae
  * *Platycheloides* cf. *P. nyasae* (de Broin, 1980)
  * *Teneremys lapparenti* (de Broin, 1980)
  TESTUDINES *incertae sedis*
  Araripemyidae
  * *Taquetochelys decorata* (de Broin, 1980)

INDIA
none

Global distribution of terrestrial and aquatic tetrapods, and its relevance to the position of continental masses

Nicholas Hotton III, Department of Paleobiology, Smithsonian Institution, Washington, D.C. 20560 USA

ABSTRACT

Families of terrestrial tetrapods of wide taxonomic diversity reflect cosmopolitan distribution from the Lower Permian to the Lower Jurassic, demonstrating a contiguity of continental masses consistent with the hypothesis of Pangea. However, consideration of endemic families indicates that faunal resemblances between specific continental areas are commonly no greater than would be expected between different continents configured as they are at present. This seeming inconsistency is probably explained by the great distances separating the areas in question, together with the climatic extremes to be expected in a landmass the size of Pangea distributed symmetrically across the paleoequator.

INTRODUCTION

Terrestrial and aquatic tetrapods of the Permian and Triassic are of particular relevance to the study of former continental contiguity for two reasons. First, because of their relatively small size, locomotor specializations, and/or persistent dependence on fresh water, they would have found seaways nearly impassible barriers to their dispersal. Second, because they were taxonomically diverse and actively evolving before and as Pangea began to break apart, their distribution provides a rich set of data for comparing continental areas.

In consequence, the distribution of fossil tetrapods of Permian and Triassic age was a primary source of evidence for former contiguity of present-day continents in the days before plate tectonics made continental drift respectable (e.g., DuToit, 1937). Paleontogical evidence is still useful in confirming former continental positions restored by the study of mechanisms of plate tectonics (Colbert, 1971; Bliek et al., 1987), and remains one of the more reliable means of dating the connection and separation of continental masses (Buffetaut and Ingevat, 1985). It provides the only record of the evolution of life on drifting continental fragments, a factor that must be taken into account in the restoration of past continental positions (e.g., Chatterjee and Hotton, 1986). Finally, during the Triassic, a few terrestrial reptiles gave rise to tetrapods that were adapted to marine environments, and the distribution of these animals provides a clue to the whereabouts of actual continental margins.

Upper Permian faunas of eastern Europe and South Africa are frequently cited as evidence for terrestrial continuity between Laurasia and Gondwana (Colbert, 1963, 1986; Romer, 1968; Charig, 1971), because of the 11 families of tetrapods they share with one another (data based mostly on Caroll, 1988). However, they share fewer families with other continental areas, and include 18 tetrapod families that are unique to eastern Europe and 22 that are unique to South Africa, arguably because they are half a world apart on opposite sides of the paleoequator. Such evidence of isolation within the supercontinent of Pangea is not commonly taken into account.

This bias arises from the perception that because of the capriciousness of terrestrial preservation, the only record that has biological significance is the presence of a taxon. The absence from one place of a taxon that is present in another is commonly assumed to reflect the faultiness of the record. In consequence, faunal similarities between areas are emphasized at the expense of differences. In many cases this procedure is unavoidable, because it must be assumed that no record is complete. An alternative possibility is offered by deposits such as those of eastern Europe and South Africa, in which most families, both shared and unique, are represented by large numbers of individuals. Such abundance justifies the working hypothesis that the absence of families from one of the records means that they never lived in that area, rather than that they have not been found yet or were not preserved. This paper treats unique families as the biological endemics that they may well be, with the aim of identifying aspects of faunas that indicate isolation as well as those that support continental contiguity.

THE RECORD

In the following account, taxonomic assignment is based entirely and distribution largely on Carroll (1988). Distribution has been modified slightly by data reported by Sues and Boy (1988), Walter and Werneburg (1988), Martens (1989), Sues and Olsen (1990), and Rubidge and Hopson (1990). Globally, the best record of terrestrial tetrapods in the time interval under study is that of the Lower and Middle Triassic. The poorest record is that of the Lower Permian, which trails in completeness, in numbers of rich areas, in the breadth of dispersal of families, and in taxonomic diversity in fauna that show a broad dispersal (Fig. 1). Except for South Africa, the record of tetrapods is not as well known in the Southern Hemisphere as in the Northern.

Chatterjee, S., and N. Hotton III, eds. *New Concepts in Global Tectonics.*
Texas Tech University Press, Lubbock, 1992, xii + 450 pp.

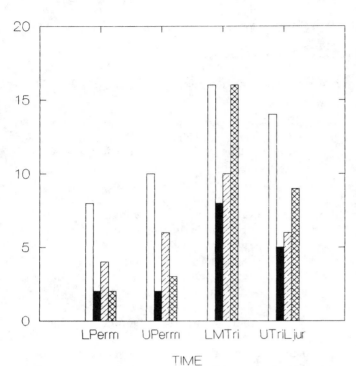

Figure 1. World record of fossil tetrapods from Lower Permian to Lower Jurassic, based on 17 fossiliferous continental areas (data mostly from Carroll, 1988). For each time interval, open bars indicate number of fossiliferous continental areas, the overall completeness of the record. Solid bars indicate number of areas with 10 or more families, the richest areas. Diagonally hatched bars indicate largest number of areas sharing one family, the breadth of dispersal. Crosshatched bars indicate number of families shared by four or more areas, the diversity of broad dispersal.

The proportion of families shared by three or more continental areas, relative to those shared by two only, increases steadily with time. It is a little more than 1:5 in the Lower Permian, 1:2 in the Upper Permian, and nearly 2:1 in the Triassic, which suggests increasing cosmopolitanism. However, this trend may be only apparent, due primarily to the overall poorer quality of the Permian record, because the faunal composition (see below) shows traces of a Permian cosmopolitanism that may have been comparable to that of the Triassic.

Each of the four time intervals reported in this paper is dominated by two continental areas that contain at least half again as many families as the richest subordinate areas (Fig. 2A–D). Dominant areas are also those that have produced the largest numbers of tetrapod fossils of all kinds, which is the primary sense that the term "richness" is used in this paper. In the Lower Permian, North America and Europe dominate with 34 and 20 families respectively; in the Upper Permian, South Africa and eastern Europe dominate with 41 and 33 families. In the Lower and Middle Triassic, South Africa and Europe dominate with 29 and 25 families respectively, and in the Upper Triassic and Lower Jurassic, Europe and North America dominate with 39 and 22 families. Thus the best of the Permian is comparable to the best of the Triassic in spite of the generally poorer Permian record.

Continued study of dominant areas over the last 20 years has not changed their apparent faunal composition significantly, although dominance has become less pronounced with continued study of subordinate areas. The richest areas, therefore, may provide the most accurate record of the past with respect to endemic as well as to shared families, a possibility that is supported by the correlations between shared and endemic families. To the extent that sharing reflects connection between two areas and endemism reflects isolation, the expected correlation should be negative. When all samples are considered, shared and endemic families show a loose positive correlation across the time interval under study (Table 1, $r = .665$, $N = 45$), which seems to have little to do with the distribution of faunas during their lives. If only the richest samples are considered, correlation between shared and endemic families, though not statistically significant, is closer to the expected (Table 1). The richer any two continental areas, therefore, the more likely that the record of a family from one area and its absence from the other indicates that the family lived at one but not at the other.

The relative importance of endemic families in the time interval under study is suggested by the contribution of shared and endemic families to their totals. In the whole sample, shared and endemics contribute comparably, as indicated by their similar high correlations with the total (Table 1). In the richest samples, however, endemics make a much larger contribution, as indicated by their higher correlation with the total (Table 1).

Table 1. Correlation among shared, endemic, and total families. Based on raw data in Appendix 1. Correlation coefficient significant at 99% level, **; significant at 95% level *; not significant (>95%) ns.

Whole sample, total > 0, N = 45

	Endemic	Total
Shared	.665 **	.922 **
Endemic		.902 **

Dominants, total > 18, N = 8

	Endemic	Total
Shared	-.090 ns	.251 ns
Endemic		.942 **

Richest areas, total > 24, N = 6

	Endemic	Total
Shared	-.362 ns	.143 ns
Endemic		.871 **

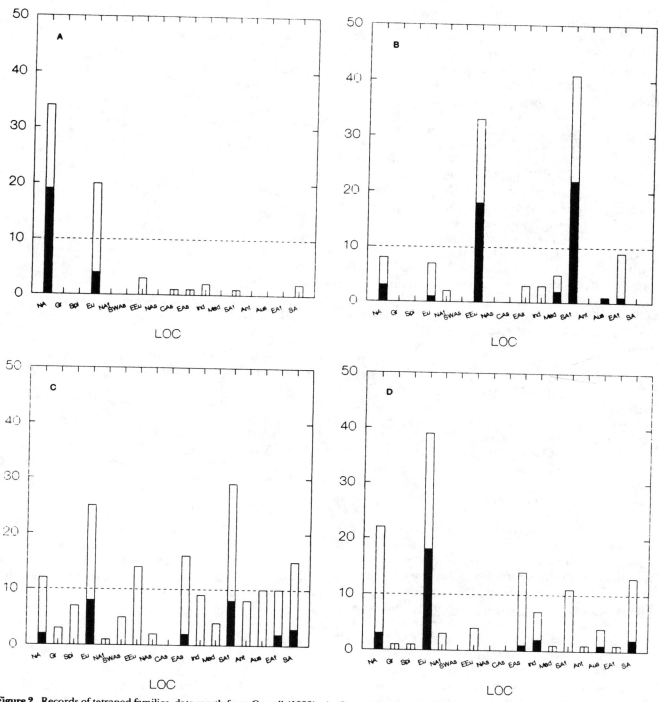

Figure 2. Records of tetrapod families, data mostly from Carroll (1988). A—Lower Permian; B—Upper Permian; C—Lower and Middle Triassic; D—Upper Triassic and Lower Jurassic. Height of bar indicates total families; solid part endemic, open part nonendemic. For geographic abbreviations see Table 2.

NATURE OF PERMIAN AND TRIASSIC TERRESTRIAL TETRAPODS

Tetrapod faunas of the Permian and Triassic consisted almost exclusively of animals of amphibian and reptilian grade. They were dominated by amphibians in the Early Permian, by synapsids in the Late Permian, and by nonsynapsid reptiles in the Triassic. Near the end of the Permian, synapsids produced a lineage that eventually evolved into true mammals, which, however, were not a conspicuous part of the fauna of the time under consideration.

Many amphibians were persistently aquatic, confined to fresh water for their entire life cycle. Others, especially in the Early Permian, enjoyed adult stages as highly terrestrialized as their reptilian contemporaries, but it must be assumed that these animals were dependent on bodies of water to accommodate their larval stages. Synapsids, which were technically of reptilian grade, though not closely related to contemporary reptiles, also may have been persistently dependent on bodies of fresh water. In synapsids the excretion of nitrogenous wastes was probably archaic and water-prodigal, more like that of amphibians and fish than like that of nonsynapsid reptiles (Hotton, 1980). Nonsynapsid reptiles had probably evolved a more effective water-conserving physiology by the Early Permian (Hotton, 1980).

Nonsynapsid reptiles of the Permian were generally small and lizardlike and comprised a relatively insignificant part of the fauna. By the beginning of the Triassic, however, they had increased greatly in numbers and diversity and included, in addition to the ancestors of living lizards and turtles, the great reptilian infraclass Archosauromorpha. Archosauromorphs evolved a variety of forms characteristic of the Triassic, including rhynchosaurs, ancestral crocodilians, and thecodonts, the latter ultimately producing dinosaurs by the Late Triassic.

Compared to post-Triassic dinosaurs and later Tertiary mammals, terrestrial tetrapods of the Permian and Triassic were not large. The largest weighed several hundred kilograms (comparable in bulk, though not in habitus, to large living crocodilians). Most were of more modest size, many with a body weight of not more than a few grams (comparable to small extant lizards).

All known Permian reptiles were quadrupedal, most were adapted to terrestrial walking, not swimming, and none were especially vagile. Their range of travel was limited by small size, robust body build, and locomotor mechanisms that suggested stability rather than speed. This state of affairs began to change in the Triassic, with the rise of archosaurs at the expense of synapsids. Some primitive archosaurs seem to have been facultatively bipedal, and the first dinosaurs were obligatory bipeds. Dinosaurs were more gracile than synapsids and primitive archosaurs, and clearly capable of covering more ground in a given amount of time. Few archosaurs were adapted to swimming and most were as dependent on a terrestrial environment for their travel as the most conservative of their predecessors.

A modest component of the Triassic fauna consisted of a variety of nonsynapsid reptiles of moderate to large size that were adapted to swimming and are preserved in marine rocks: nothosaurs, ichthyosaurs, plesiosaurs, and placodonts. All except placodonts appear to have been fish-catchers. In body form, nothosaurs were more or less lizardlike and ichthyosaurs more or less porpoiselike. Plesiosaurs and placodonts were like nothing else on earth. The body was very robust, somewhat flattened, and rather turtlelike, and the tail was short. In plesiosaurs, the neck was long and the head short (Order Plesiosauria) or the neck was short and the head long (Order Pliosauria). In placodonts, both neck and head were short, and the teeth were massive plates, apparently adapted to crushing molluscs.

Table 2. Geographic areas (and abbreviations), roughly in order of present-day proximity.

Northern Hemisphere
    NA, North America
    Gr, Greenland
    Spi, Spitsbergen
    Eu, Europe
    NAf, North Africa
    SWAs, Southwest Asia
    EEu, Eastern Europe, essentially western USSR
    NAs, Northern Asia
    CAs, Central Asia
    EAs, Eastern Asia, essentially western China

Southern Hemisphere
    Ind, peninsular India
    Mad, Madagascar
    SAf, South Africa
    Ant, Antarctica
    Aus, Australia
    EAf, East Africa
    SA, South America, essentially Brazil and Argentina

FAUNAL RESEMBLANCE

Faunal resemblance is estimated in terms of the index of similarity $(C \times 100) / N_1$, where C is the number of families shared by a pair of faunas and N_1 is the total number of families in the smaller member of the pair. Simpson (1960) recommended this index because it tends to stress the most similar parts of two faunas and is less biased for fossil samples than the alternatives. Two alternatives that he considered but rejected (Simpson, 1960) are $(C \times 100) / (N_1 + N_2 - C)$ and $(C \times 100) / N_2$, where N_2 is the total number of taxa in the larger fauna.

A useful standard for the evaluation of indices obtained from Permian and Triassic data is provided by indices of mammalian faunas reported by McKenna (1973) and by Simpson (1947). McKenna utilized $(C \times 100) / N_1$ to compare Recent faunas from the same continental mass, viz. New York/Oregon and Florida/New Mexico, and also used $(C \times 100) / N_2$ as part of a two-way comparison. Simpson, using $(C \times 100) / N_1$ alone, compared Recent faunas from different continental masses, viz. North America/Eurasia and North

America/Central East Asia, and also treated a stratigraphic range from Eocene to Pleistocene in the same areas. The difference between the indices of pairs of Recent faunas on the same continental mass (92 to 93, McKenna, 1973) and the indices of those whose members occur on different masses (52 to 63, Simpson, 1947) is striking and consistent. Indices of the fossil pairs are more variable, but do not drop below the 52 of Recent faunas on different continents (Simpson, 1947).

Pairwise comparison of Permian and Triassic data yields a total of 173 pairs in which N_1 ranges from one to 33 and the index of similarity ranges from 10 to 100 (Appendix 1). The numbers of shared families (C) in the best records of the Lower Permian, Upper Permian, Lower and Middle Triassic, and Upper Triassic and Lower Jurassic are comparable despite the overall poorer record of the Permian. In that sequence, the largest values of C are 14 (North America and Europe), 11 (South Africa and eastern Europe), 12 (South Africa and eastern Europe), and 12 (Europe and North America). Nevertheless the index of similarity varies widely, respectively 70, 33, 86, and 55 (Appendix 1), demonstrating that the value of the index is determined largely by N_1, of which endemics may be a significant component.

Values of 100 are concentrated where N_1 is no more than one or two (36% of the total), and in these the few shared families are found in relatively large numbers (four or more) of other areas. Such cases reflect the cosmopolitan distribution of many tetrapod families during the Permian and Triassic, but because of low values of N_1, their maximal index of similiarity is not a reliable indicator of faunal resemblance. Accordingly, pairs in which $N_1 < 2$ are omitted from further consideration of pairwise faunal resemblance.

These cases omitted, the index of similarity of Permo-Triassic faunas tends to be rather low; indices above 52 appear in only about 56% of the pairs of Lower Permian age, and in 31%–38% of the others (Fig. 3). Although some conform to the highest values obtained by Cox (1974), who studied Triassic faunas at coarser resolution, most are considerably lower. Cox held that his data supported Pangean land connections because the values of his indices (in the 70s and 80s) correspond to those reported by McKenna (1973) for Recent pairs within a single continental mass. However, the values of McKenna (1973) to which Cox (1974) referred are based on $(C \times 100) / N_2$ rather than $(C \times 100) / N_1$, which as noted above, are in the range of 92 to 93 (Fig. 3, interval A). Cox's indices, like those reported here, correspond more closely to the indices reported by Simpson (1947), who dealt with Recent (Fig. 3, interval B) and fossil faunas (Fig. 3, intervals C and D) on different continental masses.

Despite untidiness in detail, the data appear to reflect the reality of distance and continental configuration. Of the pairs with indices greater than 52, 47%–50% occur on a single present-day landmass or straddle adjacent Paleozoic landmasses. Forty-four percent are split between different present-day landmasses within a latitudinal hemisphere, and only 27% are split between Laurasia and Gondwana.

FAUNAL COMPOSITION AND DISTRIBUTION

As noted in the section "The Record," distribution is much narrower in the Permian. In the Lower Permian, only one family is shared by three or more continental areas for every five that are shared by two areas exclusively; but by the Triassic, two families are shared by three or more areas for every one shared by two exclusively. Because families shared by three or more areas usually occur in both Laurasia and Gondwana, their distribution is here termed "global" or "cosmopolitan," irrespective of the number of areas more than three in which they occur. They are thus distinguished from the families that are shared by only two areas, whose distribution is termed "exclusive."

Globally, Lower Permian faunas were dominated by amphibians (Appendix 2). Two groups, the Anthracosauria and Temnospondyli, included animals of moderate to large size, and the third, Lepospondyli, consisted entirely of small animals. All three groups included both persistently aquatic and terrestrial families. Among reptiles, primitive synapsids, the Pelycosauria, were the most diverse and the most common, and in size ranged from moderate to large. Nonsynapsid reptiles were of small to modest size, and though they included the ancestry of all other reptiles, they were the least conspicuous faunal components of the time. Temnospondyl (two families) and anthracosaur amphibians (one family) were the only groups to approach cosmopolitan distribution, being shared by three or four continental areas (Fig. 4). All others were shared exclusively by two areas.

By the Upper Permian, advanced synapsids, the Therapsida, had become dominant in terms of numbers, diversity, size, and general conspicuousness (Appendix 2), and four families were of cosmopolitan distribution (Fig. 5). Anapsid reptiles were more diverse than in the Lower Permian and two families were the most widely dispersed of their time. Of the amphibians, lepospondyls were extinct and anthracosaurs had undergone severe systematic and distributional contraction, although six families hung on as endemics in eastern Europe. Nine families of temnospondyls were present as endemics, two in North America, four in eastern Europe, and one each in South Africa, East Africa, and Australia. The endemic temnospondyl families of North

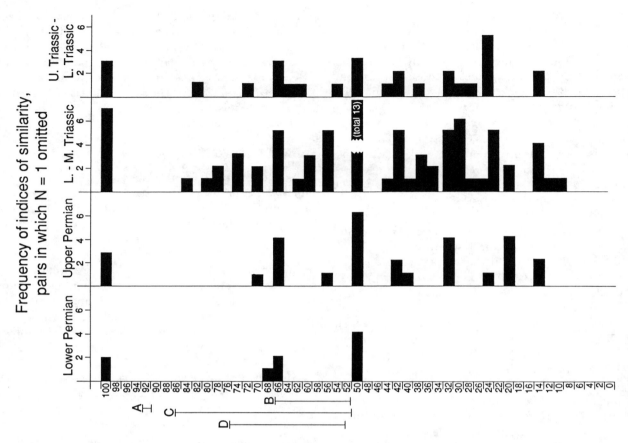

Figure 3. Frequency of indices of similarity for each time interval. Pairs in which $N_1 = 1$ omitted.

America and eastern Europe were relicts of the Lower Permian, soon to become extinct. The endemic families of South and East Africa and Australia, however, represented a secondary radiation of temnospondyls that would attain truly global dispersal in the Triassic.

Tetrapod faunas of the Lower and Middle Triassic were much more diversified than earlier (Appendix 2). Therapsids were marginally dominant, with 10 shared families, six of them of global distribution, but they were rivalled by temnospondyl amphibians, with eight shared families, seven of global distribution (Fig. 6).

Among nonsynapsid reptiles, the anapsids, which were holdovers from the Permian, were in decline, although the Procolophonidae retained global distribution. Diapsids now contributed significantly to global diversity as they included the archosaurs that were ancestral to dinosaurs and birds and were ultimately to spell the end of the therapsids. The most important diapsid groups were the Rhynchosauria with one family of global distribution, and the Thecodontia with six families, five of them of global distribution. Rhynchosaurs and thecodonts were large quadrupedal archosaurs, respectively robust herbivores and gracile predators, that commonly occur with large dicynodont therapsids such as the Kanne-meyeriidae. The latter, like rhynchosaurs, were robust herbivores of global distribution (Fig. 6), and both may have been a primary source of prey for thecodonts. The other new diapsids were marine forms, Nothosauria with four families, Placodontia with two, and Ichthyopterygia with three; all had limited distribution that centered in western Laurasia.

Upper Triassic and Lower Jurassic faunas were dominated by archosaurs, chiefly thecodonts (five families) and two orders of dinosaurs, the Ornithischia (four families) and the Saurischia (three families). Ten of these families were of global distribution (Appendix 2; Fig. 7). Marine diapsids were down to three shared families, all of limited distribution centering in western Laurasia, as in the Lower and Middle Triassic, but they were also represented by 10 families endemic to Europe. One family of lizards (North America/Europe) and two of primitive crocodylians (global) round out the diapsid picture. Relicts of earlier time are restricted to four families of temnospondyl amphibians, and five of therapsids; three families of each are of global distribution.

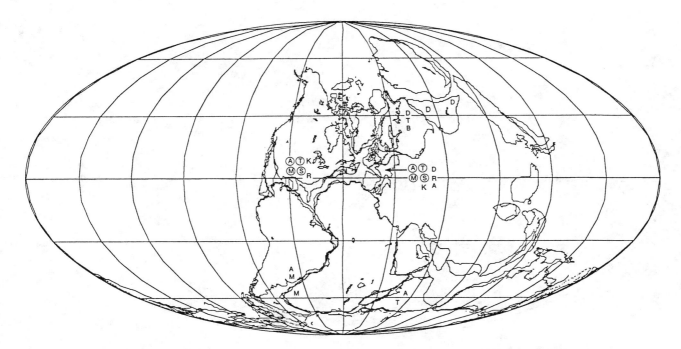

Figure 4. Global distribution of tetrapods in the Lower Permian (Sakmarian) 280 MYA. Mollweide projection, produced by the program Terra Mobilis version 2.0, for MacIntosh, by C. R. Scotese and C. R. Denham (1988). Amphibia: Anthracosauria (**A**—includes Limnoscelidae, Seymouriidae, Diadectidae, and Discosauriscidae; **D**—Discosauriscidae only); Temnospondyli (**T**—includes Eryopidae, Zatrachidae, Trematopsidae, and Archegosauridae; **T**—Eryopidae only; **A**—Archegosauridae only); Microsauria (**M**—includes Brachystelechidae and Hapsidopariontidae); Other Amphibia (**K**—Keraterpetontidae). Miscellaneous nonsynapsid reptiles: **B**—Bolosauridae; **R**—Araeoscelidae; **M**—Mesosauridae. Synapsida Pelycosauria: **S**—includes Sphenacodontidae, Edaphosauridae, and Caseidae?

More detailed treatment of distribution is concentrated primarily on pairs in which a dominant continental area is the larger member (Appendix 1, boldface). These pairs come closest to meeting the most important two conditions for the successful application of the index of similarity, which according to Simpson (1947, pp. 675, 676), are the following. First, "... the larger of the two samples (known parts of faunas) compared must be relatively large, including most or a considerable part of the whole fauna . . . ." Second, "... it must be either broadly representative of a large and regional fauna, not the strictly localized fauna of a narrow ecological niche, or must be similar in facies to or must include the facies of the smaller sample compared with it."

Lower Permian

In the Lower Permian (Fig. 4) the dominant continental areas of North America and Europe make up the only pair in which N_1 is large enough ($N_1 = 20$, Appendix 1) to warrant comparison by the index of similarity. They share 14 families for an index of 70, which falls within the range for pairs on different landmasses (Simpson, 1947) and is lower than expected if contiguity between North America and Europe obviated the only barriers to tetrapod dispersal. Faunal composition, however, shows a bias against aquatic forms; only two shared families, both amphibians, are certainly aquatic, whereas at least five of 20 North American endemics and three of four European endemics are certainly aquatic. This suggests an impediment that would have a filter effect on the dispersal of water-dependent tetrapods. Perhaps it was a matter of permanent drainage converging from west and east toward an incipient rift at the site of the future North Atlantic Ocean.

Of the 14 families shared by North America and Europe, 13 are exclusive to those areas (Appendix 2). One shared family, the Eryopidae, large amphibians that were terrestrial as adults, is of wider distribution; eryopids appear also in eastern Europe and India, the latter a Gondwana connection (Fig. 4, T).

North America shares the anapsid family Bolosauridae exclusively with eastern Europe, and South Africa shares the family Mesosauridae exclusively with South America (Fig. 4, M). Mesosaurs are anapsid reptiles of moderate size, highly adapted to aquatic or marine life, which occur in Brazil and South Africa in a distinctive carbonaceous shale that weathers white and yields very little else in the way of fossils. With a few rare exceptions still to be noted, the connections NA/Eu, NA/EEu, and SAf/SA (see Table 2 for abbreviations) are

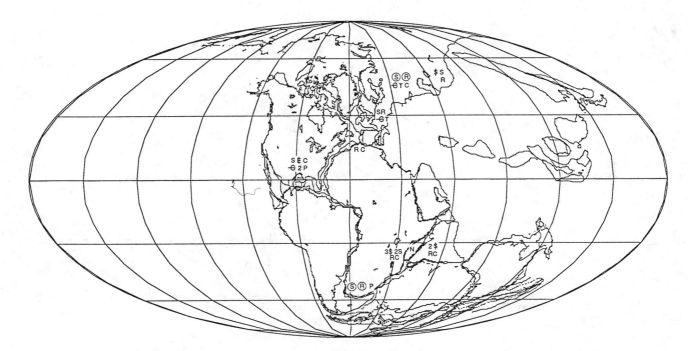

Figure 5. Global distribution of tetrapods in the Upper Permian (Tatarian) 225 MYA. Mollweide projection, produced by the program Terra Mobilis version 2.0, for MacIntosh, by C. R. Scotese and C. R. Denham (1988). Amphibia: Temnospondyli, **T**—Eryopidae. Reptilia: Procolophonoidea (**(R)**—includes Parieasauridae and Nyctephruretidae; **R**—Parieasauridae only; **N**—Nyctephruretidae only). Other nonsynapsid reptiles: **D**—Diapsida fam. inc. sed.; Captorhinida, **C**—Captorhinidae. Synapsida Pelycosauria: **C**—Caseidae; **P**—Pelycosauria fam. inc. sed. Synapsida Therapsida: **((S))**—includes Dicynodontidae, Tapinocephalidae, Whaitsiidae, Eotitanisuchidae, Burnettiidae, Moschorhinidae, Gorgonopsidae, Venjukoviidae, Galesauridae; **S**—includes Dicynodontidae, Tapinocephalidae, and Whaitsiidae only). Other Synapsida Therapsida: **E**—Brithopodidae; **$**—includes Endothiodontidae, Oudenodontidae, Kingoriidae, Cistecephalidae, Ictidosuchidae, and Diictodontidae. Therapsids marked with "**E**" also occur in eastern Europe but not in South Africa, and those marked with "**$**" also occur in South Africa but not in eastern Europe. Numbers preceding codes indicate the number of taxa found within that group at that locality.

the only ones that involve exclusively shared families. All other connections are by way of cosmopolitan families.

Europe and eastern Europe are linked by two cosmopolitan families, the Eryopidae and the Discosauriscidae, the latter small and persistently aquatic amphibians that do not occur in North America but are distributed across northwestern Laurasia (Fig 4, D). Europe is linked to Gondwana (India and South America) by a third family of cosmopolitan amphibians, the Archegosauridae, which were large, persistently aquatic fish-catchers (Fig. 4, A).

Cosmopolitan distribution is not as marked in the Lower Permian as later, but the anapsid family Captorhinidae and certain pelycosaurs suggest that it may have been comparable to that of the Triassic, but obscured by the poorer Lower Permian record. Captorhinids are listed in this paper as North American endemics during the Lower Permian (Appendix 2), but this assignment may be incorrect. They represent a lineage common to North America and Europe in the Late Carboniferous, and the family may yet turn up in Europe in the course of current work in the Rotliegendes of Germany (Martens, 1989). In addition, two Lower Permian genera of North American captorhinids are apparently identical to Upper Permian genera, respectively of East Africa (Gaffney and McKenna, 1979) and India (Kutty, 1972). The presence of these animals in the Upper Permian of Gondwana is supported by a pelycosaur endemic to South Africa in the Upper Permian, as evidence of former global distribution of a Lower Permian terrestrial fauna. Cox (1973, 1974) argued for increasing cosmopolitanism from the Permian to the Triassic, but these animals suggest that distribution was comparably cosmopolitan in the Lower Permian.

Upper Permian

In the Upper Permian, the dominant areas South Africa and eastern Europe, which lie respectively in Gondwana and Laurasia (Fig. 5), share 11 families, all of which are reptilian and fully terrestrial, and most include animals of large size. The index of similarity is only 33, because eastern Europe, the smaller member of the pair, accomodates 33 families ($N_1 = 33$) of which 18 are endemic. The endemics of South Africa are also numerous (22 out of $N_2 = 41$, Appendix 1), but of quite different composition. Eastern European endemics are dominated by 10 amphibians, seven of them persistently

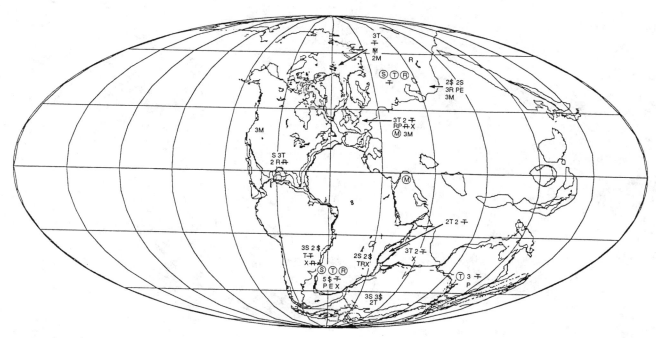

Figure 6. Global distribution of tetrapods in the Lower Triassic (Top of Scythian) 240 MYA. Mollweide projection, produced by the program Terra Mobilis version 2.0, for MacIntosh, by C. R. Scotese and C. R. Denham (1988). Amphibia: Temnospondyli (**T**—includes Capitosauridae, Brachiopidae, Trematosauridae, and Lydekkerinidae; **T**—includes one or more, but not all, families included in **T**); Other Temnospondyli (**T̄** —includes one or more of the following: Rhytidosteidae, Benthosuchidae, Indobrachyopidae, and Chigutisauridae). Reptilia: Terrestrial (**Ⓡ** —includes Procolophonidae, Sphenodontidae, Erythrosuchidae, and Proterosuchidae; **R**—Procolophonidae, Erythrosuchidae, and Proterosuchidae only); Other terrestrial (**P**—Prolacertidae; **R**—Rauisuchidae; **E**—Euparkeriidae; **X**—Rhynchosauridae); Marine (**Ⓜ**—includes Cymatosauridae, Nothosauridae, Cyamosauridae, Placodontidae, Mixosauridae; **M̄**—Mixosauridae only); Other marine (**M**—Pachycephalosauridae, Omphalosauridae, and Shastasauridae). Synapsida Therapsida: (**Ⓢ**—includes Kannemeyeridae, Lystrosauridae, Galesauridae, and Traversodontidae; **S**—includes Kannemeyeridae, Lystrosauridae, Galesauridae, and Traversodontidae only; **$**—includes Trirhachodontidae, Cynognathidae, Diademodontidae, Kingoriidae, Emydopidae, and Chiniquodontidae only. Numbers preceding codes indicate the number of taxa found within that group at that locality.

aquatic and three terrestrial; reptilian endemics include five therapsids, two diapsids, and one anapsid. South African endemics include 16 families of therapsids and one of large, persistently aquatic amphibians, three families of diapsids (including two of the earliest archosaurs), one pelycosaur, and one anapsid. Eastern European endemics are predominantly relicts of the earlier Permian, whereas South African endemics, though they include a few archaic forms such as pelycosaurs, are highly progressive in their therapsid and diapsid components. Shared families, which include six therapsids exclusive to South Africa and eastern Europe, and three therapsids and two anapsids of global distribution, likewise comprise a progressive fauna.

The pattern of South African and eastern European endemics is paralleled by the faunas shared exclusively by South Africa and East Africa, and by eastern Europe and Europe. For South Africa and East Africa there are three families of therapsids that are as progressive as South African endemics. Eastern Europe and Europe share the Eryopidae and Caseidae (Appendix 2), which are Lower Permian relicts similar to eastern European endemics. Shared faunas are rounded out by therapsid and anapsid families of global distribution, giving South and East Africa an index of similarity of 70, and eastern Europe and Europe an index of 57. Europe is linked to South Africa only by globally distributed families of therapsids (two) and anapsids (one), for an index of similarity of 43. It is linked to East Africa by the same families, also for an index of 43. Faunas of global distribution are of generally progressive aspect, similar to the endemics of South Africa.

Most of the pairs that involve one or the other of the dominant areas for an index greater than 52 are on the same or adjacent continental masses, or at least within the same latitudinal hemisphere. Pairs with one dominant that have an index of similarity of less than 52 are mostly split between Laurasia and Gondwana, like the dominants themselves. This pattern suggests that distance is a primary determinant of the level of faunal resemblance.

Three families shared by five or more areas link many Upper Permian pairs, irrespective of the index of similarity. Therogressive Dicynodontidae and Parieasauridae are

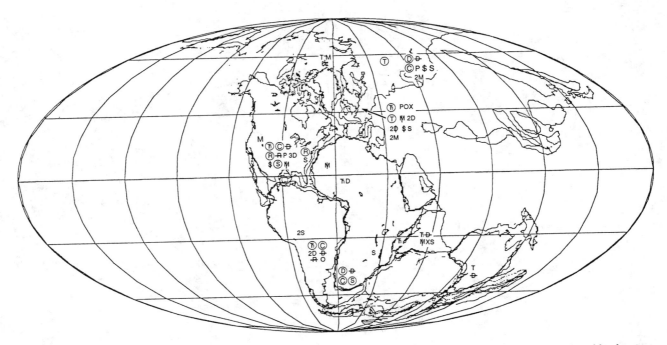

Figure 7. Global distribution of tetrapods in the Upper Triassic (Norian-Carnian) 220 MYA. Mollweide projection, produced by the program Terra Mobilis version 2.0, for MacIntosh, by C. R. Scotese and C. R. Denham (1988). Amphibia: Temnospondyli (**T**)—includes Plagiosauridae, Capitosauridae, and Mastodonsauridae; **T**—Plagiosauridae only); Other Temnospondyli, **M**—Metoposauridae. Reptilia: Thecodontia (**T̄**) —includes Phytosauridae and Stagonolepidae; **T̄**—Phytosauridae only); Other Thecodontia (**P**—Poposauridae; **O**—Ornithosuchidae; **R** —Rauisuchidae); Crocodylia (**C**)—includes Sphenosuchidae and Protosuchidae); Terrestrial miscellany (**R**)—includes Kuehneosaurida, Trilophosauridae, and Rhynchosauridae; **X**—Rhynchosauridae only); Marine (**M**—includes Pliosauridae, Ichthyosauridae, and Shastasauridae); Dinosauria (**D**) —includes Anchisauridae, Podokesauridae, Heterodontosauridae, and Fabrosauridae; **D**—includes Anchisauridae, Podokesauridae, or Heterodontosauridae only); Other dinosaurs (**Đ**—includes Melanorosauridae, Cetiosauridae, and Scelidosauridae). Synapsida Therapsida,(**S**)—includes Traversodontidae, Trithelodontidae, and Tritylodontidae; **S**—includes Traversodontidae, Trithelodontidae, or Tritylodontidae; **$**—Chiniquodontidae only. Numbers preceding codes indicate the number of taxa found within that group at that locality.

shared by eastern Europe and South Africa and also appear in Europe, eastern Asia, and East Africa (Fig. 5). The more conservative Captorhinidae appear in eastern Europe, though not in South Africa, and are shared by North America, North Africa, India, and East Africa (Fig. 5).

A therapsid-dominated fauna, of which South African families (both shared and endemic) comprise the best record, appears to have been distributed globally across Pangea in the Upper Permian. The fauna was most progressive in Gondwana, from which western Laurasia was partially isolated, as indicated by its larger component of Lower Permian relicts.

The cause of this isolation is not hard to adduce, for the major areas are separated by more than 90° of latitude and more than 60° of longitude (Fig. 5). In addition to the sheer mileage that separated these two areas, the effect of high north and south latitudes would have imposed severe climatic constraints (Robinson, 1973; Parrish et al., 1986) on dispersal of the faunas. It seems likely that the Upper Permian fauna originated from Lower Permian pelycosaur-dominated faunas of paleoequatorial distribution (Fig. 4), where the climate was at least seasonally wet. Elements dispersing to comparable climates north and south would have had to cross the arid horse latitudes, which would have isolated them from one another as if they had crossed an ocean.

Lower and Middle Triassic

In the Lower and Middle Triassic (Fig. 6), South Africa ($N_1 = 29$, Appendix 1) shares at least as many families with six subordinate areas as it does with Europe, the lesser dominant ($N_1 = 25$). The subordinate areas are eastern Europe, East Asia, India, Antarctica, Australia, and South America, and each has an index of similarity with South Africa greater than 52. The reason for such far-flung connections between South Africa and subordinate areas is that almost all of the sharing is by way of families of global distribution, which outnumber families shared only by two areas by about 2:1. Globally distributed families also occur at many more areas than in the Permian (Appendix 2). Endemics are present in only two subordinate areas, East Asia (two) and South America (three). The index of similarity with South Africa varies inversely with the number of endemics, and except for eastern Europe appears to be inversely related to distance (Appendix 1). In this scheme the value of the index of eastern Europe is anomalously high, especially as compared to its value in the Upper Permian. It will be recalled, however, that the low Upper Permian value was due to a large number of endemics. Most of these belonged to groups that had become extinct by the end of the Permian, and the survivors attained global distribution early in the Triassic.

Families of global distribution fall into three major groups (Fig. 6) that are all present in the fauna of South Africa, and occur variously at many other rich areas of both Laurasia and Gondwana (Appendix 1). The group of widest distribution consists of aquatic temnospondyl amphibians of moderate to large size and includes four families. A second group, which commonly occurs together in terrestrial environments, is a taxonomic melange of two families of thecodonts and two unrelated families of small, lizardlike reptiles. The third group is also primarily terrestrial and is represented by four families of therapsids. The aquatic amphibians of the first group and the thecodonts and other reptiles of the second are new to the Triassic, whereas the therapsids of the third group were more prominent in the Permian.

Families shared by fewer areas are mostly marine reptiles (Fig. 6), which represent considerable taxonomic diversity: nothosaurs, four families; placodonts, two families; and ichthyosaurs, three families. Nothosaurs and placodonts centered primarily in Europe, with which they were shared variously by southwest Asia, eastern Asia, and North America. Ichthyosaurs enjoyed similar distribution and also appear in the record of Spitsbergen.

Europe, the lesser dominant, shares only seven families with South Africa, for an index of similarity of 28, and its index of similarity to the subordinate areas is below 52 in all cases except India. The fauna of Europe differs from most areas in the Lower and Middle Triassic in that only about half of the fauna consists of families of global distribution and is further distinguished by three features. First, it lacks therapsids, which are present in South Africa and in the subordinate areas. Second, it includes seven families of marine reptiles of very limited distribution, which occur neither in South Africa nor in any of the six subordinate areas. Third, it has eight endemic families, four of them marine.

Thus the entire European fauna suggests isolation from South Africa and five of the six subordinate areas, irrespective of distance. The linkage of Europe to southwest Asia by marine reptiles suggests that this isolation was effected not so much distance as by proximity to the sea, presumably the western shore of Tethys (Fig. 6). The resemblance of the European fauna to the fauna of Greenland, Spitsbergen, India, and Madagascar is confined mostly to large aquatic amphibians, a cosmopolitan component; indices of similarity are inversely proportionate to distance.

Upper Triassic and Lower Jurassic

In the Upper Triassic and Lower Jurassic (Fig. 7), the dominant areas Europe and North America share 12 families for an index of similarity of 55 (Appendix 1). North America is linked to East Asia, South America, South Africa, and North Africa for indices of more than 60, and only with eastern Europe and India is its index less than 52. Europe, by contrast, is linked closely only to the adjoining areas eastern Europe and North Africa, for indices of 100. Europe's indices of similarity with East Asia and the Gondwana areas South Africa, South America, India, and Australia, are all less than 52.

Families of global distribution are in the majority, and those of widest distribution (four or more continental areas) are two families each of large aquatic temnospondyls, thecodonts, primitive crocodilians, both orders of dinosaurs, and therapsids. This pattern of distribution demonstrates global dispersal comparable to that of Lower and Middle Triassic families. Of globally distributed families, temnospondyl amphibians and thecodonts are holdovers from the Lower Triassic, and primitive crocodylians and dinosaurs are new to the Upper Triassic. Rhynchosaurs, now restricted to Europe and India, are holdovers, whereas primitive lizards and trilophosaurs, shared only by North America and Europe, are relatively new. Trilophosaurs were generalized quadrupedal archosaurs of presumably herbivorous habit that were North America endemics in the Middle Triassic. Marine reptiles are holdovers, and though they make up a smaller component of families of limited distribution than in the Middle Triassic, they remain important as 10 families endemic to Europe.

The connections of North America with East Asia, South America, South Africa, and North Africa are via groups of global distribution that include primitive crocodylians and dinosaurs, which are new to the Upper Triassic. North America and Europe are linked in part by four exclusively shared families, three of which are also new to the Upper Triassic. By contrast, the connections of Europe, both close and distant, with all areas except North America are chiefly via temnospondyls and thecodonts, both old groups of global distribution. This pattern suggests that faunas of the periphery of Pangea (Fig. 7: North America, South America, and South Africa in the west; East Asia in the east) were more conservative than those of the interior (Europe, eastern Europe, North Africa). The routes by which eastern and western peripheral faunas communicated cannot be determined on the basis of present data, but the mechanism that isolated them from the interior was probably climate (Robinson, 1971).

Europe, the larger dominant of the Upper Triassic and Lower Jurassic, contains twice as many endemic families (18) as all other continental areas combined (Appendix 1), so endemics are not important in determining indices of similarity. Nevertheless, the endemics of Europe enhance the impression of isolation of the European fauna. More than half of them are marine reptiles, and a majority of the remainder are holdovers from earlier time.

DISCUSSION

In the Permian and Triassic the evidence of cosmopolitan distribution of terrestrial tetrapods is pervasive; overall, nearly as many families are shared by three or more continental areas (mean = 4.06) as are shared by only two. This condition is most marked in the Lower and Middle Triassic, when families shared by three or more areas (mean = 4.91) are in a 2:1 majority, and most of the families shared by a given pair of continental areas are of cosmopolitan distribution. Overall, more than 40% of shared families include shared genera (Table 3).

Cosmopolitan distribution suggests a degree of continental contiguity much higher than that of the present, in the Permian as well as in the Triassic. Cox (1973) would restrict this relationship to the Triassic, but his argument is incomplete. Although it is true that global sharing of families and genera is less marked in the Permian than in the Triassic, such sharing is comparable in proportion to the number of areas in which Permian fossils are preserved. In addition, pelycosaurs and captorhinomorphs of the Upper Permian look like fragments of a "hidden" cosmopolitanism, obscured because of the scantier Permian record, which, if better preserved, would have been comparable to that of the Triassic.

Continental contiguity is supported by 39 pairs of continental areas for which the index of similarity (C × 100) / N_1 = 100, which is consistent with the indices reported by McKenna (1973) for Recent mammalian faunas from a single continent. As noted, however, 2/3 of these are of little significance as indicators of faunal resemblance because N_1 equals no more than one. In all except three of the remaining pairs, C = N_1 in the range of two to eight because N_1 consists almost entirely of families of global distribution. This condition is most spectacularly manifest in the linking of such distant areas as Eu/Ind (Lower Permian) and Gr/SAf or Gr/Aus (Lower and Middle Triassic), by the most wide-ranging families of large amphibians.

Table 3. Families represented by shared genera.

Time	Families	Shared genera	Continental areas
Lower Permian	Seymouriidae	1	NA/Eu
	Archegosauridae	1	Eu/Ind
	Eryopidae	1	Eu/Ind
	Keraterpetontidae	1	NA/Eu
	Bolosauridae	1	NA/EEu
	Mesosauridae	3	SAf/SA
	Sphenacodontidae	1	NA/Eu
	Edaphosauridae	1	NA/Eu
	Caseidae	1	NA/Eu
Upper Permian	Captorhinidae*	2	NA/Ind, NA/EAf
	Parieasauridae	2	EEu/SAf/EAf
	Caseidae	1	NA/Eu
	Dicynodontidae	1	EEu/EAs/SAf/EAf
	Tapinocephalidae	1	EEu/SAf
	Whaitsiidae	1	SAf/EAf
	Endothiodontidae	1	Ind/SAf
	Oudenodontidae	1	Ind/SAf
	Kingoriidae	1	SAf/EAf
	Diictodontidae	1	EAs/SAf
Lower-Middle Triassic	Capitosauridae	1	NA/Eu/EEu/Ind/SAf/Aus/EAf
	Brachyopidae	3	Gr/Eu, EEu/SAf, Ind/Aus
	Trematosauridae	2	NA/Gr/Aus, Eu/SAf
	Lydekkerinidae	1	Gr/EEu
	Rhytidosteidae	2	Spi/Eu, SAf/Aus
	Benthosuchidae	1	EEu/Mad
	Cymatosauridae	1	Eu/SWAs
	Nothosauridae	1	Eu/SWAs
	Erythrosuchidae	1	EEu/EAs
	Proterosuchidae	1	EAs/Ind/SAf
	Placodontidae	1	Eu/SWAs
	Mixosauridae	1	Spi/Eu/SWAs
	Kannemeyeriidae	2	Ind/SAf/EAf/SA/ (1 Ant)
	Lystrosauridae	1	EEu/EAs/Ind/SAf/Ant
	Galesauridae	1	SAf/Ant
	Cynognathidae	1	SAf/Ant/SA
	Emydopidae	1	SAf/Ant
Upper Triasic-Lower Jurassic	Metoposauridae	1	NA/Eu/NAf/Ind
	Plagiosauridae	1	Spi/Eu
	Mastodonsauridae	1	Eu/EEu
	Rhynchosauridae	1	Eu/Ind
	Phytosauridae	1	NA/NAf
	Phytosauridae	3	NA/Eu/ (2 Ind) / (1 Mad/SA)
	Anchisauridae	2	NA/SAf
	Anchisauridae	1	Eu/SA
	Ichthyosauridae	1	Gr/Eu
	Traversodontidae	1	Ind/SA
	Tritylodontidae	2	NA/Eu, Eu/SAf
	Trithelodontidae	1	NA/SAf

* North American genera Lower Permian.

The three exceptions to this pattern are quite different, the members of each pair occurring on adjacent Paleozoic landmasses. The members of Eu/SWAs (Lower and Middle Triassic) share five marine families of limited distribution, and occur at the western edge of Tethys. The members of SAf/Ant (Lower and Middle Triassic) share five families of global distribution. They also share two families that occur nowhere else and one that occurs elsewhere only in South America, so that the shared fauna is essentially a stripped-down version of the fauna of South Africa. SAf/SA (Lower Permian) is included in this lot because, although $N_1 = 1$ at the family level, the family includes three genera that are present at both areas and nowhere else.

The evidence of cosmopolitanism is sufficiently strong in the Permian and Triassic that a Pangean configuration of continental masses cannot be refuted. However, it must also be borne in mind that in 134 out of 173 pairs of continental areas, the index of similarity falls within or below the range that Simpson (1947) reported for Recent mammalian faunas on different continents (Fig. 3). The possibility that low indices of similarity are due to faulty records is discounted by the fact that some of the richest pairs have very low indices, and some of the poorest have very high ones. It is therefore probable that even with contiguous continents and potentially continuous routes of dispersal for strictly terrestrial tetrapods, many factors were in play that were as effective as seaways in isolating these faunas. Few of these factors can be sorted out with much confidence, but three possibilities are suggested here. The first is primarily distance, as indicated by the small percentage of pairs with indices above 52 that are split between Laurasia and Gondwana. The second is distance complicated by the presence of intervening belts of incompatible climate, as indicated by South Africa and eastern Europe in the Upper Permian, and South Africa and Europe in the Lower and Middle Triassic. The third isolating factor is paleodrainage patterns, possibly controlled by incipient rift zones, as indicated by North America and Europe in the Lower Permian.

REFERENCES CITED

Blieck, A., Battail, B., and Grauvogel-Stamm, L., 1987, Tetrapodes, plantes et Pangee: relance du debat sur les relations paleogeographiques Laurasie-Gondwanie: Annales de la Societe Geologique du Nord, v. 107, p. 45-56.

Buffetaut, E., and Ingevat, R., 1985, The Mesozoic vertebrates of Thailand: Scientific American, v. 253, p. 80-87.

Carroll, R.L., 1988, Vertebrate Paleontology and Evolution: New York, W. H. Freeman and Company, p. 1-698.

Charig, A.J., 1971, Faunal provinces on land: evidence based on the distribution of fossil tetrapods, with especial reference to the reptiles of the Permian and Mesozoic, in Middlemiss, F.A., Rawson, P.F., and Newall, G., eds., Faunal provinces in space and time: Liverpool, Seel House Press, Geological Journal Special Issue No. 4, p. 111-128.

Chatterjee, S., and Hotton, Nicholas III, 1986, The paleoposition of India: Journal of Southeast Asian Earth Sciences, v. 1, p. 145-189.

Colbert, E.H., 1963, Climatic zonation and terrestrial faunas, in Nairn, A.E.M., ed., Problems in palaeoclimatology: London, Interscience Publishers, p. 617-642.

———, 1971, Antarctic fossil vertebrates and Gondwanaland: Research in the Antarctic, v. 1971, p. 685-701.

———, 1986, Therapsids in Pangaea and their contemporaries and competitors, in Hotton, Nicholas III, Maclean, P.D., Roth, J.J., and Roth, E.C., eds., The ecology and biology of mammal-like reptiles: Washington, D.C., Smithsonian Institution Press, p. 133-145.

Cox, C.B., 1973, Triassic tetrapods, in Hallam, A., ed., Atlas of palaeobiogeography: Amsterdam, Elsevier Scientific Publishing Company, p. 213-223.

———, 1974, Vertebrate palaeodistributional patterns and continental drift: Journal of Biogeography, v, 1, p. 75-94.

DuToit, A.L., 1937, Our Wandering Continents. An Hypothesis of Continental Drifting: reprinted 1972, Westport, Connecticut, Greenwood Press, p. 1-366.

Gaffney, E.S., and McKenna, M.C., 1979, A Late Permian captorhinid from Rhodesia: American Museum Novitates, n. 2688, p. 1-15.

Hotton, Nicholas III, 1980, An alternative to dinosaur endothermy; the happy wanderers, in Thomas, R. D. K., and Olson, E. C., A cold look at the warm-blooded dinosaurs: Boulder, Colorado, Westview Press, AAAS Selected Symposium 28, p. 311-350.

Kutty, T.S., 1972, Permian reptilian fauna from India: Nature, v. 237(5356), p. 462-463.

Martens, T., 1989, First evidence of terrestrial tetrapods with North-American faunal elements in the red beds of Upper Rotliegendes (Lower Permian, Tambach Beds of the Thuringian Forest (G.D.R.)—first results.: Acta Musei Reginaehradecensis S. A.: Scientiae Naturales, v. 22, p. 99-104.

McKenna, M.C., 1973, Sweepstakes, filters, corridors, Noah's arks, and beached Viking funeral ships in palaeogeography, in Tarling, D.H., and Runcorn, S.K., eds., Implications of continental drift to the earth sciences: London, Academic Press, p. 295-308.

Parrish, J.M., Parrish, J.T., and Ziegler, A.M., 1986, Permian-Triassic Paleogeography and Paleoclimatology and Implications for Therapsid Distribution, in Hotton, Nicholas III, Maclean, P.D., Roth, J.J., and Roth, E.C., eds., The ecology and biology of mammal-like reptiles: Washington, D.C., Smithsonian Institution Press, p. 109-131.

Robinson, P.L., 1971, A problem of faunal replacement on Permo-Triassic continents: Palaeontology, v. 14(1), p. 131-153.

———, 1973, Palaeoclimatology and continental drift, in Tarling, D.H., and Runcorn, S.K., eds., Implications of continental drift to the earth sciences vol. 1: London, Academic Press, p. 451-475.

Romer, A.S., 1968, Fossils and Gondwanaland: Proceedings of the American Philosophical Society, v. 112(5), p. 335-343.

Rubidge, B.S., and Hopson, J.A., 1990, A new anomodont therapsid from South Africa and its bearing on the origin of Dicynodontia: South African Journal of Science, v. 86, p. 43-45.

Simpson, G.G., 1947, Holarctic mammalian faunas and continental relationships during the Cenozoic: Bulletin of the Geological Society of America, v. 58, p. 613-688.

Simpson, G.G., 1960, Notes on the measurement of faunal resemblance: American Journal of Science, v. 258-A, p. 300-311.

Sues, H.D., and Boy, J.A., 1988, A procynosuchid cynodont from central Europe: Nature, v. 331(6156), p. 523-524.

Sues, H.D., and Olsen, P.E., 1990, Triassic vertebrates of Gondwanan aspect from the Richmond Basin of Virginia: Science, v. 249, p. 1020-1023.

Walter, H., and Werneburg, R., 1988, Ueber Liegespuren (Cubichnia) aquatischer Tetrapoden (?Diplocauliden, Nectridea) aus den Rotteroeder Schichten (Rotliegendes, Thueringer Wald/DDR): Frieburger Forschungsheft, v. C419, p. 96-105.

ACKNOWLEDGMENTS

The author expresses his thanks to Lee-Ann Hayek for her cogent criticism after the reading of each of several versions of the manuscript, and for guiding him around the statistical traps implicit in the material. This, however, does not imply her agreement with opinions stated in the paper, nor is she responsible for such gaffes as may remain in it. Thanks are also due to Hans-Dieter Sues for frequent useful discussion of questions of tetrapod distribution in pre-Tertiary times, and to Mary Parrish for her persevering and ultimately successful struggle with Figs. 3-7.

APPENDIX 1

Number of families, dominant continental areas marked with *. Boldface indicates pairs in which larger member is a dominant, or which are otherwise worthy of emphasis (see text).

Lower Permian	Total Families N_1	N_2	Shared Families C	Index of Similarity	Endemic Families
NA*		34			20
Eu*	20		14	70	4
EEu	3		2	66	0
Ind	2		1	50	0
Eu*		20			4
EEu	3		2	66	0
Ind	2		2	100	0
SA	2		1	50	0
CAs	1		1	100	0
EAs	1		1	100	0
EEu		3			0
Ind	2		1	50	0
CAs	1		1	100	0
EAs	1		1	100	0
CAs	1				0
EAs	1		1	100	0
SA	2				0
Ind		2	1	50	0
SAf		1	1	100	0

Upper Permian	Total Families N_1	N_2	Shared Families C	Index of Similarity	Endemic Families
SAf*		41			22
EEu*	33		11	33	18
EAf	10		7	70	1
NA	8		2	25	3
Eu	7		3	43	1
Mad	5		1	20	2
EAs	3		3	100	0
Ind	3		2	66	0
NAf	2		1	50	0
EEu*		33			18
EAf	10		4	40	1
NA	8		4	50	3
Eu	7		4	57	1
Mad	5		1	20	2
Ind	3		1	33	0
EAs	3		2	66	0
NAf	2		2	100	0
EAf		10			1
NA	8		1	14	3
Eu	7		3	43	1
Mad	5		1	20	2
Ind	3		1	33	0
EAs	3		2	66	0
NAf	2		2	100	0
NA		8			3
Eu	7		1	14	1
Ind	3		1	33	0
NAf	2		1	50	0
Eu		7			1
Mad	5		1	20	2
EAs	3		2	66	0
NAf	2		1	50	0
EAs		3			0
NAf	2		1	50	0
Ind		3			0
NAf	2		1	50	0

Lower and Middle Triassic	Total Families N_1	N_2	Shared Families C	Index of Similarity	Endemic Families
SAf*		29			8
Eu*	25		7	28	8
EAs	16		9	56	2
SA	15		8	60	3
EEu	14		12	86	0
NA	12		6	50	2
EAf	10		5	50	2
Aus	10		8	80	0
Ind	9		7	78	0
Ant	8		8	100	0
Spi	7		4	57	0
Mad	4		2	50	0
Gr	3		3	100	0
NAs	2		2	100	0
NAf	1		1	100	0
Eu*		25			8
EAs	16		5	31	2

Appendix 1—Continued.

Lower and Middle Triassic	Total Families N_1	N_2	Shared Families C	Index of Similarity	Endemic Families
SA	15		5	33	3
EEu	14		5	36	0
NA	12		6	50	2
EAf	10		3	30	2
Aus	10		5	50	0
Ind	9		5	56	0
Ant	8		1	13	0
Spi	7		5	71	0
SWAs	5		5	100	0
Mad	4		2	50	0
Gr	3		2	66	0
NAs	2		2	100	0
NAf	1		1	100	0
EAs		16			2
SA	15		4	27	3
EEu	14		6	43	0
NA	12		5	42	2
EAf	10		3	30	2
Aus	10		2	20	0
Ind	9		3	33	0
Ant	8		2	25	0
Spi	7		1	14	0
NAs	2		1	50	0
SA		15			3
EEu	14		5	36	0
NA	12		5	42	2
EAf	10		6	60	2
Aus	10		3	30	0
Ind	9		3	33	0
Ant	8		3	38	0
Spi	7		1	14	0
NAs	2		1	50	0
NAf	1		1	100	0
EEu		14			0
NA	12		6	50	2
Aus	10		6	60	0
EAf	10		3	30	2
Ind	9		7	78	0
Ant	8		5	63	0
Spi	7		3	43	0
Mad	4		3	75	0
Gr	3		3	100	0
NAs	2		1	50	0
NAf	1		1	100	0
NA		12			2
Aus	10		4	40	0
EAf	10		3	30	2
Ind	9		4	44	0
Ant	8		2	25	0
Spi	7		5	71	0
Gr	3		2	66	0
Mad	4		1	25	0
NAs	2		1	50	0
NAf	1		1	100	0

Appendix 1—Continued.

Lower and Middle Triassic	Total Families N_1	N_2	Shared Families C	Index of Similarity	Endemic Families
Aus		10			0
EAf	10		1	10	2
Ind	9		5	56	0
Ant	8		2	25	0
Spi	7		4	57	0
Mad	4		3	75	0
Gr	3		3	100	0
NAs	2		1	50	0
NAf	1		1	100	0
EAf		10			2
Ind	9		3	33	0
Ant	8		1	13	0
Spi	7		1	14	0
NAf	1		1	100	0
Ind		9			0
Ant	8		3	38	0
Spi	7		3	43	0
Mad	4		3	75	0
Gr	3		2	66	0
NAf	1		1	100	0
Ant		8			0
Spi	7		1	14	0
Gr	3		1	33	0
Spi		7			0
SWAs	5		1	20	0
Mad	4		1	25	0
Gr	3		2	66	0
NAs	2		1	50	0
NAf	1		1	100	0
Mad		4			0
Gr	3		2	66	0

Upper Triassic Lower Jurassic	Total Families N_1	N_2	Shared Families C	Index of Similarity	Endemic Families
Eu*		39			18
NA*	22		12	55	3
EAs	14		7	50	1
SA	13		5	38	2
SAf	11		5	45	0
Ind	7		3	43	2
EEu	4		4	100	0
Aus	4		2	50	1
NAf	3		3	100	0
Mad	1		1	100	0
Gr	1		1	100	0
Spi	1		1	100	0
NA*		22			3
EAs	14		9	64	1
SA	13		8	62	2
SAf	11		9	82	0

Appendix 1—Continued.

Upper Triassic Lower Jurassic	Total Families N_1	N_2	Shared Families C	Index of Similarity	Endemic Families
Ind	7		3	43	2
EEu	4		1	25	0
NAf	3		3	100	0
Mad	1		1	100	0
EAf	1		1	100	0
EAs		14			1
SA	13		4	31	2
SAf	11		8	73	0
Ind	7		1	14	2
EEu	4		1	25	0
Aus	4		1	25	1
NAf	3		1	33	0
SA		13			2
SAf	11		7	64	0
Ind	7		2	29	2
NAf	3		2	66	0
Mad	1		1	100	0
EAf	1		1	100	0
SAf		11			0
Ind	7		1	14	2
EEu	4		1	25	0
NAf	3		1	33	0
EAf	1		1	100	0
Ind		7			2
Aus	4		1	25	1
NAf	3		2	66	0
Mad	1		1	100	0
EAf	1		1	100	0
EEu		4			0
Aus	4		2	50	1
NAf		3			0
Mad	1		1	100	0

APPENDIX 2

Faunal composition.

Lower Permian tetrapod families, shared.

Amphibia
Labyrinthodontia
Anthracosauria
    Limnoscelidae   NA,Eu
    Seymouriidae   NA,Eu
    Discosauriscidae   Eu,EEu,CAs,EAs
Temnospondyli
    Archegosauridae   Eu,Ind,SA
    Eryopidae   NA,Eu,EEu,Ind
    Zatrachidae   NA,Eu
    Trematopsidae   NA,Eu

Appendix 2—Continued.

Lower Permian tetrapod families, shared.

Lepospondyli
  Nectridea
    Keraterpetontidae   NA,Eu
  Microsauria
    Hapsidopareiontidae   NA,Eu
    Brachystelechidae   NA,Eu

Class inc. sed.
    Diadectidae   NA,Eu

Reptilia
Anapsida
  Captorhinida
    Proterothyrididae   NA,Eu
    Bolosauridae   NA,EEu
  Mesosauria
    Mesosauridae   SAf,SA
Diapsida
  Araeoscelida
    Araeoscelidae   NA,Eu
Synapsida
  Pelycosauria
    Sphenacodontidae   NA,Eu
    Edaphosauridae   NA,Eu
    Caseidae   NA,Eu

Upper Permian tetrapod families, shared.

Amphibia
Labyrinthodontia
  Temnospondyli
    Eryopidae   Eu,EEu

Reptilia
Anapsida
  Captorhinida
    Captorhinidae   NA,NAf,EEu,Ind,EAf
    Pareiasauridae   Eu,NAf,EEu,EAs,SAf,EAf
  Procolophonida
    Nyctephruretidae   EEu,Mad,SAf
Diapsida
  Diapsida inc. sed.
    Coelurosauravidae   Eu,Mad
  Lacertilia
    Eosuchia
      Tangasauridae   Mad,EAf
Synapsida
  Pelycosauria
    Pelyc. inc. sed.   NA,SAf
    Caseidae   NA,Eu,EEu,
  Therapsida
    Dicynodontidae   Eu,EEu,EAs,SAf,EAf
    Tapinocephalidae   NA,EEu,SAf
    Brithopidae   NA,EEu
    Whaitsiidae   EEu,SAf,EAf
    Procynosuchidae   Eu,SAf,EAf
    Eotitanosuchidae   EEu,SAf
    Burnetiidae   EEu,SAf
    Moschorhinidae   EEu,SAf
    Gorgonopsidae   EEu,SAf

Appendix 2—Continued.

Upper Permian tetrapod famlies, shared.

Venjukoviidae	EEu,SAf
Galesauridae	EEu,SAf
Endothiodontidae	Ind,SAf
Oudenodontidae	Ind,SAf
Kingoriidae	SAf,EAf
Cistecephalidae	SAf,EAf
Ictidosuchidae	SAf,EAf
Diictodontidae	EAs,SAf

Lower and Middle Triassic tetrapods, shared.

Amphibia
Labyrinthodontia
  Temnospondyli

Capitosauridae	NA,Spi,Eu,NAf,EEu,Ind,SAf,Aus,EAf,SA
Brachyopidae	NA,Gr,Spi,Eu,EEu,Ind,SAf,Ant,Aus
Trematosauridae	NA,Gr,Spi,Eu,EEu,Ind,Mad,SAf,Aus
Lydekkerinidae	Gr,EEu,Mad,SAf,Ant,Aus
Rhytidosteidae	Spi,Eu,NAs,SAf,Aus
Benthosuchidae	Eu,EEu,Ind,Mad
Indobrachyopidae	Ind,Mad,Aus
Chigutisauridae	Aus,SA

Reptilia
Anapsida
  Captorhinida

Procolophonidae	NA,Eu,EEu,NAs,EAs,SAf,SA

Diapsida
  Lacertilia
    Sphenodontida

Sphenodontidae	EEu,SAf

  Sauropterygia
    Nothosauria

Cymatosauridae	Eu,SWAs
Nothosauridae	Eu,SWAs
Notho. inc. sed.	NA,Eu,EAs
Pachypleurosauridae	Eu,EAs

  Archosauria
    Protorosauria

Prolacertidae	Eu,EAs,SAf,Aus

    Rhynchosauria

Rhynchosauridae	Eu,Ind,SAf,EAf,SA

    Thecodontia

Erythrosuchidae	NA,EEu,EAs,SAf,Aus,SA
Proterosuchidae	EEu,EAs,Ind,SAf,Aus
Rauisuchidae	NA,Eu,EAf,SA
Theco. inc. sed.	EEu,EAs,EAf
Ctenosauriscidae	Eu,EAs,SA
Euparkeriidae	EAs,SAf

    Placodontia

Cyamodontidae	Eu,SWAs
Placodontidae	Eu,SWAs

    Ichthyopterygia

Omphalosauridae	NA,Spi,EAs
Mixosauridae	Spi,Eu,SWAs
Shastasauridae	NA,Spi

Appendix 2—Continued.

Lower and Middle Triassic tetrapods, shared.

Synapsida
  Therapsida

Kannemeyeridae	NA,EEu,EAs,Ind,SAf,Ant,EAf,SA
Lystrosauridae	EEu,EAs,Ind,SAf,Ant
Galesauridae	EEu,SAf,Ant,SA
Traversodontidae	EEu,SAf,EAf,SA
Trirhachodontidae	EAs,SAf,EAf
Cynognathidae	SAf,Ant,SA
Diademodontidae	EAs,SAf
Kingoriidae	SAf,Ant
Emydopidae	SAf,Ant
Chiniquodontidae	EAf,SA

Upper Triassic and Lower Jurassic tetrapods, shared.

Amphibia
Labyrinthodontia
  Temnospondyli

Metoposauridae	NA,Eu,NAf,Ind
Plagiosauridae	Spi,Eu,EEu,Aus
Capitosauridae	Eu,EEu,Aus
Mastodonsauridae	Eu,EEu

Reptilia
Diapsida
  Lacertilia
    Squamata

Kuehneosauridae	NA,Eu

  Sauropterygia
    Plesiosauria

Pliosauridae	Eu,EAs

  Archosauria
    Trilophosauria

Trilophosauridae	NA,Eu

    Rhynchosauria

Rhynchosauridae	Eu,Ind

    Thecodontia

Phytosauridae	NA,Eu,NAf,Ind,Mad,SA
Theco. inc. sed.	NA,Eu,EEu,EAs,SAf
Poposauridae	NA,Eu,EAs
Stagonolepidae	NA,Eu,SA
Ornithosuchidae	Eu,SA
Rauisuchidae	NA,SA

    Crocodylia

Sphenosuchidae	NA,EAs,SAf,SA
Protosuchidae	NA,EAs,SAf,SA

    Saurischia

Anchisauridae	NA,Eu,NAf,EAs,SAf,SA
Podokesauridae	NA,Eu,EAs,SAf
Melanorosauridae	Eu,SAf,SA
Cetiosauridae	EAs,Ind,Aus

    Ornithischia

Heterodontosauridae	EAs,SAf,SA
Fabrosauridae	NA,EAs,SAf
Scelidosauridae	NA,Eu

  Ichthyopterygia

Ichthyosauridae	Gr,Eu
Shastasauridae	NA,EAs

Appendix 2—Continued.

Upper Triassic and Lower Jurassic tetrapods, shared.

Synapsida
  Therapsida
    Traversodontidae   NA,Ind,SAf,EAf,SA
    Tritylodontidae   NA,Eu,EAs,SAf
    Trithelodontidae   NA,SAf,SA
    Chiniquo. inc. sed.   Eu,EAs
    Chiniquodontidae   NA,Eu

Lower Permian tetrapod families, endemic.

Amphibia
Labyrinthodontia
  Anthracosauria
    Tseajaiidae   NA
    Archeriidae   NA
  Temnospondyli
    Dissorophidae   NA
    Doleserpetontidae   NA
    Trimerorhachidae   NA
    Parioxidae   NA
    Saurerpetontidae   NA
    Branchiosauridae   Eu
    Micromelerpetontidae   Eu
Lepospondyli
  Aistopoda
    Phlegothontiidae   NA
  Lysorophia
    Lysorophidae   NA
  Microsauria
    Goniorhynchidae   NA
    Ostodolepidae   NA
    Pantylidae   NA
    Gymnarthridae   NA
  Nectridia
    Scincosauridae   Eu
    Urocordylidae   NA

Reptilia
Anapsida
  Captorhinida
    Acleistorhinidae   NA
    Captorhinidae   NA
    Batropetidae   Eu
Synapsida
  Pelycosauria
    Ophiacodontidae   NA
    Varanopsidae   NA
    Eothyridae   NA

Upper Permian tetrapod families, endemic.

Amphibia
Labyrinthodontia
  Anthracosauria
    Chroniosuchidae   EEu
    Lanthanosuchidae   EEu
    Tokosauridae   EEu
    Nycteroletidae   EEu
    Kotlassiidae   EEu
    Seymouriidae   EEu

Appendix 2—Continued.

Upper Permian tetrapod families, endemic.

Temnospondyli
  Kourerpetontidae   NA
  Trimerorhachidae   NA
  Melosauridae   EEu
  Dvinosauridae   EEu
  Archegosauridae   EEu
  Dissorophidae   EEu
  Rhinesuchidae   SAf
  Peltobatrachidae   EAf
  Brachiopidae   Aus

Reptilia
Anapsida
  Captorhinida
    Millerettidae   SAf
    Rhipaeosauridae   EEu
Diapsida
  Diapsida inc. sed.
    Mesenosauridae   EEu
  Lacertilia
    Eosuchia
      Younginidae   SAf
      Galesphyridae   SAf
      Acerodontosauridae   Mad
  Sauropterygia
    Nothosauria
      Claudiosauridae   Mad
  Archosauria
    Proterosauria
      Protorosauridae   Eu
    Thecodontia
      Proterosuchidae   EEu
      Theco. inc. sed.   SAf
Synapsida
  Pelycosauria
    Pelyco. inc. sed.   SAf
  Therapsida
    Ictidorhinidae   SAf
    Pristerodontidae   SAf
    Eodicynodontidae   SAf
    Titanosuchidae   SAf
    Hipposauridae   SAf
    Crapartinellidae   SAf
    Scaloposauridae   SAf
    Pristerognathidae   SAf
    Hofmeyeriidae   SAf
    Galeopsidae   SAf
    Lycideopsidae   SAf
    Robertiidae   SAf
    Simorhinellidae   SAf
    Aulacephalodontidae   SAf
    Anteosauridae   SAf
    Emydopidae   SAf
    Estemmenosuchidae   EEu
    Biarmosuchidae   EEu
    Dviniidae   EEu
    Deuterosauridae   EEu
    Phthinosuchidae   EEu
    Eot. inc. sed.   NA

Appendix 2—Continued.

Lower and Middle Triassic tetrapods, endemic.

Amphibia
Labyrinthodontia
    Temnospondyli
        Sclerothoracidae   Eu
        Rhinesuchidae   SAf
        Dissorophidae   SAf

Reptilia
Anapsida
    Captorhinida
        Sclerosauridae   Eu
Diapsida
    Thalattosauria
        Thalattosauridae   NA
        Askeptosauridae   Eu
        Claraziidae   Eu
    Lacertilia
        Eosuchia
            Tangasauridae   EAf
        Squamata
            Eolacertid inc. sed.   SAf
            Paliguanidae   SAf
    Sauropterygia
        Nothosauria
            Simosauridae   Eu
    Archosauria
        Protorosauria
            Tanystropheidae   Eu
        Trilophosauria
            Trilophosauridae   NA
        Thecodontia
            Gracilisuchidae   SA
            Proterochampsidae   SA
            Erpetosuchidae   EAf
        Crocodylia
            Proterosuchid inc. sed.   Eu
        Saurischia
            Staurikosauridae   SA
    Placodontia
        Helveticosauridae   Eu
    Ichthyopterygia
        Hupehsuchidae   EAs
        Utatsuchidae   EAs
Synapsida
    Therapsida
        Bauriidae   SAf
        Ericiolacertidae   SAf
        Ictidosuchidae   SAf
        Scaloposauridae   SAf

Upper Triassic and Lower Jurassic tetrapods, endemic.

Amphibia
Labyrinthodontia
    Temnospondyli
        Trematosauridae   NA
        Brachyopidae   Aus

Appendix 2—Continued.

Upper Triassic and Lower Jurassic tetrapods, endemic.

Reptilia
Diapsida
    Order inc. sed.
        Drepanosauridae   Eu
    Lacertilia
        Sphenodontida
            Gephyrosauridae   Eu
            Sphenodontidae   Eu
            Pleurosauridae   Eu
        Squamata
            Prolacertidae   Ind
            Fulengidae   Ind
    Sauropterygia
        Nothosauria
            Nothosauridae   Eu
            Pachypleurosauridae   Eu
            Nothosaur inc. sed.   NA
        Plesiosauria
            Plesiosauridae   Eu
            Plesiosaur inc sed   Eu
    Archosauria
        Protorosauria
            Tanystropheidae   NA
        Thecodontia
            Proterochampsidae   SA
            Scleromochlidae   Eu
            Erpetosuchidae   Eu
        Crocodylia
            Saltoposuchidae   Eu
            Trialestidae   Eu
            Platyognathidae   EAs
        Saurischia
            Herrerasauridae   SA
    Placodontia
        Placodont inc. sed   Eu
        Henodontidae   Eu
        Cyamodontidae   Eu
    Ichthyopterygia
        Protoichthyosauridae   Eu
        Leptopterygiidae   Eu
        Stenopterygiidae   Eu

Alternatives to Plate Tectonics

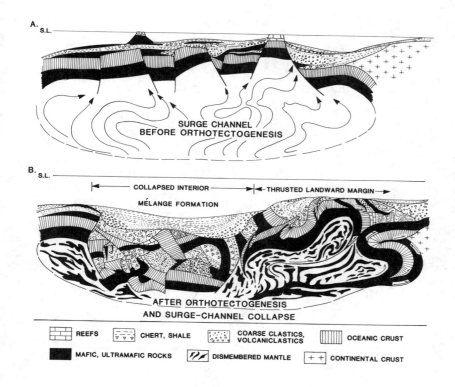

Has the Earth increased in size?

H. G. Owen, Department of Palaeontology, The Natural History Museum, London SW7 5BD, U.K.

ABSTRACT

Geological evidence of fit-together of the continents in Pangea and their displacement according to ocean-floor spreading data cannot be reconciled either with the spherical geometry of a constant modern-dimensions Earth or with the fast expansion concept of Carey. The continental margins of passive-margined oceans are extensional, yet in order to fill gaps (gores) that occur in the fit-together of the continents on a modern dimensions globe, they would need to be compressional at least in the Triassic and Early Jurassic; the opposite of reality. Carey's view that all expansion of the Earth occurred since the early Jurassic implies a closed Pacific region at that time. It is impossible, however, to refit Pangea according to the geological evidence on this fast expanding Earth model. Satellite laser ranging shows continental convergence values in the Pacific significantly smaller than is required to balance the spreading rates in the passive-margined oceans. Such balance is obligatory for a constant dimensions Earth. All of these inconsistencies are satisfied by an Earth of 80 percent of modern diameter in the Late Triassic-Early Jurassic expanding to its current size. Recent reconstructions by Weijermars disputing this view have no cartographic legitimacy.

INTRODUCTION

There is nothing more contentious in global tectonics at this time than the expanding Earth concept. Among conventional geophysicists it engenders emotions ranging from mild amusement tinged with pity for the eccentric, to positive hostility that anyone could be such an idiot as to question the incontrovertible fact that Earth has always been of its modern size. Yet these are exactly the same emotions that Taylor, Wegener, and to a lesser extent, Du Toit faced in the early days of the continental displacement hypothesis. Carey's Hobart Symposium in 1956, the published proceedings of it in 1958, and Carey's subsequent lecture tour of the United States, sparked the so-called revolution in the earth sciences. It is arguable that if paleomagnetic workers had not found divergent magnetic polar vectors in separate continents and thus provided readily understood numbers (Glen 1982), Carey would have been ignored in exactly the same manner as the earlier exponents of continental displacement had been. His contribution to the study of global geodynamics has not received the recognition that it deserves.

Most workers now accept that continental displacement has occurred, although this symposium showed that the steady-state and contracting Earth hypotheses are still alive, albeit just. I concluded (Owen 1983) from an impartial assessment of a vast amount of oceanic crustal spreading data, ocean marginal structural traverses, on-continent geological match evidence and paleontological sequence dating trends, not only that continental displacement had occurred but that the Earth has expanded during the last 200 Ma from a diameter of 80 percent of modern value to its current dimensions. If plate tectonic theory is modified to make the value of r (Earth's mean radius) a function rather than a constant in the solution of Euler's third theorem of rotations on a sphere, I find no difficulty in accepting the basic concept of plate tectonics either.

Weijermars (1989) included me in the fast expansion school of Carey, which indicates that he did not read the papers I wrote on the subject. The concept that all oceanic crustal growth occurred since the middle Jurassic, including that of the Pacific, as asserted by Carey (e.g., 1983), Shields (e.g., 1983), and Vogel (1983), and that no subduction occurred at the Pacific Ocean margins, implies an Earth of about 60 percent of modern diameter at that time. It is impossible to reassemble Pangea according to the geological evidence of fit-together of the continents on an Earth model of those dimensions (e.g., Owen 1983). Also the subsequent ocean-floor spreading data will not accord with their model. Their "fast" expansion concept has to be rejected. Unfortunately, there is a tendency among the geological and geophysical establishment to use this flawed hypothesis to reject without reason all the evidence that speaks eloquently for the slower rate of expansion determined by the author. It is the purpose of this paper to review on the one hand evidence that those adhering to current plate tectonic ideas should take heed of, and on the other to warn "fixists" against the danger of trying to develop an all-embracing mechanism for structures seen throughout the crust both oceanic and continental; the Earth's crust is heterogeneous both in bulk composition, tectonic history and age.

IMPORTANCE OF PRECISE SPHERICAL GEOMETRY

If the Earth retained its modern form and dimensions throughout geological time the following mathematical propositions can be made on the reasonable basis that, as the Earth is very close to being a spherical body, spherical geometric formulae can be applied to it. This is an observation which, despite its pedantry, needs to be made because both the constant modern dimensions

---

Chatterjee, S., and N. Hotton III, eds. *New Concepts in Global Tectonics.* Texas Tech University Press, Lubbock, 1992, xii + 450 pp.

Earth and fast expanding Earth schools ignore the mathematical implications of spherical geometric prediction.

The surface area of a sphere can be determined by the simple formula

$$4 \times 3.1416 r^2$$

In the case of a constant modern-dimensions Earth, the surface area is also constant whatever crustal generation or foreshortening has occurred and by whatever mechanism these processes have been achieved.

Proposition 1—The generation of new oceanic crustal area from midoceanic ridge systems must be balanced exactly by the subduction of oceanic crust elsewhere and, or, by orogenic foreshortening within continental regions. That is,

Generation of crustal area = Subduction + Foreshortening

Proposition 2— If little or no subduction occurs at the margins of, or within the area of, passive-margined oceans (i.e., the Arctic, North and South Atlantic, and the bulk of the Indian Ocean), the areal product of crustal generation within these oceans must be balanced largely by crustal subduction within the Pacific and northeast Indian Ocean. This requirement is independant of the rate of crustal generation within the Pacific Ocean spreading zone itself, which merely adds, albeit substantially, to the burden of subduction at or near the oceanic margins. The amount of continental crustal foreshortening is to an extent balanced by continental crustal extension and, for the purposes of this discussion, can be discounted as a factor.

If both of the above propositions can be satisfied, no Earth expansion has occurred. The ocean-floor spreading evidence of a substantial imbalance between Propositions 1 and 2 above requires for its interpretation a knowledge of cartographic methods as well as the ability to analyze the accuracy of the computer programs that are in current use for making "plate tectonic" reconstructions. Any direct measurements of relative motions between continents providing simple numbers that can be readily understood, have to be important. This underlines the importance of the LAGEOS satellite laser ranging program.

SATELLITE LASER-RANGING ACROSS THE PACIFIC OCEAN

On 4 May 1976, NASA launched the Laser Geodynamics Satellite (LAGEOS), a passive spherical-shaped satellite, as part of their Geodynamics Program, Crustal Dynamics Project. The satellite surface is covered by cube corner light reflectors by which laser beams can be reflected from one ground station to another with high accuracy of point position determinations. Measurements by satellite laser ranging (SLR) have provided data on (a) plate-tectonic motions (absolute measurements), (b) crustal deformation (absolute measurements), (c) polar motion (secular changes), and (d) Earth's gravity field (secular changes). The accuracy of these measurements can be cross-checked with an independent method of ranging using the differential reception of a quasar radio signal at the different ground stations, a technique known as very long baseline interferometry (VLBI). The preliminary report on the results obtained by laser ranging using the LAGEOS satellite was published in 1985 (Cohen et al., 1985). It is the intention here to concentrate on results to date across the Pacific Ocean, because it is the boundaries of this ocean that are required, essentially, to be the global "sink" for the bulk of the excess crust generated by ocean-floor spreading in the passive-margined oceans and the Pacific itself.

The LAGEOS laser ranging data obtained up to 1984 is plotted in Figure 1 on an oblique elliptical whole Earth map using Bartholomew's "Nordic" projection centered on the Pacific Ocean. The average annual figures in centimeters for the spreading rate at the midoceanic generating ridges are combined with the LAGEOS data derived from Christodoulidis et al. (1985) and it is immediately apparent that the accountancy of these preliminary data do not balance as they should do if the Earth is of constant dimensions. Proposition 2 above is not satisfied, therefore, and the preliminary data indicate convergence rates in the Pacific significantly less than predicted by Minster and Jordan (1978) based on a theoretical mathematical model of a non-expanding Earth. The most significant measurement is the chord length increase between Australia and South America. Bearing in mind the spreading rate in the southern South Atlantic and Indian Oceans, this chord should be decreasing in length if the Earth has maintained a constant dimensions in recent years in response to the displacement of the various continents towards each other at the expense of the area of the Pacific Ocean itself. This requirement is not significantly affected by orogenic compression of the Andean margin of South America. The period concerned for which these SLR data are available is infinitely small in comparison with geological time and, no doubt, would be dismissed as a minor short-term anomaly. However, if these preliminary measurements are combined with a detailed study of the ocean-floor spreading data back to the fit-together of the continents in Pangaea around 200 million years ago, they become significant indeed (Owen 1983).

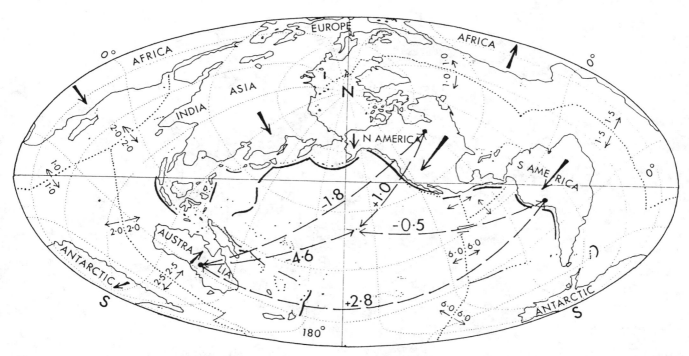

Figure 1. Outline map of the modern Earth to show the interaction of ocean-floor spreading and subduction superimposed upon which, are the average annual displacement figures in centimeters obtained by satellite laser ranging across the Pacific Ocean. Bold dotted lines are midocean spreading ridges; direction and spreading rates are shown by small arrows in centimeters per year. Broad arrows represent the relative motion of continents towards the Pacific. Heavy bold lines at the central and northern Pacific, northeast Indian Ocean, and southern South Atlantic mark deep subduction trenches. Bold pecked lines are the chords between continental stations used for satellite laser ranging; negative prefixes to bold numbers indicate convergence, positive prefixes indicate divergence (Bartholomew's "Nordic" projection).

FIT OF PANGEA AND SUBSEQUENT CONTINENTAL DISPLACEMENT

The fit of the continents together in Pangea has to accord fully with the geological match information at their common margins. Those who still maintain the concept of a steady-state Earth or even shrinking Earth, in which the ocean basins are regarded to be of great geological antiquity, disregard at their peril the detailed structural and stratigraphic information of the direct connection together of, for example, South America and Africa prior to the Early Cretaceous. In Figure 2, which represents a modern dimensions Earth in the Late Triassic-Early Jurassic, reassembly of the continents into Pangea produces relatively minor gaps between the continental margins of South America and Africa, between the Canadian margin and Greenland, and between Greenland and the Norwegian margin; major gaps occur between Canada, Alaska, and northern Russia on the one hand and on the other in the region called the Tethyan Ocean between Gondwanaland and Eurasia. These gaps (in reality spherical geometric gores), whether major or minor in extent, significantly radiate out from the center of the reconstruction—the fit of the northwest African margin into the North American east coast embayment—as would be expected if the value of surface curvature has decreased since the time of Pangea and the reconstruction is made on the modern diameter Earth. If these spherical triangular gaps in the fit existed in reality, crustal traverses across them would show either gravity lows if they consisted of deep, down-faulted marginal blocks of continental crust, or oceanic crust older than the Triassic reassembly of Pangea. Although deep, down-faulted, marginal blocks of continental crust do occur, they are of insufficient area in the passive-margined oceans to infill the various gores mentioned above. The marginal oceanic crust actually present is post-Triassic and, in certain critical areas, post-Jurassic in age. Therefore, the crust is much younger than is predicted by the constant modern-dimensions Earth model depicted in Figure 2 in which the gores would be infilled by crust older than the Triassic.

Can these problems of fit between continents be overcome by the occurrence of marginal compression subsequent to the start of continental displacement? In the case of the Tethyan Ocean, the Alpine foldbelt extending from Spain and North Africa in the west to the Burman ranges in the east is used as the convenient sink for the putative pre-Triassic Tethyan ocean crust. There is no doubt that the Himalaya between the western Ornach Nal, Chaman, Kirthar-Sulaiman wrench fault

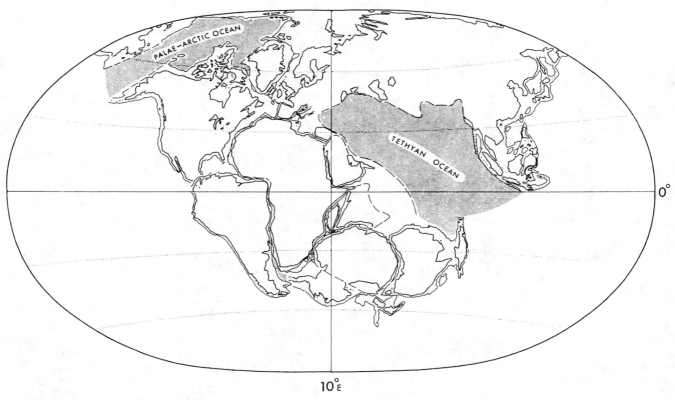

Figure 2. Pangea 180–200 Ma assuming a modern dimensions Earth. Shaded areas represent oceanic crust required to be present on such a reconstruction but for which no evidence of its presence exists in reality (Winkel's "Tripel" projection).

systems, and the corresponding system in the east between the Himalaya and Indo-Burman ranges is a major thrust zone with an infilled foredeep at its junction with the northern margin of India. However, critical analysis of the ocean-floor spreading patterns in the Indian Ocean, including a Late Jurassic to mid-Cretaceous spreading phase between greater India and Asia, indicate on a modern dimensions Earth reconstruction, a gap of 600 km at present if Antarctica is placed in its correct polar position since the Upper Cretaceous and the development of the oceanic crust of the Indian Ocean is strictly adhered to: clearly this gap is not possible. Also, there are both geological and paleontological anomalies associated with this concept of a wide Permo-Triassic Tethyan oceanic crustal area (e.g., Smith 1988). In the case of the North and South Atlantic and Arctic Ocean, there is no evidence of the necessary marginal compression of the bordering continents since their post-Triassic separation to explain these gores. Rather the reverse, for the tectonic history subsequent to break-up is essentially one of extension with some crustal thinning, with tectonic sutures in the Alaskan and corresponding Verkhoyansk-Anadyr regions marking continent-continent contact lateral displacement.

If the reconstruction of Pangea is made according to the geological evidence of fit and the pattern of subsequent break-up by ocean-floor spreading, rather than any preconceived ideas of the size of the Earth in the past, the result is that shown in Figure 3. This reconstruction produces an Earth diameter of 80 percent of modern value and there are no problems in the plotting and accountancy of the subsequent ocean-floor spreading patterns. A critical test of the ocean-floor spreading data on both the constant-dimensions Earth model and the expanding model that depends totally on the spherical geometry of the crustal growth patterns of the oceanic crust, was made in an Atlas by the author (Owen 1983). There is little to add to the results of this test other than to say that subsequent published data has continued to support the expanding Earth model described therein.

Weijermars (1989) has argued that the Earth's oceanic crustal budget between generation at the mid-ocean ridges and subduction is balanced. This view is based apparently partly upon a cartoon, largely theoretical, isochron map published by Sclater et al. (1980). This map does not agree with the strict distribution of ocean-floor magnetic anomaly data. Steiner (1977),

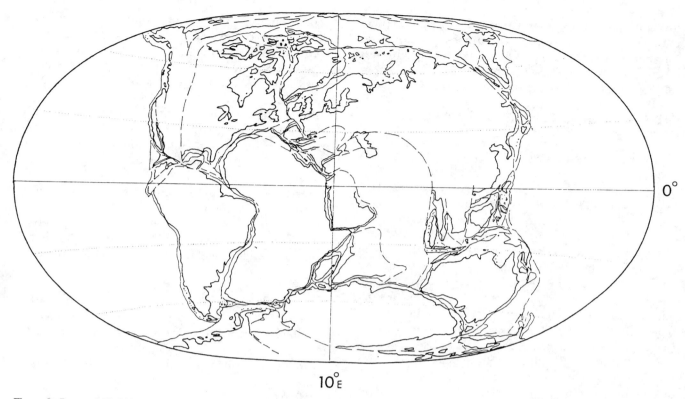

Figure 3. Pangea 180–200 Ma assuming an Earth of 80 percent of modern diameter (Winkel's "Tripel" projection).

using the older Pitman et al. (1974) map of a much higher order of accuracy, found an exponential growth of the Earth's oceanic crust suggesting a maximum radius of 89 percent in the Jurassic and possibly less. It is absolutely essential in any of these studies to go back to the original mapped data. It is particularly important to test the computer programs used in reconstructions of continental displacement against the results of conventional cartographic mensuration; a point illustrated by the author on a previous occasion (Owen 1983).

IMPORTANCE OF ACCURATE CARTOGRAPHY

The failure by Weijermars to analyze the original ocean-floor spreading data instead of depending on a largely theoretical isochron map and his construction of cartoons that have little cartographic validity, have led to the production of a series of inaccurate reconstructions (Weijermars 1989). In this line of research, there is no substitute for the accurate depiction of continental and oceanic crust according to the field data, correctly projected mathematically from a spherical surface to a flat surface (a map), however time consuming and tedious such compilations and cartographic projections might be. Crude outlines in which continental and oceanic crustal areas have been distorted in spite of the geological evidence to the contrary, such as those given by Weijermars, or in which the computer program has not corrected for changes in the amount of cartographic distortion of continental or oceanic areas in response to their changing positions in respect of the cartographic projection pole, are not acceptable.

Figure 2, herein is indeed an outline map without the ocean-floor spreading data plotted on it. However, the maps from which it is compiled (Owen 1983) are sufficiently detailed to test its accuracy. It is simply not possible on a globe to reconstruct Pangea at 180 Ma with Antarctica over the south pole as shown by Weijermars (1989) and with the north pole in the position he shows it, without disturbing the fit of Pangea together according to both the geological evidence of direct connection of the continents and the subsequent ocean-floor spreading pattern. Moreover, the continents themselves are distorted in Weijermars' reconstructions in a manner for which there is no on-continent geological evidence. The outline reconstruction of Pangea given here in Figure 2 and those of Smith et al. (1981), Barron et al. (1981), and Scotese et al. (1981), all assuming an Earth of constant modern dimensions, possess cartographic integrity, particularly so in the case of Figure 2 herein and Smith et al. Moreover, when one analyzes Weijermars' cartoons, one finds variations in continental widths and vector lengths and a failure to correct cartographically the distortion pattern of a continental outline as it changes position relative to the projection pole. One also finds that continental outlines are too crude for analytical work, let alone for use as an objective test of the writer's cartography (Owen 1983) or the cartography of the authors referred to above.

OLD CONTINENTAL CRUST AND NEW OCEANIC CRUST

It has long been recognized that, geophysically and geologically pre-Paleozoic continental rocks differ fundamentally from those of the modern ocean basins. This led Glikson (1980) to the conclusion on geochemical grounds that very little oceanic crust existed in the Archaean. Others, for example Hilgenberg (1933), Barnett (1962), and Vogel (1983), have argued on theoretical grounds that the Earth's continental crust once formed a total sialic outer shell to a lithosphere that broke up and became displaced as a result of Earth expansion. Such an Earth, if it had existed in reality, would have a diameter of around 56–60 percent of modern value. Critics of these ideas have postulated various arguments, some of which are valid, others less so, against this notion. A full review of all these various ideas and objections is not possible in the confines of this short paper, but some should be discussed here.

It has been pointed out earlier in this paper that the concept of all oceanic crustal growth having occurred since the mid-Jurassic, as postulated by Carey (e.g., 1983) and Shields (e.g., 1983), is impossible based on geological and spherical geometric fit. The fit-together of the continents in the Late Triassic–Early Jurassic on an 80 percent of modern diameter Earth is precise, but indicates the presence of a sizeable area of old Pacific Ocean crust (Fig. 3). This crustal area has been subducted during the last 180 Ma and, despite the view of the "fast expanders," the growth of ocean-floor spreading during the Mesozoic and Cenozoic supports this assertion.

The only testable evidence of Earth expansion we have is that provided by the fit-together of Pangaea and the ocean-floor spreading history and geometry of continental displacement from the Late Triassic to the present day. However, if one regards critically the various Paleozoic orogens as evidence of differential wrench motions as well as compressional belts, it is possible to achieve a total sialic crust at around 700 Ma on an Earth of around 56–60 percent of its modern diameter. The sialic plate motions, of which the orogens are the bounding sutures, will proceed kinematically from the total sialic crustal state to the configuration of Pangaea at 200–180 Ma on an 80 percent of modern diameter globe. The motions certainly do not indicate radial expansion, one of the simplistic criticisms made of Earth expansion theory as a whole. All previous reconstructions showing the continental components of a complete sialic shell in their modern configuration at diameters less than 80 percent of modern value, are invalid. It is not possible to make the required distortions implied by such reconstructions in the absence of geological evidence to support them.

Although the lithospheric crustal growth data provide evidence of increasing surface area and, therefore, Earth expansion during the last 200 Ma, the real problems in geophysics lie in our interpretation of the inner, obscure, parts of our planet. Problems arise if we regard the Earth's core structure to be in a steady state throughout the Phanerozoic. With an Earth of 80 percent of modern diameter around 180–200 Ma, the mantle would be considerably thinner than it is today. Where has the additional mantle volume of the modern Earth come from? If a 56–60 percent of modern diameter Earth was a reality, the radius of the modern core would be greater than the diameter of the Earth some 700 million years ago (Figure 4), a state clearly impossible. This is the point where the conventional geologist and geophysicist switches off, although perhaps not the astrophysicist or particle physicist. Does the above observation negate the whole concept of Earth expansion?

WHAT IS THE NATURE OF THE EARTH'S CORE

Physical theory is based on testable evidence as in any other science. Models are produced and supported by a carefully reasoned mathematical argument. However, if the basic assumptions are wrong, or need modifying in the light of new discoveries, it matters nothing how elegant the mathematics might be, the hypothesis collapses or needs radical revision. Most of our ideas about the nature of the inner parts of our planet are basically those recognized nearly a century ago and, in some cases, even earlier when the Earth was regarded as a planet essentially in a steady state. In this respect, it is still widely assumed that what holds for the present applies to the distant past also. The long-standing geological concept of a solid state nickel-iron-sulphur sphere forming the inner core to the Earth stemmed initially from differences in the computations of the planet's total specific gravity in comparison with that indicated by the continental and oceanic crustal rocks alone. Meteorite compositional evidence and seismic wave velocity and refraction characteristics were also compatible with the concept of a metallic inner core quite apart from the presence of a weak geomagnetic field. This concept might once have been satisfactory in the light of a steady-state Earth of former times. However, it is difficult to visualize in this model, the heat engine that has driven mantle convection, crustal differentiation, and con-

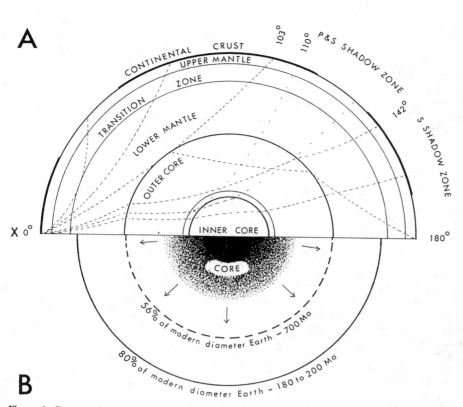

Figure 4. Cross sections through: **A**—the modern Earth, and **B**—the putative 56–60 percent and proven 80 percent diameter Earth to show relationships of core size. In A, paths of P waves only are indicated. In B, arrows indicate putative core volume growth and early shell differentiation.

tinental displacement, which we know to have been a feature of the Earth's long history.

The current scenario that one proceeds downward with an increasing value of gravity from sialic plates with very deep and old cratonic roots, probably underlain by mantle rocks of even greater age, the whole being displaced by convection currents within the highly viscous fluid of the modern mantle with its chilled oceanic lithosphere, followed abruptly inward by a molten and turbulent outer core passing into a relatively low temperature and pressure phase represented by a solid-state nickel-iron-sulphur ball, forming the putative inner core defies logic. It does not make physical sense that the temperature and pressure gradients could produce a core in a relatively low temperature solid state. It has been argued that as there is no evidence of expansion on the Moon or Mercury since the equivalent of our early Archaean, the Earth cannot have expanded either (e.g., Taylor 1983). Yet again, the uniformitarian approach obscures the fact that the Moon, Mercury, and Mars are much less dense planets than either Earth or Venus; like is not being compared with like. The concept that expansion of the Earth is due solely to the decreasing value of Universal Gravity G also reflects a uniformitarian approach that is demonstrably incorrect. However, the possibility that the potential expansion of a planet depends on its original mass and physical state after its condensation from the stellar nebula is worthy of investigation.

Halm (1935) discussed the possibility of an alternative to the iron core scenario; that the Earth's inner core approximated to the then recognized state of a white dwarf. In modern terms, this would be regarded as matter under such gravitational pressure that atomic structures could not exist (Owen 1981) (Figure 4). The potential for atomic and molecular differentiation and for planetary expansion are very real in such a model until such time as the inner core reaches a total atomic state whereupon planetary expansion would cease. In the case of Earth, which is, with Venus, relatively dense among the terrestrial planets, such a concept provides a parsimonious mechanism and sequence for shell differentiation and outgassing that the current concept of the Earth's interior and physical state does not; not only during the last 200 Ma when ocean-floor spreading data indicate global expansion, but throughout its 3.5 Ga lithospheric and atmospheric history. This concept might also explain the lack of expansion seen on the Moon, Mercury, and Mars, which are very light in weight in comparison with the Earth and Venus and probably developed an atomic state core and shells rapidly from a very much smaller amount of original matter.

POSTSCRIPT

In this brief review of the evidence of Earth expansion and the possible dynamic Earth model that might be the cause of it, I have suggested reasons why neither the steady-state concept nor the conventional plate tectonic model accord with observed data. If geologists and geophysicists had the courage to reassess their ideas in the light of new observations, major advances in geodynamics and particle physics could be achieved. At present, many workers attempt to fit their observations into a rigid framework of uniformitarian theory when an increasing amount of data shows this philosophy to be untenable.

REFERENCES CITED

Barnett, C.H., 1962, A suggested reconstruction of the land masses of the Earth as a complete crust: Nature, v. 195, p. 447-448.

Barron, E.J., Harrison, C.G.A., Sloan, J.L., and Hay, W.W., 1981, Paleogeography, 180 million years to the present: Eclogae Geologicae Helvetiae, v.74, p.443-470.

Carey, S.W., 1958, The tectonic approach to continental drift: Symposium on Continental Drift, Hobart, p. 177-355.

———, 1983, Tethys, and her Forebears: *in* Carey, S.W. (Ed.) The expanding Earth, a symposium. Sydney, p. 169-187.

Christodoulidis, D.C., Smith, D.E., Kolenkiewicz, R., Klosko, S.M., Torrence, M.H., and Dunn, P.J., 1985, Observing tectonic plate motions and deformations from a satellite laser ranging: Journal of Geophysical Research, v. 90, p. 9249-9263.

Cohen, S.C., King, R.W., Kolenkiewicz, R., Rosen, R.D., and Schutz, B.E. (Eds.) 1985, Lageos scientific results: Journal of Geophysical Research, v. 90, p. 9215-9438.

Glen, W., 1982, The road to Jaramillo, critical years of the revolution in earth sciences: Stanford University Press, p. i-xvii, 1-459.

Glikson, A., 1980, Precambrian sial-sima relations, evidence for Earth expansion: Tectonophysics, v. 63, p. 193-234.

Halm, J.K.E., 1935, An astronomical aspect of the evolution of the Earth, Presidential Address: Journal of the Astronomical Society of South Africa, v.4, p. 1-28.

Hilgenberg, O., 1933, Vom Wachsenden Erdball, Berlin, p. 1-50.

Minster, J.B., and Jordan, T.H., 1978, Present-day plate motions: Journal of Geophysical Research, v. 83, p. 5331-5354.

Owen, H.G., 1981, Constant dimensions or an expanding Earth: *in* Cocks, L.R.M. (Ed.) The Evolving Earth: London and Cambridge, p. 179-192.

———, 1983, Atlas of continental displacement, 200 million years to the Present: Cambridge Earth Sciences Series p. i-x, 1-159, 76 maps.

Pitman, W.C. III, Larson, R.L., and Herron, E.M., 1974, The age of the ocean basins: Geological Society of America Map and Chart MC-6, 2 charts.

Sclater, J.G., Parsons, B., and Jaupart, C., 1981, Oceans and continents, similarities and differences in the mechanism of heat loss: Journal of Geophysical Research, v. 86, p. 11535-11552, 2 plates.

Scotese, C.R., Snelson, S., Ross, W.C. and Dodge, L.P., 1981, A computer animation of continental drift: *in* McElhinny, M.W., Khramov, A.N., Ozima, M. and Valencio, D.A. (Eds.) Global reconstructions and the geomagnetic field during the Paleozoic: Advances in Earth and Planetary Sciences 10, p. 61-70.

Shields, O., 1983, Trans-Pacific biotic links that suggest Earth expansion: *in* Carey, S.W. (Ed.) The expanding Earth, a symposium. Sydney, p. 199-205.

Smith, A.B., 1988, Late Paleozoic biogeography of East Asia and paleontological constraints on plate tectonic reconstructions: Philosophical Transactions of the Royal Society of London v. A 326, p. 189-227.

Smith, A.G., Hurley, A.M., and Briden, J.C., 1981, Phanerozoic paleocontinental world maps: Cambridge Earth Science Series, p. 102, 88 maps.

Steiner, J., 1977, An expanding Earth on the basis of sea-floor spreading and subduction rates: Geology v. 5, p. 313-318.

Taylor, S.R., 1983, Limits to Earth expansion from the surface features of the Moon, Mercury, Mars, and Ganymese: *in* Carey, S.W. (Ed.) The expanding Earth, a symposium. Sydney, p. 343-347.

Vogel, Kl., 1983, Global models and Earth expansion: *in* Carey, S.W. (Ed.) The expanding Earth, a symposium. Sydney, p. 17-27.

Weijermars, R., 1989, Global tectonics since the breakup of Pangaea 180 million years ago, evolution maps and lithospheric budget: Earth-Science Reviews v. 26, p. 113-162.

Earth expansion theory versus statical Earth assumption

G. O. W. Kremp, Department of Geosciences, University of Arizona, Tucson, Arizona 85721 USA

ABSTRACT

There is no scientific evidence to support the ancient assumption that Earth has maintained a constant size for the past three or four billion years. Nevertheless, this assumption has been accepted, unquestioned, and unexamined. In contrast, there are many geologic indications that support the Earth expansion theory. Only a few are discussed here.

All of the Earth's continents would fit neatly together, enveloping the Earth with continental crust on a globe 60% of its present size. Atomic clocks have documented evidence that Earth's rotational speed is slowing down. Seismologists have discovered that the temperature of the outer core at the core-mantle boundary is at least 800 degrees higher than at the Seismic D" layer at the very base of the mantle, indicating that the outer core is heating up and is forcing the expansion of the Earth. This is a recent process and could not have persisted for some four billion years. It may have caused the rapid expansion of the Earth's size during the last 200 million years. Permanent magnitismus of the Indian Ocean, paleontologic research, and radiomagnetic age dating indicate that during the last 200 million years the Earth expanded from 60% to 100% of its present size. Illustrations of K. Vogel's terrella models indicate that the continental crust, fixed to its underlying asthenosphere, was pushed outward in radial direction upon the growing globe.

EARTH EXPANSION THEORY VERSUS STATICAL EARTH ASSUMPTION

Numerous arguments were presented at the discussion meeting concerning new concepts in global tectonics at the Smithsonian Institution in July 1989. Most of these arguments indicated that there seems to be something questionable in plate-tectonic theory as it is currently presented. In my opinion, the reason for this is that our geologic textbooks accept, without criticism, the ancient assumption of a statical Earth that has not changed size during the last three or four billion years of its existence.

Conventional wisdom holds that about 2,500 Ma drifting continental fragments were assembled in a supercontinent that was surrounded by the old, old Panthalassa Ocean. This permanent ocean (with a simatic crust?) already should have existed before that time, maybe 3,800 Ma. It is assumed that this mystical ocean had to be about three times as large as the supercontinent (Fig. 1).

Geologic evidence indicates that about 2,500 Ma most of the sialic (continental) crust had been formed after a major thermal event culminating in global crationization (Table 1). Paleomagnetic data presented at the 1980 International Congress in Paris confirmed the existence of a supercontinent (Fig. 2) by that time. But what about the Panthalassa Ocean?

Meager geological evidence indicates that approximately 4,000 Ma the primordial hydrosphere began to condensate. The primordial sea water was probably quite different in temperature and composition from sea water of the present and probably covered the entire globe. The uppermost crust of the primordial, consid-

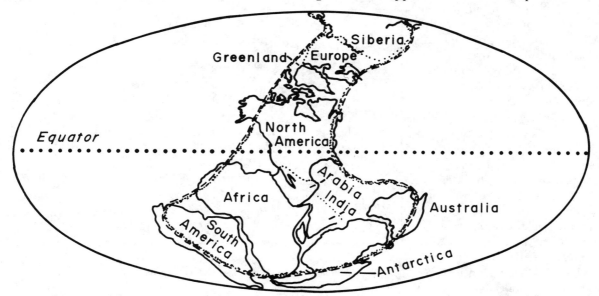

Figure 1. The Proterophytic Supercontinent Model of Piper, 1976. The dashed line represents the boundary of the Precambrian sialic (continental) crust. According to the recently outdated static Earth assumption, around 2,500 Ma the supercontinent was surrounded by an ocean of simatic (oceanic) crust that was about three times as large as the supercontinent itself. However, there are no traces of such a crust.

Chatterjee, S., and N. Hotton III, eds. *New Concepts in Global Tectonics.*
Texas Tech University Press, Lubbock, 1992, xii + 450 pp.

Table 1. Major Precambrian tectonic events (after Glikson 1980b).

Time (Ma)	Tectonic Events
Present to 1000	Operation of two-stage mantle melting processes; evidence for existence of contemporaneous sima.
1000	Major crustal tension phase involving intrasialic rifting and opening of oceanic gaps. Large-scale horizontal plate tectonics.
1700 to 1900	Peak of tectonic events within intrasialic mobile belts.
1000 to 2000	Mobile belts (inter-cratonic; of marginal or external types).
2000 to 2500	A tectonically stable interval of marked paucity.
2500 to 2700	Major thermal event culminating in global cratonization.
2600 to 3800	Sima-sial transformation.

erably smaller Earth was covered by a thin sialic crust. But the Earth's surface suffered from severe bombardment by numerous asteroidlike objects. As a result, the primordial sialic crust was to some extent destroyed and replaced by a crust of simatic (mafic-ultramafic) composition, which in the course of the next 1,500 million years, was transformed into a sialic (continental) crust due to endogene processes (Gilkson, 1980b; Kremp, 1982).

As indicated in Figure 4, the diameter of the Earth in the Cryptophytic eon may have been less than 40% of its present size. But the Earth continued to grow and the depth of the primordial hydrosphere gradually became thinner and thinner. The emergence of continental beds must have occurred before at least 1,800 million years (Cloud, 1973; Schidlowski, 1976). These continental beds, of prevailing red color, offer evidence that much of the crust had emerged from its hydrospheric cradle and was exposed to atmospheric weathering.

To summarize, the newer geologic evidence indicates that by 2,500 Ma a hydrosphere existed, covering the total globe (which was considerably smaller than at present), but it was not the Panthalassa Ocean as predicted in Figure 1. By 1975, the geologic knowledge of the Early Proterophytic eon was still very scarce and the Earth expansion theory hardly known. Before 1980 even Glikson, the most outstanding Precambrian geologist, did not understand the Earth expansion theory. Glikson's 1980b article, "Precambrian sima-sial relationship: Evidence for Earth expansion," showed his first acceptance of the theory. Based on polar wandering curves, Piper (1976) assumed that the supercontinent (Fig. 1) remained connected until 1,150± 200 Ma. Godwin (1976) was of about the same opinion (Fig. 2). Many geologists assume that the supercontinent remained more or less connected throughout the Pre-cambrian and often assume (e.g., Boucot and Gray, 1979) that the Pangea supercontinent remained in contact through all of Paleozoic time.

Nevertheless, there are many scientists who still believe in the statical Earth assumption. They assume the existence of a permanent simatic ocean, existing since 2,500 Ma. To explain their belief, they resorted to the supercontinent cycle theory. According to this idea, about every 500 million years a new incarnation of a supercontinent occurred (that is, drifting continental fragments assembled to form one body), which later broke apart until the next incarnation happened. This assumption means that Earth has experienced five supercontinental cycles in the last 2,500 million years. But where are the traces of that subducting Panthalassa Ocean (five times each!) and the traces of the bouncing-together procedure of the continental fragments (also five times each)?

As early as 1981, Carey argued that there are no traces of a simatic crust 2,500 Ma and that Piper's 1976 version cannot be supported. Carey's postulation is still valid. In

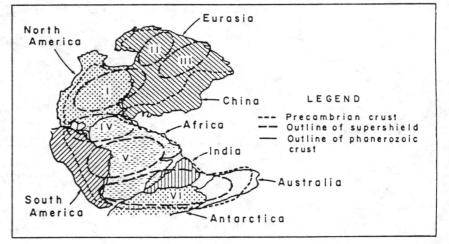

Figure 2. A Pangea-like supercontinent, existing about 2,500 Ma according to Godwin (1976). Godwin, like Piper, postulated that about 2,500 Ma a giant supercontinent of a quasi-Pangean pattern must have existed and remained connected throughout most of Precambrian time. Many geologists (e.g., Boucot and Gray, 1979) are convinced that this supercontinent remained more or less connected until the end of the Paleozoic. Their independent geologic research conforms with the Earth expansion theory (after Godwin, 1976).

contrast to the statical Earth assumption, the Earth expansion theory assumes that there existed 1,700 Ma (maybe even 2,500 Ma) a globe only 40% of its present size (Figs. 4 and 8) and we can follow with confidence Glikson's geologic research, which he concluded in 1980 (b, p. 222) with the statement "... in the absence of any other viable hypothesis it appears that the Earth's surface area must have increased with time [since early and middle Proterozoic times], in agreement with the expanding Earth theory...."

According to the investigations of Glikson (1979 to 1983), there is solid, geological evidence that sea-floor spreading began around 1,000 Ma and that the birth of the Paleo-Pacific Ocean must have happened around this time, not 2,000 or 3,000 million years earlier. Extensive ensialic riftings of tears and oceanic gaps seem to have occurred during the period of the Hudson Regime (Fig. 4). Glikson (1979–1983) and others assumed that during a continental rifting episode between 1,200 to 1,000 Ma, the Paleo-Pacific Ocean opened between the Americas and the Asia-Australia-Antarctica complex as a widening of the Cordilleran Rift. He further indicated that around 1,000 Ma, sea-floor spreadings, continental drift and modern subductionlike processes had begun. Supporters of the statical Earth assumption failed to prove that multiple sea-floor spreadings from mid-oceanic ridges as well as multiple subductions of huge oceanic crusts existed in Precambrian time. One good reason for the success of the statical Earth assumption may be peer pressure and the fact that even mentioning such a revolutionary idea as the Earth expansion theory would raise eyebrows in academic circles.

Scientists who accept the Earth expansion theory find themselves in a situation somewhat similar to that of Galileo's some 550 years ago. His colleagues refused to look into his telescope and to discuss his findings. They preferred to believe that Earth (and therefore man) was the center of the universe. To accept any other idea was heretical. Darwin and Wegener suffered the same experiences. Similar dynamics are now at work against the Earth expansion theory. To accept this theory is to contradict assumptions put forth by learned people in published papers.

The notion of an expanding Earth is not new. As Carey (1976) pointed out, it occasionally has been discussed in the scientific literature since the 19th century. In 1933, Hilgenberg constructed the first terrella models (Fig. 3). He maintained that the Earth was only about two-thirds of its present diameter when all the continents fitted neatly together and completely enveloped the globe.

Hilgenberg's experiment can be repeated easily with simple materials. One needs an accurate globe of the world, an air balloon such as children play with, and some sheer fabric. On the fabric, trace the outline of the continents along their extended shelves, then carefully cut out the shapes. Inflate the balloon to two-thirds the size of the globe and tape the "continents" to it according to their arrangement on the globe. This arrangement—the Pangaean-like supercontinent according to Godwin's construction (Fig. 2)—will cover the balloon with only narrow spaces between some of the landmasses. Thus, one has proof-in-hand of Hilgenberg's 60% terrella model.

Recently, the Earth expansion theory has attracted more attention. In 1981, its foremost advocate and spokesman, S.W. Carey organized a well-attended symposium. Of 54 papers presented, 38 favored Earth expansion; 12 were undecided and four opposed it. (It should be noted that evidence cited in the opposing four papers is easily refuted.)

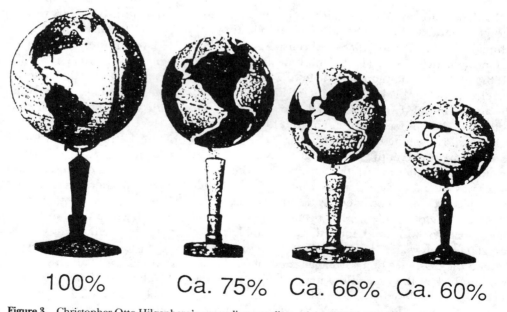

Figure 3. Christopher Otto Hilgenberg's expanding terrella models of 1933. Hilgenberg became convinced of the reality of Earth expansion when he discovered that all the Earth's continents fit neatly together, enveloping the Earth with continental crust on a globe approximately 60% of its present size. Stimulated by Wegener's ideas, he was the first to construct expanding terrella models. However, in his era, continental drift was considered a heresy, and his work was soon forgotten (courtesy Vogel, 1981).

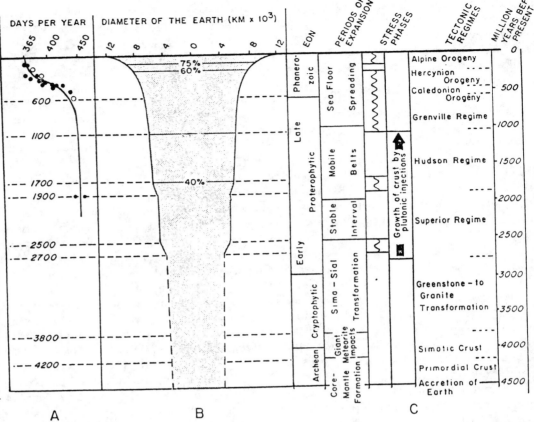

Figure 4. Proposed chart of the increase in the Earth's diameter since accretion. **A**—Variations in the number of days per year since Gunflint time (approximately 2,000–1,900 Ma). Black dots represent data obtained from growth line pattern measurements as published by Pannella (1972). Circles indicate data reported by Lambeck (1978). (See Figs. 5 and b.) **B**—Increase of the Earth's diameter since the earliest Precambrian (as tentatively assumed by the author). The geologic record indicates that Vogel's 75% terrella model might present Earth at the beginning of the Cenozoic Earth expansion phase; the 66% model at the beginning of the Mesozoic Earth expansion phase; the 60% model around or before the Jurassic expansion lull, and the 40% model perhaps around the stress phase of the Precambrian Hudson Regime (see Fig. 7). **C**—Major geodynamic events since the earliest Precambrian. (B and C from Kremp, 1984 revised from my 1981, 1983 publication.)

Evidence favoring Earth expansion has been mounting considerably since 1981 with the publications of Ahmad (1982), Carey (1988), Ehrensperger (1988), Kremp (1982-89), Noël (1989) and especially of Vogel (1983, 1990) to mention only a few. Vogel's famous terrella models provide the most convincing evidence for an expanding Earth. Included here would be the authors of the numerous arguments put forward in this volume who question details of the plate-tectonic theory as it is taught presently. In a final analysis, all these arguments would support the insight that the Earth never was a dead body of statical, unchanging size, but that it has expanded considerably during its existence.

Recently, more evidence has become apparent that our globe is growing, and, indeed, rather rapidly in the last 200 million years. In the last decades, seismologists located the existence of a particular zone, about 200 to 300 km thick at the base of the mantle. This zone was recognized and designated some 40 years ago as the D″ region. Yuen and Peltier (1980) as well Boss and Sacks (1985) postulated the existence of a substantial flow of heat across the core-mantle boundary. They concluded that if whole-mantle convection occurs in the Earth's mantle, the D″ region should be the lowest thermal boundary layer of the whole-mantle convection circulation system. The temperature of the outer core at the core-mantle boundary recently has been estimated to be about 800 degrees higher (Lay, 1989) than the thermal D″ layer of the mantle or perhaps even 1500 degrees higher (Williams et al., 1987). The statical Earth scientists cannot explain these differences by their current themes and have avoided discussing it. It can be concluded that the thermal increase in the outer core is a fairly recent process that has forced the rapid expansion of the Earth's size during the Hercynian and Alpine orogenies (Fig. 4).

Would this mean that our globe is now continuing to expand farther and farther? I almost would believe that we are again in another stress phase. It may be another cycle in which the outer core of the Earth is overheating again by atomic decay. Releasing the heat causes the expansion. When the overheating is released, expansion ceases. The stress phases that apparently started 2,700, 1,900, 1,100, 600, 150, and 80 Ma may indicate the cyclic nature of the Earth's growth. Why the Earth is expanding is not really known, and some other decompressional processes may be occurring in the core or mantle or both that affect this expansion.

Another indicator for Earth expansion is that the globe's rotation speed is slowing down. In 1972, Pannella studied the growth pattern of molluscan shells and reported that daily growth layers arranged into seasonal and tidal patterns (present in calcified structures of many modern as well as fossil organisms) provided evidence regarding the length of the lunar month and year since Ordovician times. It appears that the number of days per lunar month and per year have decreased significantly since the early Paleozoic. According to Pannella, some 1,900 Ma, the year had about 447 days and approximately 290 Ma, the year had 383 days.

The accompanying graphs (Figs. 5 and 6), redrawn from Pannella (1972), show the decrease of days per year since the Ordovician and the Precambrian, respectively. In 1980, Pannella's work received scathing criticism from a colleague who could not contemplate the concept of an expanding Earth, and caused the Pannella to discontinue his investigation on this topic. I believe that it is imperative to resurrect and continue Pannella's research on the paleontological clocks.

Atomic clocks also indicate that Earth's rotational speed is slowing down. Like the ticking of a pendulum in a grandfather clock, the rapid-fire oscillations of cesium atom's nuclei (9,192,631,770 times a second) can be used to measure time. Atomic clocks are the world's most accurate timepieces and have important applications in navigation and communication systems. These clocks have also been used to make direct measurements of continental drift, coordinate astronomical observations and test the ability of Earth's gravity to slow down time.

The U.S. Naval Observatory, based in Washington, D.C., is the official timekeeper of the nation. It is a member of a worldwide group established in 1972 to keep atomic clocks synchronized with the Earth's rotation (McCarthy, 1976). Officials at the Naval Observatory determined that the year 1987 lasted one second longer than the year 1986. The atomic clock—used as an international standard of time—had to be set back one second for the fourteenth time since 1972! Atomic time (which is constant) has gotten ahead of time measured by the Earth's rotation. This indicates that Earth's rotation is decelerating.

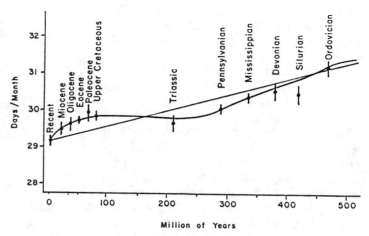

Figure 5. Graph based on growth patterns of modern and fossil invertebrates according to Pannella (1972), showing variations in the length of the synodic month since Ordovician time. The fourth-order polynomial curve was selected by the computer as having the best fit. The straight line represents the deceleration of the Earth's rotation. Vertical bars indicate standard error. Daily growth patterns of modern and fossil invertebrates provide evidence regarding the length of the lunar month and year. Pannella investigated the growth patterns of many fossil invertebrates (mollusks, brachiopods, and corals) and of stromatolites from Precambrian to Recent times and has also reexamined the reports of earlier investigators. His 1972 studies of Phanerozoic material included two nautiloids from the Silurian of Nova Scotia, and his reports on the Paleozoic generally agree with those of earlier authors. (Courtesy G. Pannella, 1972).

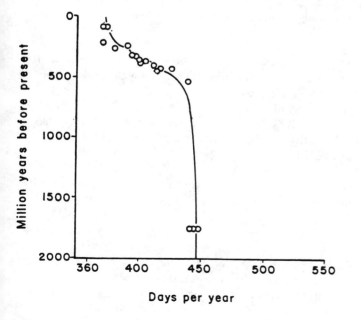

Figure 6. Graph by Pannella (1972) based on growth patterns of Phanerozoic invertebrates as well as Gunflint stromatolite laminae. As the diameter of the Earth expanded, its rotational speed decreased proportionately, resulting in fewer days per year. Pannella's interpretations of the growth patterns of fossil invertebrates and stromatolites, the so-called paleontological clocks, seem to indicate that there were about 375 days per year in Late Cretaceous time (approximately 70 Ma), 383 days per year in Late Pennsylvanian time (approximately 290 Ma), 405 days per year in Middle Devonian time (approximately 360 Ma), and 447 days per year in Early Proterozoic time (Gunflint Formation) (approximately 1,900 Ma). The data also provide evidence for the existence of the moon as far back as 2,000 Ma. Since at least that time the moon has been a faithful companion of our planet. (Modified from Schopf, 1980.)

Marshall (1987) reported that the Earth's rotation has slowed down more rapidly in the early 1980s than at present. It has not been necessary to adjust time as frequently over the last few years. Nevertheless, another leap second for the year 1989 was necessary.

According to the law of the constance of rotational impulse, extended mass decreases rotational speed. Envision an ice-skater in a fast pirouette, who, when extending the arms, slows his turns. And so it must be with Earth; expansion of its size also decreases its rotational speed. Furthermore, recent research based on the remanent magnetismus found in the Indian Ocean, paleontological evidence, and radiomagnetic age dating show compelling evidence that two Earth expansion cycles occurred during the last 170 million years (Fig. 7). Space restrictions prevent a detailed discussion about this discovery. However, it is outlined in detail in my recent paper "Paleogeography of the Last Two Cycles of Earth Expansion" to be published in Volume 25 of the Indian *Journal of Palynology*. The assumptions presented in that paper about at what time the growth of the Earth had reached the 40%, 60%, 66%, and 75% mark (Figs. 4 and 7) might have to be corrected somewhat after further research. But, it is time to discard the idea that Earth has not changed its size.

In my judgment, the most convincing evidence for Earth expansion is the terrella models that Klaus Vogel worked out since 1977 (see Fig. 8). Vogel has acknowledged the limitations and uncertainties of his terrella model constructions. There are unresolved problems concerning the history of certain geologic terranes, such as the formation and eventual accretions of Alaska, the Philippines, New Zealand and the changing of the coastal outlines of the continents due to orogenic processes, and oroclines (geotectonic mountain bending processes). Some oroclines are outlined in Carey's book of 1988, for example, the Alaskan orocline between North America and Eurasia (his figs. 15 and 18) and the Baluchistan and Punjab oroclines in the Indian region (his fig. 12). Geoscientists still have many challenging problems to solve. Yet all these questions need not discredit the overall concept of Earth expansion.

The reader will note that on the 60% and the 66% glass-terrella models, India is connected to Asia (Fig. 8). Before the new global tectonics concept was developed, India generally was considered to be an integral part of

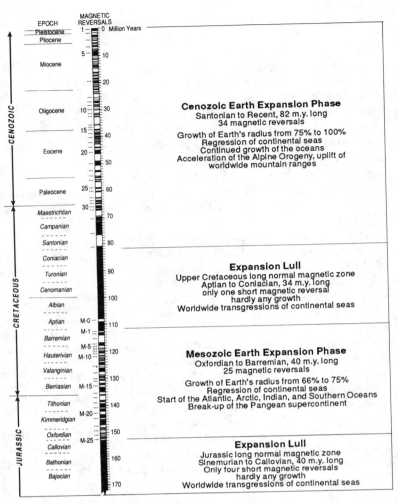

Figure 7. Geologic time scale of the last 170 million years and an interpretation of the last two Earth expansion cycles. (Geologic time scale after Heezen et al.'s (1978) geologic map concerning deep-sea drillings and magnetic anomaly lineations found in the Indian Ocean (from Kremp, in press).

Asia. Meyerhoff and Meyerhoff (1976) wrote that their work showed more than 2,000 geological and paleontological works have been published since 1883 by field geologists and paleontologists working in India and contiguous areas of southern Asia and the Soviet Union that proved the contiguity of India with Asia since middle Proterozoic time.

New ideas emerged. In 1967 the National Oceanographic Data Center in Washington, D.C. announced their plans to begin the task of exploring "the last frontier on Earth—the oceans" (*Geotimes*, November 1967, p. 16), and in January 1969, the ship, *Glomar Challenger*, completed its first 18 months of deep-sea drilling operations supported by the National Science Foundation. Most of the geoscientific community of the world now accepted the concept of sea-floor spreadings. Continental drift was finally a fact as suggested by the Polar *Forscher*, Alfred Wegener, who was killed by a glacier in 1930. Significant Carboniferous-Permian geological and paleontologic features known in South America and in Southern Africa were also found in India, that is, the Gondwana flora, coal deposits, and tillites, the remnants of glaciers. In 1972, most geoscientists recognized that the South American continent had drifted away from Africa. Therefore, one concluded, was it not possible that India also could have drifted away from the Gondwanalands, especially as paleogeographic maps showed that India could very well have nested in Paleozoic time between Southern Africa, Antarctica, and Australia? Consequently Kanawar (1972) proposed that the Indian subcontinent had drifted from Gondwana over the Indian Ocean to Southern Asia.

The deep-sea drilling continued feverishly. By August 1975, the *Glomar Challenger* had finished phase 3 of the drilling project with 43 deep-sea cores in the Atlantic, Pacific, and Indian oceans and in the Antarctic waters. More data became available about the remanent magnetismus and the paleontologic and radiometric age of the ocean floors and in early 1975, R. L. Larson published his first paleogeographic maps of assumed sea-floor spreading in the Indian Ocean.

In 1976, Johnson et al. published their paper on the "Spreading history of the eastern Indian Ocean and Greater Indian's northward flight from Antarctica and Australia." It was based on information from magnetic surveys and from deep-sea drillings that were known by that time. The authors assumed that Greater India and Antarctica-Australia dispersed from each other around 130 Ma and that after this time, sea-floor spreading began between Greater India and Antarctica-Australia, and that India had started its northward drift over the Indian Ocean to arrive at the Asian coasts about early Oligocene time. Johnson et al.'s publication found applause and complete acceptance in the scientific community of the western world. But there were two lonely oil geologists who had other thoughts. Thanks to their longtime work in Southeast Asia, the Meyerhoff's (father and son) had a good knowledge of the Russian and the Chinese geologic literature. They knew that Kalandadze (1975) published a paper concerning the first discovery of *Lystrosaurus* in the Moscow Basin of the USSR.

Lystrosaurus was an early therapsid. Most of the dozens of genera of the late Permian mammallike reptiles had been extinguished during the Permian-Triassic transition time. Only a few genera, among them *Lystrosaurus*, had survived (Stanley, 1989). At that time the western world only knew about *Lystrosaurus* finds in the Karroo Basin of South Africa and in Antarctica.

Propelled by the challenge of global tectonics, the Laboratory of Vertebrate Paleontology and Paleoanthropology in Peking had resumed their publication of *Vertebrata Palasiatica* after a lapse of six years (Gregory, 1974). The Meyerhoffs knew this also and that the Chinese paleontologist, Sun, after his expedition to

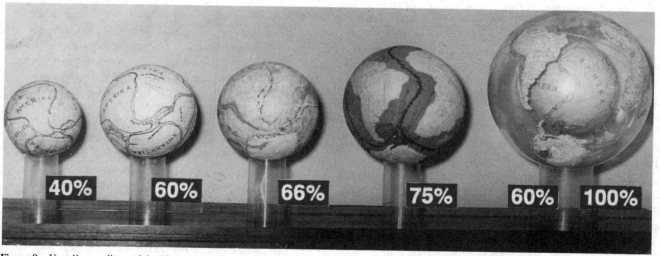

Figure 8. Vogel's terrella models. About 1977 Klaus Vogel started to construct his first terrella models. He did not know at that time of Carey's publications nor of Hilgenberg's terrella models (K. Vogel, pers. comm. 1990). But he had followed the discussion concerning the new plate-tectonic ideas, especially about the drift of India from the Gondwanalands, and he remembered Wegener's inspiring teachings about which a "modern" geography teacher had lectured, in his school when he was still a thirteen-year-old boy. "The spark hit," he wrote. He started to experiment with terrella models and discovered that on models of 55% to 60% of the Earth's present size, the continents, with shelves, completely cover the surface of the globe. He calculated that the continents, without shelves, would fit on a globe approximately 40% of the size of the present Earth. Vogel realized that this was a theoretical possibility because the continental shelves could have formed only after the brittle upper continental crust had broken into pieces. (Vogel's 1982 terrella models. Photos courtesy of K. Vogel, 1984.)

Sinkiang, had reported in 1973 about the presence of the Triassic *Lystrosaurus* in Mongolia as well as in Sinkiang. Meyerhoff and Meyerhoff (1978) had learned that Sun (1973) had written, "... it is difficult to imagine that migration could have taken place through the wide seaways [if India was still a part of the Gondwanalands in Triassic time and located about 30° South]" Understandably, the Meyerhoffs (1978) asked: How was *Lystrosaurus* able to migrate over the ocean from Godwana to Asia in the Triassic. This seems to be impossible, they said. They pointed out also that many, many papers were published documenting the intertonguing Tethys facies between India and neighboring countries.

Johnson et al. did not answer these questions in their Reply (1978), but they made the statement that "... fossils in the crystalline core of the Himalayas ... are interpreted ... as showing that the Himalayas and India are indeed part of a single plate, originally part of Gondwanaland, consistent with ... the Indus-Brahmaputra line, north of the Himalayas, as the suture between India and Asia. Hence the fossils and sedimentary rocks south of the suture are naturally related."

The Indus-Brahmaputra line (sometimes called the Indus-Tsangpo line or the Indus Suture Zone) has been claimed by paleo-tectonicists to mark the suture where thousands of kilometers of former ocean was subducted. But Carey (1988) noted that at all times the Chinese geologists have reported that "Gondwana" faunas and rock associations characteristic of India extend across the Indus-Tsangpo line far across to Tibet and China.

Ahmad (1982) stated that the so-called Indus Suture Zone is a rift valley and its eruptions might have produced some of the rocks classified as ophiolites. Numerous rift valleys and normal faults exist on the Tibetan Plateau and unequivocally bespeak of tension all over the area and not of compression that one would expect from a subduction-collision zone. Furthermore, he stated that available evidence on the distribution of flora and fauna, including vertebrates, insects, and freshwater lamellibranchs, indicates that Gondwanaland and the northern continents, particularly Angaraland and Cathaysia, were never separated by a deep ocean, as often envisaged in the past. Nor was the Tethys of geosynclinal character. Instead, it was an epicontiental sea, transgressing and regressing frequently, and in Permian time had covered the area north of India from the Pacific coast in the area of Japan and eastern China to Italy and Tunis. (Ahmad's fig. 1 presents detailed paleogeographic maps of the Tethys in the Eurasian region during the Carboniferous, Permian, Triassic, and Jurassic periods.)

In 1976 practically no one had heard anything about Hilgenberg's terrella models (Fig. 3) nor of the Earth expansion theory. Carey's 1976 voluminous book was not yet known and Vogel's terrella models (Fig. 8) not yet in existence. Despite all the enigma Johnson et al. (1976) could not see any alternative but northward flight of India from Antarctica and Australia. I made the same mistakes in my 1974 and 1977 publications. In 1984 I saw Vogel's photographs of his terrella models. These photographs convinced me that our globe was only about 66% of its present size by the Mesozoic Earth expansion phase between 150 to 110 Ma and that India was always attached to Asia and before the Mesozoic to Asia and also to Antarctica and Australia (Figs. 4, 7, and 8).

Theories, as yet unproven, surmise that slabs of subducting oceanic plates are able to penetrate deep into the lower mantle, glide along the core-mantle boundary and return as a steady stream of hot matter in the midoceanic ridges. These theories estimate that the descending slabs need only some 30 million years to reach the core, which may be also questionable.

The whole-mantle-convection theory is only a few years old. Before the investigations of Dziewonski and Woodhouse (1984, 1987), we had the theory of the two-layer-convection system. This theory assumes that upper mantle and lower mantle flows are independent of one another, purportedly caused by an impenetrable chemical and dynamic boundary at about 670 km below the surface, which would create two independent convection systems.

However, Vogel's expanding Earth terrella models indicate a different scenario. The Earth expansion theory is based on the assumption that a continental crust almost covered the surface of a smaller Earth until the Mesozoic. This assumption is well supported by research results in historical geology (Kremp, 1981–1984). The volume of Earth (not necessarily its mass) increased through an endogenic process. As a result, the "Pangaean" crust ruptured. Widening gaps between the continental pieces gave birth first to the Paleo-Pacific Ocean and later to the oceans of today.

A fascinating terrella model by Vogel is represented in Fig. 9. Here, a terrella of 55% is included within a glass model of the present-day Earth. Both terrellas indicate convincingly that the continents in general seem to be fixed more or less at their underlying asthenosphere and retain their relative positions to each other. Therefore, recent crustal movements are mainly determined by radial outward pressing of the continents and by the filling-in of the growing gaps by new oceanic crust from sea-floor spreading from the mid-oceanic ridges. The evidence, illustrated in Figure 9, convinced me that Vogel's theory is correct.

Subduction of oceanic plates is a fact. However, plate-tectonic theory is relatively new and has not yet embraced all the known data. To gain a greater understanding of

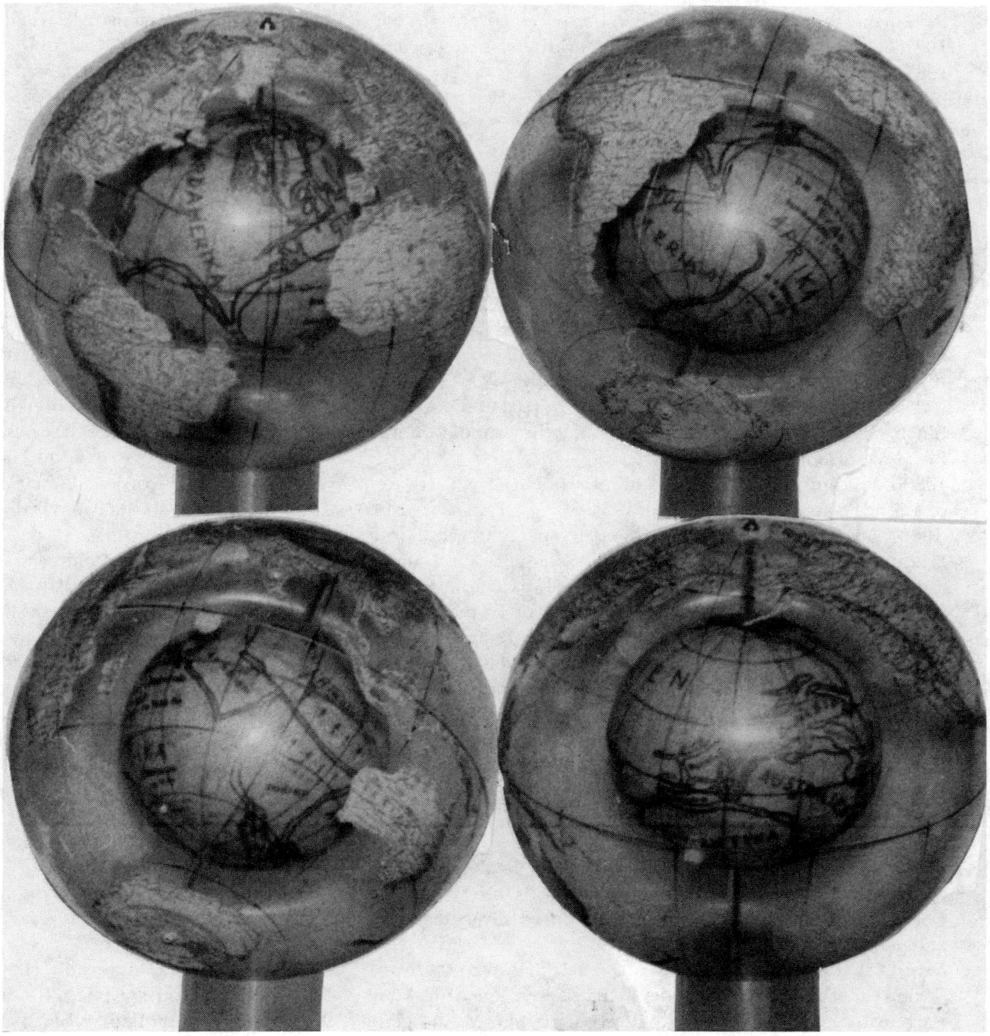

Figure 9. The rapidly changing concepts of global tectonics. Four photos of Vogel's 55% terrella model included within a transparent sphere of present-day Earth to compare the starting positions with the present situation. It appears that the continental lithospheres are more or less fixed to their underlying asthenospheres and retain their relative positions to each other. Consequently, the continental crust was pushed outward in radial direction upon the growing globe. While the oceans were growing with different speed according to seafloor spreading, the distribution pattern of the continents became more and more asymmetric. (Photos courtesy of K. Vogel, 1987.)

Earth dynamics, it is imperative to pay attention to the data of historical geology and to the Earth expansion theory. Computerized axial tomography, as taught by Dziewonski and Woodhouse, along with other seismologic techniques, will help to clarify what is occurring in the interior of our Earth.

REFERENCES CITED

Ahmad, F., 1982, The myth of oceanic Tethys: Bollittina della Sociedad Paleontologica Italiana, v. 21(2-3), p. 153-168.

———, 1983, Late Paleozoic to early Mesozoic Paleogeography of the Tethys region, *in* Carey, S.W., ed., Expanding Earth Symposium: Sydney, 1981, p. 131-145, University of Tasmania.

Boss, A.P., and Sacks, I.S., 1985, Formation and growth of deep mantle plumes: Geophysical Journal, Royal Astronomical Society, v. 80, p. 241-255.

Boucot, A.J., and Gray, J., 1979, Epilogue: A Paleozoic Pangaea?, *in* Boucot, A.J. and Gray, J., ed., Historical Biogeography, Plate Tectonics, and the Changing Environment: Oregon State University Press, p. 465-482, Corwallis.

Carey, S.W., 1976, The expanding Earth: Elsevier, Developments in Geotectonics (10), 488 pp., Amsterdam.

———, 1983a, Earth expansion and the null universe, *in* Carey, S.W., ed., Expanding Earth Symposium: Sydney (1981), p. 367-374, University of Tasmania.

———, 1983b, Evolution of beliefs on the nature and origin of the Earth, *in* Carey, S.W., ed., Expanding Earth Symposium: Sydney (1981), p. 3-7, University of Tasmania.

———, 1983c, The necessity for Earth expansion, *in* Carey, S.W., ed., Expanding Earth Symposium: Sydney (1981), p. 377-396, University of Tasmania.

———, 1983d, Tethys and her forebears, *in* Carey, S.W., ed., Expanding Earth Symposium: Sydney (1981), p. 169-187, University of Tasmania.

———, 1988, Theories of the Earth and universe; a history of dogma in the Earth sciences: Stanford University Press, 413 pp.

Cloud, P. E., 1968, Atmospheric and hydrospheric evolution on the primitive Earth: Science, v. 160, p. 729-739.

———, 1973, Paleoecological significances of the banded iron formation: Economic Geology, v. 68(7), p. 1135-1143.

———, 1976, Beginnings of biospheric evolution and their biogeographical consequences: Paleobiology, v. 2, p. 351-387.

Dziewonski, A.M., and Woodhouse, J.H., 1984, A map that gets under the Earth's skin: Science, v. 84, Currents, p. 6.

———, 1987. Global images of the Earth's interior: Science, April 3, 1987, v. 236, p. 37-48.

Ehrensperger, J., 1988, Die Expansion des Kosmos, Die Expansion der Erde: 1. Auflage, 59 pp. Verlag, W. Vogel, Winterthur, Switzerland.

Glikson, A.Y., 1972, Early Precambrian evidence of a primitive ocean crust and island nuclei of sodic granites: Geological Society of America, Bulletin, v, 83, p. 3323-3344.

———, 1976, Stratigraphy and evolution of primary and secondary greenstones; Significance of data from shields in the southern hemisphere: (NATO Advanced Study Institute, University of Leicester, 1975 report), in B.F. Windley, ed., The Early History of the Earth: p. 275-278. Wiley, London.

———, 1979, The missing Precambrian crust:. Geology, v. 7, p. 449-454.

———, 1980a, Granulite-gneiss suites of the western part of the Arunta Block, central Australia: Report, Australian Bureau of Mineral Resources, Geology and Geophysics.

———, 1980b, Precambrian sial-sima relations; evidence for Earth expansion: Tectonophysics, v. 63, p. 193-234.

———, 1983, Geochemical, isotopic, and paleomagnetic limits on Precambrian crustal surface dimensions; Evidence for a small-radius Earth (Abstract): in Carey, S.W., ed., Expanding Earth Symposium: Sydney (1981), p. 88, University of Tasmania.

Glikson, A.Y. and Baer, A.J., 1980, Comment and Reply on "The Missing Precambrian Crust": Geology, March 1980, p. 114-117.

Godwin, A.M., 1976, Giant impacting and the development of continental crust in the early history of the Earth: Nato Advanced Study Institute, University of Leicester (1975), report in B.F. Windley, ed., The early history of the Earth: Wiley: p. 77-78, John Wiley and Sons Inc, London.

Gregory, J.T., 1974, Vertebrate paleontology: Geotimes, January, 1974, p. 29-30.

Heezen, B.C., Lynde, R.P., Jr. and Fornari, D.J., 1978, Geologic map of the Indian Ocean: Lamont-Doherty Geological Observatory and Department of Geology Sciences, Columbia University, 8 pp. and geologic map, New York.

Hilgenberg, O.C., 1933, Vom wachsenden Erdball: Selbstverlag, 50 pp.

———, 1966, Die Paläogeographie der expandierenden Erde vom Karbon bis zum Tertiär nach palomagnetischen Messungen: Geologische Rundschau, v. 55(3), p. 878-924.

Johnson, B.D., Powell, C.McA., and Veevers, J.J., 1976, Spreading history of the Indian Ocean and greater India's northward flight from Antarctica and Australia: Geological Society of America Bulletin, v. 87 (November), p. 1560-1566.

———, 1978, Reply: Geological Society of America Bulletin, v. 89, p. 640.

Kalandadze, N.N., 1975. Pervaya nadhodka Lystrosaurus na territorii Evropeyskoy chasti SSSR [First discovery of Lystrosaurus in the European part of the USSR]: Paleontologicheskii Zhurnal, no. 4, p. 140-142.

Kanawar, R.C., 1972, Hilmalayas and the continental drift: Geologische Bundesanstalt Verhandlungen, no. 2, p. 265-267, Vienna.

Kremp, G.O.W., 1973, Advancing Organization; time and the orderly progression of life: Journal of Palynology, v. 9(1), p. 1-28. Lucknow.

———, 1974, A re-evaluation of global plantgeographic provinces of the Late Paleozoic: Review of Paleobotany and Palynology, v. 17, p. 113-132. Elsevier Scientific Publishing Company. Amsterdam.

———, 1977, The positions and climatic changes of Pangaea and five southeast Asian plates during Permian and Triassic times: Paleo Data Banks, no. 7(1), p. 1-21, Palynodata, Tucson, Arizona.

———, 1979, The Earth Expansion Theory and the climatic history of the Lower Permian: Proceedings, Sixth International Gondwana Symposium (Calcutta, 1977), v. 1, p. 1-20.

———, 1981, A preliminary account of some significant events in the expansion history of the Earth: Abstracts, Expanding Earth Symposium (Sydney, February 1981), p. 31.

———, 1982, The oldest traces of life and the advancing organization of the Earth (Part I: Archean and Cryptophytic): Paleo Data Banks, v. 18 (2), p. 53-128, University of Arizona, Tucson.

———, 1983, Precambrian events indicative of Earth expansion, in Carey, S.E. ed., Expanding Earth Symposium: Sydney (1981), p. 91-99, University of Tasmania.

———, 1983, The oldest traces of life and the advancing organization of Earth (Part II: Early and Late Proterophytic): Paleo Data Banks, v. 19(2), p. 65-156, University of Arizona, Tucson.

———, 1984, The oldest traces of life and the advancing organization of the Earth (Part III: Epilogue): Paleo Data Banks, v. 21(1), p. 157-396, University of Arizona, Tucson.

———, 1985, The Earth Expansion Theory. The next revolution in Earth Sciences: (Abstract) American Association of Stratigraphic Palynologists, 18th Annual Meeting, El Paso, Texas, October 16, 1985, p. 16.

———, 1989, The geodynamic situation at the Cretaceous/Tertiary Transition: (Abstract) Palynology, v. 13, p. 284.

———, in press. Paleogeography of the last two cycles of Earth expansion: Journal of Palynology, v. 25, New Delhi, India.

Lambeck, K., 1978, The Earth's variable rotation, geophysical causes and consequences: Cambridge University Press, 58 pp.

Larson, R.L., 1975, Late Jurassic sea-floor spreading in the eastern Indian Ocean: Geology, v. 3, p. 69-71.

Lay, T., 1989, Structure of the Core/Mantle Transition Zone; a chemical and thermal boundary layer: Eos, Transaction, American Geophysical Union, v. 70(4), January 24, 1989, p. 49-59.

Marshall, E., 1987, A matter of time: Science, v. 238, December, p. 1641-1642.

McCarthy, D.D., 1976, The determination of universal time at the U.S. Naval Observatory: U.S. Naval Observatory Circular No. 154, July, Washington, D.C.

Meyerhoff, A.A., and Meyerhoff, H.A., 1972, The new global tectonics: major inconsistencies: American Association of Petroleum Geologists Bulletin, v. 56(2), p. 269-336.

———, 1974, Tests of plate tectonics: American Association of Petroleum Geologists, Memoir 23, p. 43-145.

Meyerhoff, H.A., and Meyerhoff, A.A., 1978, Spreading history of the eastern Indian Ocean and India's northward flight from Antarctica and Australia: Discussion and reply/discussion: Geological Society of America Bulletin, v. 89, p. 637-640. Doc. no. 80414.

Noël, D., 1989, NUTEERIAT: Nut trees, the expanding Earth, Rottnest Island, and all that: Cornucopia Press, 200 pp. Perth, Australia.

Pannella, G., 1972, Paleontological evidence on the Earth's rotational history since early Precambrian: Astrophysics and Space Science, v. 16, p. 212-237. Elsevier.

———, 1976, Tidal growth patterns in recent and fossil mollusk bivalve shells; a tool for the reconstruction of paleotides: Die Naturwissenschaften, v. 63, p. 539-543. Springer Verlag.

Piper, J.D.A., 1976, Paleomagnetic evidence for a Proterozoic supercontinent: Philosophical Transactions, Royal Society of London, Series A, p. 469-490.

Schidlowski, M., 1971, Probleme der atmosphärischen Evolution im Präkambrium: Geologische Rundschau, v. 60, p. 1351-1384.

———, 1976, Archean atmosphere and evolution of the terrestrial oxygen budget: (NATO Advanced Study Institute, University of

Leicester, 1975 report), *in* Windley, B.F., ed., The Early History of the Earth: p. 525-534. Wiley, London.

Schopf, T.J.M., 1980, Paleooceanography: Harvard University Press, 341 pp. Cambridge, Massachusetts.

Stanley, S.M., 1989. Earth and life through times: W.H. Freeman Company, 689 pp., New York.

Sun, Ai-lin., 1973, Permo-Triassic reptiles of Sinkiang: Scientia Sinica, v. 16, no. 1, p. 152-156.

Vogel, K., 1983, Global models and Earth expansion, *in* Carey, S.W., ed., Expanding Earth Symposium: Sydney (1981), p. 17-27, University of Tasmania.

———, 1984, Beiträge zur Frage der Expansion der Erde auf Grundlage von Globenmodellen: Zeitschrift für geologische Wissenschaften, Berlin, v. 12(5), p. 563-573.

———, 1989, Recent crustal movements in the light of Earth Expansion Theory: 6th International Symposium "Geodesy and Physics of the Earth," German Democratic Republic, Potsdam, Aug. 22-27, 1988. Akademie der Wissenschaften der DDR, Zentralinstitut für Physik der Erde, p. 284-297, Potsdam, East Germany (DDR).

———, 1990, The Expansion of the Earth—an alternative model to the Plate Tectonics Theory, *in* S.S. Augustais, ed., Critical aspects of the Plate Tectonics Theory, v. II: Alternative Theories, p. 19-34: Theophrastus Publications, Athens.

Vogel, K. and Schwab, M. 1983. The position of Madagascar in Pangaea, *in* Carey, S.W. ed., Expanding Earth Symposium: Sydney (1981), p. 73-76, University of Tasmania.

Williams, Q., Jeanloz, R., Bass, J., Swendsen, B. and Ahrens, T. J., 1987,. Meeting curve of iron to 250 Gigapascals; Constraint on the temperature of Earth's center: Science, v. 230, 10 April, p. 181-182.

Yuen, D.A. and Peltier, W.R., 1980, Mantle plumes and the thermal stability of the D" Layer: Geophysical Research Letters, v. 7(9) (September), p. 625-628.

Surge tectonics: a new hypothesis of Earth dynamics

Arthur A. Meyerhoff, P. O. Box 4602, Tulsa, Oklahoma 74159 USA

Irfan Taner, 3336 East 32nd Street, Suite 210, Tulsa, Oklahoma 74135 USA

Anthony E. L. Morris, P. O. Box 64610, Los Angeles, California 90064 USA

Bruce D. Martin, P. O. Box 234, Leonardtown, Maryland 20650 USA

William B. Agocs, 968 Belford Road, Allentown, Pennsylvania 18103 USA

Howard A. Meyerhoff,* 3625 South Florence Place, Tulsa, Oklahoma 74105 USA

ABSTRACT

Intensive geotectonic research during three decades has vastly increased the database for Earth-dynamics studies. Many sets of new data have been acquired, and used mainly to support the plate-tectonics hypothesis. Among them are data on midocean ridges, continental rifts, strike-slip zones, magmatic arcs, oceanic and continental volcanism, deep-sea trenches, Benioff zones, geosynclines, orogenic plutons, ophiolites, mélanges, metamorphic core complexes, detachment fault zones, and related phenomena.

The same research has acquired many additional sets of geological and geophysical data that are unexplained by any geotectonic hypothesis. These include the presence in all active tectonic belts, whether compressive or tensile, of (1) long linear fault, fracture, and fissure systems that parallel the belts and extend for hundreds of kilometers, (2) bands of elevated heat flow (>55 mWm^{-2}) that coincide with the linear fault-fracture-fissure zones, and (3) bands of microearthquakes that coincide with both the fault-fracture-fissure zones and the elevated heat-flow zones. In addition, (4) all tectonic belts are partitioned into rather uniform segments, 50 to 300 km long. Other phenomena present in these same zones include (5) bivergent foldbelts with bilateral symmetry, (6) mantle diapirs, (7) tectonostratigraphic terranes, (8) distorted "tectonic lines" or *Verschluckungszonen*, (9) stretching lineations parallel with the tectonic belt, (10) "aseismic" oceanic ridges, (11) inverted metamorphic gradients, (12) mantled gneiss domes, (13) thermal springs, (14) lava fields, (15) dike swarms, (16) rows of anorogenic plutons, (17) kimberlite diatremes, (18) aligned evaporite basins, (19) lithosphere low-velocity zones, (20) lenses of "anomalous upper mantle" (P-wave velocities of 7.0–7.8 km/s), and (21) high-conductivity electrical and magnetotelluric anomalies. (22) Vortex structures are closely associated with the above phenomena. (23) Still another unexplained phenomenon is the eastward migration through time of many magmatic or volcanic arcs, batholiths, rifts, depocenters, and foldbelts.

All of these phenomena that are Jurassic or younger—whether explained by plate tectonics or not—form an internally consistent pattern that is totally unlike anything predicted by plate tectonics or any other hypothesis of Earth dynamics. All of these phenomena, where Jurassic or younger, lie in zones of high heat flow. Almost all have been postulated at one time or another to be underlain by shallow magma chambers, mantle diapirs, or asthenosphere upwellings. The long linear fault-fracture-fissure zones that lie in these bands of high heat flow indicate that something is moving horizontally at fairly shallow depth in these magma chambers or diapirs. The common association of lava fields suggests that the something is lava. Reflection- and refraction-seismic data reveal the presence of large lenses of anomalous upper mantle beneath all high heat-flow bands. The same data reveal the presence inside of these lenses of low-velocity zones that are acoustically transparent. We interpret these lenses to be channels containing fluid or semifluid magma.

Mapping of the phenomena listed above—particularly high heat flow and microseismicity bands—revealed a worldwide network of interconnected lithosphere channels which we call surge channels. We established nearly 40 physical criteria for locating and mapping them. The channels underlie all Jurassic and younger tectonic foldbelts, strike-slip zones, and rifts. We used most of the same criteria to identify similar worldwide networks of older surge channels, late Archean to Jurassic, that are now extinct and consequently are no longer associated with bands of high heat flow, thermal springs, young igneous rocks, and microearthquakes. Because all tectogenesis can be related to both extinct and/or extant surge channels, then surge channels must be the lithosphere's most fundamental tectonic unit.

As we interpret the data, the cause of tectogenesis is lithosphere compression. Only one Earth model, that of a cooling Earth, seems capable of producing all of the nearly 40 phenomena that we identify with surge channels. In a cooling Earth, the lithosphere is under equiplanar tangential compression. During the geotectonic cycle, the intensity of this compression increases steadily until tectogenesis takes place. Changes in the intensity of the compression cause the magma to move very slowly in pulsated surges (hence, the term surge channel). When the compression reaches an intensity sufficient to rupture the surge channel, the contents of the channel surge bilaterally upward and outward (hence the term, surge tectonics). This last process, tectogenesis, terminates the geotectonic cycle, and another begins.

Every important structure of the lithosphere, whether midocean ridge or alpinotype foldbelt, is the product of surge tectonics. Every major tectonic process, ranging from folding and thrusting to metamorphism, metallogenesis, and flood volcanism, is related directly to the presence of one or more surge channels. Surge tectonics has led us to some wholly unexpected results, not only in structural geology and tectonics, but also in geomorphology, petrology, sedimentation, paleontology, paleoclimatology, and even organic evolution. An important aspect of surge tectonics is not just that it provides a unifying mechanism for all aspects of magmatism and tectonics, but that it is internally consistent, explaining all tectonic phenomena that we have observed.

---

*Deceased, 24 March 1982. Prof. Meyerhoff began to write this manuscript in 1979 under the title, "Surge Origin of Island Arcs."

Chatterjee, S., and N. Hotton III, eds. *New Concepts in Global Tectonics.* Texas Tech University Press, Lubbock, 1992, xii + 450 pp.

INTRODUCTION

Objective and scope

The objective of this paper is to present a new, comprehensive, and internally consistent hypothesis of Earth dynamics, a hypothesis we call surge tectonics. The need for such a hypothesis has increased as important new sets of geological and geophysical data are discovered through plate-tectonics research but left unexplained. Although we have published and discussed many data unexplained by plate tectonics and other hypotheses (Meyerhoff and Meyerhoff, 1972a, 1972b, 1974), our own attempts to explain them have been unsuccessful or, at best, only partly successful (Meyerhoff et al., 1972; H. Meyerhoff and Meyerhoff, 1977). It was not until late 1987 that we discovered a single mechanism that provided a logical explanation for both the data sets explained by plate tectonics and the equally important data sets that plate tectonics does not explain or even address. This mechanism, the lateral compression and rupture of lithosphere surge channels, is the basis of our new hypothesis, surge tectonics. Before we describe this hypothesis, however, we review the data sets that no published hypothesis explains.

Data sets not explained in current Earth-dynamics hypotheses

Linear Structures—Meyerhoff et al. (this volume) reviewed recent studies of the midocean ridges that utilized side-scanning sonar. Side-scanning sonar images, or sonographs, reveal the presence everywhere on the midocean ridges of systems of linear, ridge-parallel faults, fractures, and fissures. Meyerhoff et al. (this volume) observed also that midocean ridges and their branches cross the ocean-continent boundary in many places. Because of the important implications of these observations, we decided it was necessary to verify Meyerhoff et al.'s (this volume) observations and to determine whether or not the fault, fracture, and fissure systems of the midocean ridges are present within the continents.

Such systems were found in all continental tectonic belts examined. An example of some of these systems (discussed herein) is the Western Cordillera of the United States (Fig. 33; Stewart, 1978) and, at a smaller scale within the same tectonic province, the California Coast Ranges (Fig. 61; Johnson and Page, 1976). Other examples are the East African Rift system (Fig. 55; Mohr, 1987), the Rhine graben (Fig. 59; Illies and Greiner, 1979), the Front Range of New Mexico, Colorado, and Wyoming (Fig. 38; Eaton, 1986), and the Reelfoot graben of the Mississippi embayment (Figs. 56, 57). An important fact is that such fault, fracture, and fissure systems are present in all foldbelts, rift systems, and strike-slip fault zones. They comprise a huge body of data that are unexplained in plate tectonics, and with few exceptions have not been addressed. A second important fact is that these continental fault, fracture, and fissure systems demonstrate, in accordance with Stokes's Law (see Appendix), that movements beneath them have been parallel with them, and not at right angles as required by plate tectonics (see Meyerhoff et al., this volume).

Mantle Diapirism—Since the publication of Wegmann's (1930, 1935) pioneering papers on the subject, mantle diapirism has been invoked increasingly as a mechanism for generating tectogenesis. Van Bemmelen (1933) and Glangeaud (1957), for example, favored mantle diapirism and subsequent lateral sliding and/or compression for creating the structures of the Mediterranean Sea region. Mantle-diapirism hypotheses have found favor at different times with many geologists (e.g., Maxwell, 1968) for explaining the tectonic evolution of the Alpide belt and still do (e.g., Krebs, 1975; Locardi, 1988).

Beloussov (1980, 1981, 1989) observed the evidence adduced for important mantle diapirism is formidable. Geological data, geophysical interpretations, and theoretical considerations in most of the world's mobile belts—rifts, wrench zones, and foldbelts alike—practically require large-scale, upper-mantle diapirism as a part of the tectogenetic process. Almost since the beginning of plate-tectonic theory, geophysicists such as Llibroutry (1971), Bonini et al. (1973), and many others pointed out the important role that diapirism must play in plate tectonics. Despite this, mantle diapirism and related upwelling processes received little attention as intrinsic parts of plate tectonics until Dewey (1988) recognized their possibly important roles throughout the Alpide and parts of the Circum-Pacific tectonic belts. Dewey's (1988) explanations, however, fail to account for coexisting states of compression and tension in a convergence zone, as field data require (Platt and Vissers, 1989). In contrast, surge tectonics requires the simultaneous formation of compressional and tensile regimes and structures.

Table 1, based on random sampling of the recent literature, shows how widespread the idea of mantle diapirism and upwelling is. Fully 56% of the examples listed are foldbelts; the remainder are tensile belts. Our point is that, whereas mantle diapirism may have a place in tensile regimes within the framework of plate tectonics, it has no place in both compressional and tensile regimes within the framework of that hypothesis. In contrast, surge tectonics predicts mantle diapirism in all stress regimes.

Phenomena related to mantle diapirism—Mantle diapirs and related asthenosphere upwellings are zones

Table 1. Random sample of literature proposing mantle diapirs.

Type of mobile belt / Name of mobile belt	Author(s) (Year)	Geology	Geophysics	Both	Theory
Alpinotype Foldbelt					
Yenisey-Sayan-Baykal	Sheynmann (1968)	x			
Rif, Betic Cordillera, Alboran Sea	Bonini et al. (1973)		x		
Northern Appalachians	Ramberg (1973)	x			x
California Coast Ranges	Maxwell (1974)	x			
Alps	Smith and Woodcock (1982)				x
Massif Central (France)	Nicolas et al. (1987)			x	
Rhenish Massif (Germany)	Witt and Seck (1987)			x	
Alps to the Himalaya	Dewey (1988)				x
Apennines, Tyrrhenian Sea	Locardi (1988)			x	
Carpathians, Pannonian Basin	Ádam et al. (1989)		x		
Western Cordillera of North America	Hamilton (1989)				x
Honshu (Japan)	Hirahara et al. (1989)		x		
Rif, Betic Cordillera, Alboran Sea	Platt and Vissers (1989)	x			
Pyrenees	Velasque et al. (1989)				x
Continental and Oceanic Rifts					
Mid-Atlantic Ridge	Ramberg (1973)		x		x
Zabargad Island (Red Sea)	Bonatti et al. (1981)	x			
Red Sea	Nicolas (1987)	x			
Oman, Red Sea	Boudier and Nicolas (1988)	x			
Gulf of Aden, Red Sea	Isaev (1987)			x	
Base of continental slope, west of Iberian Peninsula	Boillot et al. (1989)			x	
Rhine Graben	Clauss et al. (1989)		x		
Iceland	White (1989)		x		
East Pacific Rise	Winterer et al. (1989)		x		
Aseismic Ridges					
Hawaiian Ridge, Cape Verde Rise	White (1989)		x		
Theoretical Models					
Foldbelts and rifts	Ramberg (1973)				x
Foldbelts and rifts	Beloussov (1974)				x
Cratonal basins	Artyushkov and Baer (1986)				x
Island arcs and rifts	Turcotte (1989)				x
Flood basalt, backarc basins	White and McKenzie (1989)				x

of lowered seismic velocity that in many places are overlain at intervals by smaller zones of lowered velocity (the "Christmas Tree" effect of Corry, 1988; Fig. 70). The first (lowest) low-velocity zone above the asthenosphere generally forms a large, lenticular, diapirlike body, as shown in Figure 15. The diapirlike bodies, or lenses, range in position from the uppermost mantle (Talwani et al., 1965) to the middle to upper crust (Finlayson et al., 1989; Thompson and McCarthy, 1990; Figs. 67, 68). These features are referred to in the literature as anomalous upper mantle (Le Pichon et al., 1965), and are found beneath tensile tectonic belts, such as rifts (e.g., Mooney et al., 1983), strike-slip zones (e.g., Mooney and Weaver, 1989), and foldbelts (Taylor et al., 1980; Marillier et al., 1989). P-wave velocity values associated with them are in the 7.0- to 7.8 km/s velocity range. The literature dealing with them began with Revelle (1958) and now is huge. Yet nowhere to our knowledge have these lenses of anomalous upper mantle (Figs. 10, 15, 26, 27, 29, 31, 35, 45, 47, 51, 58, 66–68) and associated low-velocity zones (Figs. 10, 18, 19, 26, 31, 43, 47, 51, 54, 62) been integrated successfully into a plate-tectonics context, because they occur in all tectonic settings ranging from compressional through tensile (Marillier et al., 1989).

In many foldbelts, the lenses of anomalous upper mantle have been severely deformed as integral parts of the foldbelts (Figs. 45, 51, 65). Compressed and deformed lenses of anomalous upper mantle are referred to in the literature as two-sided foldbelts, bilaterally deformed foldbelts, structural fans, and other terms (Figs. 13, 28–30, 35, 37, 41, 44, 45, 48–52, 63–65, 92, 95, 96). Such structures—which we call kobergens—are wholly unexplained in plate tectonics. One flank of these great structural fans commonly is called a zone of backthrusting, but without regard for the criteria that identify backthrusts (e.g., Rich, 1951).

Examination of our figures showing bilaterally deformed foldbelts reveals at once a structure that combines the effects of compression and tension. The flanks of the structure (Fig. 13) exhibit folds, thrusts, and nappes whose vergence on one flank is opposite that of the other flank. The two flanks are separated by a zone of tension. It is only in such bilaterally deformed belts that tension and compression can act simultaneously in parallel bands for distances of hundreds of kilometers along strike. The seemingly contradictory stress regimes for which Dewey (1988), Platt and Vissers (1989), and many others have long sought an explanation are a natural consequence of bilateral deformation. The literature on bivergent foldbelts predates Suess (1885) and increases until the present (e.g., Platt and Vissers, 1989). Plate tectonics until the last six years largely ignored all of this literature. Thus, a large volume of data on backthrusting remains to be explained.

Vortex structures—These are large ovate, oval to circular areas 200 to 1,000 km in diameter. In many places, they form large depressions surrounded by major tectonic features. Some of these circular structures are what Kober (1921, 1925, 1928) called *Zwischengebirge*, and include several of the large mantle upwellings discussed by Dewey (1988). In addition to being probable areas of mantle diapirism and upwelling, field data also suggest that motions beneath them are vortical (Wellman, 1966; Searle et al., 1989). Examples include the Pannonian basin (Fig. 12), the Aegean Sea (Figs. 12, 73-75), Lake Victoria (Fig. 71; the depression between the eastern [Gregory] and western branches of the East African Rift), the Dasht-i-Lut depression of eastern Iran (Fig. 77; Wellman, 1966), the North Fiji basin (Fig. 21; Auzende et al., 1988), the Easter Island microplate (Fig. 76), and many others.

Segmentation—Meyerhoff et al. (this volume) summarized the voluminous data showing the segmentation of the midocean ridges into what White (1989) called pinch-and-swell geometry (Fig. 16). The average distance between the "pinches" ranges from 50 to 300 km, and is mainly 80 to 200 km. The pinches are underlain by structures, which in plate tectonics, have been called transform faults, devals, and overlapping spreading centers (Winterer et al., 1989). The most likely mechanism that forms pinch-and-swell geometry is ridge-parallel flow, precisely analogous to that which generates pinch-and-swell or pillow structure in lava tubes (Fig. 17; Yamagishi, 1985; Meyerhoff et al., this volume). Examples of midocean–ridge segmentation appear not only on Figures 11B, 16, and 17 of this paper, but also on figures 3, 9, 14, 16–18, and 21 of Meyerhoff et al. (this volume). Figures 22 (Walvis Ridge) and 24 (Louisville Ridge) show that segmentation in ocean basins is not limited to the midocean ridges.

Segmentation—pinch-and-swell geometry—also characterizes all continental tectonic belts. Examples in alpinotype foldbelts include the California Coast Ranges (Fig. 60; Zandt and Furlong, 1982), Oman (Boudier and Nicolas, 1988), northern Appalachians (Fig. 1; Williams, 1978), and many others. Examples of segmentation in germanotype foldbelts include the Central and Southern Rocky Mountains (Figs. 1, 40; Gries, 1983; Hamilton, 1988) and the Mesozoic Santa Maria foldbelt of the Mojave block (Tosdal et al., 1989). Segmentation is an identifying characteristic of chains of metamorphic core complexes (Coney, 1980; Tosdal et al., 1989) and mantled gneiss domes, such as the Bronson Hill anticlinorium in the New England Appalachians (Fig. 1; Thompson et al., 1968; Williams, 1978). Several examples of segmentation beneath the San Andreas fault are known (Fig. 60; Zandt and Furlong, 1982). Segmented continental rift systems include the East African Rift (Figs. 1, 55; Mohr, 1987), the Rhine graben (Fig. 59; Illies and Greiner, 1979), the Baykal Rift (Fig. 53; Logatchev and Florensov, 1978), and the Amazon Rift (Fig. 79; Nunn and Aires, 1988), among many. The remarkable geometric similarity between the structural grain of the intensely compressed Bronson Hill root zone of the northern Appalachians and the young rift valley of East Africa is shown on Figure 1 (Thompson et al., 1968; Williams, 1978; Ebinger et al., 1989).

Such segmentation in the tectonic belts of continents is reflected in their topographic features. For example, individual chains of the Central and Southern Rocky Mountains (Fig. 40) comprise several subparallel and along-strike segments. Thus, the factors that cause segmentation also determine the lengths of mountain ranges, and thereby exercise a fundamental influence on topography (e.g., the locations of mountain passes). Although we document these statements in the pages that follow, we present some examples on Figure 1. This figure illustrates the remarkably similar sinuosity and

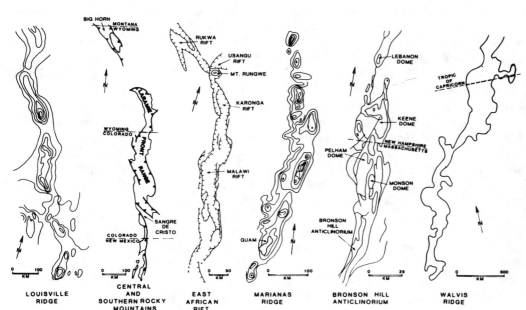

Figure 1. Similar patterns of sinuosity and segmentation from a breakout channel (Louisville Ridge), a germanotype foldbelt (Central and Southern Rocky Mountains), a continental rift (East African rift), a magmatic arc (Marianas arc), an alpinotype foldbelt (Bronson Hill anticlinorium, northern Appalachians), and a feeder channel (Walvis Ridge). See text for feeder channel definition. The close similarity among these supposedly different tectonic features suggests that they have a common origin. Sources: GEBCO (1975-1982), Williams (1978), Cheng et al. (1987), Hamilton (1988), and Ebinger (1989).

segmentation that are common to contrasting types of tectonic belts. These include oceanic tensile regimes (Louisville and Walvis ridges), continental rifts (East African Rift), magmatic arcs (Marianas Ridge), germanotype foldbelts (Central and Southern Rocky Mountains), and alpinotype foldbelts (Bronson Hill anticlinorium or root zone, of the northern Appalachians). If all of these features acquired their pinch-and-swell geometry by horizontal flow in lithosphere magma chambers (surge channels), it is clear that the whole field of geomorphology and the origin of landforms requires substantial reevaluation.

Ebinger (1989) and Ebinger et al. (1989) noted that the segmentation observed in the East African rift system is similar to that reported along the midocean ridges (Macdonald and Fox, 1983; Macdonald et al., 1988). Yet nowhere could we find comparative studies of segmentation patterns in different types of tectonic belts or large-scale landforms. Apparently the pinch-and-swell geometry that characterizes every tectonic belt that we studied has not been recognized. It is certainly a topic that has not been addressed by any Earth-dynamics hypothesis.

Rift zone to *Verschluckungszone*—A very large group of linear structures is present in foldbelts. These include such features as troughs, rifts, strike-slip fault zones, suture zones, lines, median tectonic lines, *Narbenzonen*, *Scheitelzonen*, and *Verschluckungszonen*. Examples of troughs and rifts include the Okinawa trough, the Taiwan longitudinal valley (Fig. 28), and the Kamchatka-Olyutor trough or graben. Strike-slip zones and sutures include the North Pyrenean fault (Fig. 65), the North Anatolia fault (Fig. 64), the Armorican shear zone, the Taurus-Zagros suture (Fig. 77), the Indus-Yarlung suture, the Philippine fault zone (Fig. 28), the San Andreas fault (Figs. 60–63), and the Brevard shear zone (Fig. 90). Lines and median tectonic lines, including *Verschluckungszonen*, *Narbenzonen*, and *Scheitelzonen*, are widespread (Ampferer, 1906; Ampferer and Hammer, 1911). Examples are the median tectonic lines of Japan and New Zealand (Figs. 90, 91), the Insubric line (Fig. 51), the Periadriatic line (Fig. 90), the Balanton line (Fig. 90), and many others (Fig. 90).

In our research, we found that these linear features form an unbroken genetic series from rift and graben at one end of the spectrum to *Verschluckungszone* at the other end. The literature on such linear features is large, yet attempts to interrelate them are few, especially since 1955. With rare exceptions (e.g., Mueller, 1983), the extremely useful and well-established concept of *Verschluckungszone* has all but vanished. We note that this is another great body of literature that is addressed only piecemeal in plate tectonics, most commonly with an undocumented sentence or two referring to a possible suture zone or a former subduction zone.

Hydrothermal manifestations—These are well known on land, where they form large natural geyser-steam fields in various tectonic regimes. Examples from the extensional regime include Iceland, the Great Basin, the Yellowstone area of the Western North American Cordillera, and the "black smokers" of several midocean ridges—the East Pacific Rise, the Juan de Fuca Ridge, the Galapagos Rift, and the Mid-Atlantic Ridge. Examples from transtensional and transpressional regimes include New Zealand, the Lau basin (Fig. 78), and northern California. Examples from compressive belts include the Lhasa block of Tibet, the Apennines, Japan, and to some degree New Zealand and northern California. Almost identical thermal phenomena appear in a complete spectrum of tectonic conditions, from tensional to compressional. Contrary to widespread belief (e.g., Von Damm, 1990), hydrothermal phenomena are not "... a consequence of the emplacement of hot rock at divergent plate boundaries..." (Von Damm, p. 173). Plate tectonics has explained some hydrothermal phenomena case-by-case, but has not provided a single mechanism to explain all occurrences, whatever the stress state. In contrast, surge tectonics explains all of them.

Eastward-migrating tectonic belts—Meyerhoff and Meyerhoff (1972b, 1974) described several large-scale tectonic features that suggest differential lag between the lithosphere and the strictosphere (the hard mantle below the asthenosphere; Bucher, 1956). Such differential lag, provided that the asthenosphere viscosity is sufficiently low to permit fluid movement, should be reflected in at least some eastward-directed mobility in the upper mantle and crust. Investigation of many tectonic belts over the world showed that such eastward-directed tectonic phenomena are omnipresent. Some 54 such eastward-migrating features are listed in Table 2.

Some of the major tectonic phenomena are mentioned briefly. (1) Volcanic arcs with backarc basins are concentrated in the western Pacific, and are arcuate eastward. (2) Two other eastward-facing volcanic arcs lie along the eastern margins of the Scotia and Caribbean seas respectively; they too have backarc basins. (3) Volcanic arcs lie within the continents along the eastern margin of the Pacific and deep-sea trenches associated with them are plastered against the continents. (4) Hot asthenosphere is dammed both on the western and eastern margins of the Pacific by Benioff zones which deflect their movements around them (Fig. 8). (5) North-south volcanic and plutonic belts in many foldbelts migrate eastward through time. (6) Plutonic and volcanic belts that are oriented northwest-southeast, west-east, and southwest-northeast also migrate eastward through time. In these cases, the whole belt does not move eastward, just the easternmost tip. (7) Tectogenesis within some north-south fold belts migrates eastward through time. (8) Similarly, the depocenters of some north-south basins shift eastward through time. These are just a few of the phenomena documented on Table 2. Some explanation is required of this worldwide phenomenon.

Reticulate pattern of high heat-flow bands—Plate tectonics predicts a simple heat-flow pattern around the Earth. It postulates a broad band of high heat flow beneath the full length of the midocean rift system, and parallel bands of high and low heat flow along the Benioff zones. In the Circum-Pacific, the low heat-flow band is postulated to lie seaward from the deep-sea trench, whereas the parallel high heat-flow band is postulated to underlie the adjacent volcanic arc continentward from the trench. Intraplate regions are predicted to have low heat flow.

Figure 2 shows the observed pattern. It is apparent that it is unlike the predicted pattern. At first—and even today—exceptions to the predicted pattern were explained by a phenomenon called hot spots, which are hot asthenosphere upwellings or diapirs (mantle plumes), fixed in time and space in the asthenosphere, and over which the lithosphere plates move, leaving "hot-spot trails" behind them. A given hot-spot trail shows an age progression from one end to the other. However, age progressions are very rare (Fig. 24), and dated rows of mantle diapirs that have no age progression (Fig. 80) form a large majority. (Almost all rows of mantle diapirs that do show an age progression show it to be from west to east.) Thus far plate tectonics cannot explain these contradictory data.

Figure 2 is a preliminary map subject to considerable revision in some regions, because insufficient heat-flow measurements have been made to draw a more accurate map. However, as a brief examination of the map shows, even substantial revisions in such areas as Siberia, South America, and parts of the Pacific basin will not seriously alter the reticulate pattern that is shown here.

Microearthquake bands—Modern ideas of the Earth's epicenter belts are based in large part on the classic Gutenberg and Richter (1949, 1954) hypocenter catalogs and accompanying maps. Almost all earthquakes shown by Gutenberg and Richter (1954) are in the 5.5–9.0 (Gutenberg scale) magnitude range. The belts lie along island arcs, crests of midocean ridges, continental rift zones, and wrench-fault zones. This Gutenberg and Richter image of the distribution of epicenter beds has become a cornerstone of plate tectonics.

Today, earthquakes of magnitude 1.0–2.0 can be detected, provided that proper recording instrumentation is available. Earthquakes of such small sizes can be produced by a variety of events ranging from quarry blasts to street traffic; many have no tectonic significance. Almost all earthquakes in the 3.5–5.5 range are tectonically significant, and their records (1) add important volumes of information to the Earth's seismicity data base and (2) greatly modify the Gutenberg and Richter (1954) model.

The routine recording of these smaller earthquakes in areas where good instrumentation is available (e.g., Canada, United States) reveals the presence of belts of seismicity that were essentially unknown by Gutenberg and Richter (1949, 1954). Where adequate data on the

Table 2. Evidence for eastward movements within the lithosphere (surge channels) and asthenosphere.

Phenomenon / Example(s)	Age	Comment(s)	Reference(s)
1. *Volcanic arc with backarc basin; most arcs face east*		Earth's rotation *promotes* backarc basin development	Meyerhoff and Meyerhoff (1972b, 1974)
a. Western Pacific arcs	Mainly Phanerozoic	Fig. 23	Wilson (1954)
b. Lesser Antilles arc	Late Cretaceous-present	Fig. 27	Bucher (1947); H. Meyerhoff (1954); North (1965); Meyerhoff and Meyerhoff (1972a)
c. Scotia arc	Cenozoic	—	Hamilton (1963)
d. Seychelles arc	Proterozoic-early Paleozoic	Extinct arc	Kamen-Kaye and Meyerhoff (1980)
2. *Volcanic arc without backarc basin; most arcs face west*		Earth's rotation *inhibits* backarc basin development	Meyerhoff and Meyerhoff (1972b, 1974)
a. Eastern Pacific arcs	Mainly Phanerozoic	Examples: Middle America and Peru-Chile trenches	Wilson (1954)
3. *Stack-up of tectonic phenomena only along the western sides of Benioff zones*			
a. Eastern Pacific fracture zones	?Precambrian	Examples: Clipperton, Clarion, Murray, Mendocino, Easter Island, and other FZs	Menard (1960); Meyerhoff and Meyerhoff (1972b, 1974)
b. Massive asthenosphere upwelling and/or diapirism (Fig. 8)	Precambrian	i. Backarc basins of the western Pacific; Fig. 8	This paper (see Woodhouse and Dziewonski, 1984)
		ii. East Pacific Rise; Fig. 8	Ditto
4. *Feeder channels (see text): part of the broad category of ridges called aseismic ridges*			
a. Walvis Ridge	Probably Permian-Cenozoic	These ridges branch *only* from the eastern flanks of midocean ridges; Fig. 22	This paper
b. Others: Cape Verde Rise, Cocos Ridge, Nazca Ridge, Galapagos rift, and others	—	Ditto	This paper
5. *Overflow, or breakout channels (see text): part of the broad category of submarine ridges called "aseismic ridges"*			
a. Hawaiian Ridge	Late Cenozoic	Fig. 23; these ridges branch *only* from eastern sides of volcanic arcs and continents; Fig. 20	Jackson (1976); Shaw et al. (1980)
b. Louisville Ridge	Maastrichtian to Quaternary	Fig. 24	Watts et al. (1988)
c. New England Seamount chain	Neocomian-Miocene	—	Houghton et al. (1978)
6. *North-south plutonic belt; all or part of the belt shifts eastward through time*			
a. Coast Range Batholith, British Columbia and southeastern Alaska	Cretaceous-middle Eocene	—	Gehrels and McClelland (1988)
b. Sierra Nevada and Great Basin granitic plutons	Cretaceous-Oligocene	—	Coney and Reynolds (1977); Bateman (1983)
c. Baja California and western Mexico granitic plutons	Cretaceous-middle Eocene	Fig. 35	Clark et al. (1982)
d. Granitic plutons of Colombian-Ecuadorian Andes	Cretaceous-middle Eocene	Fig. 41	Mégard (1987)
e. Peruvian Andes granites	Cretaceous-Tertiary	—	Cobbing and Pitcher (1983)
f. Chilean Andes granites	Cretaceous-Quaternary	—	Aguirre (1983)
g. Lachlan foldbelt granites of southeastern Australia	Silurian-Carboniferous	Lachlan foldbelt is in Tasman foldbelt	White and Chappell (1983)
h. Tasman foldbelt (overall)	Paleozoic-Triassic	—	Leitch and Scheibner (1987)
i. Southeastern China	Late Jurassic-Cretaceous	—	Wang Liankuis et al. (1980)
j. Korean granites	Cretaceous	—	Kim and Lee (1983)
k. Japanese granites	Late Cretaceous-Neogene	—	Utada (1980)
l. Urals granites	Paleozoic	—	Nalivkin (1973)
m. Southern Appalachians	Paleozoic	—	Fullagar and Butler (1979)
n. West Africa	Permian to recent	—	Wright et al. (1985)

Table 2. Continued.

Phenomenon Example(s)	Age	Comment(s)	Reference(s)
7. *Overall west-to-east plutonic belt; eastern tip propagates eastward*			
a. Middle and lower Yangzi Valley granites and mineral deposits	Early Jurassic-middle Cretaceous	—	Zhu Ming (1989)
8. *North-south volcanic belt; all or part of the belt shifts eastward through time*			
a. Cascade Range	Late Eocene-Recent	High Cascades are built on and east of older Cascades	Taylor (1990)
b. Argentine Andes and eastward	Triassic-middle Cretaceous	Shift is northeastward toward southern Brazil	Uliana et al. (1989)
c. New Zealand	Cenozoic	—	Stern and Davey (1989)
d. Southeastern China	Late Jurassic-Tertiary	NNE-striking Neocathaysian zone	Yang Zunyi et al. (1986)
e. Panxi rift, southwestern China	Early Permian (west) to Late Permian (east)	Figs. 9, 47	Zhang Yungxiang et al. (1990)
f. East African rift	Early Miocene-Recent	Figs. 55, 71	Baker (1987)
9. *NW to SE, W to E, and SW to NE volcanic belt; eastern tip propagates eastward*			
a. Young volcanics of British Columbia Coast Ranges	Miocene-Pliocene	—	Bevier (1989)
b. Snake River Plain-Yellowstone caldera	Miocene-Recent	—	Sears et al. (1990)
c. Southern British Columbia-northern Great Basin	Late Eocene-early Miocene	Overall NW to SE shift	Armstrong (1978)
d. Southern Great Basin	Cenozoic	Overall SW to NE shift	Leeman and Fitton (1989)
e. Colorado Plateau	Cenozoic	Overall SW to NE shift at 2.9 cm/yr	Tanaka et al. (1986)
f. Cenozoic silicic volcanics of Tasman foldbelt	Maastrichtian Recent	Mafic volcanics have no pattern	Pilger (1982)
g. Tyrrhenian Sea, Mediterranean Sea	Miocene-Quaternary	—	Locardi (1988)
10. *NW to SE, W to E, and SW to NE rift; eastern tip propagates eastward*			
a. Hetao-Yinchuan rift, Ningxia and Shaanxi, China	Late Eocene-Quaternary	Fig. 9	Zhang Buchun et al. (1985)
b. Fenwei rift, Shaanxi and Shanxi, China	Late Eocene-Quaternary	Fig. 9	Wang Jingming (1987)
c. Baykal rift, south-central Siberia	Eocene-Pliocene	Figs. 53, 54	Logatchev and Florensov (1978)
11. *North-south foldbelts migrating eastward*			
a. U. S. Western Cordillera	Devonian-middle Eocene	—	Roberts et al. (1965); Verrall (1989)
b. Lesser Antilles	Middle Cretaceous-Recent	Aves Ridge is Cretaceous, Volcanic Caribbees, younger	Meyerhoff and Meyerhoff (1972a)
12. *North-south depocenters migrating eastward*			
a. Izumi trough, Honshu, Japan	Late Cretaceous	—	Miyata (1990)
b. New Zealand	Cenozoic	—	Stern and Davey (1989)
c. Southeastern Arabian Peninsula, Oman	Late Proterozoic-Late Cretaceous	—	Ries and Shackleton (1990)
13. *Surface structures showing eastward-flow directional features*			
a. Rhine graben fault pattern, Germany	Eocene-Recent	Fig. 59. Horsetail structures point northeastward	Illies and Greiner (1979)
b. Ankara mélange, central Turkey	Late Cenozoic	Horsetail structures point northeastward	Norman (1984)

smaller magnitude earthquakes are present, they form a pattern of reticulate bands of seismicity, which coincide with the high heat-flow bands shown on Figure 2 (Stegena, 1990). These facts too are unexplained by other tectonic hypotheses, as Stegena (1990) noted.

Diffuse plate boundaries—The concept of diffuse plate boundaries was introduced in plate tectonics to

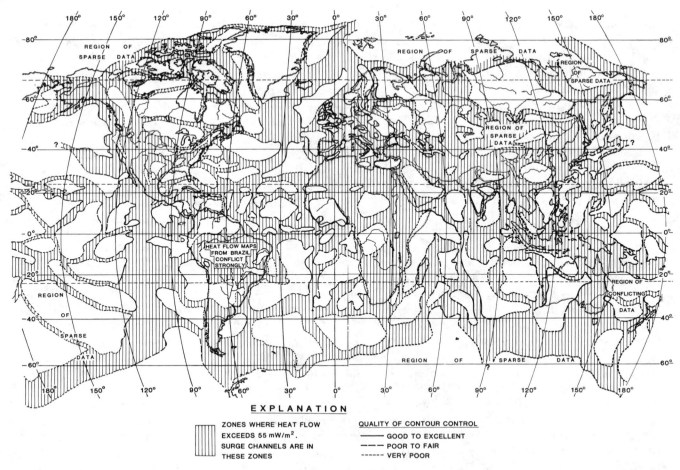

Figure 2. Preliminary world map showing bands of high heat flow (> 55 mWm^{-2}). All active surge channels lie within these bands of high heat flow. Control is weak in parts of the USSR, North America, South America, and parts of the ocean basins, especially the Pacific. The map was compiled from more than 500 sources. Some of the principal sources are: Borchert and Muir (1964), Meyerhoff (1970), Cook and Bally (1975), GEBCO (1975-1982), Erickson et al. (1977), Lachenbruch and Sass (1977), Hill (1978), Logatchev and Florentsov (1978), Smith (1978), Sykes (1978), Čermák and Rybach (1979), Cardwell et al. (1980), Basalt Volcanism Study Group (1981), KAPG (1981), Weissel (1981), Bowin et al. (1982), Hildenbrand et al. (1982), Zietz (1982), Bender (1983), Geller et al. (1983), Mooney et al. (1983), Woodhouse and Dziewonski (1984), Wright et al. (1985), Ádám et al. (1986), Gramberg and Smyslov (1986), Ma Xingyuan (1986a, 1986b), Saxena (1986), Wang Jun et al. (1986), Dziewonski and Woodhouse (1987), Geological Survey of Canada (1987), Grand (1987), Yan Zhide (1987), Bollinger and Wheeler (1988), Engdahl and Rinehart (1988), Froitzheim et al. (1988), Létouzey et al. (1988), McCartan and Architzel (1988), Newhall and Dzurisin (1988), Gough (1989), Liu Futian et al. (1989), Liu Jianhua et al. (1989), and Petroy and Wiens (1989). Unpublished data on microearthquake distribution and heat flow were obtained with the help of V.V. Beloussov, V. Čermák, Zoltán de Cserna, David Forsyth, Susan K. Goter, and the Geological Survey of Canada (R.J. Wetmiller, F. Ainglin, J. Jalpenny).

explain the presence of wide (up to several hundred kilometers across) bands of shallow seismicity. A well-known example is the San Andreas fault zone where a width of the order of 300 km has been postulated to explain (1) the broad bank of shallow diffuse seismicity of that region (Mooney and Weaver, 1989) and (2) that measured offsets along the San Andreas fault are far short of plate-tectonic predictions. Barrow et al. (1985) found that a diffuse plate boundary was essential in southern California where they showed that lateral movements along the San Andreas fault in Los Angeles County from Jurassic time to the present could not exceed 21 km, and along the whole zone of San Andreas-parallel faults, 102 km. Although the ad hoc postulate of a diffuse plate boundary proved useful in regions where plate boundaries are supposed to be present, it does not explain the many bands of shallow seismicity that are far from plate boundaries. These diffuse plate boundaries are parts of the reticulate network of microseismicity discussed in the preceding section, and underlie bands of elevated heat flow.

Stretching lineations—Many examples have been cited in the literature of stretching lineations in foldbelts that parallel the foldbelt and of similar lineations that are orthogonal to it. Brunel (1986), in studies of the Himalayas, found that stretching lineations parallel the direction of thrusting within the fault gouge of the thrust zone, and that away from gouge zones—both vertically and laterally—the stretch lineations parallel the strike of the tectonic belt. Ellis and Watkinson (1987) tried unsuccessfully to explain that apparent paradox. If Stokes's Law is considered as closely related to the strike of the lineations, the difficulty disappears: the foldbelt-parallel lineations were developed first as a consequence of flow beneath the belt and parallel with it; the lineations in the thrust gouges are younger, and are restricted to the thrust-fault gouge. As Ellis and Watkinson (1987) observed, plate tectonics does not offer a ready explanation.

Tectonostratigraphic terranes—Paleomagnetic and faunal studies gave rise to the concept that parts of various Phanerozoic foldbelts now close to or at the ocean-continent interfaces were carried thousands of kilometers (up to 9,000) from other parts of the globe to their present positions (Wilson, 1966, 1967). This proposal attracted many advocates. However, careful field work and restudies of the original—including magnetic—data showed that the concept of long-distance transport is exaggerated (e.g., Saul, 1986; Hansen, 1988; Newton, 1988; Irving and Archibald, 1990; Storetvedt, this volume). Stripped of the long-distance requirement, a tectonostratigraphic terrane is seen to be no more than an Alpine facies belt as first discovered by de Lory (1860) and as defined by many authors (e.g., Bertrand, 1987; Trümpy, 1960). These belts, if they have been transported at all, were thrust and generally lie within 200 to 300 km of their original positions. Thus, research indicates that tectonostratigraphic terranes do not comprise a unique data set, but are parts of other sets (e.g., eugeosynclines).

Other data sets—Many additional phenomena do not fit readily into plate tectonics, and therefore, have been explained case-by-case. Examples include rows of anorogenic granitic plutons, dike swarms and fissure eruptions, volcanic fields on cratons and abyssal plains, linear river courses, linear mountain divides, intracontinental basins, linear evaporite basins, and numerous others. Rather than belabor each, we present some as examples in our explanations of surge tectonics.

Summary of data sets—The most important fact brought out to this point is that the high heat-flow bands (> 55 mWm^{-2}) of Figure 2 coincide with all of the phenomena discussed thus far. These include (1) bands of microearthquakes (including diffuse plate boundaries) that do not coincide with plate-tectonic predicted locations, (2) segmented belts of linear faults, fractures, and fissures, (3) stretching lineations that parallel each tectonic belt, (4) segmented belts of identified and postulated Phanerozoic mantle upwellings and diapirs (hot-spot trails), (5) linear lenses of anomalous upper mantle (P-wave velocities in the 7.0–7.8-km/s range) that commonly are overlain by shallower, smaller, low-velocity zones, (6) the existence of bivergent deformation in all foldbelts, (7) strike-slip zones and similar tectonic lines that range from simple rifts to *Verschluckungszonen*, (8) eastward-shifting tectonic-magmatic belts that contain silicic through ultramafic magmas, which range in habitat from deep-seated plutons to extrusive volcanic fields, and (9) subaerial to submarine geothermal zones with geysers, hot springs, black smokers, and related phenomena. Plate tectonics does not explain these phenomena and their mutual association.

Data sets used in plate tectonics

Although plate tectonics does not incorporate the data sets enumerated above, it does incorporate several other fundamental data sets that are its *raison d'être*. (These are summarized in Table 5 and discussed in subsequent sections.) They include midocean and continental rifts, magmatic arcs, oceanic trenches, Benioff zones, geosynclines, preorogenic-synorogenic-postorogenic granitic plutons, ophiolite belts, zones of tectonic mélange, mantled gneiss domes and metamorphic core complexes, and a very limited number of metamorphic rock patterns (e.g., "paired" metamorphic belts). If these were the only phenomena important to the viability of plate tectonics, the hypothesis might need no reevaluation. We pointed out, however, several equally important phenomena that plate tectonics does not explain. Moreover, as we observed, the phenomena for which plate tectonics has no explanation occur in the reticulate patterns of high heat-flow belts illustrated on Figure 2. Of greater importance is that the phenomena that plate tectonics purports to explain also occur in the reticulate bands illustrated on Figure 2. This assertion is documented amply in the remainder of the paper.

Study of Figure 2 shows, however, that some foldbelts, ophiolite zones, and related phenomena do not lie in the bands shown on Figure 2. This is so because such belts are inactive. In this paper detailing the hypothesis of surge tectonics, we consider principally the active tectonic belts.

Previous tectonic hypotheses

Many of the concepts embodied in surge tectonics come from earlier hypotheses of Earth dynamics. Space restrictions, however, do not permit the careful review that these earlier hypotheses deserve. A full review will appear in *Surge Tectonics* (Meyerhoff et al., submitted).

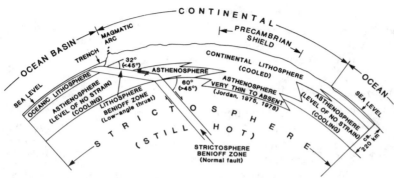

Figure 3. Idealized cross section through two ocean basins, a continent, and a Benioff zone. This figure is based on Benioff (1949, 1954), Scheidegger and Wilson (1950), Wilson (1954), Jordan (1975, 1978), Isacks and Barazangi (1977), and Lowman (1985, 1986). It shows that (1) the asthenosphere wedges out beneath continental nuclei, (2) continental nuclei are rooted to the deep mantle (strictosphere), and (3) the Benioff zone consists of two segments, a low-dipping thrust zone in the lithosphere (which is under compression) and a steeply-dipping normal fault zone in the strictosphere (which is under tension). The dip angles of the two fault segments are predicted by the Navier-Coulomb maximum shear-stress theory (Scheidegger and Wilson, 1950; Jaeger, 1962), which indicates that the strictosphere is still contracting (MacDonald, 1963).

For good summaries of all earlier hypotheses, Scheidegger (1963) and Dennis (1982) are recommended strongly.

Velocity structure of the Earth's outer shells

Basic framework—The Earth's outer shells (Fig. 3, Table 3) consist of a "hard" lithosphere above a "soft" asthenosphere (Dana, 1896; Barrell, 1914; Daly, 1940). The interpreted seismic structure of these two shells is given in Table 3, which is based on Press (1966) and Iyer and Hitchcock (1989). The asthenosphere overlies another hard shell that Bucher (1956, p. 1295) named the strictosphere. Its seismic characteristics (at least near the upper surface of the strictosphere) also are given on Table 3. Of these shells, the lithosphere is especially important because visible tectonic effects provide the principal clues to the origin of tectogenesis. It has not been too long, since the seismic structure of the crust and upper mantle became sufficiently well imaged to permit more than just educated guesses about it.

Before 1958, when Revelle (1958) discovered a layer of material with a velocity of 7.3 km/s on the southern part of the East Pacific Rise, the lithosphere was perceived as consisting of 6.6-km/s crust overlying directly the 8.1-km/s mantle. What Revelle (1958) discovered was a lens or high-velocity crust, or low-velocity mantle, with a P-wave velocity of 7.0–7.8 km/s separating the "normal" crust above from the "normal" mantle below. Similar lenses were found almost everywhere beneath the midocean ridge system (e.g., Ewing and Ewing, 1959; Talwani et al., 1965). By 1982, a similar lens of 7.0–7.8-km/s material was found beneath most of the Earth's rifts (Figs. 15, 16, 18, 31, 43, 54, 58, 79). Mooney et al. (1983) suggested that such lenses are a characteristic of extensional tectonic belts.

During the 1970s, refraction shooting in the northern Appalachians discovered a similar 7.0-km/s lens under the Acadian (Devonian) foldbelt (Taylor et al., 1980; Marrilier et al., 1989; and Fig. 44). It was not long before identical lenses beneath foldbelts were identified in many parts of the world (Figs. 10, 26, 27, 35, 44, 47, 51, 62, 64). Good images of these lenses were recorded on reflection-seismic lines used for deep continental and oceanic tectonic studies (Figs. 65–68; Nelson, 1988). Figures 67 and 68, from the English Channel and southwestern Queensland respectively, are particularly good images of two of these lenses near the base of the continental crust.

Mooney and Braile (1989), summarizing the present state of knowledge of the structure of the Earth's crust, inserted a 7.0–7.8-km/s lower crustal layer between the 6.6-km/s sialic crust above, and the 8.1-km/s mantle below. They showed the layer to be absent in places. In general, it is present beneath cratons, platforms, foldbelts, rift systems, and wrench-fault zones. Under cratonic areas the layer is 7–15 km thick; under tectonic belts it is thicker, ranging from 10–25 km. In many

Table 3. Seismic interpretation of crust and upper mantle. Principal sources are Press (1966) and Iyer and Hitchcock (1989).

Shell	Thickness (km)	Depth to base (km)	P-wave velocity range (km/s)	Comments regarding P-wave velocity range
Lithosphere	60–180	60–180	0–8.45	Includes water through upper mantle
Asthenosphere	160–40	220	7.85–8.05	Values are for the soft interior
Strictosphere	—	—	8.1–8.7	Top of shell only; upper "contact" transition a

places the 7.0–7.8-km/s velocity is gradational with mantle velocities (> 7.9 km/s). In these places, the Mohorovicic discontinuity is not a true discontinuity but a transition zone several kilometers thick (Mooney and Braile, 1989). Areas were found also beneath ancient platforms and shields where the 7.0–7.8-km/s layer is up to 30 km thick. Examples include the Baltic shield (Fig. 43; Luosto et al., 1989) and parts of the Canadian shield (Braile, 1989).

The preceding suggests that the 7.0–7.8-km/s layer is distributed rather randomly, and that its thickness in a given area is a result of random processes. These conclusions are almost unavoidable if one tries to explain the distribution and thickness with any of the Earth-dynamics hypotheses published during the last century. In surge tectonics this apparent randomness of distribution and thickness is in part predictable.

The origin of the 7.0–7.8-km/s layer almost certainly is closely related to that of the high reflectivities of the lower crust as described and illustrated by Klemperer et al. (1986), Klemperer (1987), Goodwin and Thompson (1988), Thompson and McCarthy (1990), and others (see Figs. 65–68). We emphasize the fact that a lens (or lenses) or 7.0–7.8-km/s material underlies all of the microearthquake bands that we studied, and therefore, all of the high heat-flow bands shown on Figure 2.

Continents have deep roots—An important aspect of upper mantle and crustal structure is that continental cratons have deep crustal roots (MacDonald, 1963; Jordan, 1975, 1978; Woodhouse and Dziewonski, 1984; Dziewonski and Woodhouse, 1987; Grand, 1987). Contrary to general belief (e.g., Anderson, 1987), continental roots are fixed to the strictosphere (Lowman, 1985, 1986). This conclusion is supported by large and increasing volumes of data, including neodymium and strontium studies of crustal rocks (Wasserburg and DePaolo, 1979). The absence, or near-absence, of a low-velocity asthenosphere beneath ancient cratons led Lowman (1985, 1986) to propose an Earth-dynamics hypothesis of sea-floor spreading between fixed continents. In this hypothesis, if sea-floor spreading takes place, it is restricted to suboceanic regions. Thus, the deep roots of continents are a major and very likely fatal obstacle to any hypothesis requiring continental movements (Lowman, 1985, 1986; K.B. Krauskopf, pers. comm., 1990).

An example of deep continental roots is presented in Figure 4, a seismotomographic cross section of North America. The dark shading beneath the Canadian shield shows a root extending to 400–450 km (Grand, 1987). Similar deep roots are seen beneath parts of all of the Earth's ancient cratons. In places, however, lenses of 7.0–7.8-km/s material containing low-velocity zones (Fig. 43) are present (e.g., Baltic shield; Luosto et al., 1989). Such lenses containing low-velocity layers postdate the establishment of the deep cratonic roots, as we show in subsequent sections.

Conclusion

One fundamental conclusion can be made at this point. Based on the observations summarized in these introductory sections, no Earth-dynamics hypothesis proposed to date integrates successfully all the data sets that we have mentioned. We show by examples given in subsequent sections that the surge-tectonics hypothesis accounts very successfully for all data sets.

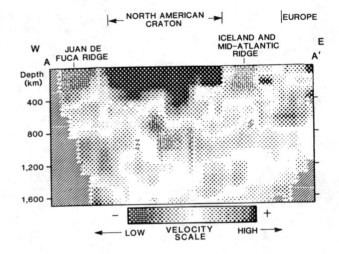

Figure 4. Southwest to northeast seismotomographic cross section showing the velocity structure across the North American craton and the North Atlantic Ocean. High-velocity (i.e., colder) lithosphere, shown in dark tones, underlies the Canadian shield to depths of 250 to 500 km. Thus, as pointed out by Jordan (1975, 1978), cratons have deep mantle roots, and the low-velocity zone beneath them is either absent or very thin. This and similar cross sections suggest that the continental nuclei are fixed to the mantle (Lowman, 1985, 1986). From Grand (1987). Note: the velocity scale is -3% to +3% above 320 km; -1.5% to +1.5% from 320 to 405 km; and -0.9% to +0.9% below 405 km.

SURGE TECTONICS

Introduction

Surge tectonics is a totally new hypothesis quite unlike previously proposed hypotheses, although many of its component parts are based on ideas long known. We believe the hypothesis provides a comprehensive and internally consistent explanation of all tectonic phenomena without the necessity of making unsupported assumptions or ad hoc explanations. We have found nothing that surge tectonics cannot explain in a simpler way than other tectonic hypotheses. Surge tectonics draws on well-known physical laws, especially those related to Newton's laws of motion and gravity. Fluid dynamics plays an important role in surge tectonics. (For more information on the laws we utilize, those mentioned in the text are defined in Appendix; those wishing more detail are referred to two standard physics textbooks by Sears et al. [1974] and Blatt [1983]. An excellent state-of-the-art fluid-dynamics text is that by Tritton [1988]).

Surge tectonics involves three separate but interdependent and interacting processes. The first process is lateral flow of fluid, or semifluid magma through a network of interconnected magma channels in the lithosphere. We call these surge channels for reasons that will become apparent. The third process is due to the Earth's rotation. This process involves differential lag between the lithosphere and the strictosphere (the hard mantle beneath the asthenosphere and lower crust), and its effects—eastward shifts—already discussed (Table 2). Because lithosphere compression caused by cooling is the mechanism that propels the lateral flow of fluid, or semifluid, magma through surge channels, we discuss the contraction (Earth cooling) hypothesis.

Contraction

General—Cooling of the Earth, or contraction, is the oldest scientific hypothesis of Earth dynamics, having been proposed late in the 16th century by Giordano Bruno, who lived from 1548 to 1600. (He was later burned at the stake for insisting that the Earth orbited the sun.) Bruno's classic analogy between a shrivelled apple and a cooling Earth was immortalized in a diagram published by Descartes (1644; see Dennis, 1982). Élie de Beaumont (1831) in Europe and Dana (1847) in North America first applied the concept to foldbelts as a general explanation of tectogenesis. Charles Davison (1887) and Sir George Darwin (1887) were the first to work out the mathematical theory of contraction, a hypothesis that remained popular until the late 1950s.

Two types of contraction hypotheses developed. The first was the elastic instability hypothesis, according to which the lithosphere, as the Earth cools, buckles by folding (the "shrivelled apple"). Jeffreys (1970) showed that the amplitude of such folds would be so great that they would be elastically unstable, and that failure would not be by folding or buckling but by fracture. Thus the second form of the contraction hypothesis was born, the fracture-contraction hypothesis (Jeffreys, 1970; Meyerhoff et al., 1972).

Cooling of the Earth, however, may not be solely the result of simple cooling of a sphere, as Jeffreys (1970) advocated. Gravitational differentiation may also be an important factor (Scheidegger, 1963). In addition, Mac-Donald (1959, 1963, 1965; pers. comms., 1981–1984) stressed that the Jeffreys (1970) contraction model failed to account for the radioactive decay that must have attended the differentiation of the Earth.

MacDonald (1959, 1963, 1965; pers. comms., 1981–1984) published several numerical models of a cooling Earth, each differing from the others by a change in the Earth's chemical composition. MacDonald (1963) showed that because of the widespread presence in the primitive Earth of radioactive elements, the Earth must have expanded during much of its history. MacDonald (1963) reasoned that during the first part of Earth's history, decay of radioactive elements caused heat production to exceed heat loss, thereby causing the Earth to expand. With continuing decay of radioactive elements, heat production and heat loss would equalize until, during the last 1.1 Ga or so, heat loss exceeded heat production and produced contraction. This contraction model differs considerably from Jeffreys' (1970) model in which contraction occurs from the Earth's inception, but it, like Jeffreys' model, is a fracture-contraction model. MacDonald (pers. comms., 1981–1984) emphasized strongly that the conclusions of his 1959–1965 papers are still valid today. Stacey (1981) and Lyttleton (1982) gave additional reasons why the Earth should still be contracting slowly.

Contraction scepticism—Many workers today either doubt that contraction is taking place or fail to see why the possibility should even be considered. Bott (1971, p. 270), expressing a common opinion, wrote that because of the success of plate tectonics in producing foldbelts, contraction now "... is irrelevant to tectonic problems." Two reviewers of an earlier version of this paper also expressed doubts that contraction can be taking place. However, one of them, K.B. Krauskopf (pers. comm., 1990), conceded that "... too little is known about what goes on in the [Earth's] interior for any definite statement to be made." He noted that MacDonald's (1959, 1963, 1965) models could easily be as sound today as they were in 1965 because "... not much more is known today

..." about the concentration of radioactive elements in the Earth's interior.

Evidence for a differentiated, cooled Earth—The evidence is straightforward. The most salient facts follow.

1. The Earth includes several concentric shells, which are explicable only if the Earth differentiated efficiently and at a much higher temperature than today.

2. The outermost of these shells may be the oceanic crust whose thickness ranges from about 4–7 km. This crust is characterized by relatively constant thickness and fairly uniform seismic properties. Both Worzel (1965) and Vogt et al. (1969) observed that if the plate-tectonic explanation of ocean-crust generation is correct, it is a truly remarkable process that produces such a uniform layer in all ocean basins regardless of the spreading rate—1.2 cm/yr or 60 cm/yr. This uniformity is explained, however, if the oceanic crust is the outermost of the Earth's concentric shells. There are other explanations, one of which is discussed later.

3. Mehnert (1969), among several, noted that the further back one looks into the geological record, the greater is the abundance of mafic rocks. This is explained if the lithosphere has been thickening through time by cooling, as Mehnert (1969) suggested.

4. Miyashiro et al. (1982), reporting on studies of the Earth's metamorphic rocks, noted that Precambrian rocks show the highest geothermal gradients and that geothermal gradients of younger rocks generally decrease to the present time.

5. A convincing evidence that huge segments of the lithosphere have been and are being engulfed by tangential compression is the existence of the previously discussed *Verschluckungszonen* (swallowing or engulfment zones) of Ampferer (1906) and Ampferer and Hammer (1911). In places along such zones, whole metamorphic and igneous belts that are characteristic of parts of a given foldbelt simply disappear for hundreds of kilometers along strike (e.g., Alps: Mueller, 1983; Kyushu-Shikoku foldbelt: Miyashiro, 1973; New Zealand Alps: Kingma, 1974; southern California Transverse Ranges: Humphreys et al., 1984). Figures 90 through 93 illustrate the characteristics of typical *Verschluckungszonen*. Although Mueller (1983), Humphreys et al. (1984), and other workers considered these features to be former subduction zones, this interpretation is difficult to defend because all of these zones, regardless of age, are near-vertical bodies that (1) reach only the top or middle or the asthenosphere (150 to 250 km deep) and (2) do not deviate more than 10° to 25° from the vertical (Mueller, 1983; Humphreys et al., 1984).

6. The antipodal positions of the continents and ocean basins mean that Earth passed through a molten or near-molten phase (Arldt, 1907; Bucher, 1933; Wilson, 1954; Harrison, 1968). Such antipodal relations are unlikely to be a matter of chance or coincidence (Harrison, 1968).

7. Theory (Jeffreys, 1970) and laboratory experiment (Bucher, 1956) showed that heated spheres cool by rupture along great circles. Remnants of two such great circles (as defined by hypocenters at the base of the asthenosphere) are active today: the Circum-Pacific and Tethys-Mediterranean fold systems. The importance of Bucher's (1956) experiment to contraction theory, in which he reproduced the great circles, is little appreciated.

8. As Earth cooled, it solidified from the surface downward. Because stress states in cooled and uncooled parts are necessarily opposite one another, compression above and tension below, the two parts must be separated by a surface or zone that Davison (1887) called the level of no strain (Fig. 3). We, as did Wilson (1954), equate the cooled layer with the lithosphere (Fig. 3). The uncooled part below is what Bucher (1956) called the strictosphere. Thus, as originally proposed by Scheidegger and Wilson (1950), Davison's (1887) level of no strain must be the asthenosphere, or a zone of no strain across which the change in stress states is gradual (Fig. 3). Only in a cooling Earth, which approximates a closed thermal system, can an asthenosphere form.

9. Continued cooling deepens the asthenosphere and the upper surface of the strictosphere. The stresses accumulated through cooling are relieved episodically by rupture along the great-circle fractures that are the Earth's cooling cracks or the Benioff zones of current literature. Because the lithosphere is being compressed and the strictosphere subjected to tension, the mechanics of rupture should follow the Navier-Coulomb maximum shear-stress theory (Jaeger, 1962). Accordingly, the lithosphere Benioff zone must dip less than 45° to a tangent to the Earth's surface (in actual fact, it dips 22° to 44°; Figs. 3, 5, 6). In contrast, the strictosphere Benioff zone must dip more than 45° to a tangent to the Earth's surface (50° to 75°; Figs. 3, 5, 6). Benioff (1949, 1954) discovered the change in Benioff-zone dip from lithosphere to strictosphere, but Scheidegger and Wilson (1950) recognized these dips as an expression of the Navier-Coulomb maximum shear-stress theory (Figs. 3, 5, 6). The dip values of the lithosphere and strictosphere Benioff zones confirm that the Earth is a cooling body. (Ritsema [1957, 1960], working independently, also discovered the abrupt dip changes in the dip of the Benioff zone with increasing depth.)

An important fact concerning the Benioff zone is that the two segments, one in the lithosphere and the other in the strictosphere, do not necessarily form a single, continuous zone as depicted in most diagrams and cross

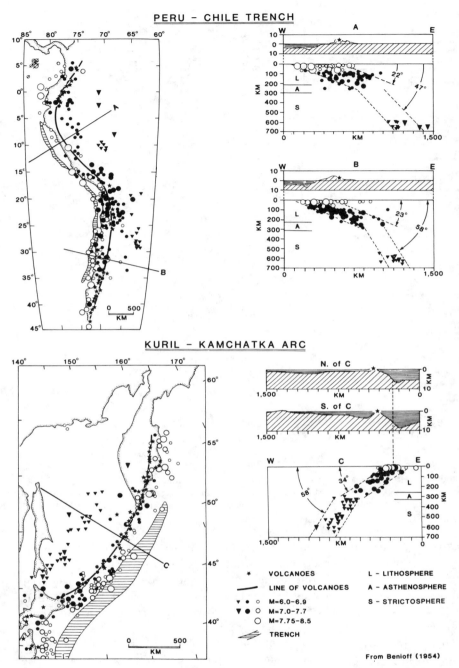

Figure 5. Sections across the Peru-Chile Trench and the Kuril-Kamchatka volcanic arc, showing the distribution of hypocenters. These sections illustrate the differences in dip between lithosphere and strictosphere Benioff zones, differences predicted by the Navier-Coulomb maximum shear-stress theory (Scheidegger and Wilson, 1950; Jaeger, 1962). Published with permission of Geological Society of America.

studies of hypocenter distribution in Benioff zones show the same clear division into a shallow, gently dipping lithosphere Benioff zone and a deeper, steeply dipping strictosphere Benioff zone (Fig. 6; Grow, 1973; Isacks and Barazangi, 1977).

Ritsema's (1957, 1960) focal-mechanism studies of shallow, intermediate, and deep earthquakes showed that the Benioff-zone dip in the lithosphere is only half of its dip in the strictosphere. An even more significant discovery made by Ritsema (1957, 1960), although he attached little importance to it, was the revelation that earthquake foci above 0.03R (approximately 180 km depth) show mainly compression, whereas those below 0.03R show mainly tension. Most earthquakes above and below 0.03R, as Scheidegger (1963) also noted, have a strike-slip component. Thus Ritsema's findings lend support to Scheidegger and Wilson's (1950) interpretation of the Benioff zone as a manifestation of the Navier-Coulomb maximum shear-stress theory.

10. Computer simulations of possible Earth thermal histories (MacDonald, 1959, 1963; Jacobs, 1961; Reynolds et al., 1966), using a broad spectrum of assumed initial temperatures and chemical compositions, show that the Earth is cooling (see also Stacey, 1981; Lyttleton, 1982). The fact that Earth's Benioff zones still are active earthquake-generating zones provides strong support for this conclusion. Perhaps the strongest support comes from the Basalt Volcanism Study Project (1981) report by 101 petrologists, mineralogists, and petrog-

sections (e.g., Figs. 3, 5, 6). Benioff (1949, 1954) found that the two segments of the Benioff zone, instead of joining near the base of the asthenosphere, may be offset for distances of 100 to 200 km (Fig. 7). In some places such as the Lesser Antilles arc, a strictosphere Benioff zone may not even be present below the lithosphere Benioff zone (e.g., Lesser Antilles and Scotia volcanic arcs). These facts are explained in surge tectonics but not by other hypotheses. In fact, all detailed modern

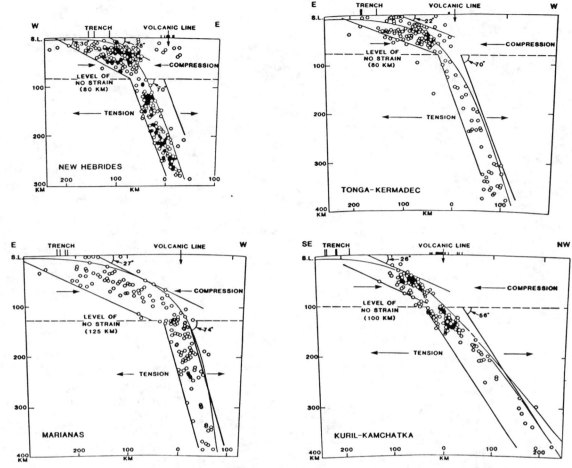

Figure 6. Sections showing hypocenter distribution beneath four western Pacific volcanic arcs (from Isacks and Barazangi, 1977). Each section displays distinct lithosphere and strictosphere Benioff zones, each with its own dip angle. For comparison, we have retained—in each section—the curved line drawn by Isacks and Barazangi (1977) to represent the shape of the Benioff zone. Compare the Kuril-Kamchatka section with Benioff's (1954) section shown on Figure 5.

raphers who wrote that the repeated extrusion of basalt to the Earth's surface through its history is proof of the Earth's long history of cooling.

11. Finally, the existence of *Verschluckungszonen* in the lithosphere and upper mantle also constitutes evidence that the Earth is actively cooling. *Verschluckungszonen* are interpreted by us to be large masses of lithosphere and upper mantle that were downbuckled into the upper mantle during tectogenesis as the lithosphere readjusted its shape to fit the underlying, cooling strictosphere (Figs. 91–93). If this interpretation is correct, the existence of *Verschluckungszonen* may constitute proof that the Earth has been cooling to the present day. We discuss *Verschluckungszonen* in a subsequent section.

Contraction as an explanation of Earth dynamics

Contraction acting alone—Despite the attraction of a cooling Earth, both Scheidegger (1963) and Bott (1971) concluded that contraction acting alone is inadequate to produce the crustal shortening measured in the Earth's many tectonic belts. For both geological and seismological reasons, this conclusion appears to be well founded. They gave several reasons; three of which and one of our own are crucial.

1. The total amount of shortening measured across the Earth's foldbelts far exceeds what can be inferred on theoretical grounds, whether one uses the contraction model of MacDonald (1963) or of Jeffreys (1970). Even if one accepts Bucher's (1955, p. 357-360; 1956, p. 1306) outstanding demonstration that apparent (measured) shortening can be and generally is four to five times true shortening, the contraction hypothesis cannot explain all apparent (measured) shortening in foldbelts. (Lyttleton's [1982] theoretical estimate of 2,000 km of shortening adequately explained the measured shortening, but his hypothesis requires cataclysmic geological events that need to be sought in the field.)

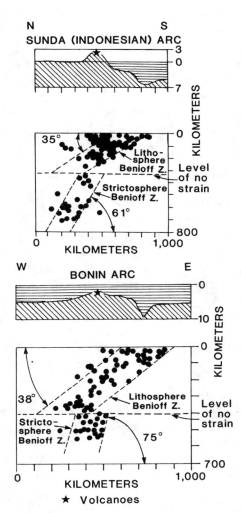

Figure 7. Two cross sections of western Pacific volcanic arcs and the distribution of the hypocenters beneath them are shown. As Benioff (1954) observed, these two sections demonstrate that the lithosphere Benioff zone need not be continuous with the strictosphere Benioff zone, a fact that is explained by contraction theory but not by plate tectonics. Published with permission of Geological Society of America.

2. Contraction alone is unable to explain the origins of all types of tectonic belts—compressional foldbelts, tensional rift zones (including midocean ridges), and strike-slip zones.

3. Ritsema (1957, 1960) and Scheidegger (1963) observed that earthquake first-motion studies show that strike-slip motions are most common in Benioff zones, not just in strike-slip and rift zones. Contraction alone cannot explain the ubiquitous strike-slip component.

4. Contraction theory requires that foldbelts are concentrated in and adjacent to oceanic trenches. This is not observed. More than 50% of the Earth's foldbelts lie at great distance from the surface trace of a Benioff zone, and all Jurassic-Cenozoic foldbelts lie within the high heat-flow bands illustrated on Figure 2. This cannot be explained by any Earth-dynamics hypothesis yet proposed. However, if contraction could lead to tectogenesis of large parts of the lithosphere far removed from Benioff zones, the preceding objections to the contraction hypothesis would be irrelevant.

Contraction as the trigger for tectogenesis—Figure 2, as we have discussed, portrays the bands of high heat flow that crisscross the Earth's surface. We believe that this reticulate network of high heat-flow bands is underlain by a network of interconnected magma chambers. In surge tectonics, these magma chambers are the mantle diapirs discussed in preceding sections and summarized in part in Table 1. Figure 2 indicates that these mantle diapirs, or magma chambers, are interconnected. The interconnected channels comprise the surge channels of surge tectonics. If the hot material in these channels is sufficiently mobile, lateral flow through them should be possible provided a pressure gradient is present. The compression already present in the lithosphere would provide the force needed to initiate and maintain flow. We emphasize that such flow would be temporally discontinuous (i.e., episodic) and, when it did occur, would be extremely slow. Flow velocities are discussed later.

We have pointed out that in a cooling Earth the lithosphere by definition is everywhere and at all times in a state of compression (Fig. 3; Davison, 1887; Scheidegger and Wilson, 1950; Bucher, 1956; MacDonald, 1959, 1963, 1965; Jeffreys, 1970; Meyerhoff et al., 1972; Stacey, 1981, Lyttleton, 1982). The compression is concentrated in planes tangent to the Earth's surface, and is equal in all directions. James C. Meyerhoff (pers. comm., 1988) called this equiplanar tangential compression, and this compression in the lithosphere is what Bucher (1956) referred to (incorrectly) as all-sided compression.

The only elements in the lithosphere that disturb this approximately equiplanar tangential stress state are the surge channels. Flow can take place in these channels wherever a pressure gradient develops. For example, the escape of lava from a channel lowers the pressure at that point, and equiplanar compression, acting at right angles to the surge-channel walls, mobilizes the fluid elements inside the channel until pressure equilibrium is restored. The presence above active channels of channel-parallel fault, fracture, and fissure systems indicates that (1) flow takes place along the full length of each tectonic belt and (2) the channels are in communication with the Earth's surface through the fault-fracture-fissure system.

Surge channels and their fault-fracture-fissure systems constitute zones of weakness in the lithosphere. Because (1) the channels are the only bodies in the lithosphere that, owing to have their potential to contain mobile fluids, they have the capacity to upset the state of equiplanar tangential compression, and because (2) they are constantly losing their contents to the surface, lithosphere compression ultimately destroys them. Their deformation and ultimate destruction are the essence of tectogenesis. Thus the cooling process in the Earth's strictosphere effectively guarantees the presence within the lithosphere of a powerful mechanism for tectogenesis.

Review of surge and related concepts in Earth-dynamic theory

Several workers proposed, on both theoretical and geological-geophysical grounds, the presence of bodies similar to surge channels in the lithosphere. Others developed concepts much like those that are the basis for surge tectonics.

Surge channels—Our study of this topic was by no means exhaustive; we may have missed important references that deal with the concept of surge-channellike edifices in the lithosphere and uppermost mantle. A particularly good example of a surge-channellike feature was proposed by Vogt (1974) for the Iceland region, including the Kolbeinsey, Reykjanes, and Faeroe-Greenland ridges. He wrote (p. 116, 118), "In the model I assume there is a pipe-like region below the spreading axis, extending subhorizontally away from a plume such as Iceland . . . . This mid-oceanic pipe extends from the base of the axial lithosphere, about 5 or 10 km deep, down to maximum depths (30 to 50 km?) from which basalt melts segregate and rise. Tholeiitic fluids would be released from the entire pipe; origin depths of 23 km for the Mid-Atlantic Ridge and 16 km for the East Pacific Rise . . . would approximate depth to the center of the pipe. The ultrabasic mush in this pipe is assumed to be flowing away from the hot spot at a rate determined by pipe diameter, viscosity, and horizontal pressure gradient . . . ." Vogt (1974) estimated that flow ranged laterally outward from Iceland from 500 to 600 km. A similar study by Gorshkov and Lukashevich (1989) suggested that channels are present beneath the full length of the midocean ridge system. According to them, flow beneath the Mid-Atlantic Ridge would be from hot spots located beneath Antarctica in the south and Iceland in the north, with the two flows converging near the equator.

Other midocean ridges where some type of axis-parallel flow and/or rift propagation have been postulated include the Juan de Fuca Ridge (Shih and Molnar, 1975) and the Galapagos Rift (Hey and Vogt, 1977). After the discovery of systematic segmentation along the midocean ridges beginning with the East Pacific Rise (Lonsdale, 1982), Lonsdale, Macdonald, Fox, and others commenced a series of investigations that led to a general postulate of ridge-parallel flow in the midocean ridges (e.g., Macdonald et al., 1988; Winterer et al., 1989). Macdonald et al. (1988) proposed that mantle diapirs rise beneath the centers of each ridge segment, thereby accounting for the greater elevations of the central parts of such segments. From the crest, ridge-parallel lateral flow commences. Such flow halts in the depressed areas between adjacent segments because of mutual impingement.

Sonographs of the midocean ridges show that the Macdonald et al. (1988) hypothesis is not tenable. Where the diapirs rise in the centers of the ridge segments, radial and/or annular structures should be present. Where the flows from adjacent segments impinge, compressional structures should be present. Neither structural form is observed. Instead, linear fractures and faults extend for hundreds of kilometers along strike, with interruptions only at transform faults, devals, and overlapping spreading centers. In several of the transform fault zones and devals, we traced individual fault traces through the fault zone from one ridge segment to the next, which negates the Macdonald et al. (1988) hypothesis.

Surge-channel, or interconnected mantle-diapir systems have been reported from small oceanic basins and continental areas. A well-known example of the former is the Tyrrhenian Sea west of Italy where geophysical techniques and very high heat flow show the presence of a very large diapirlike body at shallow depths (Locardi, 1988).

Among continental examples, the Fergana Valley in Soviet Central Asia is the best documented. Kuchay and Yeryemin (1990) discovered a very large pipelike body, or channel, below this large east-west-striking late Cenozoic structural depression nestled among the western ranges of the Tian Shan. In their summary (p. 45), Kuchay and Yeryemin concluded: "Interpretation and interpolation of geophysical data from the [Fergana Valley] lead to the conclusion that the base of the 'granitic' layer undergoes partial melting, as a consequence of which there forms a layer of lowered viscosity that has been identified seismically as a zone of reduced velocity above the Conrad discontinuity. It is possible to consider this zone as a 'granitic' asthenochannel, a subhorizontal layer that has a high strain rate and in which relative lateral displacement takes place between the contents of the asthenochannel and the surrounding rock layers. The absence of a gravity anomaly, which is an indication of variations in thickness and other properties within the zone of reduced velocity, suggests that the 'granitic' layer and the 'granitic' asthenochannel have the same density."

Use of the surge concept in tectonics—We have found three parallel and presumably independent derivations of the surge-tectonic concept. The most recent of these originated with Hollister and Crawford (1986) who used "tectonic surge" to describe rapid vertical uplift in the structural core of a bilaterally deformed foldbelt (i.e., in the center of what we call a kobergen). Hollister and Crawford (1986, p. 560) opined that "Weakening of the crust [by] anatexis and accompanying development of melt-lubricated shear zones..." is essential to rapid vertical uplift. Paterson et al. (1989, p. 116) referred to this type of tectonics as surge tectonics. Tobisch and Paterson (1990) used this term in two or three areas of southeastern Australia. Their usage is close to our own.

A second, but earlier employment of the term surge in tectonics is from Coward (1982). He noted during studies of the Moine thrust system of northwestern Scotland that the central parts of thrust sheets commonly are thrust farther than the ends of the thrust sheets, and likened the extended central part of the sheet to a surge (as in a breaking wave, the central part sweeps—surges—forward). This usage subsequently was retained and applied to the British and Irish variscides by Murphy (1985, 1990), Coward and Smallwood (1984), and Cooper et al. (1986).

The third and as far as we can determine, oldest use of the term in tectonics in recent literature was by Meyerhoff and Meyerhoff (1977). They proposed that asthenosphere surges (1) from beneath the Asian continent, (2) between North and South America, and (3) between South America and Antarctica produced the eastward-facing island arcs in the three regions. The idea was used subsequently to explain the complexities of Caribbean tectonics (Morris et al., 1990). Morris et al. (1990) used the term surge tectonics (coined by Bruce D. Martin). The paper was written during 1987–1988 and was submitted to and accepted by the Geological Society of America in 1988. Regardless, the Paterson et al. (1989) use of the term in print precedes by five months that by Meyerhoff et al. (1989). The term surge tectonics in the same sense that it is used in this paper was employed by Taner and Meyerhoff (1990).

Further evidence for surge tectonics

Seismotomographic images—Seismotomographic images of the Earth's interior were published by Dziewonski and Anderson (1984), Woodhouse and Dziewonski (1984), Dziewonski and Woodhouse (1987), and Grand (1987). These images are important in many ways, at least three of which are noteworthy.

1. Figure 8 shows the distribution of the principal hot and cold regions in the depth range 50–150 km, as determined from seismic-refraction velocities. Note that (a) hot and cold regions cross the continent-ocean boundary at will, with no obvious relation between these regions and other tectonic features; (b) only in southern Asia, the western Pacific, and (possibly) in part of western South America is some relation seen between hot and cold regions and the positions of Benioff zones; (c) hot regions, as might be expected, straddle the midocean ridges; and (d) images from the Pacific basin show a nearly continuous ring of hot upper mantle surrounding it. It should be noted (Fig. 8) that this hot upper mantle ring is west of the Benioff zone in both the western and eastern Pacific. In the former, it occupies mainly backarc basins; in the latter, hot upper mantle is mainly beneath the East Pacific Rise. Unlike the western Pacific, however, the hot upper mantle of the eastern Pacific everywhere crosses the Benioff zone except west of Peru (Fig. 8).

These observations are explained most easily by Earth rotation. We believe the Benioff zone in the western Pacific—and west of Peru in the eastern Pacific—is a barrier to eastward asthenosphere movement. It is only in these places that deep strictosphere Benioff zones are present (Benioff, 1949, 1954). They are not known elsewhere (Benioff, 1949, 1954; Isacks and Barazangi,

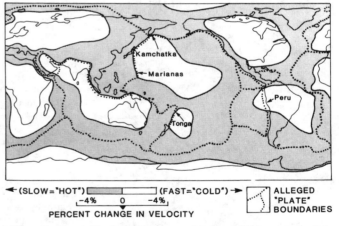

Figure 8. Generalized seismotomographic map of the upper mantle in the 50–150 km depth range (adapted from Woodhouse and Dziewonski, 1984). Low-velocity hot zones are shaded. The figure illustrates several important points: (1) ancient craton areas generally are underlain by high-velocity (cold) mantle (see Fig. 4); (2) convection-cell geometry is absent as illustrated by the facts that hot upper mantle is associated with the midocean ridge system of the eastern Pacific and with volcanic arcs in the western Pacific; (3) large areas of the ocean basins—not just continents—are underlain by cold upper mantle; (4) both hot and cold regions cross continent-ocean basin boundaries; (5) the Benioff zone in the western Pacific, especially in the Marianas, Kuril-Kamchatka, and Tonga-Kermadec arcs, is a barrier to eastward lithosphere flow (see text); and (6) cratonal areas also are barriers to eastward flow. Published with permission of J. H. Woodhouse.

1977). In most of the eastern Pacific, hot upper mantle can cross the Benioff zones except west of Peru where a strictosphere Benioff zone is present (Figs. 5, 8). In the eastern Pacific, the principal dams inhibiting eastward flow consist of the deep cratonic roots of North and South America (Figs. 4, 8). The effectiveness of the cratonic roots as dams can be judged by the fact that Figure 8 shows hot upper mantle connecting the Pacific with the Atlantic only beneath the Caribbean and Scotia seas. The deep strictosphere Benioff zones and craton roots acting as dams, have backed up hot upper mantle in parts of both the western and eastern Pacific basins. We believe that this interpretation fully explains the patterns of hot and cold material shown in Figure 8.

2. Figures 9 and 10 show seismotographic features of southwestern China. Figure 9 shows what we interpret to be a main, or trunk surge channel striking north-northwest to south-southeast beneath the Yunnan Himalayas, the Hengduan Range (Shan). Three subsidiary channels take off from the eastern side of the main Yunnan channel.

These images (Figs. 9, 10) are part of a larger project in which all of continental China to a depth of 1,100 km was surveyed seismotomographically (Liu Futian et al., 1989; Liu Jianhua et al., 1989). Liu Futian et al. (1989) discovered a network of shallow, interconnected pipelike or tubular bodies beneath the whole of China. (Figs. 9, 47). Above the strictosphere, these pipelike or tubular bodies range in width from 30 to 500 km and in

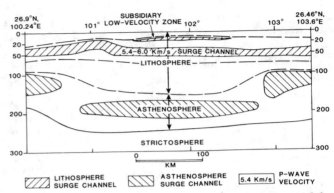

Figure 10. West-east seismotomographic cross section of part of the Yunnan surge channel, Himalaya foldbelt (Hengduan Shan foldbelt), southwestern China. The location is shown on Figure 9. Note the presence of a shallow low-velocity channel above the surge channel. This is an example of a surge-channel complex (after Liu Jianhua et al., 1989).

height from 12 to 30 km. Similar hot tubular—actually, lens-shaped—bodies were observed around the Earth by Woodhouse and Dziewonski (1984), Dziewonski and Anderson (1984), Dziewonski and Woodhouse (1987), and Grand (1987).

3. The seismotomographic images published by Dziewonski and Anderson (1984), Woodhouse and Dziewonski (1984), Dziewonski and Woodhouse (1987), Grand (1987), Liu Futian et al. (1989), and Liu Jianhua et al. (1989) demonstrate quite unmistakably that convection patterns like those predicted in plate tectonics do not exist, as inspection of Figures 4, 8, 9, and 10 shows. For example, the Juan de Fuca Ridge hot zone (Fig. 4) does not extend significantly below 300 to 400 km; the hot zone under Iceland does not extend below 350 km; a continuous band of hot mantle under the North American craton, as required in cellular convection, is absent; and so forth. A recent study of the Mid-Atlantic Ridge with long-period Rayleigh waves confirmed the seismotomographic image shown on Figure 2 (Moquet and Romanowicz, 1990). In addition, Sandwell and Renkin's (1988) studies of the geoid showed, contrary to widespread belief, that there is no reflection in the geoid of large-scale mantle convection cells.

We realize the subject of convection is broad and controversial; it is also largely theoretical insofar as the Earth's interior is concerned. As long as data did not contradict convection, it was convenient to invoke convection as a mechanism, if one was needed. Today, great volumes of data espe-

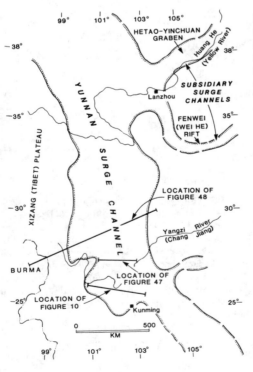

Figure 9. Seismotomographic map (at approximately 50 km depth) of surge channels in southwestern China. The principal channel (continental trunk channel) is the north-northwest-striking Yunnan surge channel that underlies the Himalaya foldbelt (Hengduan Shan foldbelt) in Sichuan Province. Three smaller channels branch eastward (branch channels) from the Yunnan channel. All three are horst-and-graben (rift) zones. All three rifts and the Himalaya foldbelt are tectonically active (see Fig. 48). Hence, this figure shows that foldbelts and rifts have a common origin. If surge channels do exist, this figure is proof that both compressive and tensile structures form above them (after Liu Jianhua et al., 1989). Locations of Figures 10, 47, and 48 are shown.

cially those from seismotomography, are becoming available. No evidence for large-scale convection is evident in these data. To do the topic justice, we reserve our detailed treatment of convection for another publication in which we can give it the attention it deserves (Meyerhoff et al., submitted).

We end this section by noting that seismic tomography provides yet another data set that is unexplained in any Earth-dynamics hypothesis proposed to date. The use of tomographic data in Earth-dynamics solutions has been extremely limited, even though the data have been available for nearly a decade. It goes almost without saying that the results of seismotomography must be incorporated in any successful tectonics hypothesis. Every discovery made with seismotomographic techniques to date has provided strong support for surge-tectonic interpretations; that is, the hot and cold zones seen in seismotomographic images are in all cases (within the upper mantle) just where surge tectonics predicts they should be.

Sonographs—We introduce the topic of deep-ocean sonography to summarize briefly the conclusions of Meyerhoff et al. (this volume) on the origin of midocean ridges. Figure 11, illustrating parts of sonographs from the Mid-Atlantic Ridge and the East Pacific Rise, shows that midocean ridges are characterized by faults, fractures, and fissures parallel with the strikes of the ridges. Comparison of Figure 11A with 11B shows this statement is true at widely different scales. The appearance of the

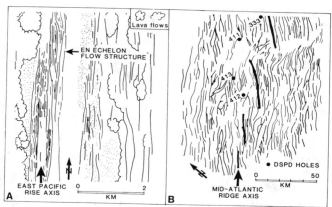

Figure 11. Faults, fractures, and fissures paralleling the crests of the **A**—East Pacific Rise (from Crane, 1987) and **B**—Mid-Atlantic Ridge (from Laughton and Searle, 1979). Note that the mechanism producing these ridge-parallel fractures operates at all scales (compare scale of A with that of B). The ridge-parallel fractures, following Stoke's Law, are analogous to streamlines in flowing liquids, suggesting that their cause is the flow of magma beneath the ridges as proposed by Meyerhoff et al. (this volume). Powerful support for this interpretation is the presence along strike of en echelon flow structures, one of which appears on Figure 11A. Fault, fracture, and fissure systems that almost certainly originated in the same way are illustrated on Figures 33, 38, 55–57, 59, 61, and 75.

fault-fracture-fissure system is nearly identical with that of fractures produced in accordance with Stokes's Law by laminar flow and viscous drag in ice streams and glaciers (Meyerhoff et al., this volume). Their orientation indicates the flow beneath the ridges parallels the ridges, and is not at right angles to them as required in plate tectonics. En echelon structures like that in Figure 11A also demonstrate ridge-parallel movement. Thus, plate tectonics's most fundamental assumption seems to be fatally flawed.

Reaching the surge-channel concept

From the preceding, one can see why we could not support any existing Earth-dynamics hypothesis, and how we reached the surge-channel concept. First, we listed or referred to some 23 data sets, or types of data that are not explained in current hypotheses. Second, the evidence for the existence in the lithosphere and uppermost mantle of pipelike, tubular, or lenslike hot bodies (e.g., diapirs) that are interconnected, is now overwhelming; most important, these lenslike tubes interconnect (Fig. 2). The Chinese data (Figs. 9, 10), in addition to proving the interconnections among lenslike bodies, demonstrate that they underlie foldbelts, rift zones, and strike-slip belts alike. The association of all features with high heat-flow belts, microearthquake zones, linear fault, fracture, and fissure systems—among many phenomena—suggested a common origin for all tectonic belts, whether foldbelt, rift zone, or strike-slip belt. The linear features indicate that motion in the lenslike bodies parallels them. If the lithosphere is under compression as it must be in a cooling earth, any mobile material within those lenslike bodies, or channels, must at times surge as the compression in the lithosphere decreases and increases. Hence, we adopted the term surge channel for these hot lenticular bodies. Because they underlie foldbelts, rift zones, and strike-slip belts alike, we adopted the term surge tectonics. One reviewer of a very early (January, 1989) version of this paper jokingly referred to the network of lithosphere surge channels as "the Earth's cardiovascular system." As a description of lithosphere structure and surge-channel function, the analogy is very apt.

Role of gravity

The principal forces acting on Earth's lithosphere are compression, rotation, and gravity. We have described the first two, and now endeavor to describe the last. Gravity controls the depth of each channel in the lithosphere, whether it is a single channel (simple surge channel) or a complex of more than one channel (surge-channel complex). Rising magma, presumably differentiated in the asthenosphere as a result of the Earth's cooling, is less dense than the rocks surrounding it. Consequently, Newton's Law of Gravity comes into play to bring the rising and lighter magma to a level where it is stable. The force that brings the lighter magma to its level of stability is the Peach-Köhler climb force (Weertman and Weertman, 1964; Weertman, 1971). This response to gravity compels the less dense magma to rise to its gravitationally stable level where the magma density equals that of the surrounding rocks (called the neutral buoyancy level [Gilbert, 1877; Gretener, 1969; Ryan, 1987; Corry, 1988; Walker, 1989]). Gilbert (1877) was the first to work out the theory in his classic Henry Mountains laccolith study. Once at its gravitationally stable level, the magma can move only laterally.

What determines the positions of surge channels in the ocean basins and continents? We have suggested elsewhere (Meyerhoff et al., this volume) that midocean ridges (which are giant surge channels) form in response to the Earth's cooling and resultant circumferential shortening. The uparching of the relatively thin ocean floors that should accompany circumferential shortening provides obvious zones of weakness that rising magma can exploit. Away from the midocean ridges, previous zones of weakness are the most likely places to accommodate the formation of new channels. Keep in mind that during the geotectonic cycle, compressive stress in the lithosphere is minimal after tectogenesis. The ruptures produced during tectogenesis and the reduced compression should facilitate surge-channel formation.

The continual rise of magma from the asthenosphere into the lithosphere has another effect, in our opinion. It provides a logical means for transporting heat from the interior, thereby carrying on the Earth's cooling process. A nearly identical conclusion was reached by the 101-member Basalt Volcanism Study Project (1981). They wrote that the Earth had most likely cooled during its entire history, and that much of the cooling was, and still is being, accomplished by the rise and release of basalt from the deep interior.

Kobergens and bilateral tectogenesis

Since our discovery of surge channels in December 1987, we have found both active and fossil ones beneath every conceivable large tectonic feature—midocean ridge, aseismic submarine ridge, oceanic rise, linear island and seamount chain, eugeosyncline and volcanic arc, foldbelt, wrench-fault system, long-linear to curvilinear valley, actively rising mountain chain and plateau, and intracratonic basin. The presence of a channel beneath such diverse structures and landforms indicates that a single geological process or mechanism produces all tectonic phenomena. This process generates compression, tension, and shear both (1) at the same time within a tectonic belt and (2) at different times in any one part of a tectonic belt. A mechanism for such diverse activities was formulated early in this century by the eminent Austrian geologist, Leopold Kober (1921, 1925, 1928).

In 1911, Kober (1921) observed that from Gibraltar to Iraq, the Alpide foldbelts are deformed bilaterally with the northern ranges vergent toward Europe and the southern toward Africa (Fig. 12). About the same time, Chamberlin and Richards (1918), Keith (1923), and Woodworth (1923) made a similar discovery about the Appalachian chain—that it verges northwestward on the west and southeastward to eastward on the east. This discovery led to Chamberlin's (1927) famous wedge theory of diastrophism. We have named each bilaterally deformed belt kobergens in honor of Kober (1921), not Chamberlin and Richards (1918), because from the beginning Kober understood the potential role that a magma chamber (equivalent to our surge channel) in the middle of the foldbelt might play in its deformation (Fig. 12). This role was not foreseen by Chamberlin and his colleagues, although hints of Kober's ideas appeared later in Chamberlin and Link's (1927) classic paper on the vertical extent and shape of batholiths.

To the best of our knowledge, the late Chester R. Longwell (Yale and Stanford universities) originated the term kobergen, a term that describes bilaterally deformed surge channels-foldbelts (Longwell's class notes by AAM, 19 April 1946). He coined it as a play on the word orogen and the surname Kober. (The term orogen was conceived and defined first by Professor Kober.)

The bilateral, or kobergenic style of tectogenesis was the basis of most of Kober's (1921, 1925, 1928) work. Although the bivergent style of the Alps-Dinarides had been known for many years (e.g., Suess, 1875, 1901), it was not regarded as a fundamental manifestation of tectogenesis. Argand (1916, 1924), Collet (1927), and others interpreted the southward-directed thrusts of the Dinarides as backthrusts that developed very late in the Alpide tectogenesis. Most later workers accepted Argand's and Collet's interpretation (e.g., King, 1951; Rich, 1951; Glangeaud, 1957, 1959, Trümpy, 1960). Rich (1951), however, noted that the bivergent style—or fan structure as he termed it—was characteristic of many foldbelts around the world. Glangeaud (1957, 1959) classified bivergent foldbelts (*chaînes biliminaires*) as one of several geosynclinal types. Aubouin (1965), following Glangeaud (1957), classified geosynclines into three categories, one of which is biliminal geosynclines, which subsequently are deformed into bivergent foldbelts.

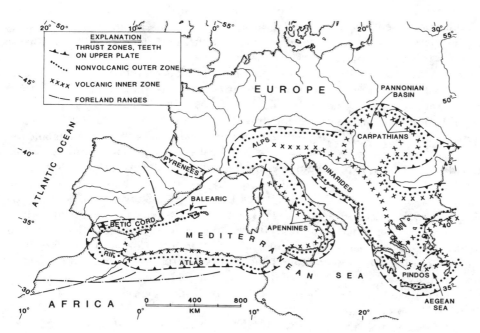

Figure 12. Bivergent Alpine foldbelts of the Mediterranean region according to van Bemmelen (1933). This figure illustrates how common bivergent foldbelts actually are and why they were so important in tectonic hypotheses published by Kober (1921, 1925, 1928), Glangeaud (1957, 1959), and many others.

Today, such foldbelts receive scant attention. In six modern textbooks on structural geology, we found only one discussion of bivergent foldbelts (Miyashiro et al., 1982) in a review of the historical importance of Kober's contributions to geotectonics.

The kobergen style of tectogenesis has been duplicated in several laboratory model studies. Among them, the most notable are those of Ramberg (1967, 1972, 1973), Emmons (1969), and Merle and Guillier (1989). Ramberg's (1973; his figs. 7, 10, 13, 15, 17) models duplicate alpinotype structures in every important detail, and without exception produce bivergent foldbelts. A compressional modeling study by Merle and Guillier (1989) produced the same results. These experiments, plus those of Emmons (1969), show the close interrelations among compressive stress, tension, and shear. In fact, the experiments by all three sets of authors show a complete intergradation between structures produced by compression, shear, and tension.

We define kobergen as a bilaterally deformed foldbelt above a surge channel (Fig. 13). Such bilaterally deformed foldbelts have been known by many names. Among them are bilaterally deformed fold-belts, biliminal foldbelts, bisymmetrical foldbelts, bivergent foldbelts, bivergent orogens, bivergent structural fans, divergent thrust belts, fan structures, flower structures, floaters, imbricate fan structures, inversion structures, mushroom structures, opposed-dip thrust complexes, pop-ups, structural fans, triangle-zone deformations, wedge complexes, and wedges (or tectonic wedges). Krebs (1975), in one of several prescient papers on tectonics, called them divergent, bilateral symmetrical mountain belts. Use of the term symmetrical, however, is not recommended, for few of them are, each flank of a kobergen having its own structural configuration.

We determined that kobergens form an evolutionary series or spectrum from continental-margin kobergens where part of the lithosphere is relatively thin, to continental-interior kobergens where the lithosphere is relatively thick. At one end of the spectrum is the alpinotype kobergen (Haug, 1900; Brunn, 1961, 1979; Aubouin, 1965; and many others). This kobergen is deformed in classic alpinotype styled with nappes, thrust sheets, metamorphic belts, flysch, molasse, ophiolite suites (Steinmann, 1905), and mélange (Fig. 13; see also Figs. 28, 29, 35, 41, 44, 45, 48, 50, 51, 63-65, 86-89, 95, and 96). At the other end of the spectrum is the germanotype kobergen exemplified by the High Atlas (Fig. 52), Sierra de Perijá (Colombia and Venezuela), Mérida Andes (Venezuela), and other ranges. Germanotype foldbelts have block faults, drape folds, and mildly to sharply compressed sedimentary sequences (of cratonal origin), which in many areas overlie more severely deformed foldbelts, some of them of alpinotype. As a rule, the alpinotype belt forms in association with island arcs or at the continent-ocean interface where the overlying lithosphere, especially on the oceanic side, is thin.

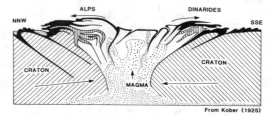

Figure 13. Kober's (1925) mechanism—compression of a magma chamber (equivalent to surge channel)—for bivergent foldbelt deformation. This fan structure is what we call kobergen and is the core of surge tectonics.

The germanotype belt forms in the much thicker lithosphere of continental interiors.

The two end members form evolutionary sequences through both time and space. The time-evolutionary sequence begins as an alpinotype foldbelt, which through repeated tectogenesis, thickens the lithosphere. As it does so, the alpinotype belt evolves into a germanotype belt. A space-evolutionary sequence involves a geosynclinal basin that in part of its length is at the continent-ocean interface (where an alpinotype belt forms during tectogenesis) and in another part of its length is in a continental interior (where a germanotype belt forms during the same tectogenesis).

We have reviewed all or parts of 95 foldbelts worldwide. These range in age from Proterozoic through the Cenozoic. Of the total, 88 are deformed bilaterally and therefore are kobergens. Of the remaining seven, all probably are kobergens, but in each case one flank is hidden by younger sedimentary rocks or the ocean. Examples of the seven include the Ouachita-Marathon foldbelt of southeastern North America, the Franklinian foldbelt of Arctic North America, the Mauretanides of West Africa, and the Cape foldbelt of southern Africa. The Urals foldbelt, commonly thought of as exhibiting only westward vergence (e.g., Gorokhov and Sharfman, 1963; Hamilton, 1970; Nalivkin, 1973) is a kobergen (Kober, 1928; Perfil'yev, 1979a, Pushcharovskiy et al., 1988).

Where bivergent structure has been mapped, the thrust system directed toward the foreland is assumed to be the dominant direction of movement. The complementary, or oppositely verging thrust system, generally is attributed to backthrusting within one-sided (monovergent) tectogenesis (e.g., Rich 1951; Elliott, 1981; Coward et al., 1986). In each foldbelt from which backthrusting has been reported, we found a bivergent foldbelt, or kobergen. The few cases of documented backthrusting that we investigated are localized areas that comprise a very small part of the total foldbelt.

Our observations suggest that identification of backthrusts must be done with the stringent use of carefully selected field criteria, some of which were outlined by Rich (1951). One criterion is that backthrusts are short—no more than a few tens of kilometers—because local factors produce them. A second criterion is that rupture in the backthrust zone is irregular, crosscutting facies and other geological boundaries (Rich, 1951). This is true because the causative factors are of local extent and rocks in the hanging wall are not pressed strongly and uniformly against the backthrust fault plane and footwall (Rich, 1951).

In plate tectonics, most zones attributed to backthrusts extend more than 90% of the length, commonly the full length of the foldbelt. They parallel the overall trend of the regional structure, including that of the foreland thrusts. This last observation indicates that the cause(s) of foreland thrusts and the so-called backthrusts is (are) the same.

Closely related is the fact that the rocks of alpinotype foldbelts are in parallel isopic (dominated by one lithologic type) facies belts (Aubouin, 1965). In the foreland, thrust faults generally form the borders of such isopic belts, a fact which shows that the belts and the thrusts are genetically related. Fault-bounded isopic belts are not limited to the foreland; they also characterize the opposite side of every kobergen, that is, the faults that have been called backthrusts also bound isopic belts. This observation, which applies in alpinotype foldbelts the world over (e.g., Trümpy, 1960; Aubouin, 1965), indicates that both the foreland thrusts and the so-called backthrusts are rooted together in the central part of the foldbelt in which they occur. If this is true, then the term backthrust as currently used is in most cases a misnomer.

From the preceding, it is evident that bivergent tectogenesis is a principal identifying characteristic of a deformed surge channel. Kobergens therefore, are one of the most important structures of surge tectonics.

Classification of surge channels

It is premature at this time to attempt an in-depth classification of surge channels, because our knowledge of them is inadequate. We know enough to permit some generalized observations. An important point to note is that although mainly active surge channels are discussed, many of the statements also apply to inactive surge channels. Several points are paramount in any consideration of these channels.

Surge channels are the principal mechanism of Earth cooling. As it cools, the asthenosphere deepens and constantly differentiates into light and heavy fractions. As we outlined, the lighter fractions rise to form surge channels in the lithosphere in response to gravity.

The positions of surge channels are controlled by (1) the position of the deep roots of continental cratons, (2) the locations of the Benioff zones, (3) proximity to nearby active surge channels, (4) lithosphere thickness, (5) accessibility of zones of lithosphere weakness, and (6) Earth's rotation.

In support of these statements, we note that active surge channels are deflected around the regions where continental roots are deepest (Figs. 4, 8; Woodhouse and Dziewonski, 1984; Dziewonski and Woodhouse, 1987; Grand, 1987). The concentration of broad bands of high heat flow west of the Benioff zone in both the western and eastern parts of the Pacific basin (Fig. 8)

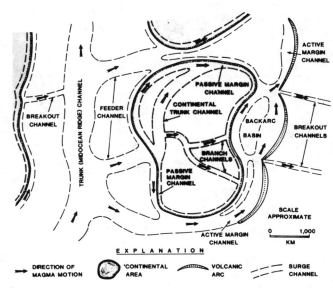

Figure 14. Tentative classification of surge channels. This figure should be used in conjunction with Table 4 and the descriptions and examples in the text.

shows how efficient the strictosphere Benioff zone is at containing the eastward flow in the asthenosphere. The concentration of active channels in Phanerozoic tectonic belts demonstrates the affinity of surge channels for regions that have thinner lithosphere and linear zones of structural weakness (Fig. 2). In general, the largest channels develop where the lithosphere is thinnest. In those places where a surge channel underlies an ancient craton, it is in a part of the craton that has been appreciably thinned by erosion (e.g., Fig. 42). Our reasons for these statements become evident as we present different surge-channel examples from around the globe.

Figure 14 and Table 4 provide a tentative classification of surge channels. Although the table gives the impression that a simple classification can be formulated, Figure 14 gives some insight into the complexity involved. As an example, Figure 14 shows that the channels can and do change from one type to another along trend.

Characteristics of surge channels

Table 5 lists the major tectonic features produced by surge channels during their histories. The list combines data sets explained by plate tectonics with those not explained by plate tectonics. Both geological and geophysical criteria are listed, not only for active surge channels (Fig. 2) but for inactive ones as well.

Figure 2 shows only active surge channels. To locate inactive ones, more information of the types outlined in Table 5 is needed. It should be stressed that a complete gradation exists from the criteria that identify active channels to those that characterize inactive ones. Rather than belabor these criteria, we present actual examples following the outline of Table 4 and discussing the criteria listed in Table 5.

Table 4. Tentative classification of surge channels (see Fig. 14).

Category	Comments
A. Ocean-basin surge channels	
1. Trunk channels	These are the midocean ridges and their principal branches. They are thousands of kilometers long and can be 1,000 to >3,000 km wide.
2. Feeder channels	These are large branches of trunk channels. They are thousands of kilometers long and are typically 200 to 600 km wide. They become the continental trunk channels. They carry magma from midocean ridges to the adjacent continent *east* of a midocean ridge. They invariably branch from the *eastern sides* of midocean ridges (Fig. 54).
B. Surge channels of continental margins	
1. Passive margin channels	All continental margins appear to be underlain by active surge channels. In fact, surge channels are one of main *raisons d'être* of continental margins.
2. Active margin channels	These are very large channels that underlie backarc basins. They are quite high and are dammed by Benioff zones (Figs. 8, 45).
3. Breakout channels	These are long linear channels, 20 to 150 km wide that typically underlie "linear island and seamount chains." They grow eastward (Figs. 42, 43), commonly showing eastward age progression. They usually are short-lived. Their western terminus typically is at the cusp where two volcanic arcs meet.
C. Continental surge channels	
1. Trunk channels	See ocean-basin feeder channels.
2. Branch channels	These are small channels 20-150 km wide, several hundred kilometers long, branching from the *eastern* sides of trunk channels.

Table 5. Characteristics of surge channels. Explanation of symbols: x = Yes, applicable; - = No, not applicable; ? = unknown, general for lack of data; * = linear valleys and basins generally form as a surge channel loses its contents through tectogenesis and other causes; + = + and - negative anomalies over tectonic belts have some significance, but we have not studied these sufficiently.

Tectonic feature generated by channel	Channel status		Tectonic regime			Figure number	Additional examples
	Active	Inactive	Oceanic	Transitional	Continental		
Linear to curvilinear lithosphere breaks							
Long linear zones of faults, fractures, and fissures	x	x	x	x	x	11	Figs. 20, 21, 33, 55-57, 61, 75-77
Horst-and-graben complexes, rifts	x	x	x	x	x	54	Figs. 1, 9, 11, 38, 40, 52, 55, 71, 78
streamline (strike-slip) fault, suture, *Verschluckungzone*, tectonic line	x	x	?	x	x	90	Figs. 28, 33, 34, 36, 51, 60, 65, 63-65, 77, 91-93
Thrust faults	x	x	x	x	x	49	Figs. 30, 36, 61
Horsetail structures	x	x	?	x	x	59	—
Eddies and vortex structures	x	x	x	x	x	76	Figs. 21, 77-77
Structures related directly to lithosphere breaks							
Magmatic arcs, eugeosynclines, alpino- and germanotype foldbelts	x	x	x	x	x	1	Figs. 5, 9, 12, 21, 30, 36, 38, 40, 49, 52, 78, 81
Tectonostratigraphic terranes	x	x	-	x	?	36	—
Fissure eruptions and volcanic fields	x	x	x	x	x	10A	—
Aligned plutons	x	x	x	x	x	49	Figs. 56, 81
Lines of thermal springs	x	-	x	x	x	78	Western United States (Eaton et al., 1978)
Kimberlite dikes, diatremes, ring complexes	-	x	?	x	x	—	South Africa
Dike swarms	x	x	x	x	x	—	Eastern United States (McHone et al., 1987)
Ophiolite belts	x	x	x	x	x	81	—
Mélange belts	x	x	x	x	x	—	Ankara mélange (Norman, 1984)
Metamorphic belts (mantled gneiss domes, core complexes, inverted gradients)	x	x	?	x	x	90	Figs. 1, 44-46
Submarine ridges (aseismic ridges)	x	x	x	-	-	22	Figs. 1, 11, 20-24, 80
Structure indirectly related to lithosphere breaks							
Fold trends	x	x	x	x	x	40	Figs. 38, 49
Stretching lineations	x	x	?	x	x	—	Himalayas (Brunel, 1986)
Mineral belts	x	x	x	x	x	—	Colorado Mineral belt (Tweto, 1975)
Morphotectonic features							
Linear river courses	*	x	-	?	x	59	Figs. 56, 57, 79
Linear topographic divides	x	x	x	x	x	38	Figs. 1, 40, 52
Basins, linear evaporite basins	*	x	?	x	x	83	Figs. 53, 56-58, 78, 79
Plateaus	x	?	x	x	x	-	Tibet Plateau
Midocean ridges	x	x	x	-	-	16	Fig. 2
Feeder channels (oceanic rises)	x	x	x	-	-	22	Figs. 1, 23
Breakout channels (linear island and seamount chains)	x	x	x	-	-	24	Figs. 1, 23, 80
Geophysical characteristics							
+ Bouguer gravity anomaly	-	x	x	x	x	79	Figs. 45, 51
- Bouguer gravity anomaly	x	-	x	x	x	15	Figs. 18, 29, 32, 51, 53
Midocean ridge magnetic anomalies[+]	x	?	x	-	-	18	Figs. 20, 21
High-conductivity, low-resistivity magnetotelluric anomaly	x	?	x	x	x	34	Fig. 54
Bands of microearthquakes	x	-	x	x	x	42	Figs. 32, 35, 57, 60, 73, 78
High heat flow bands (>55 mWm^{-2})	x	-	x	x	x	49	Figs. 2, 18, 32, 35, 42-44, 51, 54
Bands of "anomalous upper mantle" (P-wave velocity 7.0-7.8 km/s) These bands form lenses on:	x	x	x	x	x	16	Figs. 18, 19, 26, 27, 31, 43, 44, 47, 54, 58, 62, 79
(a) refraction-seismic lines	x	x	x	x	x	45	Figs. 31, 43, 48, 58, 79
(b) reflection-seismic lines	x	x	x	x	x	68	Figs. 65-67
Interior of reflection-seismic lens is transparent	x	-	x	x	x	67	Fig. 65
Interior of reflection-seismic lens is filled with reflectors	-	x	x	x	x	68	Fig. 66

Ocean-basin surge channels

Table 4 shows two types of surge channels in the ocean basins (see also Fig. 14). The principal type constitutes the gigantic trunk channels, or midocean ridges in which a great deal of magma, or magmatic "mush," must be present. The smaller channels that branch from the trunk channels only along their eastern sides seem to be conduits feeding the western margins of the continents (Figs. 2, 14). We call these feeder channels (e.g., Fig. 22). Their eastward continuations from the ocean basins comprise the principal type of surge channel beneath continental masses, where we call them continental trunk channels. In places (e.g., Gulf of California, Fig. 35), the trunk channel crosses the ocean-continent interface, and is itself a feeder channel.

A third type of surge channel occupies large parts of the ocean basins, particularly in the western Pacific basin. These are breakout channels. We have classified these as surge channels of the continental margin, because they owe their existence to features commonly associated with such margins (e.g., Benioff zones).

Mid-Atlantic Ridge—The Mid-Atlantic Ridge is the best known of the ocean's trunk channels. Shortly after Revelle's (1958) discovery of 7.0–7.8-km/s anomalous upper mantle in the southern part of the East Pacific Rise, Ewing and Ewing (1959) showed that such a layer is also present on the Mid-Atlantic Ridge and that it pinches out on either flank (Fig. 15). Where the anomalous upper mantle is absent, crust with a P-wave velocity of 6.4 to 6.8 km/s directly overlies normal mantle (8.0–8.5 km/s). The overall shape of the anomalous upper mantle bodies under the Mid-Atlantic Ridge is that of a thin, up to 30-km-thick lens in the axial zone of the ridge, up to 1,000 km wide, and thousands of kilometers long.

Figure 15, from Talwani et al. (1965), shows three interpretations of Mid-Atlantic Ridge structure based on refraction-seismic and gravity data. These sections are based on data from 27° to 47° N latitude. The models shown here are greatly simplified, as Pavlenkova's (1989) later work has shown. However, they represent some truly pioneering work that has stood the test of time.

Figure 16 from White (1989) is a seismic-refraction section parallel with the Mid-Atlantic Ridge axis between 37° and 40° N latitude, the same area as Figure 15, but at right angles to the latter. As Meyerhoff et al. (this volume) noted, Figure 16 shows the pinch-and-swell geometry that characterizes all midocean ridges. Yamagishi (1985) pointed out that the same pinch-and-swell geometry typifies all lava flows and tunnels (Fig. 17), and is produced by the surging of lava through the tube, or tunnel. The pinch-and-swell structure exhibited in Figure 16 is further evidence that ridge-parallel flow produced the midocean ridges.

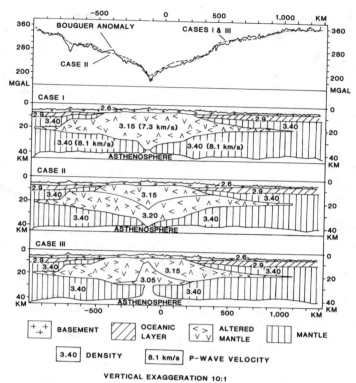

Figure 15. Three possible crustal models across the Mid-Atlantic Ridge (between 27° and 47° N latitude). All three satisfy the gravity data—at the top of the figure—and the seismic refraction data (Le Pichon et al., 1965). The numbers in the 2.6 to 3.40 range are density values; numbers 7.3 and 8.1 are P-wave velocities. We regard this figure as our "type locality," for midocean ridge surge channels (from Talwani et al., 1965).

A low-velocity layer almost certainly is present in the interior of the 7.0–7.8-km/s lens. This is supported in part by a band of microearthquakes that coincides very closely with the high heat-flow anomaly at the ridge axis. Bouguer gravity maps show a negative value at the ridge crest. The negative values are greatest between transform faults, and least at the fault zones themselves (Lin et al., 1990), thereby providing strong support for our interpretation.

Farther south along the Mid-Atlantic Ridge, at 11° S latitude, Pavlenkova (1989) and several colleagues collected detailed gravity, magnetic, heat-flow, and seismic-refraction data. Figure 18 shows a more complex internal ridge structure than Figure 15. Here, at 11° S, the ridge is 1,000 km wide and the lens of 7.7–7.8-km/s material is nearly as wide as the ridge itself at depths of 26–36 km. Above this zone are two more low-velocity zones, 450–600 km wide. A distinct heat-flow anomaly is

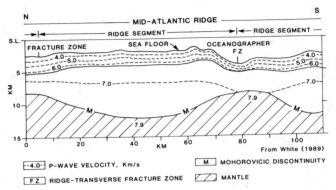

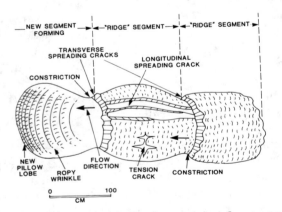

Figure 16. North-south refraction-seismic section showing the velocity structure of the Mid-Atlantic Ridge between 40° and 37° N latitude. Note how the 7.0–7.8-km/s layer thins essentially to 0 thickness where large fracture zones transect the midocean ridge. The purpose of this figure is to show the morphological similarity between the longitudinal section of a midocean ridge and that of a lava tube (Fig. 17). Figures 16 and 17, in our opinion, provide an excellent reminder that physical laws operate at all scales (another example is on Fig. 11). We suggest that the ridge segments between the major fracture zones are mechanical equivalents of the basalt pillows between transverse spreading cracks (constrictions) of Figure 17. If the analogy is correct, this constitutes further evidence that flow in the surge channels is ridge-parallel. Published with permission of the Geological Society (London).

present, and the Bouguer anomaly is nearly identical with that on Figure 15.

The lens of 7.0–7.8-km/s material shown on Figure 15 may be fatal to plate tectonics, a possibility recognized by Drake and Nafe (1968) and Vogt et al. (1969). Theoretically this material, if it is generated at the ridge crest, should be present everywhere in the ocean basins. Vogt et al. (1969, p. 595–596) discussed the significance of the lateral pinch-out, writing that if this anomalous material is produced during sea-floor spreading, "We would then require a physical process whereby the low-velocity [anomalous] mantle somehow becomes segregated into oceanic layer and normal mantle as it cools and withdraws from the axis." Drake and Nafe (1968, p. 185) expressed the opinion that the 7.0–7.8-km/s lenses were transient and their disappearance away from the ridge crests ". . . may be related to the changes in elevation associated with tectonic activity." The problem remains unsolved by plate tectonics, but is an unavoidable consequence of surge tectonics.

East Pacific Rise—The velocity structure of the East Pacific Rise is like that of the Mid-Atlantic Ridge, except the East Pacific Rise channel system is much broader than that of the Mid-Atlantic Ridge (Menard, 1960; Talwani et al., 1965). The greatest difference between the two ridges is the presence in the East Pacific Rise of very shallow low-velocity lenses with high electrical conductivity, possibly containing melt (Reid et al., 1977; Filloux, 1982). Figure 19 shows a recent interpretation of the geometry and velocity structure of one shallow lens (Harding et al., 1989). Unlike the Mid-Atlantic Ridge—at least in exploration to date—the shallowest low-velocity lens of the East Pacific Rise may extend to within 0.8 km of the sea floor (Caress et al., 1989). The presence of such shallow activity implies that magma in the principal surge channel beneath the East Pacific Rise crest is differentiating continuously. Such differentiation would explain the occurrences of extremely young andesite, dacite, rhyodacite, and other silica-rich

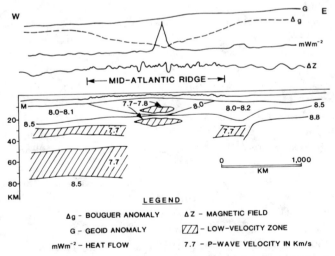

Figure 17. Schematic model of the characteristic surface structures on a lava tube. The flow direction is from right to left. Compare with Figure 16. We believe that the midocean ridges and lava tubes originate in similar ways. From Yamagishi (1985). Published with permission of Geological Society of America.

Figure 18. West-east refraction-seismic section across the Mid-Atlantic Ridge at 11° S latitude. Geoid, gravity, heat flow, and magnetic profiles also are shown. Note the close similarity between the gravity profiles of Figures 15 and 18. Figure 18 shows that the internal structure of a midocean ridge is much more complex than Figure 15 suggests. The low-velocity zone at 75–50 km is the asthenosphere; that at 37–28 km is the main surge channel; two small low-velocity channels underlie the ridge axis in the 30-12 km depth range. Together these low-velocity zones form a large surge-channel complex (from Pavlenkova, 1989).

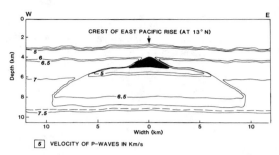

Figure 19. West-east velocity profile across a shallow magma chamber on the crest of the East Pacific Rise at 13° N latitude. This section is based on a series of expanding-spread seismic profiles and common-depth-point reflection lines (CDP). The solid black may be a zone of large melt percentage as determined from CDP lines by Harding et al. (1989). This figure illustrates a fundamental difference between the East Pacific Rise and other midocean ridges, namely, the presence of very shallow magma chambers. Another difference is that the East Pacific Rise is buoyed by an exceptionally large surge channel; it may be buoyed partly by upwelling asthenosphere as well (Fig. 8). We believe that this swell is produced by relatively unimpeded eastward flow across the broad Pacific basin. Dammed against the two American cratons, flow is forced between them, or farther poleward (Fig. 8). In contrast, in the Atlantic basin, asthenospheric swelling can take place only where a flow enters from the Pacific, mainly between 10° and 60° N latitude, but to some extent south of 30° S latitude.

volcanic products along the rise (Thompson et al., 1989).

Midocean ridge magnetic anomalies—Agocs et al. (this volume) demonstrated that magnetic anomalies of midocean ridges are explained better by magnetic-susceptibility contrasts than by magnetic field reversals, as proposed in the Morley-Vine-Matthews hypothesis (Morley and Larochelle, 1964; Vine and Matthews, 1963). Meyerhoff et al. (this volume) proposed that ridge-parallel fault, fracture, and fissure patterns of midocean ridges (Fig. 11) are the underlying cause of the linearity of magnetic anomalies. When new magnetic anomaly sets were found in the western Pacific—illustrated on Figure 20—sets that conflicted with others in the Pacific basin, an explanation had to be devised to conform with plate-tectonics concepts (Fig. 20). Larson and Chase (1972, p. 3641) concluded that the different anomaly patterns of Figure 20 ". . . must have been generated by a system of five spreading centers joined at two triple points." They stated that all correlations among the magnetic profiles were established "by eye" (p. 3631). No transformations to the pole were attempted, although the sets come from greatly different latitudes (52° N, 40° N, 5° N, Agocs et al., this volume). (That is, no adjustments were made to reduce the magnetic profiles to a common latitude so meaningful correlations might be attempted.)

In plate tectonics, an extremely complex tectonic history is required for the Pacific basins. In contrast, if the magnetic anomaly sets of the western Pacific were produced by once-active surge channels, a coherent and internally consistent pattern of flow (as shown by the magnetic anomalies) should emerge. To test this hypothesis, we chose the Phoenix anomalies (Fig. 20). We then updated the Weissel (1981) map of magnetic lineations of the western Pacific, and added the anomaly patterns discovered since 1980. The result is Figure 21, to which we added black arrows indicating our interpretation of flow directions. A coherent flow pattern is apparent. The pattern indicates flow from beneath Asia and Australia converges between the Yap and Tonga trenches, and continues east-northeastward beneath the Phoenix anomalies. The result shown on Figure 21 strongly supports surge tectonics. A carefully planned magnetic survey across the region would be a relatively inexpensive means of testing further this surge-tectonics model.

Additional data support the surge-tectonics model in this region. Furumoto et al. (1976) discovered a 20–42-km-thick 7.0–7.7-km/s layer beneath the Caroline basins, Ontong Java Plateau, and New Guinea geoidal high (Fig. 21; Table 5). This anomalous upper mantle lens falls in a band of high heat flow (Fig. 2; Gramberg and Smyslov, 1986), which also coincides with a well-defined band of microearthquakes (Hegarty and Weissel, 1988). In a study of the Ontong Java Plateau, Sandwell and MacKenzie (1989) found that the geoid-topography ratio of the Ontong Java Plateau region is

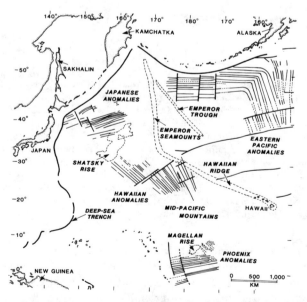

Figure 20. This figure, from Larson and Chase (1972), shows various magnetic-anomaly lineation sets in the northwestern Pacific basin. To account for these lineation sets, Larson and Chase (1972) proposed an excessively complete plate-tectonic model, which as Figure 21 shows, is unnecessary.

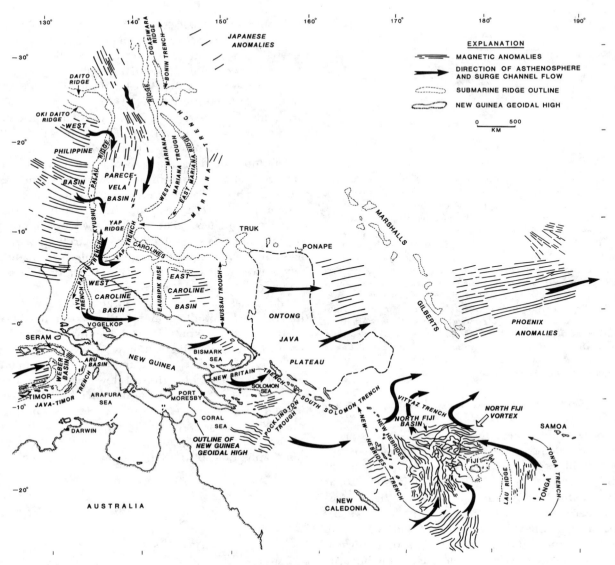

Figure 21. This figure, updated from Weissel (1981), suggests that the linear anomaly sets of Figure 20 are explained more simply by a surge-channel flow pattern. In this figure, we analyze the Phoenix set of lineations (Larson and Chase, 1972). Although magnetic lineations have not been studied in large areas of the Pacific, those that are known suggest that the Phoenix lineations overlie a large surge channel formed by the convergence of several smaller channels farther west. The Philippine basin and Parece Vela basin lineations suggest flow that converges at the Yap Ridge, and turns there toward the east. Note that, north of Yap, no material flows east, but is deflected southward. We believe the reason for this is the presence, north of Yap, of a strictosphere Benioff zone (see text and Fig. 8). Similarly, lineations in the western part of Indonesia are directed west-east, converging with the Parece Vela lineations in the West Caroline basin area. Lineations in the Bismark and Coral seas join the eastward-trending Caroline lineations, probably beneath and just east of the Ontong Java Plateau. East of Australia, north-to-south directed lineations collide in the North Fiji basin vortex structure, whence they turn eastward and northeastward in the Phoenix direction. We think that the Japanese and Hawaiian lineation sets (Fig. 20) have similar explanations. To sum up: the various magnetic lineation sets in the Pacific basin originated above different surge channels which were active at different times in the past.

too high to be explained by an isostatic compensation model. They concluded that it and the New Guinea geoidal high (Fig. 21) are sustained partly by thermal-buoyancy forces in the lower half of the lithosphere. This conclusion supports the surge-tectonics model.

Feeder channels—Feeder channels are much smaller than midocean ridges (Figs. 14, 22), typically 400–700 km wide and 2,000–4,500 km long (Fig. 22). Some once may have been branches or parts of midocean ridges. Pacific examples include the Nazca, Carnegie, Cocos, Tehuantepec, and Mendocino ridges. The East Pacific Rise itself enters North America via the Gulf of California and the Great Basin, and serves as a feeder channel. Atlantic examples are the Agulhas, Walvis (Fig. 22),

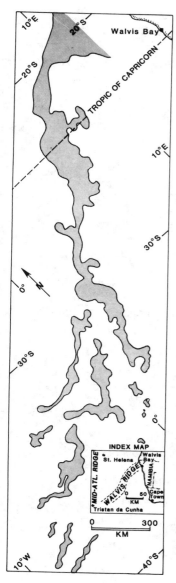

Figure 22. The Walvis Ridge feeder channel, South Atlantic Ocean. Note the sinuous, anastomosing, flow-like pattern (see Fig. 1). Eddylike features are prominent south of 30° S latitude. Compare this with Figures 24, 40, 59, 71, 72, 76, and 77.

Cameroon, Cape Verde, Azores-Gorringe, and Iceland-Faeroe ridges, rises, and plateaus. Indian Ocean examples include Broken Ridge and the Chagos-Laccadive Ridge.

Their morphology is quite complex (Fig. 22), but there is little evidence for the presence of transform faults. Many of them strike northeast-ward (Walvis, Cameroon, Nazca, Cocos, Tehuantepec). Most of them show a sinuous to anastomosing pattern. Figure 22 shows clear evidence in its sinuosity and eddylike patterns for an origin by flow (see also Fig. 1).

Feeder channels typically connect midocean ridges with the continent(s) that generally lie east of them. Feeder channels are not present between midocean ridges and the continents west of them. These ridges are little studied, and information concerning them is scarce. Plate tectonics includes them in a broad category that embraces several types of ridges, the so-called aseismic ridges, for which no unified explanation has been offered.

We selected three feeder channels from which data are abundant: the Canary Islands, Cape Verde Rise, and Walvis Ridge (Fig. 22). Typically, just like the surge channels beneath the continents, they have a 7.0–7.8-km/s anomalous lens present (Bosshard and Macfarlane, 1970; Goslin and Sibuet, 1975). They exhibit high microseismicity (Sykes, 1978) and high heat flow (Gramberg and Smyslov, 1986; V. Čermák, pers. comm., 1989). Linear and curvilinear ridge-parallel fractures are abundant (Dillon and Sougy, 1974; van der Linden, 1980). Volcanic activity, another surge-channel characteristic, continues to the present on the Cape Verde and Canary islands (Dillon and Sougy, 1974), but has not taken place along the Walvis trend since latest Cretaceous or earliest Tertiary time (Moore et al., 1984). The Walvis trend continues northeastward onshore for at least 1,500 km as a line of ring-shaped volcanic structures; the Lucapa graben and its kimberlite pipes also are on this trend (Sykes, 1978; Beloussov, 1980). We summarized these common characteristics on Table 6, which should be compared with Tables 5, 7, and 8.

Surge channels of continental margins

Breakout channels are located mainly in ocean basins. We classify them with continental-margin channels because they owe their origin to processes generated at the continental margins, not to processes generated in the ocean basins. Active margin channels are channels that are dammed behind Benioff zones (Figs. 8, 26). Undoubtedly these channel systems are structurally complex. We know the least about passive margin channels, the existence of which we did not suspect until recently. Our preliminary studies indicate that all passive continental

Table 6. Some common characteristics of feeder channels.

Feeder channel	High heat flow	High seismicity	7.0-7.8 km/s layer	Linear fracture pattern	Volcanic rocks
Canary Islands	X	X	X	X	X
Cape Verde Rise	X	X	X	X	X
Walvis Ridge	X	X	?	X	pre-Tertiary

Sources: Fúster and others (1968a, 1968b); Bosshard and Macfaflane (1970); Dillon and Sougy (1974); Goslin and Sibuet (1975); Sykes (1978); van der Linden (1980); Courtney and White (1986); Gramberg and Smyslov (1986); Newhall and Dzurisin (1988); Vladimir Čermák, written communications (1989).

Note: These ridges have been studied very little. As a consequence, the fine velocity structure, conductivity, and many other attributes are unknown.

margins may be sustained by small, deep channels. If true, surge channels are basic not only to lithosphere structure, and to continental structure in particular, but also to continental configuration (Fig. 14).

Breakout channels—Breakout channels originate along the eastern sides of continents and in magmatic arcs adjacent to them (Fig. 14; Table 4). They are included among continental-margin channels because of their clear genesis from continental-margin phenomena. In the literature, these have been called linear island and seamount chains, linear volcanic chains, and hot-spot trails (Jackson, 1976; Shaw et al., 1980; Okal and Batiza, 1987). They form another group of submarine ridges that, like feeder channels, have been lumped under the term aseismic ridges. They have several unique characteristics.

1. Their western termini, at least in the Pacific basin, are in the cusps between adjacent magmatic arcs (Fig. 23). Examples include the Obruchev Rise-Hawaiian-Emperor chain (Fig. 23), Louisville Ridge (Fig. 24), Caroline Ridge (Fig. 23), Christmas Island Ridge, Cook-Austral chain (Fig. 80), the Gilberts (Kiribati, Fig. 23), Marshalls (Fig. 23), Mid-Pacific Mountains (Fig. 23), Samoa chain (Fig. 23), Society Islands, the Tuamotus, and several others. Western Atlantic examples include the Newfoundland-Milne seamounts, New England-Corner seamounts, and the Rio Grande Rise.

Cusps are where breakout channels originate, presumably because they are the weakest links in the Benioff zones of the western Pacific. We suggest that lithosphere compression, combined with eastward asthenospheric movement produced by earth rotation, builds up stresses west of the Benioff zones (Fig. 8), stresses that are relieved from time to time by a breakout of excess magma or magmatic mush.

2. The eastern termini of these channels commonly are in the middle of nowhere (e.g., the Hawaiian Ridge), although some of them extend to the western flanks of a midocean ridge.

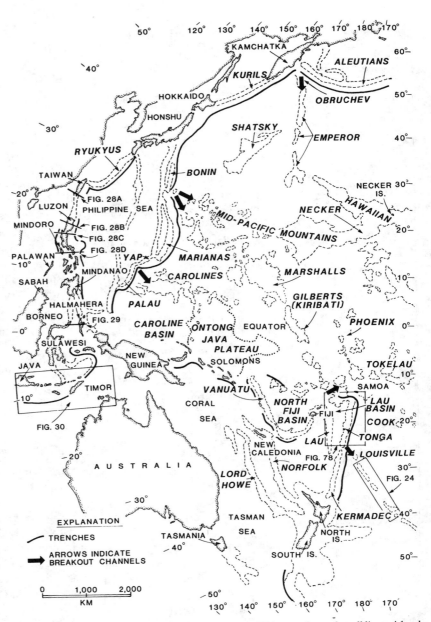

Figure 23. Sketch map of the western Pacific Ocean. This map shows that all linear island and seamount chains in this region originate (1) either at the cusp where two island arcs intersect (2) or from a linear chain that originates at the cusp between two island arcs. The locations of Figures 24, 28–30, and 78 are shown.

3. Almost all such channels are in the western Pacific. We have not found such channels in the eastern Pacific.

4. The channels are small—80–300 km across, but generally 100–200 km—and long, commonly 1,000 to 5,500 km. They are sinuous (Fig. 24), exhibiting eddy-like features (Fig. 24). They are studded with volcanoes in the form of seamounts and guyots. The Louisville

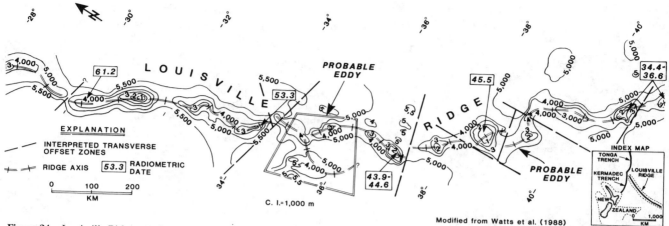

Figure 24. Louisville Ridge, southwestern Pacific, the location is shown on Figure 23. Note the sinuous pattern and the eddylike features. This is a typical breakout channel, studded with volcanoes in the forms of seamounts and guyots. This class of surge channel commonly (1) shows an age progression oldest-to-youngest from west to east, never the reverse; and (2) originates at its western end at the cusp where two volcanic arcs intersect (see Fig. 23). These ridges originate only on the western sides of ocean basins. This fact and the age progression from west to east—of some, but not all breakout channels—demonstrate that the hot spot origin claimed by plate tectonics is untenable. The location of this figure is shown on Figure 23.

Ridge of the southwestern Pacific has at least 60 volcanoes (Lonsdale, 1988).

5. They show age progressions in many areas (e.g., Louisville Ridge; Fig. 24; Hawaiian Ridge). Where age progressions are present, the youngest age is usually at the eastern terminus and the oldest age at the western (Jackson, 1976; Shaw et al., 1980; Cheng et al., 1987). Some of these west-to-east age progressions were used as proof of Morgan's (1971) hot-spot concept, but most of the island chains show no age progression at all (Fig. 80), and a few show a reverse progression (Jackson, 1976; Turner and Jarrard, 1982).

6. All geophysical data available to us indicate a surge-channel origin, in conformity with the criteria listed in Table 5. A 7.1–7.8-km/s layer has been found under the Hawaiian-Emperor chain and the Line Islands (Brune, 1969; Hill, 1969; Sutton et al., 1971; Ellsworth and Koyanagi, 1977). Ellsworth and Koyanagi (1977) found low-velocity zones beneath the Hawaiian Ridge. Von Herzen et al. (1989) found a band of elevated heat flow under the Hawaiian Ridge to Midway Island (Fig. 72). Gramberg and Smyslov (1986), on the basis of few control points, found elevated heat-flow bands beneath the Tuamotus, Marquesas, Carolines, and Mid-Pacific mountains (Figs. 23, 72). Active volcanism and microearthquake activity are known from the Hawaiian Ridge, Tuamotus, Society Islands, and the Cook-Austral volcanic chain (Talandier and Kuster, 1976; Ellsworth and Koyanagi, 1977; Talandier, 1989). Talandier (1989) pointed out that studies of active volcanism, microseismicity, and heat flow have barely begun in large areas of the Pacific basin.

7. Breakout channels are short-lived, with lifespans of 20 to 115 Ma, in contrast to 400–1,000+ Ma for other types of channels. Unlike trunk or feeder channels, they die out in the rear as their leading, forward edges progress eastward.

Active margin channels: western Pacific basin—Because of the Earth's rotation, subcontinental asthenosphere west of the Pacific basin flows outward, mainly eastward from the Australasian landmasses. This relative eastward movement is believed to produce the backarc basins that dominate continental-margin tectonics in the western Pacific (Fig. 8). Figure 25, from Meyerhoff and Meyerhoff (1977), illustrates this concept. Figure 25 also suggests that the shallow-dipping lithosphere Benioff zone can deflect the mobile hot material upward beneath the backarc basin, thereby explaining the high heat flows observed across large areas behind magmatic arcs.

Figure 26 is Rodnikov's (1988) more modern conceptualization of backarc-basin structure. It shows the velocity, density, and thermal structure of a typical western Pacific magmatic arc-backarc basin complex, and is based on refraction-seismic, gravity, and heat-flow data. We have added two shallow-level surge channels to illustrate our concept of the further complexity beneath backarc basins.

Table 7 shows some of the key characteristics of surge channels (Table 5) in relation to ten magmatic arcs, nine in the western Pacific and one in the Caribbean. Comparison with Tables 5, 6, and 8 illustrates the point that many seemingly different megastructures have similar geophysical and geological characteristics.

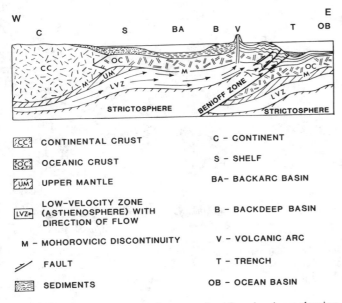

Figure 25. Schematic structural cross section of a volcanic arc showing the emplacement of asthenosphere between the lithosphere and the stereosphere. Eastward-directed movement of the asthenosphere, caused by the Earth's rotation, made it possible for the backarc basins of the western Pacific, the Caribbean, and the Scotia Sea to form, and—in our opinion—produced the transitional-type crust that characterizes backarc basins. From H. Meyerhoff and Meyerhoff (1977).

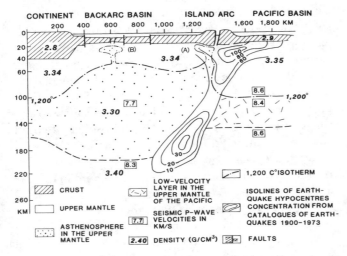

Figure 26. Velocity, density, and thermal structure of a typical western Pacific volcanic arc and backarc basin, based on refraction-seismic, gravity, and heat-flow data. From Rodnikov (1988). Note the very thick (150 km) section of asthenosphere which is dammed behind the lithosphere Benioff zone as a result of the Earth's rotation (see Fig. 8). Published with permission of Elsevier Science Publishers B. V.

Figure 27 is a southwest-to-northeast seismic-refraction section across the northeastern corner of the Lesser Antilles arc, as interpreted by Officer et al. (1959). The well-developed 7.0–7.6-km/s lens is evident. Edgar et al. (1971) have since mapped this layer westward into the Colombian and Venezuelan basins.

Figure 28 shows four west-to-east seismic-reflection profiles across the Taiwan and Philippine island arcs. Their locations appear on Figure 23. We modified Létouzey et al.'s (1988) geology slightly on the northernmost line through Taiwan (Fig. 28a), preferring instead the work of Suppe (1981, his fig. 9). The kobergenic structure on all four lines needs no elaboration. In the southern profiles (Fig. 28d), the kobergen has split into two branches. The western branch passes through Mindoro and Palawan to Sabah and Borneo. The eastern branch continues to the Molucca Sea between Sulawesi (Celebes) and Halmahera in northeastern Indonesia. Both exhibit the distinctive kobergenic style. Surge tectonics explains in a natural and unforced way the coexistence of two or more foldbelts in the same region with parallel and coeval histories of volcanism, sedimentation, and tectogenesis. Our model eliminates the need for postulating the presence of two or more coexisting Benioff zones, numerous microplates, exotic tectonostratigraphic terranes, and strongly contorted slab configurations.

Figure 29 shows three cross sections along the eastern Luzon kobergen of Figure 28d, except that these sections are farther south in the Molucca Sea between Sulawesi and Halmahera (Fig. 23). Figure 29b is a structural interpretation based on reflection and refraction seismology, and field work (Silver and Moore, 1978). These authors stated that their data provide "... clear documentation of a two-sided symmetrical collision zone" (p. 1689). Subsequent gravity modeling by McCaffrey et al. (1980) produced Figure 29c. The black prong approaching the surface on both Figures 29b and 29c appears to be a sliver of faulted mafic to ultramafic surge-channel roof. Regardless, the kobergenic style is unmistakable. The up-faulted prong (black) produces a sharp gravity high within a regional Bouguer gravity low. The close similarity with gravity profiles across the Alps (Fig. 51) and the Appalachians (Fig. 44) is apparent.

Figure 30 shows an east-west kobergen in the Banda arc (Breen et al., 1989). The north-vergent thrust faults north of Sumbawa, Flores, and Wetar are documented by seismic-reflection studies and by direct observations using SeaMARC II sonographs (Silver et al., 1986; Breen et al., 1989).

Active margin channels: North American Western Cordillera—In the eastern Pacific, backarc basins are absent for reasons discussed. Benioff zones are present in parts of the eastern Pacific rim, but not in others, presumably because, being fixed in time and space, they were overrun by lateral continental accretion.

Table 7. Some common characteristics of selected volcanic arcs.[1,2]

Volcanic arc	Rift	Wrench system	Foldbelt	Kobergen is present	High heat flow	High seismicity	7.0-7.8 km/s layer	Midcrustal low-velocity zone	Negative gravity anomaly	Linear fracture pattern	Volcanic rocks
Bonin	●				X	X	X		X	X	X
Indonesia	●	●	●	●	X	X	X		X	X	X
Japan	●	●	●	●	X	X	X	X	X	X	X
Kuril-Kamchatka	●		●	●	X	X	X	X	X	X	X
Marianas	●				X	X	X			X	X
Philippines	●	●	●	●	X	X	X	X		X	X
Ryukyu	●				X	X	X			X	X
Taiwan	●	●	●	●	X	X	X		X	X	X
Tonga-Kermadec	●				X	X	X	X	X	X	X
Lesser Antilles			●	●	X	X	X		X		X

[1]Blank spaces indicate that we found no literature on these topics. This does not mean that there is no literature. It may also mean that the topic has not yet been studied.

[2]Data sources by region are listed below: *Bonin arc:* Murauchi et al. (1986), Watanabe et al. (1977), Sychev and Sharaskin (1984), Honza and Tamaki (1985), Gramberg and Smyslov (1986). *Indonesia:* Curray et al. (1977), Milsom (1977), Silver and Moore (1978), Hamilton (1979), Bowin et al. (1980, 1982), McCaffrey et al. (1980), Silver et al. (1986), Eva et al. (1988), Mount (1988), Newhall and Dzurisin (1988), and Breen et al. (1989). *Japan:* Murauchi et al. (1968), Utsu (1971), Watanabe et al. (1977), Yoshii (1983), Sychev and Sharaskin (1984), Gramberg and Smyslov (1986), Oike and Huzita (1988), Shimazu (1988), Aihara (1989), and Hirahara et al. (1989). *Kuril-Kamchatka:* Vlasov and Belova (1964), Utsu (1971), Udintsev (1972), Fedotov (1973), Gorshkov (1973), Sychev and Sharaskin (1984), Gramberg and Smyslov (1986), and Newhall and Dzurisin (1988). *Marianas:* Pushcharovskiy (1972), Segawa and Tomoda (1976), Watanabe et al. (1977), Karig et al. (1978), Bibee et al. (1980), LaTraille and Hussong (1980), Hussong and Sinton (1983), Sinton and Hussong (1983), Eguchi (1984), Sychev and Sharaskin (1984), Gramberg and Smyslov (1986), and Bloomer et al. (1989). *Philippines:* Silver and Moore (1978), Cardwell et al. (1980), Divis (1980), McCaffrey et al. (1980), Weissel (1981), Gramberg and Smyslov (1986), Stéphan et al. (1986), Létouzey et al. (1988), Mount (1988), and Sarewitz and Lewis (1988). *Ryukyus:* Murauchi et al. (1968), Udintsev (1972), Watanabe et al. (1977), Sychev and Sharaskin (1984), Kobayashi (1985), Gramberg and Smyslov (1986), Kizaki (1986), Létouzey et al. (1988), and Ouchi et al. (1989). *Taiwan:* Ho (1982), Biq et al. (1985), Gramberg and Smyslov (1986), Richard et al. (1986), Tsai (1986), Létouzey et al. (1988), and Liu and Yu (1989). *Tonga-Kermadec:* Raitt (1956), Menard (1964), Karig (1970), Sclater et al. (1972), Watanabe et al. (1977), Gramberg and Smyslov (1986), Newhall and Dzurisin (1988), and Pelletier and Louat (1989). *Lesser Antilles:* Officer et al. (1959), Bowin (1976), Westercamp (1979), Westercamp and Tomblin (1979), and Gramberg and Smyslov (1986).

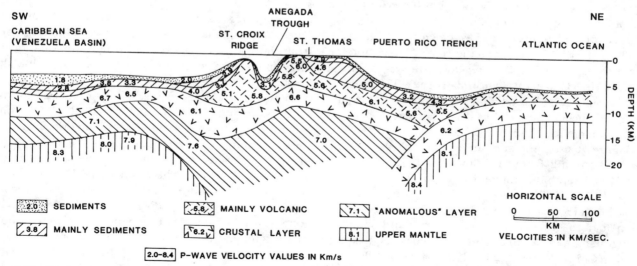

Figure 27. Southwest-to-northeast refraction-seismic cross section of the northern part of the Lesser Antilles arc (from Officer et al., 1959). This figure is included to show the similarity between the Caribbean island arc system and those of the western Pacific. See Table 7. Published with permission of Pergamon Press.

Table 8. Some common characteristics of selected tectonic belts. An "X" means that the feature is present, and "No" means that it is not. A blank space means that feature may well be present, but we found no literature on the topic.

Tectonic belt	Rift	Wrench system	Foldbelt	Kobergen is present	High heat flow	High conductivity anomaly	High seismicity	7.0-7.8 km/s layer	Midcrustal low-velocity zone	Negative gravity anomaly	Linear fracture pattern	Volcanic rocks
Alps			●	●	X	X	X	X	X	X	X	X
(North) Anatolia Fault		●	●	●	X		X				X	X
Appalachians			●	●	X		X			No	X	X
High-middle Atlas			●	●	X	X	X	X	X	X	X	X
Balkanides-Dinarides			●	●	X	X	X	X	X	X	X	X
Baltic shield					X	X	X	X	X	X	X	No
Lake Baykal	●				X	X	X	X	X	X	X	X
East Pacific Rise	●				X	X	X			X	X	X
Fenwei graben	●				X	X	X	X		X	X	X
Great Basin	●		●	●	X	X	X	X	X	X	X	X
Gulf of California	●		●	●	X	X	X	X	X	Locally in north	X	X
Hetao-Yinchuan	●				X	X	X				X	X
Himalayas (Yun-Nan; Hengduan)			●	●	X	X	X	X			X	X
Mid-Atlantic Ridge	●				X	X	X	X		X	X	X
San Andreas F.Z.		●	●	●	X	X	X	X	X	X	X	X
Canadian Cordillera			●	●	X	X	X	X		X	X	X

One area of the Western Cordillera that has a remnant lithosphere Benioff zone is the Cascade Range extending from northern California to southernmost British Columbia.

Figure 31 is a north-south seismic-refraction line across the Great Basin part of the Western Cordillera. The line extends southward from southern Idaho to Lake Mead on the Arizona-Nevada border, crossing the Great Basin surge-channel complex at an oblique angle. This channel complex underlies the entire Great Basin, is 500 to 600 km wide between the Sierra Nevada and the Wasatch Range, has a P-wave velocity of 7.3–7.8-km/s, and is 25 to 33 km deep (Landisman et al., 1971; Shurbet and Cebull, 1971; Fuis et al., 1987; Mooney and Braile, 1989). Shurbet and Cebull (1971) calculated its thickness to be not less than 10 to 12 km.

Figure 31 also shows another prominent feature of the Great Basin, a 2–6-km thick low-velocity zone (5.0–5.8 km/s) at a depth that ranges from 8 to 10 km (Prodehl, 1970, 1979; Landisman et al., 1971; Fuis et al., 1987). This low-velocity zone and the one below are nearly identical to two similar zones that underlie the Hengduan Shan (Yunnan Himalaya) channel of southwestern China (Figs. 9, 10, 47). The principal difference is that the Hengduan Shan channel underlies an active foldbelt (Fig. 48), whereas the Great Basin channel system underlies an active extensional system superimposed on a formerly active foldbelt. This seemingly contradictory situation is explained naturally in surge tectonics and is not contradictory.

Another feature portrayed on Figure 31 is a 180-km-wide lens of 7.0-km/s material centered below Mountain City, Nevada. Although we have not completed our study of the Western Cordillera, our preliminary mapping indicates that this lens is an inactive remnant of a Paleozoic surge channel—a biconvex lens—that extends from the Pacific Ocean basin through the Mendocino Escarpment to a south-to-north orientation below the Western Canada basin.

Figure 32 presents seven east-west profiles across the Great Basin at 38° N latitude. Eaton et al. (1978) compiled these profiles and found a striking overall east-to-west symmetry of several features in the Great Basin along a north-south symmetry axis located between 115° and 116° W longitude. The symmetrical elements shown on Figure 32 include heat flow (which is high), earthquake (seismicity) distribution, the low-velocity zone at 8 to 10 km depth, magnetism, gravity, topography, temperature distribution (including the location of the Curie isotherm), and depth to the Mohorovičić discontinuity. The Great Basin as a whole occupies a huge regional Bouguer gravity low (Eaton et al., 1978).

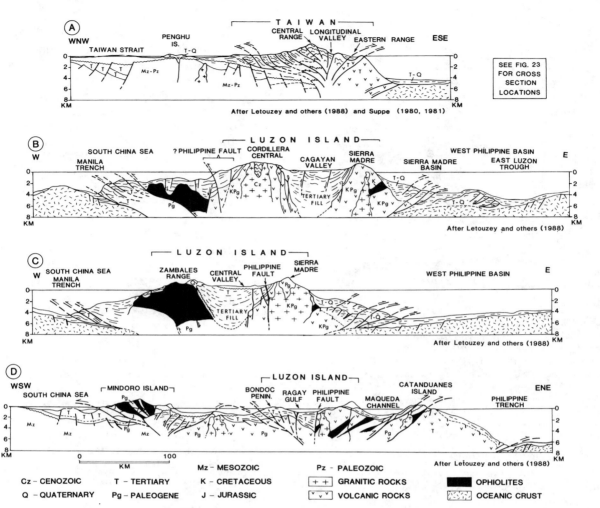

Figure 28. Four west-to-east structural cross sections of Taiwan and the Philippine arc system (from Létouzey et al., 1988). The sections are based on reflection-seismic and field-geology data. The following should be noted: (1) the Benioff zone (in each cross section, at the eastern side of Taiwan or Luzon) has a kobergen just behind it, demonstrating the influence of the Earth's rotation; (2) a major strike-slip fault system lies at the crest of the kobergens that underlie the volcanic arc—the longitudinal valley of Taiwan and the Philippine fault of Luzon; (3) the kobergen beneath northern Luzon bifurcates to form a separate kobergen beneath Mindoro and Palawan Islands (Fig. 23). Locations of cross sections appear on Figure 23.

The cause of the symmetrical features does not lie at the surface, because symmetry of the surface geology from one side of the basin to the other is totally lacking. The only geological surface phenomenon that exhibits symmetry is the distribution of Quaternary volcanic rocks, which are concentrated along the eastern and western margins of the Great Basin (Eaton et al., 1978). Thus, the cause of the symmetrical features lies at depth.

Additional phenomena characteristic of the Great Basin include a pronounced (at all scales) north-south fault, fracture, and fissure pattern (Fig. 33; Stewart, 1978); a strong low-resistivity (high conductivity) magneto-telluric anomaly (Fig. 34; Gough, 1984); metamorphic core complexes (analogous to mantled gneiss domes of the Appalachians; Coney, 1980; Wust, 1986; Davis and Lister, 1988); gravity detachment fault complexes (Wust, 1986; Davis and Lister, 1988); and kobergen structures (bivergent foldbelts). The linear fault and fracture pattern of Figure 33, as on midocean ridges, suggests (following Stokes's Law) north-south mantle flow under the Great Basin. Comparison of Figure 33 with figures 11, 9, and 10 of Meyerhoff et al. (this volume) illustrates the striking similarity between the linear structures of midocean ridges and that of the Great Basin. Moreover, as Gough (1984, p. 430) observed, ". . . in the western US the close association of electrical conductivity with the seismic low-velocity layer, and with high heat flow, makes it very probable that the conductive layer is, in fact, a layer of the uppermost mantle containing a molten mineral fraction."

Heezen (1960) considered the Great Basin to be a continuation of the East Pacific Rise-Gulf of California

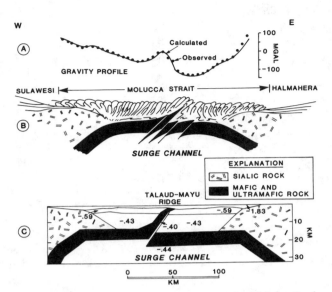

Figure 29. Three west-east cross sections across the Molucca Sea between Sulawesi and Halmahera, Indonesia (from Silver and Moore [1978] and McCaffrey et al. [1980]). The location of the cross sections is shown on Figure 23. **A**—Is a gravity profile across the line of section. **B**—is a structural interpretation—schematic—based on seismic and geologic data. **C**—Is the gravity model that satisfies both A and B. The kobergen shown here is the same kobergen that underlies Luzon (Figs. 28a-28c). The kobergen of Figure 30 is a part of the same system.

belt of seismicity and high heat flow. Plate tectonics, however, requires extensive offset along the San Andreas fault zone, which separates the Great Basin from the Gulf of California. As a consequence, Heezen's (1960) concept was discarded. We believe that Heezen was correct, and the Gulf of California belt of high seismicity and heat flow is continuous into the Great Basin. If the belt is not

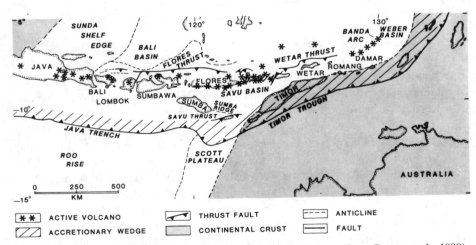

Figure 30. East-west-surge channel and kobergen of the Banda arc (from Breen et al., 1989). Together, Figures 28 through 30 illustrate a surge-channel system through a distance of 6,500 km, which is perhaps 60% of the length of this active-margin surge-channel system. Location of figure shown on Figure 23.

a continuous one, it is necessary to attribute to coincidence the juxtaposition of the Gulf of California province with the Great Basin. To illustrate the similarity, some of the Gulf of California's salient tectonic features are discussed.

Figure 35 (compare with Great Basin data on Figs. 31, 32) shows the great similarity of geological and geophysical features. These include high heat flow (Lawver and Williams, 1979), elevated seismicity (Ortlieb et al., 1989), high electrical conductivity (Fig. 35; Gough, 1984, 1989), a lens of 7.0–7.8-km/s material (Phillips, 1964; Fuis and Kohler, 1984), a lithosphere low-velocity zone (York and Helmberger, 1973), a system of north-south-striking parallel to subparallel linear faults, fractures, and fissures (Ortlieb et al., 1989), and a Late Jurassic through middle Eocene kobergen (Fig. 35; Rangin, 1984). Both belts, after undergoing a long succession of Late Jurassic through middle Eocene compressive events, became sites of tensile stress beginning in late Eocene time, a stress regime that continues in both areas today.

The kobergen tectonic styles of the Gulf of California (Fig. 35) also characterize the Great Basin, except that in the Great Basin several kobergens have developed (Fig. 36). Much of the Great Basin, however, is not adequately mapped, so that detailed reconstructions of the several kobergens there are not yet possible.

Geologists working in the Western Cordillera, despite the clear evidence in many ranges for extensive westward thrusting, have been loath to treat the Western Cordillera as anything but a monovergent, west-to-east-directed, tectonic belt. Hershey (1903, 1906) was the first to find westward-verging thrusts, these in the Klamath Mountains. Suess (1909) and Kober (1921, 1925, 1928) interpreted Hershey's (1903) discovery to mean that the Western Cordillera, like the Alpide system (Figs. 12, 13), was a bivergent system. Similar concepts were published by Burchfiel and Davis (1968) and by Yeats (1968), but as plate tectonics increasingly dominated geological thought, all aberrations from the monovergent model were attributed to backthrusting during the alleged underthrusting of North America by eastward-moving plates of oceanic lithosphere (e.g., Burchfiel and Davis, 1975). The same explanation has been used for the Western Cordillera of Mexico (Rangin, 1984, 1986) and Canada (Brown and Tippett,

1978). Proponents of bivergent thrusting are few; a notable example is Krebs (1975).

As detailed mapping continues, a very different tectonic pattern emerges (Figs. 36, 37), specifically one of linear, parallel to subparallel, bivergent foldbelts from the Sierra Nevada to the Central and Southern Rocky Mountains. For example, in western and central Nevada, Roure and Sosson (1986), and Speed et al. (1988) postulated the presence of various allochthonous terranes that have been transported eastward onto the North American block. Speed et al. (1988) suggested that some of these terranes may have rafted thousands of kilometers from some unknown part of the Pacific basin, and became attached to North America during late Paleozoic, Triassic, and possibly later times. In fact, the supposed allochthons, although tectonically disrupted and partly allochthonous, are not far from their original positions and are in their original positions with respect to each other (Meyerhoff et al., in preparation). We interpret them to be the facies belts of the Alps literature (Bertrand, 1897; Trümpy, 1960). Each group of facies belts has been deformed bivergently, as both Roure and Sosson (1986) and Speed et al. (1988) have shown (Figs. 36, 37). When mapping of the area shown on Figure 36 is completed in detail, at least five Mesozoic

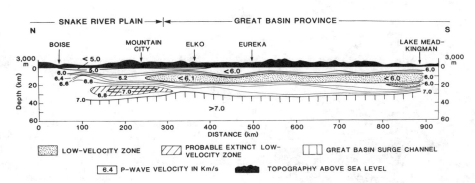

Figure 31. North-south refraction-seismic profile of the eastern Great Basin, western North American Cordillera. Location shown on Figures 34 and 36. The numbers are P-wave velocities in kilometers per second. A prominent feature of the figure is the 2–6-km thick low-velocity zone that underlies much of the Great Basin at a depth of 10 to 12 km. At the northern end of the profile is a lens of 7.0-km/s material that probably is an extinct late Proterozoic-Paleozoic lens in a surge-channel complex (see text). Its dimensions (180 x 12 km) are comparable with those of the Lake Baykal surge channel of Figure 54. Note that the "mother" surge channel, which lies under the whole of the Great Basin (with P-wave velocity of 7.0 km/s) is generally below 30 km. From Prodehl (1970, 1979).

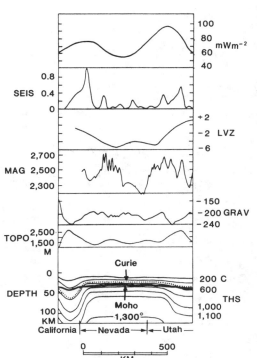

Figure 32. Seven west-to-east profiles across the Great Basin of Utah and Nevada. The location of the profiles is shown on Figures 34 and 36. Note the overall bilateral symmetry of each—heat flow, intensity of seismicity, low-velocity zone arrival-time shifts, magnetic signature, gravity expression, topography, depth to Curie point, and thermal structure. From Eaton et al. (1978). Published with permission of Geological Society of America.

Figure 33. Fault pattern of part of the Western Cordillera. Shown are normal and strike-slip faults. This pattern, following Stokes's Law, shows that the predominant flow direction beneath the Western Cordillera of the United States is north-south to north-northwest-south-southeast. compare with Figures 11, 38, 55, 56, 59, 61, 75, and 96. From Stewart (1978). Published with permission of Geological Society of America.

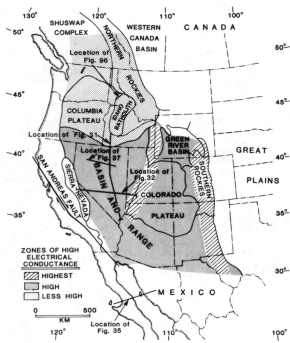

Figure 34. Zones of high electrical conductivity, Western Cordillera of the United States (from Gough, 1989). Two north-south, very high-conductivity, low-resistivity zones are present, one beneath the eastern edge of the Southern and Central Rocky Mountains, the other beneath the eastern margin of the Great Basin. Following Gough (1989), we believe that the high conductivity is related directly to the percentage of mafic melt in the surge-channel complex underlying the region. In our interpretation, the melt is in surge-channel complexes. This figure suggests that an interconnected surge-channel system underlies the whole Western Cordillera of the United States. The locations of Figures 31, 32, 35, 66, and 96, are shown.

kobergens will occupy the area of Figure 36. These kobergens have continued to be the sites of Cenozoic surge-channel activity (Figs. 31, 32).

Farther east in the Central and Southern Rocky Mountains, surge channels have been active only since Late Jurassic time (providing yet another example of eastward migration resulting from the Earth's rotation). Because the lithosphere beneath the Central and Southern Rocky Mountains is thicker than that beneath the Great Basin, the tectonic style is quite different, with germanotype structure predominating.

Recently, Eaton (1986, 1987) published two epic studies of the geological evolution of the Southern Rocky Mountains that extend 1,200 km north-south, cresting in a band 150–200 km wide, from the Medicine Bow Range in south-central Wyoming to the Franklin Mountains at El Paso, Texas (Fig. 38). He called attention to several characteristics of the Southern Rockies, to which we added a few more.

1. The Southern Rocky Mountains form a north-south unit that cuts across the northwest-southeast-trending Laramide structures (Fig. 38). They are a post-middle Eocene uplift, largely Oligocene through Quaternary (Fig. 38). The Laramide northwest-southeast trends are the trends of the Late Jurassic-early Eocene surge channels of the region.

2. The north-south, linear, Neogene to present, normal faults indicate a tensile stress perpendicular to the range (Figs. 33, 38; Stewart, 1978; Eaton, 1986). These faults occupy a belt 150 to 200 km wide, a typical width of many intracontinental surge channels (Figs. 35, 36, 42, 43, 49, 51, 54, 60).

3. The Rio Grande Rift occupies the southern two-thirds of the crest of the range. The system of faults that bound the rift extends into Wyoming (Fig. 40; Eaton, 1986, 1987).

4. An asthenosphere bulge underlies the entire Southern Rocky Mountains.

5. A band of high seismicity extends the full length of the range (Smith, 1978).

6. A belt of high heat flow is present (Lachenbruch and Sass, 1977).

7. A lens of 7.1–7.35-km/s material is well established (Prodehl, 1970, 1979; Prodehl and Pakiser, 1980).

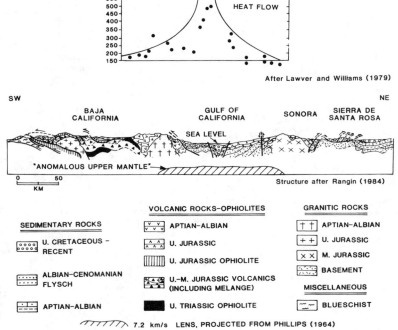

Figure 35. Southwest-northeast structural cross section across the Gulf of California. A surge channel (anomalous upper mantle) and the heat flow also are shown. The location is shown on Figure 34.

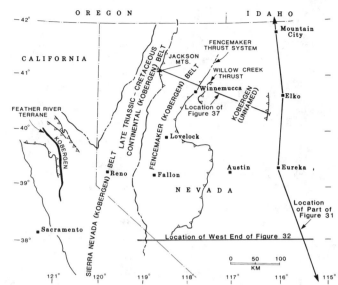

Figure 36. Index map of part of the Great Basin. This figure shows the locations of identified parallel to subparallel kobergens in this region. From west to east, these are: (1) Sierra Nevada kobergen, which bifurcates in the north into a Paleozoic-middle Cretaceous ocean-margin kobergen (Feather River belt) and a Late Triassic-middle Cretaceous kobergen in northwestern Nevada; (2) the Late Triassic-middle Cretaceous Fencemaker kobergen; and (3) a coeval unnamed kobergen. A fourth kobergen probably lies east of this kobergen. Published structural data from the region still are inadequate to outline accurately the extents of the identified kobergens. Locations of parts of Figures 31 and 32, and of Figure 37 are shown.

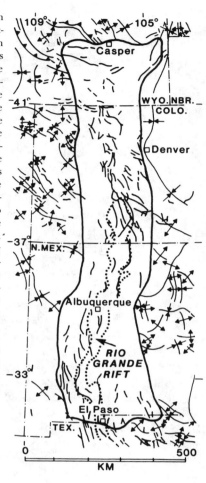

Figure 38. Southern Rocky Moutnains structures superimposed on Laramide structures (after Eaton, 1986). The Southern Rocky Mountains structures strike north-south unlike the Late Cretaceous-middle to late Eocene Laramide structures that strike northwest-southeast. The Southern Rockies uplifts are largely Oligocene and younger. One purpose of this figure is to show the important role of surge channels in creating and sustaining both uplifts and depressions. In this case, one surge channel simultaneously created (1) the north-south uplift that extends from El Paso, Texas, to Casper, Wyoming, and (2) the Rio Grande rift. Published with the permission of Elsevier Science Publishers B. V.

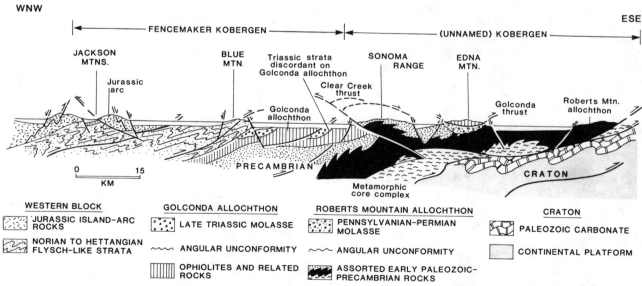

Figure 37. Northwest-southeast structural cross section across Late Triassic-middle Cretaceous kobergens, northwestern Nevada. The location is on Figures 34 and 36. Modified from Roure and Sosson (1986). In our interpretation (Meyerhoff et al., in preparation), the Fencemaker kobergen underlay a deep-marine backarc basin during Late Triassic time; the unnamed kobergen farther east underlay the continental crust east of the backarc basin. The Sierra Nevada kobergen underlay the Sierra Nevada volcanic arc west of the Fencemaker kobergen (see Fig. 36). Published with the permission of Prentice Hall.

This lens lies at a depth of 33 to 55 km (Mitchell and Landisman, 1971).

8. Mitchell and Landisman (1971) found a crustal low-velocity zone about 18 km above the 7.1–7.35-km/s lens.

9. A low-resistivity, high-conductivity magnetotelluric anomaly parallels the range, and is shown on Figure 34 (Gough, 1984, 1989).

10. A negative Bouguer anomaly is present.

11. The 550° C isotherm is sharply elevated beneath the range.

12. The range contains Cenozoic volcanic rocks in several places along the Rio Grande rift and north of it.

If our surge-tectonics concepts are valid, these data show that an active surge channel underlies the entire range, a conclusion that finds full support in a discovery by Eaton (1986). He drew five topographic profiles across the Southern Rocky Mountains at five different latitudes; he did the same for the Mid-Atlantic Ridge. He then paired them, as shown on Figure 39. He noted the great similarity between Southern Rocky Mountains and Mid-Atlantic Ridge profiles, writing (p. 175) that they ". . . demonstrate a remarkable, but coincidental, similarity in first-order morphology between a continental ridge and a slowly spreading oceanic ridge. . . . The cross-sectional dimensions of width and height of the crestal ranges, themselves, are the same, as are the topographic relief and slopes. . . . One can see that. . . the concave rises of the paired profiles track one another faithfully for hundreds of kilometers." Although Eaton (1986) considered the "remarkable similarity" to be wholly coincidental, we do not. In surge tectonics, both the Southern Rocky Mountains and the Mid-Atlantic Ridge have to be underlain by active surge channels. We regard Eaton's (1986) discovery as strong support of our interpretation. The conclusions reached here for the Southern Rocky Mountains apply equally to the Central Rockies.

Additional supporting evidence is the fact that most Central and Southern Rocky Mountain ranges have kobergenic structure. This is seen on Figure 40. Figure 40 also shows another feature of surge-channel flow that we mentioned in a preceding section. This is the segmentation of the ranges, a continental pinch-and-swell structure, a sort of texture that characterizes linear flow phenomena everywhere (Figs. 1, 16, 17, 22, 24, and many others).

Figure 41 is a structural cross section across the Colombian Andes, which we introduce here for comparison. The section extends from the Pacific basin to the continental craton. Several kobergens, parallel and partly coeval, exist side by side, just as in the North American Western Cordillera. Alpinotype structure is

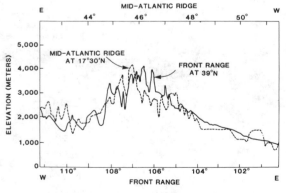

Figure 39. Superimposed topographic profiles of the Mid-Atlantic Ridge and the Southern Rocky Mountains (from Eaton, 1986). Eaton (1986) commented on the ". . . remarkable, but coincidental, similarity. . ." of the two features. The similarity, as it happens, is not coincidental, as the two features in our interpretation were produced by a surge channel. Published with the permission of Elsevier Science Publishers B. V.

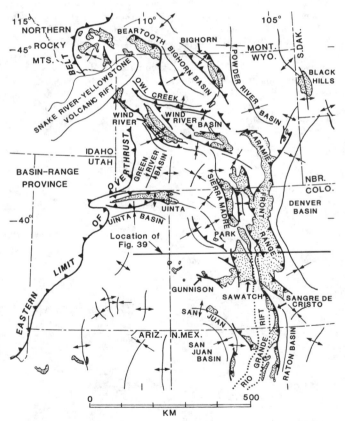

Figure 40. Central Rocky Mountains germanotype structures: uplifts (stippled), thrust faults (low and high angle), anticlines, and synclines are shown. This figure shows that each uplift, or combination of uplifts, in the Central Rocky Mountains has kobergen structure and therefore is sustained by a surge channel. This conclusion is supported not only by the geological data but also by all geophysical data—heat flow, seismicity, and so forth. Note the segmentation of the ranges and see Figure 1. Compiled mainly from Gries (1983) and Hamilton (1988).

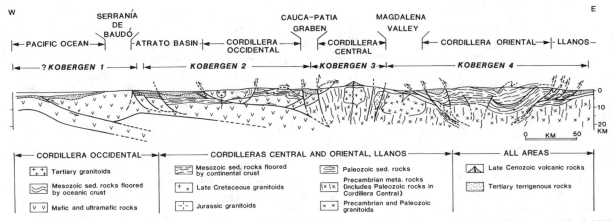

Figure 41. Andes of Ecuador and Colombia: the development side by side of alpinotype (left) and germanotype (right) kobergens (from Mégard, 1987). On the far west, just offshore and adjacent to the coastal plain, a Cenozoic kobergen may be in the process of formation (kobergen 1?). Adjacent to and east of kobergen 1? is the Late Cretaceous-Tertiary kobergen 2. Kobergen 3 also is a young feature, the site of an active volcanic arc. Kobergen 4, adjacent to the Precambrian craton, is a germanotype structure.

developed in kobergen 2 near the Pacific coast. Germanotype structure dominates in the east.

Passive margin channels: Caledonides and Appalachians—Figure 42 shows Scandinavia, Finland, the Baltic Sea, and adjacent watermasses west of northern Europe. Epicenters (Talbot and Slunga, 1989) also are shown, together with the heat-flow contours (Čermák and Rybach, 1979). The figure shows that the entire region, mainly the Scandinavian Caledonides and ancient Baltic shield, is seismogenic, a fact that is irreconcilable with other tectonic hypotheses. High concentrations of epicenters in the Skagerrat-Kateggat region (Oslo-Copenhagen), the Gulf of Bothnia, and the Norwegian coast indicate that Precambrian and Palezoic tectonic trends influence the configurations of some of the surge channels. Our conclusion regarding the basis of epicenter distribution is that one or more laterally connected low-velocity zones underlies the entire region north and west of the Baltic Sea.

Supporting data are copious. At the Gulf of Bothnia, Luosto et al. (1989) and

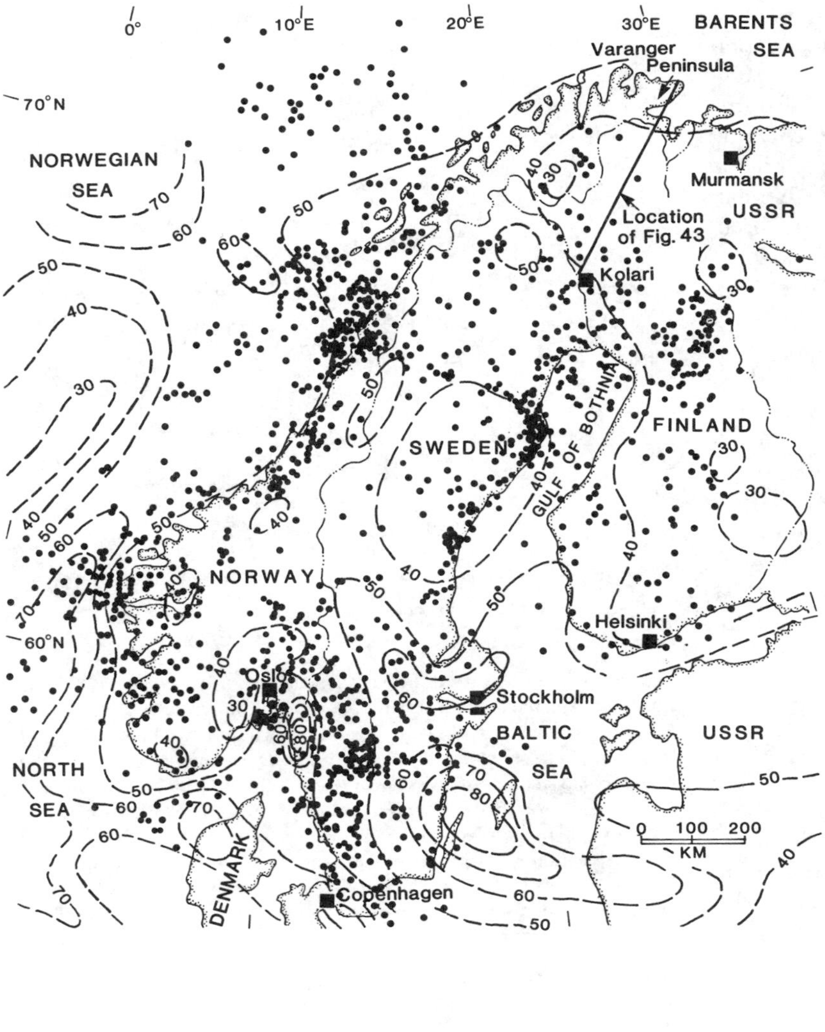

Figure 42. Epicenters and heat flow of the Scandinavian Caledonides and Baltic shield. This illustration shows several features: (1) A middle Paleozoic and older foldbelt (Caledonides) beneath Norway is underlain by an active surge channel; (2) A stable craton has been penetrated by an extensive surge-channel system; (3) stable cratons and inactive foldbelts can later become the sites of surge-channel activity; (4) Many—probably all—major linear uplifts (e.g., Caledonides) and linear depressions (e.g., Gulf of Bothnia, Gulf of Finland, central Baltic Sea) are underlain by active surge channels. The location of Figure 43 is shown. The figure was compiled from Čermák and Rybach (1979), Luosto et al. (1989), Mueller and Ansorge (1989), and Talbot and Slunga (1989).

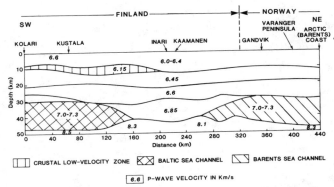

Figure 43. Southwest-northeast refraction-seismic profile across the northern margin of the Baltic shield. Two active surge channels at 30–50 km depth are shown, one in the north beneath a branch of the middle Paleozoic Caledonides and one in the south that underlies the Gulf of Bothnia. The location of Figure 43 is shown on Figure 42. The purpose of this figure is to document our statements concerning Figure 42. Modified from Luosto et al. (1989). Published with permission of Elsevier Scientific Publishers B. V.

Mueller and Ansorge (1989) found a shallow low-velocity zone (6.15 km/s) 3 to 10 km thick at a depth of 7 to 10 km (Fig. 43). About 23 to 31 km deeper is a 15-km thick lens of 7.0–7.3-km/s material that deepens from 30 km in the north (Fig. 43; Luosto et al., 1989) to 42 km near the mouth of the Gulf of Bothnia (Mueller and Ansorge, 1989). A Bouguer gravity minimum (Simonen and Mikkola, 1980) and a high-conductivity anomaly (Ádám, 1983) are present beneath the Gulf of Bothnia channel. Farther west in the mountains close to the Norwegian coast, Muir-Wood (1989) discovered a swarm of linear, seismically active, high-angle, reverse faults in zones of high heat flow and strong microseismicity.

Quite a different situation obtains in the northern Appalachians. Here, high heat-flow readings are scarce and microearthquake activity is minimal. Figure 44 shows a geological cross section from Rodgers (1987). The presence of a kobergen is at once evident, with its center or root zone at the Bronson Hill anticlinorium, an extensive north-south zone of mantled gneiss domes (analogous to metamorphic core complexes) in New England (Thompson et al., 1968). Seismic-refraction studies reveal the presence below the Bronson Hill anticlinorium of a lens of anomalous upper mantle with a P-wave velocity of 7.0-km/s (Fig. 45; Taylor et al., 1980). A narrow, sharp, positive magnetic anomaly overlies the root zone (Zietz, 1982), as does a positive Bouguer gravity anomaly (Hildenbrand et al., 1982) and a heat-flow anomaly (Birch et al., 1968; McCartan and Architzel, 1988). The highest grade of metamorphism is at the Bronson Hill anticlinorium, with decreasing grades toward both the northwest and southeast. Reverse metamorphic zonation (inverted isogradic surfaces), another kobergen characteristic, is common (Thompson and others, 1968). The winged, or kobergenic structure shows up clearly on a seismic-reflection line shot here (Fig. 46; Taylor, 1989).

We interpret the lens of 7.0-km/s material as the upper part of a surge-channel complex that was active at least from late Proterozoic through Middle Devonian time, when Acadian tectogenesis deformed the region. This surge-channel system is dead, or nearly so. We say this because the positive Bouguer gravity anomaly indicates that little or no low-density material remains in the lens of 7.0-km/s material, and the level of microearthquake activity is very low. The positive heat-flow anomaly might indicate that some hot material remains in the channel, but Jaupart et al. (1982) related the high heat-flow readings to shallow radioactivity associated

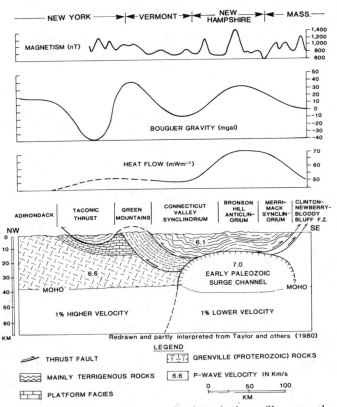

Figure 44. Northwest-southeast refraction-seismic profile across the northern Appalachians from the Adirondack mountains of New York State to the Atlantic coast near Boston, Massachusetts. The purpose of this figure is to show the Acadian (Middle Devonian) kobergen in the Appalachian system with its extinct (?) surge channel, bilateral symmetry, and central core, the Bronson Hill anticlinorium. The latter is part of a linear system of mantled gneiss domes. Heat-flow, gravity, and magnetic data are also shown. Compiled from Taylor et al. (1980), Hildebrand et al. (1982), Zietz (1982), and McCartan and Architzel (1988). The Taylor et al. (1980) data are published with the permisson of *Science* and the American Association for the Advancement of Science.

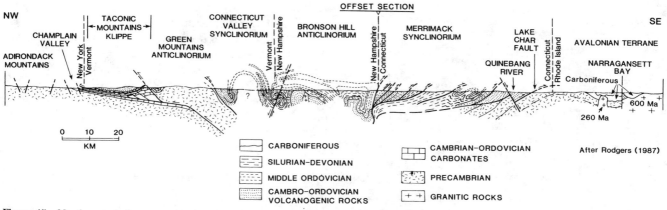

Figure 45. Northwest-southeast structural cross section of the northern Appalachians, New England. This section is located very close to the Taylor et al. (1980) section in Figure 44. Note the bivergent symmetry centered on the gneiss domes of the Bronson Hill anticlinorium. Gneiss domes, consisting of metamorphosed igneous—in large part granitic batholiths—rocks, are characteristic of the cores, or root zones, of alpinotype foldbelts. Published with the permission of Princeton University Press.

with silicic magmas of the nearby Jurassic-Cretaceous White Mountains Magma series, which formed from a post-Paleozoic surge channel that traverses the whole Paleozoic foldbelt from northwest to southeast.

Figure 1 shows the segmented Bronson Hill anticlinorium side by side with other equally segmented tectonic belts. The remarkable similarity of these supposedly diverse features, ranging from the root zone of an alpinotype foldbelt (Bronson Hill anticlinorium) to young rifts in a tensile regime, directly confirms our contention that all tectonic belts—alpinotype or rift—originate by flow parallel with the tectonic belt.

Continental surge channels

We tentatively recognize two categories of continental surge channels. (Table 4; Fig. 14). The first is the trunk channel, or continental trunk channel, which is a continuation of oceanic feeder channels. We also mentioned the East-Pacific-Rise to Great-Basin continuum, in which an oceanic trunk channel becomes a continental trunk channel. The smaller channels that branch from continental trunk channels are called branch channels. Like the feeder channels in the ocean basins, they branch only from the eastern sides of their parent trunk channels, thereby providing one more demonstration of the important role played by earth rotation. Thus the branch channels originate in a manner similar to that of breakout channels, except that a Benioff zone is not present.

Lithosphere thickness and tectonic style—The style of tectogenesis in any given tectonic belt, or part of one, reflects directly the lithosphere thickness at the location of that belt. Consequently, at this stage in our research we see no obvious way to subdivide surge channels in

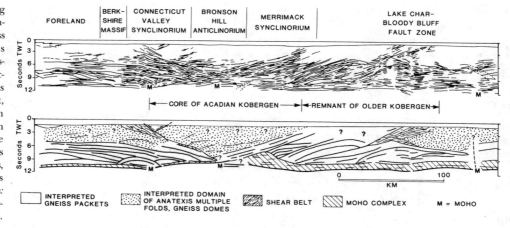

Figure 46. *Top:* Line drawing (tracing) of a northwest-southeast reflection-seismic line across the northern Appalachians from New York State to near Boston, Massachusetts. Line of section is close to those of Figures 44 and 45. Note the bivergent, or winged structure centered on the Bronson Hill anticlinorium (Fig. 1), which was the root zone of the northern Appalachians during the Acadian tectogenesis, or tectogeny. We interpret this to be a kobergen. *Bottom:* Taylor's (1989) suggested interpretation of the line drawing. The original figure is from Phinney (1986).

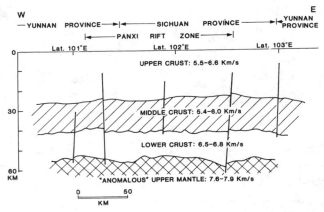

Figure 47. West-east seismotomographic section across the Permian Panxi rift southernmost Sichuan Province, southwest China. The main surge channel is below 50 km. This is the same channel which on Figure 10 is in the 60-30 km depth zone. The midcrustal low-velocity zone of Figure 19 (43-25 km) is the subsidiary low-velocity zone of Figure 10. Comparison of this figure with Figure 10 shows how much the depth of a surge channel can change in a short distance. Location of Figure 47 is shown on Figure 9. From Liu Jianhua et al. (1989).

continental environments. In general, tectonic belts closest to the continental margin, or located in thin lithosphere, deform in alpinotype style. In thicker lithosphere, germanotype style is common and, deep in the continental interiors, rift valleys of the East African type are more likely to form than any other type of tectonic belt.

In this section, we illustrate some Eurasian and North African foldbelts, followed by consideration of some rift zones around the world. The alpinotype examples probably are indistinguishable from the active margin channels of the last category of surge channels, surge channels of continental margins. In Table 8 we combined some continental margin tectonic belts, as well as belts from other regimes, with those of the continental interiors. This table, based on the criteria presented in Table 5, should be studied together with Tables 5, 6, and 7. The similarities among all the features are evident.

Yunnan Himalaya (Hengduan Shan)—Figures 9, 10, and 47 are from the seismotomographic imagery study conducted in China by Liu Futian et al. (1989) and Liu Jianhua et al. (1989). Through this study, Liu Futian and his Chinese colleagues inadvertently discovered a network of shallow interconnected lens-shaped bodies in the lithosphere of China. The great importance of this discovery cannot be overstated. Liu Futian et al. (1989) had no explanation for their discovery, and noted only that a close relationship seemed to exist between the seismotomographic images and China's gross tectonic subdivisions.

Figure 9, from Liu Jianhua et al. (1989), depicts a 1.4-million-km² area in southeastern Asia, centered in Yunnan Province. The figure portrays a large north-west-trending trunk channel in the west from which at least three much smaller branch channels branch eastward. Figure 10 shows the changes in depth that take place through short distances. Causes of the depth changes are unknown but likely include changes in lithosphere composition and density. If so, depth variations are at least partly a response to gravity requirements (Peach-Köhler climb force; Weertman and Weertman, 1964; Weertman, 1971; Ryan, 1987; Corry, 1988; Walker, 1989). Figure 47, also from Liu Jianhua et al. (1989), shows the velocity structure beneath Yunnan Province as interpreted from seismic-refraction data. The low-velocity zone at 25–42 km, just as in the East Pacific Rise, is well above the underlying anomalous upper mantle with P-wave velocities of 7.6 to 7.9 km/s, and contains lighter magma products that presumably were differentiated from the 7.6–7.9-km/s layer. The overall similarity between the velocity structure of the Yunnan compressional belt of southwestern China and that of the Great Basin (Fig. 31) should be noted.

Figures 9, 10, and 47 also illustrate differences in surge-channel sizes. The large trunk channel in the west is about 500 km wide and thousands of kilometers long, with a height of 17 to 20 km. It is associated with high heat flow (Ma Xingyuan, 1986a, 1986b), elevation of the Curie point (Chen Zongji, 1987), lines of hot springs (Ma Xingyuan, 1986a; Wang Jun et al., 1986a; Zhu Meixiang, 1986), intense seismicity (Ma Xingyuan, 1986a, 1986b; Chen Zongji, 1987; Yan Zhide, 1987), a high-conductivity (low resistivity) magnetotelluric anomaly (Liu Guodong, 1987a, 1987b), a lens of 7.5–7.9-km/s material (Ma Xingyuan, 1986a, 1986b), and numerous parallel strike-slip faults that are 1,200 km long or more (Ma Xingyuan, 1986a; Chen Zhongji, 1987; Yan Zhide, 1987). A nearly symmetrical Tertiary kobergen overlies the 7.5–7.9-km/s lens (Wang and Chu, 1988), and is illustrated on Figure 48.

In the northeastern part of the map area of Figure 9, two small surge channels branch eastward to northeastward from the large channel beneath western Yunnan Province. The northern of these (approximately 38° N latitude) underlies the active Hetao-Yinchuan rift. The surge channel is approximately 180 km wide; its associated graben is 50 to 80 km wide and at least 540 km long. Formation began in middle to late Eocene time in the southwest and worked northward and eastward. The Hetao-Yinchuan graben system is associated with high heat flow (Zhang Buchun et al., 1985; Wang Jun et al., 1986b; Ye Hong et al., 1987), intense seismicity (Zhang Buchun et al., 1985; Ma Xingyuan, 1986a), a gravity low (Zhang Buchun et al., 1985), a high-conductivity (low resistivity) magnetotelluric anomaly (Zhang Buchun et al., 1985), a linear upward bulge of the Mohorovičić

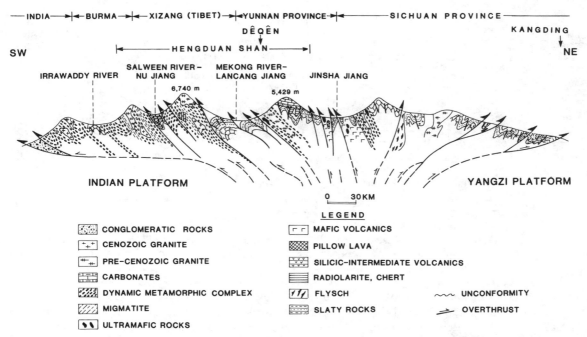

Figure 48. Southwest-northeast structural cross section of the Himalaya kobergen from northeastern Burma to central Sichuan Province, southwest China. The location is shown on Figure 9. The foldbelt is tectonically active. This figure shows an exceptionally symmetrical kobergen above a still-active surge channel (shown on Fig. 9). From Wang and Chu (1988). Published with the permission of Elsevier Science Publishers B. V.

discontinuity and the asthenosphere (Zhang Buchun et al., 1985; Ma Xingyuan, 1986a), volcanics, and strike-slip faults hundreds of kilometers long (Zhang Buchun et al., 1985; Ma Xingyuan, 1986a).

The surge channel just south of the Hetao-Yinchuan channel (approximately 35° N latitude) is overlain directly by another active graben system, the Fenwei rift (Wang Jingming, 1987). This surge channel is 30 to 100 km wide. Its associated graben is 20 to 90 km wide and 1,000 km long. The graben began to form in the southwest during middle Eocene time, and gradually extended northeastward to the present time (Zhang Buchun et al., 1985; Wang Jingming, 1987). The Fenwei rift system is associated with high heat flow (Wang Jun et al., 1986b; Chen Guoda, 1989), numerous Pliocene and Pleistocene volcanoes (Ma Xingyuan, 1986a), intense seismicity (Zhang Buchun et al., 1985; Ma Xingyuan, 1986a; Ye Hong et al., 1987), a negative gravity anomaly (Zhang Buchun et al., 1985), a high-conductivity (low resistivity) magnetotelluric anomaly (Zhang Buchun et al., 1985; Liu Guodong, 1987a), a linear upward bulge of the Mohorovičić discontinuity and the asthenosphere (Zhang Buchun et al., 1985; Ma Xingyuan, 1986a), anomalous upper mantle with a P-wave velocity of 7.8 to 7.9 km/s (Ma Xingyuan, 1986a), and numerous straight faults with a continuity of up to several hundred kilometers (Zhang Buchun et al., 1985; Ma Xingyuan, 1986a; Wang Jingming, 1987).

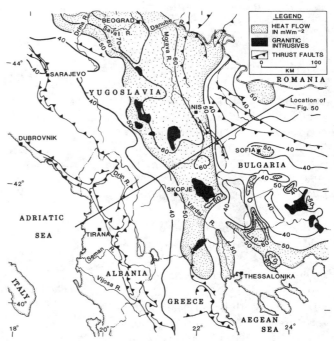

Figure 49. Kobergen structure of the Dinarides and Balkanides, Yugoslavia, Albania, Greece, and Bulgaria (from Hurtig et al., 1981). A band of elevated heat flow and a line of young intrusive granitic rocks are at the crest of this kobergen. High seismicity and a strike-slip fault zone also are present. The location of Figure 50 is shown.

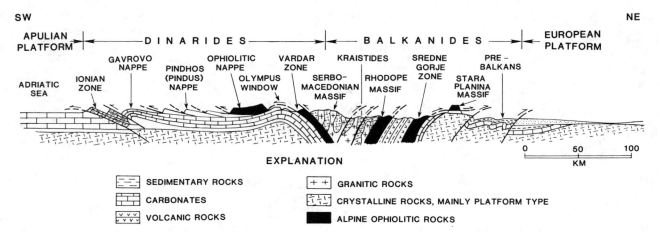

Figure 50. Southwest-northeast structural cross section of the Dinarides-Balkanides kobergen. This is another spectacularly developed symmetrical kobergen (see Fig. 48). The location of the cross section is on Figure 49. From Aubouin (1977).

From the preceding, the close genetic relations among the three structures are apparent—the Yunnan kobergen, the Hetao-Yinchuan rift, and the Fenwei rift (Figs. 9, 10, 47, 48). Figure 9 demonstrates that the three are interconnected by the same surge-channel system. This is another example demonstrating the close genetic link between tensile and compressive tectonic belts.

Dinarides-Balkanides—The foldbelts of the Balkans are associated with an unusually symmetrical kobergen (Fig. 49, 50). Figure 49, from Hurtig et al. (1981), shows westward-verging nappes in the west, and eastward-verging nappes in the east. A belt of high heat flow (up to 75 mWm^{-2}) is symmetrical with respect to the kobergen, being concentrated along its core from which nappes extend outward up to 200 km on either side (Aubouin, 1977). A high-conductivity magnetotelluric anomaly has been mapped from Hungary into the northern part of the kobergen (Gough, 1989). The core of the kobergen has been intruded by numerous young plutons (Fig. 49; Hurtig et al., 1981) and, in the Vardar zone (Fig. 50), Neogene volcanic rocks crop out (Dimitrijević, 1974). The kobergen also has high seismicity, both at the crest and in still-active thrusts faults on both flanks (Dragašević, 1974). At depth is a crustal low-velocity zone above a lens of 7.0-km/s material (Dragašević, 1974). The sparse gravity control in our possession (we have one line across the full kobergen) shows a negative Bouguer anomaly. We call attention to the great similarity between this kobergen and that of the Yunnan Himalaya, or Hengduan Shan (Figs. 48, 50).

Alps—Figure 51 is a structural cross section through the western Alps. It is based on classical field geology, to which we added data from various geophysical surveys. The kobergen structural style is evident (Fig. 13; Kober, 1925; Frei et al., 1989). The Ivrea body probably was part of the original surge channel. Its P-wave velocity is 7.2 to 7.6 km/s (Miller et al., 1982; Mueller, 1983; Frei et al.,

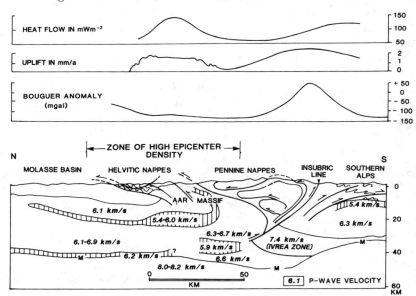

Figure 51. North-south structural cross section of the kobergen of the western Alps. Based on surface geology and several types of geophysical studies. Heat flow, rates of uplift, and Bouguer gravity also are shown. Compare with Figure 13. The low-velocity zones shown are in the crust. Note that one is believed to crop out (Miller et al., 1982; Mueller, 1983). As we interpret the section, the Ivrea zone is part of a Late Jurassic-Eocene surge channel that was partly destroyed by Late Cretaceous-early Tertiary orogeny and split into two parts, one north and one south of the Insubric Line. The latter is a *Verschluckungszone*, whose origin is discussed in the text (Figs. 90–93). This section was compiled from data published by Čermák and Rybach (1979), Miller et al. (1982), Mueller (1983), Bayer et al. (1989), and Frei et al. (1989).

1989). This body, now displaced considerably from its original position, is about 20–25 km thick, 40 km wide, and several hundred kilometers long in an east-west direction. It is very close to the root zone of the western Alps (Rutten, 1969). Its mafic to ultramafic rocks are associated closely with Caledonian granulites (Laubscher and Bernoulli, 1983). The Ivrea body, however, cannot have been in the surge-channel roof because its rocks indicate that temperatures probably did not exceed 800° C (Hunziker and Zingg, 1980). The body is closely associated with the Insubric Line (Gansser, 1968), part of an extensive wrenchlike fault system that extended from the western Alps to the Carpathians (Ádám, 1987). The Ivrea zone has been termed seismically dead (Gansser, 1983) and lies between two east-west zones of high heat flow (Čermák and Rybach, 1979; Deichmann and Rybach, 1989) associated with bands of high seismicity (Deichmann and Rybach, 1989) and very rapid uplift (Miller et al., 1982; Gansser, 1983; Mueller, 1983). In addition, at least two low-velocity zones are present (Angeheister et al., 1972; Laubscher, 1978; Miller et al., 1982; Mueller, 1983).

Parallel to and closely associated with the Ivrea zone-Insubric Line and its eastern counterparts is a row of late Alpine intrusions (late Eocene-early Miocene) that formed in an extensional belt. This same extensional belt contains linear to curvilinear wrench faults (Royden et al., 1983; Laubscher, 1988; Ratschbacher et al., 1989). The Ivrea zone-Insubric Line and its eastern continuations are underlain by a high-conductivity anomaly (Brodie and Rutter, 1987) and are known from magnetotelluric studies to continue into the Pannonian basin via the Balaton Line (Ádám et al., 1986; Ádám, 1987). These magnetotelluric, low-resistivity, high-conductivity anomalies originate in bodies 6 to 17 km deep, just below the hypocenters of the associated seismic belt (Ádám, 1987). In all likelihood, the anomaly sources are the fluid contents of one or more shallow low-velocity zones (that is, part of the surge-channel complex).

The metamorphic grades of the Alps are related closely to the position of the Ivrea zone-Insubric Line (Fig. 51). The highest metamorphic grades are closest to the Ivrea zone and its eastern equivalents, and decrease away from that zone (Mueller, 1983; Frei et al., 1989). Stretching lineations generally are parallel to subparallel with the structural strike (Ellis and Watkinson, 1987; Selverstone, 1988; Ratschbacher et al., 1989).

Figure 51 also shows, whereas a positive Bouguer gravity anomaly overlies the Ivrea zone, the two bands of high seismicity and heat flow respectively north and south of the Ivrea zone are associated with negative Bouguer gravity anomalies (Čermák and Rybach, 1979; Mueller, 1983, Bayer et al., 1989).

Figure 51 shows another important phenomenon. The 5.4-km/s low-velocity zone north of the Alps bends upward beneath the Helvetic zone, reaching the Earth's surface in the Aar massif (Miller et al., 1982; Mueller, 1983). This suggested relationship cannot be proven, but Miller et al. (1982) and Mueller (1983) highlighted the likelihood that many former low-velocity zones in the lithosphere now crop out at the surface. The rapid uplift of the entire Alps chain from the Helvetic to the southern Alps, including the Ivrea zone (Fig. 51), suggests to us that a single surge channel once underlay this part of the Alps (Ivrea zone). Subsequent tectogenesis broke up the surge channel, dividing it into at least two parts, one north of and one south of the Ivrea zone. Seismicity data suggest the two channels are linked laterally beneath the Ivrea zone, if only by a very thin and limited low-velocity zone (channel). The importance of the Alps example is apparent. It provides an example of a surge channel that through tectogenesis was deformed and segmented; yet the surge-channel system is still very active.

Middle and High Atlas of Morocco—The Middle and High Atlas of Morocco illustrate the germanotype style. In surge tectonics, germanotype foldbelts develop above surge channels well removed (approximately 200–300 km) from oceanic lithosphere and where continental lithosphere is moderately thick. Continental crust appears to have thickened appreciably during a succession of Proterozoic and Paleozoic events that are recorded in outcrops north of, south of, and within the High and Middle Atlas (Wallbrecher, 1988; Piqué and Michard, 1989). Figure 52, from Froitzheim et al. (1988), shows

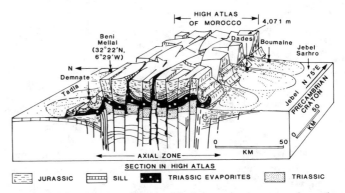

Figure 52. Germanotype block uplift above an active surge channel beneath the High Atlas of Morocco (from Froitzheim et al., 1989). This style of kobergen structure develops wherever the overlying lithosphere is between approximately 20 and 35 km thick. Note the thick development of Late Triassic evaporites (in black). These evaporites consist largely of halite. The location of this figure is shown on Figure 83. Figure 83 suggests that the Late Trisassic evaporites were precipitated directly above a surge channel. If so, the channel has remained active from at least Late Triassic time to the present. See text for detailed discussion.

the block uplift of a part of the western High Atlas. Outcrops of Proterozoic and Paleozoic rocks show that they accumulated on a sialic basement. This basement during Jurassic-Cenozoic time broke along near-vertical faults except at the margins of block uplifts. There, along the northwestern and southeastern margins of the range, small reverse faults and some low-angle thrusts developed, with thrusting northeastward in the northwest and southeastward in the southeast (Fig. 52; Froitzheim et al., 1988; Jacobshagen et al., 1988; Piqué and Michard, 1989).

Phenomena associated with the High and Middle Atlas and confirming our interpretation were described by Schwarz and Wigger (1988). They include high heat flow; thermal springs; a high-conductivity, low-resistivity, magnetotelluric anomaly; a belt of high seismicity down the crest of the western High Atlas and the Middle Atlas; a negative Bouguer gravity anomaly; Quaternary volcanic rocks; and bundles of linear faults and fractures parallel with the strikes of the ranges. A broad lens of 7.2–7.8-km/s material underlies both ranges at a depth

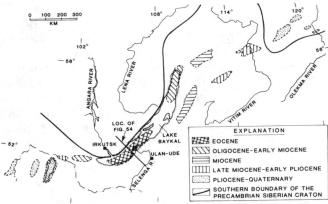

Figure 53. Lake Baykal and vicinity, south-central Siberia. The figure shows the probable chronology of development of the fault depressions in the Baykal rift system. From Logatchev and Florensov (1978). Note that, whereas the principal growth was northeastward from the vicinity of Irkutsk, some growth was westward. Published with permission of Elsevier Science Publishers B. V.

of 29 to 35 km; it is overlain near the top of the lower crust, or the base of the middle crust, by a 6.2-km/s low-velocity layer between 17 and 29 km (Schwarz and Wigger, 1988.)

Baykal rift system—Figure 53 shows the 2,500-km long Lake Baykal rift system of south-central Siberia. Mapping of the rift system indicated that it began to form about middle Eocene time in the southern part of modern Lake Baykal, and then propagated both to the east-northeast for some 1,500 km and toward the west-southwest for 500 km.

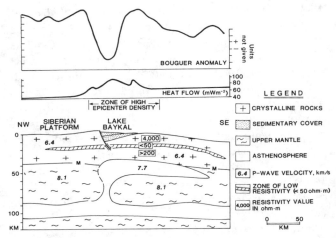

Figure 54. Northwest-southeast refraction-seismic section across the Lake Baykal surge channel, south-central Siberia. Magnetotelluric, heat-flow, and gravity data also are shown. No scale is available for the gravity curve. This is an example of a surge channel in thick continental lithosphere. A kobergen has not developed, only a rift. If the lithosphere is sufficiently thick, a kobergen may never develop. The 170-km width and 20-km height are typical of many subcontinental surge channels. Compiled from Golenetsky and Misharina (1978), Krylov et al. (1979), Zakharova (1980), Sychev (1985), and Popov (1987).

Figure 54 is a seismic-refraction profile across the rift zone (Krylov et al., 1979; Sychev, 1985). The surge channel beneath this half-graben is clearly visible at a depth of approximately 30 to 50 km. The P-wave velocity of much of the channel is 7.7 km/s (Krylov et al., 1979; Sychev, 1985) with a center low-velocity zone with velocities in the 5.4–6.0-km/s range (Krylov et al., 1979). The conduit joining this channel with the asthenosphere is at the northern margin of the depression, directly beneath the main fault zone. The lens of 7.7-km/s material is 20 km thick and 170 km wide. These dimensions are typical for large numbers of surge channels, although there is a wide range of widths (compare Figs. 9, 15, 18, 19, 22, 24, 26, 28–31.)

The Baykal rift zone overlies a band of high heat flow (Zorin and Osokina, 1984). A low-resistivity, high-conductivity, magnetotelluric anomaly develops at 10 to 20 km, and probably is a shallow low-velocity zone in which the contents are partly molten (Popov, 1987). Cenozoic volcanics are scattered along the rift zone (Genshaft and Saltykovskiy, 1989). The Bouguer gravity anomaly is strongly negative (Artemjev and Artyushkov, 1971). Lake Baykal and its associated rifts lie in a belt of intense seismicity (Golonetsky and Misharina, 1978), with hypocenters between 0 and 30.5 km, just above the top of the surge channel shown on Figure 54 (Vertlib, 1978).

A striking feature of the Baykal surge channel is its apparent extreme asymmetry with respect to the feeder conduit. Most surge channels we have studied thus far,

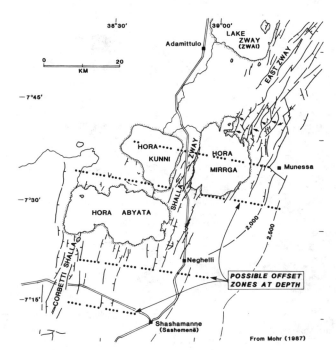

Figure 55. East African rift valley, Ethiopia. Note the linear faults, fractures, and fissures parallel with the north-northeast-striking rift. These form belts of fractures that, along strike, are offset (offsets are indicated by dotted lines at right angles to the rift valley). We call these fracture-band offsets. In our opinion, they originate in the same way as offsets of ridge-parallel fractures (Fig. 11) along the ridge-transverse fracture zones (transform faults) that cut at approximately right angles across the midocean ridges (see Meyerhoff et al., this volume), forming one end member of a series of related structures—fracture-band offsets (= offsets along transform faults), channel overlaps (overlapping spreading centers), eddylike structures, and vortex structures (these have no name in plate tectonics). Figures 55 and 71–75 illustrate in part this gradational series. Published with permission of Elsevier Science Publishers B. V.

with the exception of deformed channels (Fig. 51), are fairly symmetrical with respect to the feeder conduit.

East African rift system—Like the Baykal rift, the East African rift zone has a 7.3–7.5-km/s lens at 20 to 30 km (Mooney et al., 1983), a long history of volcanism and thermal spring activity, a shallow low-velocity zone (based on the presence of a high-conductivity layer above the anomalous lens; Wood, 1983), a negative Bouguer anomaly (Browne and Fairhead, 1983), and high heat flow (Crane and O'Connell, 1983). The high microseismicity of the rift has long been known (Tobin et al., 1969).

Figure 55 shows another characteristic, specifically, the presence of a rift-parallel fault, fracture, and fissure system (Mohr, 1987). This rift-parallel system is offset repeatedly along strike, much as the ridge-parallel fractures of the midocean ridges are offset at so-called transform fault zones. We interpret these fracture offsets as further evidence for the presence of a surge channel system beneath the rift.

Mississippi embayment—This region is known best in modern tectonic studies for generating some of North America's more spectacular earthquakes in 1811 and 1812 (Hamilton and Johnston, 1990). This earthquake zone—the New Madrid zone—lies on a band of epicenters that extends from the Ouachita Mountains to the Gulf of St. Lawrence (Sykes, 1978, fig. 14). Part of the epicenter system appears on Figure 56. The associated fault, fracture, and fissure system appears on Figure 57 (Fletcher et al., 1978; Hildenbrand, 1982;

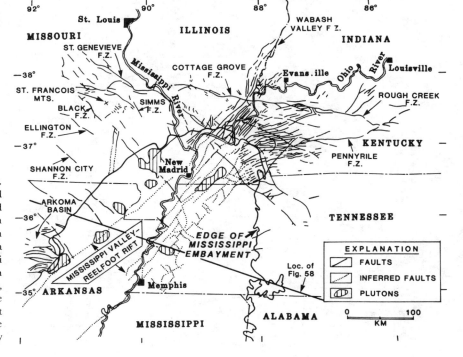

Figure 56. Upper Mississippi embayment. This map shows the linear fault, fracture, and fissure system mapped in the subsurface and at the surface. Note that Stokes's Law again applies. The principal fault zones are shown by name. The fault patterns suggest that a surge channel beneath the upper Mississippi embayment trifurcates, with one branch following the Simms-St. Genevieve fault zone, another the Wabash Valley fault zone, and the last the Pennyrile and Rough Creek fault zones. The positions of several subsurface plutons are also shown. Compiled from many sources (see Fig. 57).

Figure 57. Upper Mississippi embayment. The epicenters are plotted with relation to the principal fault zones. Site of the New Madrid earthquakes of 1811–1812 is shown. Compiled from Fletcher et al. (1978), Hildenbrand (1982), and Hinze and Braile (1988). The purpose of this illustration and Figure 56 is to show that the long linear Mississippi Valley's course and that of the associated Mississippi embayment are controlled by a surge-channel system. In short, surge channels generate topographic depressions, or basins.

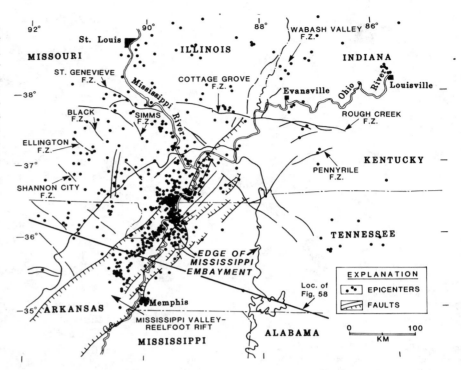

Hinze and Braile, 1988). McCartan and Architzel's (1988) heat-flow map of the eastern United States, despite the few data points, showed elevated heat flow in the New Madrid region (55–65 mWm^{-2}).

Ervin and McGinnis (1975), Fletcher et al. (1978), Hildenbrand (1982), and Mooney et al. (1983) reported the presence of a 450-km wide lens, 26 to 40 km below the surface, of anomalous low-density mantle material at the base of the crust with a P-wave velocity of 7.2 to 7.5 km/s (Fig. 58). This lens lies

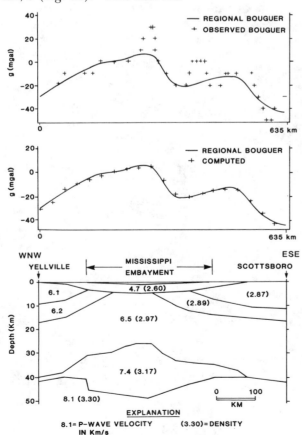

not only beneath the Mississippi embayment but also beneath a late Proterozoic graben system called the Reelfoot rift (Ervin and McGinnis, 1975). This rift contains an unusually thick section of late Proterozoic-Early Cambrian sedimentary rocks (Hinze and Braile, 1988). A positive Bouguer gravity anomaly characterizes the rift (Fig. 58). We believe the presence of a positive anomaly indicates that the surge channel present between about 36° and 38° N latitude has reached its waning stages. It has been a long-lived channel, because it is associated with numerous intrusions of different ages (Phipps, 1988). These include Early to Middle Devonian (399–383 Ma) alnöite and lamprophyre diatremes with associated dikes; Early Permian (287–275 Ma) kimberlite and lamprophyre dikes and sills; and Late Cretaceous (93–80 Ma) sills, dikes, and stocks of nepheline syenite and related rocks (Phipps, 1988). The intrusions are related closely to the fault systems flanking each side of the Reelfoot rift, and some are quite large (Hildenbrand, 1982).

Rhine graben—The Rhine graben has all of the characteristics of the Baykal, East African, and other continental rifts (e.g., Mooney et al., 1983). This graben has

Figure 58. This figure from Ervin and McGinnis (1975) shows the presence of 7.4-km/s material beneath the upper Mississippi embayment. Note the similarity between this cross section and that across the Amazon Valley on Figure 79. Both figures show intracratonic basins above a surge channel. The regional Bouguer gravity curves are shown, as well as the density values for each rock layer. Reproduced with permission of Geological Society of America.

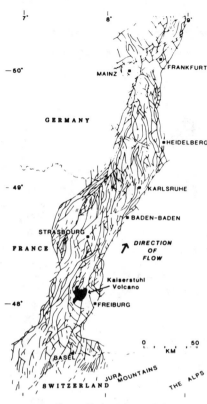

Figure 59. Fault and fracture patterns of the Rhine graben, France and Germany. From Illies and Greiner (1979). Horsetail patterns point predominantly from south-southwest toward the north-northeast in the direction of flow within the underlying surge channel. Published with permission of Elsevier Science Publishers B. V.

an unusually well-developed fault, fracture, and fissure pattern (Fig. 59; Illies and Greiner, 1979). Throughout, the fracture pattern contains horsetail structures that demonstrate flow toward the north-northeast.

Surge channels in zones of transtension-transpression

San Andreas fault—Figures 60–63 illustrate some features of the San Andreas fault zone critical to its structural interpretation. In surge tectonics, the major strike-slip zones of the Earth's lithosphere are generated at or near the crests of surge channels, and create a tectonic style that may be regarded as intermediate between that of a tensional rift zone and that of the alpinotype foldbelt.

We first concluded that strike-slip systems owe their origin to surge-channel activity after reading Lachenbruch and Sass's (1980) study of heat flow along the San Andreas fault. They found that heat flow under most of the California Coast Ranges averages 80 mWm^{-2} and is not concentrated along the fault zone. Therefore, they reasoned, friction along the fault zone does not generate the observed heat flux. If the heat is not the result of shear, they deduced further, then the observed heat flow can be explained best by "extreme mantle upwelling" (p. 6185), with asthenosphere conditions reaching a depth of only about 20 km, especially in the northern Coast Ranges. Such a zone was found by Zandt (1981) and Zandt and Furlong (1982), and is illustrated on Figure 60. We interpret the upwelled zone as evidence that the upwelling is produced by a surge channel. Note that, north of the Transverse Ranges of southern California, the surge channel becomes shallower, reaching to less than 20 km about 25 km southeast of Cape Mendocino, which is in effect the 20 km deduced by Lachenbruch and Sass (1980).

Figure 60. Structural contour map of the top of the California Coast Ranges northwest-southeast-trending surge-channel complex. Zandt and Furlong (1982), from whose work this map was taken, interpreted the contour values to represent lithosphere thickness. However, the shape, dimensions, and orientation parallel with California's fold and fault system (Fig. 61) indicate that the linear feature is a surge channel. As we point out in the text, this linear feature also is associated with Cenozoic volcanic rocks, elevated heat flow, high seismicity, a row of hot springs, 7.0–7.8-km/s material, and other phenomena generated by surge-channel activity. The depths shown are to the top of the main magma chamber, not to the top of the shallower chambers that are shown on Figure 62. The location of Figures 62 and 63 are shown. Inset map: this is a postulated eddy in the San Francisco Bay region where the zone of most intense seismic activity is transferred from the San Andreas fault to the Hayward-Calaveras fault (Smith, 1978). Published with the permission of Geological Society of America.

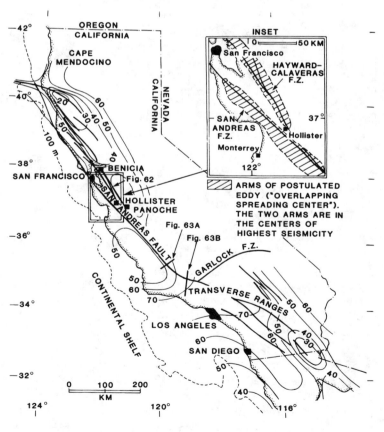

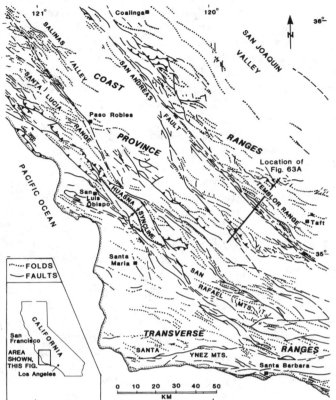

Figure 61. Map of part of the California Coast Ranges showing the north—northwest-south-southeast trends of the faults and fold axes. Following Stokes's Law, this map shows that flow beneath the region shown on the map parallels the fault and fold-axis directions. Compare this fault and fold-axis pattern with the patterns shown on Figures 11, 33, 38, 55, 56, 59, and 75. Note that the fault and flow-axis pattern directly overlies the postulated surge channel of Figure 60. From Johnson and Page (1976). Published with permission of Elsevier Science Publishers B. V.

The surge channel (as we interpret it) on Figure 60 exhibits three characteristics. These are (1) rather constant width, (2) constantly changing depth (from 20 to 70 km), and (3) a position that shifts from one side of the San Andreas fault to the other, for example, just south of Hollister where the Calaveras fault splits off (Fig. 60). The depth changes are similar to those observed elsewhere, such as western Yunnan (Figs. 9, 10).

Smith (1978) showed that most intense seismicity in central California is concentrated along the Calaveras fault east of San Francisco Bay, and on the San Andreas fault from San Francisco south. North of San Francisco, seismic activity on the San Andreas fault decreases markedly with the major activity along the Hayward-Calaveras system and its northward continuations.

The transfer of the zone of maximum seismicity from the San Andreas to the Hayward-Calaveras system—based on Smith (1978)—is shown schematically on the

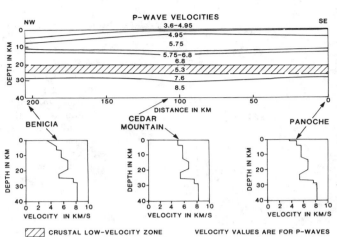

Figure 62. Velocity-depth interpretation for a northwest-southeast refraction-seismic line through the Diablo Range, central California, from Benicia to Panoche. The location of the line is shown on Figure 60. *Top:* Numbers indicate the P-wave velocities in km/s. Note the low-velocity zone associated with 7.6-km/s material below 20 km. *Bottom:* one-dimensional velocity-depth functions for selected profiles. The Benicia profile is at the northwestern end of the refraction line, the Panoche profile at the southeastern end, and the Cedar Mountain profile in the middle. The purpose of this figure is to demonstrate that a low-velocity zone associated with 7.0–7.8-km/s material at depth does underlie the Coast Ranges. From Blümling and Prodehl (1983). Published with the permission of Elsevier Science Publishers B. V.

inset map on Figure 60. The overlap pattern produced by zones of maximum seismicity is analogous, we believe, to the so-called overlapping spreading centers of midocean ridges (Macdonald and Fox, 1983). As Meyerhoff et al. (this volume) explained, such features likely are caused by eddylike motions in the upper asthenosphere. The shape and direction of this eddylike feature demonstrate that motions below the San Andreas fault zone are parallel with that zone. The fact that San Francisco Bay directly overlies one of these features is not, in our opinion, coincidental, but is related directly to the presence of this eddylike feature. In fact, depressions are everywhere found where eddy structures are present (Figs. 71–77).

Figure 61 strongly supports our interpretation of flow parallel with the San Andreas fault system. The map portrays the known faults, fractures, and fold axes of part of the San Andreas fault system (Johnson and Page, 1976). Just as the dominant pattern of faults, fractures, and fissures of midocean ridges suggests ridge-parallel flow, these faults, fractures, and fold axes in the California Coast Ranges indicate flow parallel with them. Figure 61, like Figures 11 and 33, is an illustration of Stokes's Law (Sears et al., 1974; Blatt, 1983; Appendix).

Additional facts support our contention that a surge channel underlies the San Andreas fault zone. We discussed the high heat flow (80 mWm^{-2}; Lachenbruch and

Figure 63. Two structural cross sections across the San Andreas fault, central California. In all structure sections across this fault—a statement that applies to all regional strike-slip faults—the regional strike-slip fault lies at the center of a developed, or developing kobergen. This implies that the regional master fault is at or very close to the top of a surge channel, or former surge channel. See the text for detailed discussion. The locations of these cross sections are on Figures 60 and 61. Compare this figure with Figures 64 (North Anatolia fault), and 28b and 28c (Philippines fault). Figure 51 (Insubric Line of the Alps) shows a related feature. The geometric relation between regional strike-slip faults and kobergen mechanics means that much of what has been written about strike-slip zones needs to be reevaluated. Upper diagram is published with the permission of *Science* and the American Association for the Advancement of Science; the lower diagram is published with the permission of Geological Society of America.

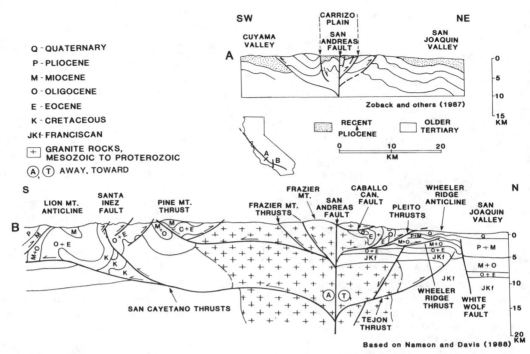

Sass, 1980), intense seismicity (Smith, 1978), and linear fault, fracture, and fold-axis pattern (Fig. 61; Johnson and Page, 1976). Of equal importance is the presence of a 7.3–7.6-km/s lens (Blümling et al., 1985) beneath the surface east (Blümling and Prodehl, 1983) and west (Walter and Sharpless, 1987) of the San Andreas fault. Figure 62 shows the layer beneath a 200-km segment of the Diablo Range, just east of the fault (Blümling and Prodehl, 1983). A low-velocity zone of 5.2–5.7-km/s material occupies the middle crust above the 7.3–7.6-km/s lens (Fig. 62), just as in midocean ridges (Fig. 18), the Himalayan foldbelt of western Yunnan (Fig. 47), the Alps (Fig. 51), Lake Baykal (Fig. 54), the Baltic shield (Fig. 43), the Great Basin (Fig. 31), and elsewhere. A gravity low is associated with the fault (Robbins, 1968), but this may reflect that a vertical low-velocity zone—presumably caused by the lower density of crushed rock in the fault zone—characterizes the fault (Feng and McEvilly, 1983). A high-conductivity magnetotelluric anomaly is associated with the San Andreas fault at both the surge-channel depth (7.3–7.6 km/s; Lienert et al., 1979) and at the shallow low-velocity zone depth (Phillips and Kuckes, 1983).

The San Andreas fault lies precisely at the crest of a developing kobergen (Fig. 63), that is, along the plane of maximum tension at the crest of the surge channel beneath it. This is illustrated clearly on the cross sections of Figure 63 (Zoback et al., 1987; Namson and Davis, 1988). The evidence of this figure has major implications for the origin of strike-slip faults. The structures east of the San Andreas fault are east-vergent; those west of it are west-vergent. This is true not only of the fault between approximately 32° and 36° N latitude where these sections are located (see Figs. 61, 62), but also of the San Andreas fault in northern and central California (Wentworth et al., 1984; Wentworth, 1987; Wentworth and Zoback, 1989).

We are aware that our interpretation of the San Andreas fault zone is unorthodox, as is our concept that lateral movements along strike-slip zones are minuscule. We document this assertion in a manuscript in preparation. Some of our reasons are presented in a companion paper by Martin (this volume).

North Anatolia fault—Our interpretation of the San Andreas fault zone applies to strike-slip faults elsewhere, as shown in Figure 64, a structural cross section through the North Anatolia fault zone in northern Turkey (Létouzey et al., 1977). The presence of a kobergen centered on the North Anatolia fault is evident. Structures verge northward north of the fault, and southward south of it. In addition, this fault is associated with high heat flow (Čermák and Rybach, 1979); numerous thermal springs (Tezcan, 1979); intense, destructive, and shallow earthquakes (Şengör, 1979); a lens of anomalous

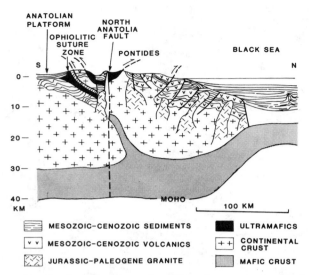

Figure 64. South-north structural cross section across the North Anatolia fault (from Létouzey et al., 1977). As in the case of the San Andreas fault, the master regional strike-slip fault lies at the center of a kobergen. This figure shows that the San Andreas is not an isolated example. Compare with Figures 28, 51, 63, and 65.

upper mantle (Fig. 64; Létouzey et al., 1977); scores of linear faults, fractures, and fissures close to and parallel with the fault (Dewey et al., 1986); and Quaternary volcanism (Pasquarè et al., 1988).

Other major strike-slip zones—Other examples of major strike-slip zones located in the geometric center of a kobergen include the North Pyrenean fault of the Pyrenees Mountains of Spain and France (Fig. 65; Choukroune, 1989), the Insubric Line of the Alps (Fig. 51; Gansser, 1968), the Alpine fault of New Zealand (Fig. 91, Kingma, 1974; Wood, 1978), the Longitudinal Valley of Taiwan (Fig. 28a; Létouzey et al., 1988), and the Philippine fault (Figs. 28c, 28d; Létouzey et al., 1988). In each example, structural vergence on opposite sides of the fault is away from the fault, that is, the structural style is that of a kobergen. The fact of opposing structural vergence centered on supposed major strike-slip zones is not addressed in any earth-dynamics hypothesis.

Surge-channel images

Most of the information on surge-channel geometry presented is based on seismic-refraction and gravity modeling studies (e.g., Figs. 15, 16, 18, 29, 31, 43, 45, 47, 51, 54, 58, 60, 64). Additional information comes from magnetotelluric (e.g., Figs. 34, 54) and seismotomographic data (e.g., Figs. 4, 9, 10, 93). In fact, the detailed seismotomographic survey of China proved to us that an interconnected network of channels, or lenses, pervades the lithosphere (Liu Futian et al., 1989; Liu Jianhua et al., 1989). Within a short time, seismotomographic imagery should become the foremost method for investigating all levels of the Earth's interior.

Beginning in March 1975, a systematic attempt (COCORP) began in the United States to study the continental crust with CDP seismic-reflection techniques (Oliver et al., 1976; Nelson, 1988). Similar programs are under way in Canada and most of Europe. The results are very good to depths up to 80 km. The principal problem is one of interpretation.

Figures 66–68 illustrate one of the main problems in interpretation. Figure 66 shows scattered reflectors to about 4.0 sTWT (4 seconds, two-way travel time), with some sections of fairly continuous data between about 2.5 and 4.0 s. The section between 4 and 6 seconds is generally transparent. Below 6.0 sTWT to about 11.0 sTWT, the reflection Moho, reflectors occur in semicontinuous bands that are separated by thin transparent zones. Many of the reflectors in the 6.0–11.0-s interval are markedly convex upward, but flatten toward the base of that interval.

Figure 67 is very similar to Figure 66. Just below 6.0 sTWT, strong convex-upward reflectors are present. They are underlain by a transparent zone at approximately 7.0–8.5 s, which in turn overlies another band of reflectors between 8.5 and 10.0 s. These reflectors are flat-lying to slightly concave upward. The principal difference between the transparent zone of Figure 66 and that of Figure 67 is that the zone is interrupted

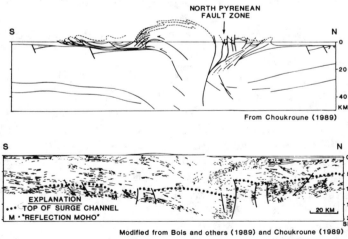

Figure 65. South-north structural cross section of the Pyrenees, border between France and Spain. This section, although based on reflection-seismic data, is supported by detailed field mapping. The lower section shows the quality of the reflections. Note that the North Pyrenean fault (upper section) lies at the crest of a kobergen structure. The line of large dots in the lower profile outlines what we believe to be the surge-channel lens beneath the Pyrenean kobergen. It has been strongly faulted (offsets up to 5 km); even the reflection Moho is offset.

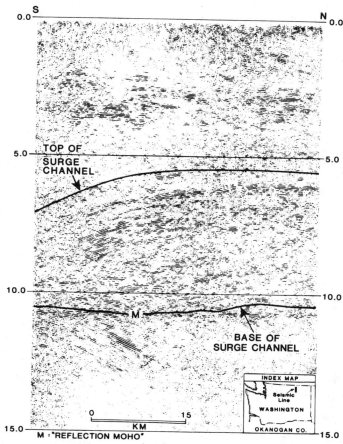

Figure 66. South-to-north reflection-seismic line through Okanogan dome, northeastern Washington (from Nelson, 1988). The Okanogan dome is a metamorphic core complex, whose origin is related closely to that of mantled gneiss domes (Figs. 44–46). The dome structure shows clearly. This structure coincides exactly with the presence of a lens-shaped, highly reflective structure in the lower crust. We show this figure to emphasize the close genetic relations among alpinotype foldbelts (Figs. 1, 35, 41, 44, 47, 51, 65), germanotype foldbelts (Figs. 1, 40, 41, 52), backarc basins and island arcs (Figs. 1, 26–28), rifts (Figs, 1, 54, 58, 67, 68, 79), strike-slip fault systems (Figs. 60, 62, 64, 65), mantled gneiss domes (Figs. 1, 44–46), and metamorphic core complexes (Fig. 66). The location of this figure is shown on Figure 34.

believe that the reflections, particularly the crosscutting ones, are the results of intracrustal shear. As a consequence, various names for the pods sprang up: rift pillows, shear pods, low-strain lozenges, anastomosing network of shear zones, and so forth. Only a few nongenetic terms are applied to these features. These include such terms as lenses, lenticles, pods, and lozenges. For the many ideas on these bodies, the papers by Klemperer (1987, 1988), Reston (1988), Finlayson et al. (1989), and Thompson and McCarthy (1990) are helpful.

Finlayson et al. (1989) found that these lenses have P-wave velocities of 7.0–7.8 km/s. Thus we equate them with the anomalous upper mantle to which we have referred repeatedly in this paper. Klemperer (1987) noted that the lenses commonly are zones of high heat flow. Hyndman and Klemperer (1989) observed that the lenses generally have very high electrical conductivity. In our own investigation, we found that where transparent zones are sandwiched in Figure 66 by minor bands of reflectors. In contrast, the transparent zone of Figure 67 is almost uninterrupted.

Figure 68 displays no transparent zone in the package of reflectors between the top of the convex-upward reflectors at 7.0 to 7.5 sTWT and the base of the flat-lying to slightly concave-upward reflectors at 13.5–14.0 sTWT. The only transparent section is above the convex-upward unit (above 7.0–8.5 s).

The main problem in interpretation is to explain the crosscutting and dipping reflectors in a region of the lithosphere where essentially flat-lying bodies are anticipated. When many of these layered masses were perceived to have a lenticular shape, speculation over the origin became even greater. Some authors believe that the reflections are from mafic bodies intruded from below, and therefore, the lower crustal layers preserve a record of extension. Some attribute the reflections to pods of subducted sedimentary rocks. Still others

Figure 67. Northwest-southeast reflection-seismic line at western end of the English Channel (Western Approaches basin); southeastern end of the line is just offshore from Brest, France. Note the lens-shaped surge channel with a transparent (low-velocity zone, possibly including some melt) layer between the roof and floor. We interpret this to be an active surge channel. From Reston (1988).

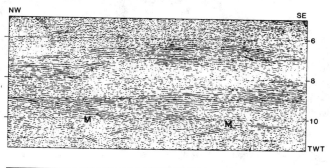

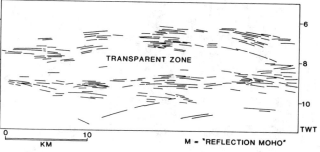

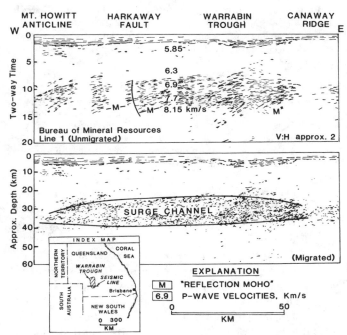

Figure 68. West-east reflection-seismic cross section of the Warrabin trough, Eromanga basin, southwestern Queensland, from Finlayson et al. (1989). P-wave velocities are shown. This lens-shaped body, which we interpret to be a surge channel, has reflectors throughout and, unlike Figure 67, has no transparent zone. This fact suggests to us that the hot contents have cooled and the surge channel (which is Paleozoic) is now inactive.

within such lenses, high heat-flow bands and belts of microearthquakes are present (Figs. 66, 67). Where a transparent layer is not sandwiched in a lens, there is neither microearthquake activity nor high heat flow (Fig. 68).

Thus, in surge tectonics, the lenses, pods, lozenges, or lenticles of these seismic-reflection lines are extremely easy to explain. They are surge channels. Surge channels, of course, may be active or inactive. An active channel should have semifluid magma or mush within it. The presence of the fluid or mush explains the transparent zone between the upper convex-upward layer and the underlying flat-lying-to-concave-upward layer. Figure 66 displays a rather young (Mesozoic-present) channel that underlies the English Channel and is still moderately active. Figure 67 displays an active channel (Jurassic-present) also, this in the North American Western Cordillera of northeastern Washington State. Figure 68 shows a channel in southwestern Queensland that no longer is active. It was active through Devonian time. Today it is essentially solidified throughout and therefore, reflective throughout.

We have defined a kobergen as a bilaterally deformed surge channel. Two examples of a kobergen that appear on seismic-reflection lines include the Pyrenees (Fig. 65) and the northern Appalachians (Fig. 46). The upper panel of Figure 46 shows the two wings of the Acadian (Devonian) kobergen of the northern Appalachians centered on the root zone of the Bronson Hill anticlinorium. Farther east is a remnant of what may have been an older surge channel centered near the Lake Char-Bloody Bluff fault. In Figure 65, the center of the deformed Pyrenees kobergen is transparent, which is consistent with the continuing activity of the Pyrenees kobergen-surge channel. It should be noted that the Pyrenees are the site of the North Pyrenean strike-slip fault discussed in the preceding section.

Surge-channel structure

Figures 46 and 65–68 provide cross-section images of what we believe are the main body of each surge channel, or surge-channel complex, before and after its initial tectogenesis. Figure 68 shows that not all surge channels undergo tectogenesis. The channels that are least likely to undergo tectogenesis are those beneath the thickest parts of the continental lithosphere.

Figures 10, 18, 26, 31, 43, 47, and 62 demonstrate that surge-channel structure is quite complex, and that the main body, or main channel is likely to be overlain by

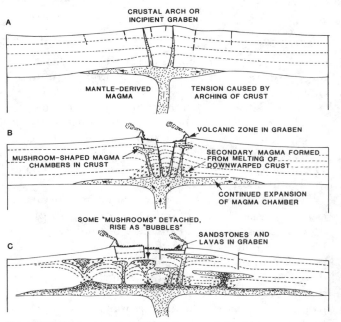

Figure 69. Bridgwater et al. (1974) presented this figure to show the nature of laccolith and sill complexes whose uparching of the overlying lithosphere causes graben formation. See the text for the physical explanation which is based on Newton's Law of Gravity, the Peach-Köhler climb force, and the level of neutral buoyancy as explained by Gilbert (1887), Weertman and Weertman (1964), Gretener (1969), Weertman (1971), Ryan (1987), Corry (1988), and Walker (1989). Bridgwater et al. (1974) referred to this laccolith-sill complex as a "nested series of mushrooms." Published with permission of Elsevier Science Publishers B. V.

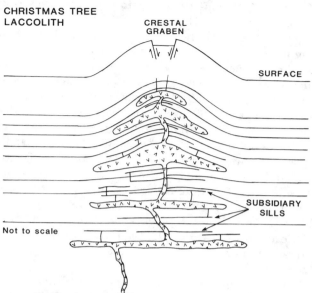

Figure 70. Corry (1988) published this figure, calling it a "Christmas tree laccolith." It differs from Bridgwater et al.'s (1974) model in that the Christmas tree structure is a series of laccoliths, one above the other. Published with permission of Geological Society of America.

smaller, more differentiated channels. It is more technically correct to use the term surge-channel complex than surge channel. However, for the sake of simplicity, we shall use the term surge channel for both a simple (single) surge channel and a surge-channel complex.

Figures 69 (from Bridgwater et al., 1974) and 70 (from Corry, 1988) present two possible surge-channel models. Both formed as a consequence of the law of gravity and the action of the Peach-Köhler climb force that we described in a preceding section (Gilbert, 1877; Weertman and Weertman, 1964; Gretener, 1969; Weertman, 1971; Ryan, 1987; Corry, 1988; Walker, 1989). The authors of Figures 69 and 70 stressed the development of rifts above laccolithlike bodies. Illies (1970) attributed great importance to large laccolithlike bodies at or near the crust-mantle boundary, and concluded that such bodies can produce continental-scale rifts. As we have shown with several examples (Figs. 35, 54, 55, 57, 59), this is also a conclusion of surge tectonics.

Eddies and vortex structures

As we prepared Figure 2, which shows the distribution of the high heat-flow bands on the Earth's surface, we repeatedly encountered several types of structures that are not generated by straight laminar flow, but by vortical flow. These structures suggest eddies, whirlpools, and vortical motions in general, and they occur at all scales, from features only a few kilometers in diameter to features 1,000 km or more across. We identified three types, but found many cases where the structure in question was intermediate between two types. We call the three types fracture band offsets, channel overlaps, and vortex structures. They seem to form a gradational series, with fracture band offsets at one end and vortex structures at the other. It is of great importance that they occur along rifts (midocean ridges, continental rifts), strike-slip zones, and compressional belts.

Fracture band offsets—An example is the East African rift. Figure 55 shows four examples from the Ethiopian (East African) rift valley. Each row of black dots indicates a transverse zone across which the rift-parallel bands of faults, fractures, and fissures are offset. At Munessa village (Fig. 55), the rift valley wall itself is offset. Mohr (1987) expressed bafflement in attempting to explain this peculiarity of rift-valley geology.

The fracture band offsets resemble similar offsets that occur at ridge-transverse fault zones (transform faults) on midocean ridges. We suggest these offset bands of fractures in continental rifts originate in the same way as they do on midocean ridges, except the original ridge-transverse faults in the continental examples now are buried by a much thicker lithosphere. Meyerhoff et al. (this volume) wrote that the formation of ridge-transverse fault zones on midocean ridges is related closely to constrictions (segmentation) in the surge channels, which interrupt straight laminar flow.

Independent studies strengthen this opinion. Ebinger (1989) and Ebinger et al. (1989) noted that the East African rift valley is segmented by constrictions every 50 to 100 km, between which rift basins are developed (see Fig. 1). Ebinger et al. (1989) concluded the megastructure of the East African rift is not greatly different from that of the midocean ridges.

We documented the great similarity between the geophysical and geological characteristics of the East African rift and the geological and geophysical characteristics of strike-slip zones and active foldbelts: a lens of 7.2–7.5-km/s material at a depth of 20 to 30 km, a shallow low-velocity zone, a high-conductivity anomaly, high-heat flow, thermal springs, active volcanism, high seismicity, and a system of linear faults, fractures, and fissures that parallel the rift (Fig. 55). In addition, the East African rift in several places occupies the center of a 950-Ma rift valley and kobergen (Villeneuve, 1983).

Channel overlaps (eddylike structures)—These structures closely resemble—and in our opinion are identical with—the eddylike features of the midocean ridges that are called overlapping spreading centers (Macdonald and Fox, 1983). These also form during channel-parallel movements, as shown by Meyerhoff et al. (this volume). Figure 71 from Dunkelman et al. (1988) shows the East African rift system. The north-south overlap of

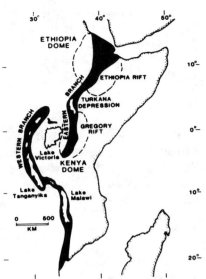

Figure 71. East African rift-valley system (from Dunkelman et al., 1988). We interpret the large-scale overlap of the Western rift (west of Lake Victoria) by the Gregory rift as an asthenosphere vortex structure that is reflected at the surface by surge-channel manifestations (horst-and-graben structures, hot springs, high heat flow, volcanics, etc.). Published with permission of Geological Society of America.

the Eastern (Gregory) rift with the Western rift is, we think, a large-scale example of an eddylike structure. Figure 72 from Legg et al. (1989) illustrates a much smaller eddylike structure. This structure lies along the San Clemente fault zone in offshore southern California. A third example, this of intermediate size, is shown on the inset map of Figure 60. Here, in the San Francisco Bay area of central California, the intense seismicity along the San Andreas fault south of San Francisco is taken up on a subparallel fault system, the Hayward-Calaveras fault zone where the intense seismicity zones of the two faults overlap.

We stated that the eddy structures portrayed on Figures 60, 71, and 72 probably originate in the same way as the so-called overlapping spreading centers of the midocean ridges. One distinguishing characteristic of an overlapping spreading center is the depression that develops between the overlapping ridge segments (Macdonald et al., 1988). In each of the three examples given here, a similar depression is evident. On Figure 60, for example, the depression of San Francisco Bay lies between the main branch of the San Andreas fault on the west and the Calaveras-Hayward fault on the east. On Figure 71, the Lake Victoria depression lies between the western and eastern (Gregory) branches of the East African rift. On Figure 72, a depression that plunges toward the northwest is evident, and is centered at 32°38′ N and 118°04′ W.

Sizable depressions everywhere are found where similar overlaps are present. Additional examples along the San Andreas fault system, for example, include the Cholame Valley and the Salton Sea. Most of these are called pull-apart basins (e.g., Legg et al., 1989), yet all that we observed are eddylike or vortical structures. The depressions that are present presumably were produced by downward vortical motions in the surge channel beneath the San Andreas fault zone (Fig. 60). Another

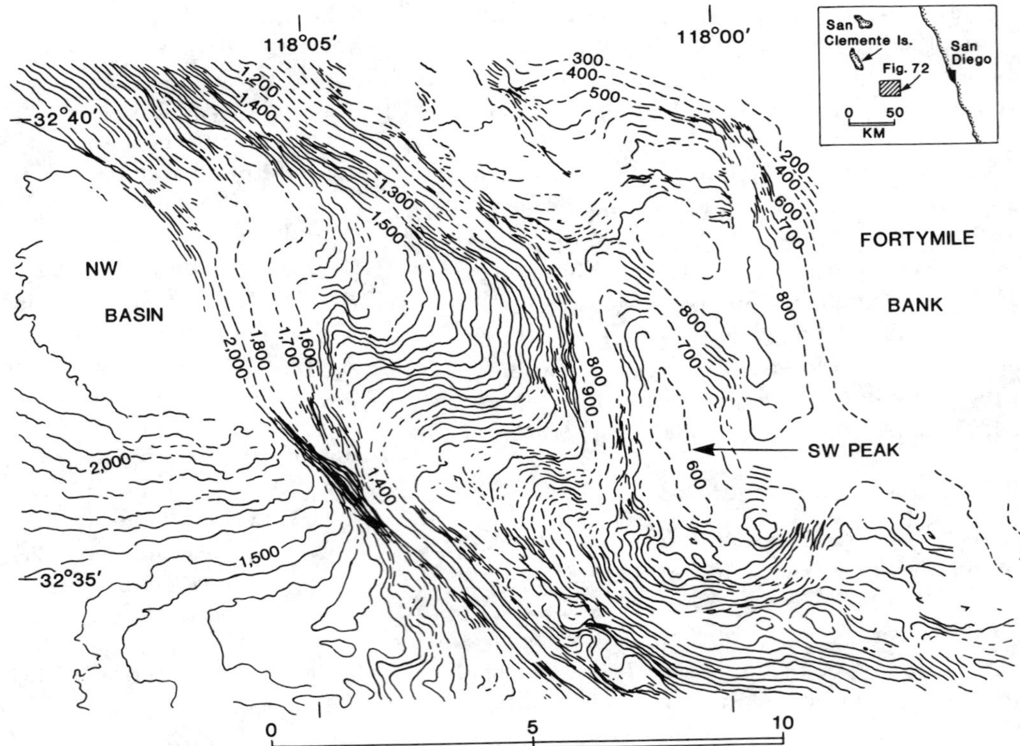

Figure 72. An outstanding example of small channel overlap (like an eddy) along the San Clemente fault zone, southern California borderland (from Legg et al., 1989). This structure resembles in every important detail the eddylike overlapping spreading centers of the midocean ridge system (Macdonald and Fox, 1983). The bathymetry clearly reflects the turbulencelike motions in the surge channel. The overlap of the channel (which is why we call them channel overlaps) is about 6 km.

example is the Dead Sea depression. Hundreds more exist around the world (e.g., Schwartz and Sibson, 1989).

The fact should be noted that eddy structures are present in widely different tectonic settings—rift zones, strike-slip zones, and compressional belts. Emmons (1969) showed in a series of laboratory model experiments how such structures develop. In all cases the eddy structure was caused by motions parallel with the strike of the tectonic belt.

Vortex structures—Vortex structures probably began as eddy structures. However, they are much more complex than eddies. Whereas fracture band offsets and eddies originate along a single surge channel, specifically where that channel is constricted or segmented (Figs. 16, 17), vortex structures originate where surge-channel flow encounters a major obstruction other than a simple along-strike constriction of the channel. One such obstruction we identified in several places is an intersection with another surge channel. However, the flow of the two intersection channels must be crosswise or even in opposite directions. Vortex structures do not appear to form where the intersecting of two channels is confluent and the two flows merge into a single larger channel.

Another type of obstruction identified is found mainly where a feeder channel from a midocean ridge passes through a continent. Here, large, deeply rooted promontories of continental lithosphere commonly protrude into the paths of continental trunk channels (Fig. 14), forcing them to change course and in places, constricting them. Large-scale eddies and vortex structures can form at these constrictions, and thereby create vortex streets (Tritton, 1988; see Appendix.)

Vortex structures are round to ovate features that range in diameter from 200 to 1,000 km, most commonly in the 300 to 800 km range. They possess several of the following characteristics: disordered topography (bathymetry) dominated by a major depression (just like eddies), fracture patterns that in places are arranged in circular spires or whorls, intense concentrations of microearthquakes, high heat flow, tensile stress in their centers, jumbled magnetic and gravity lineations, high volcanicity, abundant thermal springs and hydrothermal vents, a shallow midcrustal low-velocity layer, and a deeper lens of 7.0–7.8-km/s material.

Examples of vortex structures are numerous. Topographically and geologically, they range from rather simple ovate forms (e.g., Colombian basin of the Caribbean, Deccan Traps, Iceland, Columbia Plateau) to more complex structures (e.g., Pannonian basin: Fig. 12; Aegean Sea: Figs. 73–75; Dasht-i-Lut: Fig. 77; North Fiji basin: Fig. 21; Easter microplate: Fig. 76). They are easily recognizable because of their general shape and because they interrupt regional tectonic-structural trends.

Figures 73 through 75 show a major vortex structure beneath the Aegean Sea. Figure 73, from Papazachos and Comninakis (1977), shows the seismicity of the

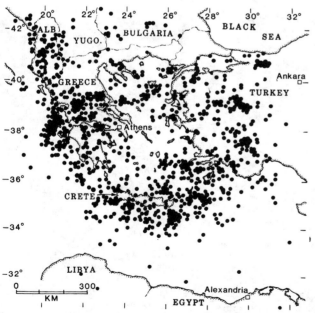

Figure 73. Aegean vortex structure: earthquake epicenter pattern from Papazachos and Comninahis (1977). Note the concentration of earthquake epicenters around the rim of the structure.

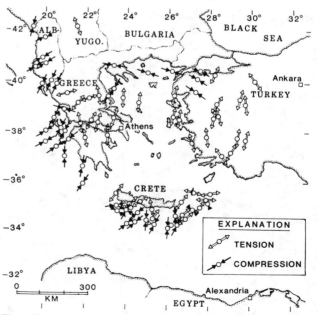

Figure 74. Aegean vortex structure: stress distribution. Tensile stress is in the center; compressive stresses are concentrated around the rim of the structure. From Papazachos and Comninahis (1977).

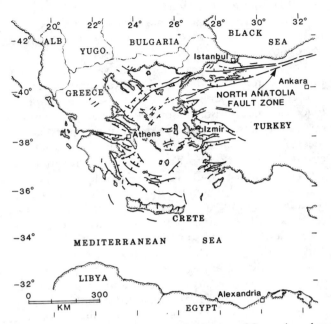

Figure 75. Aegean vortex structure: distribution of Cenozoic grabens and normal faults. Note the circular pattern. From Şengör (1979). Published with permission of the Geological Society (London).

structure. The seismicity is scattered in pockets and short bursts across the vortex structure. Figure 74, also from Papazachos and Comninakis (1977), shows stress distribution around the Aegean. The whole interior of the structure is under tension, whereas a band of compression surrounds the structure on three sides. Figure 75, from Şengör (1979), shows the tensile structures formed within the Aegean's tensile stress field. In addition, the Aegean is underlain by anomalous upper mantle with a P-wave velocity of 7.8 km/s (Berckhemer, 1977), a strongly vortical pattern of faults and fractures (Berckhemer, 1977; Lyberis, 1977), high heat flow (Erickson et al., 1977; Čermák and Rybach, 1979), a negative Bouguer gravity anomaly (Berckhemer, 1977), and a circular pattern of young volcanics (Fytikas et al., 1984).

The stress-distribution pattern shown on Figure 74 typifies the patterns encountered in active vortex structures. We interpret the pattern to mean that the structure's interior was the site of large-scale magma accumulation resulting from rupture of the intersection channels. The thickened magma accumulation presumably still has not cooled. Such concentrations of magma in the Colombian basin of the western Caribbean Sea—another vortex structure—doubled the crustal thickness there (Bowland and Rosencrantz, 1988). In any case, the presence of thickened crust beneath the Aegean forced younger channels to circumscribe it.

These channels generate compression along three sides of the Aegean (Fig. 74).

Figure 76 shows a spectacular vortex structure discovered recently in the Easter Island area of the East Pacific Rise. The fracture trends, traced from sonographs, describe a nearly 180° swirl between the two branches of the East Pacific Rise. Although Searle et al. (1989) interpreted the pattern to indicate clockwise vorticity, the fault pattern and the positions of the possible thrust faults north of the Pito fracture zone indicate that all vortical motions here have been, and continue to be, counterclockwise (Tritton, 1988). Second-order eddies can be seen along the eastern branch of the East Pacific Rise (Fig. 76).

Figure 77 shows the fault and lineament pattern of the huge Dasht-i-Lut vortex in eastern Iran. The origin of this structure has long been an unsolved geological problem. Wellman (1966) made a strain analysis of the entire region, and concluded that the Dasht-i-Lut vortex was formed by crustal material spiraling downward in a counterclockwise sense. The fact that the Dasht-i-Lut is a great regional depression surrounded by mountain ranges accords well with Wellman's (1966) interpretation. The region also is distinguished by high heat flow, active volcanism, high seismicity, ophiolite belts, kobergens, and several other phenomena that characterize surge channels (Table 5; F. Berberian and M. Berberian, 1983; M. Berberian, 1983).

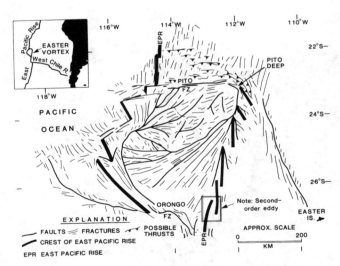

Figure 76. Easter Island vortex structure. This is a simplified tectonic map based on GLORIA and SeaMARC II sonographs. Note: (1) The East Pacific Rise axis forms a gigantic overlapping spreading center that encloses most of the vortex. (2) The orientations of the various fault sets and the presence of northward-verging thrusts north of the Pito fracture zone show that the vortex formed by counterclockwise motions and not by clockwise motions, as Searle et al., (1989) suggest. (3) Small second-order overlapping spreading centers are developed in places along the East Pacific Rise crest. From Searle et al. (1989).

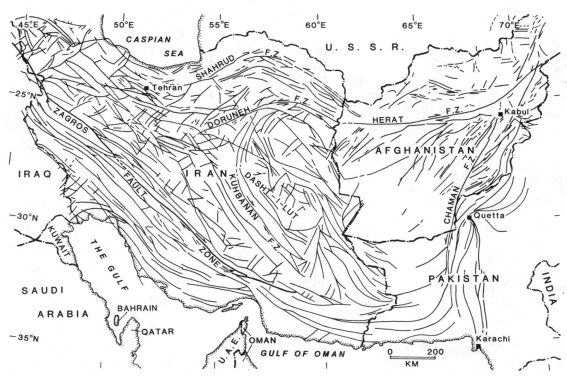

Figure 77. Dasht-i-Lut vortex structure, Iran. The orientations of the faults show that this structure is a counterclockwise vortex. From Wellman (1966).

Surge-channel flow

The evidence for horizontal flow in surge channels is overwhelming.

Geometric (structural) evidence—Several of these phenomena have been discussed. These are (1) the presence above each surge channel of long (up to thousands of kilometers), linear to curvilinear faults, fractures, and fissures that form only as a result of subcrustal flow, in accordance with Stokes's Law (Figs. 11, 20, 21, 33, 52, 55, 61, 78); (2) en echelon arrangements of the linear faults, fractures, and fissures (Fig. 11); (3) eddylike structures (Figs. 61, 71, 72); (4) vortex or whirlpool structures (Figs. 73–77); (5) pinch-and-swell structure oriented at right angles to the surge channel (Figs. 16, 17); and (6) sinuous and braided channel geometries in plan view (Figs. 1, 9, 12, 21, 22, 24, 49). Figure 22 provides a particularly dramatic illustration of point 6.

(7) Other evidence is provided by horsetail-like structures (Emmons, 1940) as shown in Figure 59, from Illies and Greiner (1979). The fault bifurcations (horsetail structure) illustrated in Figure 59 are toward the north-northeast, indicating that flow is in that direction. The north-northeastward bifurcations are analogous to a seaward-facing rill system (Schrock, 1948).

(8) Geophysical evidence consists mainly of the associations of bands of high heat flow with bands of microearthquakes of approximately the same position and width. Figure 78, from Hawkins (1974), shows

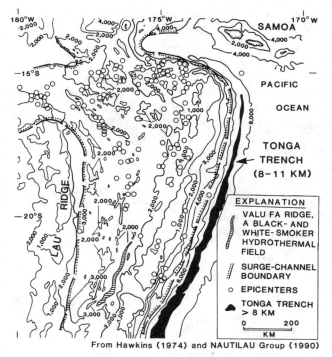

Figure 78. Tonga Ridge, Lau Ridge, and Lau basin, southwestern Pacific; epicenters are shown. Note that the epicenters tend to cluster in rows along the tops of sea-floor ridges. This fact indicates that (1) motions beneath the Lau basin parallel the basin axis, (2) streamline shears underlie these linear sea-floor ridges, and (3) subsurface flow, therefore, is laminar. Some of these ridges (e.g., Valu Fa Ridge) contain rows of black smokers (from NAUTILAU Group, 1990). Location of the figure is on Figure 23.

aligned epicenters in the Lau basin between the Tonga volcanic arc and the Lau ridge (Fig. 23). The close relation between lines of earthquake epicenters and narrow ridges is evident. The pattern is an unusually fine illustration of Stokes's Law, demonstrating that the mobile material beneath the Lau basin exhibits laminar flow. The additional fact that the basin widens, or flares toward the north also indicated that flow is north-northeastward to about 16° S latitude, where it turns westward into the North Fiji basin (Fig. 23). The North Fiji basin is a spectacular vortex structure shown in greatly generalized form on Figure 21 (Auzende et al., 1988).

Flow direction—Using the above criteria, one can determine flow direction. We summarize our findings by stating, as demonstrated with Table 2, that the preferred direction of movement is eastward. However, flow is first and foremost toward the path of least resistance, so there are exceptions (e.g., Fig. 78). It appears safe to state that where the material in the surge channel has a choice, the material will select the paths with eastward components.

Seismic anisotropy—Many workers believe that seismic anisotropy provides a reliable means of determining flow direction. The essence of the theory is that where flow takes place, the long axes of olivine (and other) crystals are oriented in the direction of flow. Because the long axes transmit seismic waves more rapidly, the seismically fast direction must be the flow direction. For example, Raitt et al. (1969) found the fast wave-propagation direction in two limited areas off the California coast is east-west, but cautioned against extrapolation of their results to the whole of the eastern Pacific or to any other part of the eastern Pacific. Despite this, the Raitt et al. (1969) results are cited repeatedly as further proof that sea-floor spreading takes place at right angles to the trend of the East Pacific Rise.

More detailed studies of anisotropy in oceanic crust have been conducted in recent years. The results are different from those predicted in plate tectonics and by Raitt et al. (1969). For example, Stephen (1985) studied the Costa Rica rift zone some 550 km south-southwest of Panama, and found that the fast seismic velocities parallel the rift. Burnett and Orcutt (1989), as a result of a study along the East Pacific Rise, found that the fast direction parallels the trend of the rise. Kuo et al. (1987) obtained the same result along part of the Mid-Atlantic Ridge. In all continental rift studies of which we know, the fast direction parallels the trend of the rift system (e.g., Rhine graben, Fuchs, 1983; Salton trough, Hearn, 1987; etc.) Studies of anisotropy, however, have scarcely begun on a global scale, and it will be many years before sufficient data are available.

Flow rates—As might be expected, these differ from one channel to the next. We summarized several examples in Table 9. In general, flow rates in ocean basins are higher—7.5 to 13.0 cm/yr—than in continental areas—1.0 to 7.2 cm/yr. One exception to oceanic flow rates—17.6 cm/yr beneath the eastern Hawaiian Islands—possibly is related to the formation of a vortex structure (Meyerhoff et al., submitted). A 24.2 cm/yr example from the Colorado Plateau seems to be related to the presence of an exceptionally shallow magma chamber above a Christmas-tree type surge-channel complex. Table 9 shows that the rate of flow is depth-re-

Table 9. Flow rates in selected surge channels.

Area or channel name	Length of dated segment (km)	Age span (Ma)	Velocity (cm/yr)	Source
Louisville Ridge	6,100	66	9.2	Watts et al. (1988)
Hawaiian-Emperor Ridge				
1. Seamount (near Suiko) to Molokai	4,620	71.7	6.4	Recalculated from Shaw et al. (1980)
2. Western Maui to Hawaii	230	1.3	17.8	Shaw et al. (1980)
Kodiak-Bowie (Pratt-Welker) Seamount Chain	1,100	21	4.4	Turner et al. (1980)
Ninetyeast Ridge				
1. Offshore	4,450	40	11.4	Luyendyk (1977)
2. On- and offshore	5,750	44	13.0	Luyendyk (1977); Saxena (1986)
Western Colorado Plateau	300	12.4	24.2	Moyer and Nealey (1989)
Lake Baykal				
1. Channel toward west	400	40	1.0	Logatchev and Florensov (1978)
2. Channel toward Pacific	1,600	40	4.0	Logatchev and Florensov (1978)
Fenwei, Shaanxi-Shanxi	1,000	44	2.5	Wang Jingming (1987)
Hetao-Yinchuan, Ninxia-Nei Monggol	700	44	1.6	Wang Jingming (1987)
Lower Yangzi Valley, Jiangxi and Zhejiang	900	77	1.2	Zhu Ming (1989)
Tasman foldbelt				
1. Part of the foldbelt	1,815	27.1	6.7	McDougall et al. (1981)
2. All of the foldbelt	2,800	38.7	7.2	Pilger (1982)

lated, for the lithosphere thickness above an oceanic surge channel should be in general less than that above a continental surge channel.

Surge-channel duration

The lifespans of surge channels are extremely variable. We have not studied a sufficient number to make many definitive statements. By using the identifying criteria in Table 5, surge channels can be mapped and dated successfully. We illustrate with a few examples.

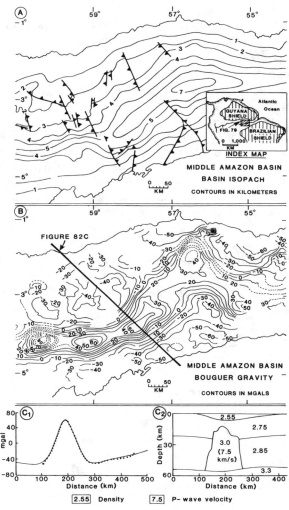

Figure 79. Middle Amazon basin, Brazil: basin isopach map, Bouguer gravity map, and cross section based on gravity data (from Nunn and Aires, 1988). Compare this figure with Figure 58. The basin depocenter and an east-west positive Bouguer anomaly coincide almost exactly. The source of the anomaly is a lens of 7.0–7.5-km/s material, probably mafic rock, the remnant of a Paleozoic surge channel (see text). The minimum lifespan (according to radiometric dates from mafic rocks in the basin) of this channel was 425 Ma. This figure shows the close genetic relationship between a surge channel and intracratonic basin formation.

Oceanic trunk channels—As shown in Table 4 and Figure 14, these are principally midocean ridges. The only ridge for which we have adequate data to provide a minimum lifespan estimate is the Mid-Atlantic Ridge. Meyerhoff et al. (this volume) summarized the age data based on dredge hauls and drilling along the ridge. St. Paul's Rocks (0°56' N latitude, 29°22' W longitude) at the ridge crest is underlain by a peridotite-hornblendite-gabbro intrusion dated at 835 Ma (Melson et al., 1972), a date confirmed independently by Umberto Cordani (*in* Melson et al., 1972). Melson and his colleagues stated that the age is not due to excess radiogenic argon. Closely associated with St. Paul's Rocks are middle Miocene sedimentary rocks and Quaternary basalt (Cifelli, 1970; Melson et al., 1972). These data suggest that the channel is at least 835 Ma.

Honnorez et al. (1975) dredged a shallow-water Middle Jurassic limestone at 10°41' N, 44°18' W. The fauna is typically Late Triassic to Middle Jurassic (P. Brönnimann, pers. comm., 1974). Ozima et al. (1976) collected 169-Ma (Middle Jurassic) basalt at 30°01' N, 42°04' W. These data show that the ridge was shallower than today and that midocean-ridge basalts were extruded during Jurassic time, thus indicating surge-channel activity during the Jurassic.

Aumento and Loncarevic (1969) found a large granitic massif of about 80 km^3 at 45°13' N, 28°52' W. The granitic rocks yielded dates of 1,690 and 1,550 Ma. They are intruded by mafic dikes or sills that are 787 Ma (Wanless et al., 1968). This last date suggests a minimum age for the surge channel. The much older granitic rocks indicate that the Mid-Atlantic Ridge surge channel may overlie an even older surge-channel complex. In surge tectonics, granites form inside surge channels (Figs. 95, 96).

At Site 334, Leg 37 (37°02.13' N, 34°24.87' W), the Shipboard Scientific Party (1977) described a gigantic gabbroic boulder in a middle Miocene nannofossil ooze. They interpreted the deposit to be a mélange atop the Mid-Atlantic Ridge. Reynolds and Clay (1977) determined the age of the boulder at 635 ± 102 Ma. They suggested the ancient date was a result of excess ^{40}Ar gas, but admitted this explanation was fraught with "difficulty" (p. 629). Thus drilling results from this location also suggest the great antiquity of the Mid-Atlantic Ridge surge channel. From the data presented here, we conclude that the Mid-Atlantic Ridge channel is not younger than 835 Ma, and probably is more than 1,690 Ma.

Feeder channels—We interpret the Carnegie Rise west of Ecuador to be a feeder channel. More specifically, we interpret it as having fed the surge channel that underlies the entire Amazon River basin from Peru and Ecuador to the Atlantic Ocean. Pertinent age data are available only from the onshore part of the channel (in the onshore area, we would call this channel a continental trunk channel; Fig. 14, Table 4).

Figure 79 shows the Middle Amazon basin segment of the Carnegie surge channel. The present shapes of the basins along the Carnegie surge channel were not acquired until Ordovician time, although a pyroxenite at the base of the sequence in the Middle Amazon basin yielded an Nd-Sm age of 565 ± 70 Ma (Early Cambrian on the Harland et al., 1982, time scale; Nunn and Aires, 1988). Another pyroxenite penetrated in a well in the basin gave a K-Ar age of 409 ± 78 Ma (latest Silurian; Szatmari, 1983), and is overlain by marine Silurian strata. The next recorded magmatic episode lasted from 293 to 140 Ma (Late Pennsylvanian to Neocomian; Bigarella, 1972; Petri and Fulfaro, 1983; de Almeida, 1986; Mosmann et al., 1986), during which an estimated 340,000 km3 of mafic dikes, sills, and flows was emplaced (Mossman et al., 1986). If we accept the radiometric dates as a measure of the minimum age of the channel, we have an age span of 425 Ma (565 less 140 Ma) for the Carnegie surge channel.

Figure 79 shows several very important characteristics of surge channels. They must be a major mechanism in the formation of intracratonic basins. Note the depocenter of the 7-km-thick Ordovician through Cretaceous section (Fig. 79a) coincides almost exactly with the positive Bouguer gravity anomaly (Fig. 79b). Because such a thick sedimentary section should generate a gravity minimum, not a maximum, Nunn and Aires (1988) postulated the presence of a thick lens of high-density rock extending well into the lower crust (Fig. 79c). The positive Bouguer anomaly indicates that little if any magma remains in the channel. The Carnegie surge channel closely resembles the Devonian channel of the northern Appalachians (Fig. 45) as well as the Devonian channel beneath the Warrabin trough (Fig. 68).

Breakout channels—Several of these are illustrated and discussed (Figs. 23, 24; Table 9). They seem to be short-lived—21 to 72 Ma for the Kodiak-Bowie, Louisville, and Hawaiian-Emperor chains. The lifespan of the New England Seamount chain was about 114 Ma, persisting from 135 Ma to about 21 Ma (Houghton et al., 1978; Duncan, 1984; Jansa and Pe-Piper, 1988).

Although the examples cited show age progressions generally from west to east, many breakout channels do not. Figure 80 shows the Cook-Austral Ridge from the southwest Pacific. The only active volcano is at the southeastern tip of the chain. The dates obtained suggest that the surge channel was active throughout its length for much of its 19 to 25 Ma of existence.

Figure 80 shows another feature, namely, the great range in the ages of rocks from Tahiti. Rocks from that island recorded events from the Pliocene to Recent, early to middle Miocene, late Oligocene, Late Jurassic, and late Proterozoic (Krummenacher and Noetzlin, 1966; Krummenacher et al., 1972; Meyerhoff and Meyerhoff, 1974).

Continental trunk channels—We described one of these, the Carnegie surge channel that passes under the Amazon Valley of South America (Fig. 79). A second example, this one a foldbelt that was deformed repeatedly, is the Appalachian surge-channel system (Figs. 1, 44, 45, 46).

The Appalachian surge-channel system came into existence after the demise of the Grenville channel system (Rankin, 1976). Data summarized by Powell et al. (1988) suggest that the shift took place between 955 and 820 to 810 Ma. The minimum age for the oldest rocks in the Appalachian geosyncline is about 850 Ma. The youngest rocks of the Appalachian surge-channel system are the Middle Triassic to Early Jurassic Newark-Eagle Mills rifts (Smoot et al., 1988) and related Late Triassic to Late Jurassic dike swarms (220–160 Ma; McHone, 1978; Mc-

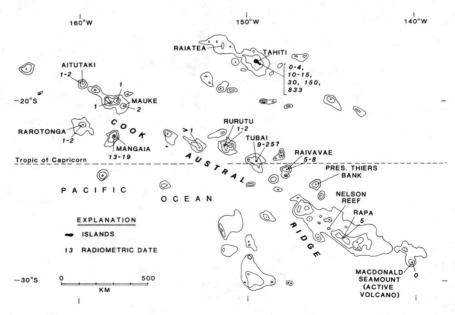

Figure 80. Radiometric dates from the linear island and seamount chains near Tahiti. Compiled from various sources, including Krummenacher and Noetzlin (1966), Krummenacher et al. (1972), and Turner and Jarrard (1982). The dates show no age progression. One age, in fact, is 833 Ma which is Proterozoic (late Riphean).

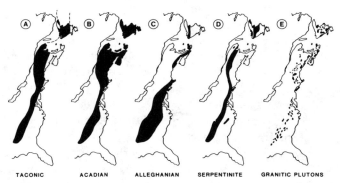

Figure 81. Appalachian foldbelt: major tectonic activity (from Williams, 1984). **A**—Area affected by Taconic orogeny; **B**—Area affected by Acadian orogeny; **C**— Area affected by Alleghanian orogeny; **D**—Distribution of Paleozoic ophiolites; and **E**—Distribution of Paleozoic granites. The figures show the complexity of a surge-channel system within a single mobile belt. The repeated shifts in the axis of maximum tectogenesis, especially in the north, are a characteristic of many long-lived surge channel systems. The figure shows clearly that multiple surge channels, probably interconnected, are likely to be present beneath a mobile belt (e.g., the Alps today; see Fig. 51 and discussion of the Alps in the text).

Hone et al., 1987). After 160 Ma, the Appalachian foldbelt system of channels became extinct. The lifespan of the Appalachian system was at least 700 Ma.

Figure 81 shows the complexity of the surge-channel history of the Appalachians. From left to right, the figure shows the areas of the Appalachian system affected by the Taconic orogeny (Fig. 81a), Acadian orogeny (Fig. 81b), and Alleghanian orogeny (Fig. 81c). Figures 81d and 81e show respectively the locations of Paleozoic ophiolites and granitic plutons (Williams, 1984).

After Late Jurassic time, a new surge-channel system, this one much smaller, established itself under the Appalachians. One channel sustains the modern elevation divide. This channel, best known along the Virginia-West Virginia border, displays microseismicity (Bollinger and Wheeler, 1988), high heat flow (Dennison and Johnson, 1971), thermal springs (Dennison and Johnson, 1971), middle Eocene dikes (Fullagar and Bottino, 1969; Dennison and Johnson, 1971), and other surge-channel phenomena.

VLBI data

Very long baseline interferometry (VLBI) measurements have been hailed by a few workers as having proved plate tectonics (e.g., Argus and Gordon, 1990). In fact, they have done no such thing, and many of the results of measurements—some taken since 1979—are confusing to say the least (Ma et al., 1989). The only fairly consistent result thus far is the growing evidence that Japan and North America are approaching one another (Ma et al., 1989; Paul D. Lowman, Jr., pers. comm., 1989). Such movement between Japan and North America is not evidence for plate motions of the sort predicted by plate tectonics unless the relative motions predicted among all plates also are observed. They are not. In fact, Japan and North America's approach (approximately 50 mm/yr) is explained by surge tectonics, as we demonstrate in Figure 82. If the surge channels beneath Japan and the East Pacific Rise (and its extension into North America) are filling, the western flank of the East Pacific Rise and the eastern flank of the Japanese surge-channel system will appear to be approaching one another as shown on Figure 82. Such a mechanism may explain the contradictory results among other continents and, in fact, within the same continent and plate.

The evidence that surge channels are filling is quite convincing. Smith (1978) studied the Rocky Mountains of northwestern Wyoming and found that they are rising at a rate of 3 mm/yr. This compares with 0.4 mm/yr in the eastern Great Basin (Naeser et al., 1983), 4 mm/yr in the Colombian Andes (Gansser, 1983), and 5 mm/yr in the Himalaya and Qinghai-Tibet Plateau (Gansser, 1983). Evidence suggests that all major Cenozoic mountain ranges are rising actively (Gansser, 1983; Nikonov, 1989). In each range mentioned by Gansser (1983) and Nikonov (1989), we found a wealth of evidence for the presence of an active surge channel at depth by using the criteria listed in Table 5.

Mapping surge channels

Almost all structures or manifestations of structures generated by surge-channel activity are linear to curvilinear, and maintain their identity for hundreds, even thousands of kilometers. The examples we have shown indicate that such structures may be as straight as the San Andreas fault zone (Figs. 33, 60) or as curved as the

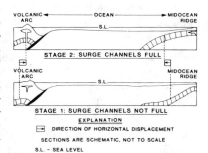

Figure 82. Schematic cross section between two surge channels. The channel on the left underlies a volcanic arc; that on the right underlies a midocean ridge. In Stage 1, the two surge channels are only half full. In Stage 2, they are full. As the two surge channels fill, the seaward-facing flank of the channel beneath the volcanic arc not only rises but also moves toward the midocean ridge. Similarly the flank of the midocean ridge that faces the volcanic arc rises and shifts toward the arc. The net effect is that distance measurements between the two facing flanks would suggest that the two are approaching one another. This is one explanation of the VLBI and other precision measurements that suggest that Japan and North America are approaching one another (Ma et al., 1989).

Dasht-i-Lut vortex (Fig. 77). The overall linearity may be produced by a few dozen megascopic folds and faults, a seemingly infinite number of stretching lineations, rows of isolated plutons (Fig. 49), or by geophysical phenomena, such as lines of epicenters (Fig. 78) and bands of elevated heat flux (Figs. 2, 49). The scale is inconsequential. Their linearity (or curvilinearity), general coherence, and great length are the important facts.

Active surge channels—In surge tectonics, the presence of an elevated mountain range is clear evidence that a surge channel is directly beneath. As we observed, many Cenozoic mountain ranges are rising. We attribute such uplift to filling of underlying surge channels. Gansser (1983) published data that show the rate of rise accelerated steadily during the last 5 to 10 Ma.

To ascertain whether an active surge-channel complex underlies a given mountain range, corroborative information as outlined in Table 5 is needed. Such information could include the presence within the mountain range of a band of high heat flow, lines of hot springs, a belt of microearthquakes, zones of recently active faults, evidence for geologically recent volcanism, a negative Bouguer gravity anomaly, and a high-conductivity magnetotelluric anomaly. Seismic refraction studies should reveal the presence of one or more shallow low-velocity zones and a lens of anomalous upper mantle with its characteristic 7.0–7.8-km/s velocities.

The existence of an actively rising plateau (e.g., Qinghai-Tibet Plateau, Colorado Plateau, the Andean Altiplano) also signifies the presence below of an active surge-channel complex. In the case of the Qinghai-Tibet Plateau, abundant field geological and geophysical data indicate at least three parallel channel complexes coexist side by side. The entire plateau is rising uniformly suggesting that the three channel complexes are laterally interconnected.

Rifts commonly develop within rising mountain ranges. An example of such a rift is the Rio Grande rift (Fig. 38). Other examples include the Great Basin-Gulf of California system (Figs. 31–35) and the East Africa rift (Figs. 55, 71). Some intermontane depressions have ground elevations substantially below sea level; examples include the Death Valley of California and the Turpan depression of western China. The unusual depths of such depressions may be expressions of graben subsidence but in some cases (e.g., the Turpan depression) vortical motions probably are involved. The great depths of the Baykal (Figs. 53, 54) and Dead Sea depressions seem to be due mainly to graben subsidence at or near the crests of surge channels.

Low-lying and deeply eroded mountain ranges such as the Appalachians (Figs. 1, 44–46) also seem to imply the presence at depth of still-active but waning surge channels. We have not studied ancient mountain ranges sufficiently to be certain of this interpretation.

Topographic depressions are almost invariably linked to surge-channel activities, as we documented for several eddy and large vortex structures (Figs. 71–77), for example the Lake Victoria basin of East Africa, the Aegean Sea, and others, including the Dasht-i-Lut depression of eastern Iran. Similarly, we documented graben-type depressions that commonly overlie centers of surge channels (Figs. 9, 28, 29, 32, 35, 38, 53–59, 71). In addition, we suggested all intracratonic basins owe their existence to the presence of surge channels, and we gave the Middle Amazon basin as an example (Fig. 79). However, unlike mountain ranges that rise as a consequence of the filling of surge channels, we believe that basins form as a consequence of the withdrawal of the fluid contents of the channels.

Our work was not directed toward the origin of intracratonic basins. However, to determine if the Middle Amazon basin is unique or unusual, we randomly selected the Williston basin for comparison. That basin overlies a lens of 7.0–7.8-km/s anomalous upper mantle (Braile, 1989). It has, among other things, a heat flow that is elevated above that of its surroundings (Kron and Stix, 1982), minor earthquake activity (Camfield and Gough, 1977; Smith, 1978), and a shallow high-conductivity magnetotelluric anomaly (Camfield and Gough, 1977; Hildenbrand et al., 1982).

Results from two intracratonic basins show they are associated with what we interpret as formerly active surge channels. Because of this, and because other basin types (e.g., grabens) are associated with a surge channel, we conclude that all basins are surge-channel related. We suggest intracratonic basins begin to form during the waning stages of a surge channel's lifespan, and fluid withdrawal from the channel is the probable cause. Study of Figure 42 which shows the Baltic Sea region suggests the sea and its arms—the Gulfs of Bothnia and Finland—may be in the process of subsiding today, especially in the areas of highest microearthquake activity.

Inactive surge channels—All gradations exist between highly active surge channels and inactive ones. A fully developed kobergen may evolve above a surge channel without destroying the channel. Examples of kobergens above active surge channels are shown on Figures 9, 28–30, 35, 38, 41, 47–52, 63–65, 73–75, and 77. For the most part the kobergens and surge channels of these illustrations are Jurassic or younger. Most older surge channels we have shown are inactive, or close to it (Figs. 36, 37, 44–46, 81).

Identification of inactive surge channels is easy provided they developed into foldbelts, either alpinotype or germanotype. Just the existence of a foldbelt, in our opinion, is sufficient to establish the presence of a kobergen with either an active or inactive surge-channel system. The many criteria listed in Table 5 are evident from field mapping and geophysical investigations: eugeosynclinal rocks, miogeosynclinal rocks, ophiolite belts, zones of mélange, tectonostratigraphic terranes, rows of granitic plutons, a central zone of metamorphic rocks, inverted metamorphic gradients, and suture zones. If the foldbelt was repeatedly tectonized and metamorphosed, metamorphic core complexes-mantled gneiss domes and median tectonic lines (*Verschluckungszonen*) will be exposed (Figs. 90–93).

In the case of a germanotype foldbelt, the rocks are mainly miogeosynclinal; linear evaporite basins are evident (Fig. 83). Volcanic rocks are silicic, some of them continental. Ophiolite is absent. Mélange may be present but is very different from that formed in analpinotype belt. The kobergenic style is evident, but faults are steeper in general than those flanking an alpinotype belt (Fig. 41, kobergen 4; Fig. 52). The undeformed inactive kobergen is the most difficult to identify (Fig. 68). Even so, the graben structure should be evident from surface mapping and its geophysical signature is quite distinctive.

Figure 84 summarizes the principal megastructures associated with a surge channel during its development. The trends of the grabens and half-grabens, fold axes, major fault-fracture-fissure system, dikes and dike

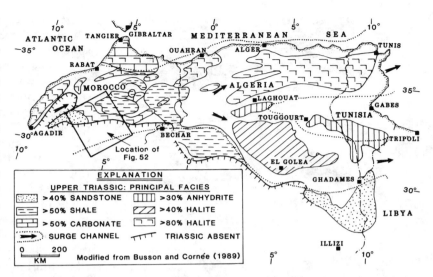

Figure 83. Principal Late Triassic lithofacies in northwestern Africa. We interpret the extensive evaporite deposits here as having been deposited above a large surge channel that extended from west to east across the region. Direction of flow is indicated. The channel may have bifurcated in western Algeria; at the very least the channel seems to have become much broader eastward. Location of Figure 52—the modern surge channel under the Middle and High Atlas—is shown.

swarms, strike-slip faults, and thrust (or gravity) fault traces are the same. In a simple surge-channel system, the tensile features are closest to the center of the kobergen and the compressive structures tend to lie closer to the flanks of the kobergen. In a complex history like that of the Appalachians (Fig. 81), the centers of tensile and compressive stresses shift during the history of the foldbelt, considerably complicating the interpretation.

SURGE TECTONICS, THE GEOTECTONIC CYCLE, AND THE GEOSYNCLINAL CONCEPT

We noted that basins form above surge channels. At the continent-ocean basin interface, the type of basin that forms is the eugeosyncline (Haug, 1900; Aubouin, 1965). We stated that in a cooling Earth, the strictosphere is contracting and in a state of tension, whereas the lithosphere above it is in a state of equiplanar tangential compression. As the strictosphere becomes smaller, the lithosphere must adjust repeatedly its inner circumference to fit it.

The adjustments by the lithosphere do not occur smoothly and continuously, but episodically, as we explain in a subsequent section. This episodicity is the basis of the geotectonic cycle, as defined by Stille (1920, 1924). When the episodic adjustments (that is, tectogenesis) occur, compression in the lithosphere is maximal and surge channels in the lithosphere are deformed. After tectogenesis, compression is minimal, and the next geotectonic cycle begins. Surge channels,

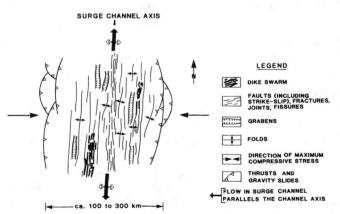

Figure 84. Plan-view sketch showing the structures produced on the roof of a surge channel.

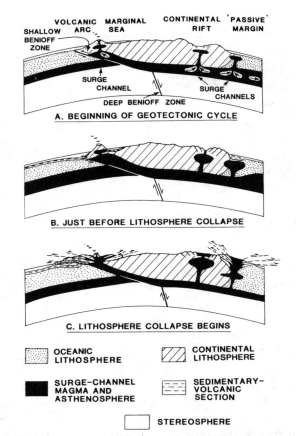

Figure 85. The geotectonic cycle (see text). The principal purpose of this diagram is to illustrate lithosphere collapse into the asthenosphere.

eugeosynclines, episodic tectogenesis, and the geotectonic cycle are inseparably linked.

How contraction triggers tectogenesis

We described the contraction process and will review only highlights. The strictosphere (Fig. 3) contracts by cooling as demonstrated by the presence in it of steep (approximately 60°–70°) Benioff zones. As a consequence, the radius the of Earth shrinks continuously. The circumference of the lithosphere, which already has cooled, must adjust to the constantly shrinking strictosphere. This is accomplished by thrusting along the less steeply inclined lithosphere Benioff zones (approximately 20°–35°). We noted these fault inclinations—20° to 35° in the lithopshere, which is under compression, and 60°–70° in the strictosphere which is under tension—are predicted by the Navier-Coulomb maximum shear stress theory (Scheidegger and Wilson, 1950; Jaeger, 1962).

The asthenosphere between the two solid spheres is equivalent to the level (or zone) of no strain where cooling and magma generation take place (Fig. 3; Davison, 1887). The asthenosphere is the source of magma for surge channels. The asthenosphere provides a structurally weak cushion between the upper and lower spheres. It alternately expands (during tectonically quiet times) and contracts (during tectogenesis). Tectogenesis is triggered by collapse of the lithosphere into the asthenosphere along the lithosphere's gently inclined Benioff zones. Lithosphere collapse compresses the underlying asthenosphere strongly, forcing magma to rise into surge channels. The sequence of events, as we interpret the process, is approximately as follows (Figs. 3, 85–87) and constitutes the geotectonic cycle.

1. The stereosphere contracts, presumably at a steady rate.

2. The overlying lithosphere, because it is already cool, does not contract, but adjusts its circumference to the shrinking strictosphere circumference by fracturing (thrusting) along the shallow Benioff zones. This fracturing is a very large-scale process that leads to tectogenesis.

3. Large-scale fracturing of the lithosphere is not a continuous process, taking place only when its dynamic support from below fails. That support is provided mainly by the soft asthenosphere and the resistance to movement along the Benioff zones. When the weight of the lithosphere overcomes the combined resistance offered by the asthenosphere and friction in the Benioff zone, lithosphere collapse ensues. Because the processes involved take place irregularly, lithosphere collapse (Fig. 85c) probably is episodic rather than cyclic. Because lithosphere collapse initiates tectogenesis, that process too is episodic.

4. During the intervals between lithosphere collapses, the asthenosphere volume increases slowly as the strictosphere radius diminishes (Figs. 85a, 85b). The increase in asthenosphere volume is accompanied by decompression in the asthenosphere.

5. Asthenosphere decompression is accompanied by rising temperature, increased magma generation, and lowered viscosity. The asthenosphere gradually becomes weaker during the interval between lithosphere collapses.

6. Rising temperature, increasing magma volume, and lowered viscosity weaken the entire suprastrictosphere zone so that, when asthenosphere viscosity is reduced to some critical value, the lithosphere collapses into it along the Benioff zones. The continentward sides of the Benioff zones override the ocean floor, thereby pushing it deeper into the asthenosphere. The lithosphere buckles, fractures, and then founders as the circumference of the lithosphere's base adjusts to that of the strictosphere's upper surface.

7. Both the lithosphere and the strictosphere fracture along great circle fractures—at the strictosphere

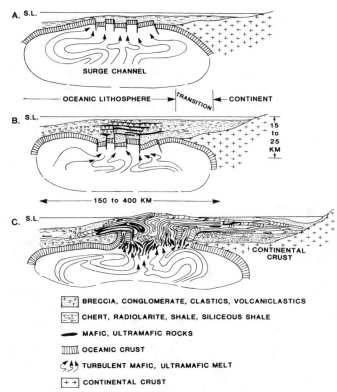

Figure 86. The eugeosyncline in the geotectonic cycle. (see tex)

level—as predicted by theory (Jeffreys, 1970) and demonstrated in the laboratory (Bucher, 1956). Only two great circle fracture zones survive in part—the Circum-Pacific and the Tethys-Mediterranean. Of these, the Tethyan barely survives and the Circum-Pacific zone is being overridden progressively by lateral continental accretion.

8. Lithosphere collapse into the asthenosphere generates major compressive stresses that are transmitted throughout both shells. Asthenosphere-derived magma in surge channels begins to surge. Wherever the volume of available magma exceeds the volumetric capacity of surge channels available to it, surge-channel rupture occurs along surge-channel roofs. In the case of eugeo-synclines, huge folds, thrust sheets, and nappes are generated (Figs. 85c, 86c, 87).

9. Lithosphere collapse into the asthenosphere combined with abrupt onset of tectogenesis is a prime example of Pascal's Law (discussed below). Sudden rupture and tectogenesis of surge channels may be likened to what happens when someone stamps a foot on a full tube of toothpaste.

10. During lithosphere collapse, some magma may be trapped in the lithosphere where it is too thick to rupture. Such magma may spread laterally, enlarging the existing surge channel or creating a new one. We call this process lithosphere wedging.

11. Once tectogenesis is completed, taphrogenesis sets in and the conditions are created for the beginning of the next geotectonic cycle.

Pascal's Law: the core of tectogenesis

Pascal's Law (or theorem)—an expression of Newton's First Law of Motion—states that pressure applied to a confined liquid at any point is transmitted undiminished through the fluid in all directions and acts upon every part of the confining vessel at right angles to its interior surfaces and equally upon equal areas. This law applies to all fluids and is the principle behind all hydraulic machines, notably the hydraulic press. A most important condition of Pascal's Law is that the pressure (force per unit area) acts equally upon equal areas. This condition lies at the very core of tectogenesis.

The Earth is, as we interpret it, a very large hydraulic press. Such a press consists of three essential parts: a closed vessel, the liquid in the vessel, and a ram, or piston. The collapse of the lithosphere into the asthenosphere is the activating ram or piston of tectogenesis. The asthenosphere and its overlying surge channels—which are everywhere connected by vertical conduits—are the vessels that enclose the

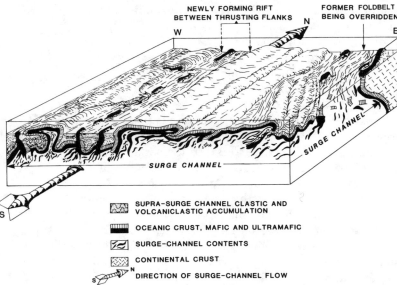

Figure 87. Our conception of a surge channel along a continent-ocean basin boundary as tectogenesis sets in. Note that, as thrusting in opposite directions takes place along the surge-channel flanks, a rift zone forms between those flanks.

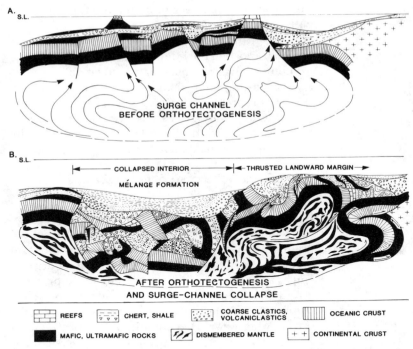

Figure 88. The origin of tectonic mélange. In surge tectonics, mélange forms as the direct result of the collapse of a surge-channel roof. In fact, the formation of mélange is a natural consequence of surge-channel tectogenesis. Recycling mechanisms, as required in plate tectonics, are unnecessary.

fluid. The fluid is magma generated in the athenosphere. This magma fills surge channels that are in the lithosphere. Because the lithosphere everywhere is under compression, one can imagine the effect of lithosphere collapse along Benioff zones at a time when the surge channels are filled, or nearly so. When the piston of lithosphere collapse drives against the asthenosphere and its magmas, great pressure is transmitted rapidly through the worldwide interconnected surge-channel network; the surge channels burst (in the eugeosynclines) and tectogenesis is in full swing.

Two objections have been voiced to us concerning the hydraulic press analogy. M.L. Keith (pers. comm., 1989) stated that the effect of a sudden application of pressure against the surge channels would be to consolidate the mafic magma (basalt) in the channels. Instead of bursting, he suggested, the channels would freeze. We agree this would be true if the channels had no communication with the surface at the time the pressure was applied suddenly. However, this is not the case. As we showed by example (this paper; Meyerhoff et al., this volume) surge channels are connected to the surface by ridge- or channel-parallel systems of faults, fractures, and fissures (Figs. 11, 33, 38, 52, 55, 59, 61, 75).

Another objection, by J.W.H. Monger and an unidentified critic, is that the magma in the surge channel is too viscous to transmit the added stress to all interconnected parts of the surge-channel system. This objection would be valid for a tectonic model in which the added stress is applied at a single point along the surge-channel system. However, in a contracting Earth, the entire lithosphere is under equiplanar tangential stress, that is, the added stress is applied everywhere along interconnected channels. During lithosphere collapse, the added stress is applied simultaneously along many thousands of kilometers of the surge-channel system. At times, worldwide lithosphere collapse and tectogenesis take place. Proof of this statement is that in every Tithonian-Eocene foldbelt in the world an angular unconformity is present between the middle part of the early Eocene and the middle part of the late Eocene (Schwan, 1980; Meyerhoff, unpubl. data). A similar worldwide tectogenesis was claimed for the Albian-Cenomanian by Croneis and Reso (1966). However, we have not studied their documentation. Regardless, lithosphere collapse occurs along thousands of kilometers of the surge-channel system, triggering tectogenesis on an extremely large scale ranging from continentwide to worldwide.

Tectogenesis versus orogenesis

To this point we used the term tectogenesis to signify what most geoscientists call orogenesis or orogeny. In our opinion, the term orogenesis is a misnomer. The word comes from the Greek words *gennao* and *oros,* which together mean "producing of mountains." There is no connotation of the intense deformation that we associate with the word tectogenesis. That word comes from two other Greek words, *tekton* (builder) and *ge* (earth), which together mean "earth builder," and which Gilluly (1973) defined as "the internal deformation of a rock mass" (p. 500). We use the term tectogenesis, to which we apply Cebull's (1973, p. 102) definition of orogenesis, namely, "... an event involving intense deformation on a scale which can be viewed in outcrop or hand specimen, which occurs over a finite period of time, and which can be recognized or reasonably inferred to extend over an area measured on the order of thousands of square kilometers." A synonym for the word orogenesis is orogeny. A synonym for the term tectogenesis is tectogeny.

Among the most fundamental principles of tectogenesis is that tectogenesis of a surge channel produces kobergens. Another principle involves the

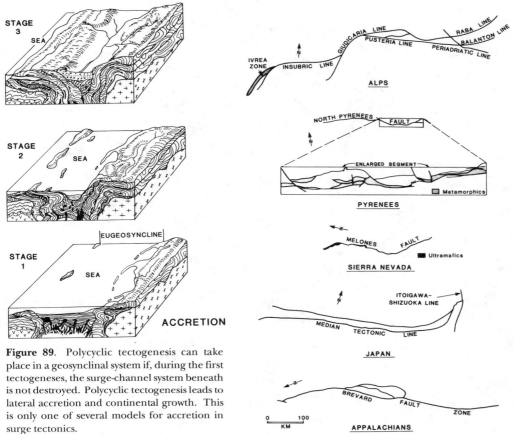

Figure 89. Polycyclic tectogenesis can take place in a geosynclinal system if, during the first tectogeneses, the surge-channel system beneath is not destroyed. Polycyclic tectogenesis leads to lateral accretion and continental growth. This is only one of several models for accretion in surge tectonics.

Figure 90. Traces of several median tectonic lines and their associated ultramafic and/or metamorphic rocks. We interpret these to be the *Narben* (scars) of *Verschluckungszonen* (engulfment zones) that formerly were straight strike-slip (streamline) fault zones but which, as a consequence of compression during circumferential shortening, have become twisted and deformed.

style of tectogenesis. That style ranges from alpinotype to germanotype (contrast Figs. 44, 48, 50, and 51 with Fig. 52). Kobergen style is a function of lithosphere thickness, with germanotype kobergens forming where the lithosphere is 20–35 km thick and alpinotype kobergens forming where the lithoshpere is 10–20 km thick. Midocean ridge-type structure forms where the lithosphere is thinner than about 10 km, and continental-type (East Africa) rifts form where the lithosphere is thicker than about 35 km.

Eugeosynclinal geotectonic cycle

From rift to *Verschluckungszone*—Figure 84 illustrates that in surge tectonics, the major structures parallel the surge-channel trend. As we show on the figure, tensile structures are concentrated toward the surge-channel crest, whereas compressional and gravity-glide structures lie at the kobergen (surge-channel) flanks. Figure 86 shows the complete geotectonic cycle of the eugeosyncline, from the beginning of sedimentation to tectogenesis. Rifts (horsts and grabens) develop early in the eugeosyncline's history (Fig. 86a), giving rise to early formation on top of the surge channel of a eugeosynclinal basin. The rifts persist in the eugeo-syncline (Fig. 86b) until tectogenesis sets in (Fig. 86c). A rift may persist during the early stages of tectogenesis, as shown on Figure 87. This figure shows lateral accretion and overriding of the next older foldbelt, the continued formation of belt-parallel structures, asymmetry of the kobergen (because of lithosphere-thickness differences between oceanic and continental margins of the surge channel), and a narrowing rift above the surge-channel center. In some cases, this narrowing rift zone collapses into the emptying magma chambers below it, forming tectonic mélange, as illustrated in Figure 88. It is through such continental-margin tectogenesis, in many examples with repeated tectogenic episodes (polycyclic tectogenesis, polycyclic tectogeny), that lateral continental accretion occurs (Fig. 89).

During the eugeosynclinal geotectonic cycle, the stress state within the surge-channel roof changes steadily from tensile to compressive. This constantly changing stress state is manifested in the degree of linearity of the faults, fractures, and fissures that form on the surge-channel roof. Such faults, fractures, and fissures, as long as there is laminar flow in the surge channel, comprise true streamlines. Laminar flow persists as long as a state of tension dominates. As the rift zone at the surge-chan-

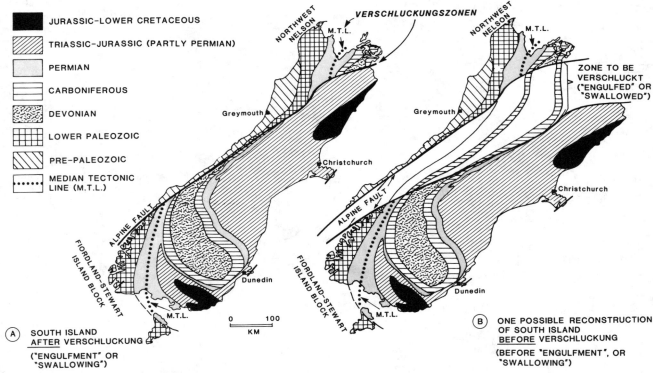

Figure 91. Alpine (and pre-Alpine) fault or *Verschluckungszone*, South Island, New Zealand. **A**—Subcrop map of pre-Campanian rocks (from Kingma, 1974). **B**—same map, but opened up at the Alpine fault. Between the two sides of the fault is shown the minimum amount of surface area that was engulfed (*verschluckert*) by the underlying surge channel during the middle Cretaceous Rangitata orogeny. We have shown the missing surface area as though the Alpine fault, or a pre-Alpine fault at the same location, as the *Verschluckungszone;* that zone originally was the Median Tectonic Line (M.T.L.). Our points are that (1) a very large surface area was engulfed along this *Verschluckungszone* in South Island; (2) much of this zone's surface trace now coincides with the Alpine fault; (3) the *Verschluckungszone* explanation obviates the need for large strike-slip movements along the Alpine fault; and (4) the *Verschluckung* process is a logical mechanism for circumferential shortening in the lithosphere. Published with permission of Elsevier Science Publishers B. V.

nel crest narrows, the zone of tension narrows and becomes concentrated into one long, straight fault trace, or fault zone. Such long, straight fault traces are the strike-slip, or wrench faults of the geologic literature (e.g., Moody and Hill, 1956). Well-known examples include the San Andreas fault (Figs. 60, 61), the North Anatolia fault (Fig. 64), and the North Pyrenees fault (Fig. 65). Because they are true streamlines in laminar flow, we propose the term wrench fault be abandoned and replaced with the term streamline fault, or streamline fault zone. Even the term strike-slip fault is a misnomer because the strike-slip takes place in the surge channel adjacent to the wall of the roof, and not in the roof itself.

As the geotectonic cycle continues, compression replaces tension at the surge-channel roof and kobergenic structures begin to form (e.g., Figs. 63–65). The streamline fault zone gradually loses its linearity, and becomes distorted, even strongly so (Fig. 90). Such distorted faults lie in the core of the kobergen, the root zone of alpinotype foldbelts. Commonly they separate rocks of totally different metamorphic or igneous facies. Many of them bound or enclosed slivers and elongate masses of mafic and ultramafic rocks. Wherever described in the literature, such distorted fault traces generally are called fossil subduction zones and suture zones. Other terms include line, median tectonic line, and steep belt (Fig. 91). Milnes (1974) proposed the term steep belt should replace the term root zone. German terms for these zones include *Narbe* (scar), *Scheitel* (vertex), and *Verschluckungszone* (engulfment zone). We prefer to use the latter because it describes accurately the principal process that takes place along such zones.

The terms *Verschluckung* and *Verschluckungszone* were introduced by Ampferer (1906) and Ampferer and Hammer (1911). *Verschluckung* is a tectonic process translated variously as swallowing, engulfment, downward sucking, and subfluence (Schwinner, 1920; de Sitter, 1964; Dennis, 1982; Mueller, 1983). It was coined specifically to explain why in some foldbelts, hundreds of kilometers of one or more facies belts that should be present are absent.

After tectogenesis, one would expect to find the following approximately symmetrical sequence of facies belts from one side of a kobergen to the other: stable cratonic or platform facies, miogeosynclinal facies, eugeosynclinal facies, miogeosynclinal facies, and stable cratonic facies (Fig. 13). The highest metamorphic grade generally coincides with the eugeosynclinal facies (Fig. 90), and lower grades appear successively away from the core of the kobergen. In many foldbelts, one or more of the anticipated facies or metamorphic belts may be absent. An extreme example would be a foldbelt in which the cratonic facies of one side of a kobergen are in juxtaposition with the cratonic facies of the other side of the same kobergen (Ampferer, 1906; Ampferer and Hammer, 1911; Schwinner, 1920; Kober, 1921, 1925, 1928; de Sitter, 1964; Mueller, 1983).

Summary and examples—Described above is the general evolution of the eugeosyncline in the geotectonic cycle. Particular emphasis is placed on our interpretation of the central, or core zone of the kobergen, where the geotectonic cycle begins with a rift and ends with a *Verschluckungszone*. A major conclusion is that the world's great strike-slip fault systems are actually streamline faults that form just as the tensile regime in the central part of the surge channel, or kobergen crest, is replaced by a compressive regime. This conclusion has major implications in tectonophysics, implications that are not discussed for lack of space. Instead we discuss them in a separate publication (Meyerhoff et al., submitted).

Examples of the early phase (when rifts predominate at the crest of the surge channel) of eugeosynclinal evolution include the Lau basin (Tonga magmatic arc, Fig. 78), Cagayan Valley (Luzon, Fig. 28b), Central Valley (Luzon, Fig. 28c), Ragay Gulf (Luzon, Fig. 28d), Molucca Strait (Indonesia, Fig. 29), and Gulf of California (Fig. 35). It should be noted (Figs. 28b-28d) that kobergenic structure forms very early. Figure 35, however, shows a polycyclic eugeosynclinal example where a modern (middle Miocene-present) eugeosyncline overlies an older one (Late Jurassic-Cretaceous) that already had been deformed into a kobergen.

The intermediate phase (when the rift zone above the surge channel is actively closing) is exemplified by one example from the Colombia Andes (kobergen 3, Fig. 41) and the longitudinal Valley of Taiwan (Fig. 28a). The valley already has narrowed to a width of only 2 to 6 km (Ho, 1982). Examples of major wrench faults (e.g., the San Andreas) previously were given. Any rifts that might have been present are gone, except for newly developed eddylike depressions in zones of channel overlap. Kobergen structure is well developed (Figs. 63-65).

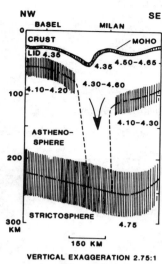

Figure 92. Northwest-southeast refraction-seismic cross section of crust and upper mantle beneath the western Alps along a line passing through Basel, Switzerland, and Milan, Italy. (1) The near-vertical slab marked with a black arrow constitutes a large *Verschluckungszone* of lithosphere that extends to the strictosphere, penetrating the entire asthenosphere. (2) Numbers are S-wave velocities, the 4.20-km/s value corresponding to the 7.7–7.9-km/s P-wave velocity. (3) The lid is the hard mantle in the lithosphere above the asthenosphere. (4) Finally, the hatched areas indicate the range of uncertainty of the locations of the bottoms of (a) the crust, (b) the lithosphere, and (c) the asthenosphere. From Mueller (1983).

Examples of the final phase range from kobergens in which compression has just begun to dominate the entire surge-channel roof to kobergens in which compression has been very strong for an extended period of time. An example of the former is the Hengduan Shan (Yunnan Himalaya) of southwestern China (Fig. 48). Examples of the latter include the Alpine chains of southern Europe (Figs. 50, 51, 90), the Brevard zone of the southern Appalachians (Fig. 90), and the Feather River zone of the northern Sierra Nevada (Figs. 36, 90).

If the concept of *Verschluckungszone* is valid, rocks that comprise facies belts that are now missing at the Earth's surface should in theory form a large nearly vertical mass in the lithosphere and possibly the asthenosphere. Should such masses exist, large geophysical anomalies should be evident. Figure 91 shows the results of a seismic-refraction study of the Alps published by Mueller (1983). A massive, nearly vertical, high-velocity slab is present to depths of 180 to 250 km. Mueller (1983) interpreted the *Verschluckungszone* (the space occupied by the slab) to be remnant of a subducted plate. As we observed in a preceding section, the subducted-plate model cannot be sustained because (1) the slab is vertical to subvertical and (2) the slab extends only to the asthenosphere and not deeper.

Nearly vertical slabs that extend to 180–250 km apparently are common. Figure 92 shows another such slab beneath the eastern Transverse Ranges of southern California. Table 10 lists several slabs that underlie different foldbelts of the world. Figure 93 presents a before-and-after model of the South Island *Verschluckungszone* in New Zealand. Surface mapping showed the

Table 10. Some interpreted *Verschluckungsonen* in the world's foldbelts.

Foldbelt	Slab depth	Reference(s)
Western Alps	180-250	Mueller (1983)
Himalaya-Nyainqentanglha Shan	ca. 230	Fei Ding et al. (1981)
Median tectonic line, Kyushu-Shikoku-southern Honshu	30+	Miyashiro (1973); Hirahara (1977)
Median tectonic line, South Island, New Zealand	131+	Landis and Coombs (1967); Kingma (1974)
Eastern Transverse Ranges, southern California	100-250	Humphreys et al. (1984)
Sierra Nevada	150-250	Jones et al. (1990)

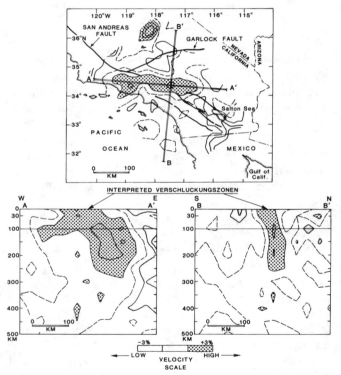

Figure 93. Results of the inversion of teleseismic P-wave delays in the southern California Transverse Ranges. *Top:* a plan-view section is shown for the depth of 100 km (superimposed on a location map for southern California). Lines of section for the lower two diagrams are shown. *Lower left:* west-east velocity cross section through the Transverse Ranges. Dotted areas show velocity deviations greater than 1.5%; shaded areas show velocity deviations less than 1.5%. (Greater than 1.5% = cold areas; less than 1.5% = hot areas.) 100-km depth line is shown. This section, parallel with the Transverse Ranges, shows a *Verschluckungszone* that reaches ca. 150-km depth in the west and 250-km depth in the east. *Lower right:* south-north velocity cross section across the Transverse Ranges. Symbols are the same as for section A-A', lower left. The high-velocity anomaly shows as a slab-like body dipping 80 to 90° toward the north that extends to approximately 250 km. From Humphreys et al. (1984).

missing facies belts in South Island occupied a zone up to 131 km wide (Kingma, 1974). Figure 93b shows how South Island might have appeared before engulfment (*Verschluckung*); Figure 93a shows South Island as it is today (with Late Cretaceous and younger rocks removed). Note that the 500-km apparent dextral offset along the Alpine fault is explained fully by the *Verschluckung* process; strike-slip movements are unnecessary.

***Verschluckungszonen* and circumferential shortening**—Given the large number of alpinotype foldbelts in the world, Precambrian to present, the amount of circumferential shortening in *Verschluckungszonen* in the lithosphere could add up to hundreds, even thousands of kilometers through geological time. Hence, the role of *Verschluckungszonen* in the Earth's circumferential shortening must be important. We emphasize that formation of such *Verschluckungszonen* is only one of the ways in which the lithosphere adjusts its shape to the constantly shrinking strictosphere beneath.

One critic of an earlier version of this manuscript commented that there would be no space in the lithosphere for such giant slabs. In fact, no space problem should exist. During tectogenesis, surge-channel rupture takes place at the channel roof (Fig. 13) and the contents of the channel move upward and outward away from the channel crest. Under a variety of conditions, the crestal zone and facies belts adjacent to it could collapse into the void space below. Equiplanar tangential compression in the lithosphere, acting at right angles to the channel walls, would press the collapsed zone into near-vertical attitudes, and provides an explanation for the origin of so-called steep belts of geological literature (Milnes, 1974).

Midocean ridges in the geotectonic cycle

Effects of tectogenesis—During the geotectonic cycle (Fig. 94a), the midocean ridges swell and contract according to the volume of magma in them. Their sizes must have changed repeatedly in the past. The

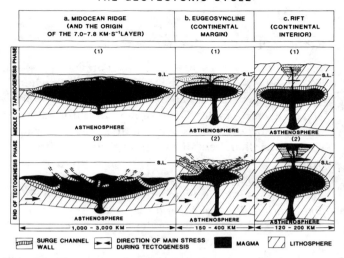

Figure 94. The geotectonic cycle in the ocean basin (midocean ridge, for example), the continent-ocean basin transition zone (eugeosyncline), and the continental interior (rift basins). *Midocean ridge:* during the taphrogenic phase of the geotectonic cycle (*a-1*), the ridge fills with magmatic products. During the tectogenesis phase (*a-2*), the ridge bursts and massive basalt floods cover the ocean floor. *Eugeosyncline:* during the taphrogenic phase, the eugeosyncline sinks, accumulating sedimentary and volcanic rocks (*b-1*). During the tectogenesis phase (*b-2*), alpinotype folding and thrusting occur. *Continental rift:* during taphrogenesis (*c-1*), continental rifting and moderate volcanism occur. During tectogenesis, rifting is accelerated and volcanism increases as the surge channel fills to near the bursting point (*c-2*). Alternatively, a continental germanotype kobergen may develop as illustrated on Figure 52.

evidence for swelling and contracting includes examples of a Middle Jurassic shallow-water limestone on the Mid-Atlantic Ridge that now lies in several hundred meters of water (Honnorez et al., 1975). Additional evidence includes the prominent north-south magnetic lineations of the eastern Pacific. The linear magnetic anomalies of midocean ridges probably formed as a result of ridge-parallel motions (Agocs et al., this volume), indicating that at some time (or times) in the past, the East Pacific Rise was broader than it is now.

During tectogenesis, midocean ridges probably rupture along their full length because, according to Pascal's Law, pressure applied to a confined liquid acts equally on equal areas of the walls of the confining vessel. This means the total force acting on the walls of a midocean-ridge surge channel is vastly greater than that acting on the roofs of the generally much smaller surge channels beneath feeder channels, breakout channels, and channels associated with continents (Fig. 14). Therefore, because the thickness of the lithosphere above the midocean-ridge surge channels is about 10 km or less, rupture of the midocean-ridge walls must take place on a massive scale. The fact that the magma beneath eugeosynclines can rupture the 15- to 25-km thickness of lithosphere above them means that rupture of the midocean ridges is much more than just a possibility.

A second reason for stating midocean ridges probably rupture along their full length during the tectogenesis phase of the geotectonic cycle is based on results of the Deep Sea Drilling Project (DSDP). During that project, 624 sites were drilled on 96 legs. The oldest basalts drilled are Callovian (Middle Jurassic; Bolli, 1980a, 1980b). As Meyerhoff et al. (this volume) pointed out, the most massive Phanerozoic continental flood volcanism took place from Late Permian through Middle Jurassic time. One can imagine the far greater volumes of basalt that burst from midocean ridges during the same time period. Huge submarine basalt floods would explain why, beneath the abyssal plains, no strata older than Middle Jurassic are known. Such floods also explain why all older, pre-Middle Jurassic rocks known from ocean basins occur on midocean ridges and other rises above the abyssal plains (Meyerhoff et al., this volume). Figure 94a illustrates our concept of what happens to midocean ridges during geotectonic cycles.

One may question whether such huge submarine flows are possible. During recent exploration of the deep sea adjacent to the Hawaiian Ridge, a huge 100,000-km^2 flow was discovered (Torresan et al., 1989) and larger flows have been sighted since (M.L. Holmes, pers. comm., 1988). The 100,000-km^2 flow described by Torresan et al. (1988) is from a very small breakout channel, much smaller than a midocean ridge and buried beneath a much thicker burden of lithosphere (Ellsworth and Koyanagi, 1977).

Such massive basalt floods as those postulated here must contribute to the construction of oceanic lithosphere. Most of the volume of each flood presumably underlies the midocean ridges, but some reaches the continental margins if the distribution of Jurassic basalts found in DSDP sites is an indication of the extent of flooding. The possibility that most of the oceanic crust of the ocean basins originates in this manner must be considered.

Kobergen structure in midocean ridges—A recent and important discovery made by Antipov et al. (1990) is that a broad zone of reverse and thrust faults borders the western flank of the Mid-Atlantic Ridge. This zone, 300 to 400 km wide, lies between the floor of the abyssal plain and the inner part of the midocean ridge. Antipov and his colleagues (1990) made no attempt to trace the zone everywhere but found it has a north-south extent of at least 1,000 km; that is, it extends north-south throughout the area they investigated. The zone shows up on single-channel reflection-seismic data obtained by

Rabinowitz et al. (1978), but only in the bathymetry; the records are not good enough to see structural detail. However, a nearly identical bathymetric zone is also present on the eastern side of the Mid-Atlantic Ridge at the junction between the abyssal plain and the ridge. If this is confirmed by multichannel reflection-seismic studies, it means that midocean ridges also have kobergenic structure, in this case two 300- to 400-km-wide bands of reverse faults separated by a 900- to 1,000-km-wide zone of block faults (Fig. 94a[2]). The importance of the Antipov et al. (1990) discovery should not be ignored, and efforts to obtain multichannel reflection (CDP) data from both ridge flanks should be made as soon as possible.

Principal cause of transgression-regression—Surge tectonics provides, for the first time, a simple mechanical link between tectonic activity and the geotectonic cycle on the one hand, and transgression-regression on the other. This link is the filling and emptying of surge channels (Figs. 85, 86, 89, 94). When midocean ridges are full (Fig. 94a[1]), transgression reaches a maximum; conversely, when midocean ridges have expelled their contents and are deflated, then regression reaches a maximum (Fig. 94a[2]). We postulate that volumetric change in the midocean-ridge system is the primary cause of transgression-regression.

To demonstrate the importance of volumetric change in the midocean-ridge system, we note the following. The midocean-ridge system and its more important branches together form a mountain chain some 70,000 km long. The average width in the Arctic, Atlantic, and western Indian oceans is about 1,100 km; its width in the eastern Indian and Pacific oceans averages more than 2,400 km and in places is 3,100 km or more. If a strip of midocean ridge just 500 km wide were uplifted an average of only 2 km, 70 million km³ of sea water would be displaced. This is equal to a worldwide sea-level rise of

$$R = \frac{V}{A} \quad (1),$$

where R is the rise in sea level, V is the volume of water in today's oceans, and A is the area of the Earth underlain by oceanic crust. Substituting numerical values in eq. (1), we obtain

$$R = \frac{7 \times 10^7 \text{km}^3}{3.54 \times 10^8 \text{km}^2} = 0.198 \text{ km},$$

or about 200 m. Even this conservative calculation shows that midocean-ridge volume changes are likely to be a major—possibly the major—cause of transgression and regression. The same mechanism accounts for the drowning of oceanic islands, atolls, and guyots, although some drowning is accommodated by differential vertical movements along oceanic fracture zones (e.g., Mendocino, Murray, and related fracture zones of the eastern Pacific Ocean).

Continental rifts in the geotectonic cycle

Examples we presented are shown on Figures 32–35, 53–55, 59, 67–71, 79, and 88. Figure 94c shows what we believe is the behavior of a continental rift system during the geotectonic cycle. The responses of continental rifts to active surge channels beneath them—mantle diapirs or magma chambers of the geological literature—are reasonably well understood. The thick (up to 50 km or more) lithosphere layer above the surge-channel complex strongly inhibits tectogenesis. As shown on Figure 94c, during the taphrogenic stage, the surge channel beneath the rift is only moderately inflated and volcanism is moderate. The most severe volcanism and rifting take place during tectogenetic stages. An important point is that, in surge tectonics, rifting in rift systems takes place at the same time that tectogenesis occurs in a eugeosyncline. Rifting (extension) and tectogenesis (compression) can take place simultaneously along the same surge-channel system; the determining factor is lithosphere thickness.

Tectonic belts intermediate between rifts and eugeosynclines are numerous. An example is the continental Cenozoic surge channel that produced the High Atlas (Fig. 52). Another example is the Late Jurassic through Quaternary Cordillera Oriental kobergen of Ecuador and Colombia (Gansser, 1973; Mégard, 1987; Fig. 41). An interesting feature of the Ecuadorian-Colombian Andes shown on Figure 41 is the presence of four adjacent and parallel kobergens. The westernmost, the Coastal Cordillera or Sierra de Baudó surge channel of middle Eocene to recent age, lies largely offshore and forms part of the channel system beneath the Panama Isthmus. It is part of the eugeosyncline complex of northwestern South America. East of it is the Cordillera Occidental, a polycyclic eugeosynclinal kobergen of the same age as the kobergen beneath the Cordillera Oriental. Between the Cordillera Occidental kobergen and that underlying the Cordillera Oriental is a fourth, which underlies the Cordillera Central and the row of modern volcanoes of the Ecuadorian-Colombian Andes (Fig. 41). Surge tectonics provides an explanation for the quadripartite division of the Colombia Andes, a puzzle whose explanation eluded geologists and geophysicists working in the region (Gansser, 1973).

Origin of granites

Classifications—Chappell and White (1974) recognized two granite types, an I-type (igneous derivation) and an S-type (sediment derivation). I-type granites are formed by the differentiation of deep mafic magmas, and S-types, by the melting of sedimentary and metamorphic rocks. Loiselle and Wone (1979) recognized A-type (anorogenic) granites, which in White and Chappell's (1983) opinion are igneous, but are derived from a zone below that from which I-type granites are derived. They observed that A-types must originate in an F- or Cl-rich nearly anhydrous source that contains abundant Ga, Nb, Sn, Ta, Zr, and rare earth elements. Hopson and Dellinger (1989), however, believed that I- and S-types are generated below A-types.

Clemens (1988, p. 445) wrote that "Many granitic magmas are generated wholly, or principally, by processes of partial fusion during high-grade metamorphism of continental crust. The high geothermal gradients necessary for production of large volumes of crustal melts ... necessitate the introduction of hot, mantle-derived material to satisfy the heat requirements. However, in granitoid genesis the crust may differentiate without the necessity for massive chemical involvement of mantle-derived magma." Basing his statements on a study of Paleozoic granites from Victoria (Australia), Clemens (1988) concluded that I-type granites may form by the partial melting of metaigneous rocks.

Other granitic types proposed include M-types, which are mantle-derived and C-types, which are crust-derived. Lyakhovich (1988) wrote a good overview of the various granitic classifications.

Compositions—The relations among coeval plutons with quite different compositions is another facet of granite genesis that only recently has begun to be studied in detail. Hopson and Dellinger (1989), reporting on a study in the northern Cascade Range in Washington, and Bruce et al. (1989), summarizing the results of a Patagonian batholith study, concluded that granitic rocks ranging from granite to diorite can form in a single magma chamber. The silicic rocks are concentrated toward the top, and the mafic rocks toward the bottom. Reintrusion and cooling can take place at different times so a composite pluton with a broad age range—more than 100 Ma—may result. It may be no longer necessary to postulate that rocks of different compositions and ages must be generated in separate magma chambers. Bruce et al. (1989) concluded further that composite batholiths may be derived from components at different levels, and even from the same level.

Vertical extents of granitic plutons—This topic has been studied increasingly in recent years. Long ago, Chamberlin and Link (1927) challenged the then-widespread acceptance of the concept that granitic plutons essentially were bottomless, and were so depicted (and still are) in almost all structural cross sections. Chamberlin and Link (1927), basing their work on published sources and their own field work, concluded that many batholiths are shaped like laccoliths and have a flat floor. Figure 95 is the conceptual model they derived for granitic intrusions in the Appalachians. They argued further that foldbelts are bilaterally deformed (Fig. 95) and, because of this, in the middle of each foldbelt there is ". . . a slight release of pressure, favoring liquefaction, and of co-operation in opening up space for advancing magma" (Chamberlin and Link, 1927, p. 348).

Many subsequent studies of granitic plutons showed that granites exhibit widely different shapes, but that Chamberlin and Link's (1927) basic premise that they have bottoms, mainly within the crust or uppermost mantle, is sound. Field studies (Chamberlin and Link, 1927; Krauskopf, 1968), gravity modeling (Thomas and Willis, 1989), and seismic-reflection studies (Lynn et al., 1981) showed the minimum thickness of many granitic plutons is on the order of 1 to 7 km. Mueller (1983) suggested that many of these tabular bodies represent the exposed parts of former middle to upper crustal low-velocity zones. Sams and Saleeby (1988), on the other hand, argued that most of the crust (at least beneath the Sierra Nevada) consists of batholithic material, with (1) mixing at depth (approximately 25 km), (2) ascent of the silicic part, and (3) lateral spreading at relatively shallow depths in the upper crust. Such a model explains both the relatively thinner tabular granitic bodies widely observed in foldbelts and the field evidence (as well as geobarometric and other data) for large plutons 10 to 20 km high (Sams and Saleeby, 1988; Hopson and Dellinger, 1989).

Most workers agree (e.g., Chamberlin and Link, 1927; Fig. 95; see also Sams and Saleeby, 1988) most granitic plutons were emplaced during times of major extension. However, these same workers either are at a loss to

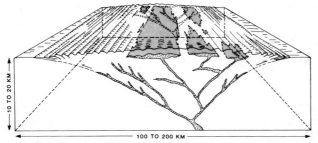

Figure 95. Chamberlin and Link's (1927) conceptual model of batholith formation. They were among the first to argue that granitic batholiths probably are fairly thin bodies that have floors. Chamberlin (1925) developed his wedge theory of tectogenesis, an hypothesis similar to Kober's (1921 and earlier). In his wedge (Fig. 95), he and Link saw that tension must characterize the central part. Consequently, the two authors postulated that granitic bodies would intrude preferentially where tension was greatest.

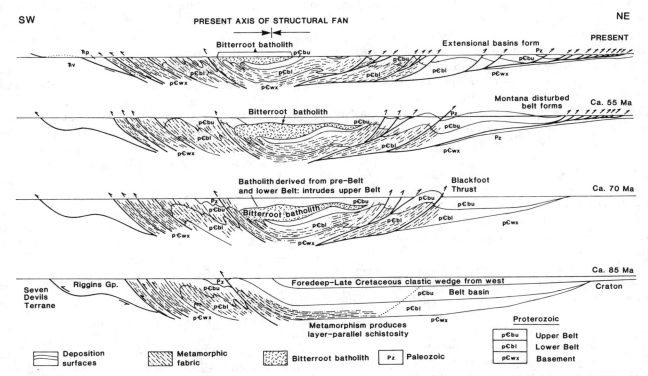

Figure 96. Bitterroot Batholith of the northern Rocky Mountains (Hyndman et al., 1988; location is on Fig. 34). Note that the batholith, of Late Cretaceous (90–70 Ma) age, is hosted in Proterozoic rocks. Hyndman et al. (1988) showed that the source rocks for the batholith were both Proterozoic sedimentary rocks and crystalline 1.7-Ga basement. Thus, the Bitterroot Batholith is the product of a Late Cretaceous to early Tertiary kobergen that developed in continental lithosphere. The exact correspondence between Chamberlin and Link's (1927) predicted locus of granite formation and the tectonic setting in which Hyndman et al. (1988) found the Bitterroot Batholith is quite a tribute to Chamberlin and Link. Published with permission of Prentice Hall.

explain how tension can exist in a deforming foldbelt, or propose mechanisms that are incompatible with foldbelt compressive stress (e.g., Hutton, 1988).

Origin in surge channels—The kobergen model (Fig. 13; Kober, 1921, 1925) in our opinion provides the logical home for the genesis of granitic plutons. Figure 13 shows a magma chamber beneath the kobergen (equivalent to our surge channel). Its position coincides with the internally generated zone of extension that provides the prerequisites—extension and space—for pluton emplacement. The close resemblance between Kober's (1925) model in Figure 13 and the Chamberlin and Link (1927) model in Figure 95 is evident.

Examples of granitic plutons in kobergen-like structures appear repeatedly in the literature. Space permits only a single example from the northern United States Western Cordillera (Fig. 96). The close resemblance between the interpreted field model (uppermost section of Fig. 96) and the theoretical Chamberlin and Link (1927) model is again evident.

We believe that the kobergen model explains most—it cannot explain all—of the problems of plutons discussed by Krauskopf (1968). He concluded that plutons, like most natural phenomena, include elements (e.g., size, shape, location, composition, internal structure, etc.) that are random and therefore unpredictable. Krauskopf (p. 17), however, noted that even " . . . the randomness, or anarchy, is in some measure the law of nature . . . ."

Ubiquity of granites—Granitic plutons are most abundant in alpinotype foldbelts (Fig. 94b). Therefore they are important indicators of the locations of former surge channels in these belts. They are generated at many times—even during a single geotectonic cycle—so are not restricted to any one part of the cycle (Miyashiro et al., 1982).

Another class of granites—usually referred to as anorogenic granites—is abundant in the cratonic or platformal regions (Fig. 94c). Examples include the Niger-Nigerian magmatic province of West Africa (Bowden et al., 1987), the Rio Grande rift of the North American Western Cordillera (Lipman, 1988), and many others. As in orogenic belts, anorogenic granites are important indicators of the locations of former surge channels, in this case continental channels (Fig. 14).

A third habitat for granites is oceanic surge channels, especially midocean ridges where granitic rocks have long been known (e.g., Ascencion Island, Mid-Atlantic Ridge; Daly, 1925). They have a considerable age range, from at least middle Proterozoic (1,690 Ma; Wanless et al., 1968) to Miocene (9 Ma; Aumento, 1969) and younger (Daly, 1925). In places, as at 45° N latitude, granite and related siliceous rocks comprise 74% of the dredge hauls (Aumento and Loncarevic, 1969). During 1988–1989, Soviet scientists studying the western side of the Mid-Atlantic Ridge from the equator to 30° S latitude found granitic and silicic metamorphic rocks to be widespread (K.M. Khudoley, pers. comm., 1989). The apparent abundance and divergent ages of granites suggest that they formed repeatedly there through geologic time. Such an interpretation explains the many puzzling geochemical data that indicate that parts of the midocean-ridge axis are continental (e.g., $^{87}Sr/^{86}Sr$ ratios of 0.704–0.723; Bonatti et al., 1971).

If our interpretation of the origin of granitic rocks is correct, future studies of their chemistry and ages should be helpful in elucidating the surge-channel history not only of the continents, but also of the oceans.

CONCLUSIONS

1. All linear to curvilinear mesoscopic and megascopic structures and landforms, and all magmatic phenomena are generated, directly or indirectly, by surge channels. The surge channel is the common denominator of geology, geophysics, and geochemistry.

2. Surge channels formed and continue to form an interconnected worldwide network in the lithosphere. They contain fluid to semifluid magma, or mush, differentiated from the Earth's asthenosphere by the cooling of Earth. All newly differentiated magma in the asthenosphere must rise into the lithosphere. The newly formed magma has a lower density and therefore, is gravitationally unstable in the asthenosphere. It rises in response to the Peach-Köhler climb force to its level of neutral buoyancy (that is, to form a surge channel).

3. Lateral movements in the Earth's upper layers are a response to the Earth's rotation. Differential lag between the (more) rigid lithosphere above and the (more) fluid asthenosphere below causes the fluid, or mushy, materials to move relatively eastward. Consequently there is a tendency for both the asthenosphere to move relatively eastward en masse, and the mushy contents of the surge channels in the lithosphere to work their way eastward.

4. Surge channels are alternately filled and emptied. A complete cycle of filling and emptying is a geotectonic cycle. During the quiescent or taphrogenic phase of this cycle, basins evolve, gravity faulting is the norm, and surge channels ultimately are filled. In contrast, during the tectogenic phase of the cycle, basins are partly to wholly destroyed, compression is the norm, and the surge channel's contents are expelled.

5. Movement in the surge channel during the taphrogenic phase of the geotectonic cycle is parallel with the channel. It is also very slow, not exceeding a few centimeters per year. Flow at the surge-channel walls is laminar as evidenced by the channel-parallel faults, fractures, and fissures observed at the Earth's surface (Stokes's Law). Such flow also produces the more or less regular segmentation observed in tectonic belts.

6. Tectogenesis has many styles. Each reflects directly the rigidity and thickness of the overlying lithosphere. In ocean basins where the lithosphere is thinnest, massive basalt flooding occurs. At ocean-continent transitions, eugeosynclines with alpinotype tectogenesis form. In continental interiors where the lithosphere is thicker, either germanotype foldbelts or continental rifts are created.

7. During the geotectonic cycle, and within the eugeosynclinal regime, the central core (crest of the surge channel) evolves from a rift basin to a tightly compressed alpinotype foldbelt. Thus a rift basin up to several hundred kilometers wide narrows through time until it is a zone no more than a few kilometers wide that is occupied by a streamline (strike-slip) fault zone (e.g., the San Andreas fault). Then as compression takes over and dominates the full width of the surge-channel crest, the streamline fault zone is distorted until it and the adjacent rocks are severely metamorphosed. If the underlying—and now deformed—surge channel still contains any void space, the overlying rocks may collapse into it, and through this process of *Verschluckung* (engulfment) become a *Verschluckungszone*.

8. In tectogenesis, surge channels deform bilaterally, imparting an overall symmetry to each deformed tectonic belt. We call such bilaterally deformed belts kobergens.

9. The cause of tectogenesis is contraction of the strictosphere induced by its cooling. The Benioff zones are the Earth's great cooling cracks. As the strictosphere shrinks, its radius and circumference decrease and the asthenosphere is filled—and thereby weakened—by magma newly differentiated from the strictosphere. Because the strictosphere retreats steadily from the lithosphere, the latter is everywhere in a state of equiplanar tangential compression, and from time to time must adjust its shape to that of the shrinking interior. It does this when the asthenosphere is fullest—and therefore weakest—by (a) lithosphere collapse along the Benioff zones and (b) the *Verschluckung* (engulfment) process within and beneath surge channels. These last two

processes account for much of the Earth's circumferential shortening.

10. In such an Earth, large-scale strike-slip motions are unlikely. The substantial strike-slip motions that seemingly have been proved along fewer than ten of the world's so-called great wrench zones (e.g., 500 km of displacement along New Zealand's Alpine fault; Fig. 91) are explained fully by the *Verschluckung*, or engulfment process.

11. Therefore, the Earth above the strictosphere resembles a giant hydraulic press that behaves according to Pascal's Law. A hydraulic press consists of a containment vessel, fluid in that vessel, and a switch or trigger mechanism. In the case of the Earth, the containment vessel is the interconnected surge-channel system; the fluid is the magma in the channels; and the trigger mechanism is worldwide lithosphere collapse into the asthenosphere when that body becomes too weak to sustain the lithosphere dynamically. Thus tectogenesis may be regarded as surge-channel response to Pascal's Law.

REFERENCES CITED

Ádám, A., 1983, EM induction in Finland and general crustal physics, *in* Hjelt, S.E., ed., The development of the deep geoelectric model of the Baltic shield, pt. 2. Proceedings of the First Project Symposium, Oulu (Finland), 15-18/11/1983: University of Oulu, Department of Geophysics, Rept. no. 8, p. 1-16.

———, 1987, Are there two types of conductive anomaly (CA) caused by fluid in the crust?: Physics of the Earth and Planetary Interiors, v. 45, p. 209-215.

Ádám, A., Duma, G., Gutdeursch, R., Verö, J., and Wallner, A., 1986, Periadriatic lineament in the Alps studied by magnetotellurics: Journal of Geophysics, v. 59, p. 103-111.

Ádám, A., Landy, K., and Nagy, Z., 1989, New evidence for the distribution of the electric conductivity in the earth's crust and upper mantle in the Pannonian basin as a "hotspot": Tectonophysics, v. 164, no. 2-4, p. 361-368.

Agocs, W. B., Meyerhoff, A.A., and Kis, K., Reykjanes Ridge: quantitative determinations from magnetic anomalies: this volume.

Aguirre, L., 1983, Granitoids in Chile, *in* Roddick, J.A., ed., Circum-Pacific plutonic terranes: Geological Society of America Memoir 159, p. 293-316.

Aihara, A., 1989, Paleogeothermal influence on organic metamorphism in the neotectonics of the Japanese Islands: Tectonophysics, v. 159, no. 3/4, p. 291-305.

Ampferer, O., 1906, Über das Bewegungsbild von Faltengebirgen: Jahrbuch der Kaiserlich-Königlichen Geologischen Reichsanstalt, Bd. 56, no. 3/4, p. 539-622.

Ampferer, O., and Hammer, W., 1911, Geologischer Querschnitt durch die Östalpen vom Allgäu zum Gardasee: Jahrbuch der Kaiserlich-Königlichen Geologischen Reichsanstalt, B. 61, no. 3/4, p. 531-710.

Anderson, D.L., 1987, Thermally induced phase changes, lateral heterogeneity of the mantle, continental roots, and deep slab anomalies: Journal of Geophysical Research, v. 92, no. B13, p. 13968-13980.

Angeheister, G., Bögel, H., Gebrande, H., Schmidt-Thomé, P., and Zeil, W., 1972, Recent investigations of surficial and deeper crustal structures of the eastern and southern Alps: Geologische Rundschau, Bd. 61, Hft. 2, p. 349-395.

Antipov, M.P., Zharkov, S.M., Kozhenov, V.Ya., and Pospelov, I.I., 1990, Structure of the Mid-Atlantic Ridge and adjacent parts of the abyssal plain at lat. 13° N: International Geology Review, v. 32, no. 5, p. 468-478.

Argand, E., 1916, Sur l'arc des Alpes occidentales: Eclogae Geologicae Helvetiae, v. 16, no. 1, p. 145-191.

———, 1924, La tectonique de l'Asie: 13th International Geological Congress, Comptes Rendus t. 5, p. 171-372. (Engl. translation by A. V. Carozzi: 1977, Tectonics of Asia: New York, Hafner Press, 218 p.)

Argus, D.F., and Gordon, R.G., 1990, Constraints from VLBI on Pacific-North America motion and deformation (Abstract): EOS, v. 71, no. 28, p. 860.

Arldt, Th., 1907, Die Entwicklung der Kontinente und ihrer Liebewelt: Leipzig, Christian Hermann Tauchnitz, 729 p.

Armstrong, R.L., 1978, Cenozoic igneous history of the U. S. Cordillera from lat 42° to 49° N, *in* Smith, R. B., and Eaton, G. P., eds., Cenozoic tectonics and regional geophysics of the Western Cordillera: Geological Society of America Memoir 152, p. 265-282.

Artemjev, M.E., and Artyushkov, E.V., 1971, Structure and isostasy of the Baikal rift and the mechanism of rifting: Journal of Geophysical Research, v. 76, no. 5, p. 1197-1211.

Artyushkov, E.V., and Baer, M.A., 1986, Mechanism of formation of hydrocarbon basins: the West Siberia, Volga-Urals, Timan-Pechora basins and the Permian basin of Texas: Tectonophysics, v. 122, no. 3-4, p. 247-281.

Aubouin, J., 1965, Geosynclines: Amsterdam, Elsevier Publishing Company, 335 p.

———, 1977, Méditerranée orientale et Méditerranée occidentale; esquisse d'une comparaison de cadre alpin: Société Géologique de France Bulletin, série 7, v. 19, no. 3, p. 421-435.

Aumento, F., 1969, Diorites from the Mid-Atlantic Ridge at 45° N: Science, v. 165, no. 3898, p. 1112-1113.

Aumento, F., and Loncarevic, B.D., 1969, The Mid-Atlantic Ridge near 45° N., III. Bald Mountain: Canadian Journal of Earth Sciences, v. 6, no. 1, p. 11-23.

Auzende, J.M., Lafoy, Y., and Marsset, B., 1988, Recent geodynamic evolution of the North Fiji basin (southwest Pacific): Geology, v. 16, no. 10, p. 925-929.

Baker, B.H., 1987, Outline of the petrology of the Kenya rift alkaline province, *in* Fitton, J.G., and Upton, B.G.J., eds., Alkaline igneous provinces Geological Society of London Special Publication no. 30, p. 293-311.

Balk, R., 1937, Structural behavior of igneous rocks: Geological Society of America Memoir 5, 177 p.

Barrell, J., 1914, The status of hypotheses of polar wanderings: Science, v. 40, n. ser., p. 333-340.

Barrows, A.G., Kahle, J.E., and Beeby, D.J., 1985, Earthquake hazards and tectonic history of the San Andreas fault zone, Los Angeles County, California: California Division of Mines and Geology Open-File Report 85-10, 139 p.

Basalt Volcanism Study Project, 1981, Basaltic volcanism on the terrestrial planets: New York, Pergamon Press, Inc., 1286 p.

Bateman, P.C., 1983, A summary of critical relations in the central part of the Sierra Nevada batholith, California, *in* Roddick, J. A., ed., Circum-Pacific plutonic terranes: Geological Society of America Memoir 159, p. 241-254.

Bayer, R., Carozzo, M.T., Lanza, R., Miletto, M., and Rey, D., 1989, Gravity modelling along the ECORS-CROP vertical seismic reflection profile through the western Alps: Tectonophysics, v. 162, no. 3/4, p. 203-218.

Beloussov, V.V., 1974, Seafloor spreading and geologic reality, in Kahle, C.F. ed., Plate tectonics—assessments and reassessments: American Association of Petroleum Geologists Memoir 23, p. 155-166.

——, 1980, Geotectonics: New York, Springer-Verlag, 330 p.

——, 1981, Continental endogenous regimes: Moscow, Mir Publishers, 296 p.

——, 1989, Osnovyy geotektonika: Moscow, Nedra, 382 p.Bender, F., 1983, Geology of Burma: Berlin, Gebrüder Borntraeger, 293 p.

Benioff, H., 1949, Seismic evidence for the fault origin of oceanic deeps: Geological Society of America Bulletin, v. 60, no. 12, p. 1837-1856.

——, 1954, Orogenesis and deep crustal structure—additional evidence from seismology: Geological Society of America Bulletin, v. 65, no. 5, p. 385-400.

Berberian, F., and Berberian, M., 1983, Tectono-plutonic episodes in Iran, in Gupta, H.K., and Delany, F.M., eds., Zagros-Hindu Kush-Himalaya geodynamic evolution: American Geophysical Union-Geological Society of America Geodynamics Series, v. 3, p. 5-32.

Berberian, M., 1983, Active faulting and tectonics of Iran, in Gupta, H. K., and Delany, F. M., eds., Zagros-Hindu-Himalaya geodynamic evolution: American Geophysical Union-Geological Society of America Geodynamics Series, v. 3, p. 33-69.

Berckhemer, H., 1977, Some aspects of the evolution of marginal seas deduced from observations in the Aegean region, in Biju-Duval, B., and Montadert, L., eds., Structural history of the Mediterranean basins: Paris, Éditions Technip, p. 303-313.

Bertrand, M., 1897, Structure des Alpes françaises et récurrence de certains facies sédimentaires: Sixth International Geological Congress, Lausanne 1894, Comptes Rendus, pt. 3, p. 161-177.

Bevier, M. L., 1989, A lead and strontium isotopic study of the Anahim volcanic belt, British Columbia: additional evidence for widespread suboceanic mantle beneath western North America: Geological Society of America Bulletin, v. 101, no. 7. p. 973-981.

Bibee, L.D., Shor, G.G., Jr., and Lu, R.S., 1980, Inter-arc spreading in the Mariana Trough: Marine Geology, v. 35, no. 1-3, p. 183-197.

Bigarella, J.J., 1973, Geology of the Amazon and Parnaiba basins, in Nairn, A. E. M., and Stehli, F. G., eds., The ocean basins and margins, v. 1. The South Atlantic: New York, Plenum Press, p. 25-86.

Biq Chingchang, Shyu, C.T., Chen, J.C., and Boggs, S., Jr., 1985, Taiwan: geology, geophysics, and marine sediments, in Nairn, A.E.M., Stehli, F.G., and Uyeda, S., eds., The ocean basins and margins, v. 7A. The Pacific Ocean: New York, Plenum Press, p. 503-550.

Birch, F., Roy, R.F., and Decker, E.R., 1968, Heat flow and thermal history in New England and New York, in Zen, E.A., White, W.S., Hadley, J.B., and Thompson, J.B., Jr., eds., Studies of Appalachian geology, northern and maritime: New Interscience Publishers, p. 437-451.

Blatt, F.J., 1983, Principles of physics: Boston, Allyn and Bacon, Inc., 815 p.

Bloomer, S.H., Stern, R.J., Fisk, E., and Geschwind, C.H., 1989, Shoshonitic volcanism in the northern Mariana arc, 1. Mineralogic and major and trace element characteristics: Journal of Geophysical Research, v. 94, no. B4, p. 4469-4496.

Blümling, P., and Prodehl, C., 1983, Crustal structure beneath the eastern part of the Coast Ranges (Diablo Range) of central California from explosion seismic and near earthquake data: Physics of the Earth and Planetary Interiors, v. 31, p. 313-326.

Blümling, P., Mooney, W.D., and Lee, W.H.K., 1985, Crustal structure of the southern Calaveras fault zone, central California, from seismic refraction investigations: Seismological Society of America Bulletin, v. 75, no. 1, p. 193-209.

Boillot, G., Girardeau, J., and Kornprobst, J., 1989, Rifting of the west Galicia continental margin: a review: Société Géologique de France Bulletin, 8e série, v. 5, no. 2, p. 393-400.

Bois, C., Pinet, B., and Roure, F., 1989, Dating lower crustal features in France and adjacent areas from deep seismic profiles, in Mereu, R.F., Mueller, S., and Fountain, D.M., eds., Properties and processes of earth's lower crust: American Geophysical Union Geophysical Monograph no. 51, p. 17-31.

Bolli, H.M., 1980a, Ages of sediments recovered from the Deep Sea Drilling Project Pacific Legs 5 through 9, 16 through 21, and 29 through 35, in Rosendahl, B.R, Hekinian, R., and others, Initial reports of the Deep Sea Drilling Project, v. 54: Washington, D. C., U. S. Government Printing Office, p. 881-886.

——, 1980b, The ages of sediments recovered from DSDP Legs 1-4, 10-15, and 36-53 (Atlantic, Gulf of Mexico, Caribbean, Mediterranean and Black Sea), in Rosendahl, B.R., Hekinian, R., and others, Initial Reports of the Deep Sea Drilling Project, v. 54: Washington D. C., U. S. Government Printing Office, p. 887-895.

Bollinger, G.A., and Wheeler, R.L., 1988, The Giles County, Virginia, seismic zone—seismological results and geological interpretations: United States Geological Survey Professional Paper 1355, 85 p.

Bonatti, E., Honnorez J., and Ferrara, G., 1971, Peridotite-gabbro-basalt complex from the equatorial Mid-Atlantic Ridge, in Bullard, Sir E., Cann, J.R., and Matthews, D.H., Organizers, A discussion on the petrology of igneous and metamorphic rocks from the ocean floor: Royal Society of London Philosophical Transactions, series A, v. 285, no. 1192, p. 385-402.

Bonatti, E., Hamlyn, P., and Ottonello, G., 1981, Upper mantle beneath young oceanic rift: peridotites from the island of Zabagard (Red Sea): Geology, v. 9, no. 10, 474-479.

Bonini, W.E., Loomis, T.P., and Robertson, J.D., 1973, Gravity anomalies, ultramafic intrusions, and the tectonics of the region around the Strait of Gibraltar: Journal of Geophysical Research, v. 78. no. 8, p. 1372-1382.

Borchert, H., and Muir, R.O., 1964, Salt deposits. The origin, metamorphism and deformation of evaporites: London, D. Van Nostrand Company, Ltd., 338 p.

Bosshard, E., and Macfarlane, D.J., 1970, Crustal structure of the western Canary Islands from seismic refraction and gravity data: Journal of Geophysical Research, v. 75, no. 26, p. 4901-4918.

Bott, M.P.H., 1971, The interior of the earth: New York, St. Martin's Press, 316 p.

Boudier, F., and Nicolas, A., eds., 1988, The ophiolites of Oman: Tectonophysics, v. 151, no. 1-4, 401 p.

Bowden, P., Black, R., Martin, R.F., Ike, E.C., Kinnaird, J.A., and Batchelor, R.A., 1987, Niger-Nigerian alkaline ring complexes: a classic example of African Phanerozoic anorogenic mid-plate magmatism, in Fitton, J.G., and Upton, B.G.J., eds., Alkaline igneous rocks: Geological Society of London Special Publication no. 30, p. 357-379.

Bowin, C., 1973, Origin of the Ninety East Ridge from studies near the equator: Journal of Geophysical Research, v. 78, no. 26, p. 6029-6043.

——, 1976, Caribbean gravity field and plate tectonics: Geological Society of America Special Paper 169, 79 p.

Bowin, C., Purdy, G.M., Johnston, C., Shor, G., Lawver, L., Hartono, H.M.S., and Jezek, P., 1980, Arc-continent collision in Banda Sea region: American Association of Petroleum Geologists Bulletin, v. 64, no. 6, p. 868-915.

Bowin, C., Warsi, W., and Milligan, J., 1982, Free-air gravity anomaly atlas of the world: Geologicl Society of America Map and Chart Series MC-46, 5 p., 74 sheets, scale, 1:4,000,000.

Bowland, C.L., and Rosencrantz, E., 1988, Upper crustal structure of the western Colombian basin, Caribbean Sea: Geological Society of America Bulletin, v. 100, no. 4, p. 534-546.

Braile, L.W., 1989, Crustal structure of the continental interior, *in* Pakiser, L.C., and Mooney, W.D., eds. Geophysical framework of the continental United States: Geological Society of America Memoir 172, p. 285-315.

Breen, N.A., Silver, E.A., and Roof, S., 1989, The Wetar back arc thrust belt, eastern Indonesia: the effect of accretion against an irregularly shaped arc: Tectonics, v. 8, no. 1, p. 85-98.

Bridgwater, D., Sutton, J., and Watterson, J., 1974, Crustal downfolding associated with igneous activity: Tectonophysics, v. 21, no. 1/2, p. 57-77.

Brodie, K.H., and Rutter, E.H., 1987, Deep crustal extensional faulting in the Ivrea zone of northern Italy: Tectonophysics, v. 140, no. 2-4, p. 193-212.

Brooks, C., James, D.E., and Hart, S.R., 1976, Ancient lithosphere: its role in young continental volcanism: Science, v. 193, no. 4258, p. 1086-1094.

Brown, R.L., and Tippett, C.R., 1978, The Selkirk fan structure of the southeastern Canadian Cordillera: Geological Society of America Bulletin, v. 89, no. 4, p. 548-558.

Browne, S.E., and Fairhead, J.D., 1983, Gravity study of the central African rift system: a model of continental disruption, 1. The Ngaoundere and Abu Gabra rifts: Tectonophysics, v. 94, no. 1-4, p. 187-203.

Bruce, R.M., Nelson, E.P., and Weaver, S.G., 1989, Effects of synchronous uplift and intrusion during magmatic arc construction: Tectonophysics, v. 161, no. 3/4, p. 317-329.

Brune, J.N., 1969, Surface waves and crustal structure, *in* Hart, P.J. ed., The earth's crust and upper mantle: American Geophysical Union Geophysical Monograph 13, p. 230-242.

Brunel, M., 1986, Ductile thrusting in the Himalayas: shear sense criteria and stretching lineations: Tectonics, v. 5, no. 2, p. 247-265.

Brunn, J.H., 1961, Les sutures ophiolitiques. Contribution a l'étude des rélations entre phénomènes magmatiques et orogéniques: Révue de Géographie Physique et de Géologie Dynamique, séries 2, v. 4, fasc. 2, p. 89-96.

———, 1979, Océans et orogènes: Société Géologique de France Bulletin, séries 7, v. 21, no. 5, p. 653-661.

Bucher, W.H., 1933, The deformation of the earth's crust: Princeton University Press, 518 p.

———, 1947, Problems of earth deformation illustrated by the Caribbean Sea basin: New York Academy of Sciences Transactions, ser. II, v. 9, no. 3, p. 98-116.

———, 1955, Deformation in orogenic belts, *in* Poldervaart, A., ed., Crust of the earth (a symposium): Geological Society of America Special Paper 62, p. 343-368.

———, 1956, Role of gravity in orogenesis: Geological Society of America Bulletin, v. 67, no. 10, p. 1295-1318.

Burchfiel, B.C., and Davis, G.A., 1968, Two-sided nature of the Cordilleran orogen and its tectonic implications: 23rd International Geological Congress, Prague 1968, Repts. v. 3, p. 175-194.

———, 1975, Nature and controls of Cordilleran orogenesis, western United States: extensions of an earlier synthesis: American Journal of Science, v. 275-A, p. 363-396.

Burnett, M.S., and Orcutt, J.A., 1989, Behavior of amplitudes and travel times of P-waves propagating about the magma chamber at 12°50' N on the East Pacific Rise (Abstract): EOS, v. 70, no. 15, p. 455.

Busson, G., and Cornée, A., 1989, Quelques données sur les antécédents climatiques du Sahara: la signification des couches détritiques rouges et des evaporites du Trias et du Lias-Dogger: Société Géologique de France Bulletin, série 8, v. 5, no. 1, p. 3-11.

Camfield, P.A., and Gough, D.I., 1977, A possible Proterozoic plate boundary in North America: Canadian Journal of Earth Sciences, v. 14, no. 6, p. 1229-1238.

Cardwell, R.K., Isacks, B.L., and Karig, D.E., 1980, The spatial distribution of earthquakes, focal mechanism solutions, and subducted lithosphere in the Philippine and northeastern Indonesian islands, *in* Hayes, D.E., ed., The tectonic and geologic evolution of southeast Asian seas and islands: American Geophysical Union Geophysical Monograph 23, p. 1-35.

Caress, D.W., Burnett, M.S., and Orcutt, J.A., 1989, Tomographic imaging of the magma chamber at 12°50' N on the East Pacific Rise (Abstract): EOS, v. 70, no. 15, p. 455-456.

Cebull, S.E., 1973, Concept of orogeny: Geology, v. 1, no. 3, p. 101-102.

Čermák, V., and Rybach, L., eds., 1979, Terrestrial heat flow in Europe: New York, Springer-Verlag, 328 p.

Chamberlin, R.T., 1925, The wedge theory of diastrophism: Journal of Geology, v. 33, no. 8, p. 755-792.

Chamberlin, R.T., and Link, T.A., 1927, The theory of laterally spreading batholiths: Journal of Geology, v. 35, no. 4, p. 319-352.

Chamberlin, R.T., and Richards, J.T., 1918, Preliminary report on experiments relating to continental deformation (Abstract): Science, new series, v. 47, p. 492.

Chappell, B.W., and White, A.J.R., 1974, Two contrasting granite types: Pacific Geology, v. 8, p. 173-174.

Chen Guoda, 1989, Tectonics of China: New York, Pergamon Press and Beijing, International Academic Publishers, 258 p.

Chen Zongji, 1987, Geodynamics and tectonic evolution of the Panxi rift: Tectonophysics, v. 133, no. 3/4, p. 287-304.

Cheng, Q., Park, K.H., Macdougall, J.D., Zindler, A., Lugmair, G.W., Staudigel, H., Hawkins, J., and Lonsdale, P., 1987, Isotopic evidence for a hotspot origin of the Louisville Seamount chain, *in* Keating, B.H., Fryer, P., Batiza, R., and Boehlert, G.W., eds., Seamounts, islands, and atolls: American Geophysical Union Geophysical Monograph 43, p. 283-296.

Choukroune, P., 1989, The ECORS Pyrenean deep seismic profile reflection data and the overall structure of an orogenic belt: Tectonics, v. 8, no. 1, p. 23-39.

Cifelli, R., 1970, Age relationships of Mid-Atlantic Ridge sediments: Geological Society of America Special Paper 124, p. 47-69.

Clark, K.F., Foster, C.T., and Damon, P.E., 1982, Cenozoic mineral deposits and subduction-related magmatic arcs in Mexico: Geological Society of America Bulletin, v. 93, no. 6, p. 533-544.

Clauss, B., Marquart, G., and Fuchs, K., 1989, Stress orientations in the North Sea and Fennoscandia, a comparison to the central European stress field, *in* Gregersen, S., and Basham, P.W., eds., Earthquakes at North-Atlantic passive margins: neotectonics and postglacial rebound: Dordrecht, Kluwer Academic Publishers, p. 277-287.

Clemens, J.D., 1988, Volume and composition relationships between granites and their lower crustal source regions: an example from central Victoria, Australia: Australian Journal of Earth Sciences, v. 35, no. 4, p. 445-449.

Cobbing, E.J., and Pitcher, W.S., 1983, Andean plutonism in Peru and its relationship to volcanism and metallogenesis at a segmented plate edge, *in* Roddick, J.A., ed., Circum-Pacific plutonic terranes: Geological Society of America Memoir 159, p. 277-291.

Collet, L.W., 1927, The structure of the Alps, 2nd ed.: London, E. Arnold, 289 p. (reprinted: 1974, The structure of the Alps, 2nd ed.: Huntington, New York, Robert E. Krieger Publishing Company, 304 p.)

Coney, P.J., 1980, Cordilleran metamorphic core complexes: an overview, *in* Crittenden, M.D., Jr., Coney, P.J., and Davis, G.H., eds., Cordilleran metamorphic core complexes: Geological Society of America Memoir 153, p. 7-31.

Coney, P.J., and Reynolds, S.J., 1977, Cordilleran Benioff zones: Nature, v. 270, p. 403-406.

Cook, T.D., and Bally, A.W., eds., 1975, Stratigraphic atlas of North and Central America: Princeton University Press, 272 p.

Cooper, M.A., Collins, D.A., Ford, M., Murphy, F.X., Trayner, P.M., and O'Sullivan, M., 1986, Structural evolution of the Irish Variscides: Geological Society of London Journal, v. 143, pt. 1, p. 53-61.

Corry, C.E., 1988, Laccoliths; mechanics of emplacement and growth: Geological Society of America Special Publication 220, 110 p.

Courtney, R.C., and White, R.S., 1986, Anomalous heat flow and geoid across the Cape Verde Rise: evidence for dynamic support from a thermal plume in the mantle: Royal Astronomical Society Geophysical Journal, v. 87, p. 815-867.

Coward, M.P., 1982, Surge zones in the Moine thrust zone of NW Scotland: Journal of Structural Geology, v. 4, p. 247-256.

Coward, M.P., and Smallwood, S., 1984, An interpretation of the Variscan tectonics of SW Britain: Geological Society of London Special Publication no. 14, p. 89-102.

Coward, M.P., Windley, B.F., Broughton, R.D., Luff, I.W., Petterson, M.G., Pudsey, C.J., Rex, D.C., and Asif Khan, M., 1986, Collision tectonics in the NW Himalayas, *in* Coward, M.P., and Ries, A.C., eds., Collision tectonics: Geological Society of London Special Publication no. 19, p. 203-219.

Crane, K., 1987, Structural evolution of the East Pacific Rise axis from 13° 10′ N to 10° 35′ N: interpretations from SeaMARC I data: Tectonophysics, v. 136, no. 1/2, p. 65-124.

Crane, K., and O'Connell, S., 1983, The distribution and implications of heat flow from the Gregory rift in Kenya: Tectonophysics, v. 94, no. 1-4, p. 253-275.

Croneis, C., and Reso, A., 1966, Physical and biological evidence for major mid-Cretaceous stratigraphic break (Abstract): American Association of Petroleum Geologists Bulletin, v. 50, no. 3, p. 609.

Curray, J.R., Shor, G.G., Jr., Raitt, R.W., and Henry, M., 1977, Seismic refraction and reflection studies of crustal structure in the eastern Sunda and western Banda arcs: Journal of Geophysical Research, v. 82, no. 17, p. 2479-2489.

Daly, R.A., 1925, The geology of Ascension Island: Philadelphia, American Academy of Arts and Sciences Proceedings, v. 60, p. 3-124.

———, 1940, Strength and structure of the earth: New York, Prentice-Hall, Inc., 434 p.

Dana, J. D., 1847, Geological results of the earth's contraction in consequence of cooling: American Journal of Science, 2nd series, v. 3, p. 176-188; v. 4, p. 88-92.

———, 1896, Manual of geology, 4th edition: New York, American Book Company, 1087 p.

Darwin, G., 1887, Note on Mr. Davison's paper on the straining of the earth's crust in cooling: Royal Society of London Philosophical Transactions, series A, v. 178, p. 242-249.

Davis, G.A., and Lister, G.S., 1988, Detachment faulting in continental extension; perspectives from the southwestern U. S. Cordillera, *in* Clark, S.P., Jr., Burchfiel, B.C., and Suppe, J., eds., Processes in continental lithospheric deformation: Geological Society of America Special Paper 218, p. 133-159.

Davison, C., 1887, On the distribution of strain in the earth's crust resulting from secular cooling; with special reference to the growth of continents and the formation of mountain chains: Royal Society of London Philosophical Transactions, series A, v. 178, p. 231-242.

de Almeida, F.F.M., 1986, Distribuição regional e relações tectônicas do magmatismo pós-paleozóico no Brasil: Revista Brasileira de Geociências, v. 16, no. 4, p. 325-349.

Deichmann, N., and Rybach, L., 1989, Earthquakes and temperatures in the lower crust below the northern Alpine foreland of Switzerland, *in* Mereu, R.F., Mueller, S., and Fountain, D.M., eds., Properties and processes of earth's lower crust: American Geophysical Union Geophysical Monograph no. 51, p. 197-213.

de Lory, Ch., 1860, Géologie du Dauphine: Paris (not seen. Mentioned in Aubouin [1965] and emended by Curt Teichert, written commun., 1988).

Dennis, J.G., ed., 1982, Orogeny: Stroudsburg, PA, Hutchinson Ross Publishing Company, 379 p.

Dennison, J.M., and Johnson, R.W., Jr., 1971, Tertiary intrusions and associated phenomena near the thirty-eighth parallel fracture zone in Virginia and West Virginia: Geological Society of America Bulletin, v. 82, no. 2, p. 501-507.

Descartes, R., 1644, Principia philosophiae: Amsterdam, Elsevier Publishing Company, 310 p.

de Sitter, L.U., 1964, Structural geology, 2nd edition: New York, McGraw-Hill Book Company, 551 p.

Dewey, J.F., 1988, Extensional collapse of orogens: Tectonics, v. 7, no. 6, p. 1123-1139.

Dewey, J.F., Hempton, M.R., Kidd, W.S.F., Saroglu, F., and Şengör, A.M.C., 1986, Shortening of continental lithosphere: the neotectonics of eastern Anatolia—a young collision zone, *in* Coward, M.P., and Ries, A.C., eds., Collision tectonics: Geological Society of London Special Publication no. 19, p. 3-36.

Dibblee, T.W., Jr., 1981, Upper Mesozoic rock units in the central Diablo Range between Hollister and New Idria and their depositional environments, *in* Frizzell, V., ed., Geology of the central and northern Diablo Range, California: Society of Economic Paleontologists and Mineralogists, Pacific Section Book 19, p. 13-20.

Dillon, W.P., and Sougy, J.M.A., 1974, Geology of West Africa and Canary and Cape Verde Islands, *in* Nairn, A.E.M., and Stehli, F.G., eds., The ocean basins and margins, v. 2. The North Atlantic: New York, Plenum Press, p. 315-390.

Dimitrijević, M.D., 1974, The Dinarides: a model based on the new global tectonics, *in* Janković, S., ed., Metallogeny and concepts of the geotectonic development of Yugoslavia: Belgrade, Faculty of Mining and Geology, Belgrade University, Department of Economic Geology, p. 141-176.

Divis, A.F., 1980, The petrology and tectonics of recent volcanism in the central Philippine Islands, *in* Hayes, D.E., ed., The tectonic and geologic evolution of southeast Asian seas and islands: American Geophysical Union Geophysical Monograph 23, p. 127-144.

Dragašević, T., 1974, Contemporary structure of the earth's crust and upper mantle on the territory of Yugoslavia, *in* Janković, S., ed., Metallogeny and concepts of the geotectonic development of Yugoslavia: Belgrade, Faculty of Mining and Geology, Belgrade University, Department of Economic Geology, p. 73-87.

Drake, C.L., and Nafe, J.E., 1968, The transition from ocean to continent from seismic refraction data, *in* Knopoff, L., Drake, C.L., and Hart, P.J., eds., The crust and upper mantle in the Pacific area: American Geophysical Union Geophysical Monograph no. 12, p. 174-186.

Duncan, R.A., 1984, Age progressive volcanism in the New England Seamounts and the opening of the central Atlantic Ocean: Journal of Geophysical Research, v. 89, no. B12, p. 9980-9990.

Dunkelman, T.J., Karson, J.A., and Rosendahl, B.R., 1988, Structural style of the Turkana rift, Kenya: Geology, v. 16, no. 3, p. 258-261.

Dziewonski, A.M., and Anderson, D.L., 1984, Seismic tomography of the earth's interior: American Scientist, v. 72, no. 5, p. 483-494.

Dziewonski, A.M., and Woodhouse, J.H., 1987, Global images of the earth's interior: Science, v. 236, no. 4797, p. 37-48.

Eaton, G.P., 1986, A tectonic redefintion of the Southern Rocky Mountains: Tectonophysics, v. 132, no. 1-3, p. 163-193.

——, 1987, Topography and origin of the Southern Rocky Mountains and Alvarado Ridge, in Coward, M.P., Dewey, J.F., and Hancock, P.L., eds., Continental extensional tectonics: Geological Society of London Special Publication no. 28, p. 355-369.

Eaton, G.P., Wahl, R.R., Prostka, H.J., Mabey, D.R., and Kleinkopf, M.D., 1978, Regional gravity and tectonic patterns: their relation to late Cenozoic epeirogeny and lateral spreading in the Western Cordillera, in Smith, R.B., and Eaton, G.P., eds., Cenozoic tectonics and regional geophysics of the Western Cordillera: Geological Society of America Memoir 152, p. 51-91.

Ebinger, C.J., 1989, Tectonic development of the western branch of the East African rift system: Geological Society of America Bulletin, v. 101, no. 7, p. 885-903.

Ebinger, C.J., Deino, A.L., Drake, R.E., and Tesha, A.L., 1989, Chronology of volcanism and rift basin propagation: Rungwe volcanic province, East Africa: Journal of Geophysical Research, v. 94, no. B11, p. 15785-15803.

Edgar, N.T., Ewing, J.I., and Hennion, J., 1971, Seismic refracton and reflection in Caribbean Sea: American Association of Petroleum Geologists Bulletin, v. 55, no. 6, p. 833-870.

Eguchi, T., 1984, Seismotectonics around the Mariana Trough: Tectonophysics, v. 102, no. 1-3, p. 33-52.

Élie de Beaumont, J.B.A.L.L., 1831, Researches on some of the revolutions which have taken place on the surface of the globe: Philosophical Magazine, v. 10, p. 241-264. (English translation of the original French article published in 1829.)

Elliot, D., 1981, The strength of rocks in thrust sheets (Abstract): EOS, v. 62, no. 17, p. 397.

Ellis, M., and Watkinson, A.J., 1987, Orogen-parallel extension and oblique tectonics: the relation between stretching lineations and relative plate motions: Geology, v. 15, no. 11, p. 1022-1026.

Ellsworth, W.L., and Koyanagi, R.Y., 1977, Three-dimensional crust and mantle structure of Kilauea Volcano, Hawaii: Journal of Geophysical Research, v. 82, no. 33, p. 5379-5394.

Emmons, R.C., 1969, Strike-slip rupture patterns in sand models: Tectonophysics, v. 7, no. 1, p. 71-87.

Emmons, W.H., 1940, The principles of economic geology, 2nd edition: New York, McGraw-Hill Book Company, Inc., 529 p.

Engdahl, E.R., and Rinehart, W.A., 1988, Seismicity map of North America: Geological Society of America Continent-Scale Map 004, scale, 1:5,000,000, 4 sheets.

Erickson, A.J., Simmons, G., and Ryan, W.B.F., 1977, Review of heatflow data from the Mediterranean and Aegean Seas, in Biju-Duval, B., and Montadert, L., eds., Structural history of the Mediterranean basins: Paris, Éditions Technip, p. 263-279.

Ervin, P., and McGinnis, L.D., 1975, Reelfoot rift: reactivated precursor to the Mississippi embayment: Geological Society of America Bulletin, v. 86, no. 9, p. 1287-1295.

Eva, C., Cattaneo, M., and Merlanti, F., 1988, Seismotectonics of the central segment of the Indonesian arc: Tectonophysics, v. 146, no. 1-4, p. 241-259.

Ewing, J., and Ewing, M., 1959, Seismic-refraction measurements in the Atlantic Ocean basins, in the Mediterranean Sea, on the Mid-Atlantic Ridge, and in the Norwegian Sea: Geological Society of America Bulletin, v. 70, no. 3, p. 291-317.

Fedotov, S.A., 1973, Deep structure under the volcanic belt of Kamchatka, in Coleman, P.J., ed., The western Pacific: island arcs, marginal seas, geochemistry: University of Western Australia Press, p. 247-254.

Fei Ding, Cheng Shixing, Hao Chunrong, Yin Shujie, Ren Rui, Dong Yun, Ma Junru, Chang Xueju, and Wang Menying, 1981, On the structural features in the central part of Xizang and obduction of the Indian plate, in Liu Dongsheng, ed.-in-chief, Geological and ecological studies of Qinghai-Xizang Plateau, v. 1, Geology, geological history and origin of Qinghai-Xizang Plateau, Proceedings of symposium on Qinghai-Xizang (Tibet) Plateau (Beijing, China): Beijing, Science Press, and New York, Gordon and Breach, Science Publishers, Inc., p. 747-756.

Feng, R., and McEvilly, T.V., Interpretation of seismic reflection profiling data for the structure of the San Andreas fault zone: Seismological Society of America Bulletin, v. 73, no. 6, p. 1701-1720.

Filloux, J.H., 1982, Magnetotelluric experiment over the ROSE area: Journal of Geophysical Research, v. 87, no. B10, p. 8364-8378.

Finlayson, D.M., Leven, J.H., and Wake-Dyster, K.D., 1989, Large-scale lenticles in the lower crust under an intra-continental basin in eastern Australia, in Mereu, R.F., Mueller, S., and Fountain, D.M., eds., Properties and processes of earth's lower crust: American Geophysical Union Geophysical Monograph no. 51, p. 3-16.

Fletcher, J.B., Sbar, M.L., and Sykes, L.R., 1978, Seismic trends and travel-time residuals in eastern North America and their tectonic implications: Geological Society of America Bulletin, v. 89, no. 11, p. 1656-1676.

Frei, W., Heitzmann, P., Lehner, P. and Valasek, P., 1989, Die drei Alpentraversen von NFP 20: Swiss Association of Petroleum Geologists and Engineers Bulletin, v. 55, no. 128, p. 13-43.

Froitzheim, N., Stets, J., and Wurster, P., 1988, Aspects of western High Atlas tectonics, in Jacobshagen, V.H., ed., The Atlas system of Morocco. Studies on its geodynamic evolution: New York, Springer-Verlag, p. 219-244.

Fuchs, K., 1983, Recently formed elastic anisotropy and petrological models for the continental subcrustal lithosphere in southern Germany: Physics of the Earth and Planetary Interiors, v. 31, p. 93-118.

Fuis, G.S., and Kohler, W.M., 1984, Crustal structure and tectonics of the Imperial Valley region, California, in Rigsby, C.A., ed., The Imperial basin—tectonics, sedimentation and thermal aspects: Society of Economic Geologists and Paleontologists, Pacific Section Book 40, p. 1-13.

Fuis, G.S., Zucca, J.J., Mooney, W.D., and Milkereit, B., 1987, A geologic interpretation of seismic-refraction results in northeastern California: Geological Society of America Bulletin, v. 98, no. 1, p. 53-65.

Fullagar, P.D., and Bottino, M.L., 1969, Tertiary felsic intrusions in the Valley and Ridge Province, Virginia: Geological Society of America Bulletin, v. 80, no. 9, p. 1853-1857.

Fullagar, P.D., and Butler, J.R., 1979, 325 to 265 m.y.-old granitic plutons in the Piedmont of the southeastern Appalachians: American Journal of Science, v. 279, no. 2, p. 161-185.

Furumoto, A.S., Webb, J.P., Odegard, M.E., and Hussong, D.M., 1976, Seismic studies on the Ontong Java Plateau, 1970: Tectonophysics, v. 34, no. 1/2, p. 71-90.

Fúster, J.M., Fernandez-Santin, S., and Sagredo, J., 1968a, Geología y vulcanología de las Islas Canarias. Lanzarote: Madrid, Consejo Superior de Investigaciones Científicas, Instituto "Lucas Mallada," 177 p.

Fúster, J.M., Cendrero, A., Gastesi, P., Ibarrola, E., and López-Ruiz, J., 1968b, Geología y vulcanología de las Islas Canarias. Fuerteventura: Madrid, Consejo Superior de Investigaciones Científicas, Instituto "Lucas Mallada," 239 p.

Fytikas, M., Innocenti, F., Manetti, P., Mazzuoli, R., Peccerillo, A., and Villari, L., 1984, Tertiary to Quaternary evolution of volcanism in the Aegean region, in Dixon, J.E., and Robertson, A.H.F., eds., The geological evolution of the eastern Mediterranean: Geological Society of London Special Publication no. 17, p. 687-699.

Gansser, A., 1968, The Insubric Line, a major geotectonic line: Schweizerische Mineralogische und Petrographische Mitteilungen, Bd. 48, p. 123-143.

———, 1973, Facts and theories on the Andes: Geological Society of London Journal, v. 129, pt. 2, p. 93-131.

———, 1983, The morphogenic phase of mountain building, in Hsü, K., ed., Mountain building processes: London, Academic Press, p. 221-228.

GEBCO, 1975-1982, General bathymetric chart of the oceans: Ottawa, The Canadian Hydrographic Service, Charts 5-1 through 5-16, scale, 1:10,000,000; Charts 5-17 and 5-18, scale, 1:6,000,000.

Gehrels, G.E., and McClelland, W.C., 1988, Early Tertiary uplift of the central and northern Coast Range batholith along west-side-up extensional shear zones: Geological Society of America, Abstracts with Programs, v. 20, no. 7, p. A111.

Geller, C.A., Weissel, J.K., and Anderson, R.N., 1983, Heat transfer and intra-plate deformation in the central Indian Ocean: Journal of Geophysical Research, v. 88, no. B2, p. 1018-1032

Genshaft, Yu.S., and Saltykovskiy, A.Ya., 1989, Kontinental'nyy vulkanizm, tsenolity i tektonika litosfernykh plit, in Beloussov, V.V., eds., Tektonosfera: yeye stroyeniye i razvitiye: Moscow, Akademiya Nauk SSSR, Mezhduvedomstvennyy Geofizicheskiy Komitet, Geodinamichskiye Issledovaniya no. 13, p. 94-105.

Geological Survey of Canada, 1987, Magnetic anomaly map of Canada, 5th ed.: Geological Survey of Canada Map 1255A, 1 sheet, scale, 1:5,000,000.

Gilbert, G.K., 1877, Geology of the Henry Mountains, Utah: United States Geographical and Geological Survey of the Rocky Mountain Region, 170 p.

Gilluly, J., 1973, Steady plate motion and episodic orogeny and magmatism: Geological Society of America Bulletin, v. 84, no. 2, p. 499-513.

Glangeaud, L., 1957, Essai de classification géodynamique des chaînes et des phénomènes orogèniques: Révue de Géographie Physique et de Géologie Dynamique, séries 2, v. 1, fasc. 4, p. 200-220.

———, 1959, Classification géodynamique des chaînes de montagnes. 9.—Chaînes intracratoniques: 1. Structure et embryologie des cratons (croute sialique): Révue de Géographie Physique et de Géologie Dynamique, séries 2, v. 2, fasc. 4, p. 197-204.

Golonetsky, S.I., and Misharina, L.A., 1978, Seismicity and earthquake focal mechanisms in the Baikal rift zone: Tectonophysics, v. 45, no. 1, p. 71-85.

Goodwin, E.B., and Thompson, G.A., 1988, The seismically reflective crust beneath highly extended terranes: evidence for its origin and extension: Geological Society of America Bulletin, v. 100, no. 10, p. 1616-1626.

Gorokhov, S.S., and Sharfman, V.A., 1963, Main Uralian fault in the southern Urals: Academy of Sciences of the U.S.S.R. Doklady, v. 149 (English trans. by American Geological Institute), p. 38-41.

Gorshkov, A.G., and Lukashevich, I.P., 1989, Computing of magma chamber temperatures in rift zones of the world ocean: Tectonophysics, v. 159, no. 3-4, p. 337-346.

Gorshkov, G.S., 1973, Petrochemistry of volcanic rocks in the Kuril Island arc with some generalizations on volcanism, in Coleman, P.J., ed., The western Pacific: island arcs, marginal seas, geochemistry: University of Western Australia Press, p. 459-467.

Goslin, J., and Sibuet, J.C., 1975, Geophysical study of the easternmost Walvis Ridge, south Atlantic: Geological Society of America Bulletin, v. 86, no. 12, p. 1713-1724.

Gough, D.I., 1984, Mantle upflow under North America and plate dynamics: Nature, v. 311, no. 5985, p. 428-433.

———, 1989, Magnetometer array studies, earth structure, and tectonic processes: Review of Geophysics, v. 27, no. 1, p. 141-157.

Gramberg, I.S., and Smyslov, A.A., 1986, The map of heat flow and hydrothermal mineralization in the world ocean: Moscow, Ministry of Geology of the USSR, 6 sheets, scale, 1:20,000,000.

Grand, S.P., 1987, Tomographic inversion for shear velocity beneath the North American plate: Journal of Geophysical Research, v. 92, no. B13, p. 14065-14090.

Gretener, P.E., 1969, On the mechanics of the intrusion of sills: Canadian Journal of Earth Sciences, v. 6, no. 6, p. 1415-1419.

Gries, R., 1983, North-south compression of Rocky Mountain foreland structures, in Lowell, J.D., and Gries, R., eds., Rocky Mountain foreland basins and uplifts: Rocky Mountain Association of Geologists, p. 9-32.

Grow, J.A., 1973, Crustal and upper mantle structure of the central Aleutian arc: Geological Society of America Bulletin, v. 84, no. 7, p. 2169-2191.

Gutenberg, B., and Richter, C.F., 1949, Seismicity of the earth and associated phenomena: Princeton University Press, 273 p.

———, 1954, Seismicity of the earth and associated phenomena, 2nd ed.: Princeton Univesity Press, 310 p.

Hamilton, R.M., and Johnston, A.C., 1990, Tecumseh's prophecy: preparing for the next New Madrid earthquake: United States Geological Survey Circular 1066, 30 p.

Hamilton, W., 1963, Tectonics of Antarctica, in Childs, O.E., and Beebe, B.W., eds., Backbone of the Americas, a symposium. Tectonic history from pole to pole: American Association of Petroleum Geologists Memoir 1, p. 4-15.

Hamilton, W.B., 1970, The Uralides and the motion of the Russian and Siberian platforms: Geological Society of America Bulletin, v. 81, no. 9, p. 2553-2576.

———, 1979, Tectonics of the Indonesian region: U.S. Geological Survey Professional Paper 1078, 345 p.

———, 1988, Laramide crustal shortening, in Schmidt, C.J., and Perry, W.J., Jr., eds., Interaction of the Rocky Mountain foreland and the Cordilleran thrust belt: Geological Society of America Memoir 171, p. 27-39.

———, 1989, Crustal geologic processes of the United States, in Pakiser, L.C., and Mooney, W.D., eds., Geophysical framework of the continental United States: Geological Society of America Memoir 172, p. 743-781.

Hansen, V.L., 1988, A model for terrane accretion: Yukon-Tanana and Slide Mountain terranes, northwest North America: Tectonics, v. 7, no. 6, p. 1167-1177.

Harding, A.J., Orcutt, J.A., Kappus, M.E., Vera, E.E., Mutter, J.C., Buhl, P., Detrick, R.S., and Brocher, T.M., 1989, Structure of young oceanic crust at 13° N on the East Pacific Rise from expanding spread profiles: Journal of Geophysical Research, v. 94, no. B9, p. 12163-12196.

Harrison, C.G.A., 1968, Antipodal locations of continents and ocean basins: Science, v. 153, no. 3741, p. 1246-1248.

Haug, E., 1900, Les géosynclinaux et les aires continentales: Société Géologique de France Bulletin, t. 28, p. 617-711.

Hawkins, J.W., Jr., 1974, Geology of the Lau basin, a marginal sea behind the Tonga arc, in Burk, C.A., and Drake, C.L. eds., The geology of continental margins: New York, Springer-Verlag, p. 505-520.

Hearn, T.A., 1987, Crustal structure and tectonics in southern California: United States Geological Survey Circular 956, p. 56-57.

Heezen, B.C., 1960, The rift in the ocean floor: Scientific American, v. 203, no. 4, p. 98-110.

Hegarty, K.A., and Weissel, J.K., 1988, Complexities in the development of the Caroline plate region, western equatorial Pacific, *in* Nairn, A.E.M., Stehli, F.G., and Uyeda, S., eds., The ocean basins and margins, v. 7B. The Pacific Ocean: New York, Plenum Press, p. 277-301.

Hershey, O.H., 1903, Structure of the southern portion of the Klamath Mountains, Cal.: American Geologist, v. 31, p. 231-245.

———, 1906, Some western Klamath stratigraphy: American Journal of Science, 4th series, v. 21, no. 1, p. 58-66.

Hey, R., and Vogt, P., 1977, Spreading center jumps and sub-axial asthenosphere flow near the Galapagos hotspot: Tectonophysics, v. 37, no. 1-3, p. 41-52.

Hildenbrand, T.G., 1982, Model of the southeastern margin of the Mississippi Valley graben near Memphis, Tennessee, from interpretation of truck-magnetometer data: Geology, v. 10, no. 9, p. 476-480.

Hildenbrand, T.G., Simpson, R.W., Godson, R.H., and Kane, M.F., 1982, Digital colored residual and regional Bouguer gravity maps of the conterminous United States with cut-off wavelengths of 250 km and 100 km: U. S. Geological Survey Geophysical Investigations Map GP-953-A, 2 sheets, scale, 1:7,500,000.

Hill, D.P., 1969, Crustal structure of the island of Hawaii from seismic-refraction measurements: Seismological Society of America Bulletin, v. 59, no. 1, p. 101-130.

———, 1978, Seismic evidence for the structure and Cenozoic tectonics of the interior of the Western Cordillera, *in* Smith, R.B., and Eaton, G.P., eds. Cenozoic tectonics and regional geophysics of the Western Cordillera: Geological Society of America Memoir 152, p. 145-174.

Hinze, W.J., and Braile, L.W., 1988, Geophysical aspects of the craton: U. S., *in* Sloss, L.L., ed., Sedimentary cover—North American craton: U. S., The geology of North America, v. D-2: Geological Society of America, Decade of North American Geology, p. 5-24.

Hirahara, K., 1977, A large-scale three-dimensional structure under the Japan islands and the Sea of Japan: Journal of the Physics of the Earth, v. 25, p. 393-417.

Hirahara, K., Ikami, A., Ishida, M., and Mikumo, T., 1989, Three-dimensional P-wave velocity structure beneath central Japan: low-velocity bodies in the wedge portion of the upper mantle above high-velocity subducting plates: Tectonophysics, v. 163, no. 1-2, p. 63-73.

Ho, C.S., 1982, Tectonic evolution of Taiwan. Explanatory text of the tectonic map of Taiwan: Taipei, Ministry of Economic Affairs, 126 p.

Hollister, L.S., and Crawford, M.L., 1986, Melt-enhanced deformation: a major tectonic process: Geology, v. 14, no. 7, p. 562-566.

Honnorez, J., Bonatti, E., Emiliani, C., Brönnimann, P., Furrer, M.A., and Meyerhoff, A.A., 1975, Mesozoic limestone from the Vema offset zone, Mid-Atlantic Ridge: Earth and Planetary Science Letters, v. 26, p. 8-12.

Honza, E., and Tamaki, K., 1985, The Bonin arc, *in* Nairn, A.E.M., Stehli, F.G., and Uyeda, S., eds., The ocean basins and margins, v. 7A. The Pacific Ocean: New York, Plenum Press, p. 459-502.

Hopson, C.A., and Dellinger, D.A., 1989, Depth dependency of granite types, illustrated by tilted plutons in the north Cascades, Washington (Abstract): Geological Society of America, Abstracts with Programs, v. 21, no. 5, p. 95.

Houghton, R.L., Thomas, J.E., Jr., Diecchio, R.J., and Tagliacozzo, A., 1978, Radiometric ages of basalts from DSDP Leg 43: sites 382 and 385 (New England Seamounts), 384 (J-anomaly), 386 and 387 (central and western Bermuda Rise), *in* Tucholke, B.E., Vogt, P.R., and others, Initial Reports of the Deep Sea Drilling Project, v. 43: Washington (DC), U. S. Government Printing Office, p. 739-753.

Humphreys, E., Clayton, R.W., and Hager, B.H., 1984, A tomographic image of mantle structure beneath southern California: Geophysical Research Letters, v. 11, no. 7, p. 625-627.

Hunziker, J.C., and Zingg, A., 1980, Lower Palaeozoic amphibolite to granulite facies metamorphism in the Ivrea zone (southern Alps, northern Italy): Schweizerische Mineralogische und Petrographische Mitteilungen, Bd. 60, p. 181-213.

Hurtig, E., Čermák, V., Haenel, R., and Zui, V., eds., 1981, Geothermal map of Europe: Gotha (East Germany), Verlag Hermann Haack, Geographisch-Kartographische Anstalt, 34 sheets (20 at scale, 1:2,500,000; 14 at scale, 5,000,000).

Hussong, D.M., and Sinton, J.B., 1983, Seismicity associated with back arc crustal spreading in the central Mariana Trough, *in* Hayes, D.E., ed., The tectonic and geologic evolution of southeast Asian seas and islands; part 2: American Geophysical Union Geophysical Monograph 27, p. 217-235.

Hutton, D.H.W., 1988, Granite emplacement mechanisms and tectonic controls: inferences from deformation studies, *in* Brown, P.E., organiser, The origin of granites: The Royal Society of Edinburgh Transactions, Earth Sciences, v. 79, pts. 2-3, p. 245-255.

Hyndman, D.W., Alt, D., and Sears, J.W., 1988, Post-Archean metamorphic and tectonic evolution of western Montana and northern Idaho, *in* Ernst, W.G., eds., Metamorphism and crustal evolution of the western United States. Rubey v. 7: Englewood Cliffs (NJ), Prentice Hall, p. 332-361.

Hyndman, R.D., and Klemperer, S.L., 1989, Lower-crustal porosity from electrical measurements and inferences about composition from seismic velocities: Geophysical Research Letters, v. 16, no. 3, p. 255-258.

Illies, J.H., and Greiner, G., 1979, Holocene movements and state of stress in the Rhinegraben rift system: Tectonophysics, v. 54, no. 1-4, p. 349-359.

Illies, J.H., 1970, Graben tectonics as related to crust-mantle interaction, *in* Illies, J.H., and Mueller, S., eds., Graben problems: Stuttgarts, Schweizerbart, p. 4-27.

Irving, E., and Archibald, D.A., 1990, Bathozonal tilt corrections to paleomagnetic data from mid-Cretaceous plutonic rocks: examples from the Omineca belt, British Columbia: Journal of Geophysical Research, v. 95, no. B4, p. 4579-4585.

Isacks, B.L., and Barazangi, M., 1977, Geometry of Benioff zones: lateral segmentation and downwards bending of the subducted lithosphere, *in* Talwani, M., and Pitman, W.C. III, Island arcs, deep sea trenches and back-arc basins: American Geophysical Union Maurice Ewing Series 1, p. 99-114.

Isaev, E.N., 1987, Structural-geophysical model of the basement complex of the Aden-Red Sea region: Tectonophysics, v. 143, no. 1-3, p. 181-192.

Iyer, H., and Hitchcock, T., 1989, Upper-mantle velocity structure in the continental U. S. and Canada, *in* Pakiser, L.C., and Mooney, W.D., eds., Geophysical framework of the continental United States: Geological Society of America Memoir 172, p. 681-710.

Jackson, E.D., 1976, Linear volcanic chains on the Pacific plate, *in* Sutton, G.H., Manghnani, M.H., Moberly, R., and McAfee, E.U., eds., The geophysics of the Pacific Ocean basin and its margin: American Geophysical Union Geophysical Monograph 19, p. 319-335.

Jacobs, J.A., 1961, Some aspects of the thermal history of the earth, *in* The earth today: A collection of papers dedicated to Sir Harold

Jeffreys: Royal Astronomical Society Geophysical Journal, v. 4, p. 267-275.

Jacobshagen, V., Brede, R., Hauptmann, M., Heinitz, W., and Zylka, R., 1988, Structure and post-Paleozoic evolution of the central High Atlas, in Jacobshagen, V.H., ed. The Atlas system of Morocco. Studies on its geodynamic evolution: New York, Springer-Verlag, p. 245-271.

Jaeger, J.C., 1962, Elasticity, fracture and flow with engineering and geological applications, 2nd ed.: London, Methuen and Company Ltd., 208 p.

Jansa, L.F., and Pe-Piper, G., 1988, Middle Jurassic to Early Cretaceous igneous rocks along eastern North American continental margin: American Association of Petroleum Geologists Bulletin, v. 72, no. 3, p. 347-366.

Jaupart, C., Mann, J.R., and Simmons, G., 1982, A detailed study of the distribution of heat flow and radioactivity in New Hampshire: Earth and Planetary Science Letters, v. 59, p. 267-287.

Jeffreys, H., 1970, The earth, 5th ed.: Cambridge University Press, 524 p.

Johnson, A.M., and Page, B.M., 1976, A theory of concentric, kink and sinusoidal folding and of monoclinal flexuring of compressible, elastic multilayers, VII. Development of folds within Huasna syncline, San Luis Obispo County, California: Tectonophysics, v. 33, no. 1-2, p. 97-143.

Jones, C.H., Kanamori, H., and Roecker, S.W., 1990, Lithospheric structure of the southern Sierra Nevada, California, from teleseismic arrival times (Abstract): Geological Society of America, Abstracts with Programs, v. 22, no. 3, p. 33.

Jordan, T.H., 1975, The continental tectosphere: Reviews of Geophysics and Space Physics, v. 13, no. 1, p. 1-12.

——, 1978, Composition and development of the continental tectosphere: Nature, v. 274, no. 5671, p. 544-548.

Kamen-Kaye, M., and Meyerhoff, A.A., 1980, Petroleum geology of the Mascarene Ridge, western Indian Ocean: Journal of Petroleum Geology, v. 3, no. 2, p. 123-138.

KAPG, 1981, Atlas of seismological maps, central and eastern Europe; Commission of the Academies of Sciences of Socialist Countries for Planetary Geophysical Research, Working Group 4.3: Prague, Czechoslovak Academy of Sciences, Geophysical Institute, 49 p. + 23 maps (various scales).

Karig, D.E., 1970, Ridges and basins of the Tonga-Kermadec island arc system: Journal of Geophysical Research, v. 75, no. 2, p. 239-254.

Karig, D.E., Anderson, R.N., and Bibee, L.D., 1978, Characteristics of back arc spreading in the Mariana Trough: Journal of Geophysical Research, v. 83, no. B3, p. 1213-1226.

Keith, A., 1923, Outlines of Appalachian structure: Geological Society of America Bulletin, v. 34, no. 2, p. 309-380.

Kim, O.J., and Lee, D.S., 1983, Summary of igneous activity in South Korea, in Roddick, J.A., ed., Circum-Pacific plutonic terranes: Geological Society of America Memoir 159, p. 87-103.

King, P.B., 1951, The tectonics of middle North America. Middle North America east of the Corilleran system: Princeton University Press, 203 p.

Kingma, J.T., 1974, The geological structure of New Zealand: New York, Wiley-Interscience, John Wiley and Sons, 407 p.

Kizaki, K., 1986, Geology and tectonics of the Ryukyu Islands: Tectonophysics, v. 125, no. 1-3, p. 193-207.

Klemperer, S.L., 1987, A relation between continental heat flow and the seismic reflectivity of the lower crust: Journal of Geophysics, v. 61, no. 1, p. 1-11.

——, 1988, Crustal thinning and nature of extension in the northern North Sea from deep seismic reflection profiling: Tectonics, v. 7, no. 4, p. 803-821.

Klemperer, S.L., Hauge, T.A., Hauser, E.C., Oliver, J.E., and Potter, C.J., 1986, The Moho in the northern Basin and Range province, Nevada, along the COCORP 40°N seismic-reflection transect: Geological Society of America Bulletin, v. 97, no. 5, p. 603-618.

Knopf, E.B., and Ingerson, E., 1938, Structural petrology: Geological Society of America Memoir 6, 270 p.

Kobayashi, K., 1985, Sea of Japan and Ryukyu Trench-back-arc system, in Nairn, A.E.M., Stehli, F.G., and Uyeda, S., eds., The ocean basins and margins, v. 7A. The Pacific Ocean: New York, Plenum Press, p. 419-458.

Kober, L., 1921, Der Bau de Erde: Berlin, Gebrüder Borntraeger, 324 p.

——, 1925, Die Gestaltungsgeschichte der Erde: Berlin, Gebrüder Borntraeger, 200 p.

——, 1928, Der Bau der Erde, 2nd ed.: Berlin, Gebrüder Borntraeger, 499 p.

Krauskopf, K.B., 1968, A tale of ten plutons: Geological Society of America Bulletin, v. 79, no. 1, p. 1-17.

Krebs, W., 1975, Formation of southwest Pacific island arc-trench and mountain systems: plate or global-vertical tectonics: American Association of Petroleum Geologists Bulletin, v. 59, no. 9, p. 1639-1666.

Kron, A., and Stix, J., 1982, Geothermal gradient map of the United States exclusive of Alaska and Hawaii: Boulder (CO), National Geophysical Data Center, National Oceanic and Atmospheric Administration, 2 sheets, scale, 1:2,500,000.

Krummenacher, D., and Noetzlin, J., 1966, Ages isotopiques K/A de roches prélevées dans les possessions françaises de Pacifique: Société Géologique de France Bulletin, série 7, v. 8, p. 173-175.

Krummenacher, D., Dowd, D.H., Duda, V.F., Cunningham, W.B., Kingery, F.L., and Speidel, W.F., 1972, Potassuim-argon ages from zenoliths [sic] and differentiates in coarse-grained rocks from the center of the island of Tahiti (French Polynesia) (Abstract): Geological Society of America, Abstracts with Programs, v. 4, no. 3, p. 186.

Krylov, S.V., Mishen'kin, B.P., Petrik, G.V., and Seleznev, V.S., 1979, O seismicheskoy modeli verkhov mantii v Baykal'skoy riftovoy zone: Novosibirsk, Geologiya i Geofizika, no. 5, p. 117-129.

Kuchay, V.K., and Yeryemin, G.G., 1990, Astenokanaly perekhodnoy zony ot Pamira k Tyan'-shanyu: Novosibirsk, Geologiya i Geofizika, no. 2, p. 37-46.

Kuo, B.Y., Forsyth, D.W., and Wysession, M., 1987, Lateral heterogeneity and azimuthal anisotropy in the North Atlantic determined from SS-S differential travel times: Journal of Geophysical Research, v. 97, no. B7, p. 6421-6436.

Lachenbruch, A.H., and Sass, J.H., 1977, Heat flow in the United States and the thermal regime of the crust, in Heacock, J.G., ed., The Earth's crust. Its nature and physical properties: American Geophysical Union Geophysical Monograph 20, p. 626-675.

Landis, C.a., and Coombs, D.S., 1967, Metamorphic belts and orogenesis in southern New Zealand: Tectonophysics, v. 4, no. 4-6, p. 501-518.

Landisman, M., Mueller, S., and Mitchell, R.J., 1971, Review of evidence for velocity inversions in continental crust, in Heacock, J.G., ed., The structure and physical properties of the earth's crust: American Geophysical Union Geophysical Monograph 14, p. 11-34.

Larson, R.L., and Chase, C.G., 1972, Late Mesozoic evolution of the western Pacific Ocean: Geological Society of America Bulletin, v. 83, no. 12, p. 3627-3643.

LaTraille, S.L., and Hussong, D.M., 1980, Crustal structure across the Mariana island arc, in Hayes, D.E., ed., The tectonic and geologic evolution of southeast Asian seas and islands: American Geophysical Union Geophysical Monograph 23, p. 209-221.

Laubscher, H.P., 1978, Foreland folding: Tectonophysics, v. 47, no. 3/4, p. 325-337.

Laubscher, H., 1988, Material balance in Alpine orogeny: Geological Society of America Bulletin, v. 100, no. 9, p. 1313-1328.

Laubscher, H., and Bernoulli, D., 1983, History and deformation of the Alps, in Hsü, K.J., ed., Mountain building processes: London, Academic Press, p. 169-180.

Laughton, A.S., and Searle, R.C., 1979, Tectonic processes on slow spreading ridges, in Talwani, M., Harrison, C.G., and Hayes, D.E., eds., Deep drilling results in the Atlantic Ocean: ocean crust: American Geophysical Union Maurice Ewing Series 2, p. 15-32.

Lawver, L.A., and Williams, D.L., 1979, Heat flow in the central Gulf of California: Journal of Geophysical Research, v. 84, no. B7, p. 3465-3478.

Leeman, W.P., and Fitton, J.G., 1989, Magmatism associated with extension: introduction: Journal of Geophysical Research, v. 94, no. B6, p. 7682-7684.

Legg, M.R., Luyendyk, B.P., Mammerickx, J., de Moustier, C., and Tyce, R.C., 1989, Sea Beam survey of an active strike-slip fault: the San Clemente fault in the California continental borderland: Journal of Geophysical Research, v. 94, no. B2, p. 1727-1744.

Leitch, E.C., and Scheibner, E., 1987, Stratotectonic terranes of the eastern Australian Tasmanides, in Leitch, E.C., and Scheibner, E., eds., Terrane accretion and orogenic belts: American Geophysical Union and Geological Society of America, Geodynamics Series 19, p. 1-19.

Le Pichon, X., Houtz, R.E., Drake, C.L., and Nafe, J.E., 1965, Crustal structure of the mid-ocean ridges, 1. Seismic refraction measurements: Journal of Geophysical Research, v. 70, no. 2, p. 319-339.

Létouzey, J., Biju-Duval, B., Dorkel, A., Gonnard, R., Kristchev, K., Montadert, L., and Sungurlu, O., 1977, The Black-Sea: a marginal basin. Geophysical and geological data, in Biju-Duval, B., and Montadert, L., eds., Structural history of the Mediterranean basins: Paris, Éditions Technip, p. 363-375.

Létouzey, J., Muller, C., and Sage, L., 1988, Structure of sedimentary basins in eastern Asia: Southeast Asia Petroleum Exploration Society Proceedings v. 8, p. 63-68.

Lienert, B.R., Whitcomb, J.H., and Phillips, R.J., 1979, Magnetotelluric measurements close to the San Andreas and Garlock faults (Abstract): EOS, v. 60, no. 18. p. 243.

Lin, J., Purdy, G.M., Schouten, H., Sempéré, J.C., and Zervas, C., 1990, Evidence from gravity data for focused magmatic accretion along the Mid-Atlantic Ridge: Nature, v. 344, 12 April, p. 627-632.

Lipman, P.W., 1988, Evolution of silicic magma in the upper crust: the mid-Tertiary latir volcanic field and its cogenetic granitic batholith, northern New Mexico, U.S.A., in The origin of granites, a symposium: Royal Society of Edinburgh Transactions, Eath Sciences, v. 79, pts. 2-3, p. 265-288.

Liu, C.C., and Yu, S.B., 1989, Fast uplifting along the plate boundary in Taiwan (Abstract): EOS, v. 70, no. 15, p. 403.

Liu Futian, Qu Kexin, Wu Hua, Li Qiang, Liu Jianhua, and Hu Ge, 1989, Seismic tomography of the Chinese continent and adjacent regions: Acta Geophysica Sinica, v. 32, no. 3, p. 281-291.

Liu Guodong, 1987a, MTS studies on the upper mantle conductivity in China: Pure and Applied Geophysics, v. 125, no. 2/3, p. 465-482.

———, 1987b, The Cenozoic rift system of the North China Plain and the deep internal process: Tectonophysics, v. 133, no. 3/4, p. 277-285.

Liu Jianhua, Liu Futian, Wu Hua, Li Qiang, and Hu Ge, 1989, Three dimensional velocity images of the crust and upper mantle beneath a north-south zone in China: Acta Geophysica Sinica, v. 32, no. 2, p. 142-152.

Lliboutry, L.A., 1971, Rheological properties of the asthenosphere from Fenno scandian data: Journal of Geophysical Research, v. 76, no. 5, p. 1433-1446.

Locardi, E., 1988, The origin of the Apenninic arcs: Tectonophysics, v. 146, no. 1-4, p. 105-123.

Logatchev, N.A., and Florensov, N.A., 1978, The Baikal system of rift valleys: Tectonophysics, v. 45, no. 1, p. 1-13.

Loiselle, M.C., and Wones, D.R., 1979, Characteristics and origin of anorogenic granites (Abstract): Geological Society of America, Abstracts with Programs, v. 11, no. 7, p. 468.

Lonsdale, P., 1982, Small offsets of the Pacific-Nazca and Pacific Cocos spreading axes (Abstract): EOS, v. 63, no. 45, p. 1108.

———, 1988, Geography and history of the Louisville hotspot chain in the southwest Pacific: Journal of Geophysical Research, v. 93, no. B4, p. 3078-3104.

Lowman, P.D., Jr., 1985, Plate tectonics with fixed continents: a testable hypothesis—I: Journal of Petroleum Geology, v. 8, no. 4, p. 373-378.

———, 1986, Plate tectonics with fixed continents: a testable hypothesis—II: Journal of Petroleum Geology, v. 9, no. 1, p. 71-87.

Luosto, U., Flueh, E.R., Lund, C.E., and Working Group, 1989, The crustal structure along the POLAR profile from seismic refraction investigations: Tectonophysics, v. 162, no. 1/2, p. 51-85.

Luyendyk, B.P., 1977, Deep sea drilling on the Ninetyeast Ridge: synthesis and a tectonic model, in Heirtzler, J.R., Bolli, H.M., Davies, T.A., Saunders, J.B., and Sclater, J.G., eds., Indian Ocean geology and biostratigraphy: American Geophysical Union, p. 165-187.

Lyakhovich, V.V., 1988, "Mantle" granitoids: International Geology Review, v. 30, no. 12, p. 1257-1271.

Lyberis, N., 1984, Tectonic evolution of the North Aegean trough, in Dixon, J.E., and Robertson, A.H.F., eds., The geological evolution of the eastern Mediterranean: Geological Society of London Special Publication no. 17, p. 709-725.

Lynn, H.B., Hale, L.D., and Thompson, G.A., 1981, Seismic reflections from the basal contacts of batholiths: Journal of Geophysical Research, v. 86, no. B11, p. 10633-10638.

Lyttleton, R.A., 1982, The earth and its mountains: New York, John Wiley and Sons, 206 p.

Ma, C., Ryan, J.W., and Caprette, D., 1989, Crustal dynamics project data analysis—1988: Greenbelt (MD), National Aeronautics and Space Administration, NASA Technical Memorandum 100723, 10 p. + 9 appendices.

Ma Xingyuan, chief comp., 1986a, Lithospheric dynamics map of China and adjacent areas: Beijing, Geological Publishing House, 2 sheets, scale, 1:4,000,000.

———, 1986b, Explanatory notes for the lithospheric dynamics map of China and adjacent areas: Beijing, Geological Publishing House, 53 p.

MacDonald, G.J.F., 1959, Calculations on the thermal history of the earth: Journal of Geophysical Research, v. 64, no. 11, p. 1967-2000.

———, 1963, The deep structure of continents: Reviews of Geophysics, v. 1, no. 4, p. 587-665.

———, 1965, Geophysical deductions from observations of heat flow, in Lee, W.H.K., ed., Terrestrial heat flow: American Geophysical Union Geophysical Monograph 8, p. 191-210.

Macdonald, K.C., and Fox, P.J., 1983, Overlapping spreading centers: a new kind of accretionary geometry on the East Pacific Rise: Nature, v. 302, no. 5903, p. 55-58.

Macdonald, K.C., Fox, P.J., Perram, L.J., Eisen, M.F., Haymon, R.M., Miller, S.P., Carbotte, S.M., Cormier, M.H., and Shor, A.N., 1988,

A new view of the mid-ocean ridge from the behaviour of ridge-axis discontinuities: Nature, v. 335, no. 6187, p. 217-225.

Macdonald, K.C., Sempère, J.C., Fox, P.J., and Tyce, R., 1987, Tectonic evolution of ridge-axis discontinuities by the meeting, linking, or self-decapitation of neighboring ridge segments: Geology, v. 15, no. 11, p. 993-997.

Marillier, F., Keen, C.E., and Stockmal, G.S., 1989, Laterally persistent seismic characteristics of the lower crust: examples from the northern Appalachians, in Mereu, R.F., Mueller, S., and Fountain, D.M., eds., Properties and processes of earth's lower crust: American Geophysical Union Geophysical Monograph no. 51, p. 45-52.

Martin, B.D., 1991, Constraints to major right-lateral movements, San Andreas fault system, central and northern California, U.S.A.: This volume.

Maxwell, J.C., 1968, The Mediterranean, ophiolites, and continental drift, in Johnson, H., and Smith, B.L., eds., The megatectonics of continents and ocean basins: Rutgers University Press, p. 167-193.

———, 1974, Anatomy of an orogen: Geological Society of America Bulletin, v. 85, no. 8, p. 1195-1204.

McCaffrey, R., Silver, E.A., and Raitt, R.W., 1980, Crustal structure of the Molucca Sea collision zone, Indonesia, in Hayes, D.E., ed., The tectonic and geologic evolution of southeast Asian seas and islands: American Geophysical Union Geophysical Monograph 23, p. 161-177.

McCartan, L., and Architzel, R.J., 1988, Heat flow map of the eastern United States: U. S. Geological Survey Miscellaneous Field Studies Map MF-2057, 2 sheets, scale, 1:2,500,000.

McDougall, I., Embleton, J.J., and Stone, D.B., 1981, Origin and evolution of Lord Howe Island, southwest Pacific Ocean: Geological Society of Australia Journal, v. 28, nos. 1-2, p. 155-176.

McHone, J.G., 1978, Distribution, orientations, and ages of mafic dikes in central New England: Geological Society of America Bulletin, v. 89, no. 11, p. 1645-1655.

McHone, J.G., Ross, M.E., and Greenough, J.D., 1987, Mesozoic dyke swarms of eastern North America, in Halls, H.C., and Fahrig, W.F., eds., Mafic dyke swarms: Geological Association of Canada Special Paper 34, p. 279-288.

Mégard, F., 1987, Structure and evolution of the Peruvian Andes, in Schaer, J.P., and Rodgers, J., eds., The anatomy of mountain ranges: Princeton University Press, p. 179-210.

Mehnert, K.R., 1969, Petrology of the Precambrian basement complex, in Hart, P.J., ed., The earth's crust and upper mantle: American Geophysical Union Geophysical Monograph 13, p. 513-518.

Melson, W.G., Hart, S.R., and Thompson, G., 1972, St. Paul's Rocks, equatorial Atlantic: petrogenesis, radiometric ages, and implications on sea-floor spreading, in Shagam, R., Hargraves, R.B., Morgan, W.J., Van Houten, F.B., Burk, C.A., Holland, H.D., and Hollister, L.C., eds., Studies in earth and space sciences: Geological Society of America Memoir 132, p. 241-272.

Menard, H.W., 1960, The East Pacific Rise: Science, v. 132, no. 3441, p. 1737-1746.

———, 1964, Marine geology of the Pacific: New York, McGraw-Hill Book Company, 271 p.

Merle, O., and Guillier, B., 1989, The building of the central Swiss Alps: an experimental approach: Tectonophysics, v. 165, no. 1-4, p. 41-56.

Meyerhoff, A.A., 1970, Continental drift: implications of paleomagnetic studies, meteorology, physical oceanography, and climatology: Journal of Geology, v. 78, no. 1, p. 1-51.

Meyerhoff, A.A., and Meyerhoff, H.A., 1972a, Continental drift, IV: the Caribbean "plate": Journal of Geology, v. 80, no. 1, p. 34-60.

———, 1972b, "The new global tectonics": major inconsistencies: American Association of Petroleum Geologists Bulletin, v. 56, no. 2, p. 269-336.

———, 1974, Tests of plate tectonics, in Kahle, C.F., ed., Plate tectonics—assessments and reassessments: American Association of Petroleum Geologists Memoir 23, p. 43-145.

Meyerhoff, A.A., Meyerhoff, H.A., and Briggs, R.S., Jr., 1972, Continental drift, V: proposed hypothesis of earth tectonics: Journal of Geology, v. 80, no. 6, p. 663-692.

Meyerhoff, A.A., Taner, I., Morris, A.E.L., and Martin, B.D., 1989, Surge tectonics, in Chatterjee, S., and Hotton, N. III, eds., New concepts in global tectonics: a discussion meeting sponsored by the Smithsonian Institution and Texas Tech University, 20-21 July, 1989, National Museum of Natural History, Smithsonian Institution, Washington, D.C., Abstracts volume: Texas Tech University Press, p. 25-26.

Meyerhoff, A.A., Agocs, W.B., Taner, I., Morris, A.E.L., and Martin, B.D., Origin of the midocean ridges: this volume.

Meyerhoff, A.A., Taner, I., Morris, A.E.L., Agocs, W.B., and Martin, B.D., in preparation, Surge tectonics: submitted.

Meyerhoff, H.A., 1954, Antillean tectonics: New York Academy of Sciences Transactions, ser. II, v. 16, no. 3, p. 149-155.

Meyerhoff, H.A., and Meyerhoff, A.A., 1977, Genesis of island arcs, in Geodynamics in south-west Pacific, international symposium, Noumea-Nouvelle Caledonie, 27 Aout-2 Septembre 1976: Paris, Éditions Technip, p. 357-370.

Miller, H., Müller, S., and Perrier, G., 1982, Structure and dynamics of the Alps: a geophysical inventory, in Berckhemer, H., and Hsü, K., eds., Alpine-Mediterranean geodynamics: America Geophysical Union and Geological Society of America Geodynamics Series, v. 7, p. 175-203.

Milnes, A.G., 1974, Structure of the Pennine zone (central Alps): a new working hypothesis: Geological Society of America Bulletin, v. 85, no. 11, p. 1727-1732.

Milsom, J., Preliminary gravity map of Seram, eastern Indonesia: Geology, v. 5, no. 10, p. 641-643.

Mitchell, B.J., and Landisman, M., 1971, Geophysical measurements in the southern Great Plains, in Heacock, J.G., ed., The structure and physical properties of the earth's crust: American Geophysical Union Geophysical Monograph 14, p. 77-93.

Miyashiro, A., 1973, Metamorphism and metamorphic belts: New York, John Wiley and Sons (Halsted Press), 492 p.

Miyashiro, A., Aki, K., and Şengör, A.M.C., 1982, Orogeny: New York, John Wiley & Sons, 242 p.

Miyata, T., 1990, Slump strain indicative of paleoslope in Cretaceous Izumi sedimentary basin along Median Tectonic Line, southwest Japan: Geology, v. 18, no. 5, p. 392-394.

Mohr, P., 1987, Patterns of faulting in the Ethiopian rift valley: Tectonophysics, v. 143, no. 1-3, p. 169-179.

Moody, J.D., and Hill, M.L., 1956, Wrench-fault tectonics: Geological Society of America Bulletin, v. 67, no. 9, p. 1207-1246.

Mooney, W.D., and Braile, L.W., 1989, The seismic structure of the continental crust and upper mantle of North America, in Bally, A.W., and Palmer, A.R., eds., The geology of North America, v. A. The geology of North America; an overview: Geological Society of America, Decade of North American Geology, p. 39-52.

Mooney, W.D., and Weaver, C.S., 1989, Regional crustal structure and tectonics of the Pacific coastal states: California, Oregon, and Washington, in Pakiser, L.C., and Mooney, W.D., eds., Geophysical framework of the continental United States: Geological Society of America Memoir 172, p. 129-161.

Mooney, W.D., Andrews, M.C., Ginzburg, A., Peters, D.A., and Hamilton, R.M., 1983, Crustal structure of the northern Mississippi embayment and a comparison with other continental rift zones: Tectonophysics, v. 94, no. 1-4, p. 327-348.

Moore, T.C., Jr., Rabinowitz, P.D., Borella, P.E., and Boersma, A., 1984, History of the Walvis Ridge, in Moore, T.C., Jr., Rabinowitz, P.D., and others, Initial reports of the Deep Sea Drilling Project, v. 74: Washington (DC), United States Government Printing Office, p. 873-894.

Moquet, A., and Romanowicz, B., 1990, Three-dimensional structure of the upper mantle beneath the Atlantic Ocean inferred from long-period Rayleigh waves, 2. Inversion: Journal of Geophysical Research, v. 95, no. B5, p. 6787-6798.

Morley, L.W., and Larochelle, A., 1964, Palaeomagnetism as a means of dating geological events, in Osborne, F.F., ed., Geochronology in Canada: Royal Society of Canada Special Publication no. 8, p. 39-51.

Morris, A.E.L., Taner, I., Meyerhoff, H.A., and Meyerhoff, A.A., 1990, Tectonic evolution of the Caribbean region, in Dengo, G., and Case, J.E., eds., The geology of North America, v. H. The Caribbean region: Geological Society of America, Decade of North American Geology, p. 423-457.

Mosmann, R., Falkenhein, F.U.H., Gonçalves, A., and Neopmuceno, F. (Filho), 1986, Oil and gas potential of the Amazon Paleozoic basins, in Halbouty, M.T., ed., Future petroleum provinces of the world: American Association of Petroleum Geologists Memoir 40, p. 207-241.

Mount, V.S., 1988, State of stress near major strike-slip faults (Abstract): Geological Society of America Abstracts with Programs, v. 20, no. 7, p. A320.

Moyer, T.C., and Nealey, L.D., 1989, Regional compositional variations of late Tertiary bimodal rhyolite lavas across the Basin and Range/Colorado Plateau boundary in western Arizona: Journal of Geophysical Research, v. 94, no. B6, p. 7799-7816.

Mueller, S., 1983, Deep Structure and recent dynamics in the Alps, in Hsü, K.J., ed., Mountain building processes: New York, Academic Press, p. 181-199.

Mueller, S., and Ansorge, J., 1989, The crustal structure of western Europe: Annual Review of Earth and Planetary Sciences, v. 17, p. 335-360.

Muir-Wood, R., 1989, Extraordinary deglaciation reverses faulting in northern Fennoscandia, in Gregersen, S., and Basham, P.W., eds., Earthquakes at North-Atlantic passive margins: neotectonics and postglacial rebound: Boston, Kluwer Academic Publishers, p. 141-173.

Murauchi, S., Den, N., Asano, S., Hotta, H., Yoshii, T., Asanuma, T., Hagiwara, K., Ichikawa, K., Sato, T., Ludwig, W.J., Ewing, J.I., Edgar, N.T., and Houtz, R.E., 1968, Crustal structure of the Philippine Sea: Journal of Geophysical Research, v. 73, no. 10, p. 3143-3171.

Murphy, F.X., 1985, The lithostratigraphy and structural geology of the Dungarvan syncline and adjacent areas, Co. Waterford, southern Ireland: unpubl. Ph. D. disser., National University of Ireland, 200 p.

Murphy, F.X., 1990, The Irish Variscides: a fold belt developed within a major surge zone: Geological Society of London Journal, v. 147, pt. 3, p. 451-460.

Naeser, C.W., Bryant, B., Crittenden, M.D., Jr., and Sorensen, M.L., 1983, Fission-track ages of apatite in the Wasatch Mountains, Utah: an uplift study, in Miller, D.M., Todd, V.R., and Howard, K.A., eds., Tectonic and stratigraphic studies in the eastern Great Basin: Geological Society of America Memoir 157, p. 29-36.

Nalivkin, D.V., 1973, Geology of the U.S.S.R.: Edinburgh, Oliver and Boyd, 855 p.

Namson, J.S., and Davis, T.L., 1988, Structural transect of the western Transverse Ranges, California: implications for lithospheric kinematics and seismic risk evaluation: Geology, v. 16, no. 8, p. 675-679.

NAUTILAU Group, 1990, Hydrothermal activity in the Lau basin: EOS, v. 71, no. 18, p. 678-679.

Nelson, K.D., 1988, The COCORP atlas, v. 1: Ithaca (NY), Cornell University, Institute for the Study of the Continents, 24 p. + ca. 65 plates.

Newhall, C.G., and Dzurisin, D., 1988, Historical unrest at large calderas of the world, v. 2: United States Geological Survey Bulletin 1855, p. 599-1108.

Newton, C.R., 1988, Significance of "Tethyan" fossils in the American Cordillera: Science, v. 242, no. 4877, p. 385-391.

Nicolas, A., 1987, Asthenosphere structure and anisotropy beneath rifts: United States Geological Survey Circular 956, p. 53-54.

Nicolas, A., Lucazeau, F., and Bayer, R., 1987, Peridotite xenoliths in Massif Central basalts: textural and geophysical evidence for asthenospheric diapirism, in Nixon, P.H., ed., Mantle xenoliths: Chichester, Wiley and Sons, p. 563-574.

Nikonov, A.A., 1989, The rate of uplift in the Alpine mobile belt: Tectonophysics, v. 163, no. 3/4, p. 267-276.

Norman, T.N., 1984, The role of the Ankara Melange in the development of Anatolia (Turkey), in Dixon, J.E., and Robertson, A.H.F., eds., The geological evolution of the eastern Mediterranean: Geological Society of London Special Publication no. 17, p. 441-447.

North, F.K., 1965, The curvature of the Antilles: Geologie en Mijnbouw, v. 44, no. 3, p. 73-86.

Nunn, J.A., and Aires, J.R., 1988, Gravity anomalies and flexure of the lithosphere at the Middle Amazon basin, Brazil: Journal of Geophysical Research, v. 93, no. B1, p. 415-428.

Officer, C.B., Ewing, J.I., Hennion, J.F., Harkrider, D.G., and Miller, D.E., 1959, Geophysical investigations in the eastern Caribbean: summary of 1955 and 1956 cruises, in Ahrens, L.H., Press, F., Rankama, K., and Runcorn, S.K., eds., Physics and chemistry of the earth, v. 3: New York, Pergamon Press, p. 17-109.

Oike, K., and Huzita, K., 1988, Relation between characteristics of seismic activity and neotectonics in Honshu, Japan: Tectonophysics, v. 148, no. 1/2, p. 115-130.

Okal, E.A., and Batiza, R., 1987, Hotspots: the first 25 years, in Keating, B.H., Fryer, P., Batiza, R., and Boehlert, G.W., eds., Seamounts, islands, and atolls: American Geophysical Union Geophysical Monograph no. 19, p. 1-11.

Oliver, J., Dobrin, M., Kaufman, S., Meyer, R., and Phinney, R., 1976, Continuous seismic reflection profiling of the deep basement, Hardeman County, Texas: Geological Society of America Bulletin, v. 87, no. 11, p. 1537-1546.

Olson, J., and Pollard, D.D., 1989, Inferring paleostresses from natural fracture patterns: a new method: Geology, v. 17, no. 4, p. 345-348.

Ortlieb, L., Ruegg, J.C., Angelier, J., Colletta, B., Kasser, M., and Lesage, P., 1989, Geodetic and tectonic analyses along an active plate boundary: the central Gulf of California: Tectonics, v. 8, no. 3, p. 429-441.

Ouchi, T., Kawakami, H., Nagumo, S., Kasahara, J., and Koresawa, S., 1989, Microseismicity in the middle Okinawa Trough: Journal of Geophysical Research, v. 94, no. B8, p. 10601-10608.

Ozima, M., Saito, K., Matsuda, J., Zashu, S., Aramaki, S., and Shido, F., 1976, Additional evidence of existence of ancient rocks in the Mid-Atlantic Ridge and the age of the opening of the Atlantic: Tectonophysics, v. 31, no. 1/2, p. 59-71.

Papazachos, B.C., and Comninakis, P.E., 1977, Modes of lithospheric interaction in the Aegean area, in Biju-Duval, B., and Montadert, L., eds., Structural history of the Mediterranean basins: Paris, Éditions Technip, p. 319-331.

Pasquarè, G., Poli, S., Vezzoli, L., and Zanchi, A., 1988, Continental arc volcanism and tectonic setting in central Anatolia, Turkey: Tectonophysics, v. 146, no. 1-4, p. 217-230.

Paterson, S.R., Tobisch, O.T., and Morand, V.J., 1989, Eastward thrusting associated with the Wyangala batholith: *surge tectonics* in the Lachlan fold belt?, *in* Australasian tectonics, structural geology and tectonics conference, Kangaroo Island, February 1989: Geological Society of Australia Abstracts, v. 24, p. 116-117.

Pavlenkova, N.I., 1989, Struktura zemnoy i verkhney mantii i tektonika plit, *in* Beloussov, V.V., ed., Tektonosfera: yeye stroyeniye i razvitiye: Moscow, Akademiya Nauk SSSR, Mezhduvedomstvennyy Geofizicheskiy Komitet, Geodinamicheskiye Issledovaniya no. 13, p. 36-45.

Pelletier, B., and Louat, R., 1989, Seismotectonics and present-day relative plate motions in the Tonga-Lau and Kermadec-Havre region: Tectonophysics, v. 165, no. 1-4, p. 237-250.

Perfil'yev, A.S., 1979a, Formirovaniye zemnoy Ural'skoy evgeosinklinali: Moscow, Nauka, 161 p.

——, 1979b, Ophiolite belt of the Urals, *in* International atlas of ophiolites: Geological Society of America Map and Chart Series MC-33, p. 9-12.

Petri, S., and Fúlfaro, V.J., 1983, Geologia do Brasil (Fanerozóico): São Paulo, Editoria da Universidade de São Paulo, 631 p.

Petroy, D.E., and Wiens, D.A., 1989, Historical seismicity and implications for diffuse plate convergence in the northeast Indian Ocean: Journal of Geophysical Research, v. 94, no. B9, p. 12301-12319.

Phillips, R.P., 1964, Seismic refraction studies in Gulf of California, *in* van Andel, T.H., and Shor, G.G., Jr., eds., Marine geology of the Gulf of California: American Association of Petroleum Geologists Memoir 3, p. 90-121.

Phillips, W.J., and Kuckes, A.F., 1983, Electrical conductivity structure of the San Andreas fault in central California: Journal of Geophysical Research, v. 88, no. B9, p. 7467-7474.

Phinney, R.A., 1986, A seismic cross section of the New England Appalachians: the orogen exposed, *in* Barazangi, M., and Brown, L., eds., Reflection seismology: the continental crust: American Geophysical Union Geodynamics Series, v. 14, p. 157-172.

Phipps, S.P., 1988, Deep rifts as sources for intraplate magmatism in eastern North America: Nature, v. 334, no. 6177, p. 27-31.

Pilger, R.H., Jr., 1982, The origin of hotspot traces: evidence from eastern Australia: Journal of Geophysical Research, v. 87, no. B3, p. 1825-1834.

Piqué, A., and Michard, A., 1989, Moroccan Hercynides: a synopsis. The Paleozoic sedimentary and tectonic evolution at the northern margin of West Africa: American Journal of Science, v. 289, no. 3, p. 286-330.

Platt, J.P., and Vissers, R.L.M., 1989, Extensional collapse of thickened continental lithosphere: a working hypothesis for the Alboran Sea and Gibraltar arc: Geology, v. 17, no. 6, p. 540-543.

Popov, A.M., 1987, O prichinakh povysheniya elektroprobodnosti v zemnoy kore: Geologiya i Geofizika, no. 12, p. 56-64.

Powell, D., Andersen, T.B., Drake, A.A., Jr., Hall, L., and Keppie, J.D., 1988, The age and distribution of basement rocks in the Caledonide orogen in the N Atlantic, *in* Harris, A.L., and Fettes, D.J., eds., The Caledonia-Appalachian orogen: Geological Society of London Special Publication no. 38, p. 63-74.

Press, F., 1966, Seismic velocities, *in* Clark, S.P., Jr., ed., Handbook of physical constants: Geological Society of America Memoir 97, p. 195-218.

Prodehl, C., 1970, Seismic refraction study of crustal structure in the western United States: Geological Society of America Bulletin, v. 81, no. 9, p. 2629-2645.

Prodehl, C., 1979, Crustal structure of the western United States: U. S. Geological Survey Professional Paper 1034, 74 p.

Prodehl, C., and Pakiser, L.C., 1980, Crustal structure of the Southern Rocky Mountains from seismic measurements: Geological Society of America Bulletin, v. 91, no. 3, p. 147-155.

Pushcharovskiy, Yu.M., 1972, Vvedeniye v tektoniku Tikhookeanskogo segmenta zemli: Akademiya Nauk SSSR, Geologicheskiy Institut Trudy, vyp. 234, 222 p.

Pushcharovskiy, Yu.M., Markov, M.S., and Perfil'yev, A.S., 1988, Tektonicheskaya evolyutsiya territorii SSSR i mobilizm, *in* Pushcharovskiy, Yu.M., ed., Aktual'nyye problemy tektoniki SSSR: Moscow, Nauka, p. 5-14.

Rabinowitz, P.D., Heirtzler, J.R., Aitken, T.D., and Purdy, G.M., 1978, Underway geophysical measurements: *Glomar Challenger* Legs 45 and 46, *in* Melson, W.G., Rabinowitz, P.D., and others, Initial reports of the Deep Sea Drilling Project, v. 45: Washington, U. S. Government Printing Office, p. 55-118.

Raitt, R.W., 1956, Seismic-refraction studies of the Pacific Ocean basin, part I: crustal thickness of the central equatorial Pacific: Geological Society of America Bulletin, v. 67, no. 12, pt. 1, p. 1623-1639.

Raitt, R.W., Shor, G.G., Jr., Fancis, T.J.G., and Morris, G.B., 1969, Anisotropy of the Pacific upper mantle: Journal of Geophysical Research, v. 74, no. 12, p. 3095-3109.

Ramberg, H., 1967, Gravity, deformation and the earth's crust: London, Academic Press, 214 p.

——, 1972, Theoretical models of density stratification and diapirism in the earth: Journal of Geophysical Research, v. 77, no. 5, p. 877-889.

——, 1973, Model studies of gravity-controlled tectonics by the centrifuge technique, *in* de Jong, K.A., and Scholten, R., eds., Gravity and tectonics: New York, Wiley and Sons, Inc., p. 49-66.

Rangin, C., 1984, Aspectos geodinámicos de la región noroccidental de México: Universidad Nacional Autónoma de México, Instituto de Geología, Revista, v. 5, no. 2, p. 186-194.

——, 1986, Contribution a l'étude géologique du système cordillérain mésozoïque du nord-ouest du Mexique: une coupe de la Basse Californie centrale à la Sierra Madre Occidentale en Sonora: Société Géologique de France, Mémoire no. 148, nouvelle série—1985, 136 p.

Rankin, D.W., 1976, Appalachian salients and recesses: late Precambrian continental breakup and the opening of the Iapetus ocean: Journal of Geophysical Research, v. 81, no. 32, p. 5605-5619.

Ratschbacher, L., Frisch, W., Neubauer, F., Schmid, S.M., and Neugebauer, J., 1989, Extension in compressional orogenic belts: the eastern Alps: Geology, v. 17, no. 5, p. 404-407.

Reid, I., Orcutt, J.A., and Prothero, W.A., 1977, Seismic evidence for a narrow zone of partial melting underlying the East Pacific Rise at 21° N: Geological Society of America Bulletin, v. 88, no. 5, p. 678-682.

Reston, T.J., 1988, Evidence for shear zones in the lower crust offshore Britain: Tectonics, v. 7, no. 5, p. 929-945.

Revelle, R., 1958, The *Downwind* Expedition to the southeast Pacific (Abstract): American Geophysical Union Transactions, v. 39, no. 3, p. 528-529.

Reynolds, P.H., and Clay, W., 1977, Leg 37 basalts and gabbro: K-Ar and ^{40}Ar-^{39}Ar dating, *in* Aumento, F., Melson, W.G., and others, Initial reports of the Deep Sea Drilling Project, v. 37: Washington, D. C., United States Government Printing Office, p. 629-630.

Reynolds, R.T., Fricker, P.E., and Summers, A.L., 1966, Effects of melting upon thermal models of the earth: Journal of Geophysical Research, v. 71, no. 2, p. 573-582.

Rich, J.L., 1951, Origin of compressional mountains and associated phenomena: Geological Society of America Bulletin, v. 62, no. 10, p. 1179-1222.

Richard, M., Bellon, H., Maury, R.C., Barrier, E., and Juang, W.S., 1986, Miocene to recent calc-alkalic volcanism in eastern Taiwan: Tectonophysics, v. 125, no. 1-3, p. 87-124.

Ries, A.C., and Shackleton, R.M., 1990, Strutures in the Huqf-Haushi uplift, east central Oman, in Robertson, A.H.F., Searle, M.P., and Ries, A.C., eds., The geology and tectonics of the Oman region: Geological Society of London Special Publication 49, p. 653-663.

Ritsema, A.R., 1957, On the focal mechanism of southeast Asian earthquakes, in Hodgson, J.H., ed., The mechanics of faulting, with special reference to the fault-plane work: Dominion Observatory (Ottawa) Publications, v. 20, no. 2, p. 341-368.

——, 1960, Further focal mechanism studies at De Bilt, in Hodgson, J.H., ed., A symposium on earthquake mechanism: Dominion Observatory (Ottawa) Publications, v. 24, no. 10, p. 355-358.

Robbins, S.L., 1968, Gravity and magnetic data over the Calaveras, Hayward, and Silver Creek faults near San Jose, California, in Dickinson, W.R., and Grantz, A., eds., Proceedings of conference on geologic problems of San Andreas fault system: Stanford University Publications, Geological Sciences, v. 11, p. 216-217.

Roberts, R.J., Crittenden, M.D., Jr., Tooker, E.W., Morris, H.T., Hose, R.K., and Cheney, T.M., 1965, Pennsylvanian and Permian basins in northwestern Utah, northeastern Nevada, and south-central Idaho: American Association of Petroleum Geologists Bulletin, v. 49, no. 11, p. 1926-1956.

Rodgers, J., 1987, The Appalachian geosyncline, in Schaer, J.P., and Rodgers, J., eds., The anatomy of mountain ranges: Princeton Unversity Press, p. 241-258.

Rodnikov, A.G., 1988, Correlation between the asthenosphere and the structure of the earth's crust in active margins of the Pacific Ocean: Tectonophysics, v. 146, no. 1-4, p. 279-289.

Roure, F., and Sosson, M., 1986, Late Jurassic collision between a composite exotic block and the North American continent: a model for the Cordillera building: Société Géologique de France Bulletin, 8e série, v. 2, no. 6, p. 945-959.

Royden, L., Horvath, F., and Rumpler, J., 1983, Evolution of the Pannonian basin system, I. Tectonics: Tectonics, v. 2, no. 1, p. 63-90.

Rutten, M.G., 1969, The geology of western Europe: Amsterdam, Elsevier Publishing Company, 520 p.

Ryan, M.P., 1987, Neutral buoyancy and the mechanical evolution of magmatic systems, in Myson, B.O., ed., Magmatic processes: physico-chemical principles: Geochemical Society Special Publication no. 1, p. 259-287.

Sams, D.B., and Saleeby, J.B., 1988, Geology and petrotectonic significance of crystalline rocks of the southernmost Sierra Nevada, California, in Ernst, W.G., ed., Metamorphism and crustal evolution of the western United States, Rubey Volume 7: Englewood Cliffs (NJ), Prentice Hall, p. 865-893.

Sandwell, D.T., and MacKenzie, K.R., 1989, Geoid height versus topography for oceanic plateaus and swells: Journal of Geophysical Research, v. 94, no. B6, p. 7403-7418.

Sandwell, D.T., and Renkin, M.L., 1988, Compensation of swells and plateaus in the north Pacific: no direct evidence for mantle convection: Journal of Geophysical Research, v. 93, no. B4, p. 2775-2783.

Sarewitz, D.R., and Lewis, S.D., 1988, Intra-arc basins in the central Philippines (II): regional extension, incipient sea-floor spreading, and implications for the creation of ophiolites (Abstract): Geological Society of America Abstracts with Programs, v. 20, no. 7, p. A127.

Saul, L.R., 1986, Pacific west coast Cretaceous molluscan faunas: time and aspect of changes, in Abbott, P.L., ed., Cretaceous stratigraphy, western North America: Society of Economic Paleontologists and Mineralogists, Pacific Section, Book 46, p. 131-135.

Saxena, M.N., 1986, Geodynamic synopsis of the Deccan Traps in relation to epochs of volcanic activity of the Indian shield, drift of the subcontinent, and the tectonic development of southern and southeastern India: Journal of Southeast Asian Earth Sciences, v. 1, no. 4, p. 205-213.

Scheidegger, A.E., 1963, Principles of geodynamics, 2nd ed.: New York, Academic Press Inc., Publishers, 362 p.

Scheidegger, A.E., and Wilson, J.T., 1950, An investigation into possible methods of failure of the earth: Geological Association of Canada Proceedings, v. 3, p. 167-190.

Schrock, R.R., 1948, Sequence in layered rocks: New York, McGraw-Hill Book Company, Inc., 507 p.

Schwan, W., 1980, The 40 ± 3 m.y. tectonic event, in Aubouin, J., Debelmas, J., and Latreille, M., coordinators, Geologie des chaînes alpines issués de la Tethys: Bureau de Recherches Géologiques et Minières Mémoire no. 115, p. 310-311.

Schwartz, D.P., and Sibson, R.H., eds., 1989, Fault segmentation and controls of rupture initiation and termination: United States Geological Survey Open-File Report 89-315, 447 p.

Schwarz, G., and Wigger, P.J., 1988 Geophysical studies of the earth's crust and upper mantle in the Atlas system of Morocco, in Jacobshagen, V.H., ed., The Atlas system of Morocco. Studies on its geodynamic evolution: New York, Springer-Verlag, p. 339-357.

Schwinner, R., 1920, Vulkanismus und Gebirgsbildung: Zeitschrift für Vulkanologie, Bd. 5, p. 175-230.

Sclater, J.G., Hawkings, J.W., Mammerickx, J., and Chase, C.G., 1972, Crustal extension between the Tonga and Lau Ridges: petrologic and geophysical evidence: Geological Society of America Bulletin, v. 83, no. 2, p. 505-517.

Searle, R.C., Rusby, R.I., Engeln, J., Hey, R.N., Zubin, J., Hunter, P.M., LeBas, T.P., Hoffman, H.J., and Livermore, R., 1989, Comprehensive sonar imaging of the Easter Island microplate: Nature, v. 341, 26 October, p. 701-705.

Sears, F.W., Zemansky, M.W., and Young, H.D., 1974, College physics, 4th ed.: Reading, MA, Addison-Wesley Publishing Company, 751 p.

Sears, J.W., Hyndman, D.W., and Alt, D., 1990, The Snake River plain, a volcanic hotspot track: Geological Society of America, Abstracts with Programs, v. 22, no. 3, p. 82.

Segawa, J., and Tomoda, Y., 1976, Gravity measurements near Japan and study of the upper mantle beneath the oceanic trench-marginal sea transition zones, in Sutton, G.H., Manghnani, M.H., Moberly, R., and McAfee, E.U., eds., The geophysics of the Pacific Ocean basin and its margins: American Geophysical Union Geophysical Monograph 19, p. 35-52.

Selverstone, J., 1988, Evidence for east-west crustal extension in the eastern Alps: implications for the unroofing history of the Tauern window: Tectonics, v. 7, no. 1, p. 87-105.

Sengör, A.M.C., 1979, The North Anatolian transform fault: its age, offset and tectonic significance: Geological Society of London Journal, v. 136, pt. 3, p. 269-282.

Shaw, H.R., Jackson, E.D., and Bargar, K.E., 1980, Volcanic periodicity along the Hawaiian-Emperor chain: American Journal of Science, v. 280-A, pt. 2, p. 667-708.

Sheynmann, Yu.M., 1968, Ocherky glubinnoy geologii: Moscow, Nedra, 231 p.

Shih, J., and Molnar, P., 1975, Analysis and implications of the sequence of ridge jumps that eliminated the Surveyor transform fault: Journal of Geophysical Research, v. 80, no. 35, p. 4815-4824.

Shimazu, M.S., 1988, Kaynozoyskiy vulkanizm i tektonicheskoye razvitiye Yaponskogo morya i yego obramleniya, in Natal'in, B.A., Tuyezov, I.K., and Uyeda, S., eds., Tektonika Vostochno-Aziatskikh okrainykh morey: Moscow, Akademiya Nauk SSSR, Mezhduvedomstvennyy Geofizicheskiy Komitet, Geodinamicheskiye Issledovaniya no. 11, p. 81-88.

Shipboard Scientific Party, 1977, Site 334, in Aumento, F., Melson, W.G., and others, Initial reports of the Deep Sea Drilling Project, v. 37: Washington, D. C., United States Government Printing Office, p. 239-287.

Shurbet, D.H., and Cebull, S.E., 1971, Crustal low-velocity layer and regional extension in Basin and Range province: Geological Society of America Bulletin, v. 82, no. 11, p. 3241-3243.

Silver, E.A., and Moore, J.C., 1978, The Molucca Sea collision zone, Indonesia: Journal of Geophysical Research, v. 83, no. B4, p. 1681-1691.

Silver, E.A., Breen, N.A., Prasetyo, H., and Hussong, D.M., 1986, Multibeam study of the Flores backarc thrust belt, Indonesia: Journal of Geophysical Research, v. 91, no. B3, p. 3491-3500.

Simonen, A., and Mikkola, A., 1980, Finland, in Geology of the European countries. Denmark, Finland, Iceland, Norway, Sweden: London, Graham and Trotman Ltd., p. 51-126.

Sinton, J.B., and Hussong, D.M., 1983, Crustal structure of a short length transform fault in the central Mariana Trough, in Hayes, D.E., ed., The tectonic and geologic evolution of southeast Asian seas and islands, part 2: American Geophysical Union Geophysical Monograph 27, p. 236-254.

Smith, A.G., and Woodcock, N.H,. 1982, Tectonic synthesis of the Alpine-Mediterranean region: a review in Berckhemer, H., and Hsü, K., eds., Alpine-Mediterranean geodynamics: American Geophysical Union and Geological Society of America, Geodynamics Series, v. 7, p. 15-38.

Smith, R.B., 1978, Seismicity, crustal structure, and intraplate tectonics of the interior of the Western Cordillera, in Smith, R.B., and Eaton, G.P., eds., Cenozoic tectonics and regional geophysics of the Western Cordillera: Geological Society of America Memoir 152, p. 111-144.

Smoot, J. P., Froelich, A.J., and Luttrell, G.W., 1988, Uniform symbols for the Newark Supergroup, in Froelich, A.J., and Robinson, G.P., Jr., eds., Studies of the early Mesozoic basins of the eastern United States: United States Geological Survey Bulletin 1776, p. 1-6.

Speed, R., Elison, M.W., and Heck, F.R., 1988, Phanerozoic tectonic evolution of the Great Basin, in Ernst, W.G., ed., Metamorphism and crustal evolution of the western United States. Rubey volume 7: Englewood Cliffs (NJ), Prentice Hall, p. 573-605.

Stacey, F.D., 1981, Cooling of the earth—a constraint on paleotectonic hypotheses, in O'Connell, R.J. and Fyfe, W.S., eds., Evolution of the earth: American Geophysical Union and Geological Society of America Geodynamics Series, v. 5, p. 272-276.

Stegena, L., 1990, Correlation between seismicity and horizontal variation of heat flow density in central Europe: Tectonophysics, v. 179, no. 1/2, p. 55-61.

Steinmann, G., 1905, Geologische Beobachtungen in den Alpen, II: die Schardt'sche Überfaltungsteorie und die geologische Bedeutung der Tiefseeabsätze und der ophiolithischen Massengesteine: Freiberger Geologische Gesellschaft Bericht, Bd. 16, p. 18-67.

Stéphan, J.F., Blanchet, R., Rangin, C., Pelletier, B., Létouzey, J., and Muller, C., 1986, Geodynamic evolution of the Taiwan-Luzon-Mindoro belt since the late Eocene: Tectonophysics, v. 125, no. 1-3, p. 245-268.

Stephen, R.A., 1985, Seismic anisotropy in the upper oceanic crust: Journal of Geophysical Research, v. 90, no. B13, p. 11,383-11,396.

Stern, T.A., and Davey, F.J., 1989, Crustal structure and origin of basins formed behind the Hikurangi subduction zone, New Zealand, in Price, R.A., ed., Origin and evolution of sedimentary basins and their energy and mineral resources: American Geophysical Union Geophysical Monograph 48, p. 73-85.

Stewart, J.H., 1978, Basin-range structure in western North America: a review, in Smith, R.B., and Eaton, G.P., eds., Cenozoic tectonics and regional geophysics of the Western Cordillera: Geological Society of America Memoir 152, p. 1-31.

Stille, H., 1920, Die Begriffe Orogenese und Epirogenese: Deutsche Geologische Gesellschaft Zeitschrift, v. 71 (1919), p. 164-240.

———, 1924, Grundfragen der vergleichenden Tektonik: Berlin, Gebrüder Borntraeger, 443 p.Suess, E., 1875, Die Entstehung der Alpen: Vienna, Braumüller, 168 p.

———, 1885, Das Antlitz der Erde, Band 1: Wien (Vienna), F. Tempsky, 779 p.

———, 1901, Das Antlitz der Erde, Band 3, erste Hälfte: Wien (Vienna), F. Tempsky, 508 p.

———, 1909, Das Antlitz der Erde, Band 3, zweite Hälfte: Wien (Vienna), F. Tempsky, 789 p.

Suppe, J., 1980, A retrodeformable cross section of northern Taiwan: Geological Society of China Proceedings no. 23, p. 46-55.

———, 1981, Mechanics of mountain building and metamorphism in Taiwan Geological Society of China Memoir 4, p. 67-89.

Sutton, G.H., Maynard, G.L., and Hussong, D.M., 1971, Widespread occurrence of a high-velocity basal layer in the Pacific crust found with repetitive sources and sonobuoys, in Heacock, J.G., ed., The structure and physical properties of the earth's crust: American Geophysical Union Geophysical Monograph 14, p. 193-209.

Sychev, P.M., 1985, Anomal'nyye zony v verkhney mantii, mekhanizm ikh obrazovaniya i rol' v razvitii struktur zemnoy kory: Akademiya Nauk SSSR, Tikhookeanskaya Geologiya, no. 6, p. 25-35.

Sychev, P.M., and Sharaskin, A.Y., 1984, Heat flow and magmatism in the NW Pacific back-arc basins, in Kokelaar, B.P., and Howells, M.F., eds., Marginal basin geology: Geological Society of London Special Publication 16, p. 173-181.

Sykes, L.R., 1978, Intraplate seismicity, reactivation of preexisting zones of weakness, alkaline magmatism, and other tectonism postdating continental fragmentation: Reviews of Geophysics and Space Physics, v. 16, no. 4, p. 621-688.

Szatmari, P., 1983, Amazon rift and Pisco-Juruá fault: their relation to the separation of North America from Gondwana: Geology, v. 11, no. 3, p. 300-304.

Talandier, J., 1989, Submarine volcanic activity. Detection, monitoring, and interpretation: EOS, v. 70, no. 18, p. 561, 568-569.

Talandier, J., and Kuster, G.T., 1976, Seismicity and submarine volcanic activity in French Polynesia: Journal of Geophysical Research, v. 81, no. 5, p. 936-948.

Talbot, C.J., and Slunga, R., 1989, Patterns of active shear in Fennoscandia, in Gregersen, S., and Basham, P.W., eds., Earthquakes at North-Atlantic passive margins: neotectonics and postglacial rebound: Dordrecht, Kluwer Academic Publishers, p. 441-466.

Talwani, M., Le Pichon, X., and Ewing, M., 1965, Crustal structure of the mid-ocean ridges, 2. Computed model from gravity and

seismic refraction data: Journal of Geophysical Research, v. 70, no. 2, p. 341-352.
Tanaka, K.L., Shoemaker, E.M., Ulrich, G.E., and Wolfe, E.W., 1986, Migration of volcanism in the San Francisco volcanic field, Arizona: Geological Society of America Bulletin, v. 97, no. 2, p. 129-141.
Taner, I., and Meyerhoff, A.A., 1990, Petroleum at the roof of the world: the geological evolution of the Tibet (Qinghai-Xizang) Plateau, Part I: Journal of Petroleum Geology, v. 13, no. 2, p. 157-178; Part II: Journal of Petroleum Geology, v. 13, pt. 3, p. 284-314.
Taylor, E.M., 1990, Volcanic history and tectonic development of the central High Cascade Range, Oregon (Abstract): EOS, v. 71, no. 9, p. 312.
Taylor, S.R., 1989, Geophysical framework of the Appalachians and the adjacent Grenville province, in Pakiser, L.C., and Mooney, W.D., eds., Geophysical framework of the continental United States: Geological Society of America Memoir 172, p. 317-348.
Taylor, S.R., Toksöz, M.N., and Chaplin, M.P., 1980, Crustal structure of the northeastern United States: contrasts between Grenville and Appalachian provinces: Science, v. 208, no. 4444, p. 595-597.
Tezcan, A.K., 1979, Geothermal studies, their present status and contribution to heat flow contouring in Turkey, in Čermák, V., and Rybach, L., eds., Terrestrial heat flow in Europe: New York, Springer-Verlag, p. 283-292.
Thomas, M.D., and Willis, C., 1989, Gravity modelling of the Saint George Batholith and adjacent terrane within the Appalachian orogen, southern New Brunswick: Canadian Journal of Earth Sciences, v. 26, no. 3, p. 561-576.
Thompson, G., Bryan, W.B., and Humphris, S.E., 1989, Axial volcanism on the East Pacific Rise, in Saunders, A.D., and Norry, M.J., eds., Magmatism in the ocean basins: Geological Society of London Special Publication no. 42, p. 181-200.
Thompson, G.A., and McCarthy, J., 1990, A gravity constraint on the origin of highly extended terranes: Tectonophysics, v. 174, no. 1-2, p. 197-206.
Thompson, J.B., Jr., Robinson, P., Clifford, T.N., and Trask, N.J., Jr., 1968, Nappes and gneiss domes in west-central New England, in Zen, E.A., White, W.S., Hadley, J.B., and Thompson, J.B., Jr., eds., Studies of Appalachian geology, northern and maritime: New York, Interscience Publishers, p. 203-218.
Tobin, D.G., Ward, P.L., and Drake, C.L., 1969, Microearthquakes in the rift valley of Kenya: Geological Society of America Bulletin, v. 80, no. 10, p. 2043-2046.
Tobisch, O.T., and Paterson, S.R., 1990, The Yarra granite: an intradeformational pluton associated with ductile thrusting, Lachlan fold belt, southeastern Australia: Geological Society of America Bulletin, v. 102, no. 6, p. 693-703.
Torresan, M.E., Shor, A.N., Wilson, J.B., and Campbell, J., 1989, Cruise report, Hawaiian GLORIA Leg 5, F5-88-HW: United States Geological Survey Open-File Report 89-198, 56 p.
Tosdal, R.M., Haxel, G.B., and Wright, J.E., 1989, Jurassic geology of the Sonoran Desert region, southern Arizona, southeastern California, and northernmost Sonora: construction of a continental-margin magmatic arc, in Jenney, J.P., and Reynolds, S.J., eds., Geologic evolution of Arizona: Arizona Geological Society Digest, v. 17, p. 397-434.
Tritton, D.J., 1988, Physical fluid dynamics, 2nd ed.: Oxford, Clarendon Press, 519 p.
Trümpy, R., 1960, Paleotectonic evolution of the central and western Alps: Geological Society of America Bulletin, v. 71, no. 6, p. 843-907.
Tsai, Y.B., 1986, Seismotectonics of Taiwan: Tectonophysics, v. 125, no. 1-3, p. 17-37.

Turcotte, D.L., 1989, Geophysical processes influencing the lower continental crust, in Mereu, R.F., Mueller, S., and Fountain, D.M., eds., Properties and processes of earth's lower crust: American Geophysical Union Geophysical Monograph 51, p. 321-329.
Turner, D.L., and Jarrard, R.D., 1982, K/Ar dating of the Cook-Austral Island chain: a test of the hotspot hypothesis: Journal of Volcanology and Geothermal Research, v. 12, p. 187-220.
Turner, D.L., Jarrard, R.D., and Forbes, R.B., 1980, Geochronology and origin of the Pratt-Welker Seamount chain, Gulf of Alaska: a new pole of rotation for the Pacific plate: Journal of Geophysical Research, v. 85, no. B11, p. 6547-6556.
Tweto, O., 1975, Laramide (Late Cretaceous-early Tertiary) orogeny in the Southern Rocky Mountains, in Curtis, B.F., ed., Cenozoic history of the Southern Rocky Mountains: Geological Society of America Memoir 144,
Udintsev, G.B., 1972, Tikhiy okean: geomorfologiya i tektonika dna Tikhogo okeana: Moscow, Nauka, 394 p.
Uliana, M.A., Biddle, K.T., and Cerdan, J., 1989, Mesozoic extension and the formation of Argentine sedimentary basins, in Tankard, A.J., and Balkwill, H.R., eds., Extensional tectonics and stratigraphy of the North Atlantic margins: American Association of Petroleum Geologists Memoir 46, p. 599-614.
Utada, M., 1980, Hydrothermal alteration related to igneous activity in Cretaceous and Neogene formations of Japan, in Ishihara, S., and Takenouchi, S., eds., Granitic magmatism and related mineralization: Society of Mining Geologists of Japan, Mining Geology Special Issue no. 8, p. 67-83.
Utsu, T., 1971, Seismological evidence for anomalous structure of island arcs with special reference to the Japanese region: Reviews of Geophysics and Space Physics, v. 9, no. 4, p. 839-890.
van Bemmelen, R.W., 1933, The undation theory of the development of the earth's crust: 16th International Geological Congress, Washington, Report 2, p. 965-982.
van der Linden, W.J.M., 1980, Walvis Ridge: a piece of Africa?: Geology, v. 8, no. 9, p. 417-421.
Velasque, P.C., Ducasse, L., Muller, J., and Scholten, R., 1989, The influence of inherited extensional structures on the tectonic evolution of an intracratonic chain: the example of the western Pyrenees: Tectonophysics, v. 162, no. 3-4, p. 243-264.
Verrall, P., 1989, Speculations on the Mesozoic-Cenozoic tectonic history of the United States, in Tankard, A.J., and Balkwill, H.R., eds., Extensional tectonics and stratigraphy of the North Atlantic margins: American Association of Petroleum Geologists Memoir 46, p. 615-631.
Vertlib, M.B., 1978, K opredeleniyu glubin ochagov zemletryaseniy v Pribaykal'ye: Geologiya i Geofizika, no. 9, p. 141-146.
Villeneuve, M., 1983, Les sillons tectoniques de Précambrien supérieur dans l'est du Zaïre; comparaisons avec les directions du rift Est-Africain, in Popoff, M., and Tiercelin, J.-J., eds., Rifts et fossés anciens: Centres de Recherches en Exploration-Production, Elf-Aquitaine, Bulletin, v. 7, no. 1, p. 163-174.
Vine, F.J., and Matthews, D.H., 1963, Magnetic anomalies over oceanic ridges: Nature, v. 199, no. 4897, p. 947-949.
Vlasov, G.M., and Belova, M.B., eds., 1964, Geologiya SSSR, t. 31. Kamchatka, Kuril'skiye i Komandorskiye ostrova, chast' I. Geologicheskoye Opisaniye: Moscow, Nedra, 728 p.
Vogt, P.R., 1974, The Iceland phenomenon: imprints of a hotspot on the ocean crust, and implications for flow below the plates, in Kristjansson, L., ed., Geodynamics of Iceland and the North Atlantic area: Dordrecht, D. Reidel Publishing Company, p. 105-126.
Vogt, P.R., Schneider, E.D., and Johnson, G.L., 1969, The crust and upper mantle beneath the sea, in Hart, P.J., ed., The earth's

crust and upper Mantle: American Geophysical Union Geophysical Monograph 13, p. 556-617.

Von Damm, K.L., 1990, Seafloor hydrothermal activity: black smoker chemistry and chimneys: Annual Review of Earth and Planetary Sciences, v. 18, p. 173-204.

Von Herzen, R.P., Cordery, M.J., Detrick, R.S., and Fang, C., 1989, Heat flow and the thermal origin of hot spot swells: the Hawaiian swell revisited: Journal of Geophysical Research, v. 94, no. B10, p. 13783-13799.

Walker, G.P.L., 1989, Gravitational (density) controls on volcanism, magma chambers and intrusions: Australian Journal of Earth Sciences, v. 36, no. 2, p. 149-165.

Wallbrecher, E., 1988, A ductile shear zone in the Panafrican basement on the northwestern margin of the West African craton (Sirwa dome, central Anti-Atlas), in Jacobshagen, V.H., ed., The Atlas system of Morocco. Studies on its geodynamic evolution: New York, Springer-Verlag, p. 19-42.

Walter, A., and Sharpless, S., 1987, Crustal velocity structure of the Sur-Obispo (Franciscan) terrane between San Simeon and Santa Maria, California (Abstract): EOS, v. 68, no. 44, p. 1366.

Wang Erchie, and Chu, J.J., 1988, Collision tectonics in the Cenozoic orogenic zone bordering China, India and Burma: Tectonophysics, v. 147, no. 1/2, p. 71-84.

Wang Jingming, 1987, The Fenwei rift and its recent periodic activity: Tectonophysics, v. 133, no. 3/4, p. 257-275.

Wang Jun, Huang Shangyao, Huang Geshan, and Wang Jiyang, 1986, Basic characteristics of the earth's temperature distribution in southern China: Acta Geological Sinica (Trial English Edition), v. 60, no. 3, p. 91-106.

Wang Liankui, Zhao Bin, Zhu Weifang, Cai Yuangji, and Li Tongjin, 1980, Characteristics and melting experiments of granites in southern China, in Ishihara, S., and Takenouchi, S., eds., Granitic magmatism and related mineralization: Society of Mining Geologists of Japan, Mining Geology Special Issue no. 8, p. 29-38.

Wanless, R.K., Stevens, R.D., Lachance, G.R., and Edmonds, C.M., 1968, Age determinations and geological studies. K-Ar isotopic ages, report 8: Geological Survey of Canada Paper 67-2, pt. A, p. 140-141.

Wasserburg, G.J., and DePaolo, D.J., 1979, Models of earth structure inferred from neodymium and strontium isotopic abundances: Washington, National Academy of Sciences Proceedings, v. 76, p. 3591-3598.

Watanabe, T., Langseth, M.G., and Anderson, R.N., 1977, Heat flow in back-arc basins of the western Pacific, in Talwani, M., and Pitman, W.C., III, eds., Island arcs, deep sea trenches, and back-arc basins: American Geophysical Union Maurice Ewing Series 1, p. 137-161.

Watts, A.B., Weissel, J.K., Duncan, R.A., and Larson, R.L., 1988, Origin of the Louisville Ridge and its relationship to the Eltanin fracture system: Journal of Geophysical Research, v. 93, no. B4, p. 3051-3077.

Weertman, J., 1971, Theory of water-filled crevasses in glaciers applied to vertical magma transport beneath the oceanic ridges: Journal of Geophysical Research, v. 76, no. 5, p. 1171-1183.

Weertman, J., and Weertman, J.R., 1964, Elementary dislocation theory: New York, Macmillan and Company, 213 p.

Wegmann, C.E., 1930, Über Diapirismus (besonders im Grundgebirge): Commission Géologique de Finlande Bulletin, v. 92, p. 58-76.

Wegmann, C.E., 1935, Zur Deutung der Migmatite: Geologische Rundschau, Bd. 26, Hft. 5, p. 305-350.

Weissel, J.K., 1981, Magnetic lineations in marginal basins of the western pacific, in Vine, F.J., and Smith, A.G., organizers, Extensional tectonics associated with convergent plate boundaries: Royal Society of London Philosophical Transactions, ser. A, v. 300, no. 1454, p. 223-245.

Wellman, H.W., 1966, Active wrench faults of Iran, Afghanistan and Pakistan: Geologische Rundschau, Bd. 55, Hft. 2, p. 716-735.

Wentworth, C.M., 1987, Implications for crustal structure in the western Coast Ranges, California, from studies along their eastern margin (Abstract): EOS, v. 68, no. 44, p. 1366.

Wentworth, C.M., and Zoback, M.D., 1989, The style of late Cenozoic deformation at the eastern front of the California Coast Ranges: Tectonics, v. 8, no. 2, p. 237-246.

Wentworth, C.M., Blake, M.C., Jr., Jones, D.L., Walter, A.W., and Zoback, M.D., 1984, Tectonic wedging associated with emplacement of the Franciscan assemblage, in Blake, M.C., Jr., ed., Franciscan geology of northern California: Society of Economic Paleontologists and Mineralogists, Pacific Section, Book 43, p. 163-173.

Westercamp, D., 1979, Diversité, contrôle structural et origines du volcanisme récent dans l'arc insulaire des Petites Antilles: Bureau de Recherches Géologiques et Minières Bulletin, deuxième série, no. 3/5, p. 211-226.

Westercamp, D., and Tomblin, J.F., 1979, Le volcanisme récent et les éruptions historiques dans la partié centrale de l'arc insulaire des Petites Antilles: Bureau de Recherches Géologiques et Minières Bulletin, deuxieme série, no. 3/5, p. 293-321.

White, A.J.R., and Chappell, B.W., 1983, Granitoid types and their distribution in the Lachlan fold belt, southeastern Australia, in Roddick, J.A., ed., Circum-Pacific plutonic terranes: Geological Society of America Memoir 159, p. 21-34.

White, R.S., 1989, Asthenospheric control on magmatism in the ocean basins, in Saunders, A.D., and Norry, M.J., eds., Magmatism in the ocean basins: Geological Society of London Special Publication no. 42, p. 17-27.

White, R.S., and McKenzie, D., 1989, Magmatism at rift zones: the generation of volcanic continental margins and flood basalts: Journal of Geophysical Research, v. 94, no. B6, p. 7685-7729.

Williams, H., 1978, Tectonic lithofacies map of the Appalachians: Memorial University of Newfoundland Map, no. 1, 2 sheets, scale, 1:1,000,000.

——, 1984, Miogeoclines and suspect terranes in the Caledonian-Appalachian orogen: tectonic patterns in the North Atlantic region: Canadian Journal of Earth Sciences, v. 21, no. 8, p. 887-901.

Wilson, J.T., 1954, The development and structure of the crust, in Kuiper, G.P., ed., The earth as a planet: The University of Chicago Press, p. 138-214.

——, 1966, Did the Atlantic close and then re-open?: Nature, v. 211, no. 5050, p. 676-681.

——, 1967, Theories of building continents, in Gaskell, T.F., ed., The earth's mantle: London, Academic Press, p. 445-473.

Winterer, E.L., Atwater, T.M., and Decker, R.W., 1989, The northeast Pacific Ocean and Hawaii, in Bally, A.W., and Palmer, A.R., eds., The geology of North America, v. A. The geology of North America; an overview: Geological Society of America, Decade of North American Geology, p. 265-297.

Witt, G., and Seck, H.A., 1987, Temperature history of sheared mantle xenoliths from the West Eifel, West Germany: evidence for mantle diapirism beneath the Rhenish massif: Journal of Petrology, v. 28, p. 475-493.

Wood, B.L., 1978, The Otago Schist megaculmination: its possible origins and tectonic significance in the Rangitata orogen of New Zealand: Tectonophysics, v. 47, no. 3/4, p. 339-368.

Wood, C., 1983, Continental rift jumps: Tectonophysics, v. 94, no. 1-4, p. 529-540.

Woodhouse, J.H. and Dziewonski, A.M., 1984, Mapping the upper mantle: three dimensional model of earth structure by inversion of seismic wave forms: Journal of Geophysical Research, v. 89, no. B7, p. 5953-5986, 6295-6297.

Woodworth, J.B., 1923, Cross section of the Appalacians in southern New England: Geological Society of America Bulletin, v. 34, no. 2, p. 253-261.

Worzel, J.L., 1965, Discussion, in Blackett, P.M.S., Bullard, E., and Runcorn, S.K., organizers, A symposium on continental drift: Royal Society of London Philosophical Transactions, ser. A, v. 258, no. 1088, p. 137-139.

Wright, J.B., Hastings, D.A., Jones, W.B., and Williams, H.R., 1985, Geology and mineral resources of West Africa: London, George Allen and Unwin, 187 p.

Wust, S.L., 1986, Regional correlation of extension directions in Cordilleran metamorphic core complexes: Geology, v. 14, no. 10, p. 828-830.

Yamagishi, H., 1985, Growth of pillow lobes—evidence from pillow lavas of Hokkaido, Japan, and North Island, New Zealand: Geology, v. 13, no. 7, p. 499-502.

Yan Zhide, 1987, Characteristics of strong earthquakes and tectonic relations in north-south seismic belt of China: Scientia, Sinica, series B, v. 30, no. 4, p. 438-448.

Yang Zunyi, Cheng Yuqi, and Wang Hongzhen, 1986, The geology of China: Oxford, Clarendon Press, 303 p.

Yeats, R.S., 1968, Rifting and rafting in the southern California borderland, in Dickinson, W.R., and Grantz, A., eds., Proceedings of conference on geologic problems of San Andreas fault system: Stanford University Publication Geological Sciences, v. 11, p. 307-322.

Ye Hong, Zhang Botao, and Mao Fungying, 1987, The Cenozoic tectonic evolution of the great North China: two types of rifting and crustal necking in the great North China and their tectonic implications; Tectonophysics, v. 133, no. 3/4, p. 217-227.

York, J.E., and Helmberger, D.V., 1973, Low-velocity zone variations in the southwestern United States: Journal of Geophysical Research, v. 78, no. 11, p. 1883-1888.

Yoshii, T., 1983, Cross sections of some geophysical data around the Japanese Islands, in Hilde, T.W.C., and Uyeda, S., eds., Geodynamics of the western Pacific-Indonesian region: American Geophysical Union and Geological Society of America Geodynamics Series no. 11, p. 343-354.

Zakharova, T.L., 1980, Izostaziya Baykal'skoy riftovoy zony: Novosibirsk, Geologiya i Geofizika, no. 5, p. 79-85.

Zandt, G., 1981, Seismic images of the deep structure of the San Andreas fault system, central Coast Ranges, California: Journal of Geophysical Research, v. 86, no. B6, p. 5039-5052.

Zandt, G., and Furlong, K.P., 1982, Evolution and thickness of the lithosphere beneath coastal California: Geology, v. 10, no. 7, p. 376-381.

Zhang Buchun, Jia Sanfa, Wang Tonghe, and Zheng Binghua, 1985, Intraplate seismotectonic features of North China: Tectonophysics, v. 117, no. 1/2, p. 177-191.

Zhang Yunxiang, Luo Yaonan, and Yang Chongxi, eds., 1990, Panxi rift and its geodynamics: Beijing, Geological Publishing House, 415 p.

Zhu Meixiang, 1986, Hydrothermal clay minerals in Rehai thermal field, Tengehong, Yunnan Province: Scientia Sinica, series B, v. 29, no. 4, p. 430-440.

Zhu Ming, 1989, Research on the isotopic ages of Mesozoic porphyries and porphyritic deposits in eastern China: Scientia Geologica Sinica, no. 2, p. 191-200.

Zietz, I., comp., 1982, Composite magnetic anomaly map of the United States. Part A: conterminous United States: United States Geological Survey, Geophysical Investigations Map GP-954-A, 2 sheets, scale, 1:2,500,000.

Zoback, M.D., Zoback, M.L., Mount, V.S., Suppe, J., Eaton, J.P., Healy, J.H., Oppenheimer, D., Reasenberg, P., Jones, L., Raleigh, C.B., Wong, I.G., Scotti, O., and Wentworth, C., 1987, New evidence on the state of stress of the San Andreas fault system: Science, v. 238, no. 4830, p. 1105-1111.

Zorin, Yu. A., and Osokina, S.V., 1984, Model of the transient temperature field of the Baikal rift lithosphere: Tectonophysics, v. 103, no. 1-4, p. 193-204.

ACKNOWLEDGMENTS

The list of persons who encouraged, contributed to, and criticized the various versions of this paper is too long to permit us to single out each person for what he or she did. We do however thank Stanley E. Cebull, Sankar Chatterjee, Jean M.S. Jenness, Stuart E. Jenness, Donna K. Meyerhoff, and James C. Meyerhoff for contributions far above the call of duty in organizing the paper, and for guidance with tables and illustrations. Kathryn L. Meyerhoff drafted, and in many cases, improved the illustrations. Ernestine R. Voyles typed the manuscript. Finally, we thank each of the following for their encouragement and critiques, which greatly improved the manuscript: D.L. Baars, M.I. Bhat, A.J. Boucot, A.L. Bowsher, S.E. Cebull, R.E. Chapman, S. Chatterjee, D.R. Choi, J.M. Dickins, Ann T. Donnelly, R.N. Donovan, A.K. Dubey, Ch. Ducloz, H. Duque Caro, A.C. Grant, P.E. Gretener, E.L. Hamilton, C.W. Hatten, N. Hotton III, M. Ilich, Jean M.S. Jenness, S.E. Jenness, M.S. Kashfi, M. Kamen-Kaye, T.T. Khoo, F.G. Koch, F.E. Kottlowski, K.B. Krauskopf, J.D. Love, W.D. Lowry, P.D. Lowman, Jr., A. and Marie Mantura, Donna K. Meyerhoff, J.C. Meyerhoff, Kathryn L. Meyerhoff, R.D. Meyerhoff, P. Miles, M.T. Moussa, W.W. Olive, T.S.M. Ranneft, D. Rigassi, V.M. Seiders, M. Schalk, W. Stannage, P.M. Sychev, B.K. Tan, Sukran Taner, Shirley S. Teitsworth, C. Teichert, W.A. Thomas, G.A. Thompson, G.B. Udintsev, Betty Whiting, C.W. Whiting, Ernestine R. Voyles, and W.L. Youngquist. The writers were inspired in part by Arthur L. Bowsher who, quite independently, developed similar concepts more than 40 years ago. Our acknowledgment of these colleagues in no way implies agreement with our views, but a surprising number of them do agree, a fact that made the writing an exceedingly pleasurable task.

All research was financed by the writers.

APPENDIX—PHYSICAL PRINCIPLES MENTIONED IN THE TEXT

Bernoulli's Theorem

In a stream of liquid, the sum of the elevation head, the pressure head, and the velocity head (1) remains constant along any line of flow provided that no work is done by or upon the liquid in the course of its flow and (2) decreases in proportion to the energy lost in viscous flow.

Hooke's Law

The stress within an elastic solid up to the elastic limit is proportional to the strain responsible for it.

Level of Neutral Buoyancy

In the lithosphere, this is the level at which a rising column of magma no longer can rise because below that level the country rock is denser than the magma, and above that level the country rock is less dense than the magma (see Walker, 1989). This level also has been called the level of gravitational stability.

Navier-Coulomb Maximum Shear-Stress Theory

We do not treat this concept in detail or with mathematics. Those interested in details are referred to Jaeger (1962, p. 75–80). The Navier-Coulomb maximum shear-stress theory states that failure (by shear) of a material occurs when the maximum shear stress equals the shear strength of the material. Compressive shear strength is always greater than tensile shear strength. Because of this, shear caused by compressive stress must be at an angle of less than 45° to the direction of that stress.

In any three-dimensional system, there are three principal stresses. If this three-dimensional system is the Earth's lithosphere, one of the three principal stresses is essentially vertical. Thus three cases—three types of faulting—arise according to whether the vertical stress direction is the greatest, intermediate, or the least (Jaeger, 1962).

Case 1, Thrust faults—As shown on Figure 97a, the vertical principal stress is the least in magnitude and the other two principal stresses are compressive. The planes of fracture pass through the direction of the intermediate principal stress and make angles of less than 45° with the direction of the greatest compressive (horizontal) stress (Fig. 97b). This case is illustrated in nature by Figures 3 and 5–7, which show the lithosphere Benioff zones dip at an angle less than 45° to the Earth's surface (see Scheidegger and Wilson, 1950).

Case 2, Streamline—strike-slip faults—As shown on Figure 97c, the vertical principal stress is the intermediate stress. Of the other two principal stresses, one will be a compression and the other will be small and may even be a tension. In this case, failure can take place on either of two vertical planes (AOB, COD) that are equally inclined at angles less than 45° to it (Fig. 97d).

Case 3, Normal faults—As shown on Figure 97e, the vertical principal stress is the greatest in magnitude. This state can be expected mainly at considerable depths (Jaeger, 1962, p. 80). Failure will take place at an angle of less than 45° to the vertical, that is, at an angle greater than 45° to the Earth's surface (Fig. 97f). Figures 3 and 5–7 illustrate natural examples in which the strictosphere Benioff zone dips at an angle greater than 45° to the Earth's surface.

Newton's Laws

All laws, theorems, and principles that involve motion are derived from Newton's three laws of motion and the law of gravity.

First Law of Motion —A body at rest remains at rest and a body in motion remains in uniform motion in a straight line unless acted upon by an external force.

Second Law of Motion —The rate of change of the velocity of a body, that is, its acceleration, is equal to the resultant of all external forces acting on the body divided by the mass of the body, and is in the direction of the straight line in which the force acts (that is, in the same direction as the resultant force).

Third Law of Motion —To every action there is always opposed an equal reaction; that is, the mutual actions of two bodies upon each other are always equal and in the opposite direction; or, for every force, there is an equal and opposite force or reaction.

Law of Gravitation —Every particle of matter in the universe attracts every other particle with a force that is directly proportional to the product of the masses of the particles and inversely proportional to the square of the distance between them.

Pascal's Law, Theorem, or Principle

Pressure applied to a confined liquid at any point is transmitted undiminished through the fluid in all directions and acts upon every part of the confining vessel at right angles to its interior surfaces and equally upon equal areas.

Peach-Köhler Climb Force

Although the following works best in mafic lavas because of their low viscosities, the statements apply to more silicic lavas as well. However, the relatively lower viscosity of mafic lavas makes it easier for them to move through the lithosphere and seek a position of gravitational stability (Walker, 1989).

In the case of a vertical crack filled with magma, the upward force exerted by the crack is a result of the lower density of the magma filling the crack (Corry, 1988). Weertman (1971, p. 1177) explained the process.

"The force of $(\rho - \rho')gV$ [where ρ is the density of the country rock at a given depth; ρ' is the magma density; g is the acceleration due to gravity; and V is the magma volume per unit of crack width] experienced by the crack is identical to the Archimedian buoyancy that would act on the crack if it were a solid of density ρ' embedded in a liquid of density ρ. There is a fundamental difference, however, between these two forces. For the true Archimedian force the body force $\rho'gV$ differs from the net force ρgV exerted on the embedded solid by the hydrostatic pressure. In the case of the liquid-filled crack, the body force $\rho'gV$ exerted on the liquid within the crack must be, and is, exactly balanced by hydrostatic forces exerted against the crack walls. "There is no net force exerted on the liquid within the crack." The Peach-Köhler climb force is produced by a gradient in elastic strain energy within the solid (rock body). The origin of the true and the pseudo-Archimedian forces is the same, namely, the gravity field. "The crack will stop propagating, and thus the magma rise will stop, when the magma reaches its level of neutral buoyancy, which is the level below which the country rock is denser than the magma, and above which the country rock is less dense than the magma"(Walker, 1989).

Poiseuille's Law

The velocity of flow of a liquid through a capillary tube varies directly as the pressure and the fourth power of the diameter of the tube, and inversely as the length of the tube and the coefficient of viscosity.

Stokes's Law and linear geologic structures

Stokes's Law, one of many expressions of Newton's Second Law of Motion, is little used by geologists, despite its fundamental importance. Because the law usually is thought of as a description of the flow of solids through air and water, it almost never is regarded as describing the interactions between any flowing medium and a solid substance (e.g., the flow of lava through lava tubes). One critic of an earlier version of this paper wrote that our use in this paper of Stokes's Law ". . . only confuses a discussion of the well-known streamline patterns of viscous flow." This critic apparently did not know that streamline patterns of viscous flow (in glaciers, for example) are commonly cited in physics texts as examples of Stokes's Law (e.g., Blatt, 1983)! Nor did

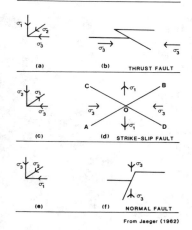

Figure 97. Navier-Coulomb maximum shear-stress theory: three examples. In each case, fracture takes place at an angle less than 45° with the direction of maximum principal stress. σ_1 in each diagram is the direction of least principal stress, σ_2 is the direction of intermediate principal stress, and σ_3 is the direction of maximum principal stress. (a) and (b) describe thrust faulting; (c) and (d) describe strike-slip faulting; and (e) and (f) describe normal faulting. For a detailed discussion, see Jaeger (1962, p. 75–80). Published with permission of Wiley and Sons, Inc. From Jaeger (1962)

he realize that such patterns of streamline flow are what led in the first place to the 1845 formulation of this law by Sir George G. Stokes (e.g., Sears et al., 1974).

Stokes's Law may be stated most simply as follows: "The force required to move a sphere through a given viscous fluid at a low uniform velocity is directly proportional to the velocity and radius of the sphere" (*Webster's Third New International Dictionary*, 1971, p. 2248). Under the heading of Stokes's Law, Sears et al. (1974, p. 225–226) wrote that "When an ideal fluid of zero viscosity flows past a sphere, or when a sphere moves through a stationary fluid, the streamlines form a perfectly symmetrical pattern around the sphere . . . ." If the fluid is viscous, however, there will be a viscous drag on the sphere. Moreover, if the surface of the sphere consists of alternate parallel bands of rough and smooth texture, differential (viscous) drag will take place between the fluid that flows over the rough-textured surfaces and that which flows over the smooth-textured surfaces. Then the streamlines are accentuated close to the sphere's surface by the difference in fluid velocities close to that surface, and are more appropriately termed sliplines. Flow close to a solid surface is called laminar flow, which has been defined as ". . . streamline flow in a viscous fluid near a solid boundary" (*Webster's Third New International Dictionary*, 1971, p. 1267). The important point is that streamlines, or sliplines are parallel with the direction of the laminar flow.

Many features related closely to laminar flow have been described by petrographers, such as Balk (1937), Knopf and Ingerson (1938), and many others. Many of the structures that we discuss in this paper are analogous to the flow and flow-line structures of Balk (1937). Thus, in geological terms, there are whole families of linear and curvilinear structures, at all scales (e.g., stretching lineations; *schlieren;* strike-slip faults; linear faults, fracture systems of the midocean ridges; several types of linear ridges), which are the products, directly or indirectly, of flow close to and parallel with them. This statement is simply another way of expressing Stokes's Law, or Newton's Second Law of Motion.

For example, the long linear faults, fractures and fissures of Figure 11 that parallel the axes of the Mid-Atlantic Ridge and East Pacific Rise respectively demonstrate that the motion—in any case, the last motion—beneath these ridges was parallel with, not orthogonal to them. Likewise, the jagged tension crack of Figure 98 clearly formed as a consequence of motions orthogonal to the overall trend of the crack. The type of fracture that forms in a solid substance is a clear reflection of the direction of the principal stress that deformed it.

Figures 11, 20, 21, 33, 38, 55, 57, 59, 61, and 75–77 from this paper provide good examples of geologic structures formed by laminar flow, in accord with Stokes's Law.

A critic of this paper wrote, "To me the linear features along a midocean ridge are more plausibly explained by the breaks and cracks of up-and-down shiftings that would result from moving apart the two sides, aided by the forcing apart by magma moving in both directions from a point of injection." This explanation, also favored by Macdonald et al. (1987), requires that a whole series of "points of injection" and "collision zones" between adjacent but opposite magma flows be present along each midocean ridge. Points of injection should be characterized by annular and/or radial structures many kilometers in diameter that alternate along strike with zones of transverse structures, including pressure ridges and like features. We have complete sonograph coverage of several lengthy segments of midocean ridges, including some 600 km of continuous coverage by the U.S. Geological Survey along the Juan de Fuca Ridge. There are no such structures; the only interruptions along strike are the ridge-transverse (transform fault) fault zones (see Figs. 9, 16–18, 21 of Meyerhoff et al. manuscript, this volume). The only other types of interruptions along strike are so-called overlapping spreading centers, which appear to be nothing more than giant eddies (see Figs. 13–15, Meyerhoff et al., this volume) that also are produced by ridge-parallel flow.

The same critic also wrote that ". . . it seems to me that *en échelon* bundles of fissures and faults could form just as well by pulling apart as the two sides move orthogonally away from the ridge." Figure 8 in Meyerhoff et al. (this volume) shows such en echelon structures on a small scale; Figure 71—an overlapping spreading center—is an en echelon structure on a large scale. As Olson and Pollard (1989) showed, such structures cannot form unless they parallel the direction of flow, as illustrated by Figures 13 and 15 of Meyerhoff et al. (this volume).

The en echelon structures are incipient vortices, in the terminology of fluid dynamics (Sears et al., 1974, p. 226; Tritton, 1988); they are the vortex structures of Meyerhoff et al. (this volume) and of this paper. As explained by Sears et al. (p. 226), "When the velocity of a fluid flowing in a tube exceeds a certain critical value (which depends on the properties of the fluid and the diameter of the tube) the nature of the flow becomes extremely complicated. Within a very thin layer adjacent to the tube walls, called the *boundary layer*, the flow is still laminar. The flow velocity in the boundary layer is zero at the tube walls and increases uniformly throughout the layer . . . . Beyond the boundary layer, the motion is highly irregular. Random circular local currents called *vortices* develop within the fluid, with a large increase in the resistance to flow. Flow of this sort is called *turbulent*." The properties of boundary layers and the vortical, eddy, and related structures of incipient and full turbulence are discussed in detail by Tritton (1988). Most of the structural features of the midocean ridges are a consequence of ridge-parallel flow, a conclusion that seems to be well documented, not just because of the linear structures, en echelon phenomena, and related structures we have mentioned, but also because of their near-identity with structures produced both in lava tubes (Yamagishi, 1985) and artificial tubes (Tritton, 1988) in which the flow, by definition, parallels the strike of the tubes. Tritton (1988), in fact, presented at least 39 photographs and line drawings of laboratory experiments in which some of the major structural features of

Figure 98. This is a photograph of a tension crack in pavement. Compare with Figure 11. Note the jagged fault trace. Tension, therefore, was at right angles to the crack, and the stress system is that shown in Figure 97e and f. If the midocean ridges are produced by tensile stresses acting at right angles to the trends of the ridges, this is the type of fault that should be observed along the ridges, not those illustrated in Figure 11.

midocean ridges, rift zones, and foldbelts are duplicated (e.g., his figs. 4.4, 17.20, 18.3, and 22.12).

Vortex street

A body towed through a fluid creates a band of oppositely flowing eddies behind it (Tritton, 1988). This band is a vortex street. (In nontechnical language, this band is a wake.) A series of immobile bodies that project into the fluid stream produce the same effect. In the case of the Earth, its rotation theoretically can cause eastward migration of the mushy part of the asthenosphere. Where mushy asthenosphere passes between two cratons rooted to the strictosphere, a vortex street might be created, as we show in a separate publication (Meyerhoff et al., submitted).

Endogenic regimes and the evolution of the tectonosphere

V. V. Beloussov,* Institute of the Physics of the Earth, USSR Academy of Sciences, Moscow, USSR

ABSTRACT

Tectonism, magmatism, and metamorphism are Earth's endogenic processes that generated the major tectonic features of the surface and underlying tectonosphere. These three processes operate everywhere together, but in different combinations; each combinations constitutes a distinct endogenic regime.

Endogenic regimes are of two types, those that construct continental crust and those that destroy it. Both constructive and destructive types include mobile and quiescent subtypes. Examples of mobil constructive types include geosynclinal, orogenic, continental-rift, and block-activization regimes; a quiescent constructive type is the platform regime. Destructive types include taphrogenic (mobile) and oceanic-basin (mobile and quiescent) regimes. The taphrogenic process of destroying continental tectonosphere is oceanization. The regime that prevails at a given time and place is determined by the heat flow and its vertical distribution of temperature.

The history of the Earth's tectonosphere consists of two stages. The first (pre-Mesozoic) involved the formation of continental crust across the Earth's entire surface. The second (post-Mesozoic), and still active, involves the destruction of the continental layer and its replacement with oceanic crust. Thus the Earth is a complex thermal machine in which physical and mechanical process interact along with the evolution of the tectonosphere.

INTRODUCTION

The endogenic development of Earth's crust involves tectonic, magmatic, and metamorphic processes. Different combinations of these processes generate different large-scale tectonic features, which are called endogenic regimes (Beloussov, 1981). Each regime is characterized by its own peculiarities of composition, structural or tectonic style, and stress state within the tectonosphere. (Tectonosphere, as used here, includes the crust and upper mantle above the asthenosphere, corresponding with the terms lithosphere and atmosphere.).

Regimes are of two types, quiescent and mobile (excited). Both types include continental platforms and the deep ocean floors. Mobile regimes include geosynclines, active foldbelts, continental rifts and other belts produced by taphrogenesis (e.g., deep depressions that are uncompensated by sedimentation present on either thinned continental or oceanic crust), plateau-basalt provinces, and midocean ridges.

FACTORS DETERMINING REGIME TYPES

Heat flow

The principal factor that determines which type of regime forms is heat flow. Quiescent regimes have low to normal heat flow (< 50 mWm^{-2}), whereas mobile regimes have higher heat flow (> 50 mWm^{-2}). The underlying asthenosphere is either thin or absent beneath quiescent regimes, but usually well developed beneath mobile regimes.

Except for the geosynclinal regime, all regimes known from the geological record exist today. Geosynclines are especially interesting, for they record a regular succession of certain endogenic processes that last for periods of tens to hundreds of million of years.

Vertical temperature distribution

The distribution of temperature within the upper mantle and crust is the second factor determining what type of regime may develop. Several different combinations of hot and cold mantle–hot and cold crust are possible. The upper mantle is considered to be hot if it generates magmatic melts; the crust is hot if it is the site of high-temperature metamorphism and granitization.

Vertical penetrability of melts

Vertical penetrability, or vertical access, to higher levels in the tectonosphere and to the Earth's surface is the third important factor that determines the type of endogenic regime that develops. The upward penetration of magma and hot fluid may be high or low, or even impossible. Most important, however, is the nature of the upward penetration—whether it is concentrated (as along a deep fault zone) or diffused (as beneath a region in which deep-penetrating faults are absent). The type of penetrability controls the intensity of thermal and physico-chemical interactions between magmas or deep fluids and the surrounding medium.

MAGMAS OF ENDOGENIC REGIMES

Tholeiitic and alkaline magmas

The magmas of each endogenic regime have distinctive compositions. Tholeiitic and alkaline magmas are the principal types, differing in the contents of incompatible (large-ion-lithosphere) elements and volatile components. Geochemical data suggest that the two are fused from mantle sources of different compositions. Tholeiitic magmas come from a "depleted" source and alkaline magmas come from an "undepleted" or "enriched" source (Gast, 1968; O'Nions et al., 1979; Hart and Allegre, 1980). The undepleted sources are deeper in the mantle than the depleted sources (Green et al.,

---

Chatterjee, S., and N. Hotton III, eds. *New Concepts in Global Tectonics.* Texas Tech University Press, Lubbock, 1992, xii + 450 pp.

*Deceased, December 1990.

1966; Yoder, 1976; Beloussov, 1983). Therefore, the upper mantle may be considered as divided roughly into two layers: an undepleted lower layer and a depleted upper layer (Jordan, 1979).

Calc-alkaline magmas

In terms of incompatible element contents, calc-alkaline magmas are intermediate between tholeiitic and alkaline magmas. Their source may be in the depleted upper layer of the upper mantle, but would be only in places where the composition of the depleted upper layer already had been altered by metamorphism generated by upward-migrating fluids (Best, 1975; Frolova et al., 1985).

SPATIAL PATTERNS AND SUCCESSIONS OF REGIMES

At any given time, the various factors enumerated above (e.g., heat flow, vertical temperature distribution, vertical penetrability, location of magma sources) control the vertical succession of regimes at any one place on the Earth's surface. These factors also control the spatial patterns of regimes at the Earth's surface. Heat escape from the interior, as seen from field data, occurs in impulses. Such impulses, or episodes of high activity, produce at any one location on the Earth's surface temporal changes in the nature of the regime, thereby imprinting a pattern of endogenic (tectonomagmatic) cyclivity.

The spatial distribution of regimes at any given time is related inseparably to the degree of penetrability of the tectonosphere. Mobile regimes (e.g., foldbelts) at the Earth's surface strike north-south (latitudinally), east-west (longitudinally), and diagonally (northwest-southeast; southwest-northeast). Such a surface pattern implies the existence in the tectonosphere of a systematic network of highly penetrable zones (Beloussov and Dmitrieva, 1984).

Magma ascent

Because heat rises from below, it first affects the lower part of the upper mantle, and melting occurs initially in the lower undepleted layer. If that layer is connected with the surfaces of the deep faults, then the alkaline magmas generated in the undepleted layer rises easily and rapidly along the fault zone to the surface. The interaction between the magmas and their surrounding medium is minor. This is an example of concentrated penetrability. If there are no available conduits to the surface, the tectonosphere plays the role of a filter, permitting only dry gaseous fluids (mainly hydrogen) to ascent. This is an example of diffuse penetrability. The dry gaseous fluids may rise to the upper depleted layer, causing some of it to melt. Whether the resulting magma is tholeiitic or calc-alkaline depends on the ability of the ascending fluid to assimilate (1) lithophile elements from the surrounding medium and (2) water needed for metasomatic processes. If lithophile elements and water are available and assimilated, the magma will be calc-alkaline; if not, the magma will be tholeiitic.

Continental versus oceanic tectonosphere

The formation of continental crust is related closely to calc-alkaline magmatism. In contrast, the formation of oceanic crust depends on tholeiitic magmatism. This difference indicates that continental upper mantle has undergone less chemical fractionation than oceanic upper mantle. If geothermal fractionation is a time-progressive process, than oceanic crust is secondary, resulting from the destruction and basification of continental crust (Beloussov, 1989).

Furthermore, the mechanisms involved in continental and oceanic fractionation entail important differences. Beneath the continents, degassing of hot ascending gaseous fluids brings a large volume of lithophile elements into the continental crust, whereas beneath the ocean basins simple selective melting is the prevailing process. This process causes oceanic crust to melt directly out from the mantle (Luts, 1980).

Some endogenic tectonic regimes build continental crust; others destroy it. I discuss the principal regimes briefly, and illustrate them on Figures 1 through 3.

Endogenic tectonic regimes that are constructive with respect to continental crust

Geosynclinal regime—Within the geosynclinal regime, the main tectonomagmatic activity is concentrated in and beneath the depressions, the so-called intrageosynclines (Fig. 1a). During the preinversion stage of geosynclinal evolution, subsidence prevails and the tectonosphere has a highly diffuse penetrability. The entire upper mantle is much heated by ascending hot gaseous fluids. Mafic and ultramafic magmas are separated, and associated with intermediate and silicic magmas. The general predominance of calc-alkaline magmas implies that hot ascending gaseous fluids have assimilated important amounts of lithophile elements. In a relative sense, the crust remains "cold." Heavy mantle material penetrates the crust during this stage, solidifying, overloading the crust, and instigating subsidence.

The mantle magma gradually heats the Earth's crust. As a consequence, the main tectonomagmatic activity migrates from the mantle into the crust, thereby initiating the inversion stage (Fig. 1b). This stage begins with folding, regional metamorphism, and granitization followed by other types of deformation. Among these other types, deep crustal diapirism is believed to be not only very important and partly responsible for the tectonic inversion and the rising of central uplifts within the intrageosyncline, but also for the crumpling of strata into folds (Fig. 1b).

In the light of most recent research, folding and related deformation should not be regarded as the crushing of plastic rocks between colliding plates. Rather, folding is the product of the same crustal heating that produces metamorphism and granitization. Even a modest crustal warming forces water to escape sediments. Although the density of geosynclinal rocks is less than that of the crust, a mixture of minerals, rock fragments, and free water provides a mass of even lower density. Such a mass of low-density material occurring between heated material below and cooler rocks above, should contribute to the inversion process and the rise of central uplifts within the intrageosyncline. (Beloussov [1981] calculated that a mixture of solid fragments and free water can decrease density and increase volume by as much as 15%.) The rise of a central uplift must lead to crumpling into folds (Beloussov, 1981, 1989; Goncharov, 1988). Thus geosynclinal folding, regional metamorphism, and granitization have a common origin in crustal heating.

During tectonic inversion, the upper mantle cools as tectonomagmatic activity migrates into the crust. This statement is supported by the observation that mantle-derived magma are not generated during the inversion stage.

Uplifts in intrageosynclines (intrageanticlines; median masses) are relatively stable cold blocks whose roots extend deep into the upper mantle. They are pierced by faults, some of which reach the undepleted layer. Consequently, magma compositions in intrageanticlines vary greatly. Even alkaline magmas occur.

During the next, or postinversion stage, all endogenous processes decline. This fact suggests that the tectonosphere is cooling during this stage.

Orogenic regime—Orogenic regimes (Fig. 1c) commence with a renewed escape of heat as a new thermal impulse occurs. The crystalline carapace formed by regional metamorphism during the preceding geosynclinal stage now hinders the penetration of melts into the crust, which remains cold. Instead, the magmas accrete to the base of crust, thickening it and causing increased isostatic uplift. Deformation of this cold crust is of the block type. Fissure intrusions and fissure eruptions are the principal magmatic phenomena. Calc-alkaline magmas dominate. The appearance of alkaline magmas—including potassic granitoids—near the end of the orogenic stage indicates that the tectonosphere is cooling from the top downward. The upper layer becomes increasingly stable and quiescent as the source of magma retreats ever deeper.

Island arcs also belong to the orogenic regime. Beloussov and Ruditch, (1960) recognize two island arc types: (1) arcs that exhibit a long previous geosynclinal history beginning from the Paleozoic or Proterozoic; and (2) arcs with no Paleozoic or older history. They are block-faulted volcanic chains, many of them recent, as exemplified by the Kuril and Aleutian arcs. They rose at the time of Holocene continental orogenesis, and have a structure and history like those of their continental counterparts—block tectonics and a lack of both granitization and regional metamorphism.

Platform regime—A platform regime (Fig. 1d) occurs when the entire tectonosphere is cold. Once established a platform regime may persist for much of geologic time or, if subjected to a new thermal pulse, may undergo tectono-magmatic reactivization and become once more the site of an orogenic regime or continental-raft regime of block tectono-magmatic activization.

Continental-rift regime—A high-temperature impulse can transform a platform regime into a continental-rift regime (Fig. 2b). This happens when thermal activity involves only the lower undepleted layer; the tectonosphere above the undepleted layer remains hard and cold. The upper layer is arched and then pierced by deep faults along which alkaline magma rises from the lower layer to the surface. The continental-rift regime is an example of a concentrated penetrability along deep faults. Volcanic pipes (e.g., kimberlite pipes, certain circular intrusions) are local manifestations of the mechanism similar to such of continental-rift formation. (Fig. 1f)

Block tectono-magmatic activization regime—Block tectono-magmatic activization is of two types: (1) activization accompanied by alkaline magmatism and (2) activization accompanied by calc-alkaline magmatism. The first type is similar to the orogenic regime, except that it usually covers a larger region and endures for a longer time. The region of the great Mesozoic-Cenozoic of Central Asia provides examples (Fig. 1g).

The second type with calc-alkaline magmatism is less common. One example is the 6,500-km long Cretaceous andesitic Chukotka-Cathaysian volcanic belt that extended from Bering Sea to Southeastern China. A second example is the Andes of South America with its Mesozoic andesitic belt of volcanic rocks that extends from northern Peru to central Chile, and which Aubouin et al. (1973) called a liminaire (incipient) geosyncline. However, there is no indication of geosynclinal development during the Mesozoic in this part of the Andes, as witnessed by the absence of pervasive folds and regional

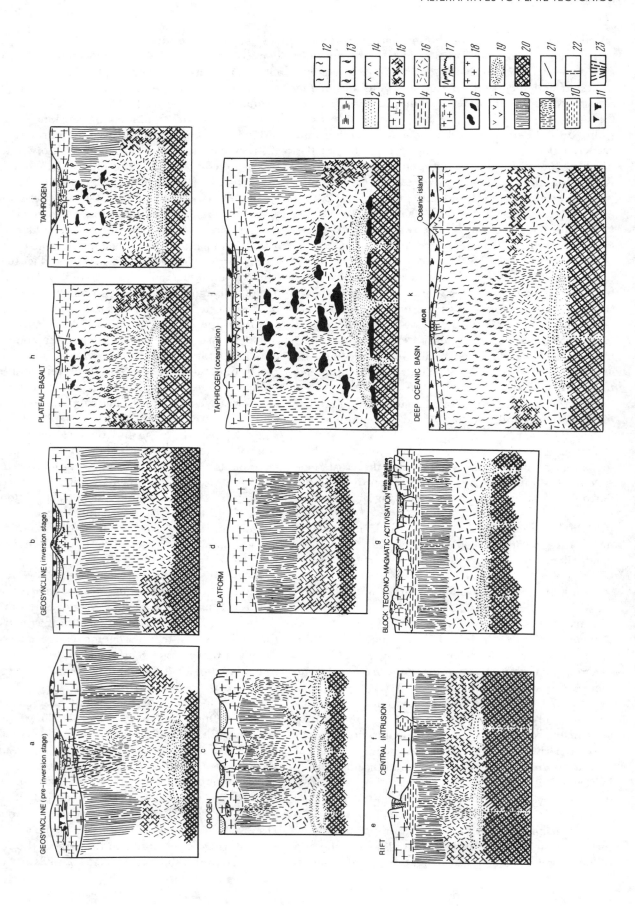

metamorphism. The deformation is of vertical block type, associated with strong outpourings of andesitic lavas (Zeil, 1979). This type of activation resembles that of an orogenic regime. It was not until Neogene time that real orogenesis began in Andes.

Endogenic tectonic regimes that are destructive to continental crust

Taphrogenic regimes—All regimes described to this point are associated either with alkaline or calc-alkaline magmatism, and are constructive with regard to the development and growth of the continental crust. A second group of regimes, all of them associated with taphrogenesis, performs the opposite role. These are regimes associated with tholeiitic magmas and are destructive to continental crust, leading ultimately to the substitution of continental crust by oceanic crust. Such replacement of continental crust is called oceanization.

Taphrogenic regimes begin in continental crust with the plateau-basalt regime (Fig. 1h). Ultimately, through susbstantial thinning, the continental crust is eliminated and replaced entirely by oceanic crust (Figs. 1i, 1j). An example is the following progressive, gradational, Neogene-Quarternary succession: (1) Pannonian basin; (2) Aegan, Levant, and Alboran seas; and (3) the Tyrrhenian Sea and Algero-Provence basin. The close correlation between basin depth and stage of evolution is apparent. The succession begins within a continental mass in places where some crustal thinning already has taken place. Continental crust is even thinner beneath the shallow Aegan, Levant, and Alboran seas. Beneath the Tyrrhenian Sea and Algero-Provence basin, the continental crust is broken into many small separate blocks, with the larger part of the basin floors being underlain by oceanic crust.

Figure 1. Structure and state of the tectonosphere during geosynclinal (a, b); orogenic (c); platform (d); rift (e); central intrusion (f); block tectono-magmatic activization with alkaline magmatism (g); plateau-basalt (h); taphrogenic (i); taphrogenic with oceanization (j); deep oceanic basin and oceanic island (k). 1—water; 2—sedimentary rocks; 3—continental crust; 4—mafic and ultramafic intrusive and extrusive rocks; 5—zone where mantle material is replacing continental crust; 6—blocks of continental crust sinking into the mantle; 7—oceanic crust; 8—depleted layer of the upper mantle, already metasomatized, in cold state; 9—depleted layer of the upper mantle undergoing metasomatism and partly melted; 10—alkaline magma; 11—calc-alkaline magma; 12—sterile depleted layer of the upper mantle, cold and hard; 13—sterile depleted layer for the upper mantle, partly melted; 14—tholeiitic basalt; 15—undepleted layer of the upper mantle, cold and hard; 16—undepleted layer of the upper mantle, partly melted; 17—folded rock; 18—granitized and/or regionally metamorphosed; 19—lens of deep-heated material (in asthenosphere); 20—middle mantle below tectonosphere; 21—faults; 22—zones of concentrated penetrability (see text); and 23—zones of diffuse penetrability.

In the Far East, the same progressive development of oceanic crust is apparent, and the same correlation between water depth and continental crust thickness is observed along the ocean-continent transition zones. The shallowest marginal seas are underlain by continental crust, which becomes gradually thinner towards areas of greater water depth. Continental crust is absent beneath areas where the water is some 2 to 3 km deep. Beneath the latter areas, oceanic crust has replaced continental crust.

The mechanism of oceanization—The thinning and eventual elimination of continental crust generally are attributed to tectonic stretching. This explanation, however, cannot apply to round basins such as the Peri-Caspian depression. The reason is that, on the continents, the stretching process would have to extend across an area hundreds of kilometers broad to account for the amount of subsidence observed (e.g., in the Peri-Caspian depression). Such areas of stretching are not observed and no structural evidence for such stretching is known (Artyushkov and Beer, 1983)

Artyushkov (1979) made a special study of the problem, concluding that the thinning of the continental crust is principally an in situ process, significantly, the ecologitization of the crust. However, for reasons given below, this hypothesis cannot be correct because all Neogene and Qarternary taphrogenic depressions are characterized by high heat flow. But high heat flow and ecologitization are incompatible processes, for ecologite is a high-pressure, low-temperature rock. Therefore, ecologitization cannot be the mechanism of oceanization.

Taphrogenic depressions are characterized not only by high heat flow, but also by high average seismic velocities. The presence of these two properties together implies the existence of a different oceanization mechanism (Beloussov, 1984, Beloussov and Pavlenkova, 1986). One may suppose, for example, that tholeiitic magma, which originates in the upper depleted layer of the upper mantle and is very low in volatiles, is capable of great overheating and of penetrating the crust just above it in a highly fluid state. The same overheating of the uppermost mantle should cause such a decrease in density that a zone of density inversion would form between the mantle and the crust. Under these conditions, the mantle would gradually absorb the crust, replacing it with oceanic crust. This mechanism would account for high heat flow and high seismic velocities. At the same time, water escaping from the destroyed continental crust could add to the volume of the oceans and fill the newly forming depressions on the Earth's surface. This hypothesis is supported by deep ocean drilling results. Deep-water drilling has discovered that (1) as a rule sedimentation on the bottoms of marginal seas

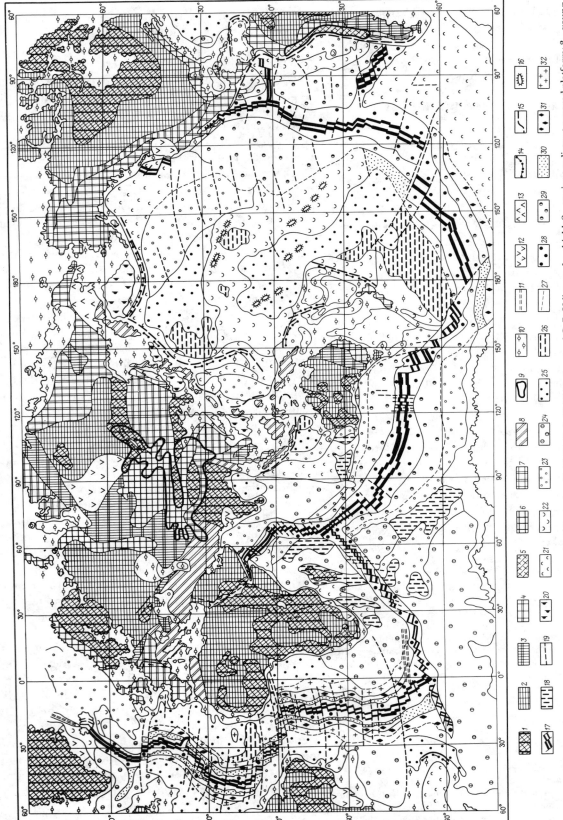

Figure 2. Tectonic sketch of the earth by V. V. Beloussov, A. V. Goriachev, E. M. Ruditch, V. N. Sholpo, and G. B. Udintsev. 1—shield; 2—ancient sediment-covered platform; 3—young platform; 4—area of block activization with alkaline magmatism; 5—area of block-activization with calc-alkaline magmatism; 6—Paleozoic foldbelt; 7—Mesozoic foldbelt; 8—Alpine foldbelt; 9—area of epiplatform orogenesis; 10—taphrogenic regime; 11—continental rift; 12—plateau basalt; 13—Chukotka-Cathaysian (Mongol-Okhotsk) andesite belt; 14—island arcs of the second type (see text) 15—deep trenches; 16—volcanic island group; 17—microcontinent; and 19—transform fault. Oceanic regions away from midocean ridge, age of oceanization: 20—Pliocene; 21—Miocene; 22—Oligocene; 23—Eocene; 24—Paleocene; 25—Late Cretaceous; 26—Early Cretaceous; 27—very approximate boundary between ocean-floor regions of different ages. Midocean ridge, age of crust in riftogenic zone: 28—Pliocene; 29—Miocene; 30—Oligicene; 31—Eocene; 32—Paleocene.

begins under either shallow-water or even continental conditions that (2) change gradually to deep-water conditions. We assume that this subsidence of the crust also is a gradual process that develops simultaneously with the destruction of continental crust and its replacement with oceanic crust.

In many regions, close coexistence of orogenic and taphrogenic regimes is observed. Examples abound in the Mediterranean area with its typical combinations of marine taphrogenic depressions surrounded by orogenic ridges (Platt and Vissers, 1989). The same is true in the Far East where marginal seas (taphrogenic) coexist with island arcs (orogenic). Both regimes are mobile, associated with high heat flow, and products of Neogene and Quarternary development. Differences between them are the consequence of different interactions between crust and upper mantle, which in turn are caused by the degree of fractionation of the upper mantle material. In some zones, incidentally, it is still possible for calc-alkaline magma to be generated. In other zones this is not possible and only tholeiitic magma can be generated.

Oceanic-basins regime (Fig. 1k)—We stress first that the deep-water drilling results show that during Late Jurassic-Early Cretaceous time large areas of the present oceans apparently were covered by shallow epicontinental seas and that the now-existing deep oceans were formed since Late Jurassic-Early Cretaceous time as a consequence of further crustal subsidence. This conclusion is supported by analyses of the characteristic of sedimentary rocks recovered in the drilled holes (Ruditch, 1983, 1984; Timofeev and Khulodov, 1984). Because with all probability, shallow epicontinental seas were underlain by continental crust, we may conclude that the formation of oceans is due to the same process of oceanization. Study of the ages of the oldest sediments in each hole shows that the change from shallow- to deep-water conditions happened in different places at different times. This means that ocean-bottom subsidence occurred differently in large separate blocks (Fig. 2).

Oceanic-basins regime and Benioff zones—Because the main event in ocean-basin formation occurred as long ago as late Mesozoic or early Cenozoic time, the oceanic tectonosphere has had sufficient time to cool. The cold tectonosphere of the Pacific is in contact with the tectonosphere of the Far East transitional zone which is still heated because the thermal events are much younger, Neogene to Quarternary. The contact between the hot and cold tectonosphere coincides with the Benioff zone. Such is the real nature of this zone, a thermal boundary between hot and cold mantle.

The inclination of the Benioff zone is everywhere toward the hotter region because the cold tectonosphere, being more dense, underflows the heated and less dense tectonosphere. This mechanism also explains the inverse inclination of the zone that is observed locally (e.g., in the Solomon and New Hebrides-Vanautu islands). Friction in the Benioff zone produces its well-known seismic activity. The energetic circulation of fluids in the zone removes lithophile elements. This process leaves behind a denser residual material which, in the plate-tectonics concept, is identified as subducted slabs of oceanic lithosphere. The writer points out that no convincing proofs of subduction in the form suggested by plate tectonics have been discovered. In contrast, the concepts presented here seem to be more realistic in light of the factual data.

Other facets of ocean-basins—The oceanization process did not reach complete fulfillment everywhere. There are oceanic plateaus and the so-called microcontinents, which differ from typical oceanic lithosphere in their structure. Typically these plateaus and microcontinents have thicker crusts and contain layers that may be considered as remnants of former continental crust. The author considers these plateaus and so-called microcontinents as remnants of former continental crust that were transformed into different degrees by the injected mantle materials during oceanization.

Also it must be pointed out that effusions of alkaline magma are common on volcanic oceanic islands. This indicates that deep faults beneath such islands extend to the lower undepleted mantle where the sources for such magma are situated.

Midocean-ridge regime (Fig. 1k)—This is a very young regime dating from near the end of the Paleogene and is superimposed on the oceanic-basins regime. Midocean ridges can be regarded as manifestations of the final stage of concentration of volcanic activity that initially was spread across the entire oceanic area and then gradually concentrated into an increasingly narrow belt.

Spreading of the oceanic lithosphere is a process that can be accepted with confidence only for the axial zone of the midocean ridge. Numerous data, newly acquired, indicate that this process is most unlikely to occur on the ridge flanks and almost certainly not within the abyssal plains. The shape and size of the linear magnetic anomalies, and the composition of the floor basalts at the crest of ridges and away from it are very different. Away from the crests, the linear anomalies become irregular splotches, and the composition of the basalts become increasingly like those of continental plateau basalts (Gordin et al., 1987).

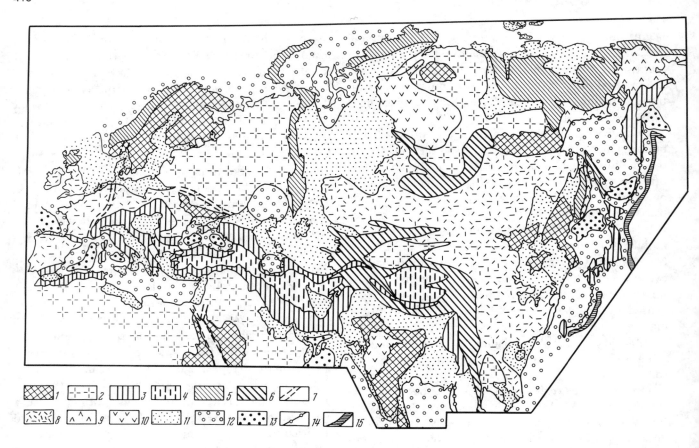

Figure 3. Endogenic regimes of Eurasia. 1—Crystalline shield; 2—sediment-covered platform; 3—very active epigeosyncline orogenic regime; 4—less active epigeosynclinal regime (on medium massifs); 5—weak orogenic regime (on Paleozoic foldbelts); 6—epiplatform orogenic regime; 7—continental-rift regime; 8—block-activization regime with alkaline magmatism; 9—block-activization regime with calc-alkaline magmatism (Chukotka-Cathaysian volcanic belt); 10—plateau-basalt; 11—taphrogenic regime on continental crust; 12—taphrogenic regime on transitional continental crust (some missing layers); 13—taphrogenic regime on oceanic crust; 14—island arc of second type; and 15—deep-water trench.

DISCUSSION AND CONCLUSIONS

Two major stages in the history of the Earth's tectonosphere are distinguished. The first was characterized by the formation of continental crust across the whole of the Earth's surface. This stage lasted until the beginning of Mesozoic time when the second stage commenced. That stage involved the destruction of the continental crustal layer and the formation of oceanic crust. The transition from the first stage to the second marks a change in the mechanism of fractionation of the upper mantle and in the mechanism of heat escape from the Earth's interior. The process of formation of the continental crust, rich in incompatible elements, is related primarily to the degassing of the upper mantle, whereas the formation of oceanic crust is related mainly to the selective melting of the depleted part of the upper mantle.

The fact has long been known that the size of the platform-covered areas of the Earth has increased gradually since the beginning of Archean time while the size of areas occupied by geosynclines has diminished. This indicates that quiescent regimes have been growing at the expense of mobile regimes, and suggests that the growth of quiescent regimes is a result of heat escape from within the earth. Initially the heat loss may have been caused by the loss of volatiles from the mantle, which was an important mechanism earlier in the Earth's history. However, deeper in the Earth, heat still accumulated and finally began to escape to the Earth's surface by means of a more energy-consuming mechanism, namely, selective melting.

When the products of selective melting reached the surface marks the time when colossal volumes of tholeiitic basalts poured out and, across a vast area, oceanic crust engulfed continental crust, substituting for the latter on a huge scale at an extremely rapid rate. Presumably the locations of the places where these fractionated depleted magmas burst forth onto the earth's surface were determined by a systematic, not a random process. This conclusion must be true if we are to explain the antipodal positions of continents and ocean basins

(Arldt, 1907; Harrison, 1968), the asymmetry of the Northern and Southern Hemispheres, and the southward-tapering tendency of all continents, as well as many other peculiar, but regular features of continents and ocean basins. Such features are hardly accidental (Harrison, 1968).

We know from the ages of basalts drilled during the Deep Sea Drilling Program (Bolli, 1980a, 1980b) that the great floods of tholeiitic magma were mainly pre-Cenozoic. In more recent geologic time, the process of the destruction of continental crust was renewed. For example, a taphrogenic regime invaded Eurasia during Neogene time, from the Atlantic Ocean in the east (marginal seas of the eastern parts of the Sino-Korean and Yangzi platforms [e.g., Bohai Gulf]). In western Europe, Central Asia, and Eastern China, the taphrogenic regime was preceded by a block-activization regime and a continental-rift regime, or both (Fig. 3). Is the function of these regimes that preparation of continental crust for the further destruction?

It appears that there are two levels of physico-chemical equilibrium, (1) continental crust formation and (2) oceanic-crust formation. The two equilibria seem to be connected through two different mantle-fractionation mechanisms. Until the Mesozoic, tectonosphere development was directed towards the first equilibrium level; afterward, tectonosphere development changed to the second equilibrium level. During the first (i.e., pre-Mesozoic), fractionation of the upper mantle and depletion of its upper layer comprised the principal physico-chemical process affecting the development of the tectonosphere. Later, after the changeover to the second equilibrium level, mantle fractionation continued, but it was combined with the very opposite process—that of homogenization, which involves the mantle's absorption of the continental crust. To date the homogenization process plays a subordinate role, fractionation and depletion being dominant.

The Earth is a complicated thermal machine in which physical and chemical processes develop along the evolution of the tectonosphere. Moreover, the tectonosphere's evolution constitutes the background against which the Earth's internal heat continually but irregularly—in both time and space—escapes. The irregular spatial and temporal escape of heat is the principal factor that determines the spatial distribution of coeval endogenic regimes and their changes through times. These changes and the thermal impulses that accompany them produce a pattern of endogenic (tectonomagmatic) cyclicity during the earth's development. All factors controlling endogenic regimes are interrelated, both spatially and temporally. The particular endogenic regime that is active in a given place at a particular time is determined by its previous state and determines its future state.

REFERENCES CITED

Arldt, Th., 1907, Die Entwicklung der Kontinente under ihrer Liebewelt: Leipzig, Christian Hermann Tauchnitz, 729 p.

Artyushkov, E.V., 1979, Geodynamics: Moscow, Nauka, 329 p. (in Russian).

Artyushkov, E.V., and Beer, M.A., 1983, About the role of vertical and horizontal movements in the genesis of a depression in folded zones on continental crust: Akademiya Nauk SSSR Izvestiya, Seriya Geologicheskaya, no. 9, p. 25-52 (in Russian).

Aubouin, J., Borello, A.V., Cecioni, G., Charrier, R., Chotin, P., Frutos, J., Thiele, R., and Vicente, J.C., 1973, Esquisse paléogéographique et structural des Indes Meridionales, Revue de Geologie Physique et de Géologie Dynamique, sér. 2, v. 15, fasc. 1-2, p. 11-72.

Beloussov, V.V., 1981, Continental endogenous regimes: Moscow, Mir Publishers, 295 p.

——, 1983, Endogenic regimes and mantle magmatism: Geotektonika, no. 6, p. 3-12 (in Russian).

——, 1984, Certain problems of the structure and evolution of transition zones between continents and oceans: Tectonophysics, v. 105, no. 1-4, p. 79-102.

——, 1989, Basics of geotectonics, 2nd edition: Moscow, Nedra, 382 p. (in Russian).

Beloussov, V.V., and Dmitrieva, B.I. 1984, Prevailing strikes in the folded structures of the Phanerozoic and Precambrian: Geotektonika, no. 5, p. 15021 (in Russian).

Beloussov, V.V., and Pavlenkova, N.I., 1986, Interrelation between the Earth's crust and upper mantle: Geotektonika, no. 6, p. 8-20 (in Russian).

Beloussov, V.V., and Ruditch, E.M., On the position of island arcs in the evolution of the Earth's structure: Sovetskaya Geologiya, no. 10, p. 23 (in Russian). (English translation, 1961: International Geology Review, v. 3, no. 7, p. 557-574.)

Best, M.C., 1975, Migration of hydrous fluids in the upper mantle and potassium variations in calc-alkaline rocks: Geology, v. 3, no. 8, p. 429-432.

Bolli, H.M., 1980a, Ages of sediments recovered from the Deep Sea Drilling Project Pacific Legs 5 through 9, 16 through 21, and 29 through 35, in Rosendahl, B.R., Hekinian, R., et al., Initial reports of the Deep Sea Drilling Project, v. 54: Washington, D.C., U.S. Governments Printing Office, p. 881-886.

——, 1980b, The ages of sediments recovered from DSDP Legs 1-4, 10-15, and 36-53 (Atlantic, Gulf of Mexico, Caribbean, Mediterranean and Black Sea), in Rosendahl, B.R., Hekinian, R., et al., Initial reports of the Deep Sea Drilling Project, v. 54: Washington, D.C., U.S. Government Printing Office, p. 887-895.

Frolova, T.I., Burikova, I.A., Gustakhin, A.V., Frolov, V.T., and Syvorotkin, V.L., 1985, Origin if the island arcs' volcanic series: Moscow, Nedra, 275 p. (in Russian).

Gast, P.W., 1968, Trace element fractionation and the origin of tholeiitic and alkaline magma types: Geochimica et Cosmochimica Acta, v. 32, p. 1057-1069.

Goncharov, M.A., 1988, Mechanism of geosynclinal folding: Moscow, Nedra, 264 p. (in Russian).

Gordin, V.M., Mikhailov, V.O., and Trebina, E.S., 1987, Methods and results of interpretation of magmatic anomalies in the central part of the South Atlantic: Akademiya Nauk SSSR Izvestiya, Fizika Zemli, no. 7, p. 69-83.

Green, D.H., Ringwood, A.E., et al., 1966, Petrology of the upper mantle: Canberra, Australian National University, Department of Geophysics and Geochemistry, Publication 444.

Harrison, C.G.A., 1968, Antipodal locations of continents and ocean basins: Science, v. 153, no. 3741, p. 1246-1248.

Hart, S.R., and Allegre, C.J., 1980, Trace elements constraints on magma genesis, *in* Hargraves, R., ed., Physics of magmatic processes: Princeton University Press, p. 121-159.

Jordan, T.H., 1979, The deep structure of the continents: Scientific American, v. 240, no. 1, p. 92-107.

Luts, B.G., 1980, Geochemistry of oceanic and continental magmatism: Moscow, Nedra, 246 p. (in Russian).

O'Nions, R.K., Evensen, N.M., and Hamilton, P.J., 1979, Geochemical modelling of mantle differentiation and crustal growth: Journal of Geophysical Research, v. 84, no. B11, p. 6091-6101.

Platt, J.P., and Vissers, R.L.M., 1989, Extensional collapse of thickened continental lithosphere: a working hypothesis for the Alboran Sea and Gibraltar Arc: Geology, v. 17, no. 6, p. 540-543.

Ruditch, E.M., 1984, Expanding Oceans: Facts and Hypotheses: Moscow, Nedra, 251 p. (in Russian).

Timofeev, P.P., and Khulodov, V.N., 1984, Evolution of sedimentary basins in the evolution of the Earth: Akademiya Naud SSSR Izvettiya, Seriya Geologicheskaya, no. 7, p. 10-34 (in Russian).

Yoder, H.S., 1976, Generation of basaltic magma: Washington D.C., National Academy of Sciences, 266 p.

Zeil, W., 1979, The Andes: a geological review: Berlin, Gebruder Borntraeger, 260 p.

Global change: shear-dominated geotectonics modulated by rhythmic Earth pulsations

Forese-Carlo Wezel, Institute of Geology, University of Urbino, I-61029 Urbino, Italy

ABSTRACT

The geological column consists of a recurrent series of depositional megasequences characterized by great lateral continuity. The lithofacies are similar regardless of the depositional setting involved. They are to be considered as almost synchronous manifestations of phases of geotectonic evolution of global importance.

Planetary compressional pulses may cause the simultaneous reactivation of many worldwide phenomena, including sea-level changes, and depositional, magmatic, tectonic, and magnetic regimes. Progressive increases in terrestrial compression have determined normal geomagnetic polarity, marine transgression, diastrophism, syn-diastrophic deposition (flysch), folding, climatic amelioration, and peaks in the diversity of genera. The "pulsatory contraction" ended in correspondence with a dramatic reorientation of lithospheric stress, followed by a shift from contraction to expansion. This geotectonic reorganization determined the reversal of geomagnetic field, regression, acid plutonism, the rise of immense topographic highs (e.g., the whole Tibetan Plateau), and a decline in global temperature, which culminated in the ice ages.

The global character of the first-order geotectonic cycle is also confirmed by the curve of the percentage of reversed polarity. The initial N-hemicycle is distinguished by break-up phases that led to oceanization, compressional suturing, and crustal shortening. The final R-hemicycle is marked by mantle diapirism, cratonization, and crustal expansion.

In the second part of the paper, a new meganeotectonic interpretation is presented, which provides information on the present tectonic state of the Earth. On the basis of morphostructures and the Seasat map of free-air gravity anomalies, one may hypothesize the existence in all the oceans of an "oblique tectonic fabric," consisting of long linear ridges and archipelagoes characterized by en echelon and sigmoidal patterns. These oblique ridges are interrupted by immense bathymetric and gravitational depressions, which are interpreted here as "master megashear belts," following the latitudinal parallels. In this way, the Earth's surface appears to be broken into megaslices, within which a powerful tectonic deformation takes place capable of producing tectonic distorsions on a continental scale. The fact that this deformation is superimposed on the midocean ridges indicates that it postdates the most recent geomagnetic inversion (i.e., of Middle to Late Quaternary age). The comparison of first-order morphostructural and tectonic stress data indicates the existence of a Labrador-Atlantic zone of oblique divergence and an Australasian zone of oblique convergence of the compressive stress, which are located at antipodes to each other. Considering their approximate correspondence with features of the geoid surface, it is hypothesized that tectonic stress in the lithosphere results from creep-flow processes occurring deep within the mantle. A double migration over the course of time of the lower mantle creep zone from the Atlantic area of divergence towards the East Asiatic zone of convergence is postulated. The mantle migrations were presumably induced by changes in the geoid shape, possibly caused by variations in the Earth's rate of rotation.

Thus, past and present-day tectonic regimes appear to be dominated by colossal strike-slip deformation occuring along vertical shear zones, which are very ancient and have periodically been reactivated. A striking similarity exists between some shear-dominated crustal structures and features produced by atmospheric circulation.

INTRODUCTION

A few first-order temporal and areal patterns of global importance are delineated herein, in order to obtain empirical information about the evolution of the Earth. These universally operative patterns basically seem to be produced by a homogeneous state of stress. As they are not constant in time, but change and fluctuate cyclically, one must assume that major changes have occurred in the state of stress in the lithosphere during Earth history. In other words, the global patterns are intrinsically dynamic and show different phases of Earth development.

A careful analysis shows that the geological column consists of recurrent series of stratigraphic units of great lateral continuity (megasequences) that reappear, but always in a slightly modified form, which differ from the preceding ones. Thus the original conditions are returned to (cyclicity), but on a higher evolutionary plane. Recurrence of this kind that follows a spiral path has been termed "helicyclic tectonics" (from the Greek "helix," meaning spiral) by the present author (Wezel, 1988). This is a new global concept that highlights an evolutionary dynamism, which orders the Earth's inorganic and organic history. The global changes represent simultaneous but independent reactions to the same worldwide process in which the general pattern of geological evolution consists of a progressive increase of complexity and interdependence (i.e., increasing frequency in time of tectonic and volcanic events). The whole tectonosphere, our planetary geosystem, is a dynamic and integrated network of morphotectonic structures that exist in a state of continual and interdependent mutation and transformation in relation to the changing environment. There are no static structures in nature: they are instead the spatial manifestation of the underlying temporal processes.

The planet as a whole has rhythms of various periodicities. The temporal cycles give rise in space to geotectonic systems and tectonic structures that are precarious and ephemeral, representing no more than transitory dynamic forms, generated by the continuous change and transformation of the lithosphere. They are structures that must be combined in their correct time sequence (relative chronology). The helicyclic concept

---

Chatterjee, S., and N. Hotton III, eds. *New Concepts in Global Tectonics*. Texas Tech University Press, Lubbock, 1992, xii + 450 pp.

highlights the fundamental role of the reactivation of long-lived, deep-seated features, crossing both continental and oceanic domains, that is the megashear mobile zones. The geotectonic cycle corresponds in fact to a strike-slip cycle (e.g., Wezel, 1988). Along megashear belts, continental dismemberment, outpouring of the magmas, regional subsidence, tectonic rotation, basin filling, crustal shortening, plutonism, vertical uplift, emergence, and erosion all take place. The horizontal shearing in vertical and mutually orthogonal planes seems to be the principal mode of deformation of the lithosphere (Hast, 1969). Thus, crustal diastrophism appears as a shear-dominated system of deformation. The Earth's crust has been, and still is, subjected to global horizontal stress fields of enormous intensity that result in shearing.

This approach contrasts with the stereotype conventional geotectonic views that favor irregularity and chance as prime factors in geodynamic evolution. Most of the basic concepts of these schemes appear as simplistic and highly unrealistic theoretical models, which I consider no longer adequate representations of geological reality.

Some of the aspects that are pertinent to global stratigraphy are discussed in the first section, and the reader is referred to Wezel (1988) for a more complete picture of the systematics of time. In the second section, however, a new first-order meganeotectonic picture of the present tectonic state of the Earth is illustrated.

PULSATING GEOTECTONIC EVOLUTION

Depositional megasequences as phases of tectonic evolution

If, while ignoring background noises, one tries to delineate the essential elements of the regional lithostratigraphy of the Tethyan realm, one is struck by its homogeneity. The main sedimentary formations of the Mesozoic appear to be characterized by lithofacies that are similar over huge geographical extensions (see Fig. 1). During the last decade, deep-sea drilling operations (DSDP) in the Atlantic and Pacific have cored different Mesozoic sequences, which show not only similarity of evolution, but also a striking synchroneity and facies correspondence with the Tethyan columns.

The spatial extension of the main formations during the Cretaceous is particularly impressive. From the synthetic data of Jansa et al. (1979) and De Graciansky et al. (1987), one may deduce a possible extension to the whole Atlantic of the following Tethyan stratigraphic units:

6 — Scaglia Rossa (Turonian-Senonian)
5 — Bonarelli Level (Cenomanian-Turonian)
4 — Scaglia Bianca (Cenomanian)
3 — Fucoid Marls (Late Aptian-Albian)
2 — Selli Level (Aptian)
1 — Maiolica (Neocomian-Early Aptian)

The Maiolica formation consists of white micritic limestones alternating with black shales that increase in number and thickness upwards, reaching a maximum of about 2 m in correspondence with a level characterized by great lateral continuity, which I have named the "Selli Level" (Wezel, 1985). An important and sharp change in sedimentary and tectonic regimes occurs with the onset of deposition of the Fucoid Marls (interbedded greenish marls and black shales) in the Late Aptian. After this the limestone formations end and more terrigenous sequences, in part hemipelagic and in part turbiditic, begin. There is a sudden increase in the accumulation rate caused by a marked terrigenous influx of clay and quartzose material. The deposition is accelerated in the Atlantic and Tethyan domains by the start of tectonic activity at the margins (see De Graciansky et al., 1987), which begin to be affected almost ubiquitously by syndiastrophic regional subsidence and foundering (see Wezel, 1985). This represents a dramatic change in environmental conditions, which is of global importance and which I call the "Aptian-Albian revolution."

The "revolution" consists of the formation of a trough and ridge topography that persists up to the sedimentation of the Bonarelli Level and is responsible for the prevalence of anaerobic conditions in diverse depositional environments. In the Central North Atlantic, beginning with the deposition of the Fucoid Marls, a ridge (west of the modern Mid-Atlantic Ridge) separated a western subbasin from an eastern subbasin (De Graciansky et al., 1987). The widespread occurrence of black shales is a clear indication of large, partially isolated basins, which were of restricted extent and which subsided. They were either euxinic or had floors that were localized below the oxygen-minimum layer. The Aptian-Albian revolution has been demonstrated by both the simultaneous tectonic mobility of a large number of the world's passive margins (e.g., Kent, 1977; Wezel, 1985), and by diapiric mobilization of the mantle on a gigantic scale, as indicated by the huge extent of mafic volcanism (flood basalts) in the Atlantic (J.-M. Auzende, pers. comm., 1989) and the western Pacific (J. Lancelot, pers. comm., 1989). According to Kent (1977) this is "a fundamental rheological modification of the crustal rocks from brittle fracture to plastic flow, which must in turn reflect a thermal event." Another example of vertical tectonics of huge proportions is represented by the formation, begining in the Aptian, of

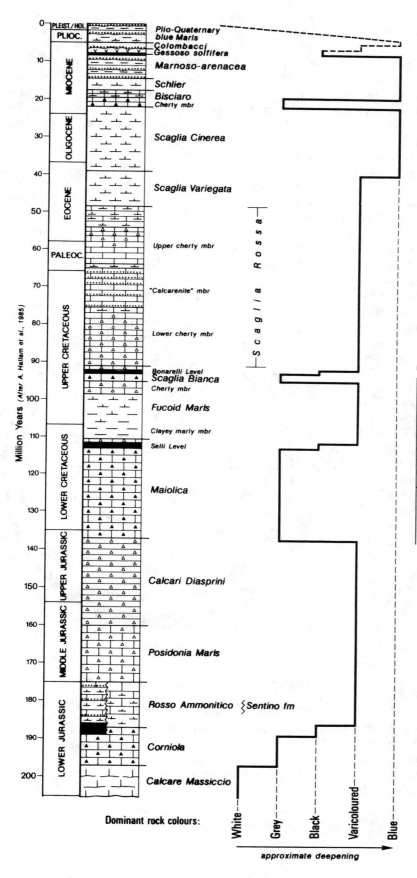

Figure 1. The main sedimentary formations of Meso-Cenozoic age, characteristic of the Tethyan realm (southern margin). The right-hand side of the diagram illustrates their correspondence with the diastrophic stages of a geotectonic cycle. The cyclic nature of both the lithofacies and the rock colors is evident. The Cretaceous stratigraphic units also show a widespread distribution throughout the Atlantic realm.

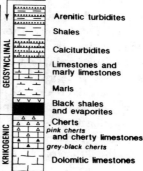

the Darwin Rise in the western Pacific, which recently has been restored to credence on the basis of DSDP data (Schlanger et al., 1981).

The lateral persistence of the stratigraphic units of the Cretaceous is enormous, but even more important, the lithofacies are similar independently of the depositional environment. The same type of deposition occurs in deep ocean and in marginal and epicontinental seas. An extraordinary example of the presence of the same facies in different basin settings is represented by the Bonarelli anoxic level of the Cenomanian-Turonian, which is found everywhere: in midocean plateaus of the Pacific Basin, cratonic interior seaways of North America, the European shelf and its interior seaways, the circum-African embayments and seaways, and the Tethyan margins (Schlanger et al., 1987), as well as in the deep Atlantic Ocean (De Graciansky et al., 1987). Evidence does not support the popular hypothesis developed over the past few years that the sediments were deposited during oceanic anoxic events, when large portions of the water column became severely or totally depleted of oxygen. Data contradicting the hypothesis of anoxia in all or most ocean waters have been presented by Waples (1983).

The surprising lateral continuity of the facies and, in part, the colors of the formations can indicate only a parallel evolution of the environment in a large number of the crustal sectors examined. Considering the variety of depositional settings involved, which go from deep marine to cratonic basin, the causative mechanism could not be other than global tectonic diastrophism. The close connection between tectonism and volcanism leads one to believe that the latter may be considered as a sort of tectonics of plastic material. I consider tectonism to be the primary active phenomenon capable of deforming portions of the mantle material (e.g., by mantle diapirism). The different tectonic movements probably cause the magma to change its composition (see below).

The large outbreaks of magmatism, high sea-level stands and high rates of oceanic openings, which accompany the Mid-Cretaceous revolution, indicate a fundamental geotectonic event. The history of the Alps provides us with pertinent information on the nature of such a catastrophic, tectonomagmatic event. It is, in fact, supposed that the Eo-Alpine phase of lithosphere subduction was initiated in Aptian-Albian time and accompanied by very severe high-P metamorphism (Winkler and Bernoulli, 1986). This is a clear indication of the existence of a powerful horizontal compressive stress field in the lithosphere. Other contemporaneous worldwide occurrences of crustal shortening during the Aptian-Albian have been summarized by Schwan (1984).

At nearly all passive margins there is evidence during the Aptian of an inversion in the direction of tilting, which may have corresponded to a change in the orientation of the maximum compressive stress. In the case of the North American Atlantic continental margin, there was a shift from horizontal compression orientated NE–SW (as at present) to compressive stress aligned N–S (Sheridan, 1976). From this point of view, the lithostratigraphic megasequences, characterized by huge lateral continuity, represent some of the indicators of temporal change in the direction and intensity of planetary stress. In fact, the lithofacies express the phases of global tectonic evolution, exemplified in schematic form by the geotectonic cycle shown in Figure

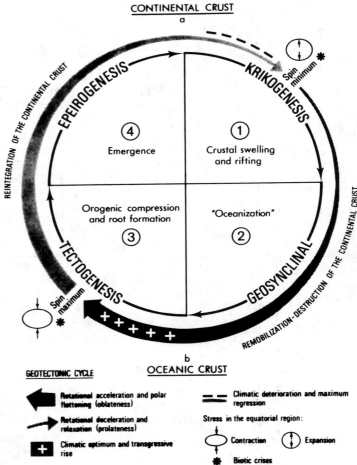

Figure 2. Idealized diagram showing the principal features and evolutive phases of a geotectonic strike-slip cycle according to Wezel (1988). All geological phenomena are manifestations of a continuous oscillation between destruction and reintegration of the continental crust (i.e., a dynamic interplay between contractive and expansive forces).

2. The Aptian-Albian episode represents the passage from a long krikogenic phase to one of geosynclinal (regional) subsidence, caused by an increase in intensity of global compressive stresses in the crust.

Another worldwide geotectonic event occurred during the Middle to Late Eocene and was marked by a set of varied geological phenomena. Large parts of the Earth's lithosphere were subjected to tangential contractional stresses. Large-scale Alpine-type orogeny and shortening were produced by this global geotectonic event, which has been dated at 40 ± 3 Ma (Schwan, 1985) and which manifested itself in many continental margins by means of a change in tilting from landward to oceanward. This change was the response to a passage from one stress system to another. In the North American Atlantic margin, for example, compression changed from N–S to NE–SW (Sheridan, 1976). In contrast, the Late Eocene stress field in the European platform was characterized by N–S compression, which replaced the NW–SE compression of the Early to Middle Eocene (Bergerat, 1987).

This event coincided with a passage in sedimentation from the deposition of marly limestone (Fucoid Marls, Scaglia Bianca, and Scaglia Rossa formations) to that of calcareous marls (Scaglia Cinerea, Bisciaro, and Schlier formations), and to the appearance of Oligo-Miocene arenaceous flysch sequences. Moreover, the Oligo-Miocene formations transgress regionally over folded and truncated Paleogene beds, which is a common feature of the Pacific margin all the way from Alaska to Indonesia (Beck and Lehner, 1974). In fact, in the Late Eocene many ocean basins underwent a hiatus in sedimentation (Moore et al., 1978).

Late Miocene and Early Pliocene compressional movements caused another catastrophic event in the form of the Messinian salinity crisis. These movements temporarily restricted the connection with the open sea and thus rapidly created conditions for the widespread deposition of evaporites in a number of partially isolated and restricted basins. The Plio-Quaternary was marked both by the rise of the Alpine chains and by phases of strong subsidence of the Mediterranean basins, which were accompanied by thick local accumulations of deltaic sediments. For the second time in the Late Miocene a maximum in the abundance of hiatuses occurred in all ocean basins except the Indian one (Moore et al., 1978).

There is evidence of a change in the stress field during the Late Miocene. In the European platform a consistent direction of compression (first NW–SE, then NNW–SSE) has dominated since the end of the Miocene (Bergerat, 1987). This stress field, which originated in the Late Miocene or subsequently (7–4 Ma), replaced the NE–SW compression that characterized the Early Miocene. On the other hand, throughout the Basin and Range Province, the stress field reorientated from NW–SE to the present-day NE–SW direction during the Late Miocene (Zoback et al., 1981). Finally, the Alpine fault system of New Zealand may also have changed to its present dominantly compressive regime during the Late Miocene (Reading, 1980).

In summarizing these results it is apparent that the global maximum stress orientation appears to have changed drastically during the course of geological history. Compressional pulses may cause the reactivation of many worldwide phenomena. Rapid sea-floor foundering (oceanization) and rapid subsidence of the cratonic basins probably correspond to periods of compressional stress in the lithosphere, as evidenced by major orogenic folding rather than tensional stress. Sea-level changes are also considered as indicators of tectonic compressional events. Thus, there is an apparent synchroneity of ocean tectonic regimes and the subsidence of cratonic basins. The cyclical application of regional stresses decreases the effective viscosity of the lithosphere. This lowering of viscosity is accompanied by an increase in heat flow, especially as viscosity is strongly related to temperature. The presence of high stress in a region of lithosphere associated with relatively high temperatures seems to be the mechanism by which this regime is produced (De Rito et al., 1983). The troughs, which are complementary to the adjacent upwarps, are related to downwarping of the lithosphere under compression.

Correlation between depositional-magnetic events and volcanic-magnetic events

In the Campanian-Maastrichtian interval of the Scaglia Rossa of Gubbio, a striking correlation has been noted between depositional rhythmic sequences and changes in the magnetic properties of rocks (Wezel, 1979). The stratigraphy of magnetic polarity roughly follows vertical fluctuations in the thickness of the rock beds, which in turn are an expression of various phases of the sedimentary cycle. The geomagnetic reversal cycle shows a lower interval of prevalent normal polarity coinciding with thinner and finer grained pink limestone beds (transgressive hemicycle), and an upper one with predominant reversed polarity typical of thicker and coarser white calciturbidites (regressive hemicycle).

The transgressive thin-bedded and finer limestones tend to correlate with normal directions of dip and higher values of intensity of magnetization (Wezel, 1979). On the contrary, the regressive thick-bedded and coarser limestones tend to correspond with negative dip and lower values of intensity.

The intervals of normal polarity also seem to be times of high heat flow, hot-spot volcanism, fast sea-floor spreading and rapid rates of the subsidence of cratonic basins (e.g., Bally, 1980). On the other hand, volcanic quiescence correlates with a dominantly reversed geomagnetic field. Moreover, volcanic and magnetic events correlate with one another: for example, Moberly and Campbell (1984) showed that Hawaiian eruptions were not random with respect to the magnetic field. They noted that most of the igneous activity along the Hawaiian-Emperor Chain occurred during times of normal polarity, even though the Earth's field polarity was reversed during one half or more of that time. They suggested that processes in the core control the magma generation in the mantle.

The concept of helicyclic tectonics indicates an apparent synchroneity between sharply defined depositional, magmatic, and magnetic regimes (Wezel, 1988). In other words, there is a correlation among processes operating respectively in the lithosphere, in the mantle and near the core-mantle boundary. The worldwide reactivation pulses that trigger the movement of lithosphere and the whole mantle up to the core, also should be capable of producing the sea-level changes associated with geotectonic cycles. The central concept is that there has been concomitant variation in a number of parameters. The transgressive time periods are distinguished (Fig. 2) by predominantly normal polarity, diastrophism (tectonic compression, high heat flow, and magmatism), accelerated basin subsidence, syndiastrophic deposition (preflysch, flysch, and postflysch), compressive folding, climatic amelioration, and peaks in the diversity of genera. On the other hand, the regressive periods are characterized by predominantly reversed geomagnetic fields, acid plutonism, epeirogenic (or morphogenic) mountain uplift, erosion, great fluxes in river sedimentation, and decline in global temperature culminating in the ice ages. Similar histories of uplift have been recorded for the Late Cenozoic in regions with quite different plate-tectonic settings. These include: continent-continent collision (Himalayan-Tibetan Plateau), ocean-continent convergence (Andes and the Altiplano), subducted ocean ridge (Colorado Plateau), and early-stage continental rifting (the East African rift valley and plateaus). Clearly, plate-tectonic theory is not a sufficient explanation for the rapid uplift of mountains (Ruddiman and Kutzbach, 1989).

Cyclic contraction and expansion and the state of the geomagnetic field

During the Alpinotype orogenic events, the lithosphere was profoundly shortened and plastically deformed. The contemporaneity of folding and thrusting in widely separated regions hardly could be explained by a local and fortuitous collision, but appears rather as a manifestation of a global and almost synchronous pulse of crustal contraction. The tangential compressive stress determined first of all the thermal remobilization of the mantle causing lithospheric weakening and, successively, a great outpouring of magmatic melts and volatiles.

When the horizontal compression is not yet intense (krikogenic stage), broad tectonospheric undulations are formed, which manifest themselves on the Earth's surface by regional arching. The anticlinal uplift and doming of the lithosphere determines a slow decompression in the sublithospheric mantle with gradual separation of its volatile constituents (bubbles). These percolate through the lithosphere, inducing the generation of acid or calcalkaline magmas (Brunn, 1986). This process occurs in island-arc systems which, in fact, develop during krikogenic stages (Wezel, 1986, 1988).

When compressional stress increases during the geosynclinal stage, strike-slip rifting is generated in the weakened arched zone, with the consequent abrupt decompression in the sublithospheric mantle. This is followed by the separation of volatiles in bubbles, their expansion, and fluidification and rise of tholeiitic-basaltic (ophiolitic) magmas (Brunn, 1986). Mafic volcanism takes place at times of basin geosynclinal subsidence and foundering, giving rise to small ocean basins. According to the concept of helicyclic geotectonic development, the island-arc (or krikogenic) phase precedes the small-ocean-basin (or geosynclinal) phase (Wezel, 1988). These tectonomagmatic processes of softening and remobilization of the lithosphere occur along continental margin zones. They are an essential prerequisite for the generation, during the successive tectogenetic phase, of the Alpino-type compressional deformation in the mobile megashear belts.

The horizontal compressional movements that generate the fold belts are substituted in time by the substantial vertical uplifts of the morphogenetic phase of mountain building (see Gansser, 1982). The morphogenetic event is triggered by substantial acid plutonism and volcanism. According to Gansser (1982), the vertical uplift of the mountain chains produced a relief that exceeded the present height of about 6,000 m in the Himalayas and the Alps.

I believe that the acid plutonism, a forerunner of the later morphogenic phase, is a putative clue to thermal expansion, which introduced heat into the upper mantle and lower crust and thus formed granites. The uplift of such voluminous plutonic melts is incompatible with a compressional stress field and, hence, requires a new state of stress that must allow the rise of immense topographic areas, for example the whole of Tibet, which measures at least 2,500,000 km² (Gansser, 1982). This represents a worldwide pulse of expansion that follows on tangential contractional stress acting during folding and shortening (Wezel, 1988).

The shifts from contraction to expansion and vice-versa reflect fundamental turning points in geotectonic history (Fig. 2). The turning points were times of crustal, climatic, and magnetic instability that may have triggered major biotic crises. They are strikingly indicated in the curve of the percentage of reversed polarity through Phanerozoic time (Fig. 3). It appears from this curve that time intervals characterized by more than 50% of normal polarity (and thus by lithospheric contraction) are the Silurian, Early Devonian, and the Mesozoic. In contrast, the Cambro-Ordovician, Permo-Carboniferous, and Paleogene are distinguished by reversed geomagnetic fields (and by lithospheric expansion). I call the former "N-periods" and the latter "R-periods."

There is also a fundamental difference in the various orogenic belts according to whether they are formed in the R or N periods. The Variscan is distinguished from the Mesozoic Alpine fold belt by the scarcity of true ophiolite suites (ultrabasites are practically absent) and the abundance of intrusive granites and migmatites (Krebs, 1976). The Variscan orogeny of Europe, for example, is mainly to be considered a widespread thermal event, which introduced heat into the upper mantle and lower crust and thus formed granites (Zwart, 1976). Major crustal shortening is a characteristic instead of the Alpine diastrophism. From this point of view, we may

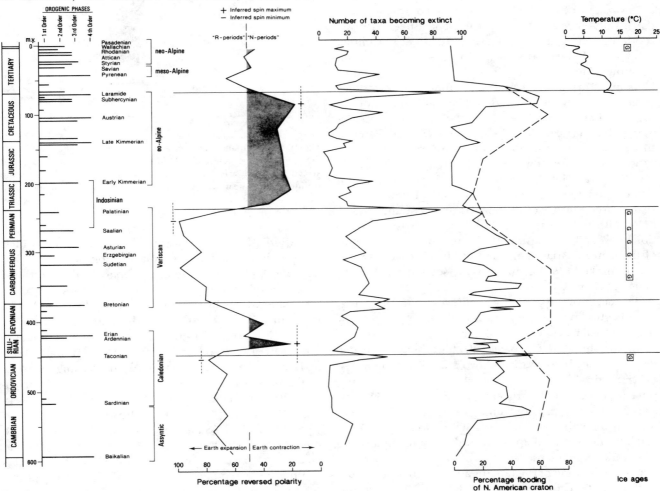

Figure 3. Some first-order global patterns of Phanerozoic history (modified from Whyte, 1977). The diagram indicates: (1) the position and intensity of compressional pulses (orogenic phases based on Schwan, 1980); (2) the percentage of reversed geomagnetic polarity; (3) the number of higher taxa af all biological groups becoming extinct; (4) the percentage flooding of the North American craton (the dashed curve is a period-by-period smoothed curve); (5) the temperature of high latitude marine waters; and (6) the occurrence of ice ages.

Note that the N-periods (in grey) are characterized by spin maximum, crustal contraction, break-up phases, and Alpino-type basement shortenings. Instead, the R-periods are distinguished by spin minimum, thermal events, uplift of voluminous plutonic melts, and gravitational spreading. The seven main biotic crises (or mass extinctions) occur at the turning points.

perhaps distinguish the Variscan Period of thermal mobilization of the mantle (migmatization and granitization) and associated vertical tectonics, from the Mesozoic Alpine Period, in which large parts of the Earth's lithosphere were subjected to tangential contraction. The passage between one regime principally characterized by processes of vertical ascending mantle diapirs and another regime mainly distinguished by horizontal compressional movements occurs at a fundamental first-order turning point, which is situated close to the Permo-Triassic boundary (Fig. 3). The major biotic crisis of the whole Phanerozoic history (i.e., the Late Permian mass extinction) is situated in close correspondence with this boundary (Fig. 3). The Permo-Triassic turning point might separate an Upper Paleozoic era characterized by pulsatory expansion from a Mesozoic one marked by pulsatory contraction. In correspondence with this, there must have been a drastic reorientation of the regional stress field. For example, the United States was subjected to N–S regional compression in Mesozoic time (Upper Jurassic and Cretaceous; Haman, 1975) after a drastic reorientation of stress that occurred during Pennsylvanian-Permian time. Following the Late Carboniferous-Early Permian suturing along the Variscan fold belt, which created Pangea, there was a fundamental geotectonic reorganization, caused by an important change in the direction of the regional stress field (Ziegler, 1988). This reorientation determined the Mesozoic compressional deformation that caused the reactivation of important megashear belts, with the consequent Mid-Jurassic crustal separation between Laurasia and Africa, and the rapid opening of the Central Atlantic in the Late Jurassic-Cretaceous, which marked the beginning of the second break-up of Pangea (Ziegler, 1988).

The curve of the percentage of reversed polarity (Fig. 3) indicates, in fact, the presence of two large cycles of the first order occurring respectively in the Siluro-Permian and the Meso-Cenozoic. Each of these first-order geomagnetic cycles consists of two pulsating hemicycles: (1) an initial cycle characterized by mainly normal polarity, continental disruption, oceanization, compressional suturing, and shortening; and (2) a final cycle marked by mainly reverse polarity, cratonization, consolidation, mantle diapirism, and crustal expansion. The initial N-hemicycle consists of break-up phases that led to the opening of the Iapetus and the Atlantic oceans and to Late Caledonian and Eo-Alpine orogenic diastrophism. The Mesozoic diastrophic period was characterized by two principal compressional pulses: the Mid-Permian to Triassic (Indosinian) folding phase, and

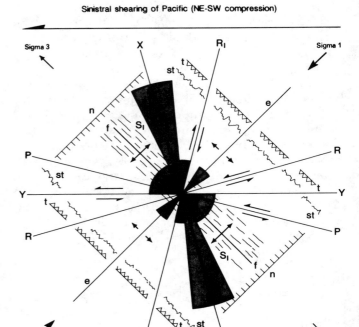

Figure 4. Orientation of the major bathymetric ridges and linear archipelagoes of the Pacific Basin. In the context of en echelon structures, which are characteristic of strike-slip fault zones evolving during simple shear (see Hancock, 1985), these correspond to X-shears, which are held by some workers to be rotated antithetic Riedel shears (or R1-shears).

the Middle-Upper Cretaceous Eo-Alpine phase, which was very intense. The first compression peaked during the Late Ladinian in the eastern part of the Southern Alps (Castellarin et al., 1988) and the folding that it created is known by various names (see Dickins, 1987): Indosinian in southeast and southern Asia; Gondwanide in South Africa, Antarctica, and South America; Sonoman in western North America; and Hunter-Bowen in Eastern Australia. The Eo-Alpine folding phase is distinguished by very intense compressional deformation with emplacement of the Austroalpine nappes, very high-P metamorphism, and lithospheric subduction (Trumpy, 1980). As has been noted, the passage to the successive R-hemicycle takes place through a fundamental change, in which regional compression ceases and is substituted by regional extension. These turning points (comprising a geotectonic reorganization) occurred during the Early Devonian and in correspondence with the K-T boundary.

A NEW MEGANEOTECTONIC PICTURE

The tectonic fabric of the ocean basins

Examination of the morphostructures of the ocean basins indicates the presence of a series of long, linear ridges and troughs, orientated obliquely with respect to the midocean ridge and fracture zones. In the Pacific Ocean, the linear ridges are represented bathymetrically by seamount chains, island archipelagoes, and islands that trend approximately northwest (Fig. 4). These long, positive features constitute a set of subparallel trends that tend to occur in the form of "elbows" (e.g., between the Emperor Seamounts and Hawaiian Ridge) or show variation in the angle at which they diverge from the parallels (e.g., between the Line Islands Ridge and Tuamotu Archipelago). The ridges are crossed approximately at right angles by northeast-trending linear troughs, particularly in the Line Islands area.

These elongated, narrow bathymetric features reveal an en echelon pattern that is present at all scales and seems to be characteristic of sea-floor topography. Multibeam bathymetric data also confirm the existence of a series of oblique lineations arranged according to an en echelon pattern (see Winterer and Sandwell, 1987), while gigantic sigmoidal patterns are also discernable (e.g., in the Marshall and Gilbert islands, at the Lord Howe Rise and on the New Caledonia-Norfolk Ridge).

A pervasive oblique tectonic fabric is thus found universally in all oceanic crust (Searle et al., 1981; Searle, 1983). It consists of multiple, closely spaced fault scarps that strike parallel to each other and are easily detectable in long-range sidescan sonographs (such as GLORIA). Sometimes these oblique tectonic lineaments display a sigmoidal grain.

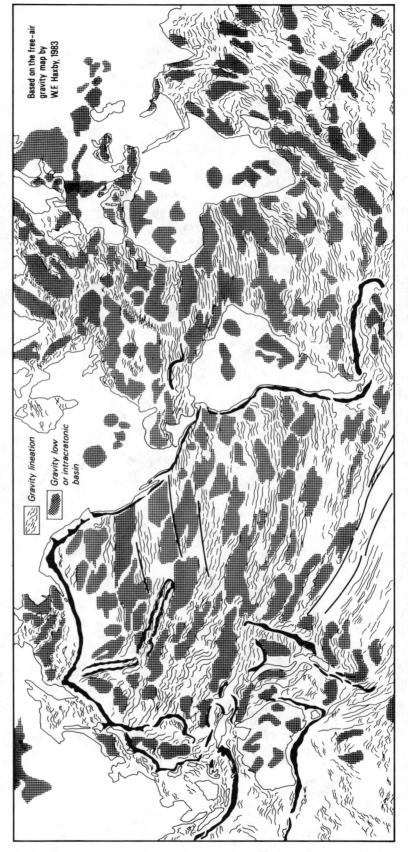

Figure 5. Sketch map of linear gravity anomalies across the world's ocean floors, based on Seasat altimeter data (after Haxby, 1983). Note the en echelon patterns formed by the swarms of gravity lineations, some of which have a sigmoidal shape.

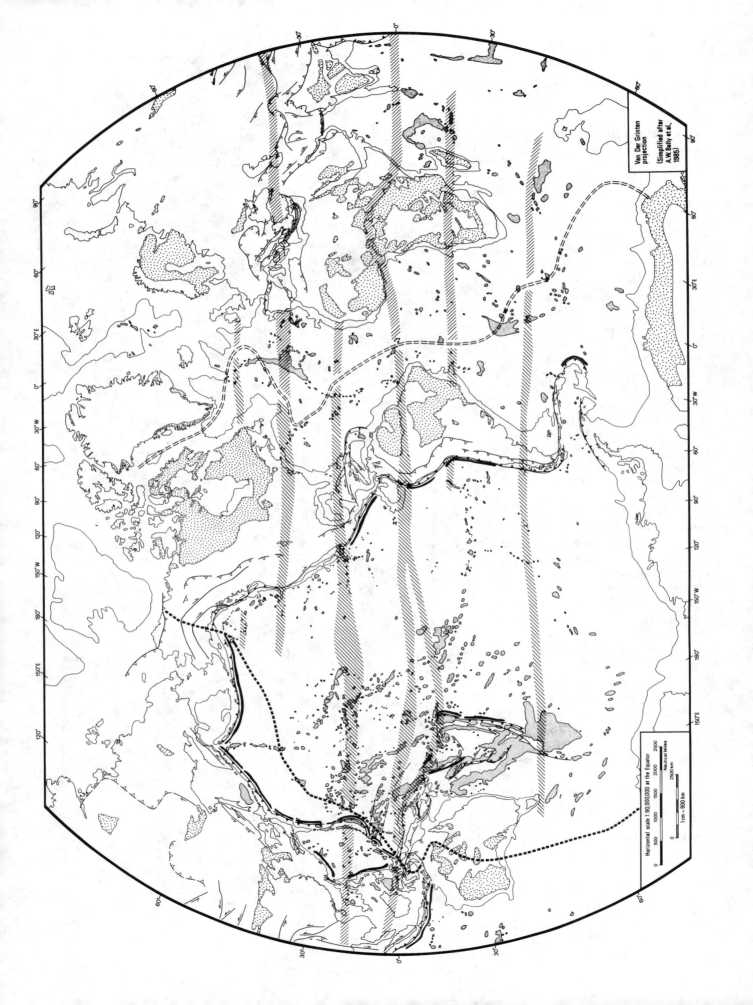

Also in the Philippine Sea, there are chains of small dome-shaped volcanoes that are orientated N45° W to N14° W approximately parallel to the direction of plate motion (Hollister et al., 1978). Their morphology suggests that they were formed recently.

The northwest-striking Pacific-type linear ridges continue also in the western part of the Atlantic Ocean (e.g., the New England Seamounts and the Rio Grande Rise). Their orientation changes completely in the eastern part of the Atlantic Basin and correspondingly in the Indian Ocean. Here, the en echelon ridges and lines of seamounts strike approximately northeast-southwest (e.g., the Reykjanes Ridge, Cameroon volcanic line, Walvis Ridge, and Southwest Indian Ridge). A spectacular en echelon pattern, composed of a series of northeast-trending segments, can be seen in both the Central Indian Ridge and the northern half of the Ninetyeast Ridge. Simple observation of a bathymetric map of the oceans shows that west of the Mid-Atlantic Ridge there is a boundary between Pacific-type lineations and those of the Atlantic-Indian type.

Many of these linear ridges and archipelagoes are volcanic island and seamount groups that are characterized by shifting volcanic activity. The linearity of these features and the presence of volcanism both indicate that the ridges correspond to major underlying fault zones that connect with the mantle.

The progressive migration of the center of volcanism along each lineation seems compatible with the interpretation of such phenom-ena as active strike-slip fault zones. This hypothesis is supported by en echelon and sigmoidal patterns of linear volcanic ridges. Where present, voluminous igneous activity should indicate periods or zones of transtension. With respect to the Pacific-type lineations, we may hypothesize that they are dextral NW–SE trending shears of regional scope (Fig. 4). The linear ridges are oblique with respect to both magnetic anomalies and fracture zones and are often found at an angle of 45° to both of these. In the case of northwest-striking seamount chains in the northeast Pacific Ocean, the linear magnetic anomalies cross the volcanic ridges with some embayments and distorsion of their trends (Barr, 1974).

Haxby's Seasat map of free-air gravity anomalies for the ocean basins (Francheteau, 1983; Haxby and Weissel, 1986) is particularly instructive regarding the analysis of tectonic deformations (Fig. 5). It shows that lineated gravity anomaly patterns occur throughout the world ocean floor. In the Pacific Ocean there are gravity lineations of small amplitude, which have wave-lengths of 150–500 km and have a trend that is clearly oblique to that of the major fracture zones and essentially parallel to the vector of absolute plate motion (i.e., N77° W). In the eastern Atlantic and Indian oceans the gravity undulations observed by Seasat show, in contrast, an approximately northeast orientation. These gravity lineations cross the East Pacific Rise, the Atlantic Ridge, and the Southeast Indian Ridge, upon which they are superimposed, indicating that they are neotectonic structures that postdate the most recent inversion of the Earth's magnetic field (i.e., they are of Middle to Upper Quaternary age).

In the Seasat image, the East Pacific Rise, the South Atlantic Ridge, the Southeast Indian Ridge, and the Pacific-Antarctic Rise are only just visible. Only the North Atlantic Ridge, the Reykjanes Ridge, and the Southwest Indian Ridge are clearly identifiable.

Master megashear belts following the latitudinal parallels

The oblique ridges on the ocean floor are interrupted and separated by immense bathymetric and gravitational depressions that follow an east-west trend, approximately in the direction of the latitudinal parallels. They have the form of corridors that widen and narrow from place to place and that at some points seem to consist of a series of depocentres, or megalenses. These extend across the ocean floors and pass through the continents in correspondence with intracratonic basins, saddles or other features that represent structural lows (Figs. 5 and 6).

Figure 6. Simplified map illustrating the present tectonic state of the Earth. The lithosphere is made up of east-west megaslices, en echelon, separated from each other by highly disturbed megashear belts (indicated by oblique lines). Inside each megaslice, powerful large-scale shearing determines: diffuse broad deformation and distortion, rapid tectonic rotation and vortices produced in the mantle, general weakening of the lithosphere and a continuously changing topography (inversion tectonics). The Labrador-Atlantic zone of oblique divergence and the East Asiatic zone of oblique convergence of the compressional stress are also outlined. They separate the Amero-Pacific (New World) first-order stress hemisphere from the Eurasian (Old World) one, which are situated at each other's antipode.

The corridor of the 40° N parallel is shown particularly clearly on the Seasat geotectonic image, in which it appears as a clean "razor slash" across the Earth's crust. The image also gives a good representation of the equatorial depression that passes across the intracratonic basin of the Amazon (the Amazonas Trough) and the Lake Turkana saddle (the equatorial break); that is, across major zones of weakness of pre-Mesozoic age.

These east-west structural troughs correspond to some of the major oceanic fracture zones of the mid-ocean ridges (e.g., the Charlie-Gibbs, Mendocino, Clarion, Kane, Easter, and Romanche fracture zones and the Romanche oceanic trench), where the axis of the ridge is composed of segments that have various orientations with respect to north, all of which are, however, oblique in relation to the corridors. Where they appear in the oceanic ridges, the troughs correspond to transform faults, whereas in the interior of the continents they follow old fault zones (some of which are of Precambrian age) that subsequently have been reactivated. For instance, the transcurrent faults of the South American equatorial zone extend from the Carnegie Ridge in the Pacific to the Romanche fracture zone in the Atlantic (De Loczy, 1974). In my opinion, they are genetically related and are of the same age. Because some of these transcurrent faults were active during the Precambrian or Early Paleozoic, it follows that the Mid-Atlantic Ridge has existed since Precambrian time. The corridors also separate various sedimentary basins that occur along the passive continental margins of America and Africa. Where they continue inland they are associated with alkaline magmatism, ring dikes, diatremes, and kimberlites, all of which are indicative of deep-seated sources of magma (Sykes, 1978).

The data taken as a whole suggest an interpretation of the east-west orientated rectilinear troughs as colossal and highly disturbed shear zones separating different intercontinental megaslices that are arranged en echelon. In this way, the surface of the Earth appears to be broken along the latitudinal lines into large, parallel segments, as a result of powerful, zonal megashears that are still active and are located in preexisting zones of weakness. Within each megaslice an intense tectonic deformation takes place that is powerful enough to produce distortions in the form of the continental margins. This is demonstrated, for example, by the parallel sigmoidal shape taken in eastern Australia by the Great Dividing Range, Lord Howe Rise, and New Caledonia-Norfolk Ridge, all of which occur within the Australian megaslice.

Large-scale shearing thus determines torsion, tectonic block rotation, fault rotation, vorticelike structures (Figs. 6 and 9), rhomboidal features, and elongation, comprising a diffuse and broadly distributed system of deformation that consists of poorly organized extension and compression and a continually changing geom-etry that inverts structures. In the opinion of Mann et al. (1983), oblique spindle-shaped features (tension fractures), S-shaped (sigmoidal) features (situated between sinistral faults) and Z-shaped features (occurring between dextral faults) all form during the early stages of strike-slip motion. During the middle stages, increased strike-slip offset produces rhomboidal features (rhomb grabens or rhomb horsts). The final development is that of long troughs, such as the active pull-apart feature represented by the Cayman Trough. The surface of the Earth shows many examples of sigmoidal structures of a large scale, including the Marshall-Gilbert island chain and the Lord Howe Rise. It also bears Z-shaped features like the Niksar Basin that occurs along the North Anatolian dextral fault zone in Turkey, and rhomb-shaped blocks such as the Manihiki Plateau and the mountain ranges situated between Peking and Sian (Siking).

The voluminous alkaline magmatism associated with megashears indicates that they are deep fractures that penetrate the entire lithosphere to tap asthenospheric sources of magma. Such translithospheric shear belts appear to guide intraoceanic and intracontinental movements of the crust.

The northwest-trending parallel linear ridges that characterize the Pacific region mark dextral strike-slip faults within megaslices that are bounded by giant east-west sinistral megashears (Fig. 6). In contrast, the northeast-striking ridges of the Atlantic-Indian Ocean region are sinistral transcurrent faults that are separated from neighboring belts by dextral megashears orientated east-west. In this interpretation the linear bathymetric ridges are considered to be pressure ridges or push-up blocks, produced by a transpressive regime.

The East Asiatic zone of convergence

Comparison of morphostructural data and gravity lineations detected by Seasat suggests that there is a zone of oblique convergence of the maximum compressional stress. This zone is located along an alignment that runs through the following areas (Fig. 6): Beaufort Sea, Gulf of Alaska, Chinook fracture zone, east of the Shatsky Plateau, Mariana Trench, Gulf of Carpentaria (probably), Great Bight of South Australia, Australia-Antarctic Basin, and Wilkesland. To the west of this belt are all of the back-arc marginal basins that occur on the periphery of Asia and also of the Westralian Superbasin (Teichert's "Westralian Geosyncline"). According to Toksoz and Bird (1977) the marginal basins correspond to a Benioff zone that is 260–680 km deep. However, the back-arc basins located on the periphery of eastern Australia in the region of the Melanesian re-entrant do not belong to this oblique convergence belt (but perhaps to another incipient zone extending from Hawaii to the Samoa Island and Kermadec-Tonga Trench).

The in situ tectonic stress data (derived from break-outs, earthquake focal solutions, and local stress measurements) indicate a stress region whose boundaries correspond to those of the convergence zone (Zoback, 1987 and Zoback et al., 1989). To the west of this zone the maximum compressive stress is orientated approximately northwest-southeast (in the Aleutian arc and Alaska, China, South China Sea, Indian Ocean, and western Australian shield), while the Pacific region, which is located to the east, undergoes compressive stress in a NE-SW direction.

In general terms, the Afro-Eurasian and Amero-Pacific stress regions come into contact along the East Asian convergence zone, and both are characterized by first-order tectonic stress of a remarkably uniform orientation. The greatest oblique compression (or transpression) occurs along the point of contact, where it causes intense tectonomagmatic activity and the formation of convergent shear basins. This is also the location of the only geotectonic features that penetrate into the mantle to any great depth (i.e., the deep-focus Benioff seismic zones and their aseismic extensions that penetrate the lower mantle to depths of more than 1,000 km, according to Creager and Jordan, 1986).

The East Asiatic convergent compressive zone corresponds approximately to the western boundary of a large geoid high that is centered on New Guinea (Fig. 7 bottom). The latter correlates with an anom-alous reduction in P-wave velocity that occurs in the depth range of 1,100-

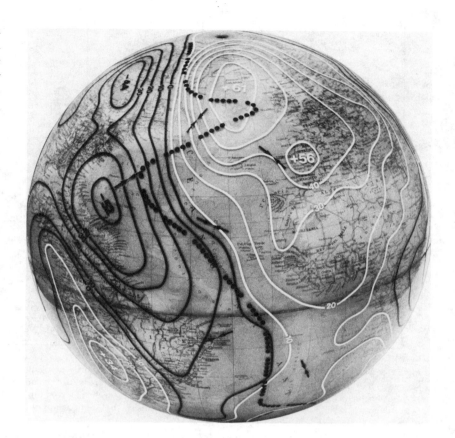

Figure 7. Photo of the globe illustrating the relationship between features of geoid undulations, comprising the Labrador-Atlantic (Top) and Australasian (Bottom) zones of divergence and convergence. The geoid highs are considered indicative of diapiric upwelling of hot mantle material.

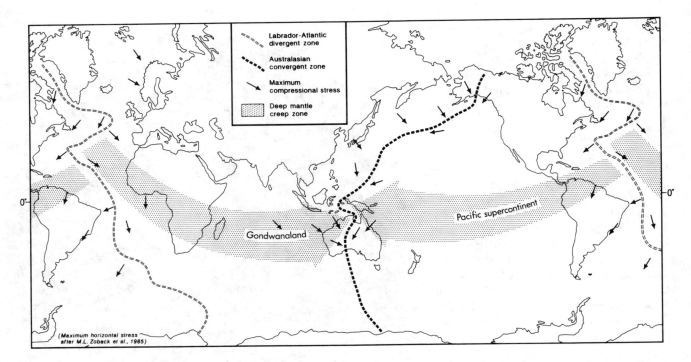

Figure 8. Schematic map showing the contemporary compressive stress orientation along the divergence and convergence zones (after Zoback et al., 1989). It is postulated that the present-day patterns of tectonic stress are determined by the double migration, beginning with the Atlantic region, of zones of the deep mantle (lower-mantle convection) towards the East Asiatic zone of convergence. During the course of their movement, beginning during the Jurassic, break-up pulses occurred, causing the oceanization of both the Pacific and Gondwana supercontinents.

1,500 km and is indicative of less dense, relatively hot material in the lower mantle (Nishimura, 1984; Pavoni, 1985). More-over, the convergence zone corresponds to the steepest drop in the geoid surface (Nishimura, 1984), which suggests that stress near the surface results from creep-flow processes occurring deep within the mantle. We may hypothesize that in the East Asiatic convergence zone the oblique compression has caused enormous diapiric upwellings in the hot mantle which, presumably, have given rise to the marginal basins (Miyashiro, 1986; Wezel, 1986). Their northward migration should indicate the existence of a transcurrent component along the convergence zone, which on the largest scale should represent an area dominated by dextral shear.

Finally, it should be emphasized that the magnetic bights of the Gulf of Alaska and the Shatsky Plateau have been produced by the proximity of two different stress regions; in fact, the line of convergence passes between them. This explains, among other things, the curvature of the Denali and Lake Clark dextral strike-slip fault systems in Alaska.

The Western Atlantic zone of divergence

Morphostructural analysis and the Seasat gravity lineations permit us to trace in the western Atlantic Ocean a line of separation between the compressive stresses. This line passes approximately through Sverdrup Basin, Baffin Bay, Davis Strait, Sea of Labrador, North American Basin, Brasil Basin, Discovery Peak, Meteor Peak, Ob Guyot, Lena Guyot, Africa-Antarctica Basin, Enderby Abyssal Plain and MacRobertson Land (Fig. 6). It separates the Amero-Pacific (New World) stress hemisphere, which is characterized by northeast-orientated horizontal compression, from the Eurasian (Old World) stress hemisphere, which undergoes northwest-trending compressive stress.

The Labrador Sea has been interpreted as the product of a sea-floor spreading process that is supposed to have commenced in Campanian time and to have ended in the Early Oligocene (see Grant, 1980). However, integrated interpretation of the multichannel reflection seismic surveys reveals that the continental crust extends into the so-called oceanic region (Grant, 1980, 1982). Moreover, in the opinion of Grant, there is sufficient evidence to argue for an increase in the rate of subsidence after the Upper Miocene. The preferred overall interpretation is one of recurrent crustal movements that consisted of uplift and foundering, and were accompanied by mafic volcanism and intrusion.

The sedimentary basins on the continental margin east of Newfoundland have also been interpreted as "formed primarily through recurring vertical displacements (inversion tectonics) rather than by crustal thin-

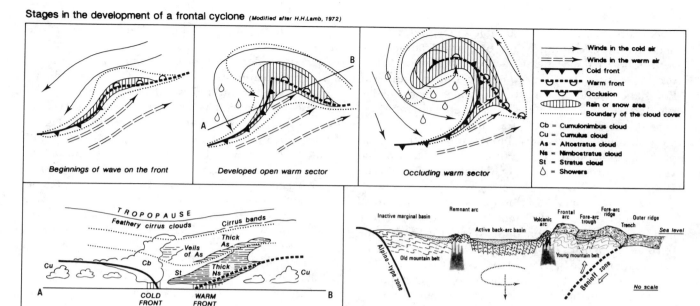

Figure 9. A striking similarity exists between some tectonic structures and features generated by atmospheric circulation. For instance, the cross section through a mature cyclone (bottom left) recalls the vertical section through a back-arc marginal basin (bottom right).

ning due to extension" (Grant, 1987). The structural process of change in which a basin becomes a structural high (eversion) or a high becomes a basin (inversion) is probably characteristic of oblique-slip shear zones. Transtension leads to the formation of subsiding basins, and transpression generates the uplift of blocks (Reading, 1980). Structures become inverted because of temporal or spatial changes in curvature or in the direction of motion.

The divergent stress zone follows the Midocean Canyon and proceeds along the lineament that connects the Corner Seamounts with the Nares Abyssal Plain. To the east it delimits the North American abyssal plains and rise. Comprehensive study of the upper rise documented the presence from the Campanian to the Pleistocene of 12 depositional sequences, each of which is bounded by erosional unconformities (Poag and Mountain, 1987). These events were sometimes accompanied by regional truncation that can be interpreted as the result of structural inversion and also may have occurred during the Upper Oligocene and Upper Miocene.

The East American divergent zone proceeds southwards as the eastern limit of the Demerara and Ceara abyssal plains and eventually reaches the Romanche fracture zone. From this point it runs to the east of the Rio Grande Rise (the Hunter Channel), beyond which it continues via Tristan Da Cunha Island, Gough Island, Meteor Seamount, and Lena Guyot to MacRobertson Land in Antarctica. Focal mechanisms also indicate an American (northeast-trending) orientation of compressive stresses in the vicinity of South Georgia Island and at the Atlantic Ridge near 50° S (see Zoback, 1987). As far as the Rio Grande Rise, this line of demarcation between different compressive stress systems also corresponds to the boundary between the geoid depression to the west and the geoid high to the east (Fig. 7 top).

CONCLUSIONS

One may speculate that over the course of time there has been a double migration of the lower mantle creep zone from the area of divergence in the Central Atlantic towards the East Asian convergence zone (Fig. 8). It is possible that this migration resulted from global reorientation (wandering) of the whole mantle relative to the Earth's spin axis (Anderson, 1984). The axis of rotation has apparently wandered about 8° in the past 60 million years and 20° in the last 200 million years (see Anderson, 1984). The large-scale tectonic stresses generated in the outer layers of the mantle and crust by the migration of the rotational bulge, as it moved towards the equator, caused rifting, fragmentation, and oceanization of both the Pacifica and Gondwana continental masses (Fig. 8). This migrating hot region may explain the correlation of tectonic activity (and polar wandering) with magnetic reversals. The mantle motion presumably reflects convection in the fluid core, offering a link between tectonic and magnetic field variations. Smooth core convection should produce few reversals, whereas most turbulent core convection should cause many magnetic reversals (see Sheridan, 1987).

I believe that the inferred double zonal motion in the deep mantle, extending from the Central Atlantic divergent zone to the western Pacific convergent region, produced both the Amero-Pacific sinistral and the Eurasiatic dextral shearing, both of which are parallel to small circles (and can be considered geolatitudinal megashears along ancient fractures). This structural pattern, which is symmetrical to the rotational axis, was probably generated by alteration of the shape of the Earth produced by cyclic rotational changes (Wezel, 1988).

The central concept of this analysis is that the global variation in sea level, climate, and state of the geomagnetic field seems to be concomitant with the Earth's rate of rotation (Whyte, 1977; Wezel, 1988). The rotational acceleration in the first half of the geotectonic cycle (Fig. 2) is accompanied by a generally rising sea level, phases of climatic amelioration, and predominantly normal polarity. The following regressive hemicycle is, instead, distinguished by an overall fall in sea level, a reversed geomagnetic field, and by climatic deterioration that culminates in ice ages. The major biotic crises (mass extinctions) occur at the turning points (Figs. 2 and 3) and may have been triggered by climatic and magnetic instabilities (Whyte, 1977).

The evidence suggests that past and present-day tectonic regimes appear to be dominated by a colossal strike-slip deformation occurring along vertical shear zones, which are very ancient and are periodically reactivated. Within these mobile shear belts, tectonic deformation was able to produce crustal fragmentation and rapid rotations, tectonothermal and mafic activity, extension and foundering, compression and uplift, wavelike migration of subsidence and uplift, inversion tectonics, and a combination of vertical and lateral movements.

A striking similarity exists between some shear-dominated crustal structures and features produced by the circulation of the atmosphere. For instance, the cross section through a cold front and a warm front in a mature cyclone recalls the vertical section through a back-arc marginal basin delimited by the Alpine-type and the Benioff Zones (Fig. 9). The cyclone also originates from shearing motions (between cold and warm air).

ADDENDA

Conventional terminology in the context of the new geotectonic concept

In order not to overburden the text by introducing new terms, I have used the conventional plate-tectonic terminology. However, to avoid confusion it will be necessary clearly to define the meaning given to each term in the context of the new conceptual model proposed.

Because a descent (underthrusting) and consumption of the oceanic lithosphere is held to be improbable, the word "subduction" refers to the large-scale rotational collapse—very much like a "mega-vortex" or a "cyclonic swirl"—of the entire region inside the island arc (Fig. 9). To describe this phenomenon of deep spinning engulfment, perhaps it would be better to use the term "sub-suction" (down-sucking). It is held that the formation of small ocean basins occurs following the postorogenic regional foundering of a block-faulted truncated mountain system, resulting from a large shear couple (Wezel, 1986). The inclined seismic surface (Benioff Zone) is postulated to be the boundary of a density contrast between the heated (thermally weakened), deep, funnel-shaped "root zone" beneath the small ocean basin, and the bordering cold oceanic lithosphere. In my opinion, there is neither descending oceanic plate nor accretion. The "accretionary prism" is, in general, nothing other than a drowned pre-Miocene mountain belt due to the young collapse tectonics (Wezel, 1986).

The obliquity and high rate of convergence are the cause, along the Pacific margins (Fig. 8), of broad, deep zones of intense tectonic deformation of the lithosphere, accompanied by anomalously high heat flow and magmatic activity (e.g., the large volumes of granite intrusions). Within these mobile megashear belts, zones of subsidence and collapse (back-arc basins) alternate with zones of rapid vertical uplift (volcanic arcs) because both compressional and extensional features may be present. In this shear-dominated system of deformation, the disruption of the Pacific continental margins into a variety of basins and uplifts occurs.

It is further held that ophiolite complexes were issued from small ocean basins, and thus are back-arc basaltic blankets. In fact, not only does the time of their fomation correspond to systaltic pulses of Earth contraction, but also the ophiolite terrains of mountain ranges are usually located on the concave side of the folded arcs. The context of a converging margin, therefore, seems more realistic than their being generated along a midocean-ridge spreading center.

In the pulsatory contraction-expansion model proposed here, the term "opening" of an ocean basin indicates a phase of crustal foundering of the continental basement and accelerated basin subsidence, simultaneous with regional compressive stress. The basin foundering is preceded by a krikogenic phase of crustal fracturing and outpouring of flood basalts (flows and mafic dikes and sills). The term "sea-floor spreading" here refers to phases of lateral expansion of the network of mafic intrusion and basalt flows ("oceanic basement" or "layer 2") with consequent broadening of basin limits. In other words, this means the broadening of the foundered, mafically-intruded old continental crust (behind a mountain arc). Mafic volcanism and intrusion occurred episodically during global pulses. In my conception, rapid sea-floor spreading corresponds to periods of horizontal compressive stress in the lithosphere, causing back-arc outpouring of flood basalts.

The model interprets the modern "midocean ridge", not as a unique quasi-continuous feature, but rather as a series of discrete rectilinear segments separated by E–W striking megashear couples (Fig. 6). The crustal rift is a broad zone of vertical shear interposed between two latitudinal megashears. It roughly represents a gigantic Riedel or conjugate anti-Riedel strike-slip fault zone evolving during simple shear.

In some regions of the South Atlantic (Jordi and Lehner, 1973) and the North Atlantic (Mutter and Buhl, 1989) within the oceanic crust, truncated older beds, which are tectonically disturbed, have been observed on reflection profiles. These inclined structures that cross the entire crustal section are not very different from the sequences of thrusts observed in the "accretionary wedges" (here interpreted as ancient mountain belts that have recently foundered). Such structures indicate to me once again the existence of foundered remnants of ancient continents ("oceanic layer 3"), and not a new generation of oceanic crust at spreading centers. In order to verify this hypothesis, the drilling of a deep hole through the oceanic basement and crust should be a primary scientific objective.

Finally, the so-called "hot-spot" traces seem to represent very long linear shear belts. These belts comprise some lines of volcanoes with en echelon arrangement (e.g., the Hawaiian Chain; Miyashiro, 1986).

These brief notes are intended to emphasize the importance of the nomenclature used in describing our conceptual models. The difficulties and impasses in the plate-tectonic theory are reflected also in the conventional terminology currently in use, which slows down the formulation of new concepts. Scientific uniformity, fixed ideas, rigid patterns, and fashionable solutions are not the best way to conduct an integrated study of the Earth system on a global scale (i.e., considering large spatial and long time scales). Only a more careful attitude, which is flexible and open to new orientations in the geosciences, will enable us to make some progress towards an understanding (factual, and not ideological) of our evolving planetary ecosystem as a whole.

REFERENCES CITED

Anderson, D. L., 1984, The Earth as a planet: paradigms and paradoxes: Science, v. 223, p. 347-355.

Bally, A. W., 1980, Basins and subsidence—A summary, in Bally, A. W., Bender, P. L., McGetchin, T. R., and Walcott, R. I., eds., Dynamics of Plate Interiors: American Geophysical Union, p. 5-20.

Bally, A. W., Catalano, R., and Oldow, J., 1985, Elementi di tettonica regionale, Pitagora Editrice, Bologna, 276 p.

Barr, S. M., 1974, Seamount chains formed near the crest of Juan de Fuca Ridge, northeast Pacific Ocean: Marine Geology, v.17, p. 1-19.

Beck, R. H., and Lehner, P., 1974, Oceans, new frontier in exploration: American Association Petroleum Geologists Bulletin, v. 58 (3), p. 376-395.

Bergerat, F., 1987, Stress fields in the European platform at the time of Africa-Eurasia collision: Tectonics, v. 6, (2), p. 99-132.

Brunn, J. H., 1986, The sublithospheric mantle, the generation of magmas and tectonics: Revue de Geologie Dynamique et de Geographie Physique, v. 27 (3-4), p. 149-161.

Castellarin, A., Lucchini, F., Rossi, P. L., Selli, L., and Simboli, G., 1988, The Middle Triassic magmatic-tectonic arc development in the Southern Alps, in Wezel, F. -C., ed., The Origin and Evolution of Arcs: Tectonophysics, v. 146, p. 79-89.

Creager, K. C., and Jordan, T. H., 1986, Slab penetration into the lower mantle beneath the Mariana and other island arcs of the Northwest Pacific: Journal of Geophysical Research, v. 91, p. 3573-3589.

De Graciansky, P. C., Brosse, E., Deroo, G., Herbin, J.-P., Montadert, L., Muller, C., Sigal, J., and Schaaf, A., 1987, Organic-rich sediments and paleoenvironmental reconstructions of the Cretaceous North Atlantic, in Brooks, J., and Fleet, A.J., eds., Marine Petroleum Source Rocks: Geological Society Special Publication No. 26, p. 317-344.

De Loczy, L., 1974, Synchronous diastrophic events in South America and Africa and their relation to phases of seafloor spreading, in Kahle, C.F., ed., Plate Tectonics—Assessments and Reassessments: American Association of Petroleum Geologists, Memoir 23, p. 246-254.

DeRito, R. F., Cozzarelli, F.A., and Hodge, D.S., 1983, Mechanism of subsidence of ancient cratonic rift basins, in Morgan, P., and Baker, B.H., eds., Processes of Continental Rifting: Tectonophysics, v. 94, p. 141-168.

Dickins, J. M., 1987, Major sea level changes, tectonism, and extinctions: Compte Rendu, 11th International Carboniferous Congress, manuscript of 14 p.

Francheteau, J., 1983, The oceanic crust: Scientific American, v. 249 (3), p. 114-129.

Gansser, A., 1982, The morphogenic phase of mountain building, in Hsu, K. J., ed., Mountain Building Processes: Academic Press, London, p. 221-228.

Grant, A. C., 1980, Problems with plate tectonics: the Labrador Sea: Bulletin Canadian Petroleum Geologists, v. 28, p. 252-278.

Grant, A. C., 1982, Problems with plate tectonic models for Baffin Bay-Nares Strait: evidence from the Labrador Sea, in Dawes, P.R., and Kerr, J. W., eds., Nares Strait and the Drift of Greenland: a Conflict in Plate Tectonics: Meddelelser om Gronland, Geoscience, v. 8, p. 313-326.

Grant, A. C., 1987, Inversion tectonics on the continental margin east of Newfoundland: Geology, v. 15, p. 845-848.

Hallam, A., Hancock, J. M., LaBrecque, J. L., Lowrie, W., and Channell, J. E. T., 1985, Jurassic to Paleogene: Part 1—Jurassic and Cretaceous geochronology and Jurassic to Paleogene magnetostratigraphy, in Snelling, N.J., ed., The Chronology of the Geological Record: Geological Society Memoir 10, p. 118-140.

Haman, P. J., 1975, Possible relationships between lineament tectonics and the dynamics of the Milky Way Galaxy, in Hodgson, R. A.,

Gray, S. P., and Benjamins, J. Y., eds., First International Conference on The New Basement Tectonics: Utah Geological Association, Publication No. 5, p. 528-536.

Hancock, P. L., 1985, Brittle microtectonics: principles and practice: Journal of Structural Geology, v. 7, p. 437-457.

Hast, N., 1969, The state of stress in the upper part of the earth's crust: Tectonophysics, v. 8, p. 169-211.

Haxby, W. F., 1983, Geotectonic imagery: a new generation of plate kinematic discoveries: Lamont Newsletter, p. 6-7.

Haxby, W. F., and Weissel, J. K., 1986, Evidence for small-scale mantle convection from Seasat altimeter data: Journal of Geophysical Research, v. 91, (B3), p. 3507-3520.

Hollister, C. D., Glenn, M. F., and Lonsdale, P. F., 1978, Morphology of seamounts in the western Pacific and Philippine Basin from multi-beam sonar data: Earth and Planetary Science Letters, v. 41, p. 405-418.

Jansa, L. F., Enos, P., Tucholke, B. E., Gradstein, F. M., and Sheridan, R. E., 1979, Mesozoic-Cenozoic sedimentary formations of the North-American Basin, western North Atlantic, in Talwani, M., Hay, W., and Ryan, W. B. F., eds., Deep drilling results in the Atlantic Ocean: American Geophysical Union, Ewing Conference Series, v. 3, p. 1-57.

Jordi, H. A., and Lehner, P., 1973, Regional seismic profiles across the Atlantic margin of South America and South Africa: Anais 27th Congresso Brasilero Geology, v. 3, p. 67-90.

Kent, P. E., 1977, The Mesozoic development of aseismic continental margins: Journal of the Geological Society, London, v. 134, p. 1-18.

Krebs, W., 1976, The tectonic evolution of Variscan Meso-Europa, in Ager, D.V., and Brooks, M., eds., Europe from Crust to Core: Wiley, London, p. 119-139.

Lamb, H. H., 1972, Climate: Present, Past and Future: Methuen Co., London, v. 1, p. 613.

Mann, P., Hempton, M. R., Bradley, D. C., and Burke, K., 1983, Development of pull-apart basins: Journal of Geology, v. 91, p. 529-554.

Miyashiro, A., 1986, Hot regions and the origin of marginal basins in the western Pacific: Tectonophysics, v. 122, p. 195-216.

Moberly, R., and Campbell, J. F., 1984, Hawaiian hotspot volcanism mainly during geomagnetic normal intervals: Geology, v. 12, p. 459-463.

Moore, T. C., van Andel, Tj. H., Sancetta, C., and Pisias, N., 1978, Cenozoic hiatuses in pelagic sediments: Micropaleontology, v. 24, p. 113-138.

Mutter, J., and Buhl, P., 1989, Segmentation of spreading ridges: MCS evidence for structural control: Lamont Newsletter, v. 22, p. 4-4A.

Nishimura, K., 1984, A schematic model of development of active continental margins as inferred from particular features of global-scale geoid undulations: Bulletin of the Disaster Prevention Research Institute, Kyoto University, v. 34, p. 187-201.

Pavoni, N., 1985, Pacific/anti-Pacific bipolarity in the structure of the earth's mantle: EOS, v. 66, No. 25, p. 512.

Poag, C. W. and Mountain, G. S., 1987, Late Cretaceous and Cenozoic evolution of the New Jersey continental slope and upper rise: an integration of borehole data with seismic reflection profiles, in Poag, C. W., Watts, A.B., et al., Initial Reports DSDP, v. 95, p. 673-724.

Reading, H. G., 1980, Characteristics and recognition of strike-slip fault systems: Special Publication of the International Association of Sedimentologists, No. 4, p. 7-26.

Ruddiman, W. F., and Kutzbach, J. E., 1989, Uplift model links tectonics to major northern hemisphere climate changes of past 20 million years: possible trigger for Plio-Pleistocene ice ages: Lamont Newsletter, v. 21, p. 4-5.

Schlanger, S. O., Jenkyns, H. C., and Premoli Silva, I., 1981, Volcanism and vertical tectonics in the Pacific Basin related to global Cretaceous transgressions: Earth and Planetary Science Letters, v. 52, p. 435-449.

Schlanger, S. O., Arthur, M. A., Jenkyns, H. C., and Scholle, P. A., 1987, The Cenomanian-Turonian Oceanic Anoxic Event, I. Stratigraphy and distribution of organic carbon-rich beds and the marine ^{13}C excursion, in Brooks, J., and Fleet, A.J., eds., Marine Petroleum Source Rocks: Geological Society Special Publication No. 26, p. 371-399.

Schwan, W., 1980, Geodynamic peaks in Alpinotype orogenies and changes in ocean-floor spreading during Late Jurassic-Late Tertiary time: American Association of Petroleum Geologists, v. 64 (3), p. 359-373.

Schwan, W., 1984, Development of the earth's lithosphere with respect to facts of plate tectonics, geosynclinal evolution and orogenic revolution: Proceedings of the 27th International Geological Congress, VNU Science Press, v. 22, p. 101-125.

Schwan, W., 1985, The worldwide active Middle/Late Eocene geodynamic episode with peaks at +45 and +37 m.yr B. P., and implications and problems of orogeny and sea-floor spreading: Tectonophysics, v. 115, p. 197-234.

Searle, R. C., 1983, Multiple, closely spaced transform faults in fast-slipping fracture zones: Geology, v. 11, p. 607-610.

Searle, R. C., Francis, J. G., Hide, T. W. C., Somers, M. L., Revie, J., Jacobs, C. L., Saunders, M. R., Barrow, B. J., and Bicknell, S. V., 1981, "GLORIA" side-scan sonar in the East Pacific: EOS, v. 62, p. 121-122.

Sheridan, R. E., 1976, Sedimentary basins of the Atlantic margin of North America, in Bott, M.H.P., ed., Sedimentary Basins of Continental Margins and Cratons: Tectonophysics, v. 36 (1-3), p. 113-132.

Sheridan, R.E., 1987, Pulsation tectonics as the control of long-term stratigraphic cycles: Paleoceanography, v. 2, (2), p. 97-118.

Sykes, L. R., 1978, Intraplate seismicity, reactivation of preexisting zones of weakness, alkaline magmatism, and other tectonism postdating continental fragmentation: Reviews of Geophysics and Space Physics, v. 16, p. 621-688.

Toksoz, M. N., and Bird, P., 1977, Formation and evolution of marginal basins and continental plateaus, in Talwani, M., and Pitman, W.C., eds., Island Arcs, Deep Sea Trenches, and Back-Arc Basins: American Geophysical Union, M. Ewing Series, v. 1, p. 379-393.

Trumpy, R., 1980, An outline of the geology of Switzerland—A guidebook: Wepf & Co. Publishers, Basel, p. 104.

Waples, D. W., 1983, Reappraisal of anoxia and organic richness, with emphasis on Cretaceous of North Atlantic: American Association Petroleum Geologists Bulletin, v. 67, p. 963-978.

Wezel, F.-C., 1979, The Scaglia Rossa Formation of Central Italy: results and problems emerging from a regional study: Ateneo Parmense, Acta Naturalia, v. 15, p. 243-259.

Wezel, F.-C., 1985, Facies anossiche ed episodi geotettonici globali: Giornale di Geologia (Selli volume), v. 47, p. 281-286.

Wezel, F.-C., 1986, The Pacific island arcs: produced by post-orogenic vertical tectonics?, in Wezel, F.-C., ed., The Origin of Arcs, Developments in Geotectonics v. 21, Elsevier, Amsterdam, p. 529-567.

Wezel, F.-C., 1988, Earth structural patterns and rhythmic tectonism, in Wezel, F.-C., ed., The Origin and Evolution of Arcs: Tectonophysics, v. 146, p. 1-45.

Whyte, M.A., 1977, Turning points in Phanerozoic history: Nature, v. 267, p. 679-682.

Winkler, W., and Bernoulli, D., 1986, Detrital high-pressure/low-temperature minerals in a late Turonian flysch sequence of the eastern Alps (western Austria): Implications for early Alpine tectonics: Geology, v. 14, p. 598-601.

Winterer, E. L., and Sandwell, D. T., 1987, Evidence from en-echelon gross-grain ridges for tensional cracks in the Pacific plate: Nature, v. 329, p. 534-537.

Ziegler, P. A., 1988, Post-Hercynian plate reorganisation in the Tethys and Arctic-North Atlantic, in Manspeized, W., ed., Triassic-Jurassic Rifting, Continental Break-up and Origin of the Atlantic Ocean and Passive Margins: Developments in Geotectonics v. 22, Elsevier, p. 1-27.

Zoback, M. L., 1987, Global pattern of tectonic intraplate stresses: IUGG, University British Columbia, Vancouver, p. 1116.

Zoback, M. L., Anderson, R. E., and Thompson, G. A., 1981, Cainozoic evolution of the state of stress and style of tectonism of the Basin and Range province of the western United States: Philosophical Transactions Royal Society of London, v. A300, p. 407-434.

Zoback, M. L., Zoback, M.D., Adams, J., et al., 1989, Global patterns of tectonic stress: Nature, v. 341: p. 291-298.

Zwart, H.J., 1976, Regional metamorphism in the Variscan orogeny of Europe: Nova Acta Leopoldina, v. 45, F. Kossmat—Symposium, p. 361-367.

ACKNOWLEDGMENTS

This paper expands a talk given by the writer at the discussion meeting on "New Concepts in Global Tectonics" (Washington, D.C., 20-21 July 1989). I wish to thank the organizers of this stimulating workshop for the invitation. Thanks are also due to J. M. Dickins, Karsten Storetvedt, and Arnoud De Feyter for helpful comments and encouragement. I thank Daniele Savelli for assistance in the preparation, Franco Paolucci for drafting the figures, and David Alexander and David Murray for polishing the English.

This is the Geological Institute of Urbino University contribution No. 130.

This research is CNR-Role of the Strike-slip and vertical tectonics in the Mediterranean Neogene contribution no. 3.

The contracting-expanding Earth and the binary system of its megacyclicity

L. S. Smirnoff,* Smirnoff Minerals Corporation, 12620 Forest Canyon, Parker, Colorado 80134 USA

ABSTRACT

Established and new geologic data are interpreted as the following Earth's parameters change: radius, gravitational acceleration, and angular velocity. The change of these parameters is related to the pulsating, contracting-primarily expanding Earth probably caused by the gravitational constant change. The empirical sequences of the sedimentary and orogeno-magmatic megacyclycity are interpreted as a finite binary multiple harmonic of the probable gravitational pulsations with a systematic decrease of the cycles' duration. The terrestrial and extraterrestrial megacyclicities are probably synchronous due to a common cause.

INTRODUCTION

It is interesting to consider the possible change through geologic time of such parameters of Earth as its radius (R), gravitational acceleration (g), and angular velocity (ω). Attempts have been made to explain some geological data as a result of a change of the above mentioned parameters (Carey, 1975; Stewart, 1977; Crawford, 1982; Weijermans, 1986; and others).

This paper not only tries to explain the regular change of R, g, and ω, but also ties them to statistical data on frequencies of orogenies, unconformities, and distribution in time and space of sedimentary and magmatic rocks. It leads to the concept of the globe's gravitational pulsations with R change, related changes of g, and ω. These parameters are probably dependent upon the gravitational constant (G) decrease estimated from geological data as $\dot{G}/G \approx 0.9 \times 10^{-10}$ per year. This concept is tied to the proposed binary multiple harmonic of geologic cyclicity, which includes tectonic, magmatic, and sedimentologic processes. This approach to cyclicity allows one to extrapolate the known data to the origin of the Earth calculated as 4.53–4.56 b.y., and probably of the Universe calculated as 13.7–14.3 b.y. The above is in agreement with Dirac's and Hoyle-Narlikar's cosmology of the expanding Universe (Smirnoff, 1989a, b, c). It should be mentioned that this concept includes some hypothetical assumptions discussed later. However, taken together, they create a consistent finite model of the contracting, and primarily expanding Earth.

THE EARTH'S RADIUS CHANGE

There is a concept of general R increase with geologic time with the main expected implication being the change of sea level and of the sea-covered areas of the continents. The rate of such average R increase for the last 0.5 b.y. has been evaluated by Egyed as $\dot{R} = 0.54$–0.76 mm/y based on old paleogeographic maps (Egyed, 1956; 1969). His calculations of a sea level that was 275–500 m higher 0.5 b.y. ago than that of today is quite close to modern data (Holland et al., 1986).

New maps (Ronov, 1983) show that the sea-covered areas of the past are quite different from data used by Egyed. Figure 1 shows that during the last 0.6 b.y. the approximated sea-covered area of continents decreased by 31×10^6 km^2. The decrease is almost half as much as that of Egyed. The periods of the prevailing sealevel changes, based on the areal distribution of seas (Fig. 1), fit independently obtained data, based on a sequence stratigraphic concept (Vail et al., 1977; Holland et al., 1986). Thus, the sea-covered area of the past reflects the sealevel changes.

If we correct Egyed's data by Ronov's calculations, then following Egyed's concept our new calculated rate of R change is:

$$R_{0.5} \approx 0.29 \text{ mm/y} \qquad (1)$$

However, Egyed's concept of relating R with sealevel changes and his way of approximating data have been the subject of debate (Armstrong, 1969; Veizer, 1971; Hallam, 1971; see also related data in Turcotte et al., 1978; Worsley et al., 1984).

There is another hypothetical way of $\dot{R}$ evaluation. Creer (1965), Egyed (1969), and others noticed that if we remove the oceanic crust from the Atlantic, Indian, and Polar oceans, the continental crust boundaries will fit together on a globe with R = 3,500–4,000 km. They assumed that the age of the continental crust is close to the age of the Earth as 4.5 b.y. and suggested average $\dot{R}$ = 0.65 mm/y. This value can be corrected also.

The most probable age of the continental crust is close to 3.8 b.y. (DePaolo, 1981; Voitkevich and Bessonov, 1986), which is close to the approximated age of the major orogeno-magmatic event dated as 3.7 b.y. (Table 2). Moreover the Pra-Pacific ocean is at least 1.0 b.y. old with its area not less than that of today (Scotese, 1987). If we assume *(assumption I)* that the Pra-Pacific (excluding the adjacent seas) is as old as the continental crust, then the initial Earth's surface is close to 0.63 of today's with corresponding $R_{3.7} \approx 5,000$ km and average:

$$R_{3.7} = 0.37 \text{ mm/y} \qquad (2)$$

which is close to (1).

Chatterjee, S., and N. Hotton III, eds. *New Concepts in Global Tectonics*. Texas Tech University Press, Lubbock, 1992, xii + 450 pp.

*Deceased, 8 May 1991

I have considered above two main $\dot{R}$ evaluations that both show a decrease from previous estimates. This should be noticed because Wesson (1978), who reviewed all data, used Egyed's (1956, 1969), and Creer's (1965, 1967) calculations and concluded that $\dot{R}_{0.3-4.5} = 0.60$ mm/y, and $\dot{R}_{0.0-0.3} = 1.0$–7.0 mm/y. The latter for the last 0.3 b.y. is based only on paleomagnetic data. Indeed, the new paleomagnetic calculations show inconclusive variations of R (Khramov, 1987, table 7.1, 7.2).

The two independent ways of $\dot{R}$ change estimation we considered above, lead us to the same conclusion. If so, the inevitable consequence of R increase is g and ω change of the expanding Earth.

GRAVITY CHANGE

Change in gravity during the Earth's history can be affected by the gravitational acceleration (g) change and by change in the universal gravitational constant (G). From my point of view, the possible approach to g change is through the study of the gravitational nature of orogenic activity. Orogenic (tectonic) events are controlled by the gravitational field (Gravity and Tectonics, 1973). Therefore, the relative change of tectonic activity should reflect the change in g. Such possible change is shown on Figure 2. It can be noted:

a. There is a systematic change of the tectonic activity from the Baikalian to the Laramian orogeny, which is expressed by minimums of the orogenic activity at the end of major geologic epochs.

b. The general decrease of unconformities development from 97% to 86% is 11% during 0.6 b.y. Let us assume *(assumption II)* that this decrease is in direct proportion to g, and with $g_{0.0} = 981$ cm sec^{-2} (86%), $g_{0.6} = 1,106$ cm sec^{-2} (97%), and:

$$\dot{g}/g = \left(\frac{1106 - 981}{1106}\right)/600 \times 10^6 = 1.88 \times 10^{-10} /\text{y}. \quad (3)$$

c. From $g = GM/R^2$, M = Const., follows $G_{0.6} = g_{0.6} \times R^2_{0.6} \times G_{0.0}/g_{0.0} \times R^2_{0.0} =$, which calculates overall G decrease:

with (1) $\dot{G}/G \approx 1.04 \times 10^{-10} /\text{y}$ (4)

with (2) $\dot{G}/G \approx 0.77 \times 10^{-10} /\text{y}$ (5)

The astronomically suggested decrease $\dot{G}/G = (0.8 \pm 0.5) \times 10^{-10}$ per year (Van Flanderen, 1975) fits (5) and (4) and Dirak's (1937) and Hoyle-Narlikar's (1971) cosmology of the expanding Universe with $\dot{G}/G \approx 1.0 \times 10^{-10}$ per year. It should be noted that $\dot{G}/G$ value depends on R. The latest data (Damour et al., 1988) is in disagreement with above suggesting $\dot{G}/G = (1.0 \pm 2.3) \times 10^{-11}$ per year. If assumption II is correct, Figure 2 also shows a pulsating g of the contracting-primarily expanding Earth.

If so, Pronin's orogenies (e.g., Taconian, Uralian, Austrian) are caused by the contractions (with probable maximums of ophiolites) and would follow the Mega-Omega extinctions of different orders, and the time of high frequency of the magnetic field polarity changes (see Piper, 1987, fig. 12.6). The above allows us to suggest that contraction is affecting the Earth's core and mantle in a way that causes the magnetic field reversals with low magnetic field intensity in between. The latter suggestion can be connected to higher levels of radiation, which with unusually strong mafic volcanism, are the main cause of the extinctions. This suggestion does not support the concept of accidental catastrophes in the Earth's history (Global Catastrophes in Earth History, 1988).

The periods of the Earth's contraction should be related to the ω change at the time of Mega orogenies. Additionally, several implications on the time scale follow from g and G general decrease.

ANGULAR VELOCITY CHANGE

Figure 3 shows the decrease of the year length probably related to ω decrease during the last 0.55

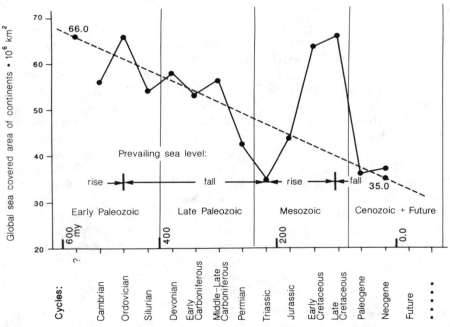

Figure 1. Change of sea-covered areas of continents with time based on Ronov's data (1983, table 11). Data averaged by us for the stratigraphic intervals close to the cycles of 2^7 order (Table 1).

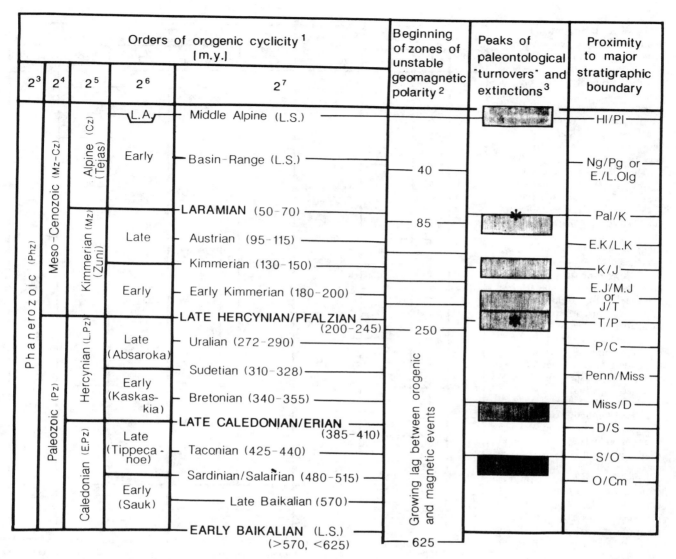

Table 1. Binary multiple harmonics of orogenic cyclicity and correlative events, Phanerozoic. **1**—Orders of orogenic cyclicity are based: a/2^3 on Table 2; b/2^4 on two Phanerozoic sea-level cycles (Vail et al., 1977, fig. 1; Holland et al., 1986; see also Fig. 1), and also on the pattern of O^{18} distribution (O'Neil, 1979), and geomagnetic field polarity bias (Piper, 1987, fig. 12.13); c/2^5 on four common tectonic epochs, and also on the boundaries of geochemical cycles of C_{org}, O^{18}, ^{34}S distribution (Kaplan, 1975; Budyko, et al., 1987); d/2^6 on four Paleozoic (Sloss, 1963), and possibly four Meso-Cenozoic megacycles as Triassic-Jurassic, Cretaceous, Paleogene, and Neogene; e/2^7 on Pronin's statistical maximums of unconformities (see Kunin and Sardonikov, 1976). **2**—after Khramov (1987). **3**—after Cutbill, Funnell (1967); Megaextinctions are shown by the asterisks. Note the correlation between the orogenies and the boundaries of sedimentary cycles.

b.y. The general trend, which can be justified by the expanding Earth, has been attributed primarily to tidal friction. In addition, it has been recognized that there is a different process responsible for periodic ω increase. As can be seen, the previously unexplained Earth's accelerations (especially between Early Paleozoic-Late Paleozoic) have happened during suggested contractions (compare Figs. 2 and 3). The quantitative estimation of ω change is complicated by the supposed side effects of R and G change on the Earth's core, mantle, and orbit. It is important to have an alternative possibility of ω change estimation (Smirnoff and Khramov, 1975).

BINARY MULTIPLE SYSTEM OF MEGACYCLICITY

The variable G-R tectonics of the contracting-primarily expanding Earth allow us to consider a multiple harmonic system of the Earth's pulsations. The simplest one would be a binary multiple harmonic (BMH) system, which can be shown for last 0.6 b.y. on the basis of the statistical maximums of orogenic activity, sedimentary megacyclicity, and other geologic data (Table 1). One

Table 2. Binary multiple harmonics of the Earth geologic megacyclicity (b.y.). **1**—The sequence of global orogeno-magmatic events and their primarily K-Ar averaged radiometric dates (b.y.): Phanerozoic as commonly accepted with the L. Baikalian age from Table 1; Riphean and older (after Salop, 1983). **2**—Stratigraphic equivalents and their ages (after Harland et al., 1982). **3**—The same after Kent and Hugo (1978). **4**—An approximation of the events' ages calculated as 0.066 + (8 × 0.02) = 0.226 b.y. **5**—Mega (Late Hercynian, Laramian) and Omega (Late Alpine) extinctions, shown by the asterisks. **6**—An age calculated from Figure 4, one of the possible ages of nucleosynthesis' event dated between 6.0–16.0 b.y. (Trimble, 1975); it is close to an age of the iron meteorites. **7**—An age calculated from Figure 4, a possible age of the Universe (T); the time of the Big Bang close to an age corresponding to an average value of the Hubble's constant, $H = 75$ km s^{-1} Mpc^{-1} (Hodge, 1981).

The cyclicity orders are based: a/ 2^1 on a hypothetical construction; b/ 2^2 after H. Stille and also two stages of endogenic mineralization (Smirnoff, 1982) on an absolute maximum of the continental drift velocity near 1.6 b.y. (Aparin and Vedenkov, 1978; Irving and McGlynn, 1981), and on correlation between maximums of the plates' movement and the orogenies (Scheinmann, 1975). It is interesting to note that Vyborgian-Hudsonian orogeny is the age of the major emplacement of the rapakivi granites; c/ 2^3 based on lithology and stromatolites (Harland et al., 1982; Semikhatov, 1974); d/ 2^4 for the Phanerozoic on Table 1, for the Riphean-Archean on lithological data and principal unconformities (Semenenko, 1970, 1976); e/ 2^5 for the Phanerozoic on Table 1; for the Riphean on the sedimentary cyclicity, unconformities, and stromatolites (Semikhatov, 1974; Harland et al., 1982; Shixing, 1982; Keller et al., 1984), and also on reversals of geomagnetic frequencies (see text); for the Aphebian-L. Archean on the same data as for 2^4, which need to be further substantiated. Note: a/ the correlation of the cycles 2^5 order with the sequence of the orogeno-magmatic events as their boundaries; b/ an assignment of order 2^6 to Katangian and Avzyan orogenies (orogenies of this order not shown on this table are absent in Salop's sequence); c/ binary two-step extrapolation in the past to the event 13.7–14.3 b.y., the further extrapolation provides big dates that are appropriate to the infinite Open Universe; d/ the possibility of extrapolation in the future from the approximated sequence "... 22 - 7" and "6 - 1" to "0 - ...." It could mean a transition from prevailing expansion to prevailing contraction which would be appropriate to "Closed Universe."

Orders of cyclicity					Global orogenic magmatic events [1]	Age [1]	Precambrian stratigraphic equivalents and their ages			Approximation by series of natural numbers [4]	
2^1	2^2	2^3	2^4	2^5			USA [2]	World [2]	South Africa [3]	Duration D/0.02	Age of events
?				Future ?	LATE ALPINE (L.S.)		* [5]			...1–6?	–0.074
	Neogean	Phanerozoic (Phz)	Meso-Cenozoic	Future & Alpine						7	0.066
				Laramian		0.066	*			8	0.226
			Paleozoic	Kimmerian		0.23–0.25	*			9	0.406
				Late Hercynian							
				Hercynian		0.41				10	0.606
				Late Caledonian							
				Caledonian	BAIKALIAN	~0.60(L.S.)	Zedian	0.59 E+M		11	0.826
		Riphean (Rph)	Late	L.Rph$_2$	Katangan	0.65–0.68		0.65 S-U	Katanga ~0.80	12	1.066
					Lufilian	0.78		0.80	Damara ~1.08	13	1.326
				L.Rph$_1$	Grenville	1.00–1.10	Yovian	~1.05		14	1.606
			Early	E.Rph$_2$	Avzyan Kibaran/Elsonian	1.30–1.40		Yurmatian ~1.35	?	15	1.906
				E.Rph$_1$	VYBORGIAN/HUDSONIAN	1.60–1.75		Burzyan ~1.65	~1.67	16	2.226
		Aphebian (Aph)	Late	L.Aph$_2$	Karelian	1.90–2.00	Xenian	~2.1	Waterberg ~2.07	17	2.566
	Protogean			L.Aph$_1$	Ladogan	~2.20		Huronian	Unnamed ~2.22	18	2.926
			Early	E.Aph$_2$	Seletskian	~2.40		~2.4	Transvaal ~2.63	19	3.306
				E.Aph$_1$	KENORAN	2.60–2.80		~2.63 ~2.8 Ran	Randian ~2.8–2.9	20	3.706
		Archean (Ar)	Late	L.Ar$_2$	Swazilandian	~3.20	Weltian	Swaziland ~3.75	Pongola ? Swaziland ~3.75	21	4.126
				L.Ar$_1$	Saamian	3.50–3.75				22	4.566
			Early	E.Ar$_2$	Godthaabian?	~4.00		Isuan ~3.9			
				E.Ar$_1$				Hadean			
					Origin of Earth	~4.53				23–30	8.806
?					Origin of elements ? [6]	~8.50				31–38	14.326
?					Origin of Universe ? [7]	~13.70					

must add one more Early Baikalian orogeny (near 0.6 b.y.) to have this BMH system complete and open to the future. Considering the status of the Phanerozoic-Riphean (Vendian) boundary and its age (Cowie and Johnson, 1985), our addition is realistic.

The BMH system is not supposed to be restricted to the Phanerozoic. It can be shown for the remaining 4.0 b.y. of the geological past with decreasing certainty for Riphean, Aphebian, and especially Archean.

Table 2 justifies, step by step, the subdivision of the entire geological sequence on 2, 4, 8, and so on, cycles. The last subdivision, the cycles of 2^5 order for the Riphean-Archean, has insufficient data for their identification and can probably be improved.

The statistical data show that Cz-Mz; Mz-L. Pz; L. Pz.-E. Pz; E.Pz-L.Rph (see Fig. 4 for abbreviations) boundaries of 2^5 order (time of contractions, Fig. 2) are the times of the most frequent reversals. Riphean boundaries of this order near 0.72; 1.10; 1.30; and 1.60 b.y. can be inferred

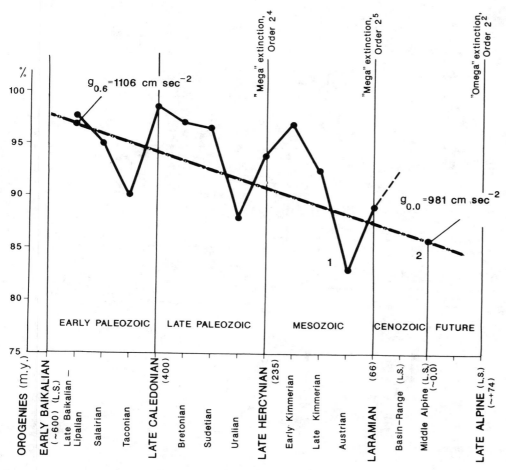

Figure 2. 1—Pronin's change of the orogenies frequency as a percentage of the stratigraphic sections with unconformities based on 1,000–5,000 analyzed sections on all continents (Kunin and Sardonikov, 1976, table 3, left column). 2—an approximation of the above. Note a binary sequence of the orogenies, the "Mega" extinctions, and an inferred "Omega" (?) extinction of the highest order in the Future from Tables 1 and 2.

from such data (Fig. 4), which are quite close to dates of Table 2. The lack of data and the uncertainty of the ages prevent doing the same for the Aphebian.

Thus, the proposed BMH system of megacyclicity is substantiated by the data for the order 2^4 (for Archean and Aphebian), 2^5 (for Riphean), 2^6 (for Meso-Cenozoic) and 2^7 (for Paleozoic), providing an inferred total BMH system of Table 2.

The proposed system reveals new properties of geologic megacyclicity and time scale.

a. The time duration (D) of the cycles 2^4 order changes systematically, which is shown on Figure 5. Figure 6 is a systematic D plot of cycles 2^5 order for the Neogean (Riphean, and Phanerozoic with future). One can check that the same plot for the Protogean shows irregular D change due to growing variability of K-Ar data caused by uncertainties of sampling and other reasons. It can be mentioned that D changes of the Cale-donian, Hercynian cycles of 2^5 order have been often attributed to the change of the galactic year (Wilson, 1981a, b).

b. The Earth's cyclical system is finite if approximated by the series of natural numbers with D of

Figure 3. The change of the number of days per year with time estimated from studies of fossil shells' growth rings (after Creer, 1975, with my approximation). The boundaries of megacycles from Tables 1 and 2.

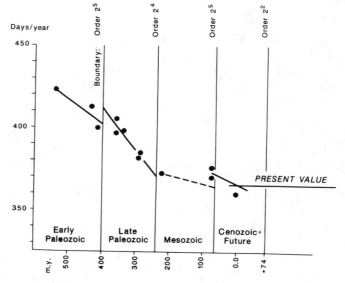

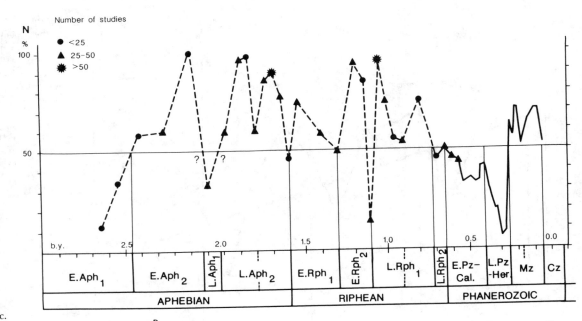

Figure 4. The bias of the geomagnetic polarity determined from the percentages of paleomagnetic results of normal polarity falling within segments of the APWP (Piper, 1987, fig. 12.6, with my approximation and stratigraphic interpretation). The abbreviations stand for: E. Aph—Early Aphebian; L. Rph—Late Riphean; E. Pz-Cal—Early Paleozoic or Caledonian; L. Pz-Her—Late Paleozoic or Hercynian; Mz—Mesozoic; Cz—Cenozoic.

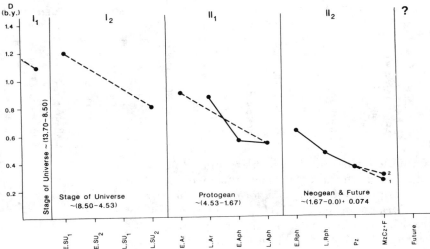

Figure 5. The change of the durations (D) of cycles 2^4 order with time calculated as the arithmetic average between minimum and maximum dates for each event from table 2. Dashed line — an approximation and extrapolation. Note: a/ two different $D_1 = 0.24$ b.y. as the duration of the Meso-Cenozoic, and $D_2 = 0.24$ b.y. + 0.074 b.y. as the duration for Meso-Cenozoic with the Future; b/ a trend for Neogean and its similarity to an approximated trend for Protogean; c/ a hypothetical approximation for *two* more stages of the Universe based on assumption III (see text).

the future as near 0.5 b.y., which is determined as: $0.074 + (6 \times 0.02) + (5 \times 0.02) + \ldots$ b.y.

c. The extrapolative age of the Earth is 4.53–4.56 b.y., which is close to an average age 4.55 b.y. of the chondrites (Wood, 1988).

d. The extrapolative age of the Universe (T) is 13.7–14.3 b.y., which is close to an age near 15.0 b.y. corresponding to the average Hubble constant as 75 km s^{-1} Mpc^{-1} (Hodge, 1981). It should be noted that an extrapolation of this kind favors *(assumption III)* a common extraterrestrial cause of megacyclycity (G change) with the synchronicity of terrestrial and extraterrestrial events, and a relative constancy of the radioactive decay rates. This would explain the puzzling solar activity (S)-earthquake energy (E) correlation (Smith, 1975; Kalinin et al., 1979), and with E-ω-R correlation (Press, Briggs, 1975; Kropotkin, 1983), it would mean G-S-E-ω-R-... correlation.

e. The sequence of the mega-cycles is probably finite biologically, due to upcoming Omega extinctions of the highest order, which provides new limitation for the existence of extraterrestrial civilizations (Hart, 1975). The nonsystematic catastrophic extraterrestrial cause of biological extinctions (Global Catastrophes in Earth History, 1988) is not necessary for this system.

f. The modern time of the Earth is near the middle of the Alpine + Future cycle—the boundary of 2^6 order (Fig. 2). It should be a time of contraction of this order, which would explain the horizontal compressive stress of the crust, ω increase, and other observed modern data (Kropotkin, 1983).

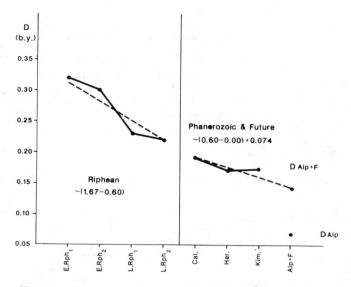

Figure 6. The change of durations (D) of cycles 2^5 order for Neogean calculated similar to Figure 5. Note D = 0.066 b.y., which does not fit the trend and D = 0.066 b.y. + 0.074 b. y., which does fit the trend.

g. The Phanerozoic time-scale change can be expressed if we calculate durations (D) change as:

$$\dot{D} = D_{cal} - D_F = (0.02 \times 10) - (0.02 \times 6)$$
$$= 0.08 \text{ b.y.};$$
$$D_{Phz + F} = 0.606 + 0.074 = 0.68 \text{ b.y., and}$$
$$\dot{D}/D^2_{Phz + F} = 1.73 \times 10^{-10} /y \quad (6)$$

which is close to the independently acquired value (3).

h. The relative time scale change between Alpine + Future and Hadean (Early Archean$_1$) cycles of 2^5 order with $\dot{D}Had = \dot{D}Aep + F = 0.02$ b.y:

$$(\dot{D}Aep + F/D^2_{Aep} + F)/(\dot{D}Had/D^2_{Had})$$
$$= (1.02 \times 10^{-9})/(1.03 \times 10^{-10}) \approx 10.0 \quad (7)$$

i. The relative change of Earth's age is:

$$D_E/(D_{Alp + F} \times 16) = 4.55/(0.14 \times 16) \approx 2.0 \quad (8)$$

One cannot exclude that simple ratios of natural numbers could mean a quantum relationship between gravity and time. Titius-Bode's law of the planets' distances to the Sun for the first eight planets is related to the sequence 2^n and is probably an unexplained precedent of the quantum meaning in macrophysics (see our addition to interpretation of this law in the Appendix, which includes all planets).

It should be taken into consideration that the binarity of geological cycles of smaller scale can be observed in relatively uninterrupted sequences. For example, the clastic sedimentary sequences are full of visible and invisible unconformities—the missing parts of geological records (Kuliamin, Smirnoff, 1973). In addition, it should be remembered that there are other causes for the cyclicity.

BRIEF DISCUSSION

The main element of the proposed system is a mechanism of R variations in time and its integration with plate tectonics. There are several possible approaches.

a. A time-space variable G-R could lead to two major scenarios of the G-R tectonics: subduction as a result of the spreading-expansion, and a process like subduction as a result of contraction (with flood basalts, ophiolites—"squeeze-outs"), or the alternation of subduction-contraction with spreading-expansion (see Kropotkin, 1983 for the last), or a combination of both.

b. Decompaction of minerals in the mantle leads to volumetric expansion of the globe with R increase and the finite R = 6,740 km (Barsukov and Urusov, 1982; Milanovsky and Mal'kov, 1985). This is close to R = 6,600 km, which can be calculated on the basis of the finite sequence 22-1 of Table 2. A finite R = 6,530 km of the pulsating and expanding Earth follows from the hypothesis of the Earth origin from the hot and high density matter with subsequent decompaction (Kuznetsov, 1984).

c. G-caused terrestrial and extraterrestrial synchronicity *(assumption III)* explaining the reasonable coincidence of geologically estimated $\dot{R}$ from equation (2) with $\dot{R}$ from another equation (McDougall et al., 1963):

$$\dot{R} = HR \approx 0.45 \text{ mm/y} \quad (9)$$

with Hubble constant H corresponding to our T = 14.3 b. y. (see Ivanenko et al., 1967; and Wesson, 1978; for the explanation of (9); T is probably to be determined soon within 10% by the upcoming Hubble telescope mission). Finally, with our assumptions I, II, III, and equations (2), and (5), we are getting, probably, the geologically justified:

$$\dot{G}/G \sim 1/T \quad (10)$$

This fits Dirac's and Hoyle/Narlikar's cosmology, and is related to the system in geology that does not fit uniformitarianism and catastrophism. The major biological extinctions become part of the system related to the contraction of the Earth and correlated processes such as the maximums of magnetic reversals (for the latter see Loper et al., 1988), and strong mafic volcanism. The proposed concept of the G-R tectonics and the BMH stratigraphy can be related to four stages of geosynclinal sedimentation, subdivided by orogenies and silicic magmatism (Khain, 1973). Moreover, the similarity of the cyclicity of each order resembles the similarity of the phylogeny and ontogeny of magmatic cycles of different orders (Rundquist, 1968).

The proposed concept has broad scientific implications and practical applications. One of those is related to sequence stratigraphy. A binary hierarchy can be clearly seen on the data for Meso-Cenozoic (Haq et al., 1988; the possibility of the binary hierarchy of Phanerozoic stratigraphy has been shown to P. Vail by the author in 1985).

REFERENCES CITED

Aparin, V.P., and Vedenkov, V.S., 1978, Chronologic conjunction between cycles and paleomagnetic polar wandering: Doklady Akademii Nauk SSSR, v. 232, p. 123-125.

Armstrong, R.L., 1969, Control of sea-level relative to the continents: Nature, v. 221, p. 1042-1043.

Barsukov, V.L., and Urusov, V.S., 1982, Phase transitions in mantle and possible Earth's radius change: Geokhimia, no. 12, p. 1729-1743.

Budyko, M.I., Ronov, A.R., and Yanshin, A.L., 1987, History of Earth's atmosphere: Springer Verlag, 139 p.

Carey, S.W., 1975, The expanding Earth—an essay review: Earth Science Reviews, v. 11, p. 105-143.

Cowie, J.W., and Johnson, M.R.W., 1985, Late Precambrian and Cambrian geologic time scale, in Snelling, N.J., ed., The chronology of the geological record: Blackwell Scientific Publication, p. 47-64.

Crawford, A.R., 1982, The Pangean paradox: where is it?: Journal of Petroleum Geology, v. 5, no. 2, p. 149-160.

Creer, K.M., 1965, An expanding Earth: Nature, v. 205, p. 539-544.

———, 1967, Earth possible expansion, in Runcorn, S.K., ed., International dictionary of geophysics: Pergamon Press, p. 383-389.

———, 1975, On a tentative correlation between changes in geomagnetic polarity bias and reversal frequency and the Earth's rotation, in Rosenberg, G.D., and Runcorn, S.K., eds., Growth rhythms and the history of the Earth's rotation: London, J. Wiley, p. 293-317.

Cutbill, J.L., and Funnell, B.M., 1967, Numerical analysis of the fossil record, in Harland, W.B., et al., eds., The fossil record: Geological Society of London, p. 791-820.

Damour, T., Gibbons, G.W., Taylor, J.H., 1988, Limits on the variability of G using binary-pulsar data: Physical Review Letters, v. 61, no. 10, p. 1151-1154.

De Jong, K.A., and Scholten, R., (eds.), 1973, Gravity and Tectonics: New York, J. Wiley, 502 p.

DePaolo, D.J., 1981, Radiogenic isotopes and crustal evolution, in O'Connor, R.J., and Fyfe, W.S., eds., Evolution of the Earth: Geological Society of America, Geodynamic Series, no. 5, p. 59-68.

Dirac, P.A.M., 1937, The cosmological constants: Nature, v. 139, p. 323.

Egyed, L., 1956, Determination of changes in the dimensions of the Earth from paleogeographical data: Nature, v. 178, p. 534.

———, 1969, A slow expansion hypothesis, in Runcorn, S.K. ed., The application of modern physics to the Earth and planetary interiors: J. Wiley–Interscience, p. 65-76.

Global Catastrophes in Earth History, 1988, Lunar and Planetary Institute and the National Academy of Sciences, Abstracts, 226 p.

Hallam, A., 1971, Reevaluation of the paleogeographic argument for an expanding Earth: Nature, v. 232, p. 180-183.

Haq, B.U., Hardenbol, J., Vail, P.R., 1988, Mesozoic and Cenozoic chronostratigraphy changes: An integrated approach: Society of Economic Paleontologists and Mineralogists, Spec. Publ., no. 42, p. 71-108.

Harland, W.B., Cox, A.W., Liewellin, P.G., Pickton, C.A., Smith, A.G., and Walters R., 1982, A geologic time scale: Cambridge University Press, 131 p.

Hart, M.H., 1975, An explanation for the absence of extraterrestrials on Earth: Quarterly Journal of Royal Astronomical Society, v. 16, p. 128-135.

Hodge, P.W., 1981, The extragalactic distance scale: Annual Review Astronomy and Astrophysics, v. 19, p. 357-372.

Holland, H.D., Laser, N., McCaffrey, M., 1986, Evolution of the atmosphere and oceans: Nature, v. 320, p. 27-33.

Hoyle, F., and Narlikar, J.V., 1971, On the nature of mass, Nature, v. 233, p. 41-44.

Irving, E., and McGlynn, J.C., 1981, On the coherence, rotation and paleolatitude of Laurentia in the Proterozoic: Developments in Precambrian Geology, v. 4., p. 561-598.

Ivanenko, D.D., Brezhnev, V.C., and Frolov, B.H., 1967, New non-statical solutions of Einstein's equations, in Sovremenniye Problemi Gravitatsii: Tbilisi University, USSR (in Russian).

Kalinin, Yu.D., and Kiselev, V.M., 1979, Solar relationship and prediction of seismic activity of the Earth, in Proceedings: Solar-Terrestrial Predictions, v. 4, p. G23-G28, Department of Commerce, USA.

Kaplan, I.R., 1975, Stable isotopes as guide to biogeochemical processes: Proceedings of Royal Society of London, Seria B, v. 189, p. 183-211.

Keller, B.M., Semikhatov, M.A., and Chumakov, N.M., 1984, Type section of the upper Erathem of the Proterozoic, in Proceedings of 27th Geological Congress, v. 5, p. 125-170, Nauka, Moscow, (in Russian).

Kent, L.E., and Hugo, P.J., 1978, Aspects of the revised South African stratigraphic classification: American Association of Petroleum Geologists, Studies in Geology, no. 6, p. 367-379.

Khain, V.Ye., 1973, Obschaya Geotektonika: Nedra, Moscow, 511 p. (in Russian).

Khramov, A.N., 1987, Paleomagnetology: Springer Verlag, 308 p.

Kropotkin, P.N., 1983, A new geodynamic model: Doklady Akademii Nauk SSSR, v. 272, p. 575-578.

Kuliamin, L.N., and Smirnoff, L.S., 1973, Tidal sedimentation cycles in Cambrian-Ordovician sands, Baltic: Doklady Akademii Nauk SSSR, v. 212, p. 697-699.

Kunin, N.Yu., and Sardonikov, N.M., 1976, Global cyclicity of tectonic movements: Bulletin Moskovskogo Obschestva Ispitateley Prirodi, Otdeleniye Geologii, v. LI, no. 3, p. 5-27 (in Russian).

Kuznetsov, V.V., 1984, Fizika Zemli i Solnechnoy Systemy: Trudy Instituta Geologii i Geopfiziki, Siberskoye Otdeleniye Akademii Nauk SSSR, vyp. 639, 92 p. (in Russian).

Loper, D.E., McCartney, K., Buzyna, G., 1988, A model of correlated episodicity in magnetic field reversals, climate and mass extinctions: Journal of Geology, v. 96, no. 1, p. 1-15.

McDougall, J., Butler, R., Kronberg, P.H., Sandquist, A., 1963, A comparison of terrestrial and universal expansion: Nature, v. 199, p. 1080.

Milanovsky, Ye.Ye., Mal'kov, B.A., 1985, Phase transitions in the core and mantle: Doklady Akademii Nauk SSSR, v. 280, p. 696-700.

O'Neil, J.R., 1979, Stable isotope geochemistry of rocks and minerals, in Jager, E., and Hunziker, J.C., eds., Lectures in isotope geology: Berlin, Springer Verlag, p. 235-263.

Piper, J.D.A., 1987, Paleomagnetism and the continental crust: John Wiley, Open University Press, 434 p.

Press, F., and Briggs, P., 1975, Chandler wobble, earthquakes, rotation and geomagnetic changes: Nature, v. 256, p. 270-273.

Ronov, A.B., 1983, The Earth's sedimentary shell: American Geophysical Institute, Reprint Series, no. 5, 120 p.

Rundquist, D.V., 1968, Metalls' accumulation and evolution of their genetic types, in Doklady Sovetskikh Geologov, 23rd Mezhdunarodny Geologichesky Congress: Nauka, Moscow, p. 212-226 (in Russian).
Salop, L.L., 1983, Geological evolution of the Earth during Precambrian: Springer Verlag, 459 p.
Scotese, C.R., 1987, Development of the circum-Pacific Panthalassic ocean during the Early Paleozoic: Geological Society of America, Geodynamic Series, v. 18, p. 49-57.
Semenenko, N.P., 1970, Intercontinental correlation of the Precambrian, in Geokhronolgiya Dokembriya: Nauka, Moscow, p. 5-22 (in Russian).
——, 1976, Hierarchy of dates and Precambrian correlation, in Aktual'niye Voprosy Sovremennoy Geokhronologii: Nauka, Moscow, p. 20-28 (in Russian).
Semikhatov, M.A., 1974, Stratigraphy of Proterozoic: Trudi Geologichesokogo Instituta Akademii Nauk SSSR, v. 256, 302 p. (in Russian).
Sheynmann, Yu.M., 1975, Certain inductive laws deriving from paleomagnetic measurements: Izvestiya Akademii Nauk SSSR, Phyzika Zemli, no. 1, p. 66-75.
Shixing, Z., 1982, An outline of studies of the Precambrian stromatolites in China: Precambrian Research, v. 18, p. 367-396.
Sloss, L.L., 1963, Sequences in the cratonic interior of North America: Bulletin of Geological Society of America, v. 74, p. 93-114.
Smirnoff, L.S., 1989a, Binary multiple harmonics of sedimentary and orogenic cyclicity, Phanerozoic: Washington, D.C., Abstracts, t. 3, p. 134.
——, 1989b, Binary multiple harmonics of sedimentary and orogeno-magmatic cyclicity, Precambrian: 28th International Geological Congress, Washington, D.C., Abstracts, t. 3, p. 134-135.
——, 1989c, Gravitational pulsations of the contracting-expanding Earth and the binary system of its tectonic megacyclicity: Smithsonian Institution and Texas Tech University Symposium, Abstracts, p. 30-31.
Smirnoff, L.S., and Khramov, A.N., 1975, Coriolis force and the texture of the sandstones vs. the paleomagnetic latitudes: Izvestiya Akademii Nauk SSSR, Phyzica Zemli, no. 3, p. 66-74; no. 10, p. 57-68.
Smirnoff, V.I., 1982, Endogenic mineralization in the Earth's history: Geologiya Rudnikh Mestorozhdeniy, no. 4, p.3-20 (in Russian).
Smith, P.J., 1975, Plate tectonics and general relativity: Nature, v. 254, p. 386.
Stewart, A.D., 1977, Quantitative limits on paleogravity: Journal of Geological Society of London, v. 133, p. 281-291.
Trimble, V., 1975, The origin and abundance of the chemical elements: Review of Modern Physics, v. 47, p. 877-976.
Turcotte, D.L., Burke, K., 1978, Global sea-level changes and the thermal structure of the Earth: Earth and Planetary Science Letters, v. 41, p. 341-346.
Vail, P.R., Mitchum, R.M., and Thompson, S., 1977, Seismic stratigraphy and global change of sea-level, part 4: American Association of Petroleum Geologists, Memoir 26, p. 83-98.
Van Flanderen, T.C., 1975, Determination of rate of change of G: Monthly Royal Astronomical Society, v. 170, p. 333-342.
Veizer, J., 1971, Do paleomagnetic data support the expanding Earth: Nature, v. 229, p. 450-481.
Voitkevich, G.B., and Bessonov, O.A., 1986, Khimicheskaya Evolutsiya Zemli: Nedra, Moscow, 213 p. (in Russian).
Weijermans, R., 1986, Slow but not fast global expansion may explain the surface dichotomy of Earth: Physics of the Earth and Planetary Interiors, v. 43, p. 67-89.
Wesson, P.C., 1978, Cosmology and Geophysics: New York, Oxford University Press, 240 p.
Wilson, G.E., 1981a, Introduction, in Wilson, G.E., ed., Megacycles: Benchmark Paper in Geology, v. 57, p. 1-14.
——, 1981b, Editor's comments, in Wilson, G.E., ed., Megacycles: Benchmark Papers in Geology, v. 57, p.162-167.
Wood, J.A., 1988, Chondritic meteorites and the solar nebula: Annual Review of Earth and Planetary Sciences, v. 16, p. 53-72.
Worsley, T.R., Nance, D., Moody, J.B., 1984, Global tectonics and eustasy for the past 2 billion years: Marine Geology, v. 58, p. 373-400.

ACKNOWLEDGMENTS

I thank Y. Victor, E. Pajon, A. Kaplan, V. Osherovich, E. Erlich, G. Ulmishek, C. deWys for their contribution and constructive criticism; G. Slonimsky, and G. Curtis for their help; and E. Smirnoff for motivation. My past association with some colleagues in the National Petroleum Institute (VNIGRI) and the National Geologic Institute (VSEGEI) of the USSR, Leningrad, have been essential for this result.

APPENDIX—Modified Titius-Bode law of the planet distances.

This controversial law is applicable only to the first eight planets: $r_1 + 0.4 + (0.3 \times 2^n)$, $n = -\infty, 0, 1, 2, 3 \ldots$. It is not excluded that the next planets belong to the second octave and could have $r_2 = 30 + (0.6 \times 2^n)$, with $\Delta\Delta r = [96 + (4 \times 2)] - [0 + (4 \times 1)] = 100$.

First								Second					Octave
1	2	3	4	5	6	7	8	1	2	3	4	5	
Mer.	Ven.	E.	Mars	Ast.	Jup.	Sat.	Ur.	Nep.	Pl.	Trans Pl-1	Trans Pl-2	......	Planets
0.39	0.72	1.00	1.52	2.9	5.20	9.54	19.18	30.07	39.67	......	......	......	Average distance to the Sun-(r)
4	7	10	16	28	52	100	196	300	396	492	684	......	~r × 10
0+(4×1)	3	3	6	12	24	48	96	96+(4×2)	96	96	192	......	Δr
		0	1	2	4	8	16		0	32	......	......	ΔΔr/3